Hydromechanics and Heat/Mass Transfer in Microgravity

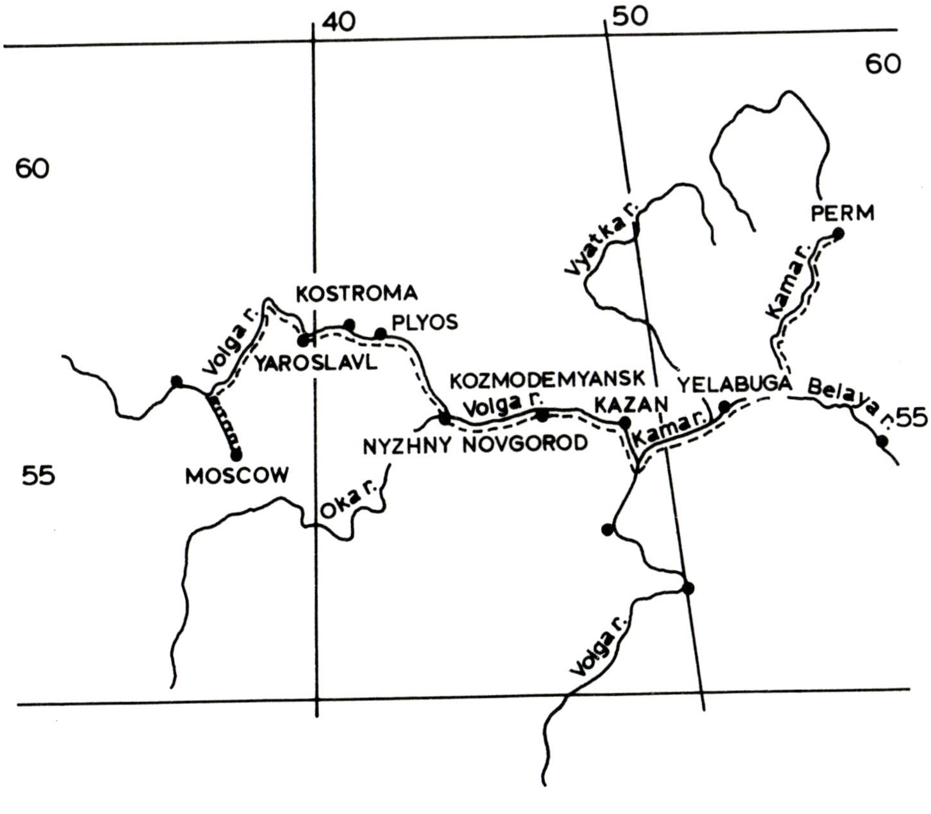

Hydromechanics and Heat/Mass Transfer in Microgravity

Reviewed Proceedings of the First International Symposium on
Hydromechanics and Heat/Mass Transfer in Microgravity

Perm - Moscow, Russia, 6-14 July 1991

GORDON AND BREACH SCIENCE PUBLISHERS
Switzerland USA Japan France Germany
Netherlands Russia Singapore Malaysia Australia

Gordon and Breach Science Publishers

5301 Tacony Street, Drawer 330
Philadelphia, Pennsylvania 19137
United States of America

Post Office Box 161
1820 Montreux 2
Switzerland

Post Office Box 90
Reading, Berkshire RG1 8JL
United Kingdom

3-14-9, Okubo
Shinjuku-ku, Tokyo 169
Japan

58, rue Lhomond
75005 Paris
France

Private Bag 8
Camberwell, Victoria 3124
Australia

Glinkastrasse 13-15
0-1086 Berlin
Germany

Emmaplein 5
1075 AW Amsterdam
Netherlands

Library of Congress Cataloging-in-Publication Data

International Symposium on Hydromechanics and Heat/Mass Transfer in Microgravity (1st : 1991 : Perm , R.S.F.S.R., and Moscow, R.S.F.S.R.)
Hydromechanics and heat/mass transfer in microgravity : reviewed proceedings of the First International Symposium on Hydromechanics and Heat/Mass Transfer in Microgravity, Perm - Moscow, Russia, 6-14 July 1991.
p. cm.
Edited by V.S. Avduevsky and others.
Includes index.
ISBN 2-88124-835-7
1. Liquids--Effect of reduced gravity on--Congresses. 2. Fluid mechanics--Congresses. 3. Heat--Transmission--Congresses. 4. Mass transfer--Congresses. I. Avduevskii, Vsevolod Sergeevich. II. Title.
TA357.5.R44I58 1991
620.1'06--dc20

92-13508
CIP

Printed in Great Britain by
Antony Rowe Ltd, Chippenham, Wiltshire

Contents

Thermocapillary Convection

Conference Organizers

The symposium was organized by:

Russian Academy of Sciences
- Institute of Continuous Media Mechanics, Urals Branch
- Institute for Problems in Mechanics
- The Gagarin Committee

Perm State University
Perm State Pedagogical Institute

With the participation of:

European Space Agency (ESA)
European Low Gravity Research Association (ELGRA)
National Aeronautics and Space Administration (NASA)
University of Aix-Marseille II

Conference Sponsors

The symposium was sponsored by:

European Space Agency (ESA)
Centre National d'Etudes Spatiales

Commercial Bank ZAPADURALBANK
Perm Branch of Perm-EVM-Fredriksson

Design Bureau "Yuzhnoye"
Central Specialized Design Bureau (Samara)

Space Research Institute, Russian Academy of Sciences
Institute of Heat/Mass Transfer, Belorussian Academy of Sciences
Institute of Hydrodynamics, Russian Academy of Sciences, Siberian Branch
Association "Astronautics - to Mankind"

Committees

Organizing Committee:

V. S. Avduevsky (Chairman)
V. A. Briskman (Vice-Chairman)
I. V. Barmin
V. G. Bar'yakhtar
P. N. Belyanin
A. I. Ivanov
D. M. Klimov
A. K. Kolesnikov
A. V. Kozlov
V. V. Malanin
Ye. V. Markov
V. V. Moshev
L. L. Regel
Ye. P. Romanov
Ye. S. Sapiro
A. A. Serebrov
M. P. Sveshnikov
V. A. Tatarchenko
Sh. Vakhidov
M. I. Yakushin

Local Organizing Committee:

V. V. Moshev (Chairman)
S. V. Mel'nikov (Vice-Chairman)
B. I. Myznikova (Vice-Chairman)
V. A. Briskman
V. I. Chernatynski
A. K. Kolesnikov
G. I. Kozhevnikova
A. Yu. Lapin
T. P. Lyubimova
N. P. Ogorodnikova
S. Ph. Onegin
L. V. Semukhina
Ye. A. Tkacheva
A. N. Vereshchaga

International Program Committee:

V. S. Avduevsky, Russia (Co-Chairman)
V. I. Polezhaev, Russia (Co-Chairman)
T. P. Lyubimova, Russia (Vice-Chairman)
X.-S. Chen, China
J. J. Favier, France
B. Feuerbacher, Germany
G. Z. Gershuni, Russia
S. Ya. Gerzenshtein, Russia
V. N. Glotov, Russia
V. L. Gryaznov, Russia
D. T. J. Hurle, UK
K. Ishida, Japan
N. A. Kulchitsky, Russia
J. C. Legros, Belgium
L. V. Leskov, Russia
O. G. Martynenko, Russia
G. Muller, Germany
A. D. Myshkis, Russia
L. G. Napolitano, Italy
T. Nishinaga, Japan
V. V. Pukhnachov, Russia
H. J. Rath, Germany
B. Roux, France
Yu. S. Ryazantsev, Russia
A. B. Sawaoka, Japan
R. F. Sekerka, USA
R. S. Sokolowski, USA
A. Triska, Czechoslovakia
M. G. Verlarde, Spain
J. P. B. Vreeburg, The Netherlands
H. U. Walter, France
B. Zappoli, France
E. M. Zhukhovitsky, Israel

Editors

V. A. Briskman, Institute of Continuous Media Mechanics, Russia
T. P. Lyubimova, Institute of Continuous Media Mechanics, Russia
B. Roux, Institut de Mécanique des Fluides, France
H. U. Walter, European Space Agency, France
V. S. Avduevsky, Institute of Machine Sciences, Russia
V. I. Polezhaev, Institute for Problems in Mechanics, Russia

I. Egry, Institute of Space Simulation, DLR, Germany
J. J. Favier, CEA/DTA/CEREM/DEM, France
B. Feuerbacher, Institute of Space Simulation, DLR, Germany
G. Z. Gershuni, Perm State University, Russia
K. Ishida, Space Technology Corporation, Japan
J. C. Legros, Université Libre de Bruxelles, Belgium
J. Meseguer, ETSI Aeronauticos, Spain
R. Peralta, Universidad Nacional Autonoma de Mexico, Mexico
P. G. Righetti, Università di Milano, Italy
J. P. B. Vreeburg, National Aerospace Laboratory, The Netherlands
B. Zappoli, Centre Spatiale de Toulouse, France

Preface by the Editors

Our conference on "Hydromechanics and Heat/Mass Transfer in Microgravity", held during a riverboat cruise between Perm and Moscow in July 1991, was the fifth in a series of such conferences. Earlier ones were held in Moscow, Perm, Chernogolovka and Novosibirsk. For the first time the conference was open to participants from all over the world, the papers were presented in English and the proceedings are also in English. This is a major contribution towards the creation of an international science community engaged in microgravity research.

The investment made by our colleagues from Russia in organizing this conference, in presenting their papers in English and delivering manuscripts in English should be praised very highly. Some of the papers reflect the difficulties of the authors in expressing themselves in that language. However, since it is the first time that these investigations are accessible to a worldwide community, the editors feel that they should be published without too much delay in their present form.

Opening Remarks

This symposium on "Hydromechanics and Heat/Mass Transfer in Microgravity" is the fifth Soviet and the first international symposium on microgravity sciences in our country, opened and democratic, which connects scientists and engineers working in this field in the former USSR and all over the world.

The scientific program of the symposium continues the topics of four previous Soviet microgravity meetings (Moscow 1979, Perm 1981, Chernogolovka 1984, Novosibirsk 1987).

The main goal of the symposium is to combine efforts in the discovery of new phenomena in microgravity sciences, their use in space and their terrestrial applications.

You are welcome to participate in this meeting and discussions.

Co-Chairmen of the Scientific Committee
Professor V. S. Avduevsky
Professor V. I. Polezhaev

Perm Address

On behalf of the Perm section of the Organizing Committee we are honored to welcome the founders of this traditional symposium, Professor V. S. Avduevsky and Professor V. I. Polezhaev, the participants of previous meetings as well as foreign guests who have accepted our invitation and have found a possibility to take part in the symposium.

The Perm scientific hydrodynamical community (Institute of Continuous Media Mechanics, Perm State University and Perm State Pedagogical Institute) began research in microgravity hydrodynamics fifteen years ago on the initiative of Professor G. I. Petrov and has contributed greatly in current advances in this area.

Previous symposia were held only for Soviet participants. But recent significant changes in home affairs has allowed the forthcoming meeting to become international. We would like to extend our most sincere thanks to our foreign colleagues not only for a valuable contribution to science but also for the tireless effort they put into preparing and organizing the symposium. We especially want to express our appreciation to the European Space Agency and personally to Dr H. U. Walter and Dr B. Roux.

We hope that this symposium will create a fruitful atmosphere of goodwill and cooperation among the scientists of the world.

Chairman of the Local Organizing Committee
Director of the Institute of Continuous Media Mechanics
V. V. Moshev

Vice-Chairman of the Organizing Committee
V. A. Briskman

Vice-Chairman of the International Program Committee
T. P. Lyubimova

Opening Session

SCIENTIFIC COOPERATION

H.U. Walter, ESA Headquarters, Microgravity Office, Paris (France)

This paper summarizes the status of cooperation between ESA and the former USSR in Microgravity Research /Fluid and Material Sciences

Science is international by its very nature, a tradition which is essential in advancing science and technology in a cost-efficient fashion and to the benefit of all.

The prerequisites for international cooperation are :

* free exchange of information and
* mobility of researchers.

The free exchange of information is normally guaranteed by publications in the open literature and conferences with international participation. The discussion of results and ideas during conferences, visits to laboratories, lecturing and working in other laboratories is highly stimulating especially for young researchers. An academic educational system without this international dimension is bound to hamper the development of young scientists and to turn in narrow spirals in research. Since science provides also the basis for technological progress, closed systems are bound to lose in the competition of the free market.

Research on gravity-dependent phenomena in Fluid and Material Sciences using microgravity carriers as a tool should be based on the free exchange of information and international cooperation as well. In fact this element should be emphasized even more since facilities and flight experiments are extremely costly and the opportunities to carry out experiments are scarce. In the past the free exchange of information and cooperation was hampered due to unrealistic expectations and overselling of industrial applications of microgravity by agencies and journalists.

It is now understood that applications can develop only following a sufficient period of fundamental research and this period may extend into the next century. The joint utilisation of the forthcoming international Space Station "FREEDOM" should therefore be planned in the spirit of international cooperation.

The cooperation between member states of ESA is defined in ESA's convention :

"The purpose of the Agency shall be to provide for and to promote, for exclusively peaceful purposes, cooperation among European States in space research and technology and their space applications, with a view to their being used for scientific purposes and for operational space applications systems :

> by elaborating and implementing a long-term European space policy, by recommending space objectives to the Member States, and by concerting the policies of the Member States with respect to other national and international organisations and institutions ;
>
> by elaborating and implementing activities and programmes in the space field ;
>
> by coordinating the European space programme and national programmes, and by integrating the latter progressively and as completely as possible into the European space programme, in particular as regards the development of applications

satellites ;

by elaborating and implementing the industrial policy appropriate to its programme and by recommending a coherent industrial policy to the Member States" (article II).

With the former USSR, ESA has signed a Framework Agreement on Cooperation in April 1990. The important terms of reference are as follows :

establishing exchanges and visits of scientists and other specialists aiming towards mutual familiarization with scientific work in progress and facilitating participation in joint research and project development work ;

exchange by each party of scientific and technical information and literature published without restriction as to its use on scientific programmes and projects, as well as exchange of assets, including equipment, where appropriate ;

cooperation on joint projects for design, development, launching and operation of payloads, and on the conduct of joint experiments in orbit ;

joint symposia and conferences ;

other cooperative activities as may be agreed by the parties.

More detailed conditions are specified in article 7 :

1. as a rule no exchange of funds is foreseen under this Agreement, unless the Parties agree otherwise. In special cases, compensation measures may be worked out for the execution of particular projects. The cooperation under this Agreement may also include projects which are implemented on a reimbursable basis.

2. Each Party shall bear the cost of dispatch of the information and the cost of the visits of its own personnel as foreseen under Article 2.2 and Article 3.

3. Detailed provisions on the abovementioned questions shall be determined within the corresponding implementing arrangement or contract.

In June 1991, the Council of ESA has decided to suspend article 7.2 during a limited period (up to December 1992) and to allow for payment of mission expenses of experts visiting ESA member states. This clause is applicable only in connection with specific projects of cooperation and the arrangements have to be negotiated in advance.

Article 5 recommends the formation of joint working groups including one on microgravity. This working group on "Fundamental Research in Microgravity" has been established in December 1990. The guidelines are concerned with joint projects and scientific cooperation ; Specific assignments to the joint working group are :

- adopt recommendations leading, where appropriate, to joint projects ;
- review and analyse the status of ongoing joint projects ;
- foster the exchange and visits of scientists and other specialists ;
- foster the exchange of scientific and technical information and literature published without restriction as to their use ;
- foster the organisation of joint symposia and conferences.

Several joint working group meetings on "Fundamental Research in Microgravity" were held during 1991 and the following projects were initiated :

(1) Experiments on protein crystallization using KASHTAN (developed by SPLAV). Flight on Photon in October 1991, 20 samples processed on the basis of cost reimbursement.

(2) Joint characterization of semiconductor crystals processed on MIR. Project underway.

(3) Joint experiment (Univ. Milano and ICMM, Perm) on the preparation of gels on sounding rockets sponsored by ESA (first experiment scheduled for MAXUS-2, 1993/1994).

(4) Experiments proposed by a team of the European Low Gravity Research Association (ELGRA) :

* Influence of the Soret effect on the Marangoni-Bénard instabilities

* Enhancement of Heat Transfer by Thermocapillary Convection.

Eight laboratories from Western Europe and two laboratories from Eastern countries are participating. It is planned to adapt a sounding rocket module (TEM-06/16) to a Photon Satellite. The scientific and cost evaluation is underway.

(5) Development of software for the modelling of 2-d and 3-d fluid flow and heat transfer in fluid configurations applicable to crystal growth (COMGA). Software package developed by IPM, Moscow.

Other projects are being discussed, including crystal growth from the vapour phase on MIR and PHOTON, furthermore the utilisation of ZONA-4 on PHOTON, ZONA-2 and ZONA-3 on MIR.

Almost all member states of ESA have signed bilateral agreements with the former USSR (see ESA/IRC(90)R/3). Several experiments have been performed on MIR and PHOTON by the French and German Space Agencies CNES and DARA, additional ones will be flown in the near future.

After several visits by ESA delegations to discuss cooperation in microgravity research with IKI, Moscow, between 1987 and 1989, a more extended visit to laboratories in Perm and Moscow was undertaken by an ESA delegation including 10 scientists from ESA member states in May 1990. Some 20 protocols on cooperation were signed during this visit. The present Conference on "Hydromechanics and Heat and Mass Transfer in Microgravity", which was sponsored and organised jointly by the USSR Academy of Sciences and ESA was one of the results of this encounter.

Presently, difficulties in cooperating with countries of the CIS are due to the changes in the civil space policy of the former USSR and the changes in the administrative structures. Flight opportunities and the use of experiment facilities can only be obtained on a commercial basis, with a small reduction possible if research institutes from CIS are involved in a given project. Since it has become virtually impossible for scientists from CIS to travel to Western Europe without paid invitations, ESA and other agencies will have to adopt their policies to this new situation.

Our Conference on "Hydromechanics and Heat and Mass Transfer in Microgravity" was a milestone in establishing cooperation. This was the first time the microgravity research communities of East and West met in an International Conference and it was a great experience for all participants. Agencies can only try to set the appropriate boundary conditions for cooperation. In reality, cooperation happens at the working level between scientists. This is usually a slow process requiring familiarisation with the mutual research projects and personal acquaintance. It was therefore most surprising to see how quickly contacts were made and joint projects were forged. In fact some 50 such agreements were elaborated during the Conference and a number of them is well underway (see Appendix). Others have stalled since the funding required is not available. ESA has presently no assignment to foster cooperation in ground-based research. Scientists interested in cooperating with their colleagues in Eastern countries will therefore have to resort to national funding agencies. The Commission of the European Communities in Bruxelles has started to provide financial support for cooperation and mobility in higher education with Eastern Europe. This Trans-European Mobility Scheme of University Students (TEMPUS) is limited to the exchange with Poland and Hungary during the pilot phase (July 1990-July 1993). It is planned to extend this programme and to include additional eastern countries. The financing of travel, fellowships, visiting scientist programmes and the possibility of placing small study contracts for the development of computer codes and numerical analysis of specific problems would be required to develop cooperation in ground-based research on gravity-dependent phenomena. The commercial procurement of flight opportunities may be very beneficial for ESA's microgravity programme as well as to Eastern Space Industries. However, International Cooperation in Science has another dimension and it is hoped that this cooperation will develop to the benefit of all.

APPENDIX

Joint Projects and agreements on cooperation identified during "Perm Conference"

Projects between Univ. Marseille and teams from the former USSR

B. Roux, Institute of Fluid Mechanics, Marseille (France)
V.I. Polezhaev, A. Fedoseyev, A. Gryoznov, Inst. Problems in Mechanics, Moscow
A.I. Feonychev, Aviation Institute, Moscow

"Mathematical and Physical Modelling of Heat and Mass Transfer Phenomena" (Numerical simulation, stability analysis)

Remarks : Coop. exists since 1981, agreement CNRS/former USSR Academy of Sciences.

B. Roux, Inst. Fluid Mechanics, Marseille (France)
G.Z. Gershuni and co-workers, Univ. Perm
T. Lyubimova, B. Myznikova, V. Briskman, Inst. Continuous Media Mechanics, Perm

"Thermo-vibrational convection"
Fundamental and Application to Crystal growth ; numerical simulation, stability analysis, model experiments)

Remarks : visit of Prof. Gershuni in Marseille in Summer 1991 (2 months, funded by French Ministry of National Education)
Joint paper presented at the July 1991 Conference

B. Roux, Inst. Fluid Mechanics, Marseille
and Scientific Centers in Novosibirsk and Krasnoyarsk
Laurentiev Hydrodynamics Institute (V.V. Pukhnachov, O. Laurentieva)
Institute of Geophysics (A.G. Kirdyashkin)
Computing Center of Krasnayarsk (Andreev)

"Fluid flows with free surfaces, thermocapillary flows"
(Numerical simulation stability analysis experiments).

B. Roux, Inst. Fluid Mechanics, Marseille
Institute of Applied Physics of Nyshni Novgorod (V. Brailovskaya)

"Mathematical and physical modelling of natural convection"

Cooperation and planned projects between LAMF/ETSIA (U.P.M., Madrid) and IMSS (PNC-UO-AN, PERM)

"Experimental and theoretical analysis of stability limits of liquid bridges between equal and unequal disks in solid-body rotation".

"Experimental and theoretical analysis of stability limits of liquid bridges in a non-axial gravity field (without oscillation)".

"Lateral oscillation of liquid bridges".

Within these three tasks, Lamf will supply the theoretical background whereas the experimental studies will be performed at IMSS.

"Development and manufacturing of a Millimetric Liquid Bridge Facility (MLBF) for Telescience applications". (Telescience Testbed) consisting of :

1. Millimetric Liquid Bridge Cell (MLBC).

   The MLBC should allow to form a small liquid bridge and apply to it at least the following stimuli :

   - Liquid injection and removal
   - Variable angle between liquid bridge axis and gravity
   - Solid body rotation, either concentric or eccentric
   - Rotation of one of the disks
   - Vibration of the liquid bridge as a whole.

2. Data Management and Image Analysis (DMIA) consisting of :

   - CCD camera and TV monitor
   - Electronics Box (to control the MLBC)
   - Data Acquisition System
   - Computer including Image Processor.

It is planned that two units of the MLBC will be manufactured at IMSS and two of the DMIA at Lamf, in order to get two complete MLBF, one in Perm and the other in Madrid. The funding for the two units of MLBC will be provided by IMSS and the fundings for the two DMIA units by Mamf, and, after mutual acceptance, the spare units will be exchanged.

To carry out the envisaged task it is suggested to have meetings every six months, alternating between Madrid and Perm.

Partners : J. Meseguer, LAMF/ETSIA, Madrid (Spain)
V.V. Moshev, Institute of Continuous Media Mechanics, Perm
L.A. Slobozhanin, Institute of High Temperature, Jarkov

Cooperation and projects with CEN, Grenoble

A.I. Fedyushkin
D. Camel, CEN, Grenoble
P. Tison, SESC/CEN, Grenoble

"Influence of magnetic field (0,15 Tesla) applied perpendiculary to the tin liquid surface"

M. Dubovik
D. Camel, CEN, Grenoble
P. Tison, SESC/CEN, Grenoble

"Influence of bismuth on thermocapillary convection of tin in liquid alloys and during Bridgman solidification".

M. Cherepanova, Center of Microelectronics, Riga
J.J. Favier, SESC/CEN, Grenoble

"Crystal growth from the melt : modelization of interfacial effects"

G.Z. Gershuni, M. Putin, Perm State University
J.J. Favier, SESC/CEN, Grenoble

"Damping of natural flows with vibration"

V.I. Polezhaev, Inst. for Problems in Mechanics, Moscow
J.J. Favier, SESC/CEN, Grenoble

"Marangoni flow in liquid metal layers"

G.A. Shvetsov, V.L. Sennitskii, Zavrentyev Inst. of Hydrodynamics, Novosibirsk, CIS
J.J. Favier, CEN, Grenoble

"Dynamics of inclusions in a vibrational liquid, vibrational cleaning of liquids"

Cooperation and planned projects with ULB 1990 - 1991

V.P. Shalimov, IKI, Moscow
J. Ryazantsev, L. Erakhin, IPM, Moscow
J.C. Legros, S. van Vaerenbergh, ULB, Bruxelles

"Investigation of the role of the Soret effect on the stability of a layer of a liquid binary mixture heated transversally"

"Marangoni-Bénard instabilities with Soret effect"

Remark : A joint paper has been presented in the Perm - Moscow meeting. The final goal is to fly on Photon.

A.A. Vedernikov, IKI, Moscow
J.C. Legros, S. van Vaerenbergh, P. Queekers, ULB, Bruxelles

"Wetting and dewetting of crytals grown from aqueous solutions"

Remark : a joint paper was presented in 1991 in San Diego.

N. Baturin, IKI, Moscow
J.C. Legros, S. van Vaerenbergh, ULB, Bruxellles

"Behaviour of a diffusion boundary layer in front of a growing crystal under high gravity conditions"

Remark : Theoretical investigation results were jointly presented at the Dubna Conference (June 1991).

Experimental investigations are under preparation on board aircraft flying circular trajectory at 5 g., for different values of the Coriolis force.

I.B. Simanovskii, A.K. Kolesnikov, Perm Pedagogical Institute
I. Vertgein, Institute Continuous Media Mechanics, Perm
J.C. Legros, Ph. Georis, M. Hennenberg, ULB, Bruxelles

"Stability of multi-layered systems"

Remark : Initiated during the ESA delegation visit (May 1990). Linear stability analysis is performed (homogeneous solution). Non linear numerical eveluation underway.

Preliminary results have been presented at different meetings. A joint paper (invited) has been presented at the Perm - Moscow meeting.

G. Gorbenko, V. Blinkov, A. Nikonov, Kharkov Aviation Institute, Kharkov, CIS

J.C. Legros, ULB, Bruxelles
J. Straub, Univ. München
W. Supper, ESTEC, Netherlands

"Flow Regimes, boiling heat and mass transfer"
"Evaporative heat exchangers and condensor networks"
"Two-phase jet-pumps, separators, pourous materials elements
"Computer modelling of multi-element two-phase
Two-phase Thermal Control Systems for large spacecraft (> 10 KW)

G.F. Putin, Perm State University
J.C. Legros, ULB, Bruxelles

"Experimental Investigation of the combined action of gravity and surface tension on the stability of a non-isothermal liquid layer"

V. Briskman, Institute of Continuous Media Mechanisms, Perm, CIS
T. Lyubimova, Inst. of Continuous Media Mechanisms, Perm, CIS
J.C. Legros, ULB, Bruxelles

"Marangoni-Bénard instabilities and Soret effect and surface deformation"

A.M. Tumaikin, V.M. Teshukov, V.V. Pukhnachov, Novosibirsk State University,
J.C. Legros, ULB, Bruxelles

"Hydrodynamics and Stability Problems"
"Phase Transitions"

Cooperation and Projects with MARS Center, Naples

V.A. Poleshaev, Inst. Problems in Mechanics, Moscow, CIS
R. Monti, Univ. Naples

"Numerical Programme to support Telescience Experiments in Fluid Dynamics"

V.G. Kozlov, Perm Pedagogical Institute, Perm, CIS
R. Monti, Univ. Naples

"G-dose validation experiments"

G.Z. Gershuni, Perm State University, Perm, CIS
R. Monti, Univ. Naples

"Thermovibrational Stability Limits in the presence of unsteady thermal field"

Cooperation and projects between various laboratories

V.A. Poleshaev, Inst. for Problems in Mechanics, Moscow, CIS
J.J. Favier, CEN, Grenoble
G. Müller, Univ. Erlangen

"Numerical codes for computation of heat transfer and convection"

V.V. Pukhnachov, Academy of Sciences, Novosibirsk, CIS
K.G. Dubonik, Inst. Problems in Mechanics, Moscow, CIS
J.C. Legros, ULB, Bruxelles

"Behaviour of liquid-gas interfaces with surface tension minimum and maximum"

G. Gouesbet, LESP, INSA, Rouen
A.T. Sukhodolsky, General Phys. Institute, Academy of Sciences, Moscow

"Instabilities in a liquid layer heated from below"
(hydrodynamical free surface instabilities)

H.F. Bauer, Inst. Space Technology, UBW, Neubiberg (Germany)
H.J. Rath, Univ. Bremen (Germany)
J.J. Favier, CEA/CEREM, Grenoble (France)
A.I. Feonychev, Aviation Institute, Moscow
I. Egry, Institute of Space Simulation, DLR, Cologne (Germany)

"Influence of Convection caused by vibrations during crystallization"

Same partners as above, plus I.S. Kalachinskaya, Moscow State University
B. Roux, Inst. of Fluid Mechanics, Marseille
J.C. Legros, Univ. Libre de Bruxelles

"Numerical simulation of oscillatory convection in floating zones and layered systems"

B. Roux, Inst. of Fluid Mechanics, Marseille
G.Z. Gershuni, Perm State University

"Stability of plane-parallel advective flows".
Investigation of different kinds of instability mechanisms, transition of instability and critical perturbation characteristics".

B. Roux, Inst. of Fluid Mechanics, Marseille
G.Z. Gershuni, Perm State University

"The investigations of thermovibrational stability and convection".

M. Gelfgat, Inst. of Physics, Riga
J.J. Favier, SESC/CEN, Grenoble
M. Moreau, Madylam/INPG, St Martin d'Hères

"Effect of constant magnetic field in the presence of thermoelectric effects" (application to metallurgy and crystal growth from the melt)

V. Kozlov, Perm Pedagocical Institute, CIS
P. Evesque, Ecole Centrale, Paris

"Dynamics of Phase Inhomogeneity in incompressible liquids under vibration"
"Convection induced by vibrations using AFPM"

A.I. Feonychev, Aviation Institute, Moscow, CIS
H.C. Kuhlmann, ZARM, Bremen

"Influence of vibrations on the onset of Taylor vortices"
"Onset of Thermocapillary flow in half-zone geometries"

A.T. Sukhodolsky, General Physics Institute, Moscow, CIS
K.S. Mironov, N.L. Bazhenov, Ioffe Phys. Tech. Institute, Leningrad, CIS
J. Schilz, DLR, Inst. Mat. Research, Cologne

"Binary Semiconductors growth and characterization (Ge-Si)

G. Ramanauskas, Institut Phys. Tech. Enerby, Kaunas, Letvia
V.E. Liusternik, Inst. High Temperature Physics, Moscow, CIS
I. Egry, Inst. Space Simulation, DLR, Cologne

"Thermal Conductivity"
"Colorimetry of levitated melts"

L.A. Slobozhanin, Phys. Tech. Inst. of Low Temperatures, Kharkov, CIS
H.F. Bauer, Univ. Bundeswehr, München

"Geometry and Stability of Liquid Bridges"

V.A. Briskman, Inst. of Continuous Media Mechanics, Perm, CIS
H.F. Bauer, Univ. Bundeswehr, München

"Dynamic Behaviour of Liquid Bridges and related thermocapillary phenomena"

V.A. Poleshaev, A. Fedyushkin, Inst. Problems in Mechanics, Moscow, CIS
T. Cherepanova, Inst. Microelectronics, Riga, Letvia
D. Henry, Lab. Fluid Mechanics, Ecole Centrale de Lyon

"Oscillatory Convection"
"Double-diffusive effects in concentrated alloys"

T. Cherepanova, Inst. Microelectronics, Riga, Letvia
D. Henry, Lab. Fluid Mechanics, Ecole Centrale de Lyon

"Oscillatory Convection"
"Convection with Soret effect"
"Hydrodynamics and solidification with magnetic field"

L.A. Slobozhanin, Inst. of Low Temperature Physics, Kharkov, CIS
J. Meseguer, Lab. of Aerodynamics, Univ. Madrid

"Hydrodynamic Instabilities"

V. Briskman, T. Lyubimova, Inst. of Continuous Media Mechanics, Perm, CIS
R. Birikh, V. Myznikov, PSPI, Perm, CIS
V. Tatarchenko, Inst. High Temperatures, Moscow, CIS
V. Zevtov, NPO Kompozit, Moscow
M.G. Velarde, Univ. Madrid
G. Müller, University Erlangen
B. Roux, University Marseille

"Instability of vibrations on thermocapillary - Stability and Flows"

A.I. Feonychev, Aviation Institute, Moscow
I. Egry, Inst. Space Simulation, DLR, Cologne
J.C. Legros, ULB, Bruxelles
H.J. Rath, Univ. Bremen
H. Kuhlmann, Univ. Bremen
J.J. Favier, CEN, Grenoble
B. Roux, University Marseille

"G-jitter on-board Spacecraft and Control of heat and mass transfer in crystallization processes"
"Transition of thermocapillary convection to oscillatory"

ZERO-GRAVITY HYDROMECHANICS: SOME RESULT AND PROBLEMS

A.D.Myshkis

Institute for Railway Transport Engineers, Moscow 103055, USSR

ABSTRACT

Results of some theoretical investigations in zero-gravity hydromechanics using modern methods of pure, applied and numerical mathematics are given and unsolved problems are indicated briefly. The main topic are: equilibrium forms; their stability; small oscillations; some other motions; convection.

Keywords: Zero-Gravity, Microgravity, Equilibrium Form, Stability of Equilibrium, Small Oscillations, Convection.

1. INTRODUCTION

Zero-gravity hydromechanics emerged as a new field in early 1960's and began to develop intensively. It was in (Ref.1) published in 1961 that this new field of science, with specific problems and methods was introduced. Since that time many hundreds of papers and several monographs have been devoted entirely or mainly to tis field. The creation of this discipline is due to the practice and the theory of space flights and necessity to investigate the behavior of fluids in very weak force fields. Zero-gravity hydromechanics is closely connected with classic theory of capillary phenomena. However, the conditions of space flights require not only study of essentially new problems but also a consideration of the simultaneous influences of many factors, e.g. heating, vibrations, electric and magnetic fields, and even self-gravitation.

This field has been studied both experimentally and theoretically. The present paper is devoted mainly to the theoretical investigations using "pure", applied and numerical mathematics. Only the topics most familiar the author will be considered.

2. EQUILIBRIUM FORMS

The equilibrium forms of liquids in immovable containers (or, which is the same, under its inertial motion) in homogeneous volume force fields and surface tension have been studied most expensively. Not only the methods of determining of such forms have been investigated but many results were derived by pure mathematical methods (solvability theorems of corresponding boundary value problems, qualitative properties and deductive estimations of solutions etc.). The most detailed exposition of these results with a large list of references is given in the monograph of R.Finn (Ref.2). One may talk about the creation of a new field of variational calculus that is connected with the theory of capillary fluids. It would be quite interesting to extend this field to zero-gravity hydromechanics considering, for example, centrifugal forces.

A summary of investigations of the equilibrium forms from the point of view of applied and numerical mathematics is given in part I of (Ref.3). This book contains some algorithms and results of their applications to construct such forms, investigations of their stability and nature of their branching near the critical values of parameters. The homogeneous volume force field, the field of centrifugal forces and surface tension were taken into consideration. If the problem being studied contains one or two parameters (for example Bond number, expressing the relation between the gravitational and capillary forces, and wetting angle), then the results of calculations may be represented visually in the form of graphs. If the number of parameters is greater, then the problem of representation of results is complicated and it may be necessary to construct approximate formulas for the main characteristics of the equilibrium forms depending on the parameters. This assertion is already valid for characterizing doubly connected equilibrium surfaces. An analogous problem arises if one tries to represent the dependence of essentially two-dimensional (i.e. non-cylindrical and nonrotational) equilibrium surface on the parameter.

In our opinion, the most important unsolved problem is to create productive mathematical models describing the emergence of hysteresis of the wetting angle. Indeed, for example, the assumption of constancy of this angle leads to the impossibility of the actually observed equilibrium of a fluid drop on an inclined plane. But this assumption now forms the basis of most of the mathematical works in zero-gravity hydromechanics. Consequently, this leads to the problem of the adequacy of these works. It may be assumed that the possibility of a small change of the wetting angle does not influence the results but until now this influence has not been investigated. Some possible approaches to the construction of the model of the above-mentioned hysteresis have been proposed. In particular, an interesting proposition made in ch.8 of (Ref. 2) was applied by its author to determine the equilibrium forms. Nevertheless it appears that none of these proposals is entirely acceptable. It is possible that a lack of systematic experimental data hampers the formulation of the corresponding relations on the phenomenological level like the relations for the dry friction.

But even if we assume that the wetting angle is constant, there still remain important unsolved problems connected with equilibrium forms. Some of them were given in (Ref. 4) and will not be discussed here. We can just mention the problem of describing a set of all equilibrium forms (in particular, stable forms) of a liquid in a giving vessel, and investigating the dependence of this set on the parameters. Of major interest are the conditions of a change of the structure of these forms from axially symmetrical into non axially symmetrical, singly connected forms into doubly connected ones etc. and also possible loss of the continuous dependence of the forms on parameters. A simple example of the latter is the behavior of a fluid in a closed vessel with a gradual increase of the rotational velocity: singly connected domain occupied by the liquid may become doubly connected or fall to two pieces etc.

3. STABILITY OF EQUILIBRIUM

The methods of investigating the stability of equilibrium forms are now well established but sometimes quite complicated, especially for essentially two-dimensional equilibrium surfaces. But we believe that a rigorous definition of such a stability corresponding to the physical intuition and also being sufficiently productive does not exist. The stability definition following the Liapunov approach excludes a priori the possibility of creating "fluid threads" that is not to be considered as a natural requirement. (It may be that this definition can be refined by imposing the rational restriction on the class of initial disturbances and external influences.) The instability definition is difficult to apply because of the necessity of non-local consideration of solutions of non-linear hydrodynamics equations with regard to surface tension. As a result one has to use the minimum potential energy criterion for the stability analysis leaving the transition from this criterion to the stability statement as a plausible postulate. (In the framework of the small disturbances theory, this transition for a rotating capillary liquid was proved in a recent paper of N.Kopachevskii and E.Vodkovich. Note that similar results about stability of capillary liquid were also derived by V.Vladimirov (Ref. 5); the important achievement here is the direct derivation of the booth upper and lower estimates of the increment of disturbances for unstable states.)

In connection with the stability of equilibrium forms one should mention the problem of gyroscopic stabilization conditions of a rotating ideal fluid in general and concrete situations, which has not been investigated.

It may be that the problem of definition of the equilibrium fluid state concept is similar to the problem of natural definition of the stability margin concept of such a state. The latter definition is not available till now. The one applied now is a formal continuation of the calculation algorithm of such margin which is known for systems with a finite number of degrees of freedom. The stability margin was not investigated for many concrete classes of problems such as for rotating fluid, for essentially two-dimensional equilibrium surfaces etc.

4. SMALL OSCILLATIONS

In the theory of small oscillations of capillary liquids the modern qualitative operator and calculation methods were applied with success. One can speak of the creation of a new branch of operator theory closely related with the zero-gravity hydrodynamics. Many results in this field can be found in part II of (Ref. 3) and in Chs. 4, 5 and 8 of (Ref.6).

The problem of boundary conditions for a mobile wetting line leads to difficulties as for the equilibrium forms. If the resistance to this motion is of dry friction type, then for ideal liquids we may apply both postulates of a stationary wetting line and of a line moving on the wetting surface with a constant wetting angle. The problem of the wetting line movement for viscous fluids is much more complicated because it follows formally from the finiteness of energy dissipation near the wetting line that the wetting angle must be 0° or 180°. Interesting results were reported by V.Pukhnachev and V.Solonnikov (Ref. 7) (see also (Ref. 8)) but the form of boundary conditions on the wetting line is not known until now. The above mentioned conclusion about the wetting angle may perhaps be valid for the oscillations of real liquids if the viscosity is sufficiently large. In the opposite case, one can use the usual value of wetting angle beyond the very narrow boundary layer which may be neglected. But this problem requires a further theoretical and experimental analysis. (See in particular the recent papers (Refs. 9, 10) of physicists.)

Important investigations of correspondent boundary value problems were carried out in the last 25 years. (It is known that in this problem the "normal movements" are investigated, the dependence of which on time is represented by the factor exp(zt); the valid values of I are called the eigenvalues, its set - the spectrum, and corresponding space factors - the eigenfunctions of the problem under consideration.) Fundamental results for the spectrum structure, the completeness of the set of eigenfunctions etc. for ideal or viscous capillary liquids with or without rotation can be found in the (Refs. 3, 6). These works describe new properties associated with the presence of surface tension in mathematical calculus, and show the physical picture of phenomena. Numerical algorithms were worked out and a number of basic asymptotic formulas were found. Operator methods showed their effectiveness in this field. It may be useful to extend the mathematical methods here, in particular to include the Sobolev weight spaces which are applied with success in some problems of mathematical physics in non-smooth domains.

Some fundamental problems and many particular ones are not solved yet. For instance, the following hypothesis confirmed by examples has not proved: there exists only a finite number of nonreal eigenvalues for viscous capillary fluids occupying the finite volume; if the viscosity is sufficiently large, such values cannot exist, i.e. then all normal movements are aperiodic.

Note the following more specific interesting problem. If the equilibrium state of a viscous liquid is stable, then all Rel $z_k < 0$, and it is natural to call $q := \max_k |\mathrm{Re}\ z_k|$ as the stability degree of the system in consideration. The value q depends continuously on the viscosity u and $q(u) \to 0$ as $u \to 0$ or $u \to \infty$ for fixed values of other parameters. Thus there exists the "optimal" value of u for which this stability degree is maximal. This value is known only for a ball (Ref. 11). One should find this characteristic for other simple systems, to construct the algorithm of its calculation for more complicate systems, and to investigate its dependence on systems parameters. (Recent results of M.Barnjaks may be useful here.)

In the above discussion, the liquid was assumed to be homogeneous (or we had a system of homogeneous liquids) occupying a finite volume. Recent papers investigate the small oscillations of continuously stratified fluids. It appears that this topic may have applications in zero-gravity hydromechanics and it should be developed. The systematic investigation of small oscillations of the fluids occupying an infinite volume is also of great interest.

Some other unsolved problems of the small oscillations theory in zero-gravity hydrodynamics can be found in (Ref. 4).

5. SOME OTHER MOTIONS

Some problems of zero-gravity hydrodynamics which are beyond the framework of the small oscillations theory and require systematic investigation on the base of advanced mathematical calculus will be pointed out here. (Some of such problem are considered A.Povitskii and L.Ljubin (Ref. 12) and in a number of papers.). Above all, these are the problems which admit the linearisation, in particular, the investigation of slow movements and forced oscillations with small amplitude and "medium" or small frequency. For oscillations with a large amplitude or high frequency the modern mathematical methods of non-linear oscillations theory can be employed. In particular, it is necessary to continue the mathematical investigation of vibration effects on the fluid in a weak volume fields. The processes involving change in the topology of domains occupied by liquids (breaking coalescence, etc.) have not been investigated in detail.

The important problem of determining the stationary flows of a capillary liquid in different situations - in particular, leading to plane-parallel or axial symmetric flows - is not considered sufficiently. In the latter case, it is of special interest to find the conditions of the emergence of a rotating vortex.

The theory of the movement of dispersion systems "fluid-gas" needs serious mathematical analysis, especially when its structure may change (from a fluid with bubbles to foam and to fog and vice versa).

6. CONVECTION

The appearance of convective movements under the action of thermo-capillary forces, vibration and self-gravitation in a heated fluid was considered in many papers; see in particular, the books of G.Gershuni and E.Zhukhovitskii (Ref. 13), the same authors with A.Nepomnjashcii (Ref. 14), and part III of (Ref. 3). However, the number of problems worked out in detail by mathematical methods is not sufficient at present. The most important general problem which has not been solved is the finding of the validity conditions of the "stability change principle" (consisting in the assertion that the moment of the emergence of convection is defined by a transition of the "critical" eigenvalue l from the left half-plane into the right one through the point l = 0, i.e. the convection appears as an aperiodical process but not as an oscillating one. This assumption which considerably simplifies the investigation is made as a work hypothesis without a mathematical substantiation.). In particular, it is plausible that this principle is valid for the emergence of convection in any vessel by heating a fluid from the bottom, if g!=0.

Some problem connected with the emergence of convection can be found in (Ref. 4). One can also mention the problem of the mathematical investigation of the thermocapillary convection weakened by the action of periodic disturbances or by addition of a thin layer of fluid taking the convection on itself (Ref. 15), and the problem of stability of movement of a dispersing medium in a temperature field.

7. ACKNOWLEDGEMENTS

I should like to thank Professors N.Kopachevskii and V.Pukhnachev who suggested to me some problems pointed out here, Professor V.Kolmanovskii and Dr.Wadhw for help in the translation of this paper, and Professor H.Walter for helpful criticism.

8. REFERENCES

1. Weightlessness - Physical Phenomena and Biological Effects. Ed. by E.T.Benedikt (Plenum Press, New York 1961).

2. R.Finn, Equilibrium Capillary Surfaces (Springer-Veriag, New York e.o. 1986).

3. A.D.Myshkis, V.G.Babskii, N.D.Kopachevskii, L.A.Slobozhanin, A.D.Tjuptsov, Low-Gravity Fluid Mechanics. Mathematical Theory of Capillary Phenomena (Springer-Verlag, Berlin e.o. 1987). This book is essentially a revised and enlarged version of the book Zero-Gravity Hydromechanics (in Russian) by the same authors (Nauka, Moscow 1976).

4. The same authors, Nonlin. Anal. Th. Meth. Appl., 4:3 (1980).

5. V.A.Vladimirov, The Stability of Rotating and Stratificiated Fluids (in Russian). Abstract of doctor thesis (Inst. of Hydrodynamics SB AS USSR, Novosibirsk 1990).

6. N.D.Kopachevskii, S.G.Krein, Ngo Zui Kan, Operator Methods in Linear Hydrodynamics. Evolutional and Spectral Problems (in Russian) (Nauka, Moscow 1989).

7. V.V.Pukhnachev, V.A.Solonnikov, Prikl. Mat. Mech. (in Russian) ,46:6 (1982).

8. K.Baiocci, V.V.Pukhnachev, Pricl. Mech. Tekhn. Fis. (in Russian), N2 (1990).

9. P.Thompson, M.Robbins, Physics Word (Nov.1990).

10. J.Koplik, J.R.Banavar, J.F.Willemsen, Phys. of Fluid, A1(5) (May 1989).

11. O.M.Lavrentieva, Prikl. Mech. Tekhn. Fis. (in Russian), N3 (1984).

12. A.S.Povitskii, L.Ja.Ljubun, Fundamentals of Dynamics and Heat- and Mass-Transfer in Liquids and Gases at Zero-Gravity (in Russian) (Mashinostrojenie, Moscow 1972).

13. G.Z.Gershuni, E.M.Zhukhovitskii, Convective Stability of Non-Compressible Liquid (in Russian) (Nauka, Moscow 1972).

14. G.Z.Gershuni, E.M.Zhukhovitskii, A.A.Nepomnjashcoii, Stability of Convective Fluids (in Russian) (Nauka, Moscow 1989).

15. J.M.Floryan, C.Chen, 6th CAS Conf. of Astronautics (Ottawa, Canada, Nov. 1990).

CONVECTIVE PROCESSES IN MICROGRAVITY

V.I. Polezhaev

Institute for Problems in Mechanics USSR Academy of Sciences, Moscow
Prospect Vernadskogo 101, 117526, Moscow USSR

Abstract

The paper presents an overview of methods and results of investigations concerned with gravity-driven convection during the last 2-3 years. Macroinhomogeneities in directionally solidified alloys and dopped semiconductors induced by different types of convection are discussed. Means to eliminate inhomogeneities and to control material processing in space and terrestrial environments are studied.

1. Introduction.

Gravity-driven convection may influence materials processing and the structure and properties of materials. Most of the first material sciences experiments in Space were based on the reduction or suppression of gravitational convection. First overviews which have been published by S. Ostrach [1,2] (see also [3,4-6]), were concerned mostly with week steady microaccelerations. During the last 5-7 years a number of new papers and overviews were devoted to different kinds of residual convection [7-9], Marangoni convection [10], stability and temperature oscillation problems [11-14]. Recent overviews and books were devoted to the problems of gravitational sensitivity and control of convection in microgravity [15-17]. This paper makes an attempt to overview the main topics of recent works by the auther and his collegues [17-25].

The focus is on gravitational sensitivity, "criteria of quality" and control of convection in view of alternatives to microgravity. The discussion is limited to simple geometries and fixed boundaries. Forced and Marangoni convection are shortly discussed for the case of coupling with gravitational convection.

2. Microaccelerations and gravitational sensitivity

Three directions of the analysis of microaccelerations for the calculation of convection in microgravity are being developed now. The first is the analysis of accelerometer measurements on the Space Station and the calculation of different kinds of "scenario" in accordance with experimental data. Most details are available from Spacelab-3 [15]. The second direction is the calculation of microaccelerations using the rotational velocity data aboard the Space Station. In this case long duration flights can be efficiently analyzed [26]. The third direction is the direct numerical simulation of microaccelerations in unmanned flights, which do not need any empirical information [17]. Only low frequency oscillations may be analyzed in this case, but, as follows from different authors [8,15,17], these are the most dangerous ones. Different flight modes (gravity-stabilised and unstable) have been analyzed. Analysis of design, trajectory, dynamics of flight is possible with the help of this method. Vibrational and rotational components of microaccelerations have been calculated [17,22]. Steady or rotational microacceleration vector cases are studied below.

3. Classifications and criteria of quality.

Different kinds of convection include forced (rotation, vibration, direct fluid flow, mixing), gravitational (thermal and concentration), Marangoni (thermocapillary and concentration), magnetic (axial, lateral, rotational).

ity for crystal growth. In crystal growth axial and lateral macro- and microinhomogeneities may be induced by convection [5,7,8,11]. Geometrical uniformity is important for thin films (epytaxial layers) [17] and steady fluid flow is needed for electrophoretic separation [17,20]. The definition of criteria and gravitational sensitivity is different for these cases. Most investigated is the lateral inhomogeneities induced by convection [5,7,8,11]. New accurate and efficient methods are needed for solving this problems.

4. Methods of analysis. PC-based system "COMGA".

Boussinesque approximations on the basis of 2D and 3D Navier-Stokes equations and different approaches (flat, axisymmetric, Hele-Shaw) for direct simulation and convective stability analysis were developed [17].

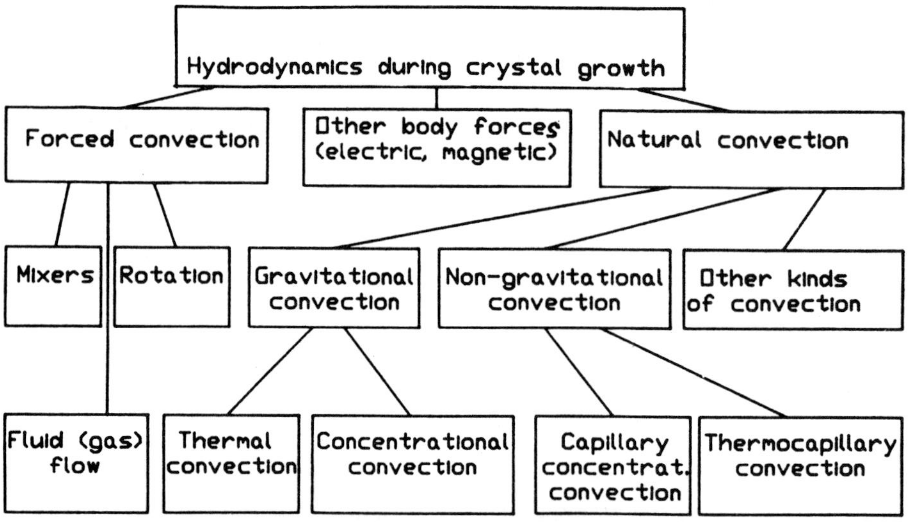

Fig 1. Classification of hydrodynamical processes

High gravitational sensitivity need more detailed classification, for example: double-diffusion gravitational and double-diffusion Marangoni (see [1,5,17]) and special attention to the coupling problem [5,17].

Gravitational sensitivity and classification of different materials processing experiments use different definition of qual-

Most efficient is a new PC-based dialog system "Convection in Microgravity" ("COMGA") [18,20,21]. Governing equations for this system are 2D unsteady Boussinesque equations [17] be written in the form:

$$\frac{\partial u}{\partial t} + u\frac{\partial u}{\partial x} + v\frac{\partial u}{\partial y} = -\frac{\partial p}{\partial x} + \nu\,\Delta u +$$

$$+g_x(t)\frac{\partial\rho}{\partial T}\frac{(T-T_0)}{\rho} \quad (1)$$

$$\frac{\partial v}{\partial t}+u\frac{\partial v}{\partial x}+v\frac{\partial v}{\partial y}=-\frac{\partial p}{\partial y}+\nu\,\Delta v+$$

$$+g_y(t)\frac{\partial\rho}{\partial T}\frac{(T-T_0)}{\rho} \quad (2)$$

$$\frac{\partial u}{\partial x}+\frac{\partial v}{\partial y}=0 \quad (3)$$

$$\frac{\partial T}{\partial t}+u\frac{\partial T}{\partial x}+v\frac{\partial T}{\partial y}=\alpha\,\Delta T \quad (4)$$

$$\frac{\partial C}{\partial t}+u\frac{\partial C}{\partial x}+v\frac{\partial C}{\partial y}=D\Delta\theta \quad (5)$$

where t is time, (u, v) is the velocity vector, p is the pressure, T is the temperature, C is the concentration of impurity, D is the diffusion coefficient, ν is the kinematic viscosity, ρ is the density and α is the thermal diffusity.

The components of the microacceleration vector $g_x(t)$ and $g_y(t)$ can be written in the next form:

$$g_x(t)=g_{x0}+(g_s+g_t\sin(\Omega_1 t))\times$$

$$\times\sin(\Omega_2 t+\phi_0) \quad (6)$$

$$g_y(t)=g_{y0}+(g_s+g_t\sin(\Omega_1 t))\times$$

$$\times\cos(\Omega_2 t+\phi_0) \quad (7)$$

where g_{x0}, g_{y0} – components of the constant microacceleration, g_s, g_l – constant and variable components due to rotation in the case that it exists, Ω_1, Ω_2 – frequencies of vibration and rotation, ϕ_0 – initial angle of inclination. This presentation includes most of the mentioned before cases of spatial and temporal variation of micracceleration vector. For example, most popular case $g_y(t) = g_t \sin\Omega t$, $g_x = 0$, $g_s = 0$ corresponds the vibration of microaccelerations in the y-direction only with zero average value, see [12,15,27-28]. In [15,17,24] more common case $g_y(t) = g_s + g_t \sin\Omega_t$ corresponds to nonzero time-average microaccelerations is studied. In the case of rotation [22,29-32] $g_x(t) = g_t \sin(\Omega t + \phi_0)$, $g_y(t) = g_t \cos(\Omega t + \phi_0)$

The boundary conditions for velocity may be:

A. The normal component for velocity is zero on every boundary and the prescribed tangent component of velocity may be either of the first kind (rigid wall) or of the second type (surface tension driven convection);

B. The normal component of the velocity is described for the part of boundary (input or output flow) and zero tangential component of the velocity. For temperature and concentration three kinds of boundary conditions (temperature, heat flux or heat transfer law) are prescribed. The principal nondimensional parameters, governing the behavior of the system and included in the menu. For mixed thermal and forced convection they are: Reynolds, Rayleigh, Prandtl numbers, aspect ratio, position and nondimensional sizes of holes, etc. (see [17,18]).

The PC-based system includes all mentioned before double- diffusive problems for gravitational-type and Marangoni convection and coupling problems as shown in Fig. 1. It was tested on several benchmarks [17,18] and may be a powerful tool for analysis of convective processes in microgravity. Recently not only ordinary but also porous media, input and output of fluid were included in this system [21].

5. Macroinhomogeneities induced by convection

Weak convection of all types transports heat and mass and induces temperature and concentration inhomogeneities in the normal direction to the flux. Reference [17,25] dis-

Table 1. **Papers dealing with maximum of lateral macroingomogeneities induced by different types of convection**

Author	Type of convection	Type of media	Theor. or Experim.	Boundary conditions	Geometry
Polezhaev 1970	Gravit., Thermal	Pr=1	Theor.	Heat flux	Square, L/H=2
Polezhaev 1974	Gravit., Thermal	Pr=1	Theor.	Heat flux	Square, Sphere Vert., Cylinder
Polezhaev, Fedyushkin 1980	Gravit., Thermal, Concent.	Pr=0.016 Sc=10	Theor.	Tw,Cw	Horisontal plane L/H=4
Nikitin, Polezhaev, Fedyushkin 1981	Gravit., Thermal	Pr=0.016 Sc=10	Theor.	Tw,Cw Moving front	Horisontal plane L/H=4
Zemskov, Belokurova, et al. 1975	Gravit., Thermal	Semicond. melt, Pr=0.016	Space Experim.	Furnace	Ampule, L/H=4
Chang, Brown 1983,1987	Gravit., Thermal	Semicond. melt	Theor.	Moving front	Axisym. Vertical Bridgman
Polezhaev Dubovik et al 1981,1987	Marangoni	Pr=1	Theor.	Tw	Plane, L/H=4
Pshenichiov Pinyagin	Gravit., Concent. Gas mixt	Pr=0.75,	Experim.	Cw	Square, L/H=4
Camel, Favier 1987	Gravit., Termal	Pr=0.016 melts	Experim.	Tw	Horisontal Bridgman
Havshu 1988	Gravit. Termal	melts	Experim.	Furnace	Horisonatal Bridgman
Alexander, Quazzani Rosenberger 1989	Gravit, Thermal	Semicond. melts	Theor.	Tw, Cw	Ideal Bridgman
Polezhaev, Nikitin 1989	Forced	Pr=1	Theor.	Tw	Plane L/H=1
Motakef 1990	Magnetic	Ge, GaAs, InSb	Theor.	Moving front	Vertical Bridgman
Nadarajah, Rosenberger Alexander 1990	Gravit., Thermal	TGS Solution Growth	Theor.	Tw, Cw	Ideal Bridgman

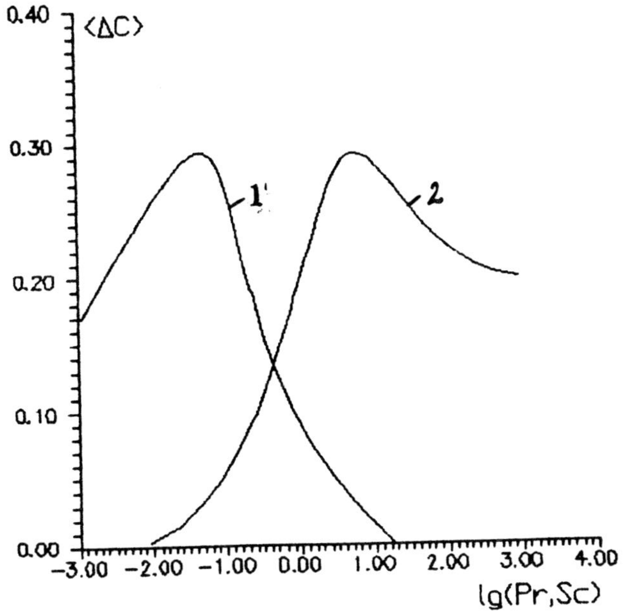

Fig.2 Lateral ingomogeneity in horisontal ampoule for different physical properties of the melts H/L = 4 [17]; 1 - Pr, 2 - Sc

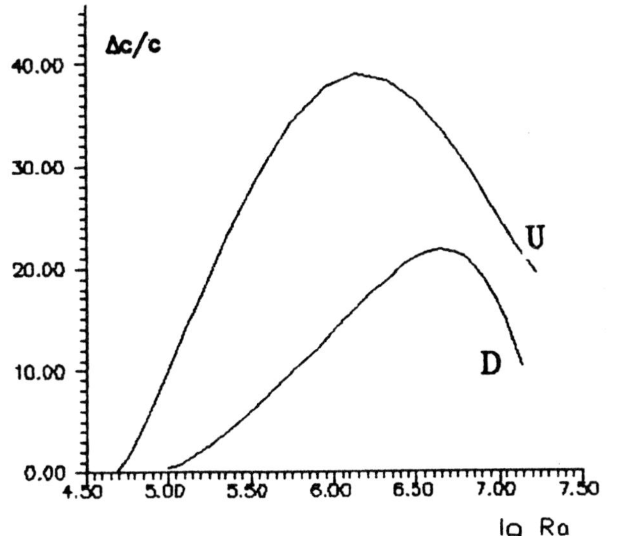

Fig.3 Geometrical ingomogeneity of the LPE [17]. U - upper, D - down substrait

cussed these phenomena for thermal, concentrational, gravity driven, Marangoni and forced type of convection. Recently S. Motakef [16] published results for the case of convection suppressed by magnetic field.

These phenomena may be the reason for lateral segregation in crystal growth. Table 1 presents a describtion of different cases, published up to now (different references see in [17,25]). The dependence of physical properties is very important for the analysis

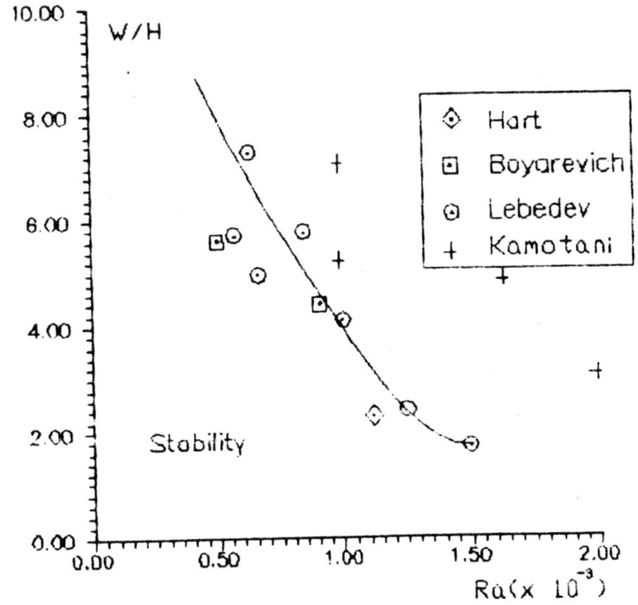

Fig 4. Experimental boundaries of temperature oscillations in horisontal layer with side heating for liquid metals [19,33,35]

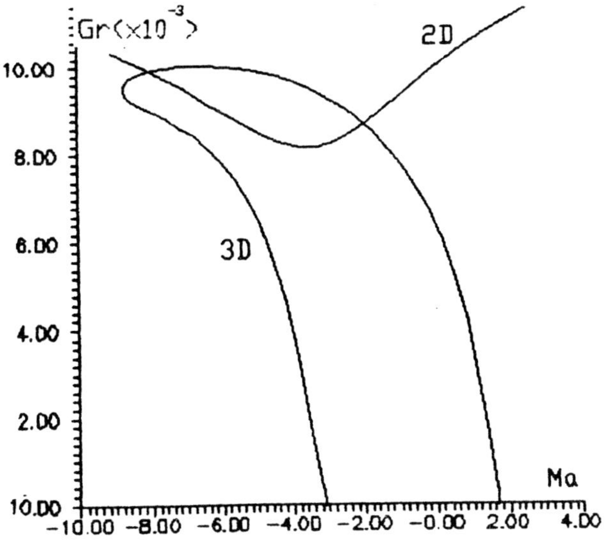

Fig 5. Stability boundaries of coupling gravity-driven and Marangoni convection in rigid-free case with side heating and adiabatic boundaries [17]

of microinhomogeneities [17]. (Fig. 2)

For the case of liquid phase epitaxy (LPE) [17] longitudinal inhomogeneities of concentrational flux also reached maximum in dependency of Ra number (Fig. 3)

6. Stability, secondary flows and temperature oscillations.

Convective instability for low Prandtl number liquids may be the main reason for

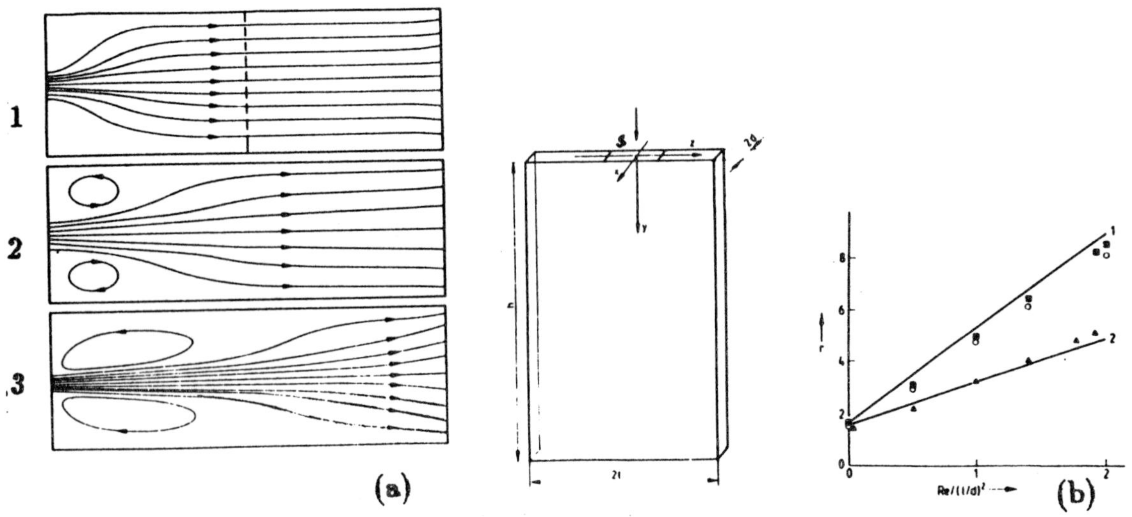

Fig. 6 Secondary flows (a) and entrance region length (b) in free flow electrophoresis chamber for zero gravity [17]. For (a): h/l=6.4, Re=200; (a1): l/d=15, S/l=0.2; (a2): l/d=10, S/l=0.2; (a3): l/d=10; S/l=0.2

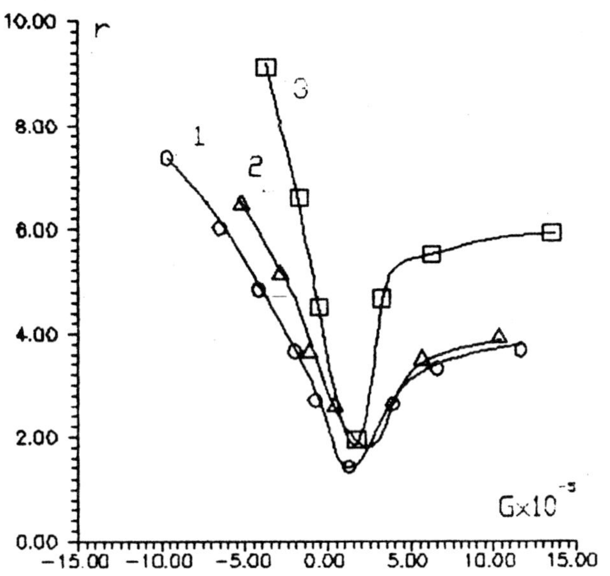

Fig. 7 Gravitational sensitivity of the free flow electrophoresis; dependency of length of entrance region from Grashof number [17] (For Re=60, 1: l/d=15, S/l=0.5, 2: l/d=10, S/l=0.5, 3: l/d=15, S/l=0.2)

the microingomogeneities in crystals [7,11]. Recently a GAMM workshop and other groups studied the problem of direct numerical simulation of the flow and temperature oscillations [13,17] and tests with experimental data and stability analysis data. Experiments show (Fig. 4) the region of stability in the case of small aspect ratio width to height. New experimental data were recently published in [19,33,35]. The difference between the data of the authors may be due to the difference in geometry and boundary conditions. More detailed analysis of experimental conditions is needed.

Coupling between thermal gravitational and Marangoni convection is an interesting phenomenon in microgravity. Fig. 5 shows new results of stability analysis for the rigid-free surface and adiabatic boundaries. The plane of Gr-Mn shows different limiting cases (zero gravity, low gravity and terrestrial conditions) and modes (2D,3D) [17].

Secondary flows in electrophoresis chamber may be reason for distortion of forced parallel flow and separation. Hele-Shaw approach have been used for analysis of this phenomenon [17,23] (Fig. 6,7)

7. Possibilities to eliminate inhomogeneities

For eliminating lateral macroinhomogeneities one needs to take into account the strong dependency from the physical prop-

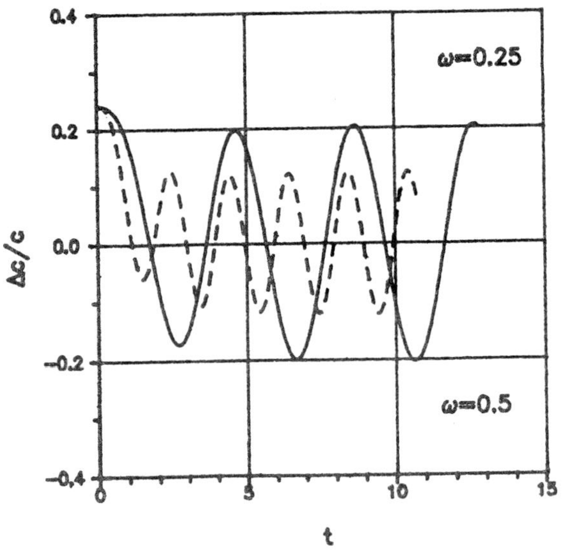

Fig. 8a

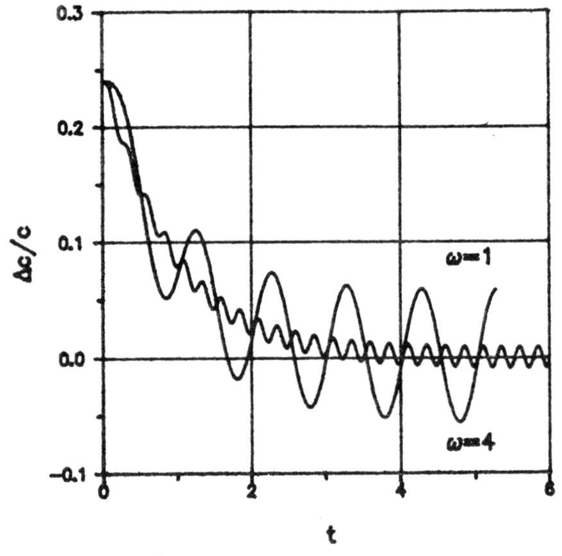

Fig 8b. Fig 8(a-b) - Time dependence of the calculated lateral ingomogeneity during a slow rotation

erties of the melts. Maximum of concentration lateral difference has also maximum for low Prandtl number (Pr=0.1) and Schmidt number 10. For Pr > 10 the lateral inhomogeneities are reduced.

Previous results show that inhomogeneities variation in dependency of geometry, level and angle of the gravity vector,

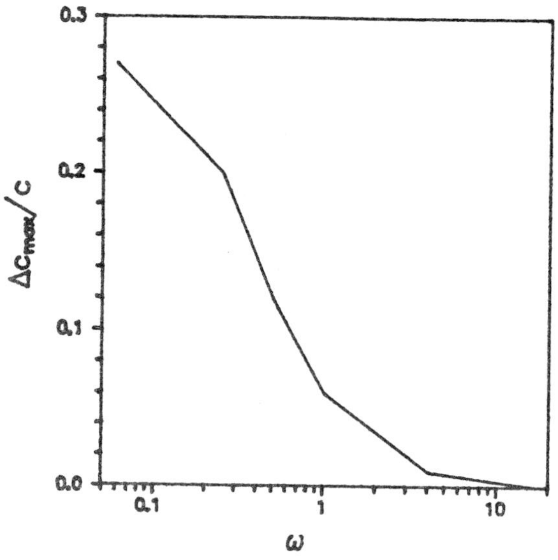

Fig. 9 Dependency of lateral ingomogeneities from the frequency of rotation

magnetic fields, temperature and dynamical control will be an alternative to microgravity [17].

Magnetic control may be efficient not only to avoid temperature oscillations.

The analysis of [16] shows, that alternative to microgravity the magnetic control for semiconductors melts (Ge,GaAs) may be efficient to avoid macroinhomogeneities for diameters of more than 2.5 cm, because for this limiting value of microgravity needs ($g/g0 = 10^{-6}$). But it is efficient only for high electroconductivity of liquid metals.

Forced convection is efficient at least for control of concentration convection during growth of crystals from solutions [36]. Forced convection control is very important in free-flow electrophoresis to reduce secondary vorticity (Fig 6). Thermal control (hot liquid in the top of the chamber (G>0) and cold (G< 0) is efficient to eliminate thermal convection secondary flows (Fig 7).

8. Rotation of the microacceleration vector

Several authors [22,29-32] discussed the problem of thermal convection due to microacceleration vector rotation. Opportunities to avoid ununiformity of the thickness of liquid phase epitaxial films were shown in [17]. The structure of temperature and flow field mechanics phenomena with the help of PC-based system "COMGA" is presented in [31]. Due to special microgravity conditions a slow rotation may be one of the cheapest and efficient way to control convection in microgravity, if the optimal rotation will be chosen.

Fig 8,9 show new results for elimination of impurity inhomogeneities during slow rotation of microacceleration vector in the x-y plane. "COMGA" system and AT386 computer have been used for this problem. The case of maximum lateral inhomogeneities, induced by thermal convection in the flat ampoule with side heating and adiabatic horizontal walls have been an initial for this calculations (Ra=10, Pr=0.016, Sc=10, L/H=4), which correspond to previous results [5,17]. For the case of rotation in x-y plane we use $g_0 = g_t = 0$, $\Omega_1 = 0$, $g_s = g$ and $\Omega_2 = \Omega = \frac{2\pi}{t_i}$, where t_i is the period of rotation. Nondimensional velocity of rotation in this case $\omega = \frac{\Omega R^2}{\nu}$ varies in the range $\omega = 0 \div 16$.

One can found from the Fig 9, that average lateral inhomogeneities of impurity in semiconductor's melt during a slow rotation in dependence of ω is eliminated. The temporal behavior illustrates more detailed this elimination for different ω. An addititional information of the temperature and flow structure can found in the recent paper [22]. Comparison with the mentioned results and [17] shows the strong dependence the conclusion upon of criteria of quality (bulk lateral or axial inhomogeneities, geometrical inhomogeneities of LPE, etc.) and governing parameters Ra, Pr, Sc, L/H, ω. For example, elimination of lateral segregation in [32] for $Ra > Ra_{cr}$ and $\frac{H}{L}$=1 needs $\omega \sim 100$.

3D calculations in [17] show, that Coriolis forces play an important role for $\omega = 5 \div 10$, so it seems, that last results for $\omega \gg 10$ need more careful 3D tests.

9. Conclusions

New results of convection studies in microgravity during the last 2-3 years show a more detailed picture of elementary convective processes. It is shown, that macroinhomogeneities for weak convection regime will be in, principle, induced by all types of weak convection (thermal gravitational, Marangoni, forced and induced by magnetic field).

Methods of mathematical modelling of convection have been developed (2D and 3D direct simulation of the basis of Navier-Stokes equations, stability analysis, new PC-based system "COMGA" for multiparametrical study of convection, axisymmetrical and Hele-Shaw approaches).

The study of convection in microgravity shows numerous opportunities to eliminate macro- and microinhomogeneities in microgravity (forced convection, magnetic field, geometric and thermal control, slow rotation) and in terrestrial environment with the use of alternative control opportunities. Multiparametrical study with the help of specialized models analysis needs for conclusions in each case of space processing.

The possibility to eliminate lateral macroinhomogeneities in semiconductor melt during a slow rotation in microgravity with the help of dialog system "COMGA" is shown.

In conclusion the author would like to thank F.Kozerev for the help and comments for ch.5, and M. Ermakov who made the calculation results in Ch.8

References

1. Ostrach S. Convection phenomena of importance for material processing in Space. In: Material sciences in Space with application to Space processing ed L. Steg, AIAA, N-Y, (1977).

2. Ostrach S. Low-gravity Fluid Flows, Annual Review of Fluid Mechanics, v.14, (1982).

3. Ostrach S. Natural convection in enclosures. J. Heat Transfer v.110, 1175, (1988)

4. Poleshaev V.I. Convective processes in microgravity In: Proc. 3th European Simp. Mater. Sci. in Space, Grenoble 24-27 April 1979, ESASP - 142 (1979), 25.

5. Poleshaev V.I. Hydrodynamics, heat and mass transfer during crystal growth In: Crystals, v.10 ed H.C. Freyhardt Springer, (1984).

6. Gebhart B., Jaluria Y., Mahajaa L.L., Sammakia B. Buoyancy-induced flows and transport N.Y., Hemisphere (1988)

7. Brown R.A. Convection and bulk transport In: Materials sciences in Space (ed. B. Feuerbacher, H. Hamacher, R.J. Naumann) Springer (1986), 55.

8. Monti R., Langbein D., Favier J.J., Influence of residual accelerations on fluid physics and Material Sciences Experiments In: Fluid Sciences and Material Sciences in Space (ed. H.U. Walter, Springer (1987), 637.

9. Legros J.C. Sanfeld A. Velarde M. Fluid Dynamics In: Fluid Sciences and Material Sciences in Space (ed.) H.U. Walter Springer (1987), 83.

10. Schwabe D. Surface-tension- driven flow in crystal growth from melts In: Crystals, v.11, Springer, (1988)

11. Müller G. Crystal growth from the melt: Convection and inhomogeneities in crystal growth from the melt In: Crystals (1988)

12. Gershuni G.Z., Zhuhovitsky E.M., Nepomniachy S.A. Stability of convective flows M. Nauka, (1989) (in Russian)

13. Numerical simulation of oscillatory convection in low-Pr flows: A GAMM workshop ed. B. Roux Braunschweig, Vieweg, (1990)

14. Favier J.J. Recent advances in Bridgman growth modelling and fluid flow J. Crystal Growth v.99, (1990), 18

15. Alexander J.I.D. Low gravity experiment sensitivity of residual acceleration: A review. Microgravity sciences and technology v.3, (1990), 52.

16. Motakef S. Magnetic field elimination of convective interference with segregation during vertical-Bridgman growth of doped semiconductors J. Crystal Growth v.104, 1990, 833.

17. Poleshaev V.I. Bello M.S. Verezub N.A. et. al Convective processes in microgravity, Moscow, Nauka, (1991), (in Russian)

18. Ermakov M.K., Grjaznov V.L., Nikitin S.A., Pavlovsky D.S. and Poleshaev V.I. Specialized software for modelling of convection in microgravity XXVIII COSPAR meeting 29 June – 6 July (1990) Hague, Netherlands

19. Lebedev A.P., Kozyrev F.V. Temperature oscillations and structure of convection in liquid metal. Preprint Inst. for problems in Mechanics, M., N.486, (1991)

20. Grjaznov V.L., Ermakov M.K., Nikitin S.A., Pavlovsky D.S. Solving convection problems on a personal computer. Prepr. Inst.. for problems in Mechanics, M., N.481, 1990

21. Grjaznov V.L., Ermakov M.K., Nikitin S.A., Pavlovsky D.S., Poleshaev V.I.

Modelling of convection in enclosure on the basis of Navier-Stokes equations: Results, applications and PC-based dialog system IV ICCBE Extended Abstract, Tokyo (1991), 162

22. Poleshaev V.I., Ermakov M.K. Thermal convection in microgravity during a slow rotation IUTAM symposium on microgravity fluid mechanics September 2-6, (1991) , Microgravity Sci. techn. IV 2(1991), 101

23. Bello M.S., Poleshaev V.I. Thermal convection and Fluid flow in continuous flow electrophoresis Microgravity sci. tech. III/4, (1991), 231

24. Lebedev A.P., Poleshaev V.I. Mechanics of wightlessness microaccelerations and gravitional convection sensitivity of the mass exchange processes in materials study in Space In: Advances in Mechanics v13, No. 1,(1990), 3

25. Poleshaev V. I., Nikitin S.A. Inhomogeneities of the temperature and concentration fields induced by connection of low gravity In: Proceedings VIII-th Europ. Symposium on material and Fluid Sciences in microgravity, Oxford UK, 1-15 Sept, 1989, ESASP-222,(June 1990), 295

26. Sarychev V.A., Beliaev M. J., Sazonov O.V. Tjak T.N. Definition of microacceleration on the orbital complex: 'Salut-6' and 'Salut-7'. Preprint of the Institute of Applied math. N.100 (1984)

27. Thevenard D., Favier J.J. Influence of g-jitter on the solidification of delute alloys In: Proceedings VII Europ. Symposium on materials and Fluid Sci. in microgravity Oxford, UK 10-15 Sept., 1989 ESASP-295 (Jan. 1990), 243

28. Polezhaev V.I. Lebedev A.P., Nikitin S.A., Mathematical simulation of disturbing forces and material sciences processes under low gravity In: Proceedings of the 5-th Symposium on material Sciences under microgravity, 5-7 November (1984)

29. Avduevsky V.S., Korolcov A.V., Kupcova V.S. Savichev V.V. Investigation of thermal gravity convection in variable fields of small acceleration vectors. PMTF, No. 1 (1987), n.1, 54

30. Ermakov S.V. Feonychev A.I. Convection under variable body forces acceleration and impurity microsegregation in crystals. In: Hydrodynamics and heat mass transfer under microgravity. Novosibirsk, (1988) , 20 (in Russian)

31. Schneider S., Straub J. Influence of the Prandtl number on a laminar natural convection in a cylinder caused by a g- jitter. J. Crystal Growth, v.97, (1989), 275

32. Mc Fadden A.B., Coriel S.R. Solutal convection during Directional Solidification In.: Proc. AIAA/ASME/SIAM/APS 15st National Fluid Dynamics Congress, Cincinnati, July 25-28, (1988), 1572

33. Kamotani Y.T., Sahraoui T. Oscillatory natural convection in rectangular enclosures filled with mercury Journ. Heat Transfer, (1990) No. 1, 253

34. Boyarevich A.V., Gorbunov L.A. Influence of magnetic fields different orientation on thermogravitational convection in electroconductive liquid with side heating MagnitoHydrodynamics No.2 (1988), 17 (in Russian)

35. Hart J., Pratt J.M. A laboratory study of oscillations in differently heating layers of mercury In: [13]

36. Braylovskaya V.A. Galushkina G.L., Zilberberg V.V., Feoktistova L.V. Modelling of free convection in solutes under the surface of growing crystal In: Hydrodynamics, heat and mass transfer during material processing Moscow, Nauka, (1990) (eds V.S. Avduevsky, V.I. Poleshaev) (in Russian).

Microaccelerations and Convection under Microgravity

MICROACCELERATIONS ON THE BOARD OF THE EARTH'S ARTIFICIAL SATELLITES

V.A. Sarychev, V.V. Sazonov, M.Yu. Belyaev*,
S.G. Zykov* and V.M. Stazhkov*

Keldysh Institute of Applied Mathematics, USSR Academy of Sciences, Moscow, USSR,
**SPA "Energiya", Kaliningrad, Moscow region, USSR*

ABSTRACT

The causes of microaccelerations arising aboard the artificial satellites of the Earth are considered: the atmosphere drag, the gravitational field gradient, the motion about the center of mass, etc. A technique is described for calculating onboard micro-accelerations on the Salyut-6, Salyut-7 and "Mir" orbital stations. In the procedure the actual motion of the station about its center of mass is first determined by the statistical processing of the onboard attitude sensor measurements; the microacceleration at any given point of the station is then calculated as a function of time for the motion obtained.

Keywords: *Microacceleration, Attitude Motion, Statistical Processing of the Measurement Data.*

1. MICROACCELERATIONS ON THE SATELLITE DURING SPACEFLIGHT.

The orbital motion of the satellite is described by

$$\ddot{\vec{R}} = \vec{g}(\vec{R}) + \frac{\vec{F}}{m}.$$

Here $\vec{R}$ is the geocentric position vector of the satellite's center of mass; the point denotes differentiation in time t ; $\vec{g}(\vec{R})$ is the strength of the earth gravity field (the gravity effect upon the satellite produced by the Moon and other bodies of the Solar System is ignored); m is the satellite mass; $\vec{F}$ is the sum vector of nongravitational forces acting on the satellite. Consider a fixed point P on the satellite body. Its position vectors with respect to the earth's center and the satellite's center of mass are denoted by $\vec{r}$ and $\vec{\rho}$, respectively: $\vec{r} = \vec{R} + \vec{\rho}$. The microaccelerations at the point P is given by the quantity $\vec{n} = \vec{g}(\vec{r}) - \ddot{\vec{r}}$, which is the difference between the strength of the earth's gravitational field at this point and its absolute acceleration. If there is a test body with mass $m_p \ll m$ at the point P then an apparent force acting on the body is equal to $m_p \vec{n}$. By taking into account the above relations the microacceleration may be written as

$$\vec{n} = -\ddot{\vec{\rho}} + \vec{g}(\vec{R}+\vec{\rho}) - \vec{g}(\vec{R}) - \frac{\vec{F}}{m}.$$

Let the satellite be a solid body. Then by denoting its absolute angular velocity by $\vec{\omega}$ we have

$$\ddot{\vec{\rho}} = \dot{\vec{\omega}} \times \vec{\rho} - (\vec{\omega} \times \vec{\rho}) \times \vec{\omega}.$$

With an accuracy sufficient for these calculations we may assume $\vec{g}(\vec{R}) = -\mu \vec{R}/R^3$, where $R = |\vec{R}|$, μ is the earth's gravitational parameter. For $\rho = |\vec{\rho}| \ll R$:

$$\vec{g}(\vec{R}+\vec{\rho}) - \vec{g}(\vec{R}) = \mu R^{-3}[3R^{-2}(\vec{R}\cdot\vec{\rho})\vec{R} - \vec{\rho}\,].$$

Now the expression for $\vec{n}$ may be written in the form

$$\vec{n} = \vec{\rho} \times \dot{\vec{\omega}} + (\vec{\omega} \times \vec{\rho}) \times \vec{\omega} + \frac{\mu}{R^3}\left[\frac{3(\vec{R}\cdot\vec{\rho})}{R^2}\vec{R} - \vec{\rho}\right] - \frac{\vec{F}}{m}. \qquad (1)$$

The expression obtained explains the causes for micro-accelerations arising on the satellite - solid body. Generally there are three of them: (1) the satellite motion about the center of mass - the summands $\vec{\rho} \times \dot{\vec{\omega}} + (\vec{\omega} \times \vec{\rho}) \times \vec{\omega}$ in expression (1); (2) inhomogeneity of the earth's gravitational field across the extent of the satellite - the summands - $-3\mu\vec{R}(\vec{R}\cdot\vec{\rho})/R^5 - \mu\vec{\rho}/R^3$; (3) nongravitational forces acting on satellite - summand $(-\vec{F}/m)$: these include various control forces, the forces due to the motions of inner parts of the satellite, the drag, the light pressure, etc. The components of microacceleration produced by the first two contributions can be calculated by rather simple formulas, but to obtain an explicit expression for $\vec{F}/m$ requires specified models, for example, the satellite - environment interaction model, the satellite control actuators operation model, etc. For satellites of the Salyut-6 and Salyut-7 type of orbital station typical values of $|\vec{F}/m|$ for some couses are given in the Table 1.

TABLE 1

Nongravitational perturbations produced by various factors upon the Salyut-6 and Salyut-7 stations during spaceflight	Typical values of $\lvert\vec{F}/m\rvert$, m/s^2
Orbital corrections by the control system	10 - 1
Attitude control system operation	10^{-2} - 10^{-1}
Fans, compressors and other service device	10^{-5} - 10^{-2}
Physical exercises by a crew	10^{-3} - 10^{-2}
Aerodynamic drag	10^{-5}
Sunlight pressure	10^{-7}
Magnetic field of the Earth	10^{-11}
Micrometeorites	10^{-15}

The following example helps to access these perturbations. We consider a satellite in circular orbit with altitude of about 340 km. A satellite is fixed in the orbital system of coordinates. The vector $\vec{\rho}$ is assumed to be parallel to the vector $\vec{R}$ and has the length of 2 m. Then

$$\left|\vec{\rho}\times\dot{\vec{\omega}}+(\vec{\omega}\times\vec{\rho})\times\vec{\omega}+\frac{\mu}{R^3}\left[\frac{3(\vec{R}\cdot\vec{\rho})}{R^2}\vec{R}-\vec{\rho}\right]\right| =$$

$$=\frac{3\mu\rho}{R^3} = 7\cdot 10^{-6}\ m/s^2 .$$

Since it is impossible to remove completely the microaccelerations on the satellite we need information about their values to analyse the results of flight experiments. Microaccelerations may be determined either by accelerometers or by using equation (1) and appropriate information on the satellite motion. These two approaches are not equivalent, each has its own application. Analysing the accelerometer measurements requires a more complicated mathematical model of microaccelerations in comparison with formula (1). This problem will be discussed below. However, first, we shall illustrate the second approach with an example of determining the microaccelerations that occured on the Salyut-6 and Salyut-7 orbital stations during their uncontrolled flight.

2. CALCULATION OF THE MICROACCELERATIONS ON THE SALYUT-6 and SALYUT-7 ORBITAL STATIONS.

This problem is interesting for two reasons. First, technological experiments on the Salyut-6 and Salyut-7 stations were usually performed during a special type of uncontrolled flight - under gravity gradient stabilization conditions that provided low perturbing microaccelerations. Second, there is an efficient technique for determining the uncontrolled rotational motion of these stations by using onboard sensor measurements (Refs.1,2). From the nongravitational forces we shall take into account only the aerodynamic drag, approximated by the formula

$$\frac{\vec{F}}{m} = -b\rho_a|\dot{\vec{R}}|\dot{\vec{R}}. \tag{2}$$

Here b = **const** is the ballistic coefficient, ρ_a = **const** is the free stream density.

We shall give a brief description of the technique (Refs.1,2) only for the part that is interesting for the problem under consideration. The station (together with the docked Soyuz or Progress spacecraft) is assumed to be a solid body whose center of mass moves in a fixed circular orbit (R and $|\dot{\vec{R}}|$ are constant, $\vec{R}\cdot\dot{\vec{R}}=0$). To write down the equations of the station motion about the center of mass we introduce two right hand Cartesian coordinate systems: the system formed by principal central axes of inertia $Ox_1x_2x_3$ and the orbital system $OX_1X_2X_3$. The axes OX_3 and OX_1 are directed along the vectors $\vec{R}$ and $\dot{\vec{R}}$, the axis Ox_1 is close to the longitudinal axis of the station and directed to its service modul, the axis Ox_2 is parallel to the rotation axis of nonsymmetric solar array, the latter being in half-space $x_2>0$. The attitude of the coordinate system $Ox_1x_2x_3$ with respect to the system $OX_1X_2X_3$ is given by the transformation matrix $\|a_{ij}\|_{i,j=1}^{3}$, where a_{ij} is cosine of angle between axes OX_i and Ox_j . For an arbitrary vector $\vec{a}$ the expressions $\vec{a}=(a_1,a_2,a_3)_x$ or $\vec{a}=(a_1,a_2,a_3)_X$ describe its components in the coordinate systems $Ox_1x_2x_3$ or $OX_1X_2X_3$ respectively. If

$$\vec{a}=(a_1,a_2,a_3)_x=(A_1,A_2,A_3)_X$$

then

$$a_i = A_1a_{1i}+A_2a_{2i}+A_3a_{3i} \quad (i=1,2,3).$$

The gravitational and aerodynamic torques are taken into account in the rotational motion equations of the station (Refs.1,2):

$$\begin{aligned}
\dot{\omega}_i &= \frac{I_j-I_k}{I_i}(\omega_j\omega_k-3\omega_0^2a_{3j}a_{3k})+\\
&+\frac{\omega_0^2I_1}{I_i}(a_{1j}p_k-a_{1k}p_j),\\
\dot{a}_{1i} &= a_{1j}\omega_k-a_{1k}\omega_j-\omega_0a_{3i},\\
\dot{a}_{3i} &= a_{3j}\omega_k-a_{3k}\omega_j+\omega_0a_{1i}.
\end{aligned} \tag{3}$$

Here i , j , k form all even permutations of numbers 1,2,3; $(\omega_1, \omega_2, \omega_3)_x=\vec{\omega}$; $\omega_0=\sqrt{\mu R^{-3}}$ is the orbital frequency; I_i are the station's moments of inertia with respect to the axes Ox_i , p_i are dimensionless functions describing the aerodynamic torque. The functions p_i depend on direction cosines a_{ij} and the parameters that specify the direction to the Sun in the coordinate system $OX_1X_2X_3$. It is necessary to take into account the position of the Sun because the solar arrays of the station are pointing to the Sun. During integration of equations (3), the elements a_{2i} of the transform matrix are calculated by the formula $a_{2i}=a_{3j}a_{1k}-a_{3k}a_{1j}$.

Initial conditions on the station motion and unknown parameters in the equations of motion are determined by statistical processing of the measurement data from solar and magnetic sensors. The apriori unknown parameters are dimensionless combinations of the free stream density and the geometrical and inertia characteristics of the station. They are contained in the expressions for the functions p_i . The sensors measure the earth's magnetic field strength $\vec{H}=(h_1,h_2,h_3)_x$ and the unit vector $\vec{S}=(s_1,s_2,s_3)_x$, which indicates the direction to the Sun, at known times (the measurements of $\vec{S}$ are possible only on the sunlit part of orbit). The time interval for data processing is approximately equal to the orbital period and contains a few tens of points where the measurements are available. The data processing is performed in several steps. First, for the actual orbit of the station and the times at which the vectors $\vec{S}$ and $\vec{H}$ have been measured, the components of these vectors are calculated in the orbital system of coordinates: $\vec{S}=(S_1,S_2,S_3)_X$, $\vec{H}=(H_1,H_2,H_3)_X$.

Then the measured and calculated values of $\vec{H}$ are normalized and the following function is composed:

$$\Phi = w_S \sum_{\vec{S}} \sum_{i=1}^{3} \left(s_i - \sum_{j=1}^{3} S_j a_{ji}\right)^2 + w_H \sum_{\vec{H}} \sum_{i=1}^{3} \left(h_i' - \sum_{j=1}^{3} H_j' a_{ji}\right)^2 .$$

Here w_H and w_S are positive weights; $(h_1', h_2', h_3')_x$ and $(H_1', H_2', H_3')_x$ are measured and calculated values of the vector $\vec{H}/|\vec{H}|$ respectively; the outer summation extends to all measurements of $\vec{S}$ and $\vec{H}$ included in the data processing. Minimizing Φ with respect to initial conditions and parameters of equations (3), the real motion of the station may be determined in the time interval under consideration. Then, knowing this motion one can determine the microacceleration at any point of the station as a function of time.

In order to make a detailed analysis of microaccelerations we divide the total microacceleration (1) and (2) into several components. First, taking into account the above assumptions on the station orbit, we write

$$\vec{n} = \vec{n}^a + \vec{n}^0, \quad \vec{n}^a = K_a (a_{11}, a_{12}, a_{13})_x,$$
$$\vec{n}^0 = \vec{\rho} \times \dot{\vec{\omega}} + (\vec{\omega} \times \vec{\rho}) \times \vec{\omega} + \omega_0^2 [3(\vec{e}_R \cdot \vec{\rho})\vec{e}_R - \vec{\rho}], \quad (4)$$
$$K_a = c \rho_a \omega_0^2 R^2, \quad \vec{e}_R = (a_{31}, a_{32}, a_{33})_x .$$

Here the component $\vec{n}^a$ is due to the aerodynamic drag, while $\vec{n}^0$ results from inertial and gravitational forces. Second, from $\vec{n}^0$ we single out the component

$$\vec{n}^r = \vec{\rho} \times \dot{\vec{\omega}} + (\vec{\omega} \times \vec{\rho}) \times \vec{\omega} - (\vec{\omega}_0 \times \vec{\rho}) \times \vec{\omega}_0 ,$$
$$\vec{\omega}_0 = \omega_0 (a_{21}, a_{22}, a_{23})_x , \quad (5)$$

which results from the station motion with respect to the orbital system of coordinates. If the station is fixed in this system, $\vec{\omega} = \omega_0 =$ **const** and $\vec{n}^r = 0$. The component $\vec{n}^0 - \vec{n}^r$ results from the gradients of inertial and gravitational forces.

Below we give the results of calculations of microaccelerations on the Salyut-7 orbital station by using the measurements of $\vec{S}$ and $\vec{H}$ taken during revolution 1595 on July 29, 1982 (Refs.3,4). The station motion was constructed as a result of measurement data processing and is shown in Figure 1 by the curves $\gamma(t)$, $\delta(t)$, $\beta(t)$ and $\omega_i(t)$ (i = 1,2,3).

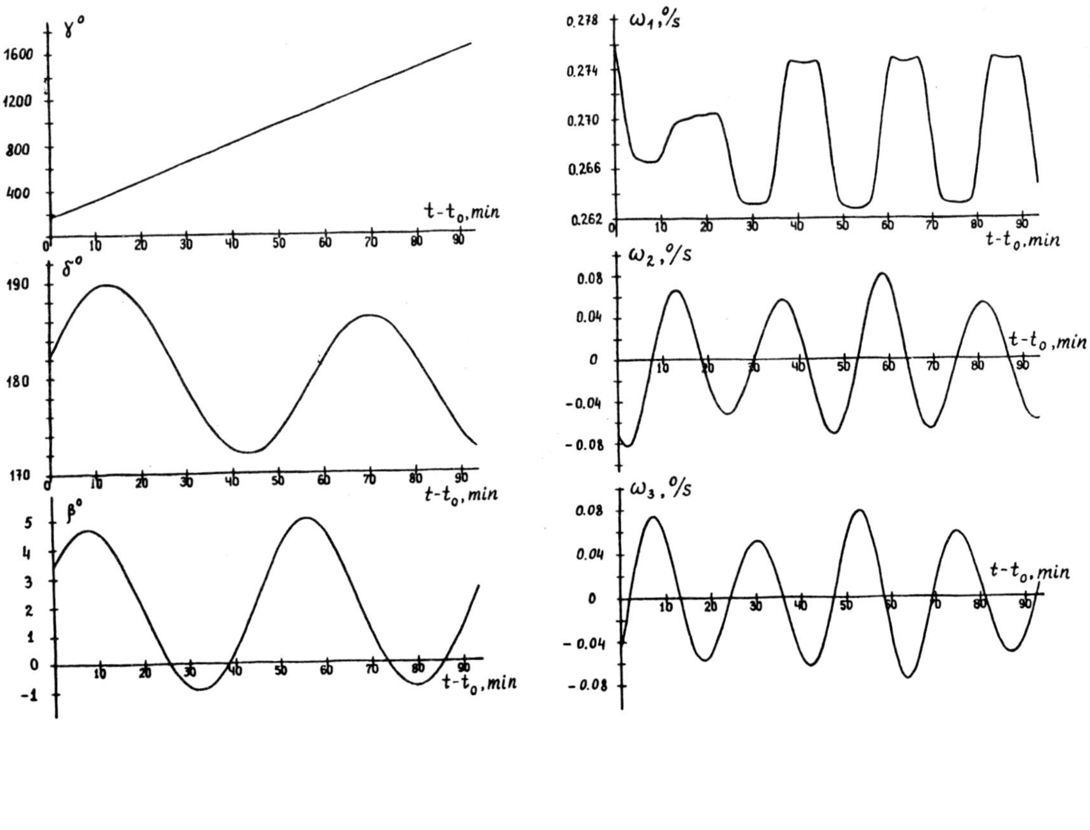

Figure 1: The attitude motion of the Salyut-7 orbital stations on July 29, 1982.

Here $t_0 = 8^h 26^m 26^s$ corresponds to the decree Moscow time, the angles γ, δ and β are determined by the relations

$$\beta = \arcsin a_{21}, \quad tg\gamma = -\frac{a_{23}}{a_{22}}, \quad tg\delta = \frac{a_{11}}{a_{31}},$$

$$sgn\frac{\cos\gamma}{a_{22}} = -sgn\frac{\cos\delta}{a_{31}} = 1.$$

The angles δ and β assign the position of axis Ox_1 in the orbital coordinate system and the angle γ specifies the station rotation about this axis. From the figure and the determined angles follows that the station rotates uniformly about the axis Ox_1, which performs small oscillations with respect to the axis OX_3. The rotation period is about 23 min.; the oscillation periods of Ox_1 in angles δ and β are approximately 55 and 48 min respectively. This motion is a particular case of the gravity gradient stabilization mode.

The results of calculations of the microacceleration $\vec{n} = (n_1, n_2, n_3)_x$ and its components $\vec{n}^z$, $\vec{n}^o$, $\vec{n}^a$ for the obtained motion are shown in Figure 2 and 3.

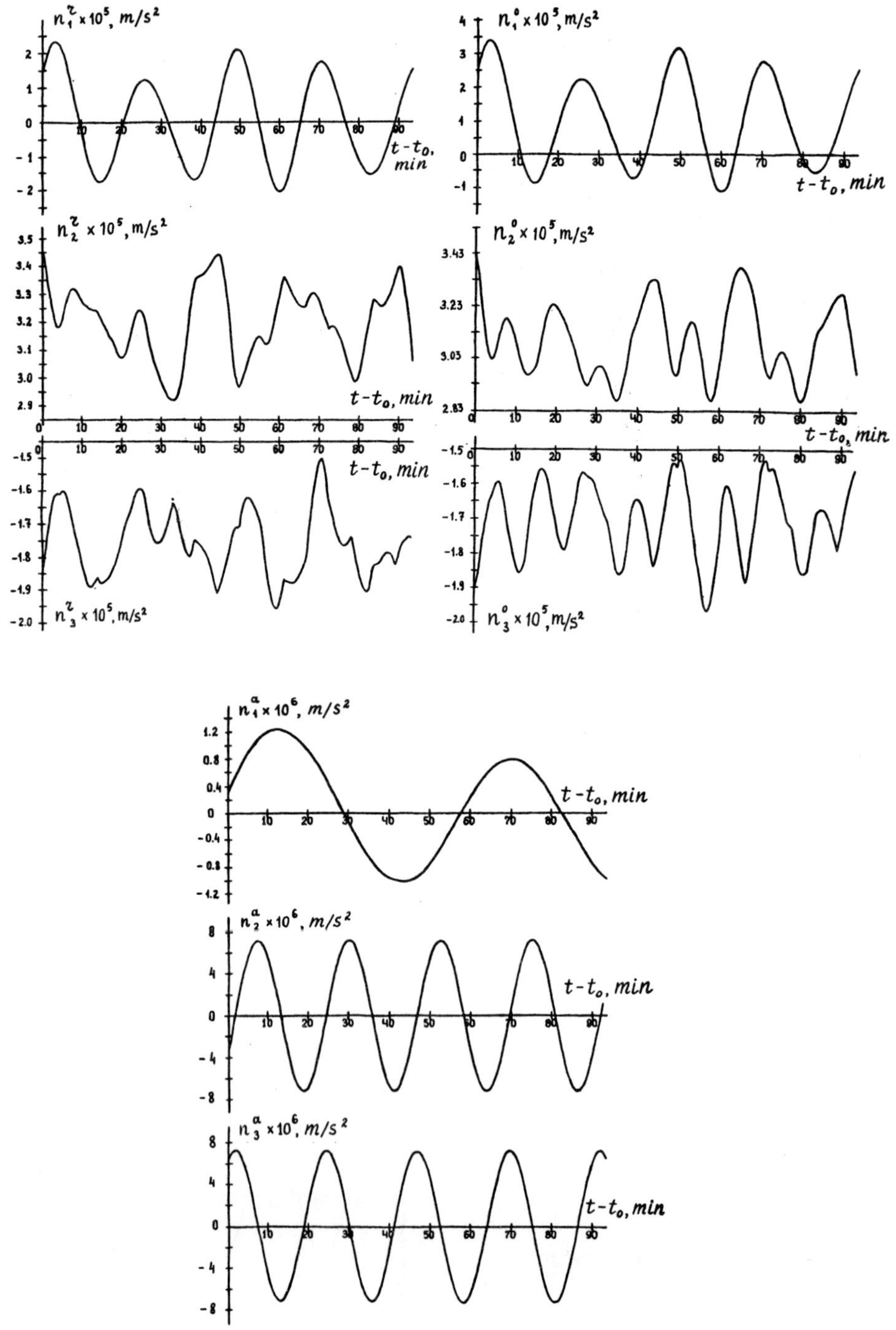

Figure 2: The results of calculations of the microacceleration components $\vec{n}^z, \vec{n}^o, \vec{n}^a$.

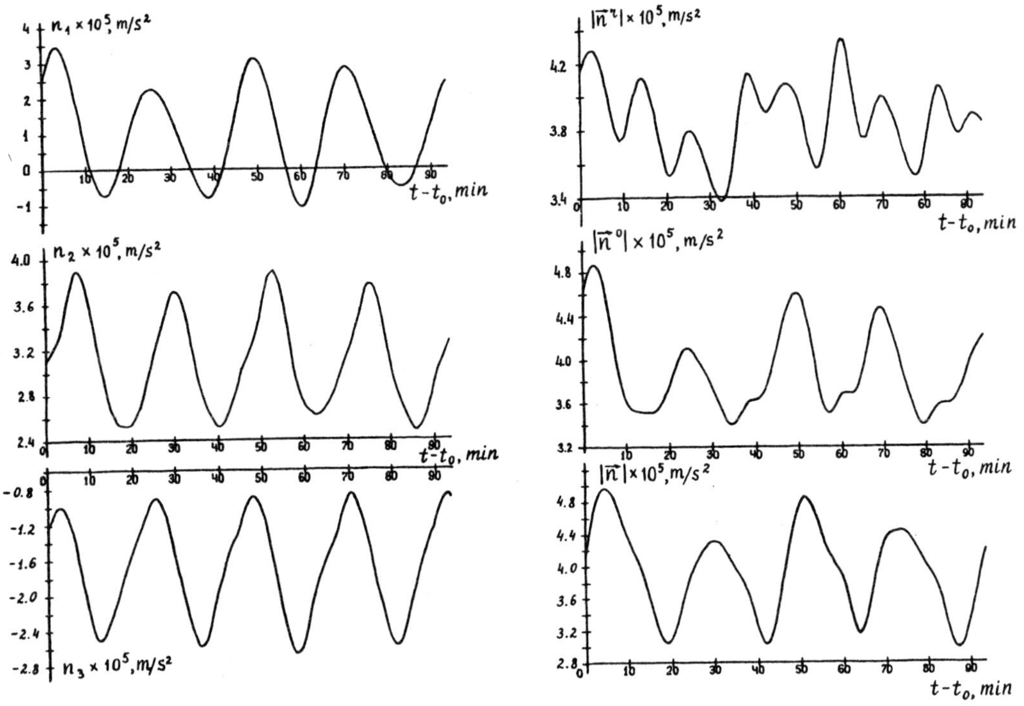

Figure 3: The results of calculations of the full microacceleration $\vec{n}$ and the values of $|\vec{n}^z|$, $|\vec{n}^o|$, $|\vec{n}|$.

These figures are drawn for $\vec{\rho}$ = (2.5m, 1.45m, -0.8m)$_x$(the technological device "Splav") and $K_a = 7.257\cdot 10^{-6}$ m/s^2. As is seen from the figures, the microacceleration is subjected to rather regular oscillations. For example, the components n_1^z, n_1^o, n_2^α, n_3^α, n_1, n_2, n_3 perform oscillations with the period of about 23 min, while the components n_2^o and n_3^o have a period of about 12 (≈0.5·23) min. An analysis of formulas (4) and (5) shows that these oscillations are due to the rotation of the station about the axis Ox_1. It is very illustrative to compare the curves of components n_1^z, n_1^o, n_2^α, n_3^α, n_1, n_2 and n_3 with the curves of $\omega_2(t)$ and $\omega_3(t)$ in Figure 1 (under the gravity gradient stabilization conditions (Ref.2) $\omega_2 \approx \omega_0 \cos\gamma$, $\omega_3 \approx -\omega_0 \sin\gamma$). The component n_1^α performs oscilations with the period 55 min. Such oscillations may be associated with the station oscillations in angle δ (compare Figure 1) because for the motion under considerations $n_1^\alpha \sim \sin\delta$. Many more similar calculations of microaccelerations on the Salyut-7 orbital station are considered in (Refs.4,5). This calculation method is, with slight changes, applied also for the determination of microaccelerations on the "Mir" station during its uncontrolled motion.

3. MICROACCELERATIONS CAUSED BY VIBRATIONS OF THE SATELLITE.

The microacceleration calculated by formulas (1), (2) or (4), (5) represents a background quantity. In reality, it is affected by oscillations and peaks produced by various shocks, body vibrations, etc. We cannot take into account all these effects if the satellite is assumed to be a solid body. Therefore, the calculations in section 2 only give a lower bound on the real microaccelerations. Results of similar calculations were used for planning and interpretations of technological experiments performed on the Salyut-6 and Salyut-7 stations. Such calculations are expedient for assessment of technological satellite projects, to predict a minimum level of microaccelerations which may be provided by the satellite.

In order to have true information about arising microaccelerations one should use accelerometers. The question how accurately the available accelerometers can measure background values of $\vec{n}$ requires special consideration. Here we will study only microaccelerations caused by short-period relative oscillations of the satellite within a simple problem formulation. Frequently the accelerometers allow to measure only this type of microaccelerations (Refs.6,7).

Consider a satellite as a perfect rigid body with a mass inside this body (Figure 4).

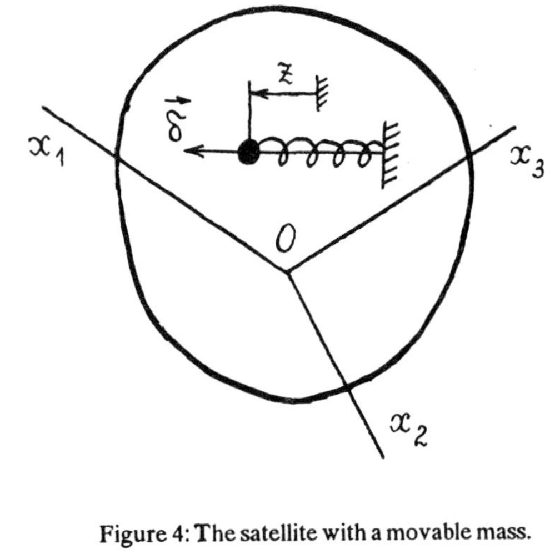

Figure 4: The satellite with a movable mass.

The mass is connected to the body by a spring and has a single degree of freedom with respect to the body, i.e. it may move along a straight line. We introduce the notations: z is the spring deformation, $Ox_1x_2x_3$ is the body coordinate system formed by principal central axes of the satellite's inertia at $z=0$, I_i (i = 1,2,3) are the inertia moments of the satellite with the movable mass relative to the axes Ox_i at $z=0$, m_1 is the body mass, m_2 is the mass inside the satellite, c is the spring constant, $\vec{\omega}=(\omega_1,\omega_2,\omega_3)_x$ is an absolute angular velocity of the body, $\vec{\delta}=(\delta_1,\delta_2,\delta_3)_x$ is a unit vector along the motion trajectory of the mass ($|\vec{\delta}|=1$),
$(a_1+\delta_1 z,\ a_2+\delta_2 z,\ a_3+\delta_3 z)_x$
are coordinates of the mass.

The effect of external torques upon the satellite being ignored, the motion equations have the solution $\omega_1=\omega_2=\omega_3=0$, $z=0$ corresponding to a fixed orientation of the satellite in inertial space. Small oscillations of the satellite near its equilibrium position are described by the equations (Ref.8)

$$I_i\dot{\omega}_i+MD_i\ddot{z}=0 \quad (i=1,2,3),$$

$$M(\ddot{z}+D_1\dot{\omega}_1+D_2\dot{\omega}_2+D_3\dot{\omega}_3)+cz=0, \tag{6}$$

where $D_i=a_j\delta_k-a_k\delta_j$; i,j and k is an even permutation of numbers 1,2 and 3; $M=m_1m_2/(m_1+m_2)$. By writing down the expression for $\dot{\omega}_i$ from the first three equations (6) and substituting it into the last equation we obtain

$$M\left(1-\frac{MD_1^2}{I_1}-\frac{MD_2^2}{I_2}-\frac{MD_3^2}{I_3}\right)\ddot{z}+cz=0. \tag{7}$$

It is the equation of small oscillations of the mass with respect to the body. If the point $P=(b_1,b_2,b_3)_x$ is fixed in the system $Ox_1x_2x_3$ the microacceleration at this point may be calculated by using the linearized formula (compare (1) at $\mu=0$, $\vec{F}=0$)

$$\vec{n}=\vec{OP}\times\dot{\vec{\omega}}+\frac{m_2\ddot{z}\vec{\delta}}{m_1+m_2}=(n_1,n_2,n_3)_x,$$

$$n_i=\ddot{z}\left[\frac{m_2\delta_i}{m_1+m_2}+M\left(\frac{D_kb_j}{I_k}-\frac{D_jb_k}{I_j}\right)\right]. \tag{8}$$

Here the new term $\sim\vec{\delta}\ddot{z}$ takes into account the displacement of the satellite's center of mass relative to the point O at $z\neq 0$.

The relations (6) - (8) are derived for a particular case of the satellite motion, however, they are valid also more generally under certain conditions. Thus, if the frequency in (7) is significantly higher than both the average angular velocity and the reciprocal value of typical time during which external perturbations act, then for such a motion

$$\vec{\omega}(t)=\vec{\omega}^{(0)}(t)+\vec{\omega}^{(1)}(t),\quad z(t)=z^{(0)}(t)+z^{(1)}(t),$$

$$\vec{n}(t)=\vec{n}^{(0)}(t)+\vec{n}^{(1)}(t),$$

where $\vec{\omega}^{(0)}(t)$, $z^{(0)}(t)$, $\vec{n}^{(0)}(t)$ are the slow time functions (for $m_2\ll m_1$ the functions $\vec{\omega}^{(0)}(t)$ and $\vec{n}^{(0)}(t)$ may be calculated by assuming the satellite to be a solid body and $z^{(0)}(t)\equiv 0$), while $\vec{\omega}^{(1)}(t)$, $z^{(1)}(t)$, $\vec{n}^{(1)}(t)$ are determinated by the relations (6) - (8). In typical situations $|\vec{n}^{(0)}|\ll|\vec{n}^{(1)}|$. We give an example. The considered problem may be used as a model of many processes but in the pure form it arises when processing the measurements of a cosmonaut's mass taken by a massmeter (Ref. 8) - the device on which the cosmonaut performs oscillations relative to the manned space station body. Taking the numerical data from (Ref.8), specifically, m_1 = 25000 kg, m_2 = 100 kg, c = 460 kg/s^2 and z_{max} = 0.1 m we obtain $|\vec{n}^{(1)}|\approx 2\cdot 10^{-3}$ m/s^2.

4. REFERENCES.

1. V.A. Sarychev, M.Yu. Belyaev, S.P. Kuzmin, V.V. Sazonov and T.N. Tyan, *Cosmic Research* 26 (1988) 390.

2. V.V. Sazonov, M.Yu. Belyaev, S.P. Kuzmin and T.N. Tyan, *Cosmic Research* 26 (1988) 675.

3. V.A. Sarychev, M.Yu. Belyaev, V.V. Sazonov, T.N. Tyan, *Cosmic Research* 24 (1986) 337.

4. V.A. Sarychev, M.Yu. Belyaev, V.V. Sazonov, T.N. Tyan, *Preprint, Keldysh Inst. Appl. Math. USSR Ac. Sci* (1984) N 100.

5. V.A. Sarychev, M.Yu. Belyaev, S.P. Kuzmin, V.V. Sazonov and T.N. Tyan, *Preprint, Keldysh Inst. Appl. Math. USSR Ac. Sci* (1985) N 129.

6. S.D. Grishin, V.B. Dubovsky, L.V. Leskov, S.S. Obydennicov et al., *Cosmic Research* 20 (1982) 479.

7. H. Hamacher, R. Jilg, U. Merbold, *Acta Astonautica* 16 (1987) 241.

8. V.A. Sarychev, V.V. Sazonov et al., *Cosmic Research* 18 (1980) 536.

INVESTIGATIONS PREPARATORY TO THE WET SATELLITE MODEL EXPERIMENT

J.P.B. Vreeburg
National Aerospace Laboratory NLR
P.B. 90502, NL-1006 BM AMSTERDAM

ABSTRACT

The Wet Satellite Model experiment is to be executed with a small spacecraft that has a tank partially filled with liquid. The spacecraft is instrumented with a distributed arrangement of linear accelerometers in order to determine its angular and translational motions. From these data the forces and moments exerted by the liquid on the rigid part of the spacecraft are to be calculated.

Investigations on different subjects have been conducted in preparation of the experiment. These concern the formulation of the dynamic behaviour of the spacecraft, the stability of various spacecraft configurations, the arrangement of the accelerometers, and some practical difficulties.

For numerical simulation of the dynamics of the spacecraft use is made of a computer programme developed at NLR. Part of this programme is a routine that simulates the behaviour of liquid in microgravity and under capillary forces. A breadboard model of the spacecraft has been operated during aircraft parabolic flight.

INTRODUCTION

Experimental investigations on scaled mechanical models are difficult to perform in terrestrial gravity. The problems originate from the necessity for support forces and can be distinguished in two types. The first type is related to local support; in a liquid this becomes apparent in the hydrostatic pressure gradient which drives convective flow and flattens surfaces. The second type is connected to the support of the weight of the body. In some cases the experiment is performed with constant centre of mass location (e.g. gyrostat behaviour) and the support can be provided almost frictionless by an airbearing. Also, centre of mass excursions may sometimes be compensated by balancing techniques. Nevertheless, if the experiment depends on the determination of small forces, or torques, of unknown orientation, microgravity could be the indicated environment.

Dynamics research with microgravity relevance finds application in space technology. For spacecraft control it is important to have reliable models for the behaviour of liquids and of flexible or jointed spacecraft. Another use of microgravity is the verification and test of complicated manoeuvres (Ref. 1). The investigations at NLR are designed for the development and test of models for behaviour of spacecraft with a large liquid fraction. Primarily are studied liquid configurations with a free surface and motions with non-negligible capillary effects. An important objective is also to develop instrumentation for microgravity dynamics experiments.

In order to generate data for model construction, for validation of simulations and for measurement technique development, NLR is preparing non-supported, i.e. free-floating, experiments for different launch opportunities. One of these, the Wet Satellite Model experiment, is to be put in a suborbital trajectory by sounding rocket and so become a satellite for approximately seven minutes. A sketch with some typical features is given in figure 1.

Precursor investigations are performed during aircraft parabolic flight, for system test and instrumentation development. The applicable theory, various difficulties and possible applications are discussed in the sequel.

EQUATIONS OF MOTION

Consider a rigid body with mass m and inertia tensor $\underline{\underline{I}}$. The inertia tensor is defined with respect to three orthogonal directions and relative to the centre of mass of the body. The three directions define a Cartesian coordinate system with origin A (system A for short) that is fixed in the body.

The location of the centre of mass in system A is $\underline{\rho}$. The inertia tensor $\underline{\underline{I}}$ is symmetrical and so has six components. The mass, centre of mass location coordinates and inertia tensor components constitute the ten inertial parameters of the body with respect to system A.

The inertial parameters are related to the forces and torques on the body and to the angular rotation rate and the linear and angular accelerations, via the equations of Newton and Euler:

$$\underline{F} = m\,\{\underline{a} + \underline{\Omega} \times (\underline{\Omega} \times \underline{\rho}) + \dot{\underline{\Omega}} \times \underline{\rho}\} = \text{total force on the body} \qquad (1)$$

$$\underline{T} = \underline{\underline{I}}\,\dot{\underline{\Omega}} + \underline{\Omega} \times \underline{\underline{I}}\,\underline{\Omega} = \text{total torque on the body} \qquad (2)$$

$$\underline{T} = \underline{R} \times \underline{F} + \underline{T}_T \qquad (3)$$

where $\underline{a}$ = linear acceleration of origin A
$\underline{\Omega}$ = angular rate
$\dot{\underline{\Omega}}$ = angular acceleration
$\underline{R}$ = distance from centre of mass to $\underline{F}$
$\underline{T}_T$ = sum of pure torques on the body.

With reference to equations (1) to (3), the objectives of the investigation can be restated as:

i. develop an experimental technique for the determination of the R.H.S. of equations (1) and (2) as a function of time.

ii. develop models and perform simulations to predict the L.H.S. of equations (1) and (2), and distance $\underline{R}$ in equation (3).

Objective ii. is to be accomplished firstly to assess the effect of liquid on spacecraft but is equally applicable to elastic, articulated or compound bodies.

The force and torque from liquid motions on a rigid tank in the absence of body forces are (Ref. 2):

$$\underline{F}_L = \int_{S_w} \{p\underline{n} - \mu(\underline{n}.\nabla)\underline{u}\}dS + \oint_C (\sigma\underline{\nu} - \mu\underline{u} \times \underline{s})\, dc \qquad (4)$$

$$\underline{T}_L = \int_{S_w} [\underline{r} \times \{p\underline{n} - \mu\,(\underline{n}.\nabla)\,\underline{u}\} + \mu\underline{n} \times \underline{u}]dS + \oint_C \underline{r} \times (\sigma\underline{\nu} - \mu\underline{u} \times \underline{s})dc \qquad (5)$$

where $\underline{r}$ = location of liquid element in system A
$\underline{u}$ = velocity of liquid element in system A
μ = viscosity
p = pressure

p = pressure
σ = surface tension
S_w = wetted tank wall, with outward normal $\underline{n}$
C = liquid-solid contact line, with tangent $\underline{s}$
$\underline{\nu}$ = $-\,\underline{n}_L \times \underline{s}$, where $\underline{n}_L$ is outward normal on the liquid free surface
$\underline{s}$ is related to $\underline{n}_L$ by a right-handed screw

Equations (4) and (5) hold for Newtonian fluid with scalar viscosity μ. The determination of $\underline{u}$ and p is to be accomplished by solution of the Navier-Stokes equations with appropriate field terms, boundary and initial conditions. It is noted that the force at the moving contact line is different from the static force. Equations can be derived for liquids with any type of stress tensor (Ref. 3). It seems reasonable to expect that the experimental determination of $\underline{F}_L$ and $\underline{T}_L$ will then allow to determine the values of parameters in the stress tensor, of interest for rheology.

Equations (1) to (5) represent only one way to model a compound body with a rigid part. If the body is largely liquid, it becomes necessary to follow a different approach (Ref. 4). Here the body is considered rigid ("frozen") at every instance and acted upon by the differences between the momenta in the rigidized and the actual part. The consequent numerical model has much better stability properties (Ref. 5). However, the frozen body does no longer possess constant valued inertial parameters.

The "frozen liquid" formulation is the customary one for the investigation of the stability of motion of rotating spacecraft. A large bibliography exists on this subject (Refs. 6, 7) for both single and dual-spin satellites. The stability of the nominal motion sequence of the Wet Satellite Model cannot be tested conclusively on earth and will be investigated by numerical simulation as exemplified in reference 8.

MEASUREMENT METHOD

The acceleration field in a rigid body is parametrized as:

$$\underline{\sigma}\,(\underline{r}) = \underline{a} + \underline{\Omega} \times (\underline{\Omega} \times \underline{r}) + \dot{\underline{\Omega}} \times \underline{r} \qquad (6)$$

where $\underline{a}$ is the acceleration of the origin, as before, and $\underline{r}$ is the location of the field point in system A.
In consequence of objective i, the vectors $\underline{a}$, $\underline{\Omega}$ and $\dot{\underline{\Omega}}$ are to be measured.
This can be achieved in a number of ways but the selected method is based on an instrument indicated by "ballistometer". It consists of a rigid arrangement of linear accelerometers and is used on earth for motion studies, notably crash impact on anthropomorphic dummies (Ref. 9). Typical arrangements, as discussed in reference 9, are shown in figure 2. Each accelerometer, in a simple model, is subject to nine error sources, viz. four instrument errors, three location and two attitude errors (Ref. 2). The calibration of a complete instrument with, say, nine sensors is a demanding task and will not be discussed here.

Once rotation and acceleration vectors are determined, they can be used to determine inertial properties of the rigid spacecraft. For example consider the motion of a rigid body under no torques, whence

$$\underline{\underline{I}}\,\dot{\underline{\Omega}} + \underline{\Omega} \times \underline{\underline{I}}\,\underline{\Omega} = 0 \qquad (7)$$

In component form, referred to system A, one has

$$\underline{\underline{I}} = \begin{bmatrix} P & -W & -V \\ -W & Q & -U \\ -V & -U & R \end{bmatrix}, \quad \underline{\Omega} = \begin{bmatrix} \Omega_1 \\ \Omega_2 \\ \Omega_3 \end{bmatrix}, \quad \dot{\underline{\Omega}} = \begin{bmatrix} \dot\Omega_1 \\ \dot\Omega_2 \\ \dot\Omega_3 \end{bmatrix}$$

Substitution in equation (7) and reordering with respect to vectors

$\begin{bmatrix} P \\ Q \\ R \end{bmatrix}$ and $\begin{bmatrix} U \\ V \\ W \end{bmatrix}$ yields:

$$\left(\dot\Omega_1\dot\Omega_2\dot\Omega_3 + \frac{1}{2}\,\Omega_1\Omega_2\Omega_3\,\frac{d\Omega^2}{dt}\right)\begin{bmatrix} P \\ Q \\ R \end{bmatrix} = \left\{\begin{bmatrix} 0 & -\Omega_3\dot\Omega_3 & \Omega_2\dot\Omega_2 \\ \Omega_3\dot\Omega_3 & 0 & -\Omega_1\dot\Omega_1 \\ -\Omega_2\dot\Omega_2 & \Omega_1\dot\Omega_1 & 0 \end{bmatrix}\left(\begin{bmatrix} \dot\Omega_1 & 0 & 0 \\ 0 & \dot\Omega_2 & 0 \\ 0 & 0 & \dot\Omega_3 \end{bmatrix}\begin{bmatrix} 0 & \Omega_3 & \Omega_2 \\ \Omega_3 & 0 & \Omega_1 \\ \Omega_2 & \Omega_1 & 0 \end{bmatrix} + \right.\right.$$

$$\left.- \begin{bmatrix} \Omega_1 & 0 & 0 \\ 0 & \Omega_2 & 0 \\ 0 & 0 & \Omega_3 \end{bmatrix}\begin{bmatrix} 0 & \dot\Omega_3 & \dot\Omega_2 \\ \dot\Omega_3 & 0 & \dot\Omega_1 \\ \dot\Omega_2 & \dot\Omega_1 & 0 \end{bmatrix}\right) + \frac{1}{2}\,\frac{d\Omega^2}{dt}\begin{bmatrix} 0 & \Omega_3^2 & \Omega_2^2 \\ \Omega_3^2 & 0 & \Omega_1^2 \\ \Omega_2^2 & \Omega_1^2 & 0 \end{bmatrix}\begin{bmatrix} \Omega_1 & 0 & 0 \\ 0 & \Omega_2 & 0 \\ 0 & 0 & \Omega_3 \end{bmatrix} +$$

$$\left. + \begin{bmatrix} 0 & \dot\Omega_3^2 & \dot\Omega_2^2 \\ \dot\Omega_3^2 & 0 & \dot\Omega_1^2 \\ \dot\Omega_2^2 & \dot\Omega_1^2 & 0 \end{bmatrix}\begin{bmatrix} \dot\Omega_1 & 0 & 0 \\ 0 & \dot\Omega_2 & 0 \\ 0 & 0 & \dot\Omega_3 \end{bmatrix}\right\}\begin{bmatrix} U \\ V \\ W \end{bmatrix} \qquad (8)$$

where $\Omega^2 = \Omega_1^2 + \Omega_2^2 + \Omega_3^2$

If numerical data of $\underline{\Omega}$ and $\dot{\underline{\Omega}}$ are given at two instances, of nonsteady rotation, a set of homogeneous equations can be constructed and solved for U, V, W, up to a constant. Resubstitution provides values of P, Q and R. For nonzero, known torques at two instances a similar calculation can be performed. Additional linear acceleration data $\underline{a}$ allow determination of $\underline{\rho}$ from equation (1), provided force $\underline{F}$ is known. The mass m can be determined for non-zero forces but also when a known configuration change of the rigid body is effected (Ref. 10).

Conversely, if the inertial properties are known, the data $\underline{a}$, $\underline{\Omega}$ and $\dot{\underline{\Omega}}$ can be used for calibration of the ballistometer, since their consistency can be verified with the (exact or numerical) solutions of equations (7) and (8).

EXPERIMENTS AND TESTS

Preliminary investigations with a model spacecraft have been performed during aircraft parabolic flight. The test article is a straight annular cylinder with overall external size of 30 cm and an annular width of 1.5 cm (Fig. 3).
The mass of the article varied between 5 and 10 kg, depending on the instrumentation inside and the liquid loading. Tests were performed in free-float mode.

One flight result was that the air in the aircraft cabin induced accelerations O (10^{-4} g) on the test object. This is due to displacement of the air by the hull about the free tank and from drafts in the cabin. Air velocities may easily reach 0.5-1 ms^{-1}. Note that an average value for seismic noise on earth is often quoted as 10^{-6} g.

The nominal operations sequence starts with motionless release of the test article. Fixed to the tank structure is a rotating momentum wheel. After release, a brake engages and so induces transfer of the angular momentum in the wheel to the whole article. In this way the interaction with the liquid in the tank is initiated.

It was found out soon that the nominal motionless initial state of the liquid could not be realized (Ref. 8), whence experiments with liquid are useful mainly for observation of liquid distribution features in the (transparent) tank (Fig. 4). During the early flights recording of the motions of the tank and of the liquid inside was performed by video and cine cameras.

At a later stage the tank was fitted with three linear accelerometers that were placed on the three axes of a Cartesian coordinate system with origin at the centre of mass of the empty test article. The accelerometers have their sensitive axis aligned with the coordinate direction and are at a distance $d \approx 17$ cm from the centre of mass. It is straightforward to show that for no forces on the (empty) tank the linear acceleration $\underline{a}$ of the origin is zero and only centrifugal acceleration is measured. Thus, the ideal output of the sensor on axis i equals

$$s_i = d\ (\Omega_i^2 - \Omega^2) \tag{9}$$

The recorded output of all three accelerometers for a motion under no forces of the empty test article is presented in figure 5. This record was selected from many more because it covers a long period of free float and shows strong motion. The angular momentum is not known a priori since the motion was induced by hand.

The values of $\underline{\Omega}$ and $\dot{\underline{\Omega}}$ were determined from the data in the graph at the beginning and end of the free float. Accelerometer output was corrected for bias as determined from output for no motion. The angular data were used to determine U, V and W as the eigenvector corresponding to the zero eigenvalue of the coefficient matrix that resulted from the analysis which is outlined following equation (8). It was found that the smallest eigenvalue of the coefficient matrix was almost an order of magnitude smaller than the other eigenvalues but not (nearly) zero. Hence, sensor errors are still influential. It was decided to first conduct additional, similar tests with more extensive data collection before the gravity of the errors will be investigated in greater detail. These tests, in parabolic flight, are planned for the end of June 1991.

REFERENCES

1. Vreeburg, J.P.B. STOF Complementary Analysis Study. Part I: Diagnostic method and applications. NLR CR 91201 L, Jun. 1991.
2. Vreeburg, J.P.B. Free motion of an unsupported tank that is partially filled with liquid. IUTAM Symp. Microgr. Fl. Mech. Bremen, FRG, Sep. 1991.
3. Dahler, J.S., Scriven, L.E. Theory of structured continua. I. General consideration of angular momentum and polarisation. Proc. Roy. Soc. London A275 (1963) 504-527.
4. Gantmakher, F.R., Levin, L.M. The flight of uncontrolled rockets. Pergamon, 1964.
5. Vogels, M.E.S. A numerical method for the simulation of liquid-solid body dynamics. NLR MP 87030, presented at IMACS '88.
6. McIntyre, J.E., Miyagi, M.I. A general Stability Principle for Spinning Flexible Bodies with Application to the Propellant Migration - Wobble Amplification Effect. ESA SP-117, May 1976, 159-175.
7. Magnus, K. Kreiselmechanik. ZAMM 58 (1978) T56-T65.
8. Vreeburg, J.P.B., Dam, R.F. van den. Effects of low Bond number liquid motions on spacecraft attitude. AGARD-CP-489 (1990) 30.
9. Mital, N.K., King, A.I. Computation of Rigid-Body Rotation in Three-Dimensional Space from Body-Fixed Linear Acceleration Measurements. J. Appl. Mech. 46 (1979) 925-930.
10. Bergmann, E., Dzielski, J. Spacecraft mass property identification with torque-generating control. J. Guid. Contr. Dyn. 13 (1990) 99-103.

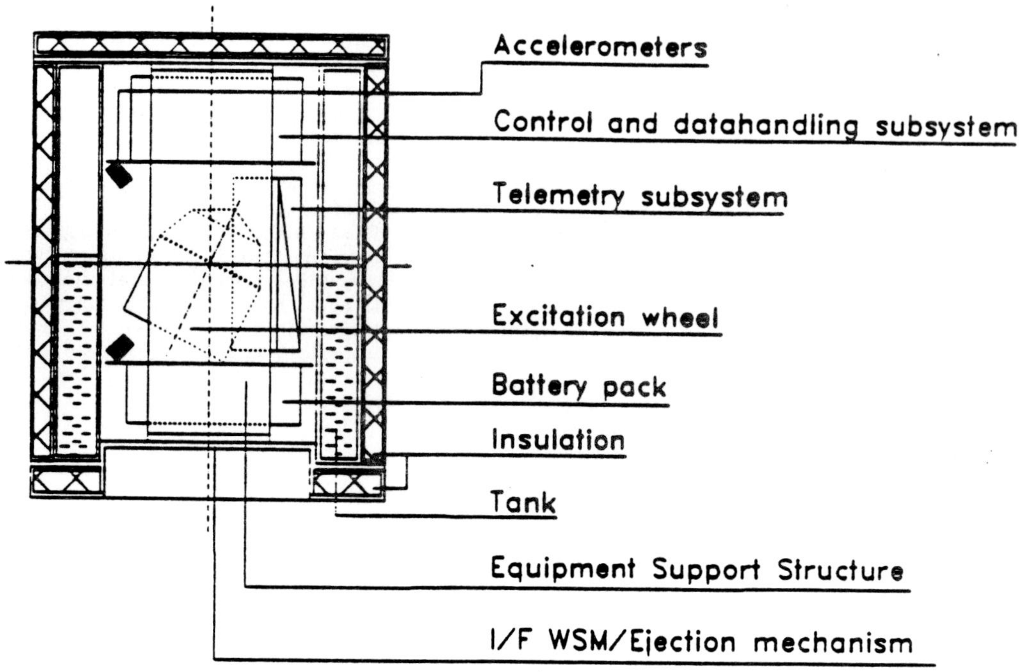

Figure 1: Tentative layout of WSM.

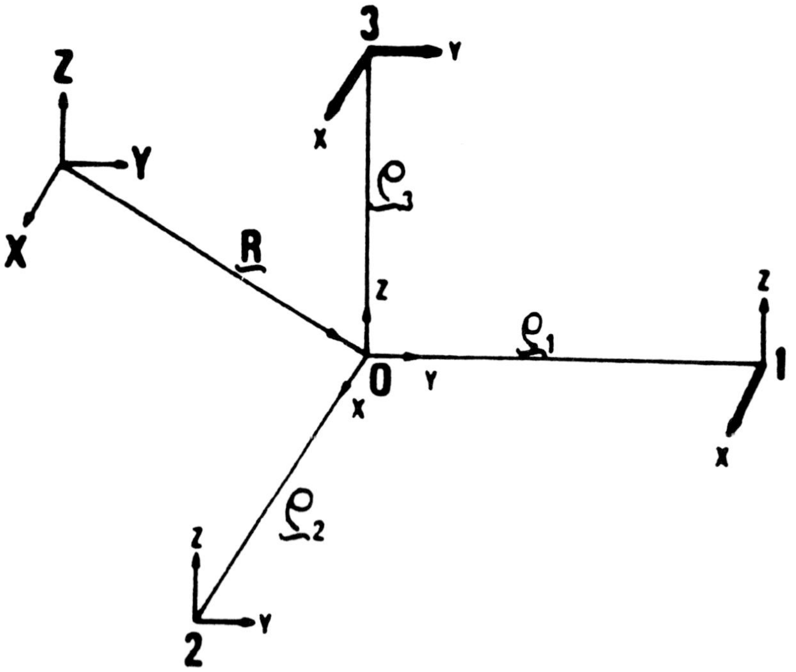

Figure 2: Six and nine-accelerometer configurations.

Figure 3: Parabolic flight test article.

Figure 4: During weightlessness in the KC-135 aircraft.

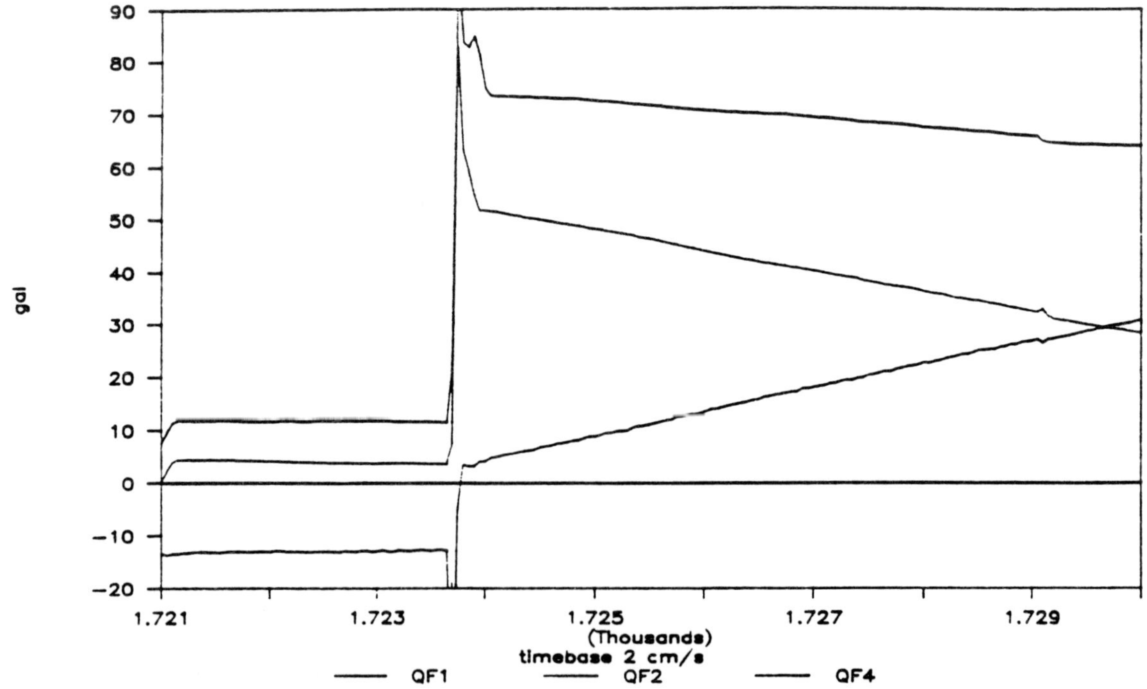

Figure 5: Accelerometer output following impulsive spin-up of the empty WSM breadboard.

STABILITY OF FLOWS WITH INTERACTION BETWEEN DIFFERENT CONVECTIVE MECHANISMS IN MICROGRAVITY

D.S.Pavlovsky
Institute for Problems in Mechanics USSR Academy of Science
prospect Vernadskogo 101, 117526 Moscow, USSR

ABSTRACT

The paper concerns two problems of stability of flows in thin layers. The first one deals with thermal convection in a horizontal layer with longitudinal temperature gradient. The boundaries of the layer are adiabatic, upper surface is free and surface tension is a function of the temperature. This problem is modelling hydrodynamic processes in unidirectional crystallization and it is necessary to take account of either gravitational and Marangoni convections. The second problem deals with forced flows in a vertical layer with internal heat sources. It is simulating free flow electrophoresis processes. The infinite layer approximation is used and critical values of parameters are found by analysis of spectra of linearized Navier-Stokes equations. Spatial and temporal characteristics of secondary flows are found.

Keywords: thermal convection, thermocapillary convection, linear stability analysis.

1. INTRODUCTION

In numerous technological processes a liquid phase is present. Various forces act on the liquid and their intensity and interaction determines the quality of a produced material. It is therefore of utmost importance to have a clear understanding of the regimes of convection under various conditions. In the space technologies gravitation convection is sufficiently week and loss of flow homogeneity can be dependent on the initial bifurcations. Beginning of the unsteady flow regimes in thin layers can be study by the linear stability theory.

Combined thermal and thermocapillary convection occurs in numerous situations, where a temperature gradient is applied along liquid-gas interface. For example, such kind of convection is observed during crystal growth from the melt in horizontal open boat (Bridgman configuration). It is well known that when a critical temperature gradient is reached an unsteady flow regime results. The temperature oscillations threshold for a flow in a long cavity is connected with loss of stability of the one-cell circulation. It may be obtained with help of linear stability analysis of the 1D steady flow for infinite layer. This analysis has been done by Hart[1], Laure and Roux[2], Kuo and Korpela[3] for thermal convection and Smith and Davis[4] for thermocapillary convection. The study of combined effects for conducting boundary conditions for the temperature (thermocapillary forces don't depend on the flow field) has been done by Myznikov[5]. In the present paper we describe some results of linear analysis of problem with insulating horizontal boundaries. In this case it must take into account influence of the temperature fluctuations on the free surface.

The second problem that we consider here is stability analysis of flow under coupling action two different mechanisms: forced convection induced by pressure gradient applied along the layer and thermal convection caused by internal heat sources. This problem arises in free-flow electrophoresis modelling. In this case the stability of the flow under each of mechanisms mentioned above has been studied too. It is either problem of stability of the Poiseuille flow[6] or one for vertical layer with internal heat sources[7].

2. MATHEMATICAL MODEL

Both problems mentioned above may be described in the similar way. Let us consider convection in the infinite layer with height H. The character of hydrodynamic processes depends on different dimensional parameters of liquid: the kinematic viscosity (ν), the dynamic viscosity (μ), the thermal conductivity ($\varkappa$), the thermal diffusivity (a), the coefficient of volumetric expansion (β), the surface tension (σ), and the characteristic of gravitation forces - the gravity acceleration (g). The governing system is given by the 3D Navier-Stokes and energy equations with Boussinesq approximation, written in Cartesian coordinates (origin is in the middle of the layer and z-axis is across the layer). These equations in non-dimensional form in the region $-1<z<1$ are (we use corresponding subscripts for partial derivatives)

$$V_t+Gr(V\cdot\nabla)V = -\nabla p+\Delta V+Te$$

$$divV=0 \qquad (1)$$

$$T_t+Gr(V\cdot\nabla T) = \Delta T/Pr+S$$

Here we use a common notation, e is the unit vector against gravity and S denotes the heat sources. The Prandtl number is defined as $Pr=\nu/a$.

Let

$$Q = \int_{-1}^{1} udz$$

denotes a flow rate along the layer (along x-coordinate).

The boundary conditions for the system (1) are determined by the considered problem.

If the system (1) have state 1D solution (main flow) we can study its stability. For this aim we use the usual theory of small perturbations. Let the disturbances depend on x and y coordinates and time as

$$exp(i(kx+hy)+\tau t)$$

so we obtain the linear boundary value problem for eigenvalue τ. The real part of τ is an amplification rate and the imaginary part is an oscillation frequency.

For fixed parameters, the critical Grashof number is calculated as

$$Gr_c = \inf Gr_0(k,h),$$

where Gr_0 is the value of Gr such that largest eigenvalue of the linearized system is purely imaginary.

We solved this eigenvalue problem by a method of K.I.Babenko. An application of this method to convective stability problems may be found in the preprint[8]. The unknown functions were approximated by interpolating polynomials with the interpolation nodal points placed in the zeros of the Chebyshev polynomials of first kind. The order of approximation is the number of the nodes and it determines the degree of the approximating polynomials. Fundamental interpolating polynomials are chosen in such a way, that they satisfy demanded boundary conditions. After substitution of the approximating polynomials in the equations, we consider the studied system in the nodes of interpolation, which are also the collocation points. Thus an algebraic eigenvalue problem is obtained.

3. CONVECTION IN A HORIZONTAL TEMPERATURE GRADIENT

Let us consider the convection in the horizontal infinite layer (the vector e is along positive z-direction) induced by the longitudinal gradient of temperature A. As we simulate the flow in an enclosure we must set

$$Q = 0.$$

When $H/2$, $H^2/4\nu$, $2g\beta AH^3/\nu$ and $8AH$ are taken as scaling factors for length, time, velocity and temperature respectively governing system is (1) with

$$S = 0$$

(internal heat sources are absent) and Grashof number is defined as

$$Gr = g\beta AH^4/\nu^2$$

The boundaries of the layer ($z=-1$, $z=1$) are insulated:

$$T_z = 0,$$

the bottom plane ($z=-1$) is rigid:

$$u = v = w = 0,$$

and the upper boundary ($z=1$) is assumed to be free and planar

$$w = 0,\ u_z = -4MT_x,\ v_z = -4MT_y$$

where

$$M = -\sigma_t/(g\beta\mu H^2).$$

The strength of buoyancy forces is characterized by the Grashof number Gr, while the number M characterizes the buoyancy - thermocapillary relation. This number M is equal to Marangoni number divided by Rayleigh number.

The stability analysis has been made for a wide range of the numbers Pr and M. Three families of eigenva-

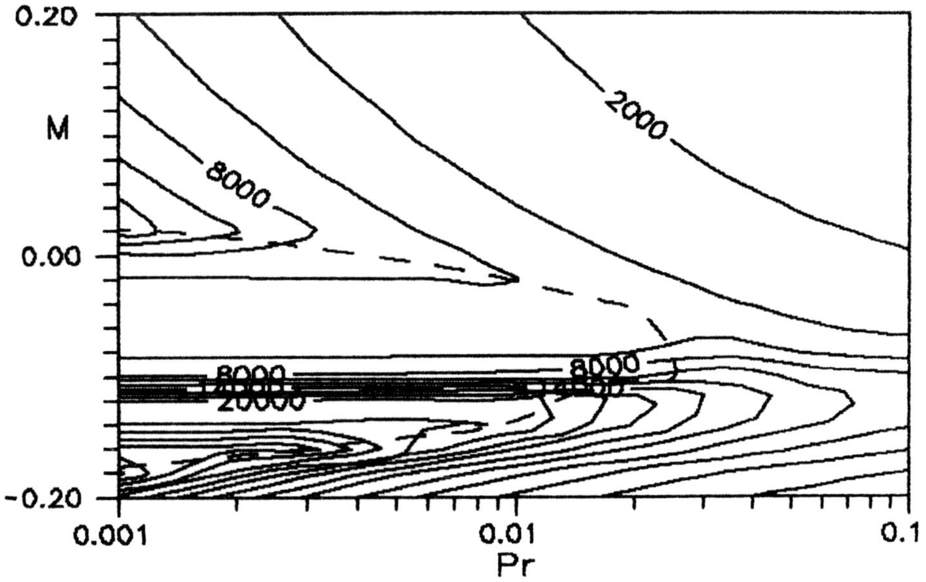

Figure 1. Isolines of critical Grashof number for low Prandtl numbers

lues have been obtained. One corresponds to solutions with the wavevector (k,h), with $k \neq 0$, $h \neq 0$; these perturbations are three - dimensional (3D mode). The other family corresponds to solutions of the form $(k,0)$ which are two - dimensional (2D mode). The third one corresponds to solutions of the form $(0,h)$ which are named spiral (Spr mode).

Our results for low *Pr* numbers are shown in Figure 1. A dashed line is a boundary between 2D and 3D modes. Inside this curve the 2D mode dominates the 3D mode. Thus for small *Pr* and *M* numbers oscillatory 2D perturbations are most dangerous and when *Gr* becomes more than critical one the temperature oscillations appear. The wavelength of the neutral perturbations of 2D mode depends only on the number *M* and is varied from 4 to 12.

For almost all other values of determining parameters oscillatory 3D mode is most dangerous. Its wavelength depends on Prandtl number, equals 180 for $Pr=0.001$ and is decreased with increase of *Pr*.

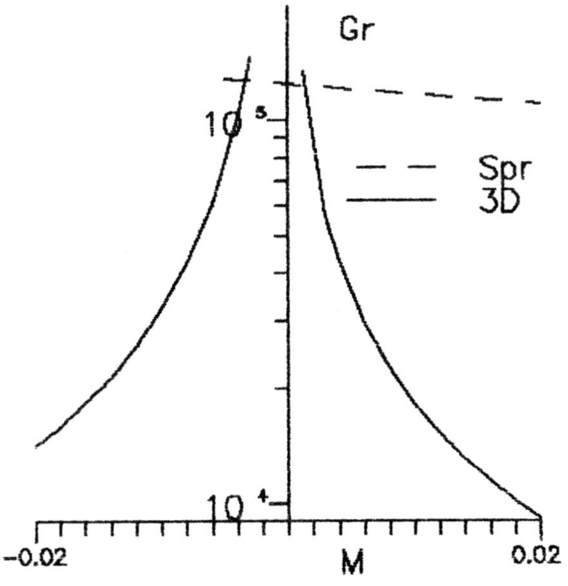

Figure 2. Neutral curves for Pr=0.56

For $M=0$ and $Pr=0.4$ unstable 3D mode disappears and for $0.4<Pr<1.2$ monotonic spiral mode determines the flow instability. The neutral curve for $Pr=0.56$ are shown in Figure 2. For $Pr>1.2$ and vanishingly small *M* main flow is always stable. The isolines of the critical Grashof number

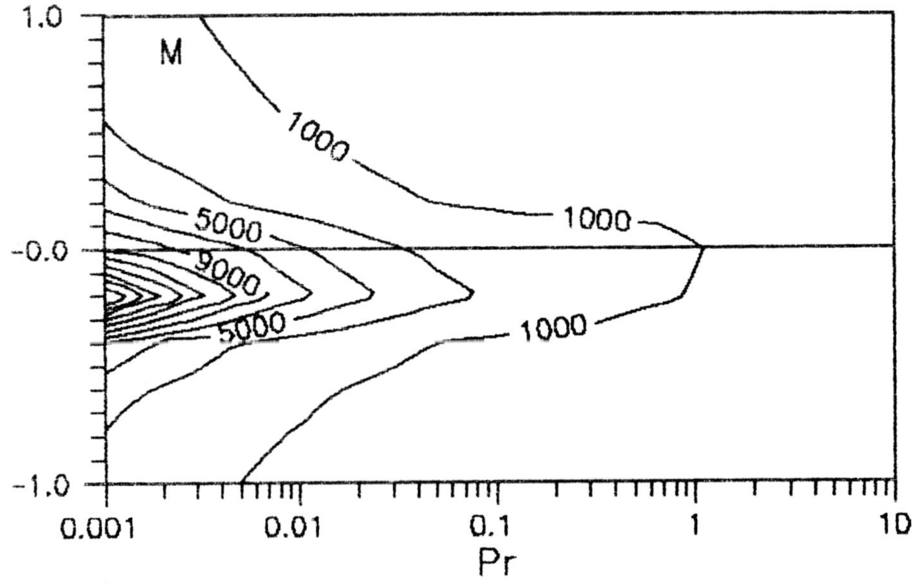

Figure 3. Isolines of critical Grashof number for 3D mode

for 3D mode in the wide range of the parameters are shown in Figure 3.

4. FORCED CONVECTION WITH INTERNAL HEAT SOURCES

Let us consider the convection in the vertical infinite layer (the vector e is along positive *x*-direction) induced by the longitudinal gradient of pressure. We assume that liquid is heated by the internal heat sources with intensity *I*.

When $H/2$, $H^2/4\nu$, $g\beta IH^4/(32\nu æ)$ and $IH^2/(8æ)$ are taken as scaling factors for length, time, velocity and temperature respectively governing system is (1) with

$$S = 2/Pr$$

and Grashof number is defined as

$$Gr = g\beta IH^5/(64\nu^2 æ)$$

The boundaries of the layer ($z=-1$, $z=1$) are isothermal:

$$T = 0,$$

and rigid:

$$u = v = w = 0.$$

The strength of buoyancy forces is characterized by the Grashof number *Gr*, while the number *Q* characterizes the relative intensity of the forced flow. This number *Q* is equal to Reynolds number *Re* divided by Grashof number, where

$$Re = u_a H/\nu$$

and u_a is average velocity along the layer.

There is heating of liquid inside the layer. Hence there is upward flow near a middle of the layer. For this problem the main flow can be described as follows:

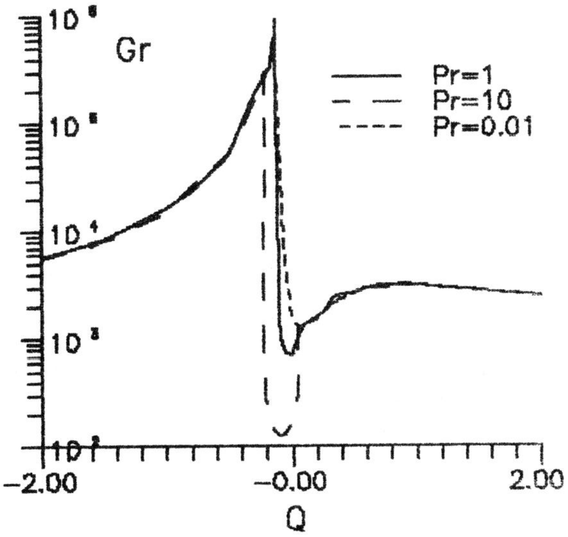

Figure 4. Neutral curves for different Prandtl numbers

$$u_0=(5z^4-6z^2+1)/60+0.75Q(1-z^2)$$

$$T_0 = 1-z^2.$$

There is the inflexion point of the basic velocity profile for

$$-2/15<Q<8/15.$$

The stability analysis has been considered for several values of the Prandtl number

$$Pr = 0.1,\ 1,\ 10.$$

For this problem 2D mode is more dangerous then 3D one. Hence only two dimensional perturbations can be considered.

Our results are shown in Figures 4,5. The calculations show that there are two different modes depending on number *Q*. In the region where the inflexion point exists the most dangerous mode is defined by the convection induced by the internal heat sources. This mode is continuation of the one which has been found in cited above[7].

For

$$|Q| \to \infty$$

this problem becomes like the stability problem for the Poiseuille flow and relation

$$Re = Q/Gr_c = 7679$$

is hold. Moreover the critical wavenumber tends to $k=1.02$

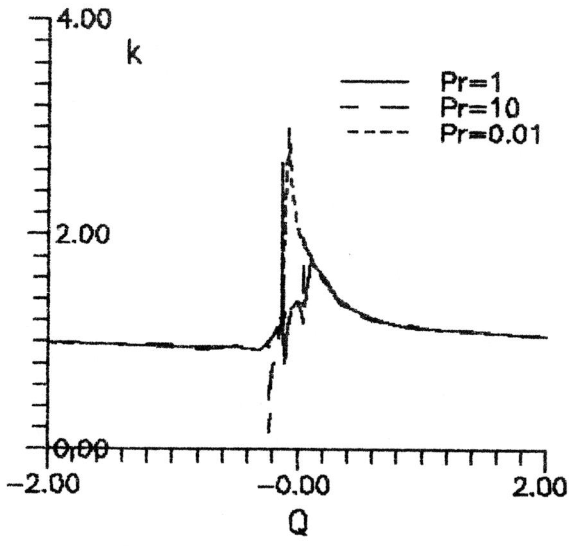

Figure 5. Critical wavenumber for different Prandtl numbers

5. REFERENCES

1. J.E.Hart, *J.Fluid Mech.*, 132, 271-281, (1983)

2. P.Laure and B.Roux, *CRAS, Ser.II*, 305, 1137-1143, (1987)

3. H.P.Kuo and S.A.Korpela, *Phys. Fluids*, 31, N1, 33-42, (1988)

4. M.K.Smith and S.H.Davis, *J.Fluid Mech.*, 132, 119-144 (1983)

5. V.M.Myznikov, in: *Convective Flows*, (Perm, USSR 1981), p.76-82, in Russian

6. C.C.Lin, *Quart. Appl. Math.*, 3, N2, 117-142, (1945)

7. G.Z.Gershuni, E.M.Zhukhovitsky and A.A.Yakimov, *J.Appl. Math. Mech.*, 37, N3, 546-568, (1973), in Russian

8. D.S.Pavlovsky, preprint of the Institute of Problems in Mechanics N416 (1989)

INFLUENCE OF ELECTRIC FIELDS ON FLUID UNDER CONDITIONS OF WEIGHTLESSNESS.

M.K.Bologa, I.A.Kojukhari, N.S.Alexeeva

Institute of Applied Physics of the RM Academy of Sci., 5,Grosul str., 277028 Kishinev, Republic Moldova.

Keywords: electrohydrodynamic, experiments in the space, heat transfer in electrical field.

Recently, studies have been widely performed in electrohydrodynamics (EHD), investigating the fluids and gases movement in an electric field /1.2/. The report is concerned with the potential programmes of space experiments, the applied aspects essence of which is described below, while their realization importance doesn't need any reasoning.

The general expression for the force, causing the EHD flows is

$$\vec{f}=\rho\vec{E}-\frac{\varepsilon_o}{2}E^2\nabla\varepsilon+\frac{\varepsilon_o}{2}\nabla\left[E^2\gamma\left(\frac{\partial\varepsilon}{\partial\gamma}\right)_T\right], \quad (1)$$

where

$$\rho = \sigma\,\vec{E}\,\nabla\,\tau, \qquad \tau = \varepsilon_o\varepsilon/\sigma \quad (2)$$

from which one may see that the field influence on the medium is a consequence of its electrophysical parameters ε and σ inhomogeneities.

The ε - and σ - inhomogeneities of the medium, as a rule, manifest themselves, also as a density inhomogeneity (heterogeneous systems, nonuniformly heated media). That is why the EHD-flows, seen in electrical and gravitational fields, are deformed by the latter, especially at low intensities of the electric fields, which complicate the interpretation and generalization of the results. This calls for the necessity to study the EHD-flows in the "pure state", under weightlessness conditions. The EHD-effects study in weightlessness has also practical aims namely the thermocarrier transport and turbulization conditions determination, the heat exchange acceleration, models desingning and EHD-evaporating - condensing systems computing.

It is essential to study the electrothermal convection stability and structure in order to determine the beginning in the liquid under the influence of its appearance. The well - known results, confirming the threshold existence /1/, are theoretical works, while the experimental data cannot be considered correct, as the beginning of the convection is observed at Coulomb forces values, commensurable with the gravitation ones /1/.

It is desirable to study the deformation, dynamics, coagulation, destruction of the gaseous inclusions (bubbles) the electric field, depending on the carrier phase electrophysical parameters (viscosity, electrical conductance, dielectric permeability) and field parameters (direction, gradient, frequency).

In the studies of the medium electrization in the electric field the probe method is used, which is known to bring in certain distortions in the studied object initial state. At the same time the bubbles in the liquid, situated in an electric field, are rather sensitive sensors of the field intensity gradient, provided the gravitational influences are excluded. Information on the electric field parameters and the free charges may be obtained by solving the bubble movement equation

$$0.5\ \varepsilon_0(\varepsilon-1)V\ \nabla E^2 = 6\pi\eta v a \qquad (3)$$

where a is the bubble radius, η - the carrier phase viscosity, V - the bubble volume, E - the electric field strength, v - the experimentally determined speed of the bubble.

A number of works is known /1/ on the charged or polarized fluid drops stability. These studies have a great practical importance for the atmospheric electricity phenomena description, electron-ion technology, etc., but they are mainly theoretical, while the experimental investigations in this field are impeded by the gravitational force influence. Weightlessness may provide favourable conditions for such experiments. Thereby, considering the possibility of using big size drops, one may greatly enlarge the range of factors, essential for the given phenomena.

Some EHD-effects, observed in the heterogeneous systems, are explained by the electric field influence on the surface tension σ_n, however, in this case the obtained data are not quite correct because of the gravitational forces influence. The bubbles, freely flying in the gas in the external field may also serve as observations objects for the E - influence on σ_n besides the above-mentioned drops in the gas, bubbles in the fluid This phenomenon must show some peculiar distortion and dynamics of the film.

The electric field heat exchange experimental data are generalized by a dependence of the type

$$Nu = C\ \Pi_1^m\ \Pi_{\nabla\tau}^n\ \Pi_2^p\ P_r^q \qquad (4)$$

where Π_1 is the quantity, determining the characteristic forces rations (Coulomb forces to viscosity forces in the case of convective heat exchange or to surface tension forces at evaporation and condensation), $\Pi_{\nabla\tau}$ - is the quantity, describing the ingomogeneity of the medium for the electrical relaxation time τ, Π_2 - is the quantity, describing the characteristic times (time of mechanical relaxation l^2/τ, electrical relaxation, the time of film renewal for the evaporation or condensation case). Therein the constant C and the degree indices q, m, n, p values are determined for the mentioned similarity numbers because of the gravitational forces influences. Particularly, for small Π_1 - values, some heat emission decrease is observed in the presence of the gravitational field; it may be explained by the negative influence of the latter on the electroconvection which has already begun. For the widening of these numbers range, in which the constants C , m , n, p and q experimental determination is possible and for their correction it is desirable to study the heat exchange in weightlessness conditions.

Tne investigations of the two-phase systems EHD-flows accompanied by drops and bubbles interaction with each other and with the field, by fluid dispersion, by films forming on the walls of the cell and on electrodes, were carried out at high electric field intensities, in order to exclude to a certain degree the influence of the gravitational field. Moreover, in some cases the transport processes of the condensate film in the EHD-evaporation systems at the gravitational forces presence cannot be achieved in principle.

The conduction of the above experiments with single objects (drops and gaseous inclusions), with their assemblies in different factors complex influences conditions and the thereby achieved data in weightlessness will be useful for the regularities study, models designing and the EHD-systems computational methods working out for the flying machines temperature regimes provision.

The experiments are desirable to be and temperature. The experimental reperformed on an universal device, consisting of a closed contour with electrodes of different geometrical forms to provide a certain field configuration, the electrodes being connected with controlled high voltage source. The device also must consist of local heaters and refrigerators, dampers and it must be partly filled with fluid and gas.

The necessary objects, the medium concentration and the regimes in the given parts of the contour are to be subjected to the complex actions of the electric field, inertial forces, corresponding actions of tne dampers and temperature. The experimental results may be recorded mainly by well-known methods, including temperature measurements, filming and photographing.

REFERENCES

1. M.K.Bologa, F.P.Grosu and I.A.Kojukhari, Electroconvection and Heat Transfer (Shtiintza, Kishinev, 1977) (in Russian).
2. I.B.Rubashov, Yu.S.Bortnikov, Electrogasodynamics (Atomizdat, Moscov, 1971) (in Russian).
3. L.D.Landau, E.M.Lifshitz, Electrodynamics of Continuous Media (GIITL, Moscow, 1957) (in Russian).

CONVECTIVE HEAT AND MASS TRANSFER IN THE PRODUCTION OF MATERIALS IN MICROGRAVITY.

A. I. Feonychev[1], G. A. Dolgikh[1], I. S. Kalachinskaya[2]

[1] *Research Institute of Applied Mechanics and Electrodynamics of Moscow Aviation Institute, Volokolamskoe sh. 4, 125871,*

[2] *Moscow State University, Leninskie gory, 117234*

ABSTRACT

The thermocapillary and vibrational convection and its influence on heat and mass transfer processes in production of materials are investigated numerically for space flight conditions. Effects of macro- and microsegregation caused by convection are considered. Transition to the regime of oscillatory thermocapillary convection is analyzed. Possibility of regulation of heat and mass transfer processes in application to two- layer liquid systems and vibration of crystal is shown.

Keywords: *Microgravity, Hydromechanics, Convection, Heat and Mass Transfer, Vibration, Marangoni Effect, Crystallization, Macro- and Microsegregation of Impurity.*

1. INTRODUCTION

Absence of total zero-gravity state on board of spacecraft and availability of insufficiently investigated ungravitational forms of convection influencing the heat and mass transfer processes in materials production define interest of theoretical hydromechanics to space technology. An insignificance of buoyancy force and sedimentation effects allows to study fine physical processes occurrent near the phase boundary and their effect on quality of obtained materials.

In this work some new effects of thermocapillary convection and crystal vibration, reducing to the infringement of diffusion processes of heat and mass transfer, are investigated. A study of these phenomena permit to pass to their reasonable use of crystal vibration and two-layer fluid systems for regulation of the obtained materials quality.

2. THERMOCAPILLARY CONVECTION AND MACROSEGREGATION OF IMPURITY

In series of experiments conducted in zero-gravity conditions by *Bridgman* method the availability of radial inhomogeneity of impurity distribution was connected with action of thermocapillary convection under partial separation of the melt from the ampoule's wall. It has been shown earlier (Refs. 1-2) how the modification of separation zone length influences the character of impurity distribution. These data were received under boundary condition of I kind for impurity concentration on phase boundary (C_S=*const*). Mass balance on phase boundary corresponds to experimental conditions more exactly. It can be expressed in dimensionless form as follows:

$$\left(\frac{\partial C}{\partial z} \right)_S = -Sc\,(1-k_0)\;Re_c\;C_S \qquad (1)$$

Here, z is the distance from the phase boundary along the axis of cylindrical ampoule, k_0 is the equilibrium coefficient of impurity distribution. $Re_c=V_c\,\delta/\nu$. V_c is the movement velocity of crystallization front. r_o is the radius of the ampoule (linear scale), ν is the kinematic viscousity. $Sc=\nu/D$ is the *Smidt* number. D is the diffusion coefficient of impurity. $C=c/c_0$. index "s" refers to the phase boundary. index "o" to process beginning.

On figure 1 we show the concentration distribution of impurity in the melt on the phase boundary. The parameters of calculations were the ampoule length, h , separation zone length near the crystallization front, h_1, the *Marangoni* number $Ma_T=-(\partial\sigma/\partial T)\ \Delta T\ r_o(\rho\nu a)^{-1}$. the *Prandtl* number, $Pr = \nu/a$. Here σ is the surface tension, ρ density, a is the thermal diffusivity, ΔT typical temperature drop.

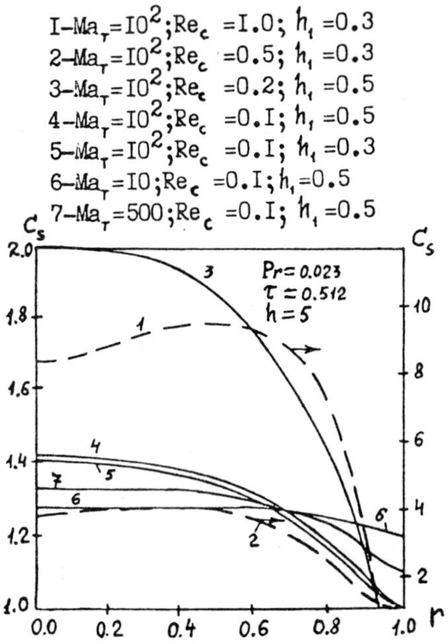

Figure 1: Distribution of impurity concentration in liquid.

Calculations conditions (Pr= 0.023, Sc= 10 and k_0= 0.087) correspond to the case of germanium melt with gallium impurity. Maximum of impurity concentration is near the ampoule axis and it is displaced from it under increase of crystallization velocity.

The radial inhomogeneity of impurity distribution at the phase boundary ($\Delta C_s = C_s^{max} - C_s^{min}$) as a function of *Marangoni* number is given in the figure 2 by solid line for the three values of Re_c.

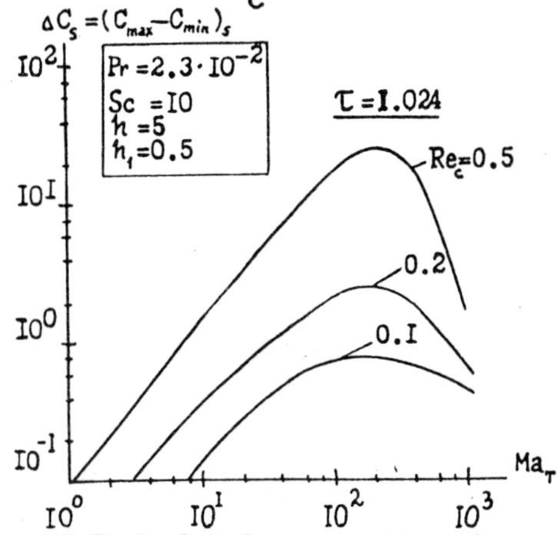

Figure 2: Radial inhomogeneity of impurity.

It is interesting, that the curve under $Ma_T < 10$ is very gently sloping. It is an evidence of the fact that the influence of thermocapillary convection on impurity distribution is difficult to exclude, as the molecular impurity transfer taking place under condition Pr_d = Re Sc < 1 is achieved in view of typical *Shmidt* number (Sc= 5-15) at convective flows velocities corresponding to Re $\leq 10^{-1}$ or $Ma_T \leq 1$.

3. STABILITY AND OSCILLATION REGIMES OF THERMOCAPILLARY CONVECTION IN ONE-AND TWO-LAYER SYSTEMS

Two-layer liquid systems are suggested for elimination of unfavorable effect of thermocapillary convection.

Numerical investigation of the thermocapillary convection at raised values Ma_T has been made for the model of floating-zone process. Three types of boundary thermal condition on the free cylinder surface were discussed: a) the exponential distribution of temperature with maximum in the middle of free surface (z=L/2); b) the exponential distribution of heat flux with maximum in the middle of free surface (z =L/2); c) the constant heat exchange on the free surface (*Biot* number $Bi=\alpha\ r_o/\lambda$ = const, where α is the heat exchange coefficient, λ is the thermal conductivity coefficient, r_o is the radius of cylinder). The temperature of crystallization T_o was defined on the top and bottom of cylinder in the first two cases. For the third case, which models the bottom half of the actual zone melting, the upper surface is maintained at the temperature T_1 and the lower surface T_0. The two-dimensional *Navier-Stokes* equations are solved by finite-different method (Ref.5). The nonlinear terms are approximated by scheme of third order of accuracy with upward finite differences.

The transition to the oscillatory regime of convection was not found in the first case of boundary condition in the range of *Marangoni* number up to Ma= 10^5 by Pr= 10^{-2} and L/r_0=4. In two other cases the oscillatory regime was found at $Ma_T \sim 5 \cdot 10^4$, that is in agreement with experimental data (Ref.6). The variation of face surface temperature in the middle point is presented in fig. 3.

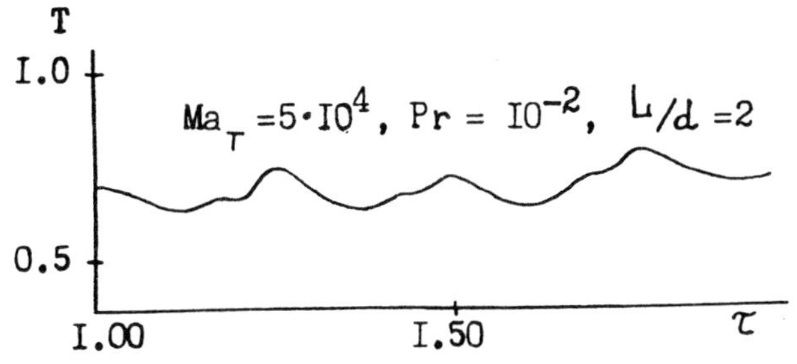

Figure 3: Oscillation of temperature at super-critical thermocapillary convection.

The mean period of oscillations $\Delta\tau = \Delta t\ \nu / r_0 \cong 0.25$ is connected with periodical changes of flow structures. The oscillation, when coupled with the solidification process in a real floating-zone, is believed to be responsible for the existence of undesirable striations in the material.

The state of equilibrium stability in two flat layers (thickness h_1 and h_2) of unmixing liquids with thermocapillary convection at the interfacial surface was studied under linearity assumption. At the solid surfaces, limiting the system of infinite layers, the temperature was maintained at T_0 and T_1 value ($T_1 > T_0$).

On figure 4 we show the diagrams of stability. In a plane β-δ the curves corresponding to maximum wave numbers for the

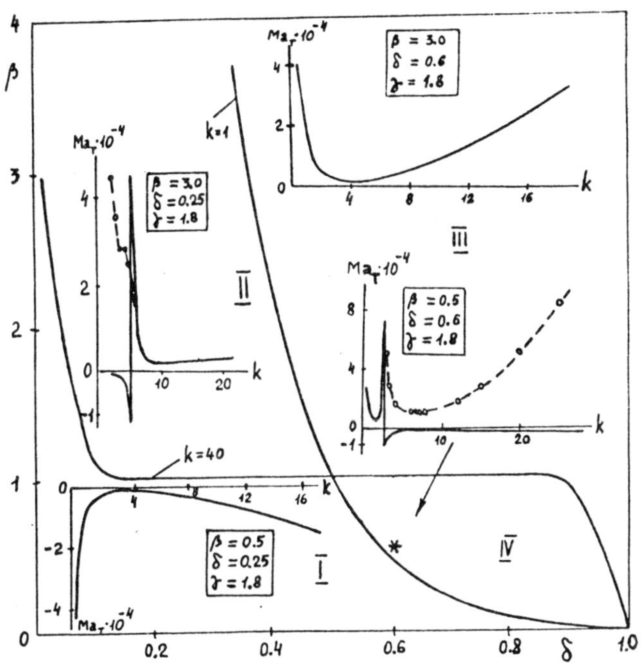

Figure 4: Diagrams of stability.

disturbances (k=1 and k=40) form four zones (I-IV). The neutral curves of stability $Ma_T(k)$ were plotted at the same fixed parameters δ,β,γ. Here, $\delta = h_2/(h_1+h_2)$, $\beta = \alpha_2/\alpha_1$, $\gamma = \rho_2\nu_2/\rho_1\nu_1$, index 1 refers to the thick layer (the basic fluid) and 2 to the thin layer (additional fluid). In regions II and IV there exists an oscillatory instability (broken curves-oscillatory neutral curves). The typical frequency of temperature oscillations for point marked by "*" in figure 4 is $\tau_1 = t_1 \nu/(h_1+h_2)^2 = 0.01$

Further, we consider two layer system, in which the main floating-zone at its free surface is bounded by a thin layer of viscous liquid. We deal with vertical section of the half of floating-zone due to the problem symmetry. The boundaries of such region are the symmetry axis of the main liquid with admixture, the free surface of the cover which is considered to be thermo-insulated and two other boundaries simulate the frontier of crystallization and symmetry plane of liquid zone with temperature kept at T_0 and T_1 ($T_1 > T_0$). The admixture concentration at the frontier of crystallization is given by (1). The computer simulation was carried out for *Marangoni* number from 0 to 10^2 at each interfacial surface. It was found out that this system has a wide range of oscillatory regimes for thermocapillary convection. Some of them are self-exited when the oscillatory amplitude is maximum (figures 5-6).

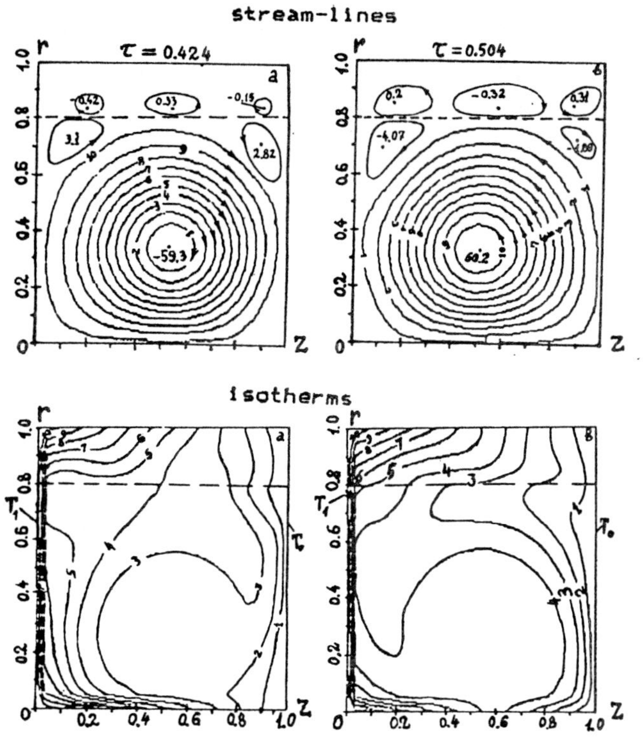

Figure 5: Flow structure and temperature

Numerical research was carried out under following parameters $\rho_2/\rho_1=0.27$, $\nu_2/\nu_1=8.34$, $\lambda_2/\lambda_1=0.26$, $Pr_2=7$, $Pr_1=0.02$, $Sc=7$, *Marangoni* number at the liquid interfacial surface Ma_{T_1} equals 10^2, and at the free external surface of the thin liquid layer $Ma_{T_2}=10$. Stream lines and isotherms describe in figure 5 different phases of oscillation. They are marked as "a" and "b" at the time-axis in figure 6. The figure 6a reveals the maximum positive and negative values of stream function in the basic liquid. The change of maximum difference of admixture concentration at the boundary of crystallization in the basic liquid for velocity $Re_c=0.1$ is shown in figure 6b for one-layer system (broken line) and two-layer system (solid line). It should be pointed out that the process of admixture oscillation for two layer systems with period of the basic oscillation mode $\tau_*=t\nu_1/(h+h_2)^2\approx0.16$ is rather complex. This period corresponds to the stream function oscillation in figure 6a. The amplitude and period of oscillation can be controlled by changing of natural properties of the thin layer and its thickness for different basic materials. In results described above we have studied the Ge(Ga)-B_2O_3 composition.

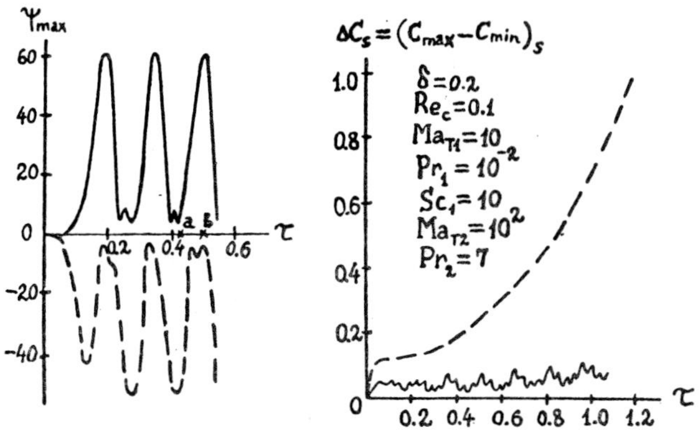

Figure 6: Variation of Ψ_{max} and radial segregation of impurity.

4. VIBRATION EFFECTS

Vibration is a feature of dynamic conditions on board of spacecraft. However, vibrational frequencies and amplitude can not cause visible fluid flows, if this effect is considered as variational buoyancy force. In (Ref. 3) another vibrational way of influence based on fluids flow near vibrating wall is suggested. It is the so called second *Stokes* problem (Ref. 7).

In simulations we have employed two variants of boundary motion. In the first case equations in moving coordinates were used and at solid vibrating planes the vortex, determined by (Refs. 7-8), in accordance with the direction of boundary motion, was defined. In the second case equations in rest coordinate system were used and velocity of the walls was defined. At the free liquid surface the vortex and stream function determined by the model of vibrational surface deformation were set.

The vibration of cylindrical crystal with radius r_1 is considered in the direction of z-axis. The deformation of free liquid surface near vibrating crystal ($r_1<r<r_0$) leads to appearance of normal speed component on this surface. Three types of free liquid surface geometry are considered: 1) on the lateral surface of cylinder at the length of h_1, ($r_1=r_0$); 2) on the lateral surface of cylinder at the length of h_1 and on the face between the crystal ($r=r_1$) and the ampoule wall ($r=r_0$); 3) on the face between r_1 and r_0 ($r_1\le r\le r_0$) only.

Since $z_0 \ll r_1$, one may consider a problem by the invariable configuration of calculated field giving a velocity $V=-z_0\omega \sin \omega t$ on the crystallization boundary and writing down on the free fluid boundary the harmonic expression for the normal component of velocity with the condition of constant liquid volume too. The thermal conditions on the boundaries are usual for the crystallizati n by the *Bridgman*, *Czochralski* methods and the zone melting.

In figure 7 on the left we show the time-averaged flow structure influenced by the thermocapillary effect on the face between the crystal ($r=r_1$) and the wall of ampoule ($r=r_0$) also. On the right we show the flow,

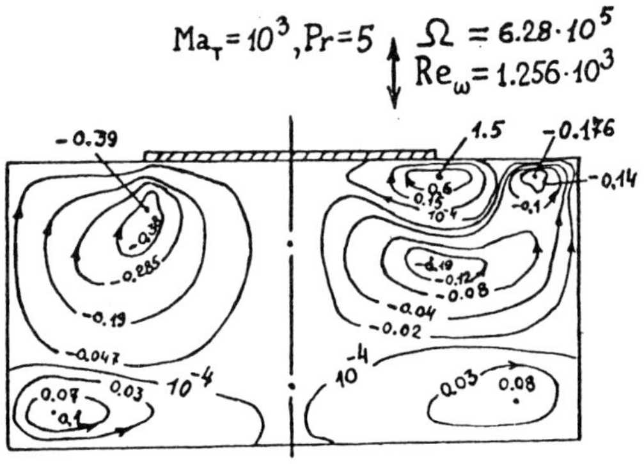

Figure 7: Stream-lines with (on the right) and without (on the left) vibration of crystal.

structure in view of the vibration effect by the scheme (3) at the dimensionless frequency $\Omega=\omega\, r_0^2/\nu$ = $6.28\cdot10^5$ and the amplitude of vibration z_0 corresponding to *Reynolds* number $Re_\omega=\Omega z_0/r_0= 1.256\cdot10^3$. The vibrational convection prevail over thermocapillary motion near the crystal.

The time-averaged flow leads to macrosegregation of impurity that can be self regulated by means of Ω and Re_ω. There is a momentary flow too. The harmonic oscillations of all parameters of liquid are typical for this flow.

On the crystallization boundary a thermal balance takes place

$$-\lambda_f\left(\frac{\partial T_f}{\partial z}\right)_s + q_s = -\lambda_{cr}\left(\frac{\partial T_{cr}}{\partial z}\right)_s \qquad (2)$$

Here, $q_c = r_{ph} V_c \rho_f$, r_{ph} is the phase transition heat. Periodic variations of the first component on the left in (2) due to the crystal vibration cause the corresponding changes of q_c up to the reverse sign (the "remelting" effect) at conservation of the expression on the right. It leads to the change of crystallization velocity, impurity flow and impurity concentration on the crystallization boundary according to (1).

On figure 8a we show the momentary values of the heat flow on the oscillational crystal under $r=r_0=1$ and $r=r_1=0.4$ for the three different regimes of vibration. *Prandtl* number, Pr = 5, corresponds to the rare-earth scandium garnets and thermal conditions on the boundary are usual for the *Bridgman* method.

On figure 8b we show the momentary values of heat flow in liquid on the oscillating crystal (germanium with gallium) for the three schemes of free fluid boundary described above. It should be noted that the differences of these schemes influence the amplitude of flow variation weakly.

In figure 9, we have plotted the impurity concentration in liquid on the crystallization boundary corresponding to two natural frequencies of vibration for the hull of *Salyut-Soyuz-Progress* space complex and the amplitude of vibration z_0 of 30 $m\mu$. The other parameters of calculations of calculations for figure 8b and 9

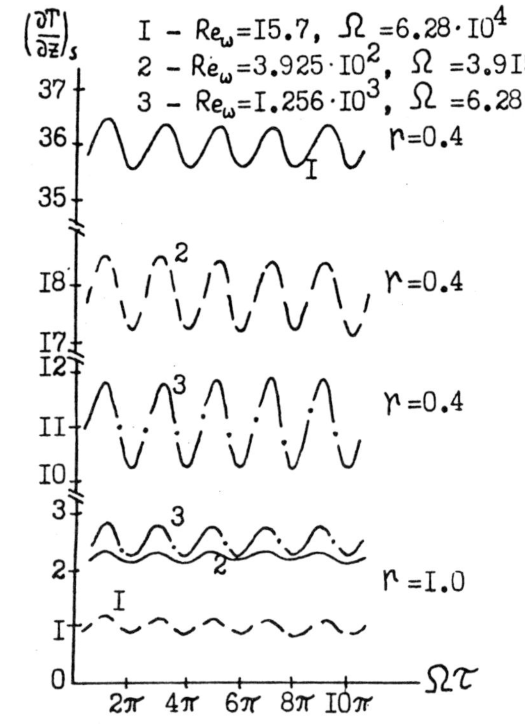

a) Pr = 5

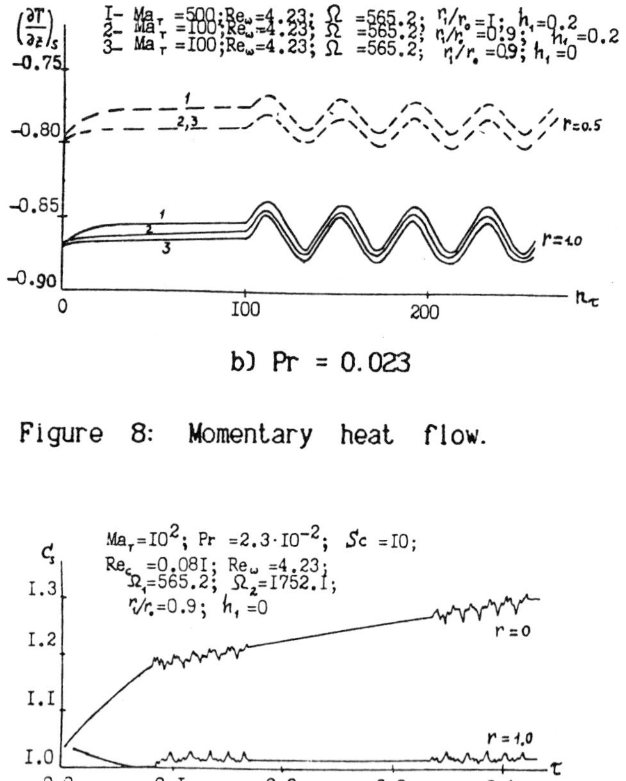

b) Pr = 0.023

Figure 8: Momentary heat flow.

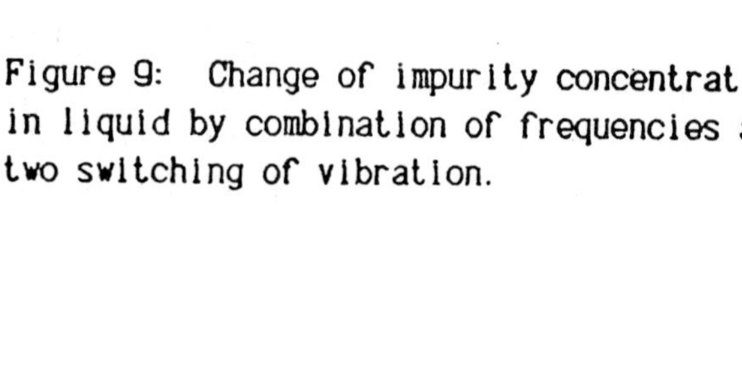

Figure 9: Change of impurity concentration in liquid by combination of frequencies and two switching of vibration.

correspond to the melt of germanium with gallium (Pr = 0.023, Sc =10, k_0=0.087) and the characteristics of *Kristall* furnace. The conducted calculations witness that crystal vibration can be the reason of appearance of impurity microsegregation during crystallization on board of spacecraft.

5. CONCLUSION.

The compulsory vibration of crystal and the use of two- layer liquid systems can be efficient methods of the control of heat and mass processes on the crystallization boundary. The unfavorable effect of impurity microsegregation may be excluded by means of choosing of the thickness, the physical properties of additional fluid in two-layer systems and vibration frequency of crystal. The problem of stability of thin liquid film may be solved by the selection of materials and its thickness too.

6. REFERENCES

1. A. I. Feonychev, *Fluid Dynamics* 16 (1981) 155.
2. G. A. Dolgikh and A. I. Feonychev, in: *The Problems of Mechan. and Heat Exchange in Rocket Engineering*, O. M. Belozerkovski Ed., Mashinostroenie, Moscow (1982) 224 (in Russian).
3. S. V. Ermakov and A. I. Feonychev, in: *Hydromech. and Heat/Mass Exchange in Weightlessness*, Novosibirsk (1988) 20 (in Russian).
4. G. A. Dolgikh, A. I. Feonychev, N. R. Storozhev and E. V. Zharikov, in: *Intern. Conf. of Solidific. and Microgravity*, Miscolc, Hungary. Abstracts (1991) 17.
5. T. Kavamura, H. Takami and K. Kuvahara, in: *Proc. Intern. Conf. of Hydrodyn.* Tokyo (1986) 291.
6. S. Ostrach and Y. Kamotani, in: *Proc. AIAA/IKI Microgr. Sci. Symp.*, Moscow, USSR, May (1991) 25.
7. G. G. Stokes, *Trans. Cambr. Phil. Soc.* 9 (11) (1851) 8.
8. M. B. Glauert, *J. Fluid Mech.* 1(1956) 97

ON THE VIBRATIONAL CONVECTION IN A PLANE LAYER WITH A LONGITUDINAL TEMPERATURE GRADIENT

Rudolf.V.Birikh
Perm State Pedagogical Institute, Perm, 614600, USSR

ABSTRACT

The exact solution of the vibrational convection equations for the arbitrarily oriented plane fluid layer with a longitudinal component of the temperature gradient is given. A structure of the convective flow are determined by two dimensionless measures of the problem. A plane parallel convective flow is shown to exist in the absence of a static field at inclined vibrations. If gravity is present, a stable averaged mechanic equilibrium state may always be reached by a suitable choice of the vibrations direction and amplitude.

Keywords: Natural convection, Vibrational convection, Longitudinal temperature gradient, Exact solution.

RESULTS AND DISCUSSION

Technological experiments with non-uniformly heated fluid under low gravity conditions stimulated interest in different mechanisms for stimulation of convection. The conditions for the onset of convection by high frequency vibrations in a completely filled cavity are discussed by Gershuni and Zhukhovitsky (Refs.1,2). Also the closed vibroconvective flow in a plane layer with a longitudinal component of temperature gradient has been considered. It exists only if there is a temperature difference between the layer boundaries. The occurence of the vibroconvective flow under vibrations at some arbitrary angle has not yet been addressed. In another work (Ref.3) they have studied the influence of a vibrational field on advective motion in the horizontal layer (Ref.4). The discussed cases of longitudinal and cross vibrations have no limit of purely vibrational convection.

In the present paper the exact solution of the equations of vibrational convection in the arbitrarily oriented plane fluid layer with a longitudinal component of the temperature gradient is given. A plane parallel convective flow is shown to exist in the absence of a static field at inclined vibrations. On the other hand gravitational plane parallel convective flow may always be suppressed by a suitable choice of the direction and amplitude of vibration.

1. A plane layer of viscous, incompressible fluid subject to harmonic high-frequency oscillations with frequency $\boldsymbol{\Omega}$ and amplitude b along the vector n is investigated. There is the static gravity field $-\mathbf{k}g$, and a constant longitudinal component A of the temperature gradient is produced in the fluid. Equations for the averaged velocity field $\mathbf{V}$, temperature T, pressure p, and complementary function $\mathbf{w}$ (the amplitude of the oscillating component of the velocity) may be written in the form

$$\frac{\partial \mathbf{V}}{\partial t}+\frac{1}{P}(\mathbf{V}\nabla)\mathbf{V}=-\nabla p-\Delta\mathbf{V}+\mathbf{k}RT+R_v(\mathbf{w}\nabla)(T\mathbf{n}-\mathbf{w}) \tag{1}$$

$$P\frac{\partial T}{\partial t}+(\mathbf{V}\nabla)T=\Delta T \tag{2}$$

$$\text{rot}\,\mathbf{w}=\text{rot}(T\mathbf{n}),\ \text{div}\,\mathbf{V}=0,\ \text{div}\,\mathbf{w}=0 \qquad (3)$$

$$R=\frac{g\beta Ah^4}{\nu\chi},\quad R_V=\frac{(b\Omega\beta Ah^2)^2}{2\nu\chi},\quad P=\frac{\nu}{\chi}$$

The following units of measurement are used here: distance - half width of the layer h, time - h^2/ν, velocity - χ/h, temperature and w - Ah, pressure - $\rho\nu\chi/h^2$, (ν,χ,β - coefficients of kinematic viscosity, thermal diffusivity and thermal expansion). The system (1)-(3) contains three dimensionless parameters: the Rayleigh number R, the vibrational analog of the Rayleigh number R_V, the Prandtl number P - and two unit vectors $\mathbf{n}$ and $\mathbf{l}$ which give the directions of the external forces. One more direction in space is given by the constant component of the temperature gradient - vector $\mathbf{k}$.

We discuss the case when vectors $\mathbf{n}$, $\mathbf{l}$ and $\mathbf{k}$ are situated in one plane. This plane may be taken as the plane xz of the Cartesian coordinate system. The axis x is directed along the normal to the layer, the axis z is along the vector $\mathbf{k}$. The origin of coordinates is chosen in the middle of the layer. Equations of motion (1)-(3) allow a solution in the form of plane-parallel flow:

$$\mathbf{V} = \mathbf{k}V(x),\ T= -z+\tau(x),\ \mathbf{w}=\mathbf{k}w(x)$$

$$\mathbf{n} = \mathbf{i}n_x + \mathbf{k}n_z,\ \mathbf{l} = \mathbf{i}l_x + \mathbf{k}l_z \qquad (4)$$

where $\mathbf{i}$ and $\mathbf{k}$ are unit vectors of coordinate axes x and z. Substitution of (4) in the system of equations (1)-(3) results in the system of ordinary differential equations:

$$V''' -(n_z^2R_V- l_zR)\tau' =n_xn_zR_V- l_xR,$$
$$\tau'' +V=0,\qquad w' =n_x+n_z\tau' \qquad (5)$$

where the prime indicates differentiation with respect to x. Dimensionless measures of the problem together with components of unit vectors $\mathbf{n}$ and $\mathbf{l}$ are combined in two parameters

$$q=n_xn_zR_V- l_xR,\qquad Q=n_z^2R_V- l_zR$$

which determine the structure of the convective flow.

For homogeneous boundary conditions on V and τ, parameter q is the only element that makes the problem nonhomogeneous and consequently determines the amplitude of the convective flow. The value $q=0$ corresponds to the condition of mechanic equilibrium. One possibility is that the static and vibration fields satisfy the condition

$$R_V=(l_x/n_xn_z)R \qquad (6)$$

In the absence of a static field, equilibrium is obtained either at cross vibrations ($n_z=0$) or at longitudinal vibrations ($n_x=0$). At other vibration directions equilibrium is impossible.

The parameter Q determines velocity and temperature distributions in the layer. For $Q=0$ a flow with a cubic velocity profile and a polynomial temperature distribution is obtained. The parameter Q is function of orientation and a static gravity field and may have any value. Positive and negative values of this parameter correspond to different convection regimes.

2. To analyze the convection regimes the boundary conditions must be formulated and the solutions of the system (5) determined. Consider zero-transport flow between solid planes. This causes the following conditions:

$$\int_{-1}^{1} V(x)dx=0,\quad x=\pm1:\ V=0$$

The velocity profile form is not sensitive to the boundary conditions for temperature and at $Q>0$ the function $V(x)$ takes the form

$$V=V_0\frac{\cosh(rx)\sin(rx)\tanh(r)\cos(r)-\sinh(rx)\cos(rx)\sin(r)}{2Br^2} \quad (7)$$

$$r=\sqrt[4]{\frac{Q}{4}}, \quad B=(\tanh^2 r\cos^2 r+\sin^2 r)\cosh(r)$$

The same velocity distribution occurs in the case of gravitational convection between the heated vertical planes at different temperatures and a longitudinal temperature gradient directed upward (Ref.5), and for the advective flow in a longitudinal vibration field (Ref.3). The analytic expression (7) demonstrates that flow intensity decreases with the increase of parameter r; at $r>3$ Gill-boundary layers are formed on the walls.

The velocity amplitude V_0 is determined by the temperature distribution in the layer and is different for different temperature boundary conditions. When formulating the boundary conditions for temperature two extreme cases may be discussed: i) ideally heat conducting boundaries ($x=\pm1:\tau=0$), ii) adiabatic boundaries($x=\pm1:\tau'=0$). In either case the temperature distribution is of the form

$$T=-z-\frac{q}{Q}x+\frac{V_0}{BQ}(\sinh(rx)\cos(rx)\tanh(r)\cos(r)+\cosh(rx)\sin(rx)\sin(r)), \quad (8)$$

$$V_0^1=q, \quad V_0^2=\frac{2qB\cosh(r)}{r(\sinh(2r)+\sin(2r))}$$

where V_0^1 and V_0^2 are velocity amplitudes at boundary conditions i) and ii).

The comparison of amplitudes V_0^1 and V_0^2 shows that the velocity of the convective flow in the case of adiabatic boundaries decreases more quickly with the increase of r than for conducting walls (at $r=0$ the amplitude proportion is equal to one and at $r\to\infty$ is equal to $1/r$).

In the case of $Q<0$ the distributions of velocity and temperature in the discussed variants of boundary conditions are of the form

$$V=\frac{V_0}{2\mu^2}\left(\frac{\sinh(\mu x)}{\sinh(\mu)}-\frac{\sin(\mu x)}{\sin(\mu)}\right) \quad (9)$$

$$T=-z-\frac{q}{Q}x+\frac{V_0}{2Q}\left(\frac{\sinh(\mu x)}{\sinh(\mu)}+\frac{\sin(\mu x)}{\sin(\mu)}\right) \quad (10)$$

$$V_0^1=q, \quad V_0^2=\frac{2q}{\coth(\mu)+\cot(\mu)}, \quad \mu=\sqrt[4]{-Q}$$

Velocity distribution (9) is also well known in the theory of free convection. For instance it arises in a vertical canal at inclined temperature gradient corresponding to heating from below (Ref.5). This flow is peculiar for the presence of critical values of parameter μ ($\mu=m\pi$, m=1,2...) at which the velocity and temperature amplitudes turn into the infinity. In the vicinity of the parameter $Q=-(m\pi)^4$ the postulated plane-parallel flow cannot be realized.

3.The stability of the equilibrium state established by condition (6) is discussed with respect to small disturbances of exponential time dependence. The behavior of disturbances with space structure of the kind (4) (wave number is equal to zero) is investigated, because long-wave disturbances are the most dangerous ones in the analogous problem of the theory of free convection. Functions $V_1(x)$, $\tau_1(x)$, $w_1(x)$ of these disturbances for the decrement corresponding to zero decay satisfy the following equations and boundary conditions

$$V_1''' - Q\tau_1' = 0, \quad \tau_1'' + V_1 = 0, \quad w_1' = n_z \tau_1' \quad (11)$$

$$x = \pm 1: \quad V_1 = \tau_1 = 0 \quad (12)$$

$$x = \pm 1: \quad V_1 = \tau_1' = 0 \quad (13)$$

Here (12) is for heat conducting and (13) for adiabatic boundaries.

Boundary problems (11)-(13) coincide with the corresponding problems of equilibrium stability in a plane vertical layer with vertical temperature gradient (Ref.5). Solutions different from zero exist only for $Q<0$ (heating from below). Critical value of the parameter $Q=Q_*$ at which the most dangerous disturbance becomes neutral is equal to $-\pi^4$ in case (12). In case (13) it is equal to $-(2.365)^4$. At $Q<Q_*$ the equilibrium is unstable and leads to plane parallel convective flow.

Hence, the averaged equilibrium in a plane layer with a longitudinal temperature gradient exists for the following correlations of gravitational and vibrational Rayleigh numbers and the vibrational direction

$$n_x n_z R_V - l_x R = 0, \quad n_z^2 R_V - l_z R > Q_* \quad (14)$$

The analysis of correlations (14) shows that convection flow may always be suppressed by the corresponding choice of the vector **n** and vibration intensity (R_V).

REFERENCES

1 G.Z.Gershuni and E.M.Zhukhovitsky, Dokl. AN SSSR 249(3) (1979)580.

2 G.Z.Gershuni and E.M.Zhukhovitsky, in Fluid Mechanics and Transport Processes in Zero Gravity, (Izd. Ural. Nauch. Tsentra AN SSSR Sverdlovsk 1983) 86.

3 G.Z.Gershuni and E.M.Zhukhovitsky, J.of Engineering Physics 56 (Minsk, Nauka i Tekhnika 1989) 238.

4 R.V.Birikh, J.Appl. Tech. Phys. 7 (1966) 43.

5 G.Z.Gershuni and E.M.Zhukhovitsky, Convective Stability of Incompressible Fluid. Izd. Nauka, Moscow (1972).

EXPERIMENTAL INVESTIGATION OF VIBRATIONAL CONVECTION IN PSEUDOLIQUID LAYER

V.G.Kozlov

Perm State Pedagogical Institute, Perm 614000, Russia

ABSTRACT

The dynamics of sand and water in horizontal layer subject to high frequency horizontal linear vibration was experimentally investigated. The vibrational convection of sand giving birth to stationary wave relief on the borderline was found. The nondimensional vibrational parameter determining the form of relief was ascertained. Presentation of sand and water as two liquids of different density gives satisfactory conformity of experimental results with theoretical ones.

Key Words: Pseudoliquid layer, Vibrational convection, Stationary relief, Transport, Two liquid theory.

1. INTRODUCTION

Pseudo-liquefied layer is the layer of fine disperse medium possessing fluidity due to some exposure and is found in many technical problems. Let discuss closed cavity with incompressible fluid partly filled by fine disperse medium. If medium and fluid differ from each other by density, gravitational distribution of these phases will occur. The behavior of this system in high frequency vibrational field is of great interest especially in the case of full fluidization of disperse medium.

2. EXPERIMENTAL PROCEDURE

The partition boundary of quartz sand and water in the horizontal cavity of the form of the rectangular parallelepiped performing high-frequency horizontal reciprocating vibrations is investigated. Without vibrations, sand uniformly covers the lower boundary of the cavity. The cavity height (Figure 1) is h = 2.0 cm, basis area is 14.6 x 3.5 cm^2. Vibrations occur along the long side. The cavity is done from the plexiglas and is partly filled by sand (layer depth h_2) and partly by water. Gas phase inside the cavity is absent.

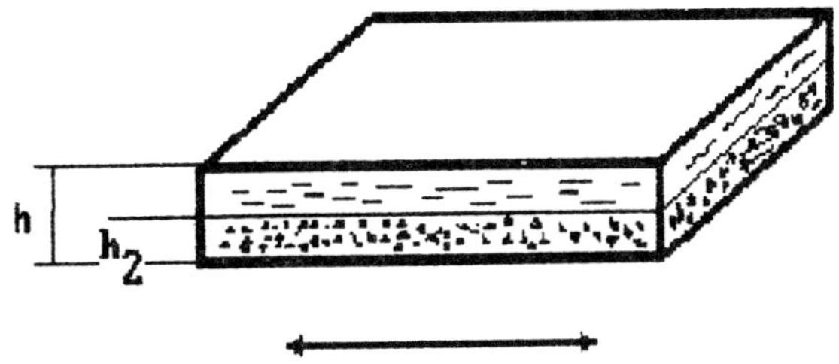

Figure 1: Experimental cavity scheme.

Experiments were carried on the sand layer with relative gauge h_2/h = 0.42 (absence of vibrations). As measurements showed, the density of the wetted sand in this case was ρ_2 =

2.0 g/cm^3. The size of sand grains varied in the interval from 0.1 up to 0.5 mm. Due to vibrations especially at high amplitudes partial loosening of the sand layer occurred. In this case its average density may easily be calculated by the effective average layer thickness obtained from photographs. The cavity was set in motion by the horizontal mechanical vibrator maintaining harmonic linear vibrations with frequency f = 1-30 Hz and amplitude b = 0-6 cm.

The surface form of the sand layer and the motion of quartz sand grains inside it were observed by the method of filming and photography. Observations were carried at stroboscopic lighting. Flash lamp was used for photographing.

3. EXPERIMENTAL RESULTS

Experiments showed that under the influence of even relatively weak vibrations pseudo-liquefying of the sand layer occurs, it obtains the fluidity properties. During this the stationary periodic relief in the form of two-dimensional hills and valleys oriented perpendicular to the vibration axis forms on the borderline.

Investigations showed that at high frequencies the form of the sand layer relief does not depend on frequency and is determined only by the amplitude of the linear velocity of vibrations. Due to the growth of the velocity amplitude the wave length of the surface relief increases. It`s worth mentioning that the observed phenomenon is of a non shock-wave character: the finest relief with wave length $\lambda \sim 1$ mm, i.e. comparable with the size of grains arises at very weak vibrations.

In Figure 2 photographs of the typical surface relief are given. Fragments a,b,c,d are obtained at vibration frequency f = 17.7 Hz, at amplitudes b = 1.5; 1.8; 2.5 and 3.4 mm correspondingly. Fragment e corresponds to f = 7.0 Hz and b = 10.8 mm.

As will be shown below at the given relative thickness and relative density of pseudo-liquefied layer the dimensionless parameter characterizing the surface form is the vibrational parameter $W = (b\omega)^2/gh$. Here ω is the cyclic frequency, g - free fall acceleration. Experimental correlations between the dimensionless wave number $K = 2\pi h / \lambda$ and vibrational parameter W obtained at different frequencies of vibrations are given in Figure 3. Symbols 1, 2 and 3 correspond to the fixed frequencies f = 7.0; 11.4 and 17.7 Hz. As the diagram shows, the stationary relief of the very small wave length occurs in a not shock-wave way at very faint vibrational exposure ($W \sim 0.1$). Along with W growth, the wave number decreases (wave length increases). Experiments revealed that the results obtained at f > 10 Hz correspond to the field of high frequency

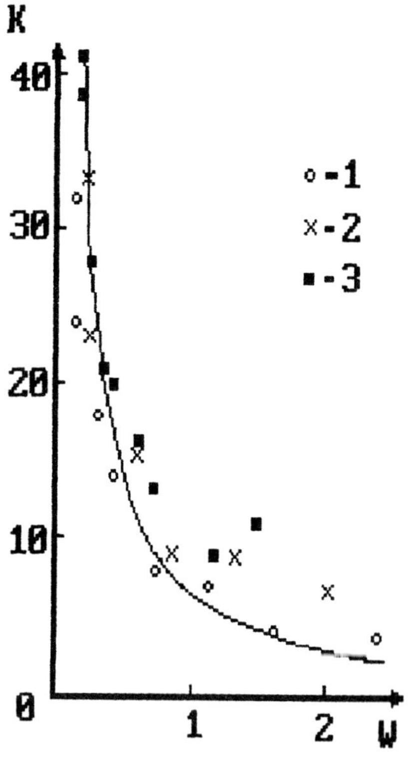

Figure 3: Variation of nondimensional wave number with vibrational parameter.

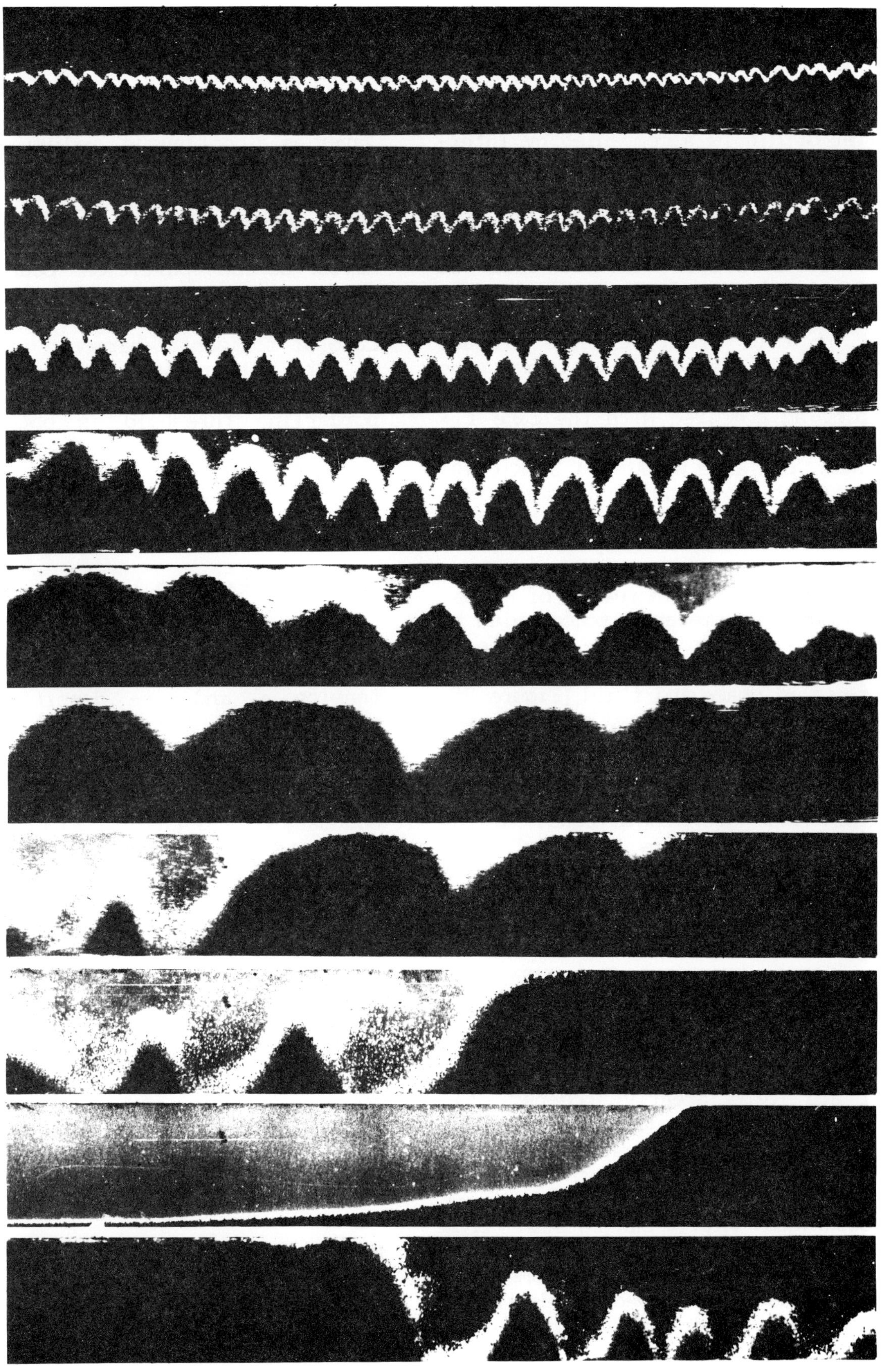

Figure 2: Vibroconvective relief photos.

asymptotic. Points obtained at frequencies f = 17.7 and f = 11.4 Hz arrange in one correlation and at f = 7.0 Hz they are somewhat lower.

Observations showed that the relief origin is accompanied by intermixing of sand grains within the layer. The depth of intermixing is comparable with the wave length of the relief. At moderate wave lengths (K > 10) relief origin do not lead to sufficient layer loosening. Measurement of photographs (Figure2) reveal that in this field relative average depth of the sand layer grows (the relative density lowers) not more than 20 p.c. as compared with the case f = 0. Significant difference may be found in case of large W. At W > 2 (K < 10) significant loosening of the sand layer is observed. One can see that borderline is washed out (Figure 2d , W = 1.1).

The further increase of the vibrational parameter is accompanied by the wave length growth and leads to the full intermixing of fine disperse solid phase and fluid (Figure 2f, W = 6.1). At such vibrational parameter values the graduate movement of the fine disperse component to one part of the layer is observed, i.e. the relief with wave length equal to the horizontal cavity size is formed. Such process of sand transport is depicted in photographs f-h obtained at one value W = 6.1 with time interval of about one minute. As it is shown in the figure transport and compaction of fine disperse admixture in one end of the cavity occurs at some restricted period of time. Only insignificant part of sand remains in the second half and forms greatly loosened stationary hills under the conditions of vibration. Figure 2i shows the sand distribution after vibration.The choice of direction of sand transport occurs in occasional way.It can be seen from Figure 2j obtained at the same value W=6.1. The decrease of the parameter W (W < 6.0) causes the graduate destroing of sand packed in one end of the cavity.

4.DISCUSSION

Discuss the pseudo-liquefied sand layer as a continuous medium with density different from that of water (at least in case of insignificant loosening). Taking such an approach we may use theoretical descriptions of the behavior of the partition boundary of non-mixing uncompressed fluids in the vibrational field (Ref.1). For the case of horizontal harmonic vibrations of the layer with relative filling x = 0.5 the next correlation for borderline threshold is obtained in this work

$$(b\omega)^2=((\rho_1+\rho_2)^3/2\rho_1\rho_2(\rho_2-\rho_1)^2)*$$

$$(\alpha k+(\rho_2-\rho_1)g/k)\,th(kh)$$

Here α is the coefficient of surface tension, ρ_1 and ρ_2 - densities of non mixing fluids, g - free fall acceleration, k - wave number.

Two fluid description model may obviously be used until the partition boundary of two phases (water and wetted sand medium) exists. But the question remains open concerning the coefficient of surface tension on the borderline of water and pseudo-liquid. Taking into account that sand grains are not linked with each other and the distance between them is filled with water the coefficient of surface tension may be taken as zero. Considering α = 0 the equation obtains the form:

$$W=((\rho+1)^3/2\rho(\rho-1))\,thK/K$$

Here $W = (b\omega)^2/gh$ is the vibrational parameter, $\rho = \rho_2/\rho_1$ - relative density, $K = 2\pi h/\lambda$ - dimensionless wave number. Plotting this correlation in Figure 3 (solid line) for the case ρ =

2, x = 0.5 we may be convinced of satisfactory agreement of the experimental results with the obtained curve observed in the section W < 1. The chosen ρ = 2 corresponds to the case of insignificant loosening (W < 1) of the sand layer. Relative layer thickness in the experiment in this section is close to the calculated one ,x = 0.48 +0.05 . The disagreement of the theoretical and experimental results in the section W > 1 is explained by alteration of relative thickness and density of pseudo-liquied layer.

5.CONCLUSION

i. Experimental study showed great influence of horizontal vibration on the form of borderline of fluidized sand layer and water. In high frequency limit the form of the surface is determined by one nondimensional parameter W. This effect is of vibroconvective nature.

ii. The two-liquid theoretical model with the coefficient of surface tension α = 0 may be applied to describe vibroconvective influence on the partition boundary of the pseudo- liquefied layer - fluid. Further, the agreement between the wave number of the experimentally observed relief and the theoretically calculated long wave boundary of instability area furnishes the supposition that the most "dangerous" are long wave disturbances.

iii. Analyzing the theoretically obtained expression of the long wave boundary of the spectrum of unstable disturbances we may mark out that at W > 6.75 the origin of the relief with endless wave length (K = 0) comes to be possible. This conclusion conforms with the above described results of the experiments (Figure 2h). At large enough value of vibrational parameter W, graduate movement of sand to one end of layer is observed.

6.REFERENCES

1. D.V.Lyubimov, A.A.Cherepanov, Izv.AN SSSR, Mech. Zhidk. Gaza, N6 (1986) 8.

ON THERMOVIBRATIONAL CONVECTION IN AN EXOTHERMAL LIQUID IN WEIGHTLESSNESS

G.Z. Gershuni[1)], E.M. Zhukhovitsky[2)], A.K. Kolesnikov[2)], B.I. Myznikova[3)] and Yu.S. Yurkov[2)].

1) *Perm State University, 15, Bukirev Str., 614600, Perm, USSR*
2) *Perm Pedagogical Institute, 24, K.Marx Str., 614600, Perm, USSR*
3) *Institute of Continuous Media Mechanics UB Acad. Sci. USSR, 1, Acad. Korolyov Str., 614061, Perm, USSR*

ABSTRACT

The non-linear regimes of thermovibrational convection in a plane layer of exothermally reacting liquid in the presence of longitudinal vibration are investigated. The characteristics of dynamics and heat transfer are determined. It is shown that there is substantial enhancement of thermal explosion threshold as a result of non-linear convection.

Keywords: *Plane Layer, Exothermal Liquid, Longitudinal Vibration, Thermovibrational Convection, Non-linear Regimes.*

1. INTRODUCTION

Vibration can be a factor with a substantial effect on the normal thermogravitational convection in a static gravity field. Moreover, under certain conditions, vibration acting on a cavity filled with fluid can itself provoke the appearance of regular convective flow (the "thermovibrational convection" phenomenon, TVC). In a static gravity field the superposition of thermogravitational and thermovibrational convection occurs. It is important to note that TVC can exist in conditions of pure weightlessness.

There are a number publications that are devoted to the TVC-problem for which temperature inhomogeneity is caused by internal heat production. In (Ref.1) the stability of mechanical equilibrium in a fluid with homogeneously distributed internal heat sources is considered. The configuration is a plane horizontal layer of fluid subject to gravitational and vibrational fields. The axis of vibration is arbitrarily oriented with respect to the layer. Two different boundary conditions for temperature are considered: a) both boundaries are isothermal and their temperatures are the same; b)one of the boundaries is isothermal, the other is insulated. The boundaries of stability and the characteristics of critical perturbations are determined. The case of Biot-like boundary conditions is investigated in (Ref.2).(Ref. 3-5) are devoted to investigations of non-linear regimes of TVC in a plane horizontal layer of internally heated fluid which result from loss of stability of mechanical equilibrium. The conditions of excitation are analysed as well as the characteristics of the dynamics and heat transfer in non-linear regime. In (Ref.6) the more complicated case is investigated when heat generation is caused by an exothermal chemical reaction. The linear stability of the equilibrium is studied in the presence of longitudinal or transversal vibrations.

In the present paper we investigate the non-linear regimes of TVC in exothermal fluid which arise in a plane layer as a result of bifurcation from an equilibrium state in pure weightlessness under the action of longitudinal vibration. The main attention is concentrated on the problem of the effect of vibration on the threshold of thermal explosion.

2. THE BASIC EQUATIONS

A plane infinite fluid layer is confined between two parallel isothermal plates $z = 0$ and $z = h$, which are held at constant absolute temperature T_o. The layer consists of exothermically reacting fluid; the heat generation depends on the temperature by the Arrenius law $Q = Q_o k_o \exp(-E/RT)$. Here Q - heat generation per unity of volume, T - absolute temperature, E - activation energy, R - gas constant. In the abscence of gravity the convection is caused by vibration of the system. The frequency Ω as high as required for asymptotic analysis and the amplitude of the displacement b is similarly low. The mean fields of velocity, pressure and temperature may be described (in appropriate system of coordinates) by the system of equations (the usual Boussinesq approximations are accepted):

$$\frac{\partial \vec{v}}{\partial t} + (\vec{v}\,\nabla)\vec{v} = -\frac{1}{\rho}\nabla p + \nu\Delta\vec{v} + \varepsilon(\vec{w}\,\nabla)(T\vec{n} - \vec{w}), \qquad (1)$$

$$\frac{\partial T}{\partial t} + \vec{v}\,\nabla T = \chi\Delta T + (k_o Q_o/\rho c_p)\cdot \exp(-E/R(T_o+T)), \qquad (2)$$

$$\mathrm{div}\vec{v} = 0, \quad \mathrm{div}\vec{w} = 0, \quad \mathrm{rot}\vec{w} = \nabla T\mathrm{x}\vec{n}. \qquad (3)$$

Here $\varepsilon = (\beta b\Omega)^2/2$ - the vibrational parameter, β - coefficient of thermal expansion, T - temperature with respect to T_o, $\vec{n}$ - the unit vector along axis of vibration, $\vec{w}$ - additional variable - solenoidal part of vector field $T\vec{n}$ (at the same time it is the amplitude of the oscillational part of the velocity); the other nomenclature is as usual. We choose h, h^2/ν, χ/h, RT_o^2/E, $\rho\nu\chi/h^2$ as the units of length, time, velocity, temperature and pressure respectively. Then the system may be written in nondimensional form as follows:

$$\frac{\partial \vec{v}}{\partial t} + \frac{1}{Pr}(\vec{v}\,\nabla)\vec{v} = -\nabla p + \Delta\vec{v} + Ra_v(\vec{w}\,\nabla)(T\vec{n} - \vec{w}), \qquad (4)$$

$$Pr\frac{\partial T}{\partial t} + \vec{v}\,\nabla T = \Delta T + Fk\exp(T), \qquad (5)$$

$$\mathrm{div}\vec{v} = 0, \quad \mathrm{div}\vec{w} = 0, \quad \mathrm{rot}\vec{w} = \nabla T\mathrm{x}\vec{n}. \qquad (6)$$

Here nondimensional parameters are introduced : the numbers of Prandtl, Frank-Kamenetsky and the vibrational Rayleigh number:

$$Pr = \frac{\nu}{\chi}, \quad Ra_v = \frac{\varepsilon h^2}{\nu\chi}\left(\frac{RT_o^2}{E}\right)^2, \quad Fk = \frac{k_o Q_o E h^2}{\rho c_p \chi R T_o^2}\exp\left(-\frac{E}{RT_o}\right). \qquad (7)$$

In (5) the insignificance of RT_o/E is taken into account. On the solid boundaries the following conditions are to be imposed:

$$z = 0,\ z = 1:\ \vec{v} = 0,\ T = 0,\ w_z = 0 \qquad (8)$$

Moreover the closure conditions for the longitudinal parts of $\vec{v}$ and $\vec{w}$ must be taken into account.

3. MECHANICAL EQUILIBRIUM

Under the described conditions, the mechanical quasi-equilibrium, i.e. the state in which the mean velocity is zero ($\vec{v} = 0$), is possible. The equilibrium fields of temperature, pressure and $\vec{w}$ depend on the transversal coordinate $\theta_o(z)$, $p_o(z)$, $\vec{w}_o(z)$. For θ_o we have the classical problem of Frank-Kamenetsky:

$$\theta_o'' + Fk\exp(\theta_o) = 0; \quad \theta_o(0) = \theta_o(1) = 0. \qquad (9)$$

As is well known (Ref.7) the solution of this problem exists only in the region $0 \le Fk \le Fk_* = 3.514$. If $Fk < Fk_*$ the steady regime of heat diffusion exists; if $Fk > Fk_*$ the initial temperature perturbation continues to grow and a steady regime is not possible ("thermal explosion" phenomenon). But in a fluid or gas the regime of diffusion may become convectively unstable even if $Fk < Fk_*$ under the action of the TVC-mechanism.

4. LINEAR STABILITY

To investigate the stability of the mechanical equilibrium we consider the disturbed fields $\theta_o + T'$, $p_o + p'$, $\vec{w}_o + \vec{w}'$, $\vec{v}$. After linearization the system for perturbations becomes:

$$\frac{\partial \vec{v}}{\partial t} = -\nabla p' + \Delta \vec{v} + Ra_v [(\vec{w}_o \nabla)(T'\vec{n} - \vec{w}') + (\vec{w}' \nabla)(\theta_o \vec{n} - \vec{w}_o)] \qquad (10)$$

$$Pr \frac{\partial T'}{\partial t} + \vec{v}\, \nabla \theta_o = \Delta T' + Fk \exp(\theta_o) T', \qquad (11)$$

$$div\vec{v} = 0, \quad div\vec{w}' = 0, \quad rot\vec{w}' = \nabla T' x \vec{n}. \qquad (12)$$

We shall consider only the case of longitudinal vibrations; and limit the consideration to 2D-perturbations: $v_y = w'_y = 0$, $\partial/\partial y = 0$. The problem may be formulated in terms of transversal components v_z and w'_z. We consider normal modes of type $\exp(-\lambda t + ikx)$ where λ is decrement and k wave number. Now the spectral problem for amplitudes v, w and θ may be written:

$$-\lambda \Delta v = \Delta^2 v - Ra_v D\theta_o (ikDw - k^2\theta), \qquad (13)$$

$$-\lambda Pr \theta = \Delta\theta - D\theta_o v + Fk \exp(\theta_o)\theta, \qquad (14)$$

$$\Delta w = -ikD\theta; \qquad (15)$$

$$z=0,\ z=1: \quad v = Dv = 0, \quad w = 0, \quad \theta = 0. \qquad (16)$$

Here $D = d/dz$, $\Delta = D^2 - k^2$. Decrement $\lambda = \lambda_r + i\lambda_i$ is the eigenvalue of our problem. The problem was integrated numerically by a Runge-Kutta-Merson method. The boundary of stability is connected with monotonous perturbations ($\lambda_i = 0$) and may be determined by the condition $\lambda = 0$. So we determine the critical Rayleigh number Ra_v as a function of parameters Fk and k.

Some results of calculations are presented in Figure 1. The critical Rayleigh number Ra_{vm}, minimized with respect k, and the corresponding wave number k_m are plotted as functions of Fk. It is apparent that both are monotonous functions of Fk decreasing while Fk is increasing. When $Fk \to Fk_*$ then $Ra_{vm} \to 1489$ and $k_m \to 1.8$. The behaviour of functions $Ra_{vm}(Fk)$ and $k_m(Fk)$ for $Fk \ll 1$ can be determined

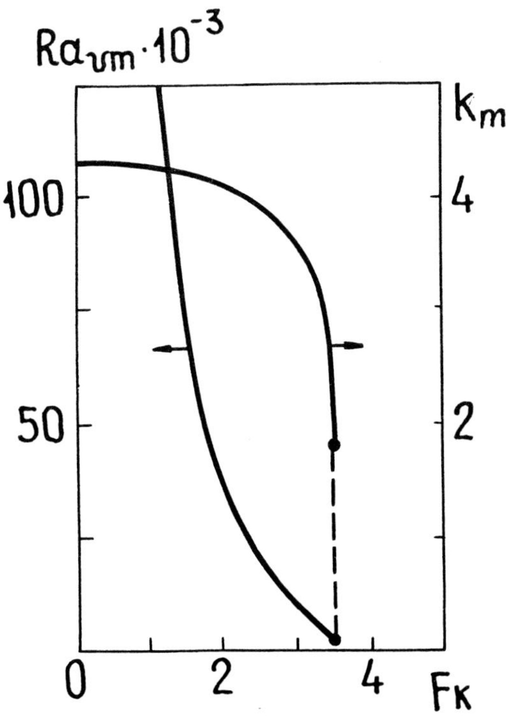

Figure 1: The critical Rayleigh number and wave number as functions of Fk.

from the consideration that in this region the Frank-Kamenetsky problem (9) reduces to the case where heat generation is spatially homogeneous with $Q = RT_o^2\kappa Fk/Eh^2$, where κ is coefficient of heat conductivity. That means that (on the base of results of (Ref.1-3)) we have as asymptotics $Fk \to 0$ as $k_m \to 4.35$; $Ra_{vm} \to 269.3\times10^3/Fk^2$.

5. FINITE-AMPLITUDE REGIMES: FORMULATION OF THE PROBLEM

It is necessary to use the full system of non-linear equations (4)-(6) to describe non-linear regimes of TVC-flows bifurcating from equilibrium. A 2D-variant of this system may be written by means of two stream functions ψ and φ for the mean flow velocity $\vec{v}$ and the oscillating part $\vec{w}$. The system of non-linear equations in terms ψ, φ, T is as follows:

$$\frac{\partial}{\partial t}\Delta\psi + \frac{1}{Pr}\left(\frac{\partial\psi}{\partial z}\frac{\partial\Delta\psi}{\partial x} - \frac{\partial\psi}{\partial x}\frac{\partial\Delta\psi}{\partial z}\right) = \Delta^2\psi + Ra_v\left(\frac{\partial^2\varphi}{\partial z^2}\frac{\partial T}{\partial x} - \frac{\partial^2\varphi}{\partial x\,\partial z}\frac{\partial T}{\partial z}\right), \qquad (17)$$

$$Pr\frac{\partial T}{\partial t} + \left(\frac{\partial\psi}{\partial z}\frac{\partial T}{\partial x} - \frac{\partial\psi}{\partial x}\frac{\partial T}{\partial z}\right) = \Delta T + Fk\exp(T), \qquad (18)$$

$$\Delta\varphi = \frac{\partial T}{\partial z}. \qquad (19)$$

The boundary conditions are

$$z=0,\ z=1:\quad \psi=0,\ \frac{\partial\psi}{\partial z}=0,\ T=0,\ \varphi=0. \qquad (20)$$

The solution of the problem has been found in a rectangular region $0\le z\le 1$, $0\le x\le L$ with periodity conditions

$$f(L,z) = f(0,z) \qquad (21)$$

where f is any of the variables. For a numerical solution the finite-difference method has been applied. The ADI-scheme of fractional steps was used. The main calculations were carried out on a 31x31 grid. The Prandtl number and the spatial period were fixed: $Pr = 1$, $L = 1.5$. Value $L = 1.5$ corresponds to wave number $k = 4.19$ which is close to the critical value k_m of linear theory in a wide region of Fk (except in the vicinity of the Fk_* - point).

6. FINITE-AMPLITUDE REGIMES: SOME NUMERICAL RESULTS

In region of stability $Fk<Fk_*$, $Ra_v<Ra_{vm}$ any initial perturbation is damped. There is only one attracting regime - that of heat diffusion. If $Fk<Fk_*$ and $Ra_v>Ra_{vm}$ initial perturbations grow and a secondary stationary regime of TVC is established. It is interesting to emphasize that in the supercritical region a steady stationary state may be observed even in the region $Fk>Fk_*$. If $Ra_v>Ra_{vm}$ and held fixed while Fk is increasing, then the stationary regime has an end-point which is connected with thermal explosion. So we may construct the map of regimes in the (Fk - Ra_v)- plane - Figure 2. Symbols I, II and III mark regions of mechanical equilibrium, steady TVC-flows and thermal explosion respectively. The left boundary of the TVC-region practically coincides with the threshold of instability as determined by linear theory. In the TVC-region the flows have a symmetrical four vortices structure. The intensity of vortices may be characterized by the extremal value ψ_m of mean stream function. The dependence of ψ_m on Fk for fixed values of Ra_{vm} is presented in Figure 3.

The most interesting result of the analysis we find the substantial enhancement of the thermal explosion threshold in the convective regime. This result is quite clear from a physical point of view: developed supercritical TVC-flows intensify substantially the heat transfer from fluid layer.

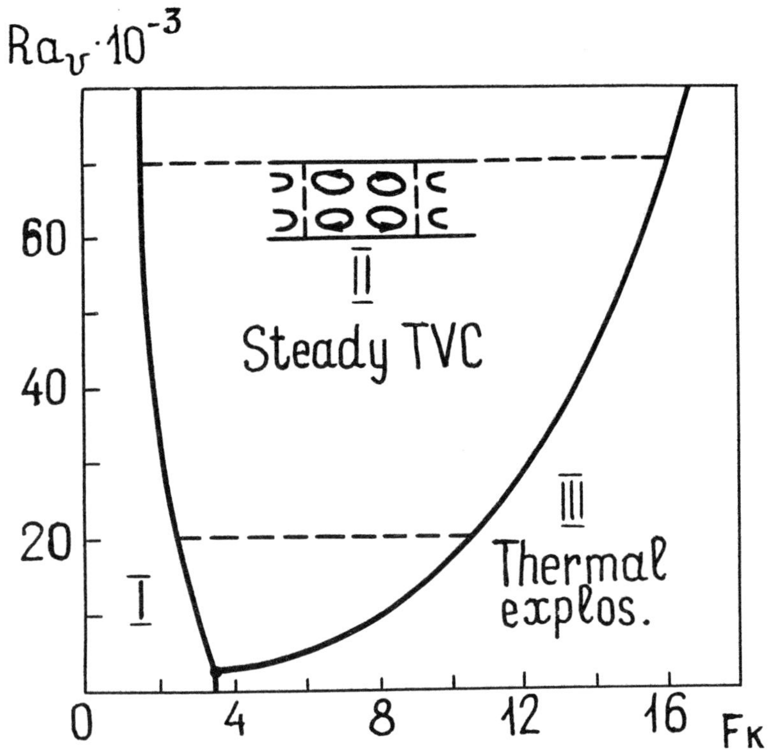

Figure 2: The map of regimes

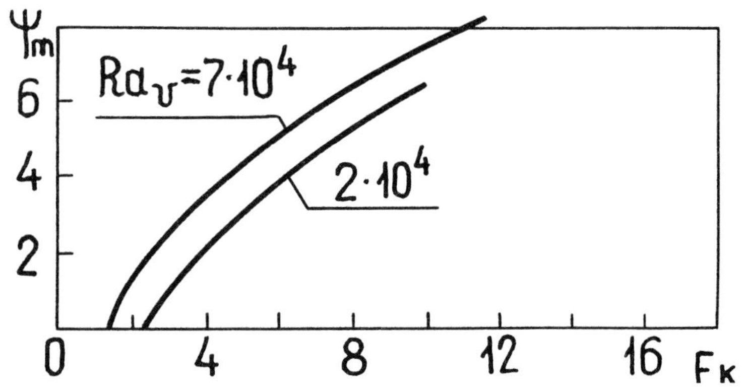

Figure 3: The extremal value of stream function plotted against Fk .

7. REFERENCES

1. G.Z. Gershuni, E.M. Zhukhovitsky and A.K. Kolesnikov, *Izv. AN SSSR, Mech. Zhidk. Gaza* 5(1985)3.

2. G.Z. Gershuni and E.M. Zhukhovitsky in *Numerical and Experimental Modelling of Hydrodynamical Phenomena in Nongravity,* Sverdlovsk (1988),p.72.

3. G.Z. Gershuni, E.M. Zhukhovitsky, A.K. Kolesnikov and Yu.S. Yurkov, *Int. J. Heat Mass Trans.* 32(1989) 2319.

4. G.Z. Gershuni, E.M. Zhukhovitsky and Yu.S. Yurkov in *Convective Flows,* Perm (1989),p.45.

5. G.Z. Gershuni, E.M. Zhukhovitsky and Yu.S. Yurkov, *Model in Mech.*, Novosibirsk 4/21(1990)103.

6. G.Z. Gershuni, E.M. Zhukhovitsky and A.K. Kolesnikov, *The Phys. of Combust. and Explos.* 5(1990)91.

7. D.A. Frank-Kamenetsky, *Diffusion and Heat Transfer in Chemical Kinetics* (Nauka, Moscow 1967).

THREE-DIMENSIONAL OSCILLATORY CONVECTION OF THE LOW PRANDTL NUMBER FLUID IN A RECTANGULAR CAVITY WITH DIFFERENTIAL HEATING FROM THE SIDE (EXPERIMENT)

A.Bojarevičs, Yu.M.Gelfgat, L.A.Gorbunov
Institute of Physics, Latvian Academy of Sciences
Salaspils-1, Latvia, 229021.

ABSTRACT *An experimental investigation of natural convection of an eutectic melt In-Ga-Sn (Pr = 0.019) in a square cavity with temperature difference between the end walls was carried out for the range of Gr numbers in the vicinity of* Gr_{osc} *when onsets the oscillatory convection. It is shown that the type of oscillatory convection strongly depends on the aspect ratios* $A_1 = L/H$ *- length to depth - and* $A_2 = W/H$ *- width to depth. At* Gr_{osc} *and* $A_1 = 4$, $2.5 \leq A_2 \leq 5.44$ *the quasi-two-dimensional steady natural convection gave place to oscillatory three-dimensional one with two longitudinal rolls. The values of Grosc and nondimensional frequences of temperature oscillations were found for several values of aspect ratios* A_1 *and* A_2*. Characteristic distributions of instantaneous temperature in the melt were measured. The flow visualisation on the upper boundary was carried out for* $Gr > Gr_{osc}$.

Key Words: oscillatory natural convection, low Prandtl number, transition to three-dimensional flow, heating from side

The problem of the low Prandtl number fluid convection in a rectangular cavity with differentially heated end walls are widely studied numerically and experimentally. In 1988 there was announced an international workshop [1] where such a problem served as a test to compare different numerical techniques. The problem also has an applied value - e.g., as a model of heat transfer at different metal or semiconductor crystal growth processes from melts. Experiments [2,3] have shown that at Grashof numbers higher than critical value $Gr \geq Gr_{osc}$ ($Gr = g\beta\Delta T H^4/(L\nu^2)$) onsets the oscillatory regime of natural convection. Often the problem is considered in the two dimensional approximation. It is supposed that at aspect ratio of width to depth of the enclosure W/H higher than unity the two dimensional approximation is valid on plane of symmetry z = 0 (Fig.1). But there are experimental results demonstrating the three-dimensional character of such convection after transition to the oscillatory convection at $Gr > Gr_{osc}$ [2].

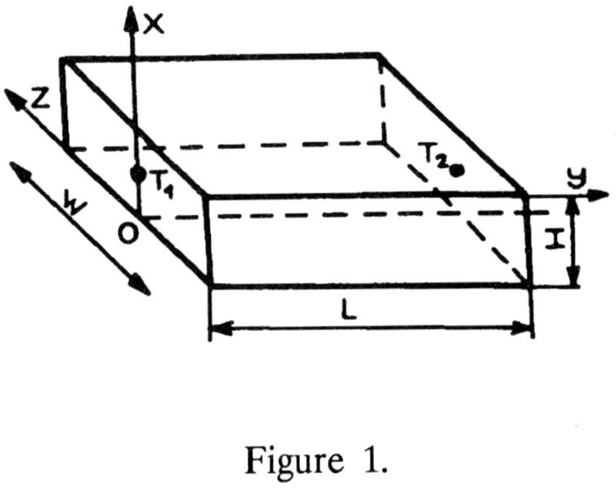

Figure 1.

The present experiment dealt with an investigation of oscillatory natural convection in a liquid with $Pr << 1$ in a rectangular cavity with differentially heated end walls. The aspect ratio $A_1 = L/H$ was usually fixed at the value 4, which corresponded to that announced for the GAMM workshop [1]. The second aspect ratio - $A_2 = W/H$ - varied from 2.5 to 5.44. As a working fluid an eutectic melt In-Ga-Sn was used: melting temperature $t^o = 10.4\ ^oC$, viscosity $\nu = 3.1 \cdot 10^{-7}\ m^2/s$, density $\rho = 6.4 \cdot 10^3\ kg/m^3$, thermal expansion $\beta = 1.4 \cdot 10^{-4}\ K^{-1}$, heat conductivity $\lambda = 31\ W/m \cdot K$ and the Prandtl number $Pr = 1.9 \cdot 10^{-2}$. Bottom ($x = 0$) and side walls ($z = W/2$) of the test cell where made of plexiglas, heat conductivity of which is much smaller than that of the melt - $\lambda_w/\lambda \approx 5 \cdot 10^{-3}$. Copper end walls - $y = 0$ and $y = L$ - where kept at temperatures T_1 and T_2 by thermostated water to the precision of temperature difference $T_2 - T_1 = \Delta T = 0.05\,K$. An upper boundary of the melt - $x = H$ - where covered by a thin oxide film. During experimental runs the mean temperature of the melt $(T_1 + T_2)/2$ where kept at the room temperature. The whole test cell where enclosed by an additional thermal isolation. Upper boundary where covered by a lid so that only a thin layer of air remained between the oxide film and the lid, minimizing the heat exchange on the upper boundary by the air convection. To reasonable degree it could be regarded that the end walls where isothermal and other boundaries adiabatic for steady stationary heat flow in the test cell.

The flatness of the upper boundary were ensured by a sharp drop in the wetting conditions melt-wall at the height $x = H$, which were created to eliminate regions of menisci at the walls.

Temperature drop between the end walls were controlled by two thermocouples situated on the wetted surfaces at the points [H/2;0;0] and [H/2;L;0]. The local temperature of the melt were measured by differential thermocouples copper-melt-constantan. The temperature measurements were registered with precision not less than 0.01 K.

The steady stationary heat transfer in this test cell was usually reached after more than two hours. The critical Grashof number for the transition to the oscillatory heat transfer Gr_{osc} were found by increasing temperature drop ΔT by discrete steps 0.1 K, or by decreasing ΔT from the values $Gr > Gr_{osc}$. While increasing Gr the oscillatory convection usually did not onset at $Gr > Gr_{osc}$ without finite outside perturbation and stationary convection were registered even at $Gr \approx 2Gr_{osc}$ for more than a hour. Gr_{osc} were determined while decreasing Gr from the developed oscillatory regime or by giving a perturbation to the convection by thermocouple (diameter - < 0.2 mm) motion and registering the increment of the temperature oscillations. The perturbation to the heat transfer by the stationary thermocouple were found negligible in respect to the value of Gr_{osc}. This were demonstrated by determining the Gr_{osc} in the absence of the measuring thermocouple by registering the onset of temperature oscillations on the end walls, where ΔT control thermocouples were situated so that they did not disrupt the end wall flatness. For all three methods of determining the Gr_{osc} its value was the same to the precision of the temperature step 0.1 K.

The Gr_{osc} were determined for three values of aspect ratio A_2 at $A_1 = 4$. In all cases the width of the cell was the same $W = 60$ mm. The results are shown on the table in comparison with values found in similar experiments. It was found that the Gr_{osc} strongly depends on aspect ratio A_2 and decreases with relative width of the test cell, but the nondimensional frequency $f^* = (H/\nu)f$ increases with the relative width (in the investigated range of aspect ratio A_2).

There were completely periodic oscillations of the local melt temperature at steady developed regimes of the oscillatory convection. The main frequency of the oscillations at all points in melt were the same, but the amplitude and the form of the oscillations varied in space - Fig.2. The

Melt	$Pr \cdot 10^2$	A_1	A_2	$Gr_{osc} \cdot 10^{-4}$	f*	Reference
In-Ga-Sn	1.9	4.0	5.44	3.8 ± 0.8	------	[7]
In-Ga-Sn	1.9	4.0	5.44	3.7 ± 0.5	------	[3]
In-Ga-Sn	1.9	4.0	3.64	5.2 ± 0.8	19.4	[7]
In-Ga-Sn	1.9	4.0	2.50	6.6 ± 0.8	15.5	------
Hg	2.6	4.0	3.8	3.65	12.8	[6]
Ga	2.0	4.76	2.0	2.5	18.0	[2]
Ga	2.0	3.45	1.48	7.0	39.4	[2]
In-Ga-Sn	1.9	3.37	2.11	7.0 ± 0.8	19.3	------
In-Ga-Sn	1.9	3.20	2.00	40 ± 1	88.0	------

analysis of the registered temperature oscillations demonstrated that at any point in the melt there are the main frequency f and its double $2f$. The relative phase of the double frequency in respect to that of the main and their amplitudes vary in space. An analysis of synchronously registered oscillations by two thermocouples demonstrated that at any moment in time the local temperature in points symmetrical to the central longitudinal crossection z = 0 coincide. The temperature field at the regular oscillatory regimes of convection in this test cell has a plane of mirror symmetry at z = 0.

Exploiting the regularity of temperature field in time, it was possible to find the instantaneous temperature distributions in the melt at $Gr \geq Gr_{osc}$. The signals of two thermocouples were registered simultaneously. The position of one was fixed for all the runs and served as a time mark, but the position of other was varied to register the temperature at the arbitrary point in the melt. Thus the instantaneous temperature distributions were found by substituting the values of temperature at different space points in one moment of time t by the values at similar moments $(t + n\tau)$, where τ was the period of the temperature oscillations and n = 0; 1; 2.... The Fig.2 illustrates the technique. Here signals 1 and 5 were registered simultaneously, but the signals 2, 3 and 4 - with the shift in time $n\tau$ in respect to the signal 1. The detailed investigation of the time periodic temperature field in the melt were carried out for $Gr = 2.3 Gr_{osc}$, $A_1 = 4$ and $A_2 = 3.64$.

The peak to peak amplitude of the temperature oscillations at the center of the test cell (H/2;L/2;0) reached 20% of the longitudinal temperature drop between the end walls ΔT. The period of temperature oscillations in this regime was τ = (32,5 ± 0.5) s, or nondimensional frequency f^* = 27.0.

The Fig 3. shows the instantaneous temperature distributions in four crossections in the moments shifted by a half of the oscillation period τ. The distributions

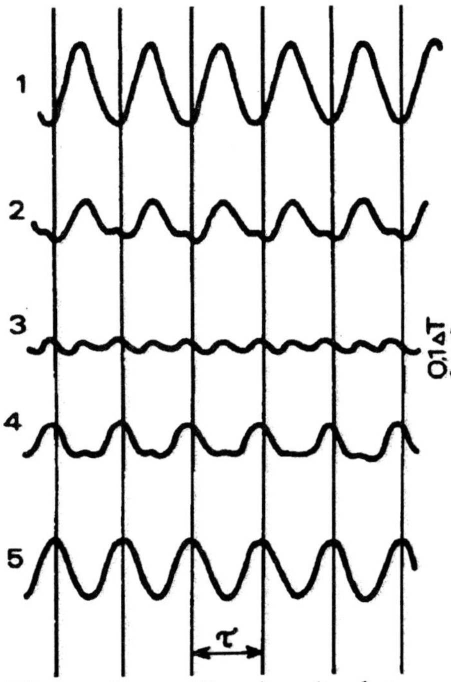

Figure 2. *Synchronized temperature oscillatory components at five points of lateral crossection x = 0.1 H and y = 0.5 L: 1) z = 0; 2) z = 0.125 W; 3) z = 0.25 W; 4) z = 0.375 W; 5) z = 0.48 W.*

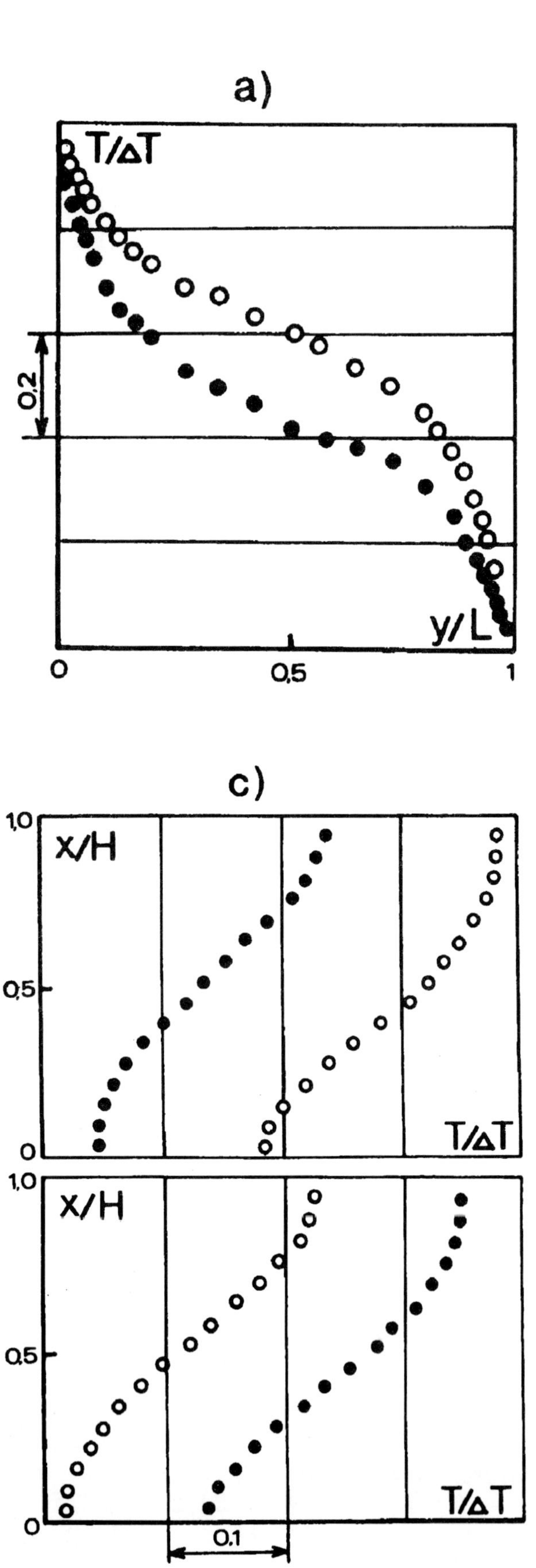

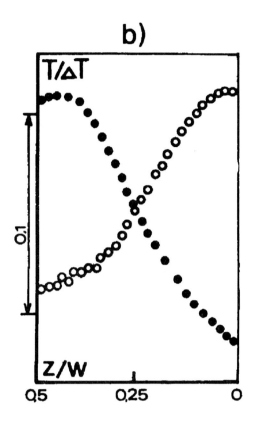

Figure 3. *Temperature distributions at two moments shifted in time by a half of the oscillation period. Full circles correspond to the moment in time , when at the point x = 0.5 H, y = 0.5 L and z = 0 there is a minimum of temperature, open circles - when there is a maximum.*
a) longitudinal temperature distributions at x = 0.5 H, z = 0;
b) lateral at x = 0.9 H, y = 0.5 L;
c) along the depth at y = 0.5 L, z = 0;
d) along the depth at y = 0.5 L, z = 0.48 W.

shown by full circles correspond to the moment in time when in the center of test cell (H/2;L/2;0) there were a minimum of local temperature, and by an open circles - a maximum. For the longitudinal cross-ection - Fig.3a - approaching the end walls the relative phase of the frequencies f and $2f$ shifted together for an approximately one fourth of the period in respect to the signal at the center of the melt and at the same time the amplitude of the frequency $2f$ were growing in respect to that of the f. It must be noted that the amplitude of the oscillations on the end walls did not reached the zero value, what can be explained by the finite thickness of copper walls and high but finite conductivity. It follows that at oscillatory convection the boundary conditions on the end wall can not be considered as isothermal.

On the Fig.3b there are shown the lateral instantaneous temperature distributions and to those correspond the time traces from the Fig.2 of the oscillating components registered simultaneously at five points of this crossection. Peak to peak amplitude of the lateral temperature drop, averaged in time being very small, reached $\pm 0{,}1\Delta T$. This clearly demonstrated that the oscillatory convection in the cell were three-dimensional.

Fig.3c,3d shows instantaneous vertical distributions of temperature in the plane $y = L/2$ at two points, corresponding to the lateral crossection on the Fig.3b. It follows from these that the vertical temperature drop changed little during the oscillation period and at any moment in time remained positive in the vicinity of the central lateral crossection. At this region both frequencies f and $2f$ changed their amplitude with the depth very little, so that the oscillations are nearly synphase in all the melt depth.

It follows from the temperature measurements that, along with a longitudinal heat transfer between the end walls, there existed an intensive lateral heat flow. However, the time averaged lateral heat flow were small. It is obvious that such a periodic temperature field in the melt could exist only at fully three-dimensional convective motion. It follows that the numerical simulations of the problem given in [1] should be solved in three-dimensional formulation.

To obtain some, even if not very correct, information about the flow velocity field, the visualization of the motion at the upper boundary were realized. Instead of the oxide film the upper boundary were covered by a thin layer of the HCl solution in pure alcohol and the melt surface under this solution became mobile. The temperature measurements in the volume of the melt demonstrated that the oscillatory convection were changed little. The frequency of the oscillations remained practically the same and the formerly described features of the oscillatory convection did not changed qualitatively. The motion on the melt surface was visualized by small particles of graphite.

The pattern seen on the surface is sketched on Fig.4. The motion on the melt surface was periodic and correlated well with the temperature measurements in the melt done while the surface was covered by the oxide film. While at the plane of the symmetry $z = 0$ the maximum of the velocity in the direction from the hot wall to the cold one was seen in the region at $y \geq L/2$, the central jet impinged on the cold end wall and diverged to the lateral walls and further along those in the opposite direction. But it should be remembered that the flow is essentially three-dimensional and the observed pattern on the surface gives only incomplete picture of flow projected on the plane.The temperature maximum near the surface at $y = L/2$ and $z = 0$ were delayed in respect to that of velocity for roughly a fourth of the period τ. After a fourth of a period the motion on the surface slowed down nearly to standstill, and afterwards developed the pattern of motion sketched on the Fig.4b.

The reason for the phase shift for practically 180^0 of the main frequency between the planes $z = 0$ and $z = \pm W/2$ became obvious. The double frequency are probably connected with the existence of free shear layer of velocity in the regions

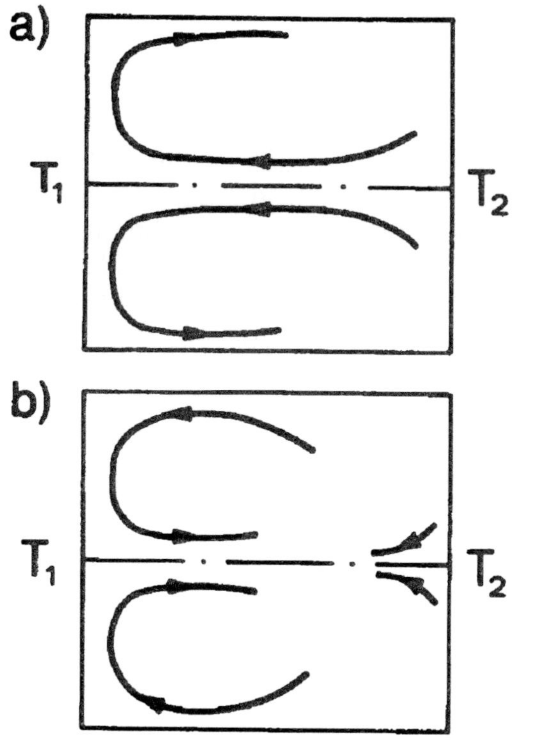

Figure 4. *Sketch of flow patterns on the melt surface in two moments shifted in time by a half of the oscillation period.*

at the planes z ≈ W/4 sufficiently far away from the end walls.

It could be proposed that the strong dependence of Gr_{osc} on the nondimensional width of the cell A_2 is not so much an effect of viscous drag on the side walls but an effect of geometrical restraint on the wavelength of the three-dimensional oscillatory motion by the fixed length scale in the lateral direction.

The experiment demonstrated that for this test cell at the fixed dimensional width W the investigated oscillatory type of convection exists for a wide range of parameters *Gr*, A_1 and A_2. It is limited in the range of *Gr* from the upper side by the transition to more complicated flows and in the range of A_2 from the downside due to restrictions on possible wavelengths in lateral direction. The few test runs at A_1 at values < 4 showed that the described oscillatory regimes of convection does not exist at nondimensional length $A_1 < 3$ - notice the sharp increase in the value of Gr_{osc} at the $A_1 = 3.2$ (table). It could be proposed that the vertical profiles of temperature and velocity are affected by the proximity of the end walls and are stable to such three-dimensional perturbations.

REFERENCES:

1. GAMM-Commitee on Numerical Methods in Fluid Mechanics. Benchmark for the GAMM-Workshop "Numerical Simulation of Oscillatory Convection in Low Prandtl Fluids". Marseille, 12-14 October, 1988.

2. D.T.J. Hurle, E. Jakeman, C.P. Johnson. Convective temperature oscillations in molten gallium. *J.Fluid Mech.*, 1974, **v.64, part 3**, pp.565-576.

3. A.Bojarevitsh, L.A.Gorbunov, F.V.Kozyrev, A.P.Lebedev. An experimental study of critical regimes of flow in liquid metal layers. *5th EPS Liquid State Conference on Turbulence*. Moscow 1989. Abstracts, pp.298-301.

4. A.E.Gill. A theory of thermal oscillations in liquid metals. *J.Fl.Mech.*, 1974, **v. 64, part 3**, pp.577 -588.

5. A.Bojarevičs. Oscillating convective flows at Pr << 1 in a rectangular cavity with horizontal temperature drop (experiment). *Matematicheskoye modelirovaniye. Prikladniye zadachi matematicheskoi fiziki*, **v.1**. Riga: Latvian University, 1990, pp.111-120.

MODELLING OF THERMOCONCENTRATIONAL CONVECTION UNDER MICROGRAVITY CONDITIONS ON THE PERSONAL COMPUTER

M.K. Ermakov, V.L. Grjaznov, S.A. Nikitin, D.S. Pavlovsky, V.I. Polezhaev
Institute for Problems in Mechanics, USSR Academy of Sciences
prospect Vernadskogo 101, 117526 Moscow, USSR

ABSTRACT

The paper contains a short description of special dialog system for study of convective processes on the basis of the unsteady Navier-Stokes equations. This is a desktop scientific analysis program for both beginning and experienced scientists. The presented system is used for the modelling of the thermoconcentrational convection (including Marangoni effects) in the flat rectangular region. Different kinds of boundary conditions may be used. Body force acceleration is a function of the time. Several test examples and convection in an enclosure for the case of slow rotation of the microacceleration vector are discussed.

Keywords: *thermal convection, microgravity, numerical modelling, numerical codes.*

1. INTRODUCTION

The development of high-speed personal computers with convenient periphery systems and effective numerical methods for Navier-Stokes equations and long-time experience of creation and analysis of the models for convective processes for common and special use (see for example [1]) revealed a new possibilities for numerous users in material and fluid sciences. This paper is devoted to the results of calculations and new PC-based system COMGA version developed during 1990-91 from the time of the first presentation [2].

2. MATHEMATICAL MODEL

The presented dialog system is used for the modelling of the thermoconcentrational convection on the basis of the unsteady Navier-Stokes equations with Boussinesq approximation in the flat rectangular region length L and height H. The governed equations are

$$\frac{\partial \mathbf{V}}{\partial t}+(\mathbf{V}\cdot\nabla)\mathbf{V} = -\frac{1}{\rho}\nabla p+\frac{\mu}{\rho}\Delta u+\mathbf{g}(t)(\beta_T T+\beta_C C)$$

$$\operatorname{div} \mathbf{V} = 0$$

$$\frac{\partial T}{\partial t} + \mathbf{V}\cdot\nabla T = \alpha\, \Delta T$$

$$\frac{\partial C}{\partial t} + \mathbf{V}\cdot\nabla C = D\, \Delta C$$

Several kinds of boundary conditions may be used. The normal component of velocity is zero on every boundary, and for the tangent component of velocity u_s the condition may be either of the first kind $u_s = 0$ (rigid wall) or of the second kind

$$\mu \frac{\partial u_s}{\partial n} = \frac{\partial \sigma}{\partial T}\frac{\partial T}{\partial s} + \frac{\partial \sigma}{\partial C}\frac{\partial C}{\partial s}$$

(free surface). For the temperature and concentration the boundary condition of first, second or third kind may be used.

The following notation are used: **V**-velocity vector, p-presure, ρ-density, T-temperature, C-impurity concentration, t-time, μ-dynamical viscosity, α-thermal diffusivity, D-diffusivity, σ-surface tension, β_T, β_C-volumetric expansion coefficients, **g**-body force acceleration.

The density of the binary liquid and the surface tension are given, as linear functions of concentration and temperature. The components of the microacceleration vector $g_x(t)$ and $g_y(t)$ can be written in the next form

$$g_x(t)=g_{x0}+(g_s+g_t\sin(\Omega_1 t))\sin(\Omega_2 t+\varphi_0)$$

$$g_y(t)=g_{y0}+(g_s+g_t\sin(\Omega_1 t))\cos(\Omega_2 t+\varphi_0)$$

where g_{x0}, g_{y0} - components of the constant microacceleration, g_s, g_t - constant and variable components due to rotation in the case that it exists, Ω_1, Ω_2 - frequencies of vibration and rotation, φ_0 - initial angle of inclination. Such kind of presentation includes principal cases of space and temporal variation of microacceleration vector.

The problem is defined by the following undimensional parameters:

L/H - aspect ratio;
$Pr=\mu/(\rho\alpha)$ - Prandtl number;
$Sc=\mu/(\rho D)$ - Schmidt number;
$Gr=g\beta_T\Delta TH^3/\nu^2$ - thermal Grashof number;
$Gr_C=g\beta_C\Delta CH^3/\nu^2$ - concentrational Grashof number;
$Ma=(\partial\sigma/\partial T)\Delta TH/(\mu\alpha)$ - thermal Marangoni number;
$Ma_C=(\partial\sigma/\partial T)\Delta CH/(\mu\alpha)$ - concentrational Marangoni number.

This time-dependent problem is solved numerically using the finite difference technique with "upwind" or central convective approximations on the staggered mesh with the use of the A.J.Chorin's method for velocity-pressure variables [3] and ADI method for vorticity-stream function variables [1].

3. STRUCTURE OF SOFTWARE

The program consists of three components: pre-processor, solver with fast processing capabilities and post-processor. The pre-processor has following important possibilities:

- formulation of the physical problem including the size of region, the initial and boundary conditions, the physical properties of the fluid; also it is possible to impose the time-dependence of the body force value and direction;
- assigning parameters of the finite-difference solver including number of grid nodes along each direction, the approximation method for the convective members of the equations, the method for time step calculation, the scales of physical quantities, time of solving;
- assigning fast processing parameters including position of sensor for indicating current values of the physical fields, point (subregion) and function for displayed flow history, time step for updating the isolines of the fields;
- assigning post-processing parameters including points, subregion and time step for saving values of the fields, moments of the time for saving whole fields for full analysis;
- grid generation;
- task control and several service functions including loading and saving task status, parameters input from a file, system setup, parameters check.

While the problem is solved the following possibilities may be used:

- saving the values of the field in some points on the disk;
- saving the problem status including all the fields for restart or post-processing;
- displaying the name and current time of the task, the current values in some points and subregion;
- displaying the isolines of temperature, impurity concentration or stream function;
- plotting flow history in some point/subregion.

Post-processing capabilities of the program include all the mentioned functions and some other ones for example drawing the sections of the fields in any direction and the visualization of the flow structure by "particles". There are interfaces with common purpose graphical systems.

The dialog system has easy-to-use interface, commands and parameters input is realized by means of pull-down menu, window interface, hot keys. The system represent friendly software with error control, menu prompts and detailed on-line help panels explaining the possible input selections and defaults which ease user input requirements. The described program works on any IBM-compatible hardware with 640 KB RAM, coprocessor and EGA/VGA monitor. The grid may contain up to 5000 nodes. The structure of the program allows to include new numerical and processing methods.

4. TEST EXAMPLES

The program and numerical methods have been tested (see [4]) with the help of well-known test-problems [5,6].

First one is natural convection of air (Pr = 0.71) in a square enclosure with isothermal side boundaries and insulated horizontal walls. Our solution of this problem was obtained at Rayleigh number Ra = Gr Pr = 10^5 using uniform mesh 21 21. Figure 1 shows well-known isolines of temperature (a) and stream function (b) and particles tracks (c). The maximum value of the stream function is ψ_{max} = 9.907, test result of [5] is ψ_{max} = 9.612, and discount is about 3%.

Second one is convection problem in the melt horizontal layer with side heating. Aspect ratio is 4:1, Pr = 0, mesh is 101 26 at two Grashof number, Gr = 2 10^4 and Gr = 4 10^4 in accordance with the GAMM-test [6]. The values of ψ_{max} are 0.4066 (test result ψ_{max} = 0.4155) and 0.4146 (test result ψ_{max} = 0.4167), and discount for above two cases are 0.5% and 2%. Flow structure for these cases are shown in Figure 2 by isolines of stream function.

Let us consider some other examples of problem solutions. Figure 3 shows a result of modelling of ther-

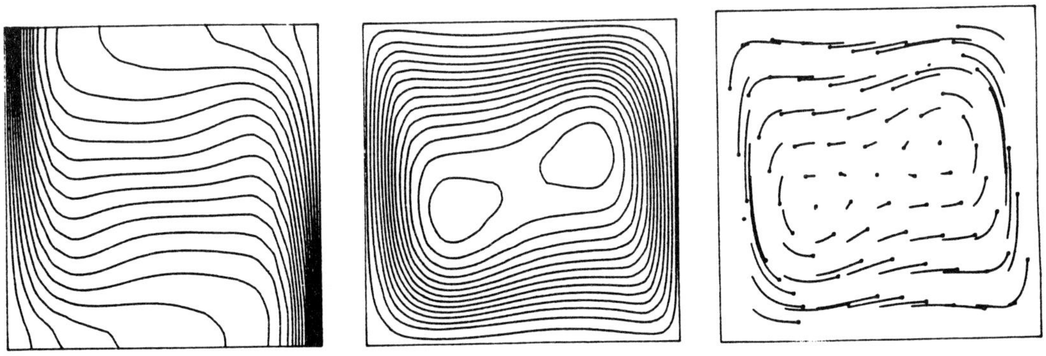

Figure 1. Convection in air due to side heating, Ra=10^5.
Stream function, temperature and particles tracks

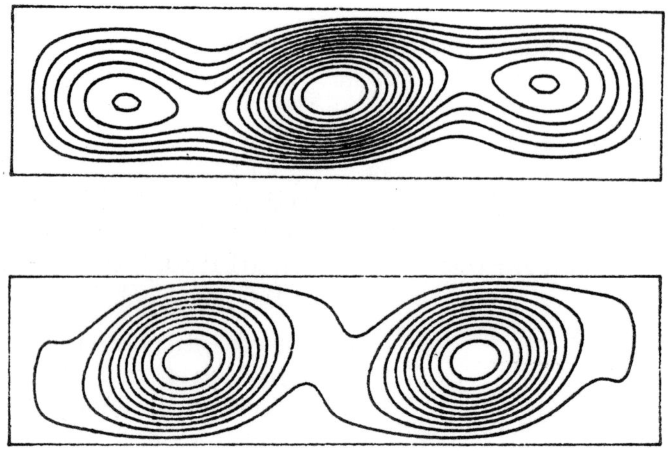

Figure 2. GAMM test. Side heating, RR-case, L/H=4, Pr=0. Straem function. a - Gr=2 10^4, b - Gr=4 10^4

mocapillary and capillary-concentrational convections interaction in a cavity with aspect ratio L/H=6 without body forces. Temperature and impurity concentration at the side wall are constant and different, bottom and free upper surface are insulated. Boundary conditions at the free surface include thermal and concentrational Marangoni numbers, Ma=1000, Ma_c=1100. Periodical moving of rolls in region are obtained like in [1].

Figure 4 shows results of modelling of double diffusion processes induced by interaction between thermal and concentrational convection for case of stable vertical impurity stratification induced by constant concentrational gradient and side heating. In [1] it was shown that layer structure appeared and stepped impurity distribution was formed for some relations between undimensional criteria (thermal and concentrational

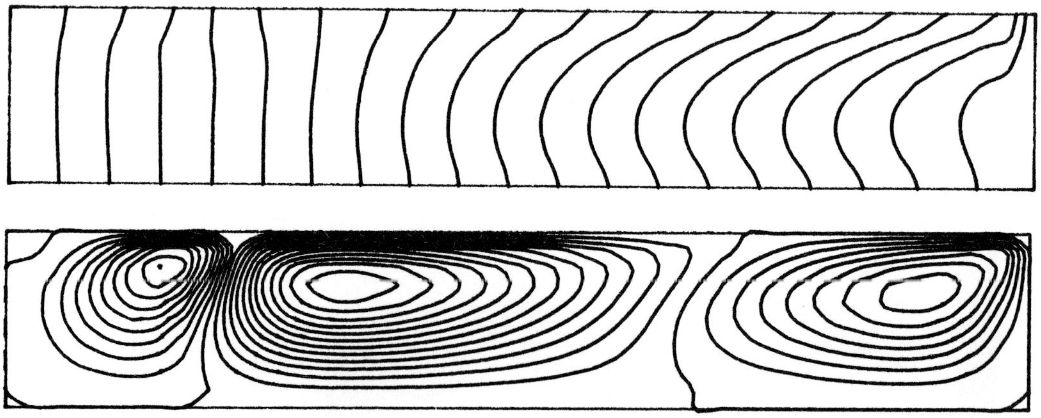

Figure 3. Thermocapillary and capillary-concentrational convecrions interaction in a cavity with aspect ratio L/H=6 without body forces. Ma=1000, Ma_C=1100; temperature, stream function

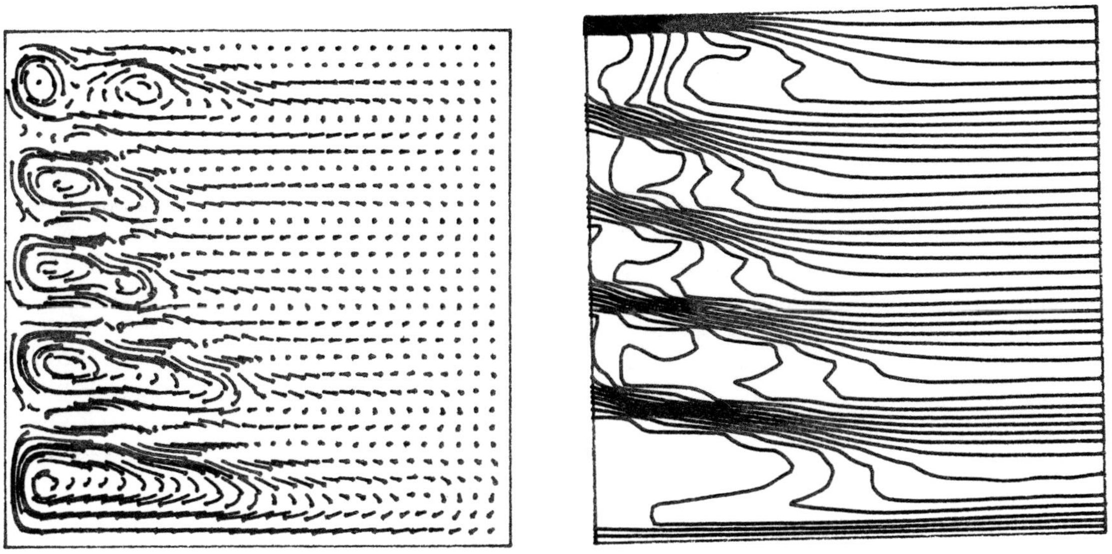

Figure 4. Convection for the case of stable vertical impurity stratification induced by concentrational gradient and side heating, Pr=10, Sc=1000, $Gr=10^6$, $Gr_C=4\ 10^6$; particles tracks, concentration

Grashof numbers, Prandtl and Schmidt numbers). PC results were obtained on 31 31 grid and are in a good agreement with [1], where it was used 129 129 grid.

Calculations and test examples for porous media and convection in enclosure with input and output fliud presented in [7].

One of features of convection in microgravity is residual gravitation convection due to time and space dependency of a microaccelerations induced by dynamics of spacecraft and systems inside (gravity gradient phenomena, vibrations, rotations etc., see, for example [8-11]). We used discribed PC-based system for study convection in a square enclosure with a slow rotation of gravity vector direction. The temporal behavior of Nusselt number for different initial inclination of enclosure and frequency of rotation gravity vector direction, frequency dependence of parameters, "resonance" phenomenon, structures of convection and temperature fields was studied [12]. Slow rotation may be efficient for elimination of temperature (concentration) inhomogeneities.

5. CONCLUSIONS

Presented version of PC-based system COMGA includes gravity-driven thermo-solutal convection, thermo-solutal Marangoni convection, coupling of these types of convection,steady and unsteady(vibration, rotation etc.) microaccelerations for ordinary and porous media. Fixed 2D rectangular geometry is restriction of this system. This version of COMGA includes more friendly software, two types of finite-difference schemes (u-p and ω-ψ) with colour graphics windows for temperature and concentration fields presentation, profiles, animation. For typical microgravity conditions ($Gr=10^4-10^5$) calculations for PC AT/386 may take several minutes.

Cases of convection with side heating, bottom heating, thermal and double-diffusion convection, convection in porous media, convection during a slow rotation and convection in enclosure with input and output fluid have been calculated and tested. This system may be efficient for telescience systems for short and long time duration (drop tower, sound rockets, orbital space stations).

6. REFERENCES

1. V.I.Poleshaev, A.V.Bune, N.A.Verezub, G.S.Glushko, V.L.Grjaznov, K.G.Dubovik, S.A.Nikitin, A.I.Prostomolotov, A.I.Fedoseev, S.G.Cherkasov, Mathematical modelling of convective heat and mass transfer on the basis of Navier-Stokes equations, Nauka, Moscow, 1987 (in Russian).
2. Ermakov M.K., Grjaznov V.L., Nikitin S.A., Pavlovsky D.S., Polezhaev V.I. Specialized software for modelling of convection in microgravity. XXVII COSPAR Meeting, 1990, Adv. Space Res. (in press).
3. A.J.Chorin, Numerical solution of Navier-Stokes equations, Math. Comput. 22, 104, 745-762 (1968).
4. Grjaznov V.L., Ermakov M.K., Nikitin S.A., Pavlovsky D.S. Solving convection problems on a personal computer. Prepr. Inst. for Problems in Mechanics, M., 1990, N 481.
5. G de Vahl Davis, Natural convection of air in a square cavity: a bench mark numerical solution, Int. J. Numer. Methods Fluids, 3, 249-264 (1983).
6. Numerical simulation of oscillatory convection in low - Pr fluids, ed. B. Roux, Notes on Numerical fluid mechanics, 27, 1990.
7. Grjaznov V.L., Ermakov M.K., Nikitin S.A., Pavlovsky D.S., Poleshaev V.I. Modelling of convection in enclosures on the basis of Navier-Stokes equations: results, applications and PC-based dialog system. Extended abstracts of the IV-ICCCBE ´91 Conference, Tokyo.
8. Polezhaev V.I. Influence of gravity gradient on the temperature stratification in a cylinder vessel. "Cosmicheskie issledovania", 1972, XII, No.6 (In Russian).
9. Kamotani Y., Prasad A., Ostrach S. Thermal convection in an enclosure due to vibrations aboard spacecraft. AIAA J., 1981, v.19, No.4, pp.511-516.
10. Polezhaev V.I., Lebedev A.P., Nikitin S.A. Mathematical simulation of disturbing forces and material science processes under low gravity. Proc. 5th European Symp. on Material Sciences under microgravity, 1984.(ESA SP-222).
11. Ermakov S.V., Feonychev A.I. Convection under variable body forces acceleration and impurity microsegregation in crystals. In: Hydromechanics and heat masstransfer under microgravity. Novosibirsk, 1988, pp.20-34.(In Russian).
12. Polezhaev V.I., Ermakov M.K. Thermal convection in microgravity during a slow rotation. Microgravity Sci. Technol., 4, 101, 1991.

OSCILLATORY PENETRATIVE CONVECTION

Aza Azouni[1)], Christiane Normand[2)]

1) Laboratoire d'Aérothermique du CNRS, 4 ter Route des Gardes, 92190 Meudon, France
2) Service de Physique Théorique, Direction des Sciences de la Matière, C.E. Saclay, 91191 Gif-sur-Yvette Cedex, France

ABSTRACT

Convection in an internally heated water layer near the maximum density point is investigated. It is shown that the combined effect of a constant internal heating and the nonlinear density state equation leads to a vertical density stratification which is not uniformly unstable. In fact two unstable sublayers interact through a stable region lying between them. Depending on the respective extent of these three sublayers, the convective perturbations which develop in this system will be either in the form of stationary or oscillatory modes.

Keywords: *Penetrative convection, internal heating, density extremum.*

1. INTRODUCTION

Penetrative convection refers to a configuration in which an unstably stratified layer of fluid is bounded either above or below by a stable layer[1,2]. The present study generalizes this situation by including the case where two unstably stratified layers can interact through a stable layer lying between them. Systems in which two distinct regions are unstable have common features with other systems in which several driving instability mechanisms are present. Here we are mainly interested in how the interaction between different kinds of instability can lead to an oscillatory behaviour at onset (Hopf bifurcation). A short review of the various fields in which oscillatory instability has been shown to occur is presented below.

In homogeneous binary fluid mixtures, the behaviour at the onset of instability depends on the direction in which thermodiffusion is acting (Soret effect). For instance, in a fluid layer heated from above, oscillatory instability is predicted to occur if the denser component migrates towards the upper hot wall[3].

Directional solidification of a binary alloy is another field where an oscillatory instability has been reported to occur when the solidification speed exceeds a specific value[4]. Davis[5] argued that this new oscillatory mode is generated by the coupling of two pure stationary modes respectively induced by a hydrodynamical and a morphological instability.

Instability mechanisms that occur in surface tension driven flows are either attributed to purely hydrodynamical disturbances of isothermal shear flows or to purely thermoconvective disturbances of the Marangoni type. In fact, the coupling of these two pure modes can lead to propagating hydrothermal waves[6].

The convective stability of two superimposed layers of immiscible fluids have shown that, depending on the thicknesses and the physical properties of the two layers, instability sets in either as a stationary or an oscillatory mode[7].

The model of penetrative convection considered in the present study can be viewed as a simplified version of the last example in which the role of the planar interface between two immiscible fluids layers is played by a stable layer of finite height which has the same composition as the two neighbouring unstable layers. Such a multi-layered configuration can be obtained when a nonlinear conductive temperature profile is established in a fluid layer whose density is a non monotonic function of the temperature.

Our original motivation comes from experimental observations in a cylindrical column of water subjected to a vertical temperature difference such that the top boundary is maintained at the temperature $T_m=4°C$ for which the density is extremum[8]. It is known that for closed cavities, the mechanical equilibrium condition is only realized for specific boundary conditions. It has been verified in our experiment, where the lateral walls are not quite perfectly insulating, that it is difficult to establish a linear temperature distribution in the conductive regime and that a better approximation is a quadratic distribution. Since we know that a quadratic distribution is an exact analytical solution for the conductive temperature profile in a horizontal layer where a constant internal heating is present, we decide to investigate this problem which is easier to solve.

The multi-layering of a homogeneous water layer, which results from the combined effect of a quadratic temperature profile and a quadratic variation of the density is considered in the next section.

2. STABILITY ANALYSIS

A water layer of height h is bounded by two rigid surfaces maintained at different temperatures: $T=T_m=4°C$ at the top boundary, and $T=T_0$ at the bottom boundary. The vertical temperature profile

$$T(z) = T_0 + A z^2 + B z \quad (1)$$

is maintained by a constant heat source proportional to A. It is also assumed that the temperature reaches an extremum value $T=T_{ex}$ at z=d where dT/dz=0. Thus, together with the expression of the boundary condition at z=h, this yields an algebraic system for the three unknown quantities : A, B and d. The solutions are expressed in terms of the height ratio $\lambda=h/d$ and the temperature difference: $\Delta T=T_{ex}-T_0$. When h is taken as the length scale, we get

$$T - T_0 = \lambda\ \Delta T\,(2 - \lambda z)\,z. \qquad (2)$$

The state equation for water is approximated by the relation

$$\rho = \rho_m\,[1 - \beta\,(T - T_m)^2], \qquad (3)$$

where ρ_m is the maximum density and β the thermal expansion coefficient. To express the temperature difference $T-T_m$ in a simple way it is convenient to take as a temperature scale the quantity

$$\Delta T_1 \equiv T_m - T_0 = \Delta T\,[1 - (1 - \lambda)^2\,], \qquad (4)$$

and to introduce a new parameter μ related to λ by

$$\mu = \frac{2\lambda}{1 - (1 - \lambda)^2}. \qquad (5)$$

Thus, we get: $T-T_m= \Delta T_1\,\Theta(z)$, with

$$\Theta(z) = (1 - \mu)\,z^2 + \mu z - 1. \qquad (6)$$

The vertical temperature profile is drawn on Figure 1 for different values of μ. The respective positions, z_m and z_{ex}, of the isothermal lines $T=T_m$ and $T=T_{ex}$ are given on Figure 1 as functions of μ. The whole layer divides in three sublayers: the bottom region where $T_0<T<T_m$ and the top region where $T_{ex}<T<T_m$ are unstable owing to the density maximum at $T=T_m$ while the intermediate region $T_m<T<T_{ex}$ is stable.

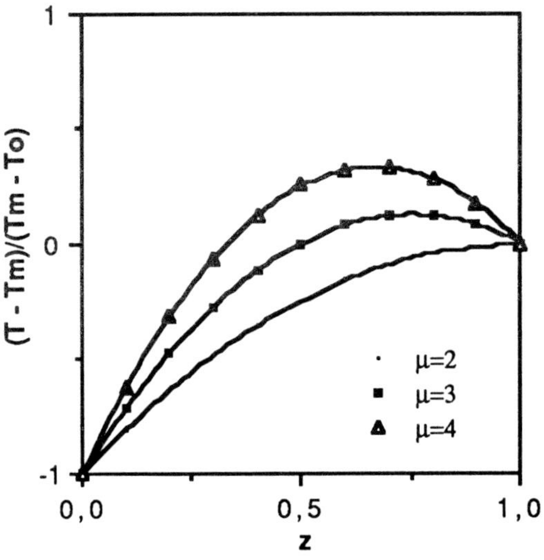

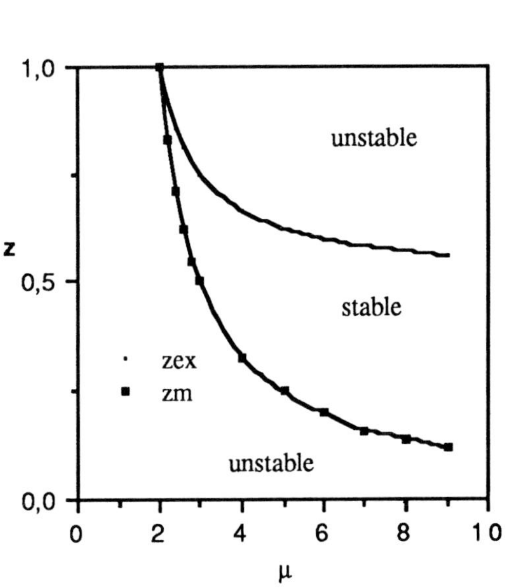

Figure 1: vertical temperature profile (upper graph) and position of the isotherms $T=T_m$ and $T=T_{ex}$ (lower graph).

The stability analysis of this system has been done[9] for perturbations of the form

$$\{w, \theta\} = e^{st}\, h(x, y)\, \{f(z), g(z)\}, \quad (7)$$

where w and θ are respectively the vertical velocity and the temperature disturbances; the function h(x, y) is periodic with a horizontal wavenumber a.
Therefore the governing equations for the perturbations are

$$\left(\frac{s}{Pr} - £\right) £\, f = Ra\, \Theta(z)\, a^2\, g \quad (8a)$$

$$[s - £]\, g = f\, D\Theta(z) \quad (8b)$$

where $D \equiv \frac{d}{dz}$ and $£ \equiv D^2 - a^2$.
The associated boundary conditions at z = 0 and z =1 are

$$f = Df = g = 0. \quad (8c)$$

The two parameters Pr and Ra are respectively the Prandtl number ($Pr=\nu/\kappa$) and the Rayleigh number defined as

$$Ra = \frac{2\, g\, \beta\, h^3\, \Delta T_1{}^2}{\nu\, \kappa}. \quad (9)$$

The set of equations (8) has been solved numerically[9] and the results are presented below.

3. RESULTS

The neutral stability curves have been obtained for different values of μ. Starting with the value μ=2 for which the whole layer is unstable we get a neutral curve similar to that of ordinary convection. When the value of μ is increased, two distinct unstable sublayers are present. Owing to penetration, convective motions do not remain confined in the unstable sublayers but interact through the stable intermediate layer and complex neutral curves are found. Representative examples are shown in Figure 2.

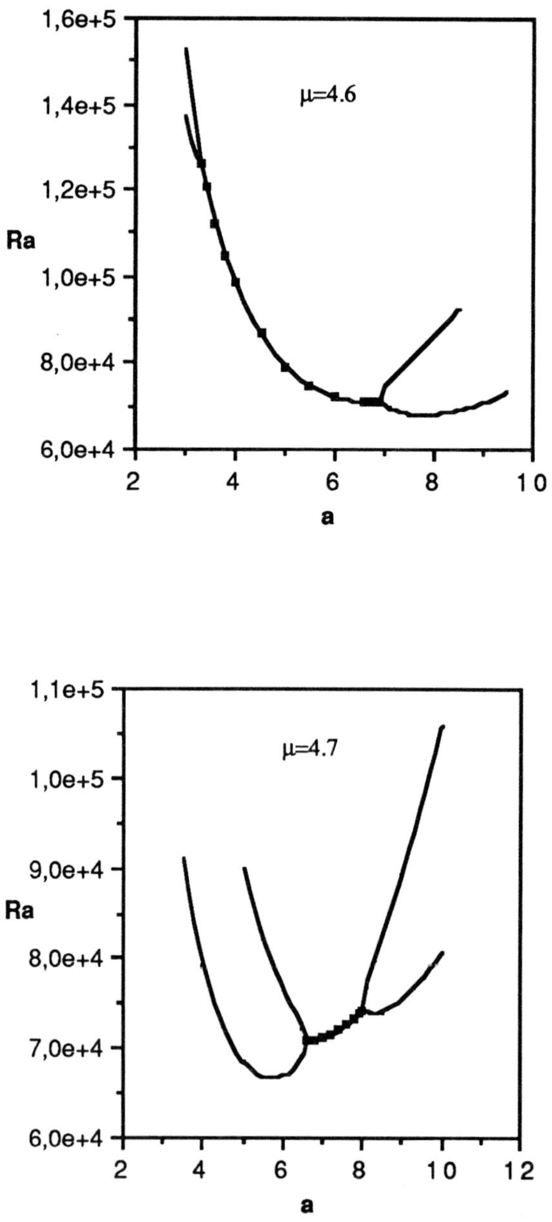

Figure 2: Neutral stability curves for stationary modes (full line) and oscillatory modes (square dotted line).

The neutral curve for stationary modes now consists of two disconnected branches, and in the gap between them oscillatory modes are found. For $\mu=4.6$, the critical point is located on the right stationary branch, while for $\mu=4.7$, the critical point is located on the left stationary branch. For a specific value of $\mu\equiv\mu*=4.65$, the minimum on the left branch and the minimum on the right branch occur for the same value of the Rayleigh number but for two different wavenumbers, leading to a degenerated situation. In fact, around this value, $\mu=\mu*$, the critical point is located on the oscillatory curve[9].

The vertical distribution of the velocity and temperature disturbances are shown on Figure 3 for the critical stationary modes corresponding to $\mu=4.6$ and $\mu=4.7$. In the first case, the convective disturbance takes the form of two superposed counter-rotating rolls and the temperature keeps a constant sign through the whole layer with a maximum located in the bottom unstable layer. In the second case, convection takes the form of two vertical corotating rolls, the most intensive being at the top of the layer.

The critical oscillatory modes are shown on Figure 4 for two times, t_1 and t_2, separated by a quarter of a period and for $\mu=4.65$. One can see that the time variation of the vertical profiles corresponds to an oscillation between the two stationary modes we have previously identified (Figure 3).

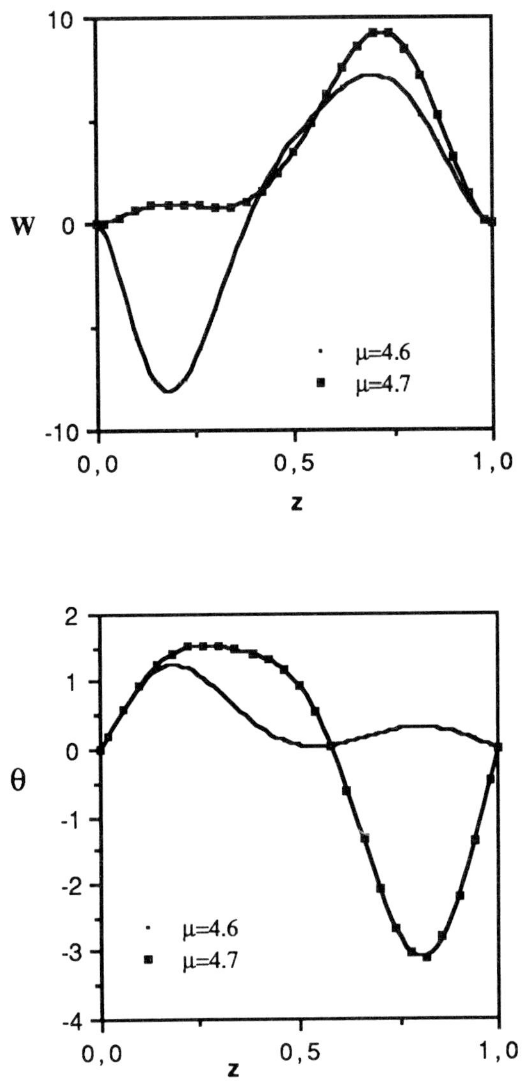

Figure 3: velocity (upper graph) and temperature (lower graph) profiles in the stationary regime.

Thus, around the value $\mu=\mu*=4.65$, instabilities appear simultaneously in the two unstable layers and interact through the stable layer, giving rise to oscillatory modes. Moreover near $\mu=\mu*$ the critical values of the Rayleigh number for the two stationary modes and the oscillatory mode are close to each other and one can expect that the nonlinear dynamics near this tricritical point will provide a complex behaviour.

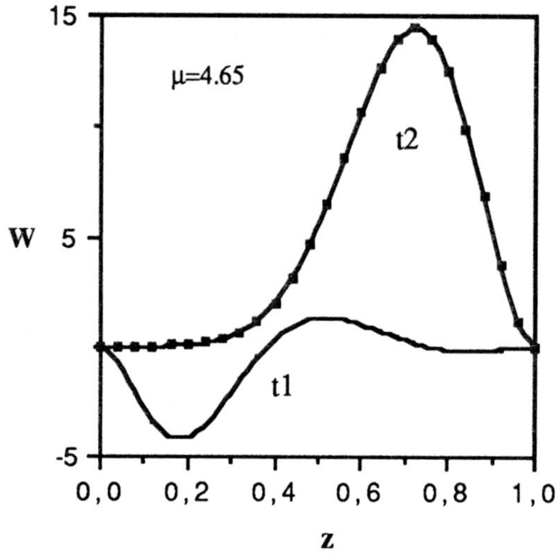

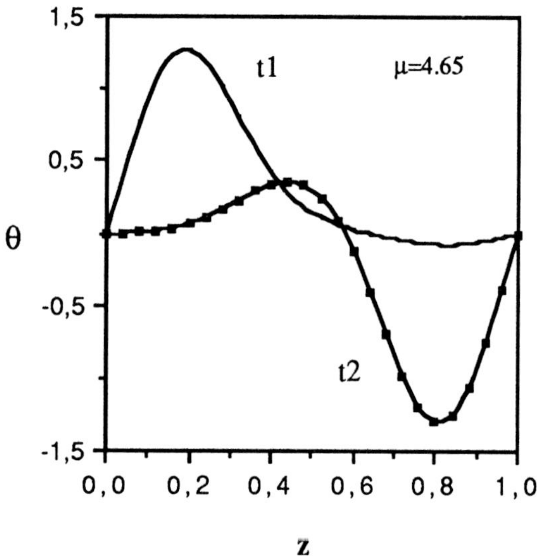

Figure 4: Velocity (upper graph) and temperature (lower graph) profiles in the oscillatory regime.

4. REFERENCES

1. G. Veronis, *Astrophys. J.* **137**, 641 (1963).
2. J. A. Whitehead, M. M. Chen, *J. Fluid Mech.* **40**, 549 (1970).
3. D. T. J. Hurle, E. Jakeman, *J. Fluid Mech.* **47**, 667 (1971).
4. S. R. Coriell, M. R. Cordes, W. S. Boettinger, R. F. Sekerka, *J. Crystal Growth* **49**, 13 (1980).
5. S. H. Davis, *J. Fluid Mech.* **212**, 241 (1990).
6. M. K. Smith, S. H. Davis, *J. Fluid Mech.* **132**, 119 (1983).
7. G. Z. Gershuni, E. M. Zhukhovitskii, *Sov. Phys. Dokl.* **27**, 531 (1982).
8. M. A. Azouni, C. Normand, *Geophys. Astrophys. Fluid Dynamics.* **23**, 209 (1983).
9. C. Normand, A. Azouni, to be published in Phys. Fluids A (1992).

CONTROL OF TWO-PHASE FLOW HEAT TRANSFER AND HYDRODYNAMICS BY ELECTRIC FORCES.

M.K.Bologa, S.M.Klimov and S.I.Chuchkalov

Institute of Applied Physics of the RM Academy of Sci.,
5, Grosul str., 277028 Kishinev, Republic Moldova, USSR

ABSTRACT

Results of experimental investigation of DC electric field influence on refrigerant-113 two-phase flow heat transfer and hydrodynamics in narrow vertical annulus are presented. Boiling heat transfer intensification in an electric field is considered to be caused by the liquid microlayer thinning resulting from a velocity increase of the interfacial boundary movement and is accompanied by a reduction of the "dry spots" due to splitting of large vapour bubbles and the development of intensive electromechanical convection. Condensation heat transfer intensification is provided by augmentation of vapour flow effect on the condensate film and by additional interaction of vapour flow and the transverse droplet flow of liquid dispersed by the field.

Keywords: two-phase (boiling and condensation) heat transfer, hydraulic resistance, electrohydrodynamic (EHD) effects, electromechanical convection.

1.INTRODUCTION

The investigations of boiling and condensation heat transfer in organic liquids in an electric field (Refs.1,2) show that under certain conditions an electric field effect on two-phase media results in considerable quantative and qualitative changes in thermal and hydrodynamic phenomena. Heat transfer in microgravity could be accomplished by a closed-cycle electrohydrodynamic evaporation-condensation system (EHD ECS). This requires study of the EHD effect on boiling and condensation in special structures such as narrow ducts or porous structures. An objective is to determine methods for initiation and control of stable two-phase flow in equipment with these elements; especially the range of flow regimes for effective drainage of condensate and departure of vapour from heat-exchange surfaces. In additional, there are essential requirements for compactness, low system weight, reliability and long life expectancy.

These considerations shaped our investigations on the influence of a DC electric field on the heat transfer and hydrodynamics of two-phase flow. The chosen configuration is a narrow vertical annulus in a closed forced circulation loop.

2.EXPERIMENTAL SET-UPS AND PROCEDURES.

The research on condensation, with the description and perfomance of the experimental set-up and the measurement techniques, is described in Reference 3. The configuration for boiling research is given in Figure 1. In both cases the test fluid flow through the system is supported by a pump. The flow rate, when boiling is studied, is determined by a pressure drop measurement from a calibrated Venturi tube placed in the single-phase flow part of the loop before the test section. The halogenated refrigerant R-113 ($C_2Cl_3F_3$) was used as test fluid.

The test section is made up of annulus with small annular clearance. The in-

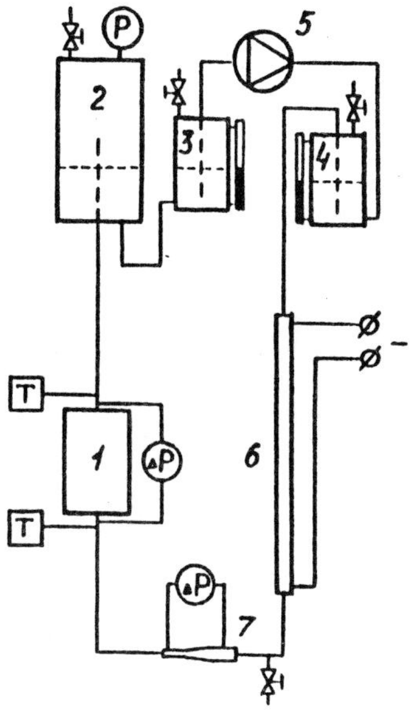

Figure 1: Scheme of the experimental set-up used to research flow boiling. 1 - Test section; 2 - Condensator; 3,4 - Buffer tanks for condensate; 5 - Pump; 6 - Electroheater; 7 - Venturi tube.

ner tube of the annulus is used as HV electrode. The outer thin-walled stainless steel tube (inner diameter 13,4 mm) is grounded and is heated or is cooled by water from a termostate. The annulus for the boiling set-up is 120 mm long with annular clearance S=0.7 mm, and that for the condensation set-up is 220 mm long with S=1,7 mm (in an other case S=2,7 mm).

out at atmospheric pressure (10^5 Pa) measured at the test section entrance. The boundary condition at the heat-exchange surface was taken as $\overline{T}_w \approx$ const. The operating conditions of the experimental runs are given in the Table 1 (where $\Delta\overline{T}_{sw}$ - the mean difference between the temperature of saturation and that at the heat-exchange surface; W - flow velocity at the annulus entrance; ΔP_{Vt} - pressure drop over the Venturi tube; ΔT_{subc}- temperature of subcooling; E - electric field strength).

In this study the influence of DC electric field, of the flow rate, of $\Delta\overline{T}_{sw}$ on the heat transfer and the pressure drop over the annulus during flow boiling and flow condensation are investigated.

2. EXPERIMENTAL REZULTS AND DISCUSSION

The most typical experimental results for heat transfer and pressure drop during subcooled R-113 flow boiling in the annulus are presented in Figures 2-4. It can be seen from Figure 2 that the increase of the flow rate at the annulus entrance (Curves 1-3) and the subcooling temperature growth (Curves 2,2') brings down the heat transfer coefficient and leads to an increase of the critical heat flux. The explanation is aided by a previous study of boiling in annuli submerged in a pool of refrigerant with a visualization

TABLE 1 The operating conditions of the experimental runs

process	$\Delta\overline{T}_{sw}$,°C	W, m/s	ΔP_{Vt} , Pa	ΔT_{subc},°C	E, kV/sm
BOILING	0,1-35,0		40 360	8,0-20,0	0-85
CONDENSATION	0,1-20,0	0-10,0	-	-	0-85

The refrigerant moving upwards (the boiling set-up) or downwards (the condensation set-up) in the annulus is boiled (or is condensed) on the inner surface of the outer tube. The heat-exchange surface mean temperature is measured with a thermistor, the total pressure drop ΔP over the annulus-with a U-shaped liquid differential manometer. All the experiments are carried

system (Ref.1). Boiling heat transfer had been determined earlier to have much in common with the slug-churn regime of two-phase flow in the investigated range of geometrical and operating conditions. It is observed that the increase of both the flow rate and the subcooling temperature is succeeded by replacement of flow regimes along the annulus. This prevents the inten-

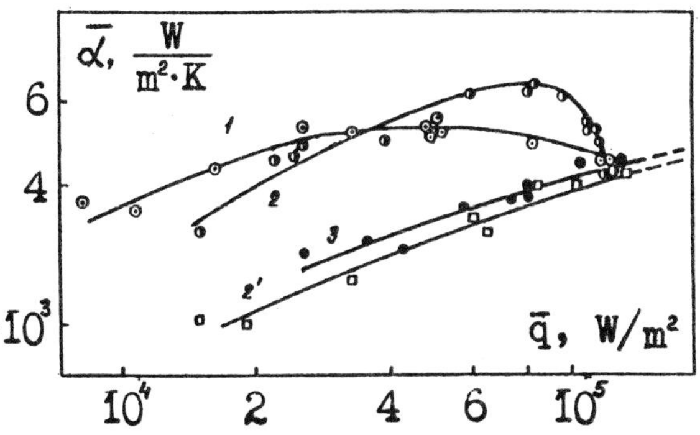

Figure 2: Mean heat transfer coefficient versus mean heat flux density curves during subcooled R-113 flow boiling inside the annulus.
W(1): W(2,2'): W(3)= 1,0 : 1,4 : 2,8;
ΔT_{subc},°C: 11,3(1-3), 17,0(2').

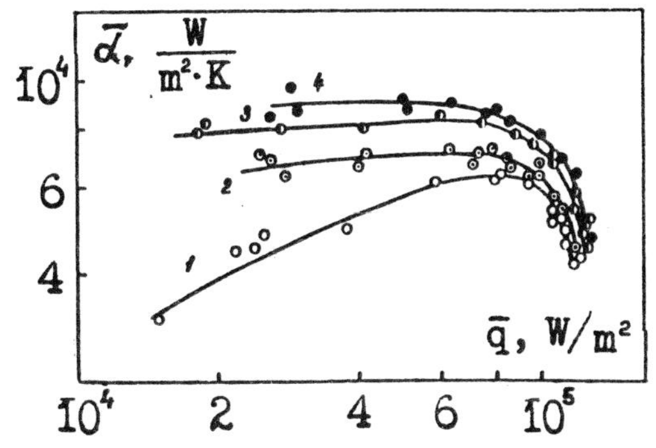

Figure 3: Boiling mean heat transfer coefficient of subcooled R-113 under the influence of an electric field for ΔT_{subc} = 11,3°C.
E, kV/sm: 0(1), 50(2), 71(3), 85(4).
(Value of W is the very same as for Curves 2.2' in Figure 2).

sive generation of large-scale vapour volume. Because of this the void fraction is decreased which is followed by a reduction of the heating surface area occupied by the liquid microlayer that appears under vapour bubbles. The latter phenomenon is responsible for the reduction of the microlayer evaporation contribution to the total heat transfer.

At this instance the heat flux increase results in slug-churn regime renewal or even annular regime formation, and so the microlayer evaporation becomes the principal contribution to the total heat transfer again. Thus is explained the independence of heat transfer intensity from subcooling and flow rate under high heat flux.

In the flow rate and subcooling temperature ranges that were investigated, the electric field strongly enhances heat transfer but changes the critical flux value hardly at all (Figure 3). In our previous work (Ref.1) the augmentation of boiling heat transfer by an electric field was stated to be caused by microlayer thinning resulting from a velocity increase of the interfacial boundary movement. The intensification of heat transfer had been found to be accompanied by a reduction of the "dry spots" (which are formed on the heating surface) due to splitting of large vapour bubbles and the development of the intensive electromechanical convection. (It should be mentioned that the liquid jet transport from the microlayer by electric forces thins the microlayer also. The role of jet transport of liquid in microlayer thinning and in the total heat transfer had been considered to be negligible (Ref.1). However, the obtained results on condensation heat transfer in an electric field (Figures 5,6) show that it can be significant). The described phenomena appear to hold true for forced flow boiling heat transfer. For this case the boiling features, the two-phase flow regimes and its ranging along the annulus are markedly affected by the cross section of the annulus, the heat flux density, heating conditions and subcooling of refrigerant.

A more detailed explanation for the descovered relations requires closer study of two-phase flow hydrodynamics and structures, microlayer dynamics and interior boiling characteristics for channels of various geometry and orientation.

Figure 4 presents data that demonstrate the influence of flow rate and electric field on the hydraulic resistance during flow boiling. In the large range of mean heat flux density the flow rate influence on hydraulic resistance is more markedly than the void fraction influence is. Analogous results have been obtained for flow boiling of subcooled water in pipes with small diameter (Refs.4,5).

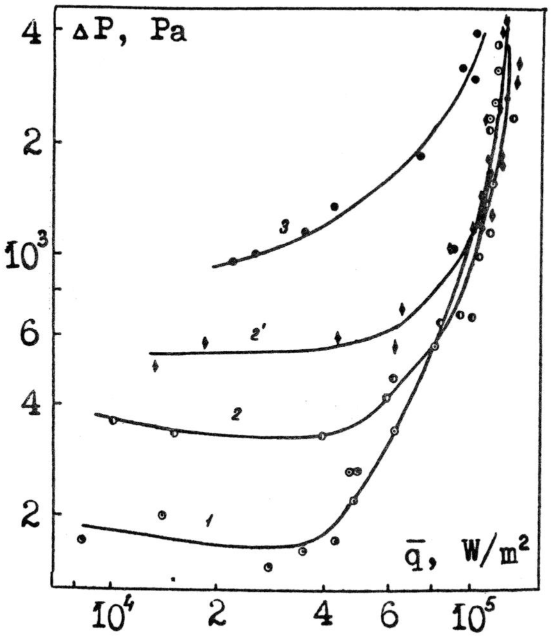

Figure 4: Variation of total pressure drop with boiling mean heat flux density and flow rate for ΔT_{SUBC} = 11,3°C. W(1): W(2,2'): W(3) = 1,0 : 1,4 : 2,8; E, kV/sm: 0(1-3), 71(2').

The application of an electric field was found to raise hydraulic resistance. Depending on flow regime it can be explained by electroreological phenomena (the effective viscosity increase of a two-phase medium under electric field effect), electromechanical convection in the annulus and the interaction of a two-phase flow vapour core and a jet-droplet cross-flow of liquid dispersed by the field.

Thus, high flow rate furthers stability of boiling heat transfer under high heat flux only. At the same time the hydraulic resistance is larger than in an electric field. Therefore, high rate of two-phase flow with boiling inside annulus may be reasonable to grow the critical heat flux only. Under moderate heat flux density and electric field, modest flow rate and small subcooling are more preferable to optimize the heat transfer and hydraulic perfomances. In weightlessness the two-phase flow rate should be sufficient to transport vapour from a heating surface.

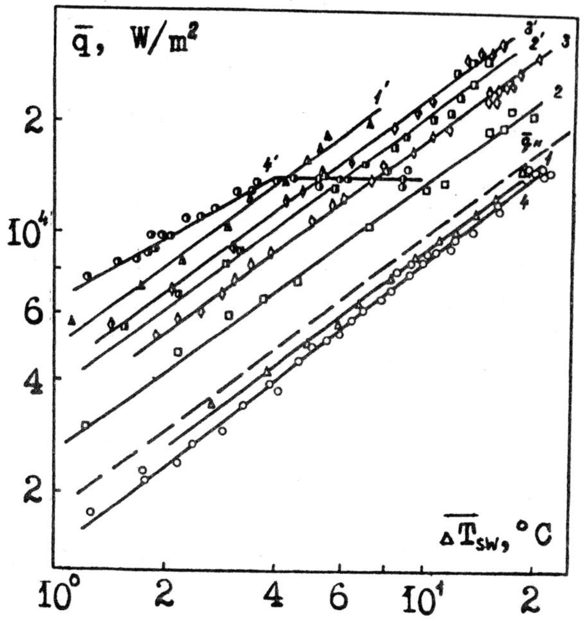

Figure 5: Condensation mean heat flux density versus the mean temperature difference curves.
S, mm: 1,7(4,4'), 2,7(1-3,1'-3');
W, m/s: 2,2(1,1',4,4'), 6,9(2,2'), 8,8 (3,3'); E, kV/sm: 0(1-4), 82(1'-4').

In Figure 5 are plotted experimental relations of heat flux density $\bar{q}_E$ during flow condensation versus the temperature difference. The curves are for the annulus geometry and have as parameters the steam velocity at the entrance W and the electric field strength. The values of $\bar{q}_E$ are generally larger than predicted from Nusselt's theory for quiescent vapour condensation ($\bar{q}_N$) or measured during flow condensation without an electric field ($\bar{q}_0$).

During flow condensation of freons shear stress effect on the condensate film and on the heat transfer is considerable for not great velocity of vapour flow even (Ref.6), that was observed in our experiments as well (Curves A,B in Figure 6).

The electric field induces EHD forces (owing to the presence of free and bound charges (Ref.7)) that have an effect upon the vapour-condensate interface. As a result, the interface is destabilized, the condensate film that covers the heat-exchange surface is disrupted and is dispersed with the

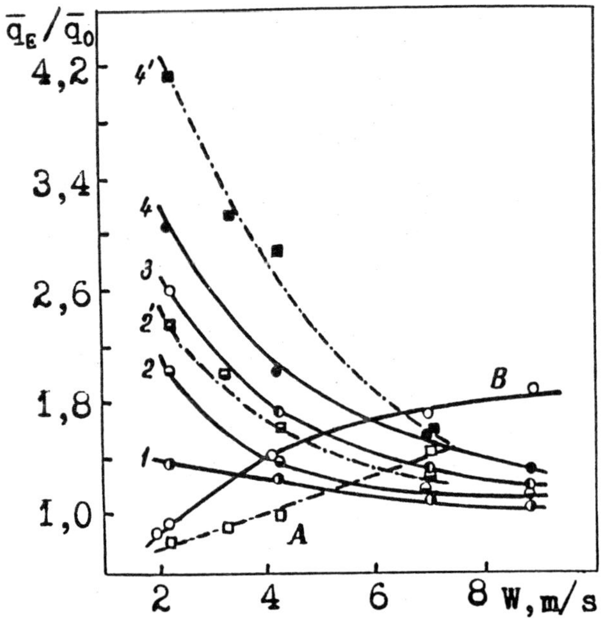

Figure 6: Variation of condensation heat transfer enhancement by an electric field effect with entrance steam velocity (Curves A,B are relations $\bar{q}_0/\bar{q}_N=f(W)$).
S, mm: 1,7 (A,2',4'), 2,7 (B,1-4);
E, kV/sm: 49(1), 60(2,2'), 74(3), 82 (4,4').

formation of jets and droplets into the inter-electrode gap. The interaction of vapour flow and jet-droplet cross-flow of liquid furthers more effective condensate removal from the annulus. As a consequence of these phenomena, the liquid film becomes thinner and thereby condensation heat transfer is intensified. Thus EHD phenomena, as develope in a channel, give rise to heat transfer augmentation that have been related to the electric field effect on condensate film. The greatest heat transfer enhancement (as much as 4,1 times) was recorded for minimum steam velocity at the annulus entrance (Figure 6). This is explained by the long period of existence of the liquid film inside the annulus and hence from the maximum value of electric charge induced in the liquid volume by the field. On that account the Coulumb interaction of the electric field and the medium is largest. The decrease of heat transfer intensification, as the steam velocity at the annulus entrance is increased, is produced partly by the reduced life time of the liquid film in the electric field, and partly from the fact that the high-speed vapour flow spreads the liquid film at the walls and so diminishes the deformation of the interface by electric forces, thereby hindering dispersion of liquid into the vapour flow core.

In smaller annular clearances, the growth of EHD effects results in greater intensification of heat transfer (Curves 4, 4' in Figure 5,- the gently sloping part of the Curve 4' corresponds to complete condensation in coming vapour flow; Curves 2',4' in Figure 6).

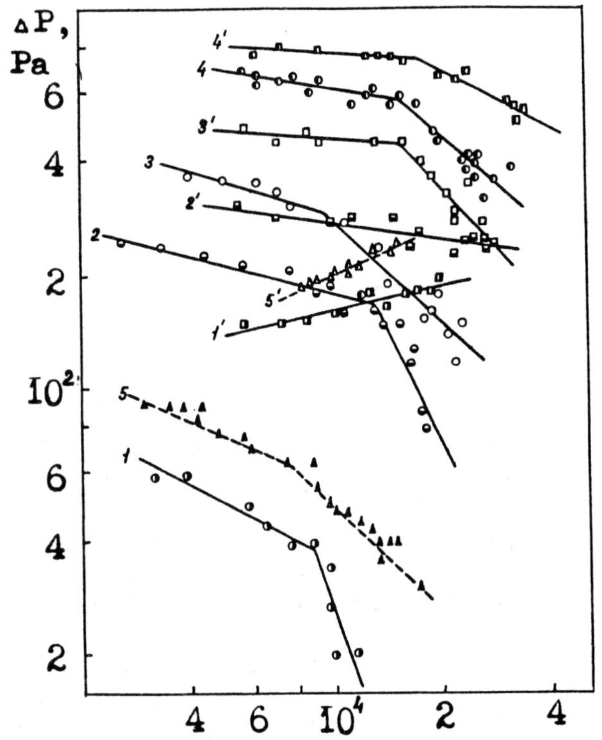

Figure 7: Variation of total pressure drop with condensation mean heat flux density.
S, mm: 1,7 (5,5'), 2,7 (1-4,1'-4');
W, m/s: 2,2(1,1',5,5'), 4,2(2,2'), 6,9 (3,3'), 8,8(4,4'); E, kV/sm: 0(1-5), 82(1'-5').

The notions developed above are in qualitative agreement with the experimental data for pressure drop at the annulus during flow condensation (Figure 7). Electric field strength increment and annular clearance decrement lead to more intensive interaction of vapour flow and jet-droplet cross-flow of liquid, and so raises resistance to the two-phase flow. As mean heat flux density increases, the "suction" of vapour along the annulus augments. This decreases the mean velocity of vapour flow inside the annulus and so reduces the total pressure drop over the annulus. The pronounced break in most curves in Figure 7 appears to have been caused by a change in the character of interaction between vapour flow and liquid film at part of the annulus or along its full length.

4. CONCLUSIONS

The previous observations allow to draw the following conclusions:

i. The DC electric field effect considerably enhances heat transfer during both flow boiling and flow condensation inside narrow annuli. This allows effective control of the thermal regime of heat transfer equipment and so reduction of equipment size. The energy expenditure for the creation of the electric field is insignificant in comparison with the heat flux increase (0,1-1% only). Since liquid is evacuated out of condensator by the joint action of vapour flow and electric field, this method may be recommended for application in microgravity.

ii. Via the EHD effect one can control the hydrodynamic resistance of the annulus during two-phase flow, and so control the mass flow rate in the circulation loop. Measured total pressure drop values in annuli show that the necessary hydraulic pressure head for fluid transport in the loop can be provided with a EHD pump which is more reliable and produces less vibrations than a mechanical pump.

Thus, one may assume the practical possibility for the development of closed EHD ECS, with advantages in compactness, high heat transfer characteristics, convenience and immediate control of operating regime.

5. REFERENCES

1. M.K.Bologa, G.F.Smirnov, A.B.Didkovsky and S.M.Klimov, Boiling and Condensation Heat Transfer in an Electric Field (Shtiintza, Kishinev, 1987) (in Russian).

2. P.G.H.Allen, in Proc. 2nd UK Nat. Conference on Heat Transfer, vol.1 (Glasgow, 1988) 861.

3. M.K.Bologa, S.I.Chuchkalov, S.M. Klimov and A.B.Didkovsky, Izv. SO AN SSSR, Ser.tekhn.nauki, 2 (1990) 126 (in Russian).

4. P.Saha and N.Zuber, in Proc. 5th Int. Heat Trans. Conf., vol. IY (Tokyo, 1974) 175.

5. Yu.A.Zeigarnic, I.V.Kirillova, A.I. Klimov and E.G.Smirnova, in Teplofizika i gidrogazodinamika protzessov kip. i kond.. Mat.Vses.konf., vol. IY, part 3 (Riga, 1986) 108 (in Russian).

6. O.P.Ivanov and V.O.Mamchenko, Kholodilnaya tekhnika, 6 (1973) 23 (in Russian).

7. M.K.Bologa, F.P.Grosu and I.A.Kozhukhar, Electroconvection and Heat Transfer (Shtiintza, Kishinev, 1977) (in Russian).

THE INFLUENCE OF THERMOMAGNETIC CONVECTION ON HYDRODYNAMICAL DRAG IN A WEIGHTLESSNESS STATE

Viktor K. Polevikov

Byelorussian State University, Minsk, 220080, USSR

ABSTRACT

The hydrodynamical drag of the cylinder, coated with a layer of magnetic fluid which is kept on the solid surface by magnetic field, is studied numerically. It was found earlier that the low-viscous magnetofluid coating reduces a drag 2-3 times. The subject of present investigation is how to decrease the drag at zero-g by means of the thermomagnetic convection phenomenon in the coating layer. Drag reduction in this case is attained due to flow intensification at the magnetic-nonmagnetic fluid interface by the convective mechanism. It is shown that the drag value may fall off to zero with increasing magnetic Grashof number.

Keywords: Cylinder, Magnetofluid Coating, Drag Reduction, Zero-g, Thermomagnetic Convection, Numerical Modelling.

1. INTRODUCTION

Works (Refs.1-3) deal with numerical modelling of separated flow past a solid cylinder coated with a magnetic fluid layer kept on the solid surface by a nonuniform magnetic field. It is found that the low-viscous magnetofluid coating reduces a cylinder drag 2-3 times at the Reynolds numbers of 50-100 (Ref.1). If such a coating has high thermal conductivity, then heat transfer from a heated cylinder is increased as many as 2-3 times (Ref.2). These effects manifest themselves the stronger, the higher is the Reynolds number.

In (Ref.3), it is shown that fluid magnetization as a function of temperature in a nonuniform magnetic field may induce, in the coating layer, the so-called thermomagnetic convection capable to change substantially the structure of separated flow around a heated cylinder at zero-g.

This study is an attempt to use the thermomagnetic convection phenomenon for reducing the drag at zero-g. Thermal conditions have been developed so that the convective motion direction at the magnetic-nonmagnetic fluid interface would coincide with the external flow one. Drag reduction in this case is attained due to interface flow augmentation caused by the convective mechanism, but not to decreasing of coating viscosity.

2. MATHEMATICAL MODEL

Consider an infinite viscous nonmagnetic fluid flow, having a constant temperature T_* and moving in the straight direction with a constant velocity U , past a long transverse cylindrical conductor of radius R which is kept at a constant temperature $T_c > T_*$ and coated with a magnetic fluid layer immiscible with the flow liquid. A conductor current strength I is assumed to be sufficiently high so that the magnetic-nonmagnetic fluid interface may be considered to be a circumference a > R in radius (Ref.1).

Under these conditions and with no gravitation, in the coating layer there appears circulation due to thermomagnetic convection, whose direction at the interface is opposite to the external flow (Ref.3). In this situation shear stresses grow, so the drag increases. To change an unfavourable convective flow direction, a thin nonconducting plate-fin is placed at the cylinder leading edge. A plate width is equal to a coating thickness and its temperature, with respect to cylinder one, is T_c.

Introduce dimensionless variables by choosing, as measuring units of a distance, velocity, and temperature, a cylinder radius R , undisturbed flow velocity U , and a temperature drop $\Delta T = T_c - T_*$, respectively. Use a polar coordinate system r , φ with a pole on the cylinder axis. If necessary, denote by subscript 1 the quantities referring to a magnetic fluid and by subscript 2, external flow.

To simplify the mathematical model, we use the inductionless approximation which enables us to neglect the effect of magnetic fluid nonisothermity on a magnetic field. In such a case, there is no need to solve the Maxwell equations as their solution for a magnetic field of a cylindrical conductor is known. Applicability of the inductionless approximation for thermomagnetic convection problems is shown elsewhere (Refs.4,5).

Under the assumptions made, two-dimensional steady-state thermomagnetic convection in the layer $1 \le r \le \delta = a/R$ and nonmagnetic fluid flow outside it ($r \ge \delta$) are described by the dimensionless equations (Refs.2-4):

$$\frac{1}{r}\,\frac{\partial(\psi_i, T_i)}{\partial(\varphi, r)} = A_i \nabla^2 T_i,$$

$$\frac{1}{r}\,\frac{\partial(\psi_i, \omega_i)}{\partial(\varphi, r)} = B_i \nabla^2 \omega_i + F_i,$$

$$\nabla^2 \psi_i + \omega_i = 0; \quad i = 1, 2;$$

where

$$A_1 = \gamma / \mathrm{Re}\,\mathrm{Pr}\,\sigma\lambda, \quad A_2 = 1/\mathrm{Re}\,\mathrm{Pr},$$

$$B_1 = \mu/\mathrm{Re}\,\lambda, \quad B_2 = 1/\mathrm{Re},$$

$$F_1 = -B_1^2\,\mathrm{Gr}_m \frac{1}{r^3}\,\frac{\partial T_1}{\partial\varphi}, \quad F_2 = 0;$$

T, ψ , and ω are a dimensionless temperature, stream-function, and vorticity;

$\gamma = k_1 : k_2$, $\sigma = c_1 : c_2$, $\lambda = \rho_1 : \rho_2$, $\mu = \eta_1 : \eta_2$ are the ratios of the magnetic and nonmagnetic fluid thermal conductivity, specific heat, density, and dynamic viscosity, respectively;

$\mathrm{Re} = U R \rho_2 / \eta_2$ is the Reynolds number; $\mathrm{Pr} = \eta_2 c_2 / k_2$ is the Prandtl number;

$\mathrm{Gr}_m = \mu_0 K_1 I \Delta T \rho_1 R / 2\pi \eta_1^2$ is the magnetic Grashof number, μ_0 is the magnetic permeability of vacuum, K is the pyromagnetic coefficient.

In the case i=1 these equations are determined on the coating ring region $1 \leq r \leq \delta$, and in the case i=2, on the outer region $r \geq \delta$. The numbers Re, Pr, Gr_m and ratios δ, γ, σ, λ, μ are determining parameters of the problem.

When a numerical method is applied the determination domain must be assigned finite, and the flow is therefore considered undisturbed at rather a great distance from the cylinder $r \geq r_* >> \delta$. Moreover, let us assume a solution to be symmetric by confining to the angular coordinate range $0 \leq \varphi \leq \pi$. On the solid surface of a cylinder and a fin, the attachment conditions are stated and at the interface, continuity conditions for a velocity and temperature, the conditions for surface permeability, shear stress and heat flux balances. The mathematical formulation of the boundary conditions is of the form:

$$r=1: \ T=1, \ \psi=0, \ \partial\psi/\partial r = 0;$$

$$r=r_*: \ T=0, \ \psi=r_*\sin\varphi, \ \omega=0;$$

$$\begin{matrix} r\geq 1, \ \varphi=0 \\ r>\delta, \ \varphi=\pi \end{matrix} : \ \partial T/\partial\varphi=0, \ \psi=0, \ \omega=0;$$

$$1\leq r\leq\delta, \ \varphi=\pi: \ T=1, \ \psi=0, \ \partial\psi/\partial\varphi=0;$$

$$r=\delta: \ T_1=T_2, \ \gamma\,\partial T_1/\partial r=\partial T_2/\partial r, \ \psi=0, \ \partial\psi_1/\partial r=\partial\psi_2/\partial r,$$

$$\mu\left(\omega_1+\frac{2}{\delta}\frac{\partial\psi_1}{\partial r}\right)= \omega_2+\frac{2}{\delta}\frac{\partial\psi_2}{\partial r}.$$

A dimensionless magnetofluid-coated cylinder drag is calculated by the formula from (Ref.1):

$$W = \delta\,\mathrm{Re}\int_0^{\pi}\left(\mathrm{Re}\left(\frac{\partial\psi}{\partial r}\right)^2\cos\varphi + 2\,\delta^2\frac{\partial}{\partial r}(\omega_2/r)\sin\varphi\right)\Big|_{r=\delta}d\varphi.$$

3. COMPUTATIONAL ALGORITHM

The problem was solved by the finite-difference method. Constructing of a grid in the coating layer and in the outer semi-ring region was made in the same manner as in Refs.1-3. A dimensionless "infinity" radius r_* was chosen equal to 20 radii of a conductor, i.e. $r_*=20$. In terms of a polar angle the grid step was assigned equal to $\pi/40$ and in terms of a radius the uniform grid was constructed in the coating layer with 10 partitions in number, and within the range $\delta \leq r \leq r_*$ the nonuniform grid was constructed with a step growing in the geometrical progression and with 20 partitions in number. In this case, the first step of the nonuniform grid was equal to the radial one of the internal uniform grid.

The differential equations were approximated by a monotonic conservative difference scheme having, on a regular grid pattern, the second approximation order and on an irregular one, the first one (Ref.6). Solid wall vorticity was predetermined by the approximate Woods condition. Boundary condition derivatives were second-order approximated on a minimum grid pattern. A solution to the obtained difference problem was found using the iteration-relaxation method (Ref.6).

Specific features of the problem are such that the accuracy of its difference solution depends not only on the net construction but also on the "infinity" radius r_* and boundary condition at $r = r_*$. Their choice was based on the methodological conclusions of Ref.7. The comparison of calculated data for a non-coated cylinder (Ref.1) with the known results (Refs.7-10) points to the reliability of the mathematical model and computational algorithm.

4. SOME RESULTS AND CONCLUSIONS

Calculations were made at fixed $\delta = 1.1$, $\lambda = 1$, $\sigma = 1$. Figure 1 plots the drag vs viscosity of a magnetofluid coating in a smooth cylinder, i.e. when a fin is absent. The ratio W/W_o where W_o is the non-coated cylinder drag evidences, how much the drag changes due to coating. As we see, decreasing the drag may be attained due to coating viscosity reduction.

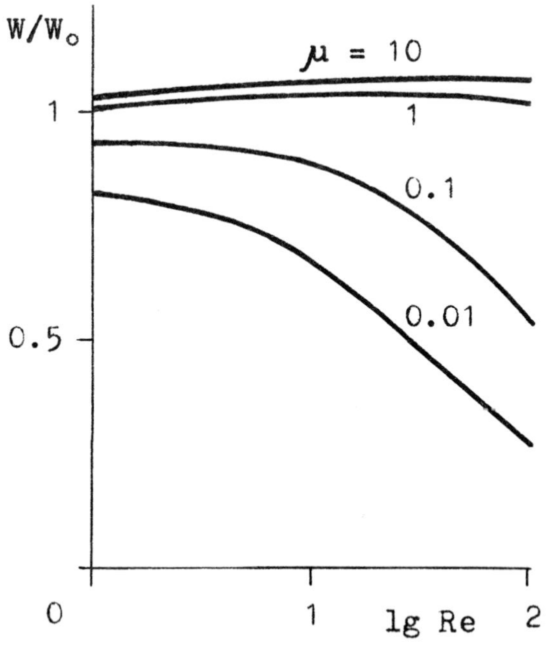

Figure 1: Coated-cylinder drag vs the Reynolds number at different viscosity ratios μ. $Gr_m = 0$. A fin is absent.

The existing magnetic fluids possess rather high viscosity: it is not lower than that of water. So, small values of μ which would provide a substantial drag reduction may be practically obtained only at high flow viscosity. In experiment (Ref.11), e.g., to demonstrate the efficiency of a magnetofluid coating, glycerine was chosen as a flow liquid.

Calculation results show that a drag may be reduced not only by decreasing μ but also by thermomagnetic convection.

Figure 2 enables one to judge, how much the convective mechanism may affect the drag. Owing to the presence of a fin two convective cells are formed on the leading edge of a coating which enhance interface motion, thereby reducing a drag. With increasing Gr_m, the drag value falls up to zero, and at rather great Gr_m the inversion of this drag force occurs, i.e. it converts to a thrust force.

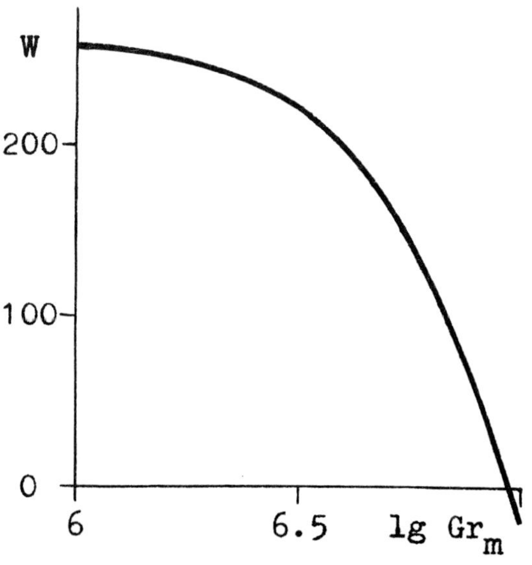

Figure 2: Magnetic Grashof number effect on a drag of a finned magnetic fluid-coated cylinder. $Re=10$, $Pr=1$, $\mu=1$, $\gamma=1$.

Probably, the thermomagnetic convection effect may be more significant in the case of a nonuniform temperature distribution over the cylinder surface, at which the temperature would decrease monotonically from a maximum value on the fin to a minimum one on the rear cylinder edge. The convective motion in this case is formed either in the entire region of a semi-ring or over its greater part.

5. REFERENCES

1. V.K.Polevikov, Izv. AN SSSR. Mekh.Zhid. i Gaza,3(1986)11.
2. V.K.Polevikov, Izv. AN SSSR. Mekh.Zhid. i Gaza,6(1988)11.
3. V.K.Polevikov, Magnitnaya Gidrodinamika, 1(1989)41.
4. V.E.Fertman, Magnetic Fluids - Natural Convection and Heat Transfer (Nauka i tekhnika, Minsk 1978).
5. V.G.Bashtovoi, B.M.Berkovsky and A.N.Vislovich, An Introduction to Thermomechanics of Magnetic Fluids (Press of High-Temperature Institute, Moscow 1985).
6. B.M.Berkovsky and V.K.Polevikov, Computational Experiment in Convection (Universitetskoye, Minsk 1988).
7. V.A.Gushchin, Zh.Vychisl. Matem. i Matem. Fiziki, 20(1980)1333.
8. S.C.R.Dennis and G.-Z.Chang, J.Fluid Mech., 42(1970)471.
9. A.E.Hamielec and J.D.Raal, Phys.Fluids, 12(1969)11.
10. D.J.Tritton, J. Fluid Mech., 6(1959)547.
11. B.M.Berkovsky, V.F.Medvedev and M.S.Krakov, Magnetic Fluids (Khimiya,Moscow 1989).

MAGNETIC BÉNARD CONVECTION UNDER MICROGRAVITY

W.v. Hörsten, S. Odenbach

University of Munich, Sektion Physik, Schellingstraße 4, D-8000 München 40, FRG

ABSTRACT

Thermal convection can be induced in a magnetizable fluid even without gravity when suitable magnetic fields are applied. Our experiment consists of a container with cylindrical gap filled with ferrofluid (a suspension of small magnetite particles in an appropriate solute). The gap is heated from its inner wall and cooled on the outer surface. The geometrie used in our experiment favours periodically boundary conditions not feasible in a ground based experiment. The flow profile has been investigated by small thermistors placed at the inner surface of the outer wall of the container. The magnetic field required for the onset of convection and the wavelength of the resulting flow pattern have been measured in the experiment.

Magnetic convection was clearly observed in a first microgravity experiment flown with a TEXUS sounding rocket flight in spring 1990. The measured flow profile does not fully agree with what had been expected.

1.INTRODUCTION

The scope of our experiment is the investigation of magnetic Bénard convection during the microgravity time of a sounding rocket flight. The magnetic Bénard convection is a form of thermal convection under microgravity. A TEXUS sounding rocket flight was used as flight facility.

Normal Bénard convection (Ref.1,2) is observed in a horizontal fluid layer heated from the bottom and cooled at the upper surface under the influence of gravity. In this case the force leading to the convective flow is given by an interaction between gravity and a temperature induced density gradient.

In contrast to this an interaction between a gradient in magnetization and a magnetic force leads to a convective flow in a ferrofluid layer. This form of convection is called magnetic Bénard convection.

Under microgravity it is possible to establish periodic boundary conditions which are the most favourable for theoretical investigations. We realised these boundary conditions by a ferrofluid in the gap between two concentric cylinders under the influence of an azimuthal magnetic field with radial gradient.

A ferrofluid is a suspension of very small magnetite particles (typical diameter 10 nm) in an appropriate carrier fluid. The magnetite particles are covered with a surfactant to prevent agglomeration. Such fluids show liquid behaviour coupled with superparamagnetic properties (Ref.3).

2.ORIGIN OF THE RADIAL FORCE

The origin of a radial magnetic force in a layer of ferrofluid between two concentric cylinders is explained in figure 1. The inner cylinder is heated to a temperature T_1 while the outer one is cooled to $T_2 < T_1$. This temperature difference gives rise to a radial gradient of magnetization in the ferrofluid. In addition a circular magnetic field is applied by a currentleading wire in the cylinder axis. Such a field has a radial gradient of azimuthal magnetic fieldstrength H_φ. The gradient of azimuthal magnetization M_φ in this field caused by the temperature difference is antiparallel to the fieldgradient.

If a volume element ΔV with large magnetization M_φ is displaced adiabatically in the direction of the fieldgradient it will be surrounded by hotter fluid with lower magnetization $M_\varphi - \Delta M_\varphi$ (see figure 1). This gives rise to a force $\vec{F}_r$ on the volume element ΔV due to the gradient of the field H_φ. The resulting force $\vec{F}_r$ is given by the difference between the force $\vec{F}_p$ on ΔV and the force $\vec{F}_s$ on the fluid that was there before the displacement of ΔV:

$$\vec{F}_r = \vec{F}_p - \vec{F}_s = \mu_0 \, \Delta M_\varphi \, (\partial H_\varphi / \partial r) \, \Delta V \, \vec{e}_r . \quad (1)$$

The gradient of the magnetic field is antiparallel to the unit vector in radial direction in cylindrical coordinates $\vec{e_r}$; $\partial H_\varphi/\partial r < 0$. Equation (1) gives a force in the direction of the fieldgradient. Alternatively one may displace the volume element antiparallel to the fieldgradient and get a resulting force also antiparallel to it. The resulting force always acts in the same direction as the displacement and causes magnetic buoyancy.

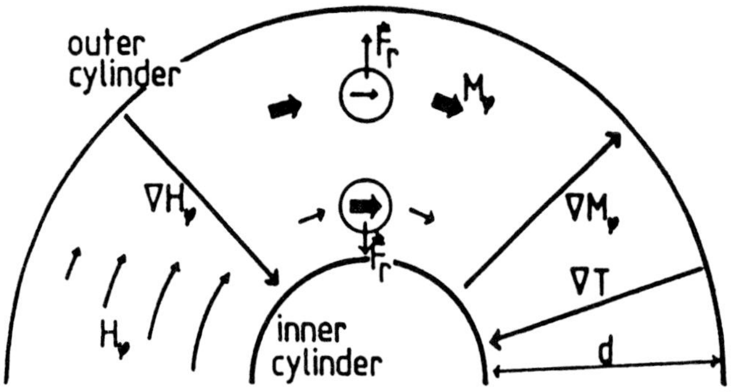

Figure 1: The origin of the destabilizing magnetic force.

Due to this destabilizing force a disturbation in the fluid can grow but it is opposed by viscous effects and by thermal diffusivity. For a real destabilization of the fluid the temperature difference or the fieldgradient must exceed certain critical values. The situation of the system is characterized by the magnetic Rayleigh number R_m (Ref.4,5).

$$R_m = \frac{\mu_0 \, K \, G \, \Delta T \, d^3}{\kappa \, \eta}. \qquad (2)$$

Here K is the negative derivative $-\partial M_\varphi/\partial T$ of the magnetization by the temperature, G the magnetic fieldgradient, μ_0 the permeability of free space, ΔT the temperature difference, d the thickness of the fluid layer, η the dynamic viscosity and κ the thermometric conductivity of the ferrofluid.
If the dimensionless ratio R_m exceeds a critical value depending on the boundary conditions the fluid layer becomes unstable.

3.THE TEXUS SOUNDING ROCKET FLIGHT FACILITY

It was mentioned before, that a TEXUS sounding rocket was used as flight facility for our experiment. Those sounding rockets offer 360 seconds of microgravity time with a microgravity level of approximately $10^{-5} g_0$. During the measuring time it is possible to change experiment parameters by telemetry. The experimental data are stored in a memory cassette in the module. This memory cassette is read out after recovery. Data necessary for the control of the experiment can be given on line to the ground station on several channels. The TEXUS rockets are launched near Kiruna in North Sweden twice a year.

4.EXPERIMENTAL SETUP

Our apparatus was designed to investigate the onset of convection in the absence of gravity and the resulting flow pattern during a 360 seconds measuring period. The effects are studied by variation of the azimuthal magnetic fieldstrength. The experiment cells have the form of two hollow-cylindrical containers with different outer diameters (see figure 2).

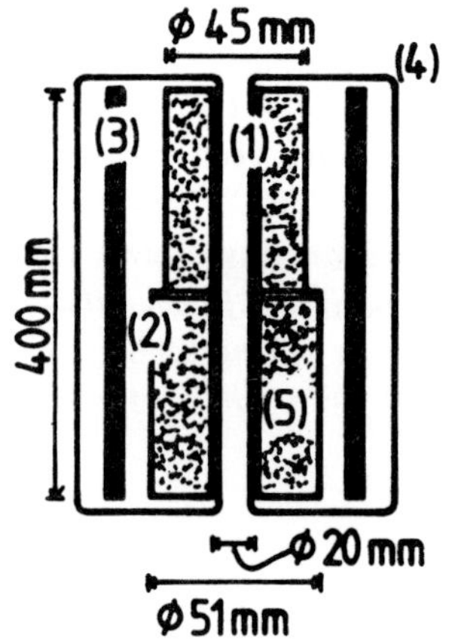

Figure 2: Experiment cells with inner (1) and outer (2) cylinder; axial (3) and azimuthal (4) field coils and the ferrofluid (5) between the cylinders.

It is possible to apply a magnetic field in the direction of the axis and another one in circular direction. The axial field is used to align the expected convection rolls along the cylinder axis (see Ref.4). The circular field yields the radial fieldgradient and the azimuthal component of magnetization. In this geometry we expected to find a flow profile like that shown in figure 3.
The investigation of the flow profile is realised by

measuring the temperature distribution on the inner surface of the outer cylinder by means of 224 microthermistors. The thermistors are embedded in small glass tubes of about 0.5 millimeter diameter.

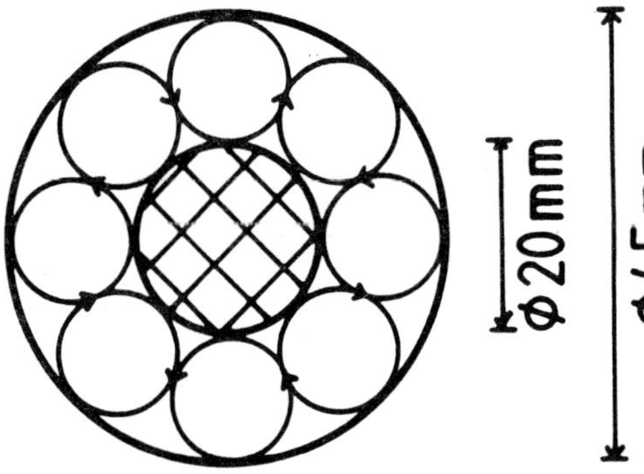

Figure 3: Cross section with the expected convective pattern.

They are arranged along two circular lines and one axial line on the outer wall of each container. The dimensions of the two containers are given in table 1.

TABLE 1 Dimensions of the containers

length of each container	200 mm
radius of inner cylinder	10 mm
Container 1: radius of outer cylinder	25.5 mm
gap width	15.5 mm
Container 2: radius of outer cylinder	22.5 mm
gap width	12.5 mm

5.RESULTS OF THE 1G TESTS

In several 1g tests the method of measurement and the whole setup were controlled. During one of those tests the module was positioned with the cylinder axis horizontal, i.e. perpendicular to gravity. In this case the expected flow profile is shown in figure 4 a. Hot fluid rises from the inner to the outer cylinder at the top while cold fluid streams towards the inner cylinder at the bottom. The measured temperature profile is shown in figure 4 b. The temperature resolution is better than 0.01 K.

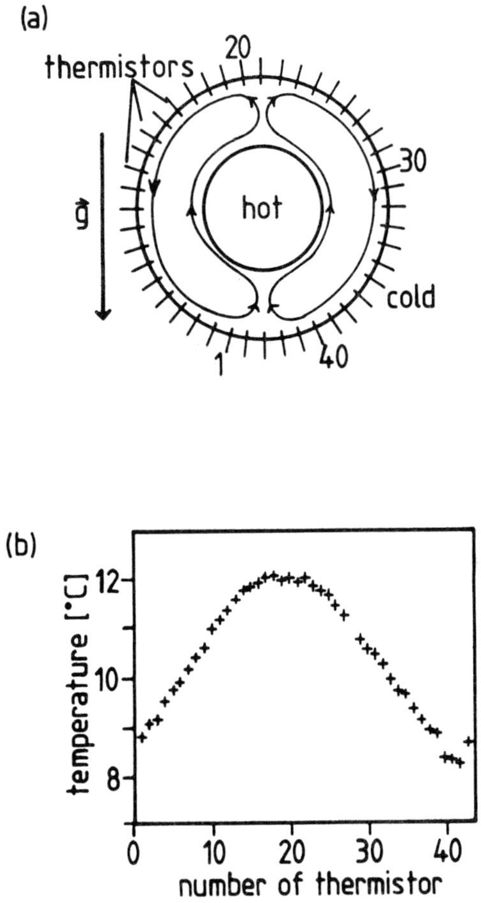

Figure 4: a) Expected flow profile and b) measured temperature distribution during the 1 g test.

6.RESULTS FROM THE μG EXPERIMENT

The TEXUS-25 campaign offered 351 seconds of μg time. During this time the azimuthal magnetic field was increased from 0 to 1.9 mT in 15 steps. Data acquisition for all microthermistors was taken one times at each step. After changing the fieldstrength a delay time of about 4 seconds to the next data acquisition was observed to allow the flow pattern to stabilize.
An example of the results is shown in figure 4, the temperature distributions of the axial line and one of the two cirular lines of thermistors of the larger container (see figure 2). The axial fieldstrength is about 7.5 mT while the circular field was 0.75 mT in the middle between the cylinders with a gradient of about $43 \cdot 10^{-3}$ T/m. The temperature difference is 20 K.

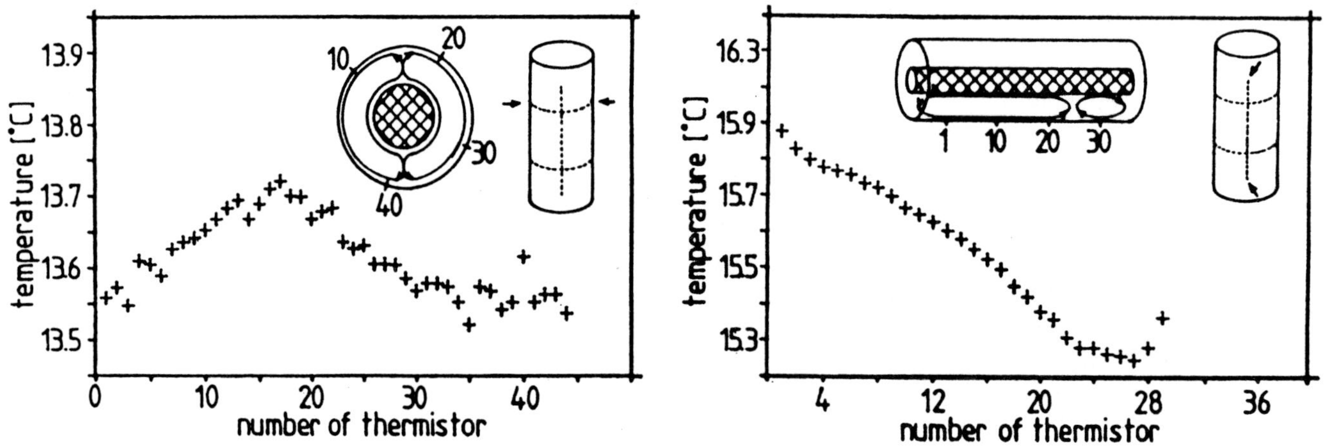

Figure 5: Temperature distribution along a circular line (a) and the axial line (b) of thermistors under μg. Insets: possible flow patterns corresponding to the measured temperature profiles and distribution of the thermistors on the cylinder surface.

Only two longitudinal convection rolls are established instead of eight expected. The reason for this difference is due to the maximal available magnetic fieldstrength which was too low to obtain a Rayleigh number significantly higher than the critical Rayleigh number. But it was observed that the amplitudes of the rolls increase proportionally to the azimuthal magnetic fieldstrength (figure 6).

This dependence of the amplitude of convection on the magnetic field is a strong indication for magnetic buoyancy.

Unexpectedly one pair of azimuthal rolls is observed in the cylinder (figure 4b). These rolls are caused by a gradient of the axial magnetic field. Due to spatial limitations the axial magnetic field coil had the same length as the convection cell (see figure 2). Therefore the axial magnetic field decreases at the outer end of the convection cell. The resulting axial gradient in fieldstrength leads to a resulting force on the fluid similar to that described in chapter 2. The convection caused by this axial magnetic field gradient gives rise to the two azimuthal convection rolls.

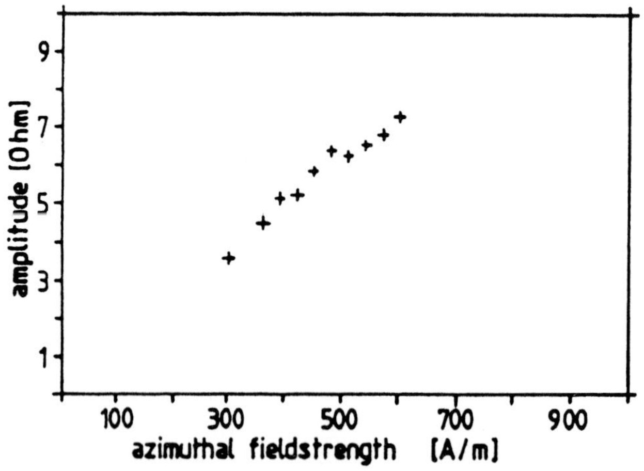

Figure 6: Dependence of the amplitude of the axial convection rolls shown in figure 5 a as a function of the azimuthal magnetic fieldstrength.

7.ACKNOWLEDGEMENTS

The TEXUS microgravity experiment was supported by BMFT and DLR. The experiment module was built by the MBB ERNO TEXUS team. Dr. Wellner from Bayer AG Leverkusen supplied a special varnish to protect the inner surface of the containers from the ferrofluid. Dr. L. Schwab of the University of Munich has proposed this experiment and has made many valuable remarks. We thank all institutions and their representatives for their valuable support which made this research possible.

8.REFERENCES

1. M. Velarde, C. Normand, Convection, *Scientific American* 243, 78 (1980)

2. F. H. Busse, Trans. in Turb. in Rayleigh-Bénard Conv., in*Hydrodynamic Inst. and the Trans. to Turb.*, ed. H.L.Swinney, and J.P.Gollub, Topics in appl. phys. 45; Springer 1981

3. R. E. Rosensweig, *Ferrohydrodynamics*, Cambridge University Press, Cambridge 1985

4. L. Schwab, *Konvektion in Ferrofluiden*, Dissertation LMU München, München 1989

5. B. A. Finlayson, Convective Instab. of Ferromagnetic Fluids, *J.Fluid Mech.* 40, 753 (1970)

Thermocapillary Convection

THERMOCAPILLARY INSTABILITY IN A MULTILAYER SYSTEM

Ph. Géoris, M. Hennenberg, J.C. Legros, A.A. Nepomnyashchy*, I.B. Simanovskii**, I.I. Wertgeim***.

Université Libre de Bruxelles (Brussels, Belgium)
**Technion (Haifa, Israel)*
***Perm State Pedagogical Institute (Perm, USSR)*
****Institute of Continuous Media Mechanics of Ural's Branch USSR Academy of Science (Perm, USSR)*

ABSTRACT

The Marangoni-Bénard Instability for a three-layer system under microgravity conditions is theoretically examined by analytical and numerical calculations. The marginal stability analysis performed for monotonous disturbances shows that convection is driven by one destabilizing interface, the other one being stabilizing. Numerical studies by finite differences method confirm the linear calculations showing that strong convective motions are concentrated in the layers bounding the destabilizing interface. Stationary stream functions, isotherms and finite amplitude curves for two real systems (air-water-air and Fluorinert-silicone oil-Fluorinert) have been built illustrating the influence of the parameters on the convective pattern.

Key-Words : *Thermocapillary, Microgravity, Instability, Interface, Multilayer.*

1. INTRODUCTION

This paper dealing with the Marangoni-Benard Instability of systems with two free interfaces (three layers) is a preparation to a microgravity experiment which will fly during the IML2 Spacelab mission.

The specific feature of thermocapillary convection is defined by the interaction of hydrodynamic and heat transfer at interfaces. These interactions lead to the appearance of qualitatively new mechanisms of disturbances evolution.

Thermocapillary convection in systems with single interface (two layers systems) has been investigated in linear [1-3] and non-linear [4-6] formulations.

In these works, the two-layer approach is used : the conservation laws and the boundary conditions are written for both media. Depending on the physical parameters of the fluids, on ratio of the layer's thicknesses and on thermal boundary conditions, the thermocapillary convection may be described both by the monotonous and oscillatory disturbances.

The interest of the three layer configuration lies in the presence of two interfaces influencing differently the stability of the system. The stability of a single layer bounded by two free interfaces has been examined by T. Funada [7] whereas Wahal and Bose [8] have studied the case of a two-layer configuration when the upper layer has two free interfaces.

In this work, the full three-layer system is considered analytically for the linear part and numerically for the non- linear one.

2. STATEMENT OF THE PROBLEM

Consider three superposed layers of immiscible viscous fluids between two rigid conducting plates. A constant temperature difference ΔT is held between the two plates (Figure 1.).

We assume that the interfacial tensions vary linearly with the temperature. Indexes 1 and 2 are related to the exteriors layers and index 3 is related to the cental one.

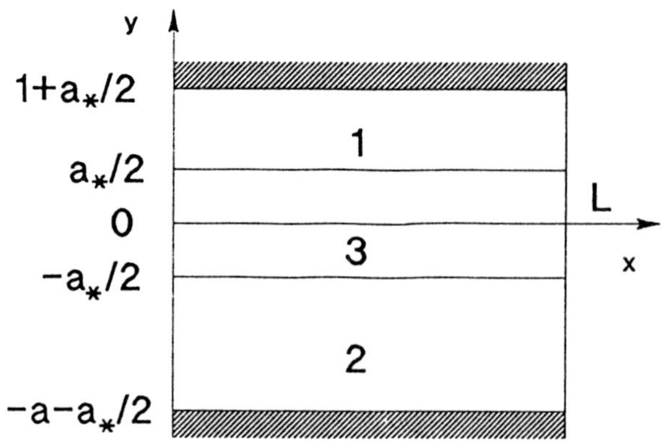

Figure 1. : Three-layer system

We use the following denominations for the physical parameters ratios:

$\nu_* = \frac{\nu_1}{\nu_2}$, $\nu = \frac{\nu_1}{\nu_3}$: *kinematic viscosities ratios*

$\eta_* = \frac{\eta_1}{\eta_2}$, $\eta = \frac{\eta_1}{\eta_3}$: *dynamical viscosities ratios*

$\kappa_* = \frac{\kappa_1}{\kappa_2}$, $\kappa = \frac{\kappa_1}{\kappa_3}$: *heat conductivities ratios*

$\chi_* = \frac{\chi_1}{\chi_2}$, $\chi = \frac{\chi_1}{\chi_3}$: *thermal diffusivities ratios*

$a_* = \frac{a_1}{a_2}$, $a = \frac{a_1}{a_3}$: *thicknesses ratios*

At the interfaces, normal components of velocity are vanishing, the continuity conditions for tangential components of velocity, viscous stresses, temperature and heat flux hold.

3. LINEAR STABILITY ANALYSIS

We have performed the linear stability analysis for the configuration described above when the layers of infinite extension in the OXY plane are of the same thickness and when the same fluid composes layers 1 and 2.

The system of linearized dimensionless differential equations is the following :

$$(D^2-k^2)^2 W_i = 0 \qquad (i=1,2,3) \qquad 2.1$$

$$(D^2-k^2)\theta_i = -W_i \qquad (i=1,2) \qquad 2.2$$

$$(D^2-k^2)\theta_3 = -\chi\kappa W_3 \qquad 2.3$$

with : $D \equiv \frac{d}{dz}$,
W = *vertical velocity disturbance,*
k = *wave number,*
θ = *temperature disturbance.*

If we consider that each layer has its own vertical coordinate system, the 18 dimensionless boundary conditions are :

$$W_i(0) = W_i(1) = 0 \qquad (i=1,2,3) \qquad 2.4$$

$$\theta_1(0) = \theta_3(1) = 0 \qquad 2.5$$

$$\theta_2(1) = \theta_3(0) \qquad 2.6$$

$$DW_1(1) = DW_2(0) = 0 \qquad 2.7$$

$$DW_2(1) = DW_2(0) \qquad 2.8$$

$$\theta_1(1) = \theta_2(0) = 0 \qquad 2.9$$

$$\kappa D\theta_2(1) = D\theta_3(0) \qquad 2.10$$

$$\kappa D\theta_1(0) = D\theta_3(1) \qquad 2.11$$

$$D^2 W_2(1) - \frac{1}{\eta} D^2 W_3(0) = -\frac{Mk^2\theta_2(1)}{2+\kappa} \qquad 2.12$$

$$\frac{1}{\eta} D^2 W_3(1) - D^2 W_1(0) = -\frac{Mk^2\theta_3(0)}{2+\kappa} \qquad 2.13$$

The expression for marginal Marangoni numbers M, obtained from the analytical solution of system (2.1-2.3) by Fourier method is :

$$M = \pm(2+\kappa)\kappa\chi\sqrt{\frac{A}{B}} \qquad 2.14$$

with :

$$A = -(c_{4m}^2 - c_{3m}^2)\left(\left(a_1 + \frac{a_1}{\eta}\right)^2 - a_2\right)$$

$$B = (a_1^2 - a_2^2)(c_3^2 - c_4^2) + (a_1 c_3 - a_2 c_4)^2 \left(\frac{1}{\chi^2} - \frac{2}{\chi}\right)$$

$$c_{4m} = \frac{1}{k\sinh k}$$

$$a_1 = \frac{1}{2\sinh^2 k} - \frac{\cosh k}{2k\sinh k}$$

$$a_2 = -\frac{\cosh k}{2\sinh^2 k} + \frac{1}{2k\sinh k}$$

$$c3 = -\frac{1}{4k\cosh k\sinh^3 k} + \frac{1}{8k^2\sinh^2 k} - \frac{1}{4k\cosh k\sinh k} + \frac{\cosh k}{8k^3\sinh k}$$

$$c_{3m} = -\frac{\cosh k}{k \sinh k}$$

$$c_4 = -\frac{\cosh^2 k}{4k \cosh k \sinh^3 k} + \frac{\cosh k}{8k^2 \sinh^2 k} + \frac{1}{8k^3 \sinh k} + \frac{1}{8k \sinh k}$$

The critical Marangoni number for the air-water-air system ($\varkappa=0.0396, \chi=138.42$, $\eta=0.0182$) is M_c = 1925 with $k_c=1.92$. The critical Marangoni number for the second studied system, Fluorinert FC70 - silicone oil 50 Cs - Fluorinert FC70, ($\varkappa=0.467$, $\chi=0.319$, $\eta=0.568$) is M_c = $\pm$ 2312.7 with k_c = 1.92.

4. NUMERICAL SIMULATION

The boundary value problem for finite amplitude disturbances is approached using 2D finite difference method. The governing 2D non-linear equations are :

$$\frac{\partial T_i}{\partial t} + \frac{\partial \psi_i}{\partial y}\frac{\partial \varphi}{\partial x} - \frac{\partial \psi_i}{\partial x}\frac{\partial \varphi_i}{\partial y} = \frac{c_m}{P}\Delta T_i \qquad (i=1,2,3) \qquad 2.15$$

$$\frac{\partial \varphi_i}{\partial t} + \frac{\partial \psi_i}{\partial y}\frac{\partial \varphi_i}{\partial x} - \frac{\partial \psi_i}{\partial x}\frac{\partial \varphi_i}{\partial y} = d_i \qquad (i=1,2,3) \qquad 2.16$$

$$\Delta \psi_i = -\varphi_i \qquad (i=1,2,3) \qquad 2.17$$

with : $d_1 = c_1 = 1$, $d_2 = \frac{1}{\nu_*}$, $c_2 = \frac{1}{\chi_*}$,

$d_3 = \frac{1}{\nu}$, $c3 = \frac{1}{\chi}$, $P = \frac{\nu_1}{\chi_1}$.

Equations 2.12, 2.13 and 2.14 are approximated on an uniform mesh with a second order precision on the space coordinates. The Poisson equations are solved by an iterative method on each time step : the accuracy is assumed to be 10^{-4}. Thom [9] and Kuskova & Chudov [10] formulae are used to approximate the vorticity on the solid boundaries. At the interfaces of layers 1 and 2, the expressions for the vorticity write :

$$\varphi_1(x,\frac{a_*}{2}) = -\frac{2(\psi_3(x,\frac{a_*}{2}-\Delta y) + \psi_1(x,\frac{a_*}{2}+\Delta y))}{\Delta y^2(1+\eta)} - \frac{M_r}{1+\eta}\frac{\partial T_1(x,\frac{a_*}{2})}{\partial x} \qquad 2.18$$

$$\varphi_2(x,-\frac{a_*}{2}) = -\frac{2(\psi_3(x,-\frac{a_*}{2}-\Delta y) + \psi_1(x,-\frac{a_*}{2}+\Delta y))}{\Delta y^2(1+\eta_*)} - \frac{Mr_*}{1+\eta}\frac{\partial T_1(x,-\frac{a_*}{2})}{\partial x} \qquad 2.19$$

where : $M_r = \frac{\eta M}{P}$, $M_{r*} = \frac{\eta_* a_* M}{P}$

Analogous expressions are used to express the vorticity at the two free interfaces of layer 3. The temperatures on the interfaces were calculated by the second-order approximation formulae :

$$T_1(x,\frac{a_*}{2}) = T_3(x,\frac{a_*}{2}) = \frac{4T_3(x,\frac{a_*}{2}-\Delta y) - T_3(x,\frac{a_*}{2}-2\Delta y)}{3(1+\kappa)} + \frac{\kappa(4T_1(x,\frac{a_*}{2}+\Delta y) - T_1(x,\frac{a_*}{2}+2\Delta y))}{3(1+\kappa)}$$

5. RESULTS AND DISCUSSIONS

5.1. Air-Water-Air system

This system has been solved numerically in a cavity (l/h=2/3) using different time steps for air and water because of the considerable difference between parameters. This procedure is valid only when the stationary solution is looked for. Stream function and isotherms are presented in Figure 2, finite amplitude curve is given in Figure 3.

As expected, motions appear mainlyly at the upper water-air interface because the high thermal diffusivity of air rends the lower interface stabilizing. This reasoning is confirmed by an analogy with the two layers configuration air-water [11] which is stable when heated from above. Isothermal lines are concentrated in the insulating air layers.

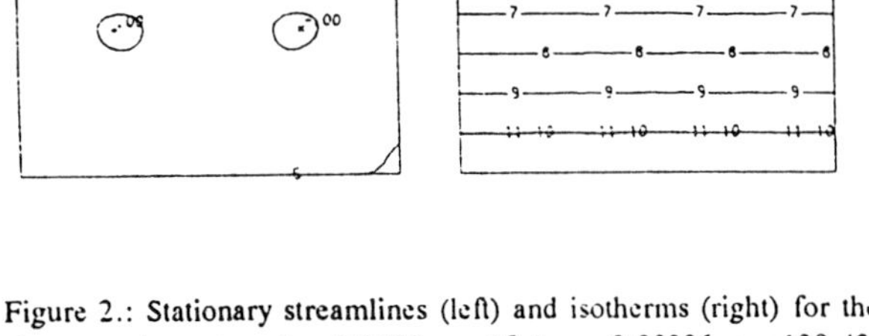

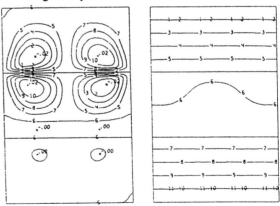

Figure 2.: Stationary streamlines (left) and isotherms (right) for the air-water-air system ($\eta=0.0182$, $\nu=15.1$, $\kappa=0.00396$, $\chi=138.42$, $P=0.758$), $M=8000$.

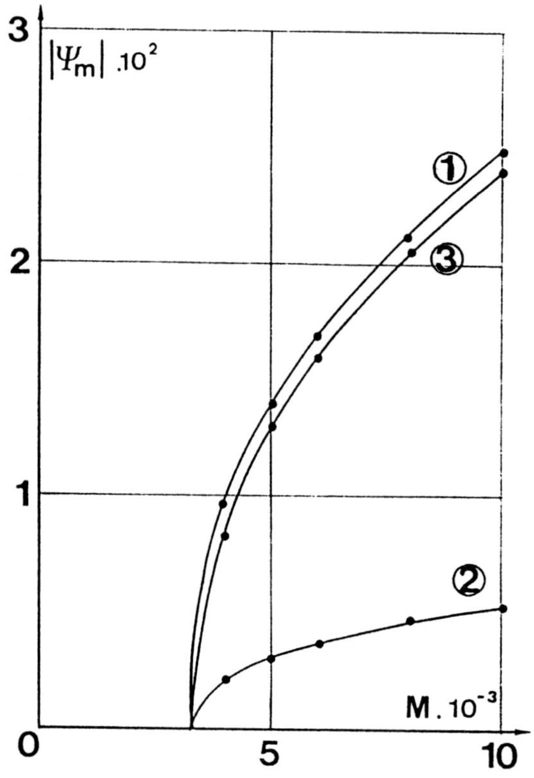

Figure 3. : Bifurcation diagram for the air-water-air system. Numbers refer to the corresponding layers.

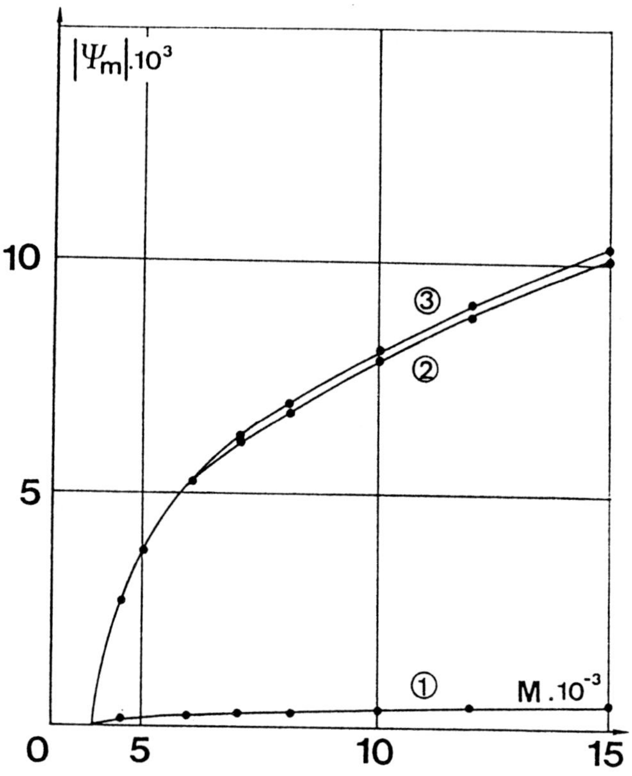

Figure 5. : Bifurcation diagram for the Fluorinert-silicone oil-Fluorinert system. Numbers refer to the corresponding layers.

5.2. Fluorinert-silicone oil-Fluorinert system

The second system considered has been solved in the same conditions as the previous one. Streamlines and isotherms are showed in Figure 4, bifurcation diagram is given in Figure 5.

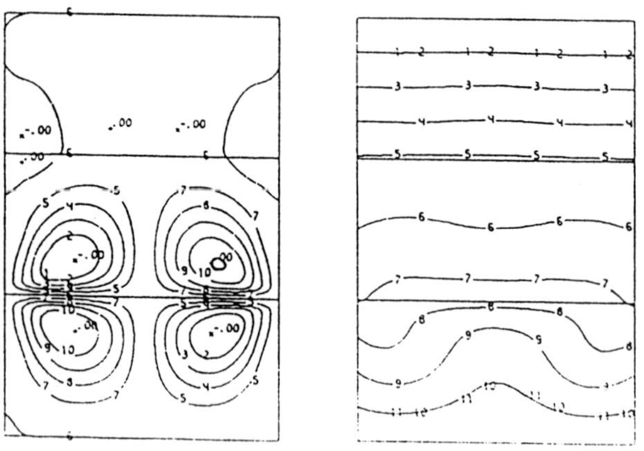

Figure 4.: Stationary streamlines (left) and isotherms (right) for the Fluorinert-silicone oil-Fluorinert system ($\eta=0.568$, $\nu=0.28$, $\kappa=0.467$, $\chi=0.319$, $P=406$), $M=8000$.

Here the convection has a higher amplitude at the lower interface. This result can be foreseen using linear stability analysis. Indeed, due to the symmetry of the configuration, only one interface can be destabilizing, the other being destabilizing. Linear analysis shows that the fluorinert layer is more stable when it is bounded by two silicone layers, one stabilizing and one destabilizing boundary, than when it is bounded by a rigid plate and a silicone layer. This indicates that the system Fluorinert-silicone oil is unstable when heated from the Fluorinert side and stable when the thermal gradient is applied in the opposite direction.

6. CONCLUSION

Depending on the parameters, the instability is generated by the 2-3 interface or the 3-1 interface for the three-layer system; however, no oscillating behaviour has been detected using linear theory or non-linear simulation. The symmetry of the problem (fluid 1 is identical to fluid 2, the layers have the same thickness) imposes that only one interface can be destabilizing at the same time. It is expected that non symmetrical configurations, like Fluorinert-silicone oil-air, may show more varied behaviour.

7. ACKNOWLEDGMENTS

We are indebted to SPPS for supporting this research through a PAI. This work is performed in the framework of an agreement signed between ESA and USSR.

8. REFERENCES

1. C.V. Sternling, L.E. Scriven, *AIChE j.*,**5**,n°4, 51-523,(1959).

2. K.A. Smith, *J. Fluid Mechanic*, **24**, 401-404, (1966).

3. A.A. Nepomnyashchy, I.B. Simanovskii, *Pricl. Mech. i Techn. Fiz.*, **1**, 62-65, (1985).

4. A.A. Nepomnyashchy, I.B. Simanovskii, *Docl. Akad. Nauk SSSR*, **272**, 825-827, (1983).

5. A.A. Nepomnyashchy, I.B. Simanovskii, *Izv. Akad. Nauk SSSR. Mekh. Zh. i*, Gaza **4**, 158-163, (1983).

6. A.A. Nepomnyashchy, I.B. Simanovskii, *Docl. Akad. Nauk SSSR. Mekh. Zh .i*, Gaza **3**, 178-182, (1984).

7. T. Funada, *J. Phys. Soc. Jap.*, **55**, n° 7, 2191-2202, July, (1986).

8. S. Wahal and A. Bose, *Phys. Fluids*, **31**, 12, (1988).

9. A. Thom, K. Eiplt, *Numerical calculations of fields in technics and physics*, Moscow : Energy, (1964).

10. T.V. Kuskova, L.A. Chudov, *Computational methods and programing*, 11,27-31, Moscow State University Computing center (1968).

11. A.Y. Gilev, A.A. Nepomnyashchy and I.B. Simanovskii, *Fluid Mechanics - Soviet Research*, **16**, 3, 44-48, (1987)

VIBRATIONAL THERMOCAPILLARY CONVECTION AND STABILITY

V.A.Briskman

Institute of Continuous Media Mechanics, UB Acad.Sci.USSR, Perm, USSR.

ABSTRACT

The present paper deals with the problems of flows and stability of nonisothermal fluid with free surface under joint action of thermocapillary and vibrational forces. The system of equations and boundary conditions are formulated for high-frequency approximation. In the framework of this approximation the analysis is carried out of transversal vibration influence on equilibrium stability of a plane fluid layer with transversal temperature gradient (Marangoni stability). It has been shown that vibrations efficiently increase the stability. Consideration is given to thermocapillary flow in a plane layer under the influence of longitudinal temperature gradient. An exact solution has been found which allows to describe plane-parallel flows. It is shown that the amplitude of velocity decreases with increased vibrational parameter and the flow is displaced in a boundary layer. The possibilities of vibrational controlling of thermocapillary convection are discussed.

KEY WORDS

Microgravity, vibrations, thermocapillary convection, Marangoni stability, high-frequency approximation.

1. INTRODUCTION

Thermocapillary and vibrational forces are the main factors governing heat and mass transfer in microgravity. However, until recently no sistematical study has been attempted to examine the joint action of these forces. There exist few studies dealing with calculations of forced motions having the same frequency as vibration. Meanwhile, the most interesting results should be expected in situations, when the low frequency vibrations induce resonance effects e.g. parametric resonance, or when the high- frequency vibrations give rise to quasi-stationary flows.

For the sake of clarity it will be better to illustrate this point by examples from some other fields of hydrodynamics. Theoretical and experimental investigations have demonstrated that the high-frequency vibrations normal to the isothermal media interface prevent development of Rayleigh-Tailor (Ref.1,2,3), Kelvin-Helmholz (Ref.2) and Tonks-Frenkel (Ref.2,4) instabilities. To our knowledge there is also an evidence for preventing Rayleigh- Bernard thermoconvective instability (Ref.5) related to the action of volumetric buoyancy forces in the nonisothermal fluid. In contrast to the above mentioned cases tangential high-frequency vibrations may lead to destabilization of initial plane interface , e.g. to generation of interface quasi-stationary relief (Refs.6,7,8). The present paper is concerned with two simple cases of high-frequency effect on thermocapillary phenomena in microgravity-namely, on stability of nonisothermal equilibrium in a fluid with free surface, and on plane-parallel thermocapillary flow. Preliminary results have been published in Refs. 9,10,11.

1. THE AVERAGED EQUATIONS AND BOUNDARY CONDITIONS FOR THE HIGH-FREQUENCY APPROXIMATION

Let us assume that the cavity is partially filled with viscous, incompressible nonisothermal liquid, which in zero gravity experiences as a unit harmonic oscillations with amplitude b and frequency ω. In the following these oscillations will be refereed to as vibrations. In the noninertial frame of reference related to the vessel, the equations of heat and mass transfer can be obtained from usual equation of heat convection by substituting vibration acceleration $b\omega^2 \cos(\omega t)$ for the static acceleration of gravity (Ref.5):

$$\frac{\partial \vec{V}}{\partial t}+(\vec{V}\cdot\nabla)\vec{V} = -\frac{1}{\rho}\cdot\nabla P+\nu\Delta\vec{V}+\beta b\omega^2 T\vec{n}\cos(\omega t)$$

$$\frac{\partial T}{\partial t}+(\vec{V}\cdot\nabla)T = \chi\Delta T \;; \qquad div\,\vec{V} = 0 \qquad (1)$$

here $\vec{V}$ is the fluid velocity, T is temperature, P - pressure, β, ν and χ are coefficients of volume expansion, kinematic viscosity and temperature conductivity; vector $\vec{n}$ has direction of vibrations.

We shall consider high-frequency effects, assuming thus that vibration period is small as compared to all characteristic hydrodynamic times t_H: viscous - h^2/ν, thermal - h^2/χ, capillary - $(\rho h^3/\sigma)^{1/2}$ and thermocapillary $(\rho h^2/(\nabla T|\partial\sigma/\partial T|))^{1/2}$, where σ is coefficient of surface tension, $\partial\sigma/\partial T$ is negative. It is then reasonable to divide the velocity, temperature and pressure into slow - changing and rapid - pulsating components:

$$\vec{V}=\vec{V}_0+\vec{V}_1\;;\; T=T_0+T_1\;;\; P=P_0+P_1\;;\; (t_H \gg 1/\omega)$$

Let us further assume that the amplitudes of rapid components are small over that of slow ones. Having the pulsative terms of equations singled out, one can distinguish between the orders of smallness with respect to the small parameter $1/\omega$. Using the first-order smallness approximation for the equation under consideration suggests that nonlinear terms and dissipative terms $\nu\,\Delta\vec{V}$, $\chi\,\Delta T$ are neglected. Subsequent integrating with respect to time indicates that the amplitudes of pulsative components may be adequately represented in terms of the slow components. Then we have for velocity and temperature:

$$\vec{V}_1 = \beta\, b\,\omega\,\vec{W}\sin(\omega t)$$
$$T_1 = \beta\, b\,(\vec{W}\nabla)T_0\cos(\omega t) \qquad (2)$$

Here $\vec{W}$ is a new variable, defined as a divergence - free part of the field $T_0\vec{n}$ ($curl\,\vec{W} = curl\,T_0\vec{n}$).

Substituting (2) into the system (1) and averaging the equations over the vibration period gives a closed system for slow motions:

$$\frac{\partial \vec{V}}{\partial t}+\frac{1}{Pr}(\vec{V}\cdot\nabla)\vec{V} = -\nabla P+\Delta\vec{V}+R_v(\vec{W}\cdot\nabla)(T\vec{n}-\vec{W})$$

$$Pr\,\frac{\partial T}{\partial t}+(\vec{V}\cdot\nabla)\,T = \Delta T \qquad (3)$$

$$div\,\vec{V} = div\,\vec{W} = 0 \;; \quad curl\,\vec{W} = curl(T\vec{n})$$

$$R_v = (b\omega\,\beta A h^2)^2/2\nu\chi \;; \quad Pr = \nu/\chi \qquad (4)$$

Here index 0 is omitted.

The existence in equations (1) of terms which are nonlinear with respect to time functions is responsible for the fact that harmonic vibrations lead to occurrence of the time average thermovibrational forces in (3).

Dimensionless quantities in (3) are expressed in terms of the distance h, time h^2/ν, velocity χ/h, temperature Ah and pressure $\rho\nu\chi/h^2$, where A is characteristic temperature gradient. Dimensionless parameter R_v represents the vibrational analog of the Rayleigh number, Pr is conventional Prandtl number.

To proceed further it is necessary to define boundary conditions. As it has been shown earlier, pulsative temperatures and velocities obey the equations with dissipative terms being omitted. Thus, when describing pulsative flows it is appropriate to use boundary conditions commonly accepted for the ideal fluid. From (2) it is readily seen that the boundary conditions for W and V_1 coincide. As regards the velocity of the averaged flow it has to go to zero at the solid surfaces. From the assumption of undeformable free surfaces it follows that the normal components of all velocity fields at these surfaces will become zero. The temperature is assumed to satisfy normal conditions. Finally, the free surface is desired to meet the requirement of balance of all stresses. Temperature dependence of the surface tension, which is considered to be linear, gives rise to a thermocapillary force tangential to the surface. The latter is balanced by the viscous force. However with nonviscous approximation accepted for the pulsative flow it is obvious that thermocapillary forces are uncapable of exerting direct influence on the pulsative velocities. From this follows that thermocapillary forces should be taken into account only by the boundary conditions adopted for the averaged motion. However, since the pulsative velocities and temperatures are determined by the averaged components based on the expressions (2), the pulsative fields will be finally dependent on thermocapillary forces.

As there are no nonlinear terms in the boundary conditions, they are averaged in a trivial manner and for the averaged components they are similar by form to the normal conditions. In expressions for the balance of tangential stresses which will be given below for particular problems, thermocapillary forces are represented by Marangoni criterion

$$M = A h^2 |\partial\sigma/\partial T| / \eta\chi \qquad (5)$$

Therefore the problems of vibrational-thermocapillary convection in zero gravity are defined by three criteria R_v, M and Pr.

It should be noted that the system of equations (3) coincides with the system, describing the ordinary vibrational thermal convection (Ref.12) Vibration effects are exhibited through the volumetric thermovibrational force. What is new in the problem under consideration is, that apart from the volume forces, it takes into account the surface thermocapillary forces.

Interaction of thermovibrational and thermocapillary forces may give nontrivial results which are qualitatively different from the results, obtained in the absence of the surface forces. In particular, some states of quasi-equilibrium are found nonfeasible. Discussion of particular problems will be given below.

2. THE INFLUENCE OF HIGH-FREQUENCY VIBRATIONS ON MARANGONI STABILITY

Consider a fluid layer in zero gravity confined between parallel planes, which have different temperatures. One of the planes (which is more cold) or both of them are considered to be free. In the absence of vibrations this is the classical problem of Marangoni stability, in which the equilibrium becomes unstable when the Marangoni numbers reach their critical values.

We shall consider the influence of transversal vibrations on Marangoni stability. Z-axis of Cartesian coordinates is directed perpendicular to the layer, so that $n_x = n_y = 0$.

The origin of the coordinates is chosen in the middle of the layer; the layer half-width has been chosen as characteristic size.

The solid boundary is considered thermally insulated while the free surface is assumed to have a heat flux which is proportional to temperature disturbances with coefficient expressed in terms of dimensionless Biot number B. If both of the surfaces are free, the conditions imposed on them are assumed equal.

As it is readily seen, the system (3) has quite an obvious solution:

$$\vec{V} = \vec{W} = 0 \;; \quad (\nabla T)_z = const = -A$$

which describes strict equilibrium. This implies that both slow and

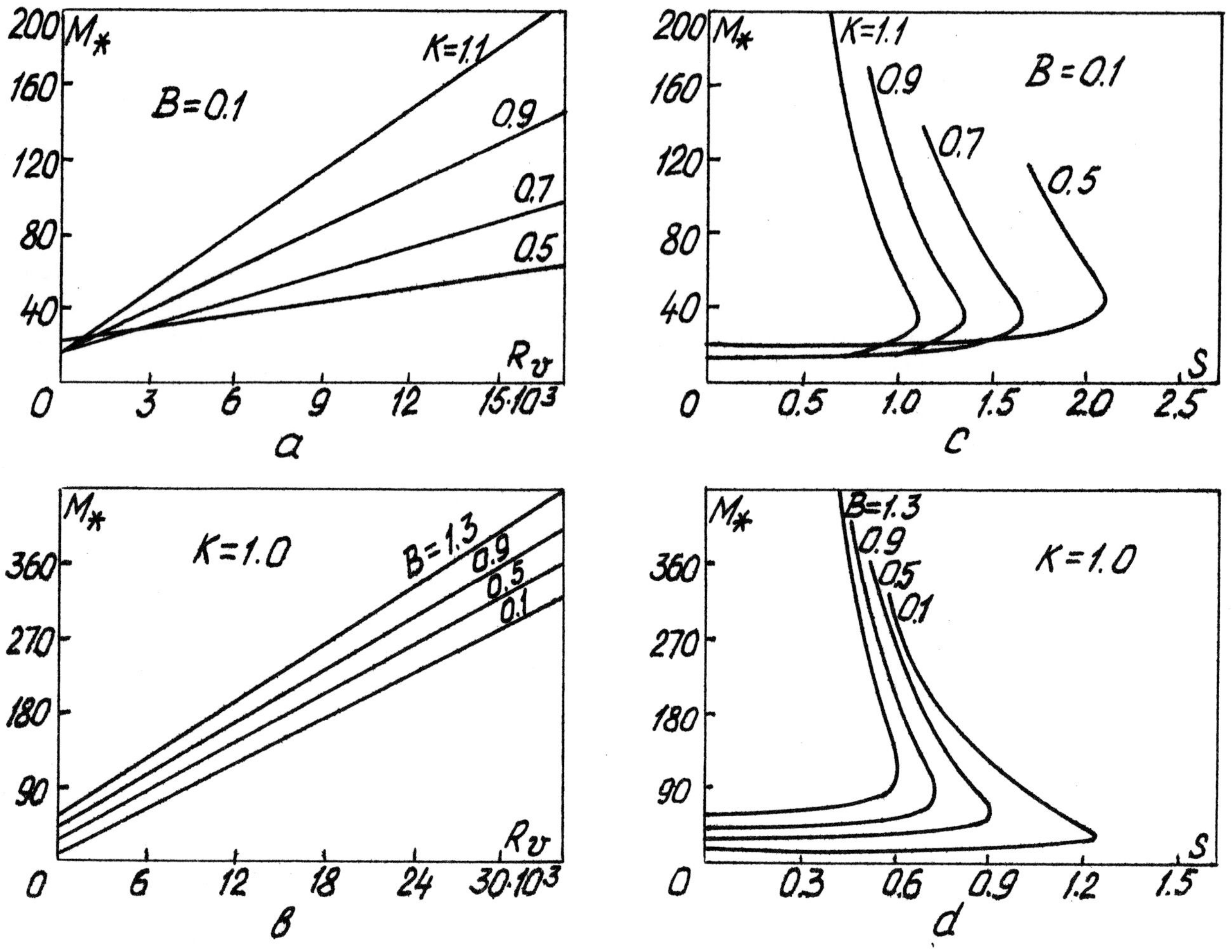

Fig.1. Neutral curves for the fixed values of the wave number k and Biot number B

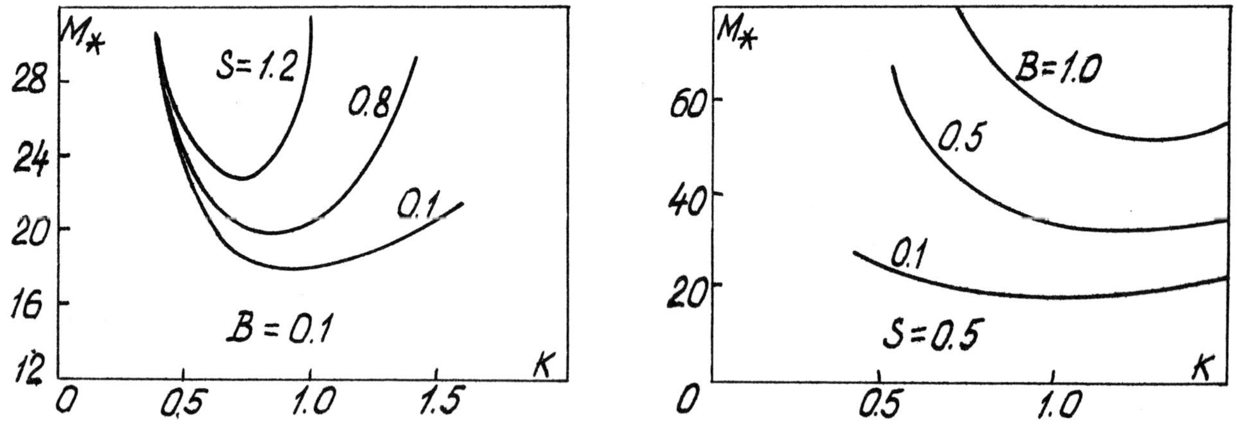

Fig.2. Critical values of Marangoni number M vs. wave number k at fixed values of B and S

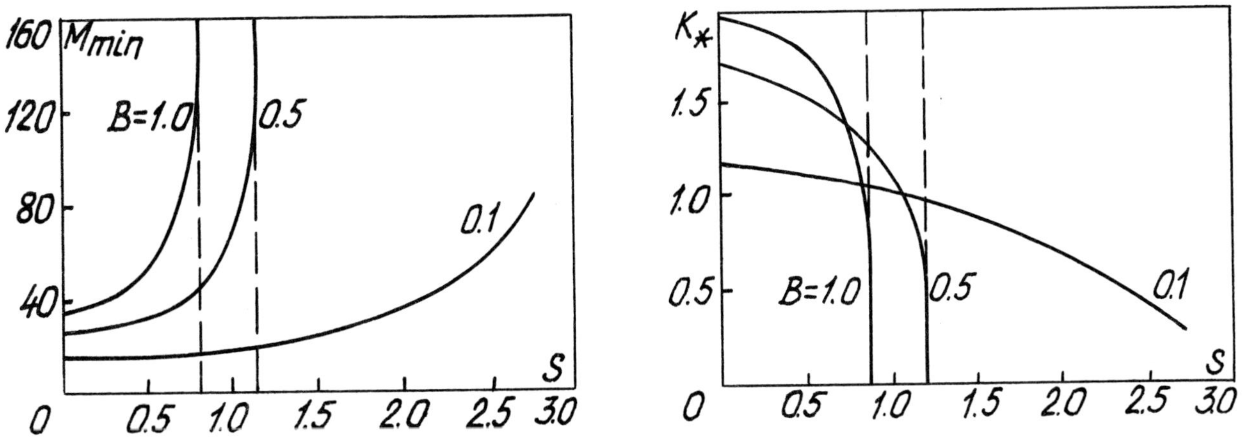

Fig.3. Variation of minimal critical values of Marangoni number M_{MIN} and corresponding wave number k_* with vibrational parameter S

rapid velocities are zero and the temperature gradient is constant. Oscillation is experienced by the pressure only.

Having linearized the averaged equations in the vicinity of equilibrium and accomplished usual transformations we obtain the system of equations for transversal components of the velocities V_z and W_z and the temperature disturbances T:

$$\frac{\partial}{\partial t}\Delta V_z = \Delta\Delta V_z - R_v \Delta_\perp W_z$$

$$\frac{\partial}{\partial t}\vartheta = \frac{1}{Pr}(\Delta\vartheta + V_z) \qquad (6)$$

$$\Delta W_z = \Delta_\perp \vartheta$$

Boundary conditions are written in the form:

$$z = -1:\quad V_z = W_z = 0;\quad \frac{\partial V}{\partial z} = \frac{\partial \vartheta}{\partial z} = 0$$

$$z = +1:\quad V_z = W_z = 0;\quad \frac{\partial \vartheta}{\partial z} = -B\vartheta \qquad (7)$$

$$\partial^2 V_z / \partial z^2 = M \Delta_\perp \vartheta;\quad \Delta_\perp = (\partial^2/\partial x^2 + \partial^2/\partial y^2)$$

The last condition represents the transformed expression for balance of viscous and thermocapillary forces, which was obtained taking into account continuity equation.

Stability is analyzed with respect to small normal disturbances which take the form:

$$V_z(x,y,z,t) = V(z)\exp(\lambda t + i K_1 x + i K_2 y)$$

The sign of λ defines stability. The boundaries of stability region are determined assuming $\lambda=0$. Then the dependence of disturbances on the transverse coordinate is described by the system of ordinary differential equations of the 8th order, for which $V = \exp(\delta z)$ are partial solutions, where

$$\delta = (k^2 + j\, k\, R^{1/4})^{1/2};\qquad j^4 = -1$$

Substitution of these solutions into boundary conditions (7) gives the system of algebraic equations, the solvability condition of which requires that the complex determinant of the 8th order should go to zero. The latter leads to a disperse relation which correlates the Marangoni number with vibrational Rayleigh R_v and wave k numbers. The Figures 1-3 illustrate predictions of disperse relation for the case of microgravity, when both surfaces are free. Similar results have been obtained for cases, using different boundary conditions and allowing for gravity.

Fig.1 shows neutral curves at some fixed values of Biot and wave numbers. Instability region lies to the left of and higher than neutral curves. The most significant result is that with increased intensity of vibrations the stability grows (Marangoni critical number increases with the increased Rayleigh vibrational number (Fig.1a and 1b). This effect is more pronounced at the plots of Fig.1c and 1d, plotted in coordinates M, S. Here $S = (R_v)^{1/2}/M$ is vibrational parameter including no temperature gradient and representing dimensionless velocity of vibrations $b\,\omega$.

As it is seen in Fig.1b and 1d, with

the increase of the Biot number (i.e. heat conduction properties of the free surface) the thermocapillary effect is restricted and neutral curves are displaced into the region of large Marangoni numbers.

The procedure and results of critical Marangoni number minimization with respect to the wave number are illustrated in Fig.2 and Fig.3. From the plots of Fig.3 it follows that starting with some values of vibrational parameter S, stability is maintained at arbitrary values of temperature gradient.

3. THE INFLUENCE OF VIBRATIONS ON PLANE-PARALLEL THERMOCAPILLARY FLOW

Consider the plane layer of fluid which displays high-frequency harmonic vibrations tangential to the layer surface. The longitudinal temperature gradient is maintained constant in the fluid. The system of coordinates is the same as in section 2. X-axis is directed along the temperature gradient, so that $(\nabla T)_Y = (\nabla T)_Z = 0$; $n_Y = n_Z = 0$. Both surfaces of the layer are considered to be free. The layer is assumed infinite in X-direction, though consideration is a case of closed flows.

Thermovibrational convection between solid planes has been discussed in analogous formulation in Ref.13. In that study the vibrations produce no average motion, i.e. the state of mechanical equilibrium holds. In case, when there are no vibrations and the fluid surfaces are free, the layer gives rise to plane - parallel thermocapillary flows. It will be shown below, that under joint action of vibrational and thermocapillary forces the equation will have solutions corresponding to the plane - parallel flow: (8)

$$\vec{V} = \vec{i}\, V(z); \quad T = x + \theta(z); \quad \vec{W} = \vec{i}\, W(z)$$

where $\vec{i}$ is the unit vector of coordinate axes X.

The system of equations (3) with account of (8) may be reduced to ordinary differential equations:

$$V''' - R_V \theta' = 0 \qquad (9)$$

$$\theta'' + V = 0\,; \qquad W' = \theta'\,;$$

where the prime corresponds to differentiating by z.

The boundary conditions are formulated as follows. The condition of vanishing tangential stresses at free surfaces with account of thermocapillary forces gives:

$$z = \pm 1: \quad V'(z) = M \qquad (10)$$

Both surfaces will be assumed to be thermally insulated:

$$z = \pm 1: \quad \theta'(z) = 0 \qquad (11)$$

The flow should be closed:

$$\int_{-1}^{+1} V(z)\, dz = 0 \qquad (12)$$

The boundary problem (9-12) has exact solution: (13)

$$V = \frac{M}{2\,F_1(\mu)}\,[F_2(\mu)\ ch\,\mu z\,\cos\mu z - F_3(\mu)\ sh\,\mu z\,\sin\mu z]$$

$$T = x - \frac{M\mu^{-2}}{8\,F_1(\mu)}\,[F_3(\mu)\ ch\,\mu z\,\cos\mu z + F_2(\mu)\ sh\,\mu z\,\sin\mu z]$$

where: $\mu = (R_V/4)^{1/4}$

$$F_1(\mu) = \mu\,(ch^2\mu - \cos^2\mu)$$

$$F_2(\mu) = (ch\,\mu\,\sin\mu - sh\,\mu\,\cos\mu)$$

$$F_3(\mu) = (ch\,\mu\,\sin\mu + sh\,\mu\,\cos\mu)$$

It may be shown that the exact solution also exists at vibrations inclined to the layer plane.

The distribution of velocity and temperature across the layer are presented at Fig.4. As can be seen from the expressions (13) and the plots, the profiles of velocity and temperature are governed only by vibrational Rayleigh number R_V. The amplitudes of velocity and temperature increase proportionally to the growth of the Marangoni number and decrease with the growth of the vibrational Rayleigh number. This means that thermocapillary convection may be reduced with the help of vibrations. From comparison of velocity profiles, plotted for various values of μ (Fig.4a) it is readily seen that when the intensity of vibrations is high, the convective motion is located in thin boundary layers whereas the most part of the fluid is practically at rest.

Let us estimate with the help of relations (4,5,13,) the possibilities of experimental investigation

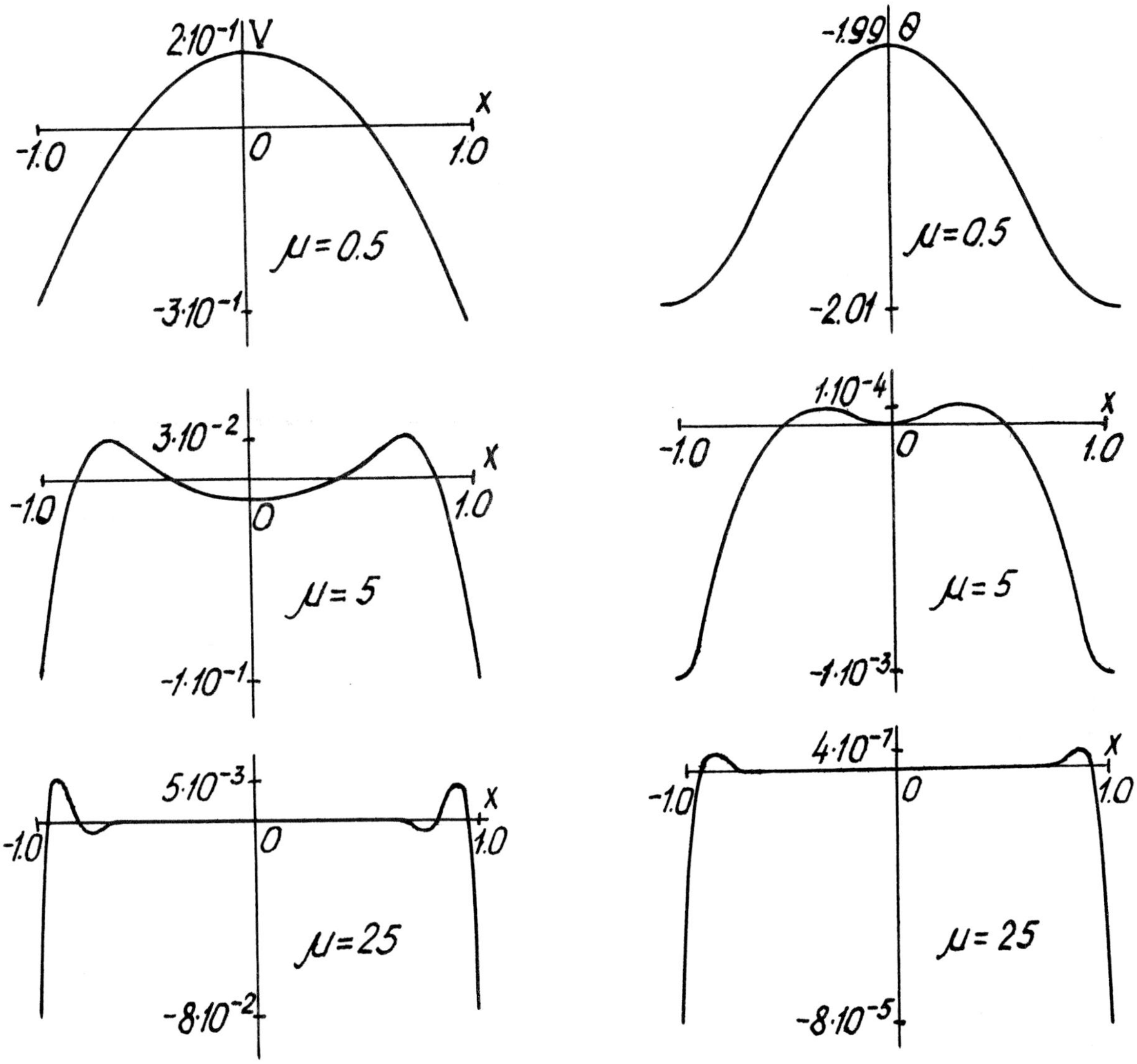

Fig.4. Velocity V and temperature θ profiles at various values of vibrational parameter μ ($M = 1$)

of vibration effects on thermocapillary flows, e.g. in 5 cm layer of ethanol at the temperature gradient of 1 K/cm. Under assumption of nondeformable fluid surface the capillary forces are insignificant, so that the shortest characteristic time is the thermocapillary one, which constitutes about 15 s. This implies that the high-frequency approximation holds true already at frequencies of order of 1 Hz. In order to estimate the force action of vibrations we shall take for example such values of vibrational parameter $\mu = 2.5$, for which the velocity at the layer axis will go to zero (the velocity profile is of the intermediate type as compared to velocity profiles of Fig.4 at $\mu = 0.5$ and $\mu = 5$). This requires the vibration rate $b\omega = 1.3$ cm/s, and at the frequency 10 Hz the amplitude $b = 0.02$ cm and acceleration 0.1 g. It should be borne in mind that conducting such experiments in terrestrial conditions presents additional difficulties. In order to avoid thermogravitational convection, the experiments should be carried out for layers the thickness of which is less than 0.3 cm. Then vibrational rates, corresponding to the same values of μ will increase in inverse proportion to h.

CONCLUSION

The two examples considered above present evidence for existence of vibrational thermocapillary phenomena. However, in both cases the structure of vibrations and temperature fields and the problem geometry are such that the fluid maintains equilibrium in the absence of thermocapillary forces. However, there are the other situations, in which vibrations are the immediate reasons of fluid motions. It is readily seen from calculations that these situations provide higher possibilities of controlling thermocapillary effects due to competition between vibrational and thermocapillary forces. Such is the case of thermocapillary vibrational flow in a plane layer under vibrations inclined to the layer.

Let us point out some other results of joint action of vibrational and thermocapillary forces, which have been not described in this paper. Based on the analysis of complete unaveraged system of equations (1), it may be shown that under vibrations the stability loss in Marangoni problem can no longer be related solely to quasi-monotone disturbances, as it has been observed in the high-frequency case, discussed in Section 2 of the present paper. Vibrations may also give rise to parametric instability, leading to occurrence of flows oscillating with half vibration frequencies (the main region of parametric resonance). The similar results are obtained for the low-frequency modulation of the uniform temperature gradient. If, in addition, one takes into account the surface deformations, consideration of the problem will then involve parametric excitation of thermocapillary surface waves.

Generally speaking in all vibration-thermocapillary problems one should take into account both the possibility of forced motion and surface deformations, and initiation of surface instabilities. In particular, since the spectrum of the surface disturbance wavelengths has no lower limit, there always exists a wave which may be adjusted to a resonance or near-resonance state with respect to vibration. However, in the analysis of vibrational suppression of slowly developing monotone instabilities (e.g. convective ones) the employment of the average approximation is still valid.

As already stated (Section 2) the threshold values of vibration rate S, at which thermocapillary instability is suppressed, are independent of frequency. At the same time the threshold values for excitation of parametric waves increase with the frequency ω (Ref.2). Therefore, starting with some sufficiently large value ω_* there exists a parameter range in which the equilibrium is absolutely stable. In this range the vibrations are high enough to prevent the occurrence of Marangoni instability, but are too low to excite waves. This is an ordinary situation which has been observed in our experiments (Refs.14,15) on vibrational suppression of large-scale deformations and ruptures, caused by thermocapillary convection in very thin fluid layers at longitudinal temperature gradient. The suppressing effect appeared to be so strong that it can be observed even at the background of the surface waves (Refs. 14,15). However an increase of vibration frequency (more than 400 Hz) in sufficiently viscous fluid (silicone oil PMS-200) allows to avoid the excitation of waves and simultaneously suppress the initial deformation.

Finally it is to be stated that vibrations may effectively increase or decrease the stability of fluid equilibrium and strongly affect the intensity and structure of thermocapillary flows. Thermocapillary convection is known to be the main reason of defects initiated in materials created in microgravity. Therefore the possibilities of vibrational control of thermocapillary effects are of essential application concern for space technologies.

REFERENCES

1. G.H.Wolf, The dynamic stabilization of the Rayleigh-Taylor instability and the corresponding dynamic equilibrium, *Z. Phys.*, 227 (1969) 291-300.

2. V.A.Briskman, Parametrical stabilization of fluid interface, *Dokl. Akad. Nauk SSSR*, 226/5 (1976) 1041-1044.

3. N.A.Bezdenezhnykh, V.A.Briskman, A.A.Cherepanov and M.T.Sharov, Control of the stability of liquid by means of variable fields, *Fluid Mechanics - Soviet research* (SCRIPTA TECHNICA, USA), 15/1 (1981) 11-32.

4. V.A.Briskman and M.T.Sharov, Study of the vibrational stabilization of liquid free surface in microgravity with the help of magnetic liquid, In: *Numerical and experimental simulation of hydrodynamical phenomena in microgravity* (Sverdlovsk 1988).

5. G.Z.Gershuni and E.M.Zhukovitsky, *Convective stability of incompressible fluid* (Nauka, Moscow 1972).

6. D.V.Lyubimov and A.A.Cherepanov, On occurrence of wave relief at media interface under vibration, *Izv.AN SSSR Mekh.Zhidk i Gasa*, 6 (1986) 8-13.

7. V.A.Briskman, D.V.Lyubimov and A.A.Cherepanov, Stability of the rotating liquids interface under axial vibrational field, In: *Numerical and experimental simulation of hydrodynamical phenomena in microgravity* (Sverdlovsk 1988), 18-26.

8. N.A.Bezdenezhnykh, V.A.Briskman, A.Yu.Lapin, D.M.Lyubimov, T.P.Luibimova, A.A.Cherepanov and I.V.Zakharov, The influence of high-frequency tangential vibrations on the stability of the fluid interfaces in microgravity, *Microgravity science and technology*, IV/2 (1991) 96-97.

9. V.A.Briskman, Vibrational-thermocapillary convection and stability, *Abstracts of International symposium of hydromechanics and heat/ mass transfer in microgravity* (Perm-Moscow 1991), 18-19.

10. M.A.Zaks, On the action of high-frequency vibrations on the Marangoni instability in a plane fluid layer, *Abstracts of International symposium of hydromechanics and heat/ mass transfer in microgravity* (Perm-Moscow 1991), 116.

11. V.A.Briskman, M.A.Zaks and A.L.Zuev, On the action of high-frequency vibrations on thermocapillary flow and stability, *Abstracts of 1st European Fluid Mechanics Conference of EUROMECH* (Cambridge 1991).

12 G.Z.Gershuni, E.M.Zhukovitsky and A.A.Nepomnjashchy, *Stability of convective flows* (Nauka, Moscow 1972).

13. R.F.Birikh, On the vibrational convection in a plane layer with a longitudinal temperature gradient, *In the present proceedings*.

14. A.L.Zuev, Influence of different factors on the thermocapillary deformation of a thin liquid layer, *Abstracts of International symposium of hydromechanics and heat/ mass transfer in microgravity* (Perm-Moscow 1991), 24.

15. V.A.Briskman and A.L.Zuev, Influence of different factors on the thermocapillary deformation of a thin liquid layer, *In the present proceedings*.

EXPERIMENTAL AND THEORETICAL STUDY OF MARANGONI FLOWS IN LIQUID METALLIC LAYERS

P. Tison, D. Camel, I. Tosello, J-J Favier
CEA/DTA/CEREM/DEM, Section d'Etudes de la Solidification et de la Cristallogenèse
C.E.N.G., 85 X, 38041 Grenoble Cedex, France

ABSTRACT

To study capillary flows in liquid metals on the ground, parallelepipedic geometries with high length/height (A= L/H) ratios and horizontal applied thermal gradients G are used. This method is applied to liquid tin-based alloys. Experiments are performed under high vacuum (< 10^{-7} mbar). Surface velocities are measured by recording the motion of thin oxide particles used as tracers. For pure tin, capillary flows dominate when H ≤ 2 mm. Velocity measurements are in agreement with order of magnitude analysis and numerical simulations which predict a transition to a boundary layer flow when Re_M/A increases. For low values of Re_M where a viscous unidirectional flow is obtained, significant damping (factor 2) is observed when a magnetic field of 0.15 Tesla (Ha ≈ 14) is applied perpendicularly to the surface. In the case of Sn-Bi alloys (0.5 to 3 at % Bi), it is shown that surface tension goes through a maximum as a function of temperature: two contra-rotating cells then appear, separated by a stagnant line. Unsteady thermocapillary flows are observed with a threshold of $1000 < Re^c_M < 1500$ for l/H = 20. Convective motions and related thermal instabilities are also observed during the directional solidification of the liquid layers.

Keywords: Marangoni flows, liquid metals.

1. INTRODUCTION

Marangoni convection is likely to play a significant role in the solidification process when the liquid includes a free surface close to the liquid-solid interface. Research into the effects of this type of convection and on ways of overcoming them has been stimulated by the enhanced possibilities offered by microgravity experimentation [1].

However, most experimental studies are related to transparent systems where convection can be directly visualised. Relatively few studies have been carried out on metallic liquids despite the fact that specific hydrodynamic regimes are predicted for low values of the Prandtl number, Pr, and that possible methods of blocking the convection are also specific to these systems.

Most of these studies concern floating zone crystal growth under microgravity. For example, it has been shown that Marangoni convection was responsible for striations observed and evidence was given of the contribution of this convection to macrosegregation phenomena [2]. A more quantitative study of macrosegregation was carried out in a somewhat different configuration (semi-confined Bridgman) [3, 4]. All these results are based on the post-experimental characterisation of solidified crystals. To the authors' knowledge, only one study [5] conducted on the ground has made use of in-situ diagnosis equipment (micropyrometers) to measure the occurrence thresholds of unsteady thermocapillary flow regimes and the resulting temperature fluctuations, in floating zone growth of Mo and Nb crystals of 4 and 6 mm diameter.

Various ways of eliminating Marangoni convection, or at least attenuating its effects in metallic systems, have been proposed. For example, experimentation has been initiated in order to determine whether controlled oxidation would enable a thin oxide film to be maintained on the liquid metal surface, thereby eliminating surface movements [6]. Another method proposed consists in adding a surfactant element to the melt in order to reduce the surface tension temperature coefficient [7] but, to the authors' knowledge, no attempt has been made to test this idea. Finally, numerical simulations have already shown that the application of a magnetic field in floating zone crystal growth should enable the convection speed to be reduced and push back the instability thresholds [8, 9]. Experiments on doped germanium crystal growth on board the Photon satellite have indeed demonstrated that the degree of microsegregation was effectively reduced by the application of an axial magnetic field of 0.03 Tesla [10].

In the work described in this paper, Marangoni convection in liquid metals and its effects on solidification have been studied using a laterally-heated flat layer configuration. This configuration is representative of horizontal Bridgman crystal growth. Moreover, a large quantity of theoretical results are available for making detailed comparisons with the experiment.

2. TEST PRINCIPLE

The tests are based on the four following main points:

a) The thickness of the liquid layer depends on the relative importance to be given to the two types of convection (thermocapillary or thermogravity).
b) The liquid is placed in a horizontal temperature gradient.
c) The upper surface is kept under high vacuum ($\approx$ 1 to 3 x 10^{-8} mbar $\approx$ 1 to 3 x 10^{-6} Pa). This surface can be unbounded, with or without thermocapillary convection, or rigid (oxidised).
d) The surface velocity field is known by recording the displacement of fine oxide particles on the free surface with a video camera.

Notations for all symbols and parameters used in this work are given in appendix.

Many theoretical studies have been carried out on the configuration considered here. The essential, non-exhaustive, conclusions of these studies are summarised below as they are indispensable for the discussion of the experimental results. Consideration is given to the theoretical predictions concerning, first of all, the case of an infinite, two-dimensional flat layer, followed by the influence of the presence of side walls.

2.1 Infinite layer

For an infinite layer, the movement equations are based on a conventional solution (cf. Levich [11] for the case of Gr = 0, and Birikh [12] for the general case). The velocity u is parallel to the temperature gradient and depends only on y. On the surface:

$$u = -\frac{Re_M}{4} + \frac{Gr}{48} \qquad (1)$$

In the expression (1), the velocities are, by convention, taken as positive when flow is from hot to cold on the surface. Note that for pure metals, σ' is negative so that the two contributions are cumulative.

The ratio of the two contributions is as follows:

$$\frac{u_{Re_M}}{u_{Gr}} = 12\,\frac{Re_M}{Gr} \qquad (2)$$

For example, for a 2 mm thick layer of liquid tin, $u_{Re_M}/u_{Gr} \simeq 25$ which means that the flow is thermocapillary. Levich [11] has already reported that this solution should be considered valid only if $u \ll 1$. Instabilities of this flow have been studied in [3].

2.2 Finite extension layer

In the case of a real layer of finite area, the question is to know under what conditions the following transitions will be observed:

* flow transitions predicted for an infinite layer,
* modified transitions,
* new transitions.

The possible answers given by available theoretical work will be examined for transitions in steady flow conditions, and then for transitions from steady to unsteady flow regimes.

For low values of Gr and Re and for high aspect ratios, the flow is identical to Birikh's flow in the central part, with the areas affected by shear along the side walls and recirculation at the ends being limited to the order of the thickness of the layer. The only steady transition predicted for Birikh's flow is dynamic in nature and corresponds to the case where the upper surface is rigid (or where the two opposing terms in equation (1) almost cancel each other out).

The influence of side walls on steady flows in the limiting case Gr = 0 has been analysed in detail in [14] and [15 - 17] using bifurcation analysis and direct simulation with extensive correlation of numerical data. These authors find that, at high Re_M values and high aspect ratios A, the following regions can be distinguished, starting from the hot end of the cavity:

- a hot end region for $x \leq 1$
- an establishment zone, where u/Re_M is an increasing function of x/Re_M ; this zone can be subdivided into:
  a) a zone where a developed boundary layer regime occurs with $u/Re_M \simeq 1.2 \times (x/Re_M)^{1/3}$ for $1 < x < 2.5\text{x}10^{-3}\ Re_M$
  b) a transition zone towards the Couette flow, the limiting velocity $u/Re_M = 1/4$ being reached at $x \simeq 0.025\ Re_M$ (3)
- a cold end region where vortices develop, their number increasing with Re_M and A.

The physical meaning of the boundary layer region is that the thickness δ of the layer driven by viscosity increases more and more slowly with x when Re_M increases. The corresponding scaling laws can be recovered through simple scaling arguments as follows.

The boundary condition at the free surface imposes the following relation between u and δ:

$$\frac{u}{\delta} \simeq Re_M \tag{4}$$

Another relation is given by the condition that viscous forces have to be of the same order as inertia forces within the boundary layer, which is written:

$$\frac{1}{\delta^2} \simeq \frac{u}{x}$$

It follows, therefore, that:

$$\delta \simeq \left(\frac{x}{Re_M}\right)^{1/3} \tag{5}$$

$$\text{and} \quad \frac{u}{Re_M} \simeq \left(\frac{x}{Re_M}\right)^{1/3}$$

Note that numerical simulations give this law with a constant which is close to unity, as is often found in boundary layer regimes.

From the above-mentioned results, the surface velocity in the middle of the layer is given by an expression of the following form:

$$\frac{u}{A} = f\left(\frac{Re_M}{A}\right)$$

with: $\frac{u}{A} = \frac{1}{4}\frac{Re_M}{A}$ (Birikh's solution) for: $\frac{Re_M}{A} < 20$ or equivalently: $\frac{u}{A} < 5$ (6a)

and $\frac{u}{A} = 0.95\left(\frac{Re_M}{A}\right)^{2/3}$ (boundary layer regime) for: $\frac{Re_M}{A} > 200$ (6b)

3. EXPERIMENTAL SET-UP (Figure 1)

A pure iron crucible is used to give an oblong shape to a fine liquid layer of pure tin or tin alloyed with bismuth.

Two independent Joule-effect heating devices located at each end of the crucible are used to control the temperature difference to within ± 0.1°C. One end of the crucible can be cooled by gas or water circulation.

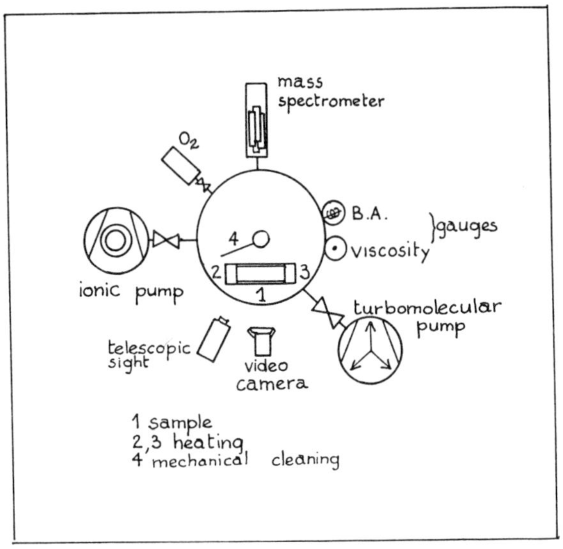

Figure 1: Experimental set-up.

The temperature is controlled by K-type thermocouples (Chromel-Alumel) inserted in the crucible. The real temperature of the liquid is determined by movable thermocouples of the same type of 0.5 mm diameter in 1 mm dia. iron sheaths inserted in the bath via the free surface. The azimuthal position of these sensors is determined by video enlargement. Their depth in the melt is known to within ± 0.01 mm by using a telescopic sight fitted to a precision measurement support.

Special attention is paid to the real conditions at the liquid-gas interface. This explains why the sample preparation and work chamber technology are compatible with ultra-vacuum techniques (only the seals of the transparent pyrex wall restrict the working pressure to 10^{-9} mbar). Tin in its liquid state can be placed under vacuum as its vapour pressure is still only about 7×10^{-10} mbar at 600°C. In order to protect the work chamber from contamination (especially from hydrocarbons), the backing pressure is obtained by a turbomolecular pump. During this operation, the pressure is measured by a viscosity ball gauge and a Bayard and Alpert gauge. A 200 l/s ionic pump associated with titanium sublimation is used to obtain and maintain the very low pressures ($\simeq$ 1 to 3×10^{-8} mbar). At these pressures, 30 to 100 seconds are needed to form a mono-layer on the liquid free surface. The gaseous constituents of the residual atmosphere above the liquid are analysed by a four-pole mass spectrometer (masses 1 to 200) equipped with an electron multiplier.

Before taking surface velocity measurements, the liquid surface is mechanically cleaned using brushes made of very fine thermally insulated pure iron or stainless steel meshes. These brushes are also used to deposit extremely fine particles of natural oxide which will serve as tracers for velocity measurements.

This process is also used to calculate the oxidation kinetics of the free surface under experimental conditions. Even when in movement, the liquid-gas interface may become partially or totally oxidised as a result of molecular bombardment of the gas phase or by degassing of the dissolved oxygen in the liquid tin [20], or possibly from the tracers. Several minutes after cleaning the surface, a fine transparent oxide film may develop from stagnant regions of the free surface (edges of the crucible or in the immediate proximity of tracers accumulated on the downstream side of the surface movement). The progressive growth of this transparent oxide can be illustrated by depositing particles which move to the limits of the oxide film. These measurements will serve to determine the maximum time that can elapse between two cleanings of the work surface. Depending on the vacuum and degassing conditions of the tin used, this time may vary between ten minutes or so and several hours.

When surface oxidation is to be induced in order to study the behaviour of a liquid with a surface that is no longer free but rigid, a micro-leak is arranged in order to adjust the oxygen partial pressure in the work chamber.

A samarium-cobalt permanent magnet can be placed above the melt in order to apply a magnetic field of 0.15 Tesla perpendicular to the surface.

The vacuum established above the free surface and the low maximum temperature values (420°C) reached in the tin are such that the the surface can be considered to be adiabatic.

Measurement of surface velocities

The displacement of tracers on the liquid surface is recorded by a video camera which films at a speed of 25 frames/second. The recording is then played back frame by frame. Determination of the position of the tracing particle each 1/25th of a second gives the fluid surface velocity in the time interval considered. the accuracy of the measurement depends on the accuracy of the particle position determination, and thus on the fidelity of the frame-on-frame playback of the video film. In most cases, the video document is "processed" by an image analyser in order to improve results.

Figure 2 gives an example of the unprocessed results obtained on velocities calculated as a function of particle position measured with respect to a starting point. The acceleration stage predicted by the calculations of H. Ben Hadid and B. Roux [16] for high Re_M numbers was clearly apparent.

The crucible dimensions determine the geometrical parameters (H, L, l). Consequently, the experimental variables are the absolute temperature (which determines the values of ρ and ν) and ΔT.

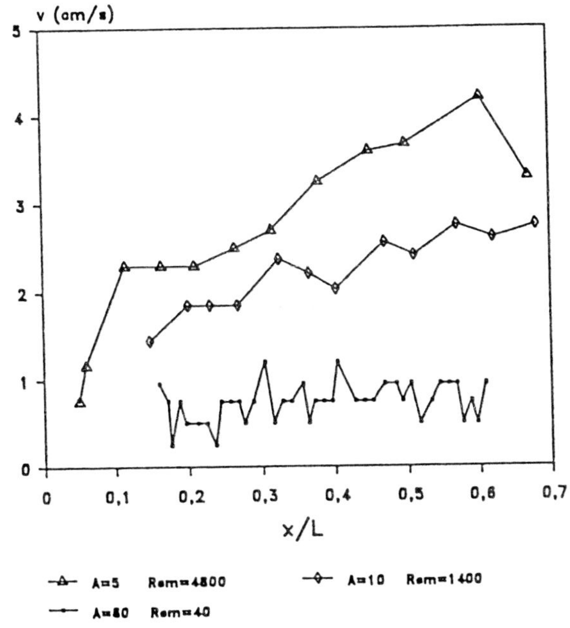

Figure 2: Particle velocity measured at the surface of the pure liquid tin layer as a function of distance to hot end for different Re_M values.

The experimentation gives access to surface velocity values as a function of temperature gradient (Figure 5). When the velocities are measured under conditions satisfying the criterion $u/A < 5$, the experimental Reynolds number Re_M is known from equation (6a). This value can be compared to the theoretical Re_M value by taking $\sigma' = 0.07$ mN/m.K. This value is derived from the work of Ibragimov [18] who also studied the surface tension of Sn-Bi alloys. Note, however, that the values given in the literature range from -0.07 to -0.22 mN/m.K (see for example [19] or [20] for a discussion of these data). This high degree of inaccuracy in the knowledge of $d\sigma/dT$ is fairly general for metallic systems. Apart from the difficulty of obtaining clean surfaces in such systems, this discrepancy results from the fact that this parameter is deduced from the absolute measurement of σ using methods which have an accuracy limit of the order of 1% and that σ varies only slightly with temperature (typically $1/\sigma \cdot d\sigma/dT \simeq 10^{-4}$ K^{-1}). The value selected here should not, therefore, be considered as a definitive value for pure tin. The choice made can be explained simply by the fact that the data presented by the author for Sn-Bi alloys have a very low dispersion and give a change in sign of $d\sigma/dT$ at low temperature, a result that, as will be seen, is confirmed by the experiments described here.

4. EXPERIMENTAL RESULTS AND DISCUSSION

4.1 Surface velocities in the case of pure tin

Figure 3 shows the variation in the ratio u/A as a function of the ratio of theoretical Re_M number to aspect ratio,Re_M/A. This figure also gives plots of the straight lines corresponding to the theoretical predictions in the viscous regime and the boundary layer regime.

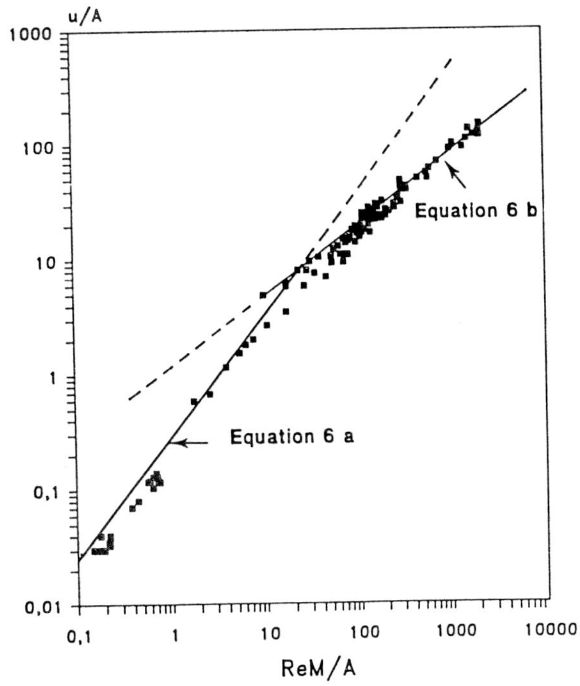

Figure 3: $\frac{u}{A} = f\left(\frac{Re_M}{A}\right)$

The predictions of H. Ben Hadid and B. Roux [16] (see expressions 6a and 6b) are effectively observed, the experimental transition zone between the two regimes is indeed between the Re_M/A values of 20 and 200 given by these authors.

For each aspect ratio, the results with $u/A < 5$ were used to deduce an experimental value of $\partial\sigma/\partial T$. The results are given in Table 2.

The excessively low values of σ' result from surface velocities which are themselves lower than the ideal case. This deviation may be induced by contamination of the melt: oxygen in equilibrium with the tracer oxide; partial removal of the crucible cleaning products used to ensure perfect wetting of the liquid metal or perhaps excessively long Sn-Fe contact time giving rise to $Fe\text{-}Sn_2$ precipitates in the liquid.

A	σ' (mJ m^{-2}K^{-1})
83	3.9 to 5 x 10^{-2}
63	4.4 to 5.9 x 10^{-2}
30	6.5 to 7.5 x 10^{-2}
12.5	4.3 to 5 x 10^{-2}
12.5	5.3 to 5.9 x 10^{-2}
12.5	3.9 to 4.3 x 10^{-2}
10	6.3 to 9 x 10^{-2}

Table 1: Experimental values of σ' for pure tin deduced from velocity measurements for the case of u/A < 5

4.2 Influence of a magnetic field on thermo-capillary convection

One of the methods considered for eliminating defects in crystals caused by liquid melt instabilities is magnetic damping [8], [9], [10].

Uni-directional thermocapillary convection is a configuration which should be well-suited to magnetic damping [21] [22].

In order to check the theoretical predictions of J. Ochiai et al. [21], a magnetic field of about 0.15 Tesla was applied perpendicularly to the liquid surface. The corresponding Hartmann number is Ha = 14. The magnetic source consists of two small samarium-cobalt magnets. The soft iron pole pieces channel and apply the field above a liquid bath measuring 2.5 x 1.5 x 0.2 cm. The entire assembly can be withdrawn to allow reference surface velocity measurements to be taken in a nearly zero magnetic field.

Figure 4 gives the overall results. The application of the magnetic field halves the velocity. The calculations of J. Ochiai et al. [21] predicted a velocity $u \simeq Re_M/Ha$ for values of Ha > 10; with $u_o = Re_M/4$ (velocity with zero magnetic field), a ratio of 0.29 should be obtained between velocities measured with and without a magnetic field.

This preliminary confirmation of the effect of a magnetic field on Marangoni flow was made on a liquid layer bounded by a pure iron crucible. When the magnetic source is applied, the flow is no longer uni-directional. This may be due to the channelling effect of the crucible on the field lines and to the presence of thermoelectric currents.

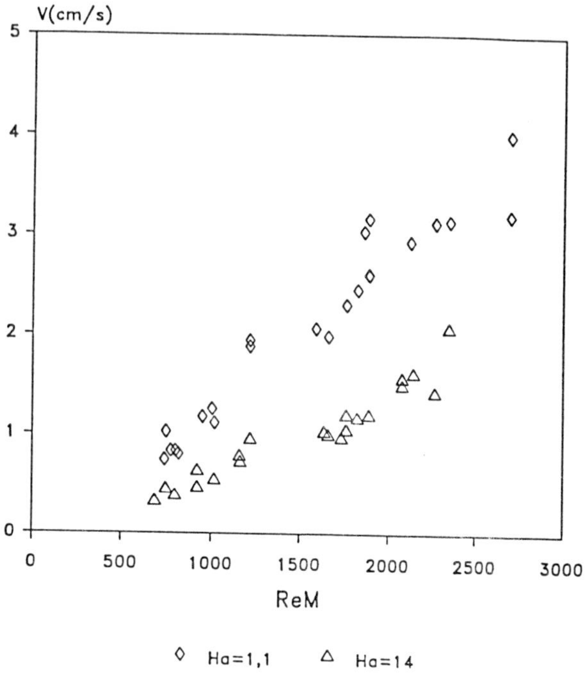

Figure 4: Influence of a magnetic field normal to the flow on thermocapillary convection. H = 2 mm Cavity 1 x 12.5 x 7.5

4.3 Influence of bismuth on thermocapillary convection of tin

It is a known fact that adding bismuth to tin reduces the surface tension. The experimental results of Y.A. Klyachko and L.L. Kuniv [23] and Kh.I. Ibragimov et al. [18] even show that the sign of the temperature coefficient changes at low temperatures.

The influence of bismuth on thermocapillary convection was studied in alloys of low bismuth content (0.2 to 3.1 at%) in 2 mm thick liquid layers.

Figure 5 shows the variation in surface velocity as a function of temperature gradient for alloys with 0.2 and 0.5 at% bismuth at temperatures of the order of 290°C.

For bismuth contents ≥ 0.5 at%, the direction of movement on the surface is reversed at the lowest temperatures. At intermediate temperatures, two counter-rotary rolls are present in the cell. Depending on the temperature gradient, these opposing surface shear forces give rise to a neutral zone on the surface of variable size in which movements are practically zero.

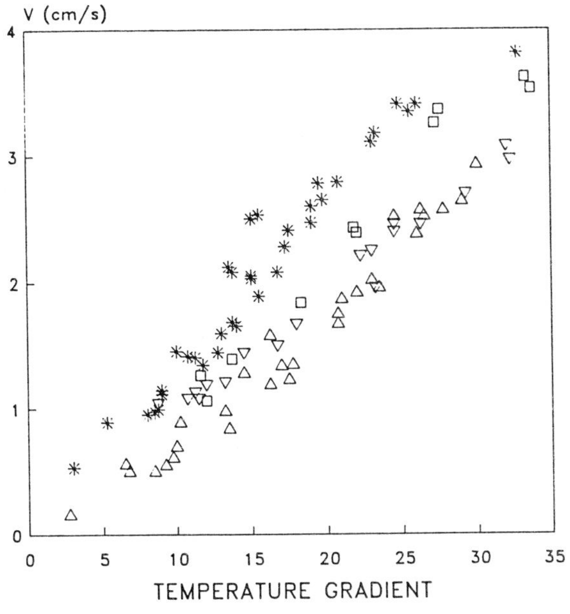

Figure 5: Influence of bismuth on thermocapillary convection in tin (mean temperature ≃ 290°C).

Figure 6 represents the variations in mean temperature in this zone as a function of Bi content for temperature gradients of the order of 30 K/cm. However, this temperature must not be mistaken for that at which surface tension is maximum as the velocity field is not always symmetrical with respect to the changeover point.

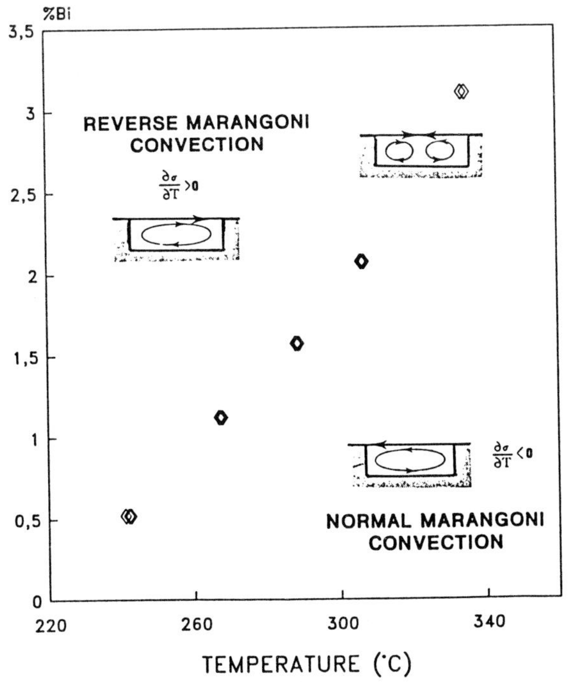

Figure 6: Variation in mean temperature of the movement reversal zone as a function of Bi content. G = 30 K/cm.

In order to measure the variation in $\partial\sigma/\partial T$ around the changeover temperature, experiments are currently being conducted with a gentle temperature gradient.

4.4 Thermocapillary-induced instabilities

The linear stability analysis developed by Smith and Davis [13] indicates that, for a fluid with Pr = 0.015, hydrothermal instabilities appear at a Reynolds-Marangoni number of 600.

The first attempts of the present authors to detect thermal instabilities of thermocapillary origin were performed on 2 mm thick liquid layers with 1 x 10 x 5 and 1 x 12.5 x 5 cavities. No instability greater than or equal to 0.05°C was evidenced for these aspect ratios up to $Re_M \leq 3000$.

However, as noted by H. Ben Hadid and B. Roux [16], these disturbances can be detected only in cavities with large lateral dimension as they propagate at an angle from ± 77° to the direction of movement. Also, the critical wavelength in the z direction is ℓ_z = 19H for Pr = 0.015 (λ_z = 3.8 cm).

Consequently, a cavity of 1 x 30 x 20 (with H = 2 mm) was used. Under these conditions, instabilities appear for Re_M values of the order of 1400. Figure 7 illustrates the aspect of these instabilities for several Re_M values. For Re_M = 2100, temperature variations may reach ± 0.5°C. The frequency of these oscillations varies from 0.19 to 0.23 Hz. Smith and Davis's calculations [13] predict $F = 2.4\ \nu / H^2 = 0.15$ Hz at 280°C

These instabilities are of thermocapillary origin since they are cancelled out when the surface is left for several hours under a total pressure of 2 to 6 x 10^{-8} mbar (the melt temperature remains stable). When the surface is cleaned (Figure 7), these instabilities are detected once more. In addition, when the tin is alloyed with 0.5 at. % of bismuth, the instabilities are not detected under conditions which would make a pure tin melt unstable. For example, for a temperature difference of 390 - 270°C, the melt temperature is stable. In pure tin, this temperature difference would give rise to a Re_M value of 1790. The presence of bismuth leads to a significant drop in the experimental Re_M number which, in the example mentioned, was about 300 in the cold part of the melt and ≃ 750 in the hot part.

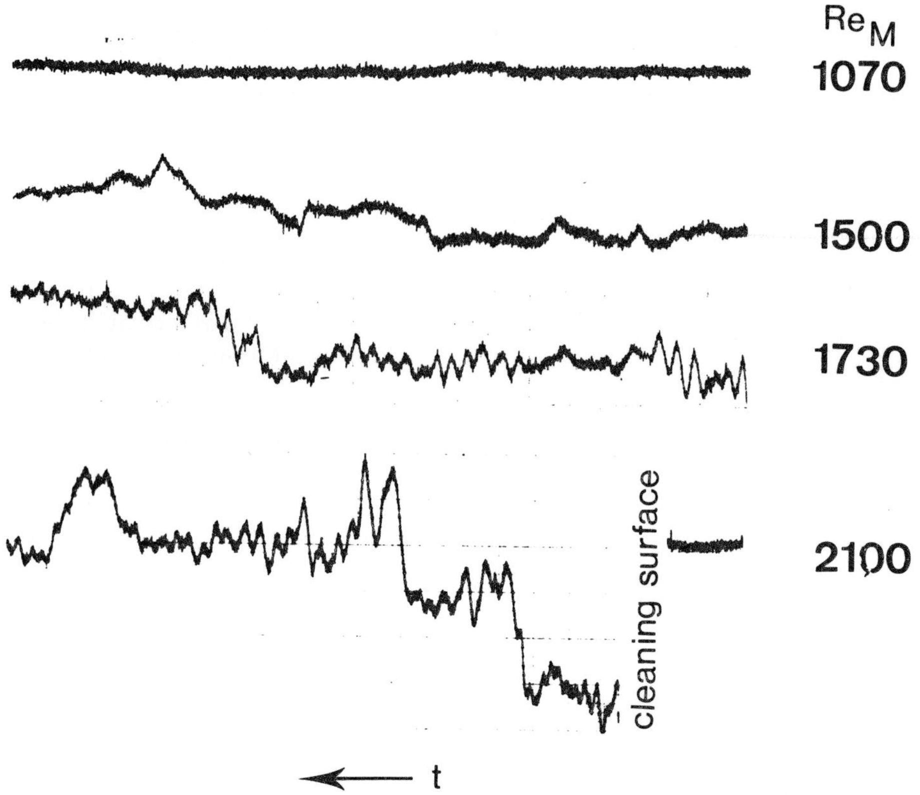

Figure 7: Time-based signal of thermocapillary-induced thermal instabilities for a few values of Re_M. (Cavity: 1 x 30 x 20; H = 2 mm)

4.5 Soluto-capillary convection caused by directional solidification of Sn-Bi alloys

The partition coefficient of the Sn-Bi system is equal to 0.25. The solidification of such an alloy leads to solute rejection ahead of the liquid-solid front and thus a bismuth concentration gradient in the liquid.

The variation in surface tension ahead of the solidification front is therefore given by:

$$\frac{d\sigma}{dx} = \frac{\partial \sigma}{\partial T} \cdot G + \frac{\partial \sigma}{\partial c} \cdot \frac{\partial c}{\partial x} \qquad (8)$$

In the present case, bismuth lowers the surface tension of tin and, aided by the partition coefficient, the second term promotes a cold-to-hot movement. To illustrate the relative importance of this term, some values given by K.I. Ibragimov et al. [18] were used to draw up figure 8. This figure shows that a 0.6 at% bismuth content at 250°C has just as much effect on the surface tension as a temperature difference of 200°C in pure tin. Thus for Bi contents of around one atomic percent, the soluto-capillary effect predominates if:

$$\frac{1}{G} \cdot \frac{\partial c}{\partial x} > \frac{0.6\%}{200} = 3 \times 10^{-3}\ \%/K$$

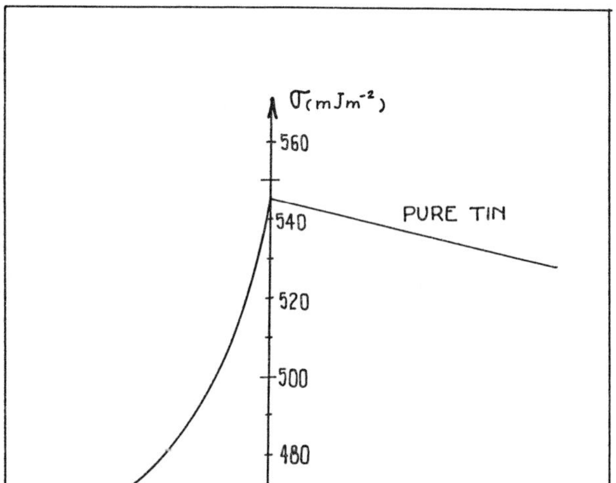

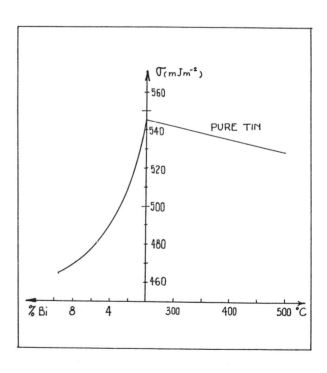

Figure 8: Comparison of the temperature- and concentration-dependence of the surface tension of Sn-Bi alloys (after [18])

By comparison, in front of a destabilised solidification front, assuming zero constitutional undercooling, the following value is obtained:

$$\frac{1}{G}\cdot\frac{\partial c}{\partial x} > \frac{1}{m_L} \simeq 0.46\%/K$$

Table 2 gives a few examples encountered during directional solidification of alloys with low bismuth content. The surface velocities measured ahead of the L/S front during solidification are compared with the velocities obtained in the homogeneous liquid alloy subjected to the same temperature gradient.

Three cases are presented where the temperature of movement reversal at the surface of the homogeneous alloy is respectively below, in the region of and above the solidification temperature.

In the first case, a zone in which the movement is reversed appears ahead of the front. This zone is bigger in the second case and the flow velocities are much higher than in the homogeneous alloy. In the third case, the movement has been reversed in the homogeneous alloy and is considerably higher in velocity during solidification.

Table2: Comparison between movements observed at the surface of Sn-Bi alloys, respectively before and during directional solidification. Cavity: 1 x 12.5 x 5 H = 2 mm

Homogeneous Liquid	Solidification
$V = 2\ cm.s^{-1}$ COLD 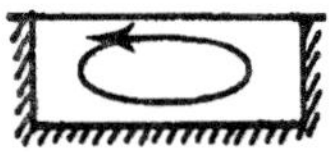 HOT	$V = 2\ cm.s^{-1}$ COLD 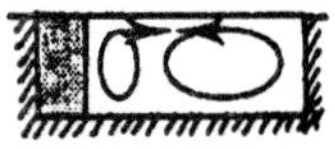 HOT
0.2 at% Bi $G = 20°K.cm^{-1}$	
$V = 0.8\ cm.s^{-1}$ COLD 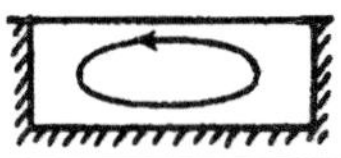 HOT	$V = 5\ cm.s^{-1}$ COLD 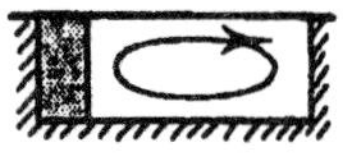 HOT
0.5 at% Bi $G = 12.5°K.cm^{-1}$	
$V = 0.1\ cm.s^{-1}$ COLD 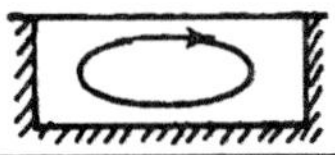 HOT	$V = 2\ cm.s^{-1}$ COLD 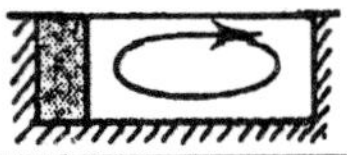 HOT
1.12 at% Bi $G = 3°K.cm^{-1}$	

Growth direction

——————>

In conclusion, a surfactant can slow down thermocapillarity-induced movements in a totally liquid alloy. On the other hand, it may prove to be a source of intense convection during solidification. The thermal consequences of this type of convection are discussed below.

The thermocouples placed at the melt surface check that the temperature is stable ahead before the appearance of the first solid seed. When the solid starts to form, the submerged thermocouples detect instabilities associated with surface movements. Figure 9 shows a characteristic example obtained on an Sn-1.5 at% Bi alloy. Temperature fluctuations are of the order of one degree.

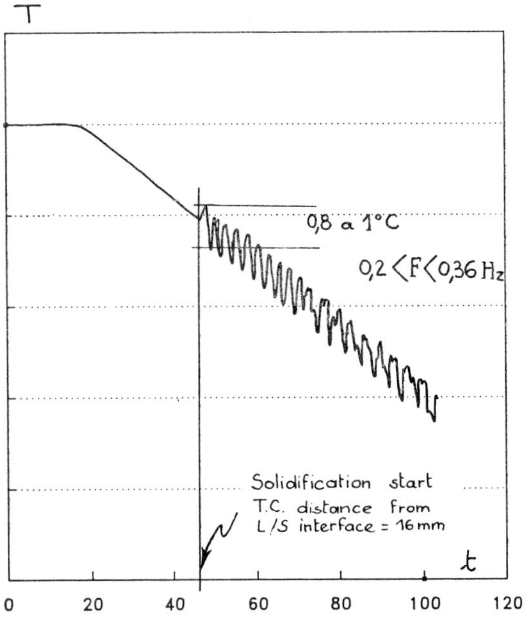

Figure 9: Thermal instabilities induced by soluto-capillary convection ahead of the solidification interface. Alloy: Sn-1.5 at% Bi Cavity: 1 x 30 x 20 H = 2 mm $G = 2\ K/mm$ Solidification rate: 1.33×10^{-2} mm/s

CONCLUSIONS

The experimental configuration, consisting of a laterally-heated flat liquid layer of small thickness, was used to study capillary convection in tin and Sn-Bi alloys.

By recording the displacement of the oxide particles used as tracers, it was possible to visualise the surface flow and to measure the corresponding velocities. For low-velocity flows, the velocity measurement was used as a basis for evaluating σ' and thus Re_M for the experimental conditions considered. The higher flow velocities were measured to verify - as predicted in numerical simulations - that the movement is accelerated at the surface and that, concurrently, the velocity deviates by default from the linear variation curve of Re_M/A.

A reduction in surface velocities (by a factor of two) was obtained by application of a magnetic field of 0.15 Tesla (Ha $\simeq$ 14) normal to the surface. It is suggested that the efficiency of such a magnetic field could be increased by using a non magnetic and electrically insulated crucible.

The slow-down in movement - and even a reversal in movement direction - was observed in alloys containing 0.5 to 3 at% of bismuth. The temperatures at which the movement is reversed are close to those at which σ goes through a maximum according to Ibragimov's data.

Evidence of hydrodynamic instabilities of purely thermocapillary origin was given for 1 x 30 x 20 cavities (thickness 2 mm). The obtained results indicate that the critical threshold, Re^c_M, at which these instabilities appear must depend on the aspect ratio. Thus, a value of $Re^c_M \simeq 1400$ was obtained in this type of cavity whereas no instability whatsoever was observed up to a value of $Re_M \simeq 3000$ for 1 x 10 x 5 cavities with a thickness of 2 mm. These instabilities are not predicted in two-dimensional numerical simulations. It was checked that these instabilities disappear when the surface is oxidised or when σ' is reduced by the addition of bismuth. In addition, these instabilities were observed to attenuate and finally disappear after holding for a prolonged time in vacuums of the order of 2 to 6 x 10^{-8} mbar. This result illustrates that a very minute amount of oxygen contamination is enough to reduce the effects of thermocapillary convection. This process might therefore be used to eliminate microsegregations during solidification, as suggested by J.D. Verhoeven et al. [6].

Finally, the method described here provides a means of studying the convective phenomena involved in directional solidification. The presence of a surfactant element with partition coefficient < 1 gives rise to soluto-capillary movements. For the cases studied (destabilised solidification front), the induced surface velocities are higher than those resulting from thermocapillary effects. Thermal instabilities of amplitude approaching one degree are associated with these movements. Further experiments are therefore necessary in order to determine whether the addition of a surfactant element can reduce capillary convection in the case of plane-front solidification.

ACKNOWLEDGEMENTS

The authors would like to extend particular thanks to Messrs. B. Roux, H. Ben Hadid and M. Papoular for the extremely fruitful exchanges which provided essential guidance throughout the study. They are also most gratefule to Mr Brunet-Manquat for his valuable participation in the experimental work. This work was carried out within the context of the GRAMME agreement between the CEA and CNES.

REFERENCES

[1] D. SCHWABE, *Crystals: Growth, Properties and Applications* (Springer-Verlag 1988), pp 75-112.

[2] A. CROELL, W. MUELLER and R. NITSCHE, *Proc. 6th European Symp. on Material Sciences under Microgravity Conditions*, Bordeaux, France, 1987, ESA SP-256, pp 87-94.

[3] E. TILLBERG and T. CARLBERG, *J. Crystal Growth* 99 (1990), 1265-1272.

[4] D. CAMEL and P; TISON, *Proc. 7th European Symp. on Materials and fluid Sciences in Microgravity*, Oxford, U.K., 10-15 Sept. 1989, ESA SP-295, pp 63-68.

[5] M. JURISCH, *J. Crystal Growth*, 102 (1990), pp 223-232.

[6] J.D. VERHOEVEN, M.A. NOAK, W.N. GILL and R.M. GINDE, *Principles of Solidification and Materials Processing* (Ed. R. TRIVEDI, Vol. 1, Trans. Tech. Pub. 1990), pp 331-346.

[7] P. DESRE and J.C. JOUD, *Acta Astronautica* 8 (81) pp 407-415.

[8] J. BAUMGARTL, M. GEWALD, R. RUPP, J. STIERLEN and G. MUELLER, *Proc. 7th European Symp. on Materials and Fluid Sciences in Microgravity*, Oxford, U.K., 10-15 Sept. 1989, ESA SP-295.

[9] A.S. SENCHENKOV, pers. communication.

[10] J.V. BARMINE and A.S. SENCHENKOV, personal communication.

[11] V.G. LEVICH, *Physico-chemical Hydrodynamics* (Prentice Hall Inc., Engleword Clips, N.J. 1962).

[12] R.V. BIRIKH, *Journal of Applied Mechanics and Technical Physics*, January 1966, pp 43-44.

[13] M.K. SMITH and S.H. DAVIS, *J. Fluid Mech.* **132** (1983) pp 119-144 and 145-162.

[14] P. LAURE, B. ROUX and H. BEN HADID, *Phys. Fluids* **A2 (4)** (1990) 516-524.

[15] H. BEN HADID, Thesis, Marseille (1989).

[16] H. BEN HADID and B. ROUX, submitted to *J. Fluid Mech.*

[17] H. BEN HADID, B. ROUX, P. LAURE, P. TISON, D. CAMEL and J.J. FAVIER, *Adv. Space Res.* **8 (12)** (1988) pp 293-304.

[18] Kh. I. IBRAGIMOV, N.L. POKROVSKII, P.P. PUGACHEVICH and V.K. SEMENCHENKO, *Soviet Physics-Doklady* **9 (3)** (1964) pp 227-229.

[19] N. EUSTATHOPOULOS and J.C. JOUD, *Current topics in Materials Science* (Ed. E. KALDIS, Vol. 4, North Holland 1980), pp 281-360.

[20] R. SANGIORGI, C. SENILLOU and J.C. JOUD, *Surface Science* **202** (1988), pp 509-520.

[21] J. OCHIAI et al, T. MAEKAWA and I. TANASAWA, *Proc. 5th European Symp. on Material Sciences under Microgravity*, Schloss Elmau, 5-7 November 1984, ESA SP-222, pp 291-295.

[22] M. PAPOULAR, P. TISON , D. CAMEL and J.J. FAVIER, CEA Internal Report, Note Technique DEM N° 14/90, Grenoble, April 1990.

[23] Yu A. KLYACHKO and L.L. KUNIN, *DAN* **64 (1)** (1949) 85.

[24] L.D. LUCAS. *Mem. Sci. Rev. Met.* **69** (1972) 479.

[25] L.D. LUCAS, *Techniques de l'Ingénieur* **M 66**.

APPENDIX

Nomenclature

- Cavity 1 x A x B

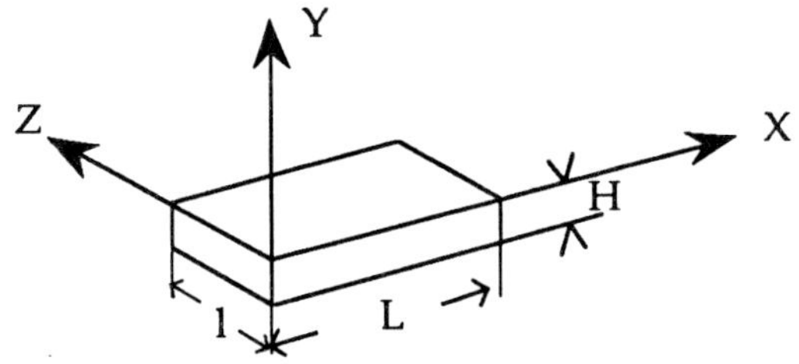

Aspect ratios: $A = \frac{L}{H}$ $B = \frac{l}{H}$

H height of the liquid layer

L length of the liquid layer

l width of the liquid layer

$x = \frac{X}{H}$

T temperature

t time

ν kinematic viscosity

ρ density
for Sn, $1/\rho = 0.14315 + 14.9 \times 10^{-6} (T - T_f)$ [24]

$\rho\nu = 5.05 \times 10^{-4} \exp \frac{5356}{8.3144\, T}$ [25]

σ surface tension

$\sigma' = \frac{\partial \sigma}{\partial T} = 7 \times 10^{-2}$ mJ m^{-2} K^{-1} [18]

v dimensioned velocity

u dimensionless velocity $= \frac{v \times H}{\nu}$

β $-\frac{1}{\rho_o} \cdot \frac{\partial \rho}{\partial T} = 1 \times 10^{-4}$

$$Re_M = \left(\frac{\partial \sigma}{\partial T}\right) \cdot \frac{\Delta T}{L} \cdot \frac{H^2}{\rho \nu^2} = \frac{Ma}{Pr}$$

$Pr = \frac{\nu}{K} = 0.011$ to 0.015 for Sn

K thermal diffusivity

$$Gr = g\beta \left(\frac{\Delta T}{L}\right) \frac{H^4}{\nu^2}$$

$\ell = \frac{\lambda}{H}, f = \frac{F H^2}{\nu}$: dimensionless wavelength and frequency of the instabilities

k partition coefficient = 0.25 for Sn-Bi

H_a Hartmann's $= B H \sqrt{\frac{C}{\rho \nu}}$

B magnetic field

C electrical conductivity
for Sn, $1/C = 0.49 \times 10^{-6}\ \Omega$m

$\mathcal{L}$ Langmuir's number = 10^{-6} torr.s equivalent to the deposition of a monolayer.

NON-ISOTHERMAL SPREADING OF LIQUID DROPS ON HORIZONTAL PLATES

P. Ehrhard

Institut für Angewandte Thermo- und Fluiddynamik
Kernforschungszentrum Karlsruhe GmbH.
Postfach 3640, W-7500 Karlsruhe 1, Germany

ABSTRACT

A viscous drop spreads on a smooth horizontal surface, which is uniformly heated or cooled. Experimental and theoretical results on thin drops are discussed. It is found for isothermal drops that gravity is very important at large times and determines the power law for unlimited spreading. Predictions compare well with the experimental data on isothermal spreading of axisymmetric drops. It is found that heating (cooling) of the plate retards (augments) the spreading process. Thus, heat transfer may serve as a sensitive control of the spreading.

Keywords: *Spreading, Wetting, Contact line, Themocapillarity.*

1. DEFINITION OF THE PROBLEM

The axisymmetric spreading of a drop on a horizontal plate is considered. The plate is uniformly heated/cooled with respect to a surrounding, passive gas (see figure 1). The viscous liquid, forming the drop, is bounded from the bottom ($z=0$) by a solid smooth plate at temperature T_W. The top boundary of the liquid region ($z=h$) is formed by a free liquid/gas interface, while the gas temperature at some distance is T_∞ and constant. At a position ($r=a$, $z=0$) a contact line is established; the angle between the liquid/gas interface and the plate is termed contact angle θ.

The following mechanics is present in the problem: Due to a hydrostatic head, gravity will promote the spreading. Mean surface tension couples the pressure jump across the liquid/gas interface ($z=h$) with mean curvature of this interface. Due to non-isothermal conditions ($T_W \neq T_\infty$),

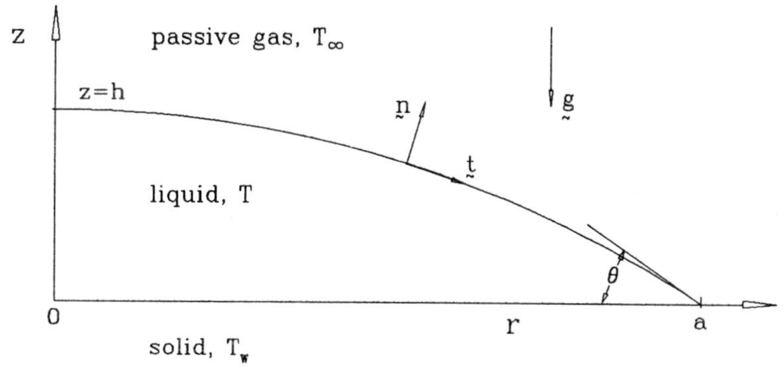

Figure 1: Problem

gradients of surface tension cause shear stresses on the liquid/gas interface (Marangoni effect). Viscosity has a dissipating effect onto fluid motion. Besides the above physical effects, chemistry enters the problem via the wetting characteristics. This is strongly dependent on the selection of solids and liquids and is usually encoded in the dependency $\dot{a} = f(\Theta)$.

2. THEORY

On tackling the above problem theoretically, one has to solve Navier-Stokes equations, continuity equation and heat transport equation within the liquid domain. That is done time-dependently within cylindrical coordinates (r,z). The boundary conditions have to be formulated at the solid plate $(z=0)$ and on the free liquid/gas interface $(z=h)$ together with appropriate initial conditions $(t=0)$. One can approximately solve this so-called "free boundary value problem" using adequate modelling of surface tension and wetting chemistry. Complete details of the formulation are given in Ehrhard & Davis [1].

The set of dimensionless groups involves the Bond number G, which represents a ratio of gravitational forces and surface tension forces. The Marangoni number M measures the relative intensity of thermally-caused shear stresses and the capillary number C relates viscous forces to surface tension forces. C is considered to be small, i.e. $C \ll 1$.

Figure 2 shows a result obtained with the theoretical model of Ehrhard & Davis [1]. It can be seen both the spreading of a drop on a isothermal $(M=0)$ and on a heated $(M=0.2)$ plate. Note that h^*, r^* are scaled versions of h, r. Thus figure 2 shows "flat" drops, stretched vertically due to separate scaling. For both cases initial contours $(t^*=0)$ are given together with evolution in time steps $\Delta t^* = 0.4$. As $t^* \to \infty$ we see a final steady state given by the dotted contours. For the heated plate one can clearly recognize a retarded spreading. The final drop contour is steeper and therefore less extended on the plate. A careful analysis of the flow pattern demonstrates this effect caused by thermally-induced shear stresses at the liquid/gas interface "pulling" the interface towards the drop summit. Therefore the transport of fluid from the drop centre towards the outer region is prevented.

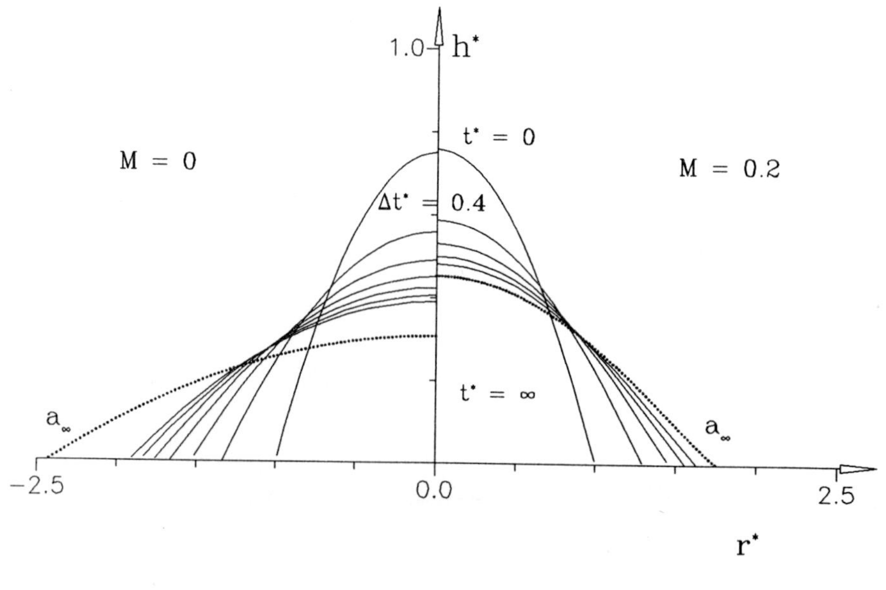

Figure 2: Effect of heated plate

3. EXPERIMENT

Figure 3 shows the experimental setup. Within a Helium-rinsed plexiglas cylinder (Helium temperature T_∞) a copper/glass mirror at temperature T_W is situated. A silicon oil drop (M100) spreads on the glass surface. The wetted region is visualized by means of a Schlieren setup: parallel light enters the test section via a beam splitter and is reflected by the related interfaces. The reflected light is spatially filtered in a Fourier plane and recorded by a CCD camera.

The above optical setup is based on the following: The unwetted mirror and the middle region of the drop reflect the light in a parallel manner. This parallel light passes the low pass filter (pin hole) and is recorded as bright by the CCD camera. In contrast, non-parallel reflected light from the outer drop region is stopped by the filter and appears dark. Thus a sharply contrasted picture of the drop is obtained which is likewise shown in figure 3. Using an image processing system together with an initial calibration a highly accurate measurement of the wetted area can be inferred. Besides $a(t)$ all temperatures, e.g. T_∞, T_W, are monitored using Pt-100 resistance thermometers. Silicone oil on glass exhibits complete wetting, which is in contrast to the partially-wetting behaviour of the materials in figure 2.

In figure 4 a comparison between our isothermal measurements ($T_\infty = T_W$) and experiments of other authors is drawn. Additionally, theoretical curves following the model of Ehrhard & Davis [1] for isothermal conditions are included. The dimensionless radius $a^*(t^*)$ is plotted in a logarithmic diagram using the scaling laws of Ehrhard & Davis [1]. By incorporating the experiments of Chen [2] and Cazabat & Cohen Stuart [3] the experimental data involve silicone oil drops of vastly different volumes and viscosities.

Following the model of Ehrhard & Davis [1] we can infer a different importance of gravity for different drop volumes. For this reason two theoretical curves are given in figure 4, limiting the experimental range of Bond numbers $0.5 \le G \le 16.4$. Most of the experimental data $a^*(t^*)$, given by symbols, are confined within a band spanned by the two theoretical curves (thin lines). From analysing time laws of individual drops one can, moreover, infer a small slope for short times developing towards a larger slope as $t^* \to \infty$. This indicates a transition

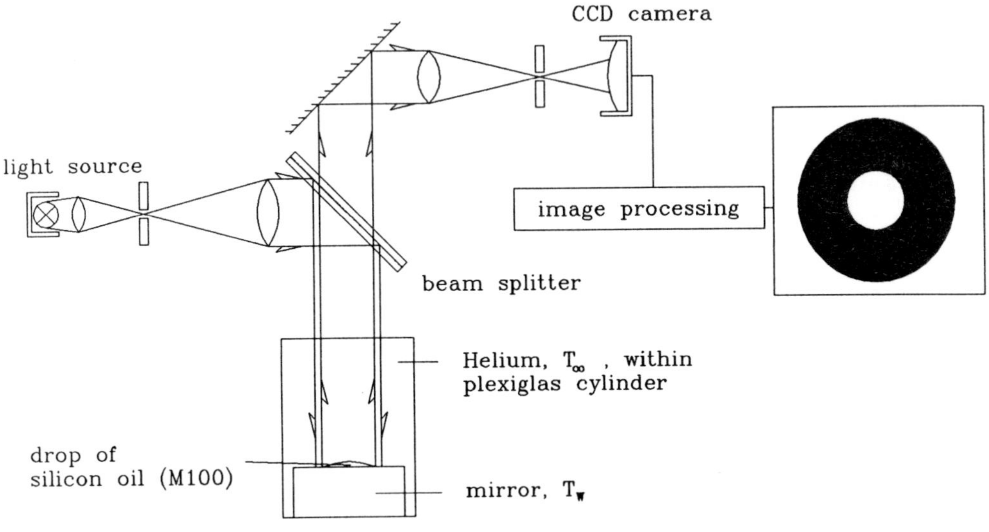

Figure 3: Experimental setup

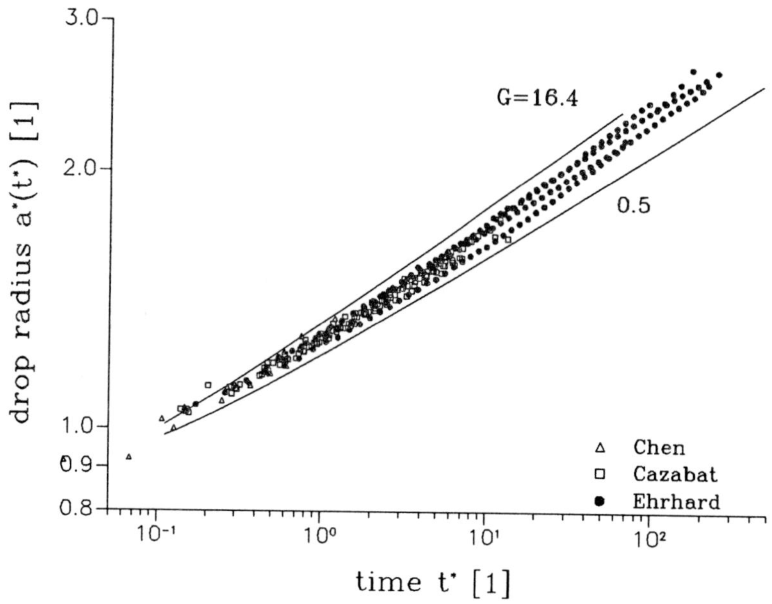

Figure 4: Isothermal spreading.

from surface tension controlled spreading towards gravity controlled spreading (see e.g. Ehrhard & Davis [1]).

Our experimental results for the spreading on a heated/cooled plate are encoded in figure 5. As reference three theoretical curves (following [1]) for $G = 10$ and various Marangoni numbers M are employed. The curve $M = 0$ represents the isothermal spreading results, as experimentally observed at an average Bond number of $\bar{G} \cong 10$ (cf. figure 4). According to the theoretical predictions for the heated plate ($M=0.04$) a retarded spreading within the experiments is observed. In contrast, the cooled plate results in a substantially augmented spreading.

4. SUMMARY

From experiments and theory one finds accordingly for the isothermal case a surface tension controlled spreading for short times and a gravity controlled spreading for larger times. The time laws for this process can be determined at reasonable accuracy by the model of Ehrhard & Davis [1]. This model holds for "flat" drops (lubrication approximation) and small capillary numbers C. Our experiments for the heated/cooled plate demonstrate, moreover, that the effect of thermocapillarity is correctly obtained by this model. A retarded/augmented spreading is observed for those cases.

5. REFERENCES

1. P. Ehrhard and S.H. Davis, *Journ. Fluid Mech.* **229**, pp. 365-388 (1991)

2. J.D. Chen, *Journ. Coll. Interf. Sci.* **122**, pp. 60-72 (1988)

3. A.M. Cazabat and M.A. Cohen Stuart, *Journ. Phys. Chem.* **90**, pp. 5845-5849 (1986)

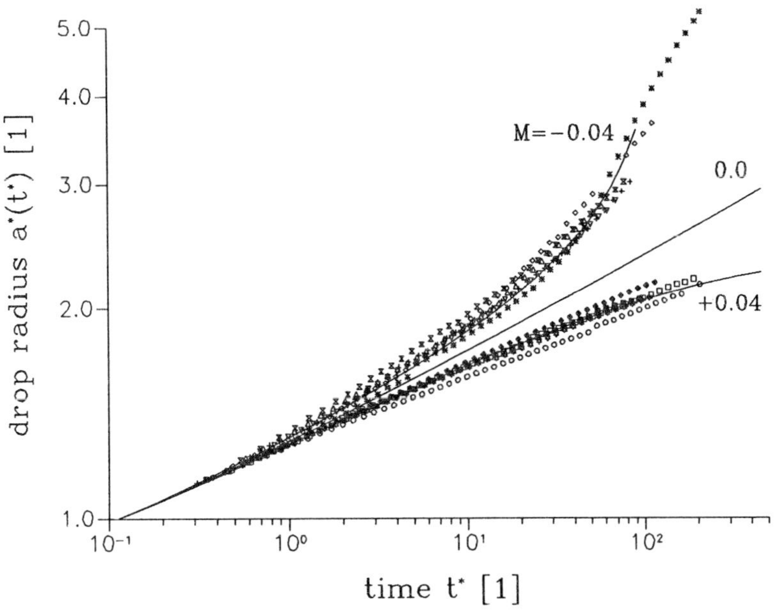

Figure 5: Non-isothermal spreading.

INFLUENCE OF DIFFERENT FACTORS ON THE THERMOCAPILLARY DEFORMATION OF A THIN LIQUID LAYER

V.A.Briskman and A.L.Zuev

Institute of Continuous Media Mechanics, UB Acad.Sci.USSR, Perm, USSR.

ABSTRACT

In microgravity conditions the thermocapillary convection is known to become a powerful mechanism of heat and mass transfer. It may exert significant influence upon those processes of space technology which possess non-isothermal free fluid surfaces and cause defects in material production. Therefore the search for methods of acting on the thermocapillary convection in order to suppress and weaken it is of interest. The intensity of thermocapillary flows essentially depends on different physical factors, such as the geometry and dimensions of the liquid volume or the presence of surfactants or a thin layer of an other immissible fluid on the free surface. Another factor, also capable to effect the thermocapillary convection, is vibrations. The paper is devoted to the experimental investigations of the suppression of thermocapillary deformations of thin liquid layers.

KEY WORDS

Experiment, thermocapillary convection, thin liquid layer, surface deformation, vibrations, microgravity.

1. THERMOCAPILLARY DEFORMATION AND BREAKDOWN OF A THIN LIQUID LAYER

When modelling thermocapillary phenomena in the laboratory, one has to secure the prevalence of thermocapillary convection over the thermogravitational one. This can be done in systems with relatively small vertical dimension, for example, in a thin horizontal fluid layers with the free upper surface, which is subject to a longitudinal temperature gradient. The thermocapillary convection in such a layer was studied by us earlier both experimentally and analytically [1,2]. It was shown, in particular, that when the layer depth is small enough (less than 1-2 mm), its surface becomes deformable. The decrease of the layer depth leads to the increase in the horizontal component of the pressure gradient. The latter generates the deformation, and, as a result, the depth over the hot regions becomes less than over the colder ones. The value of this deformation is limited by capillary and gravitational forces. Vibrations coincided with the direction of gravity create the average additional mass force. Therefore we expected that transversal vibrations can essentially alter the layer surface. The shape of the upper surface of a thin liquid layer of one fluid or two-fluid system with a horizontal thermal gradient was studied with the help of the apparatus shown in Figure 1. A circular or rectangular thin fluid layer was placed on a metallic plate of the corresponding shape. Its sharp edges prevented the liquid from flowing over. The rectangular liquid layer was 70x74 mm in size, the diameter of the circular layer 90 mm. The plate was heated at the center and cooled at the periphery. In order to prevent the non-uniform evaporation of a fluid the maximal temperature difference ΔT did not exceed 30 K. The cavity with the liquid layer was placed on the vibrating platform. This allows vertical vibrations with a frequency of 10-100 Hz and amplitudes of 0-8 millimeters. Experiments were performed with organic fluids of different viscosity: heptane, decane, hexadecane and ethanol. The liquid layer thickness h varied from 0.3 to 1.5 mm. The usual thermogravitational convection in such thin layers is negligible (small Rayleigh numbers Ra). In the experiment with a two-fluid system a thin layer (0.1-0.4 mm) of hexane was placed on a thicker water layer. The location of isolines of the surface slope was determined using the schlieren technique. The subsequent graphical integration yielded the shape of the surface. When the thermocapillary convection begins, the layers adjacent to the free surface move to the colder side whereas the near-bottom layers display reversed motion to the heater. This flow was shown to generate a deformation, which grows with the increase of temperature gradient. The graph in Figure 2 shows the dependence of the local layer

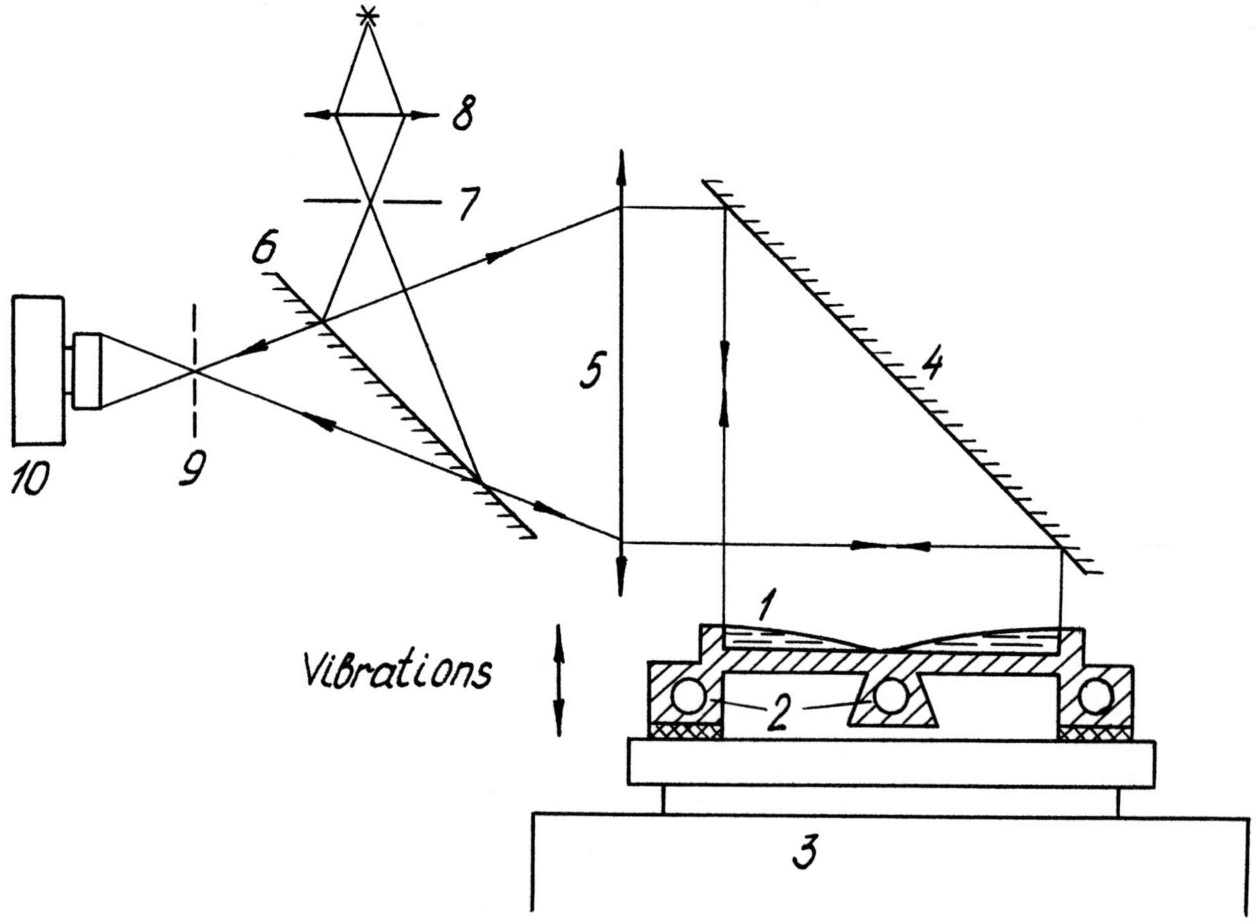

Figure 1. Apparatus for surface shape investigation of vibrating liquid layer
1 - fluid layer, 2 - heat exchanger, 3 - vibrator, 4 - mirror, 5 - objective
6 - mirror, 7 - diaphragm, 8 - microobjective, 9 - grid, 10 - photocamera

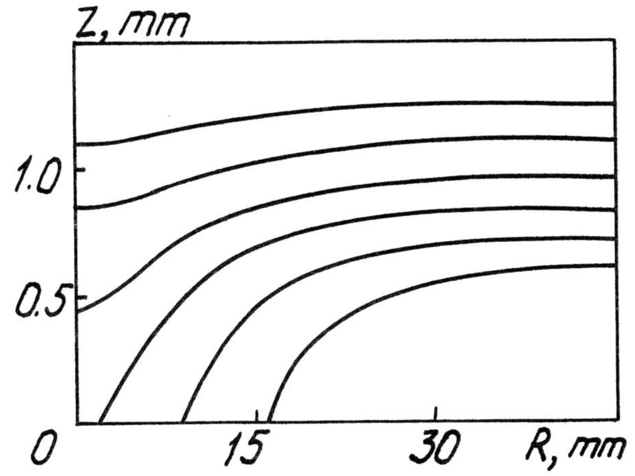

Figure 2. Fluid layer surface shape

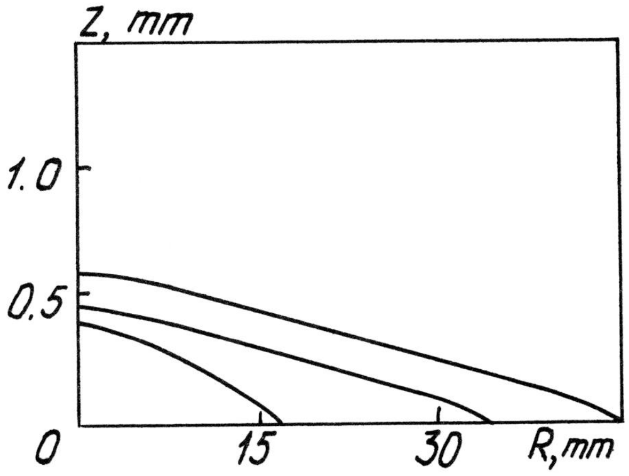

Figure 3. Fluid layer surface shape

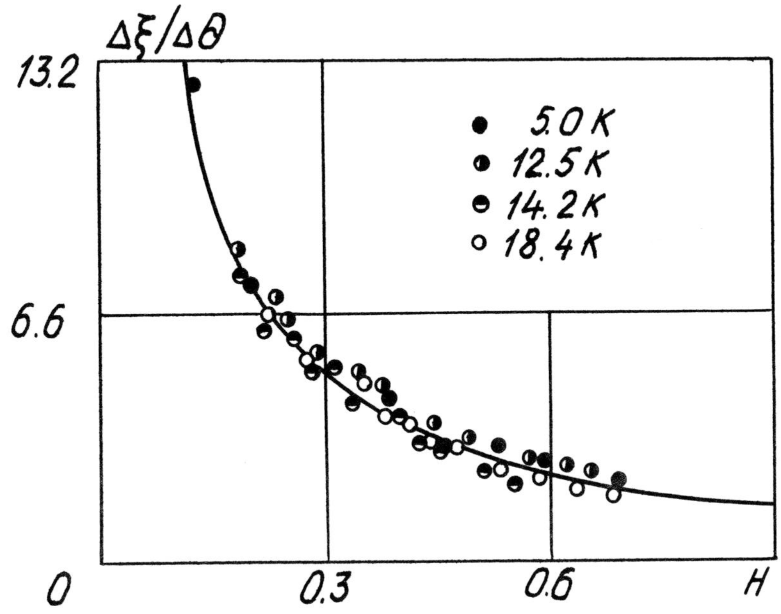

Figure 4. Scaled maximal depth difference between hot and cold regions

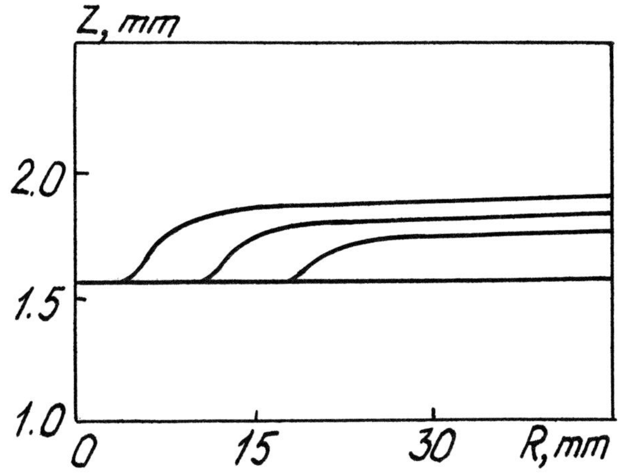

Figure 5. Fluid layer surface shape

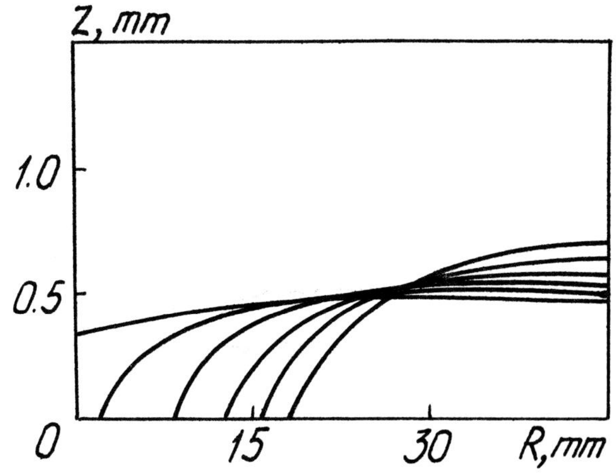

Figure 6. Fluid layer surface shape

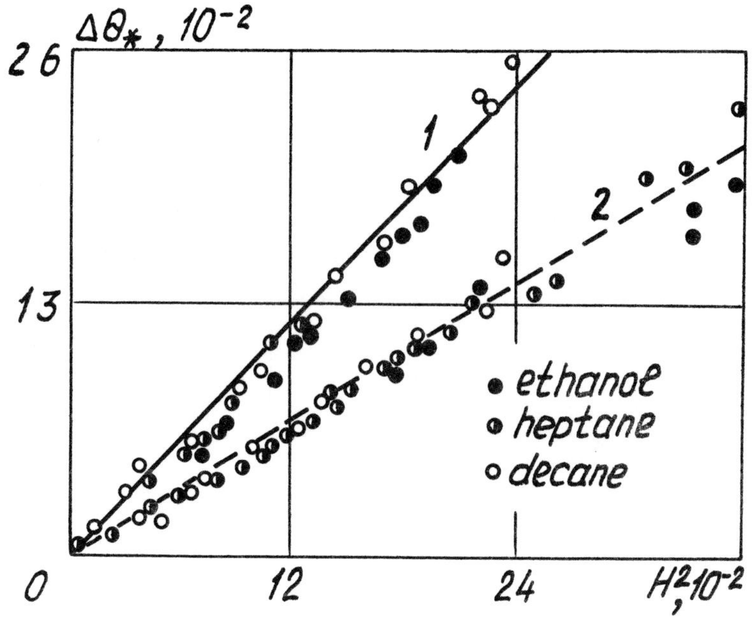

Figure 7. Scaled critical temperature difference needed to destroy fluid layer thickness on the horizontal coordinate measured from the heater in the case of the circular container. Different curves correspond to various total amounts of fluid (decane, $\Delta T = 14$ K, $h = 0.47$; 0.63; 0.80; 0.94; 1.10; 1.26 mm). The change in orientation of the temperature gradient results in changing the direction of thermocapillary motion and in the center we now observe the axially symmetrical fluid hill (Figure 3, decane, $\Delta T = 14$ K, $h = 0.07$; 0.45; 0.75 mm). The same results were obtained for the rectangular layer. For this geometry our experiments [1,2] with various fluids using different temperature conditions have shown good agreement with analytical predictions [1,3]. The maximal difference of depths between the most heated and the most cooled regions is proportional to the temperature difference and inversely proportional to the initial thickness of the layer (Figure 4). The experimental points for different temperature conditions agree well with the theoretical curve. Here the initial layer thickness H and depths difference $\Delta\xi$ are measured in units of capillary radius $(\sigma/\rho g_0)^{1/2}$; $\Delta\vartheta$ is the dimensionless temperature difference $\Delta\vartheta = \Delta T\,(\partial\sigma/\partial T)/\sigma$.

The same situation was obtained also in two-layered systems. The graph in Figure 5 ($\Delta T = 2.5$ K, $h = 0.16$; 0.24; 0.30 mm) presents the shape of a thin hexane layer placed on a water surface. As the water is usually contaminated with surfactants, the thermocapillary motion on its surface is practically absent. The effect is due to the thermocapillary forces on the upper hexane surface. As distinguished from the case of the hexane layer on the solid surface, the significant deformation is observed by now under very small temperature gradients. Another difference is that the dynamic contact angle between the upper fluid and underlayer equals to zero.

For sufficiently thin layers or large temperature gradients the deformation may grow so strong, that it reaches the bottom of the layer, and, thus, one may observe breaking of the integrity of the layer and creation of a dry spot above the heater (Figure 6). Various curves correspond to different temperature differences (decane, $\Delta T = 3.5$; 6.1; 10.2; 16.1; 22.5; 30.0 K; $h = 0.47$ mm). Our experimental results with various fluids to obtain the conditions, where the thermocapillary flow lead to the rupture of a thin liquid layer, are shown in Figure 7. The measured critical temperature differences $\Delta\vartheta_*$, which destroy layers of various thicknesses in the rectangular

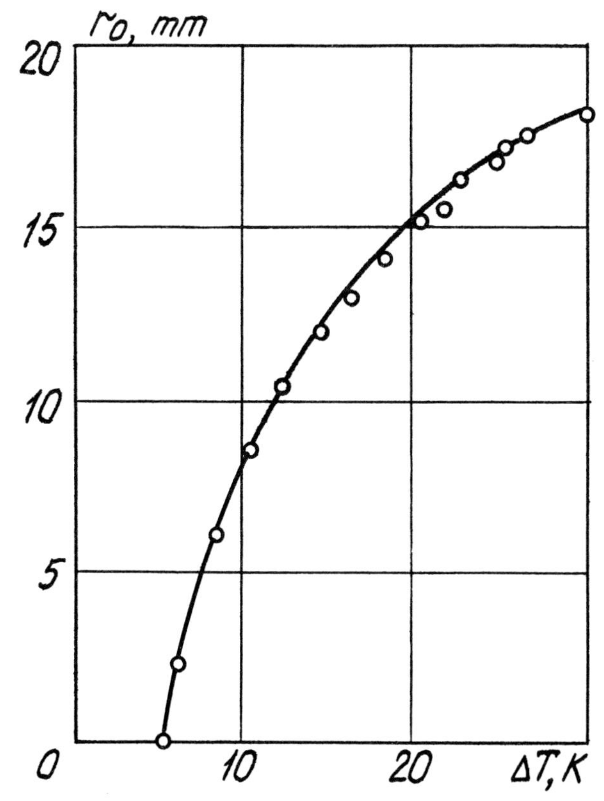

Figure 8. Radius of the dry zone on different temperature gradients

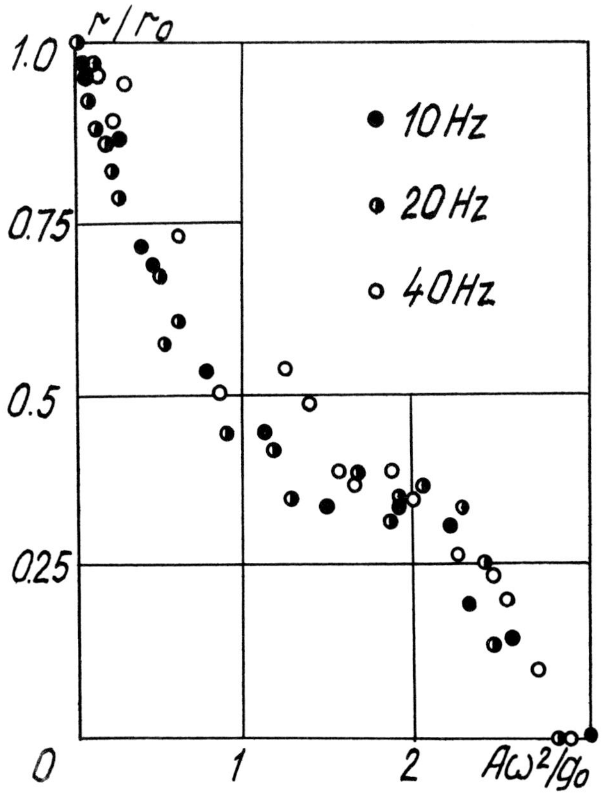

Figure 9. Reduction of the dry zone radius under the action of vibrations

container agree with the theoretical results of [2] (line 1): $\Delta\theta_* = H^2$ In the circular container (dashed line 2) the deformation is even stronger than in the rectangular one. With further increase of the temperature gradient the deformation grows and the radius r_o of the heated bare region of the bottom displays monotonous extension (Figure 8, decane, h = 0.5 mm).

2. ACTION OF VERTICAL VIBRATIONS ON THE THERMOCAPILLARY DEFORMATION

The presence of transversal vibrations was established to suppress deformations and to decrease the radius of the dry zone. Under sufficiently high amplitudes of vibrations the integrity of the fluid layer could be restored. The restoring archimedean force due to the deformation of free surface is proportional to the gravity acceleration. In the high-frequency vibrational field this force has the average component, which acts on thermocapillary deformations similarly to the gravity force and leads to the leveling of the surface.

At the moment of vibrations the layer surface turned out to be not enough stationary (besides some surface trembling) to use the sensible schlieren method even with stroboscopic light. Simultaneously another coexisting effect - the appearance of parametric waves on the free surface - arises even for viscous fluids. Therefore the influence of vibrations only on the conditions of the thermocapillary breakdown of the layer was studied. In experiment the thermocapillary deformation of the decane layer of 0.5 mm height with a temperature difference of 24 K was observed. At the absence of vibrations the radius of the dry zone in the center was equal to 16 mm, but under the action of vibrations it began to reduce. Figure 9 presents the ratio of this radius r to the initial one as a function of the ratio of vibrational acceleration to the gravitational one. To answer the question which one of two vibrational parameters, vibration velocity or acceleration is responsible for

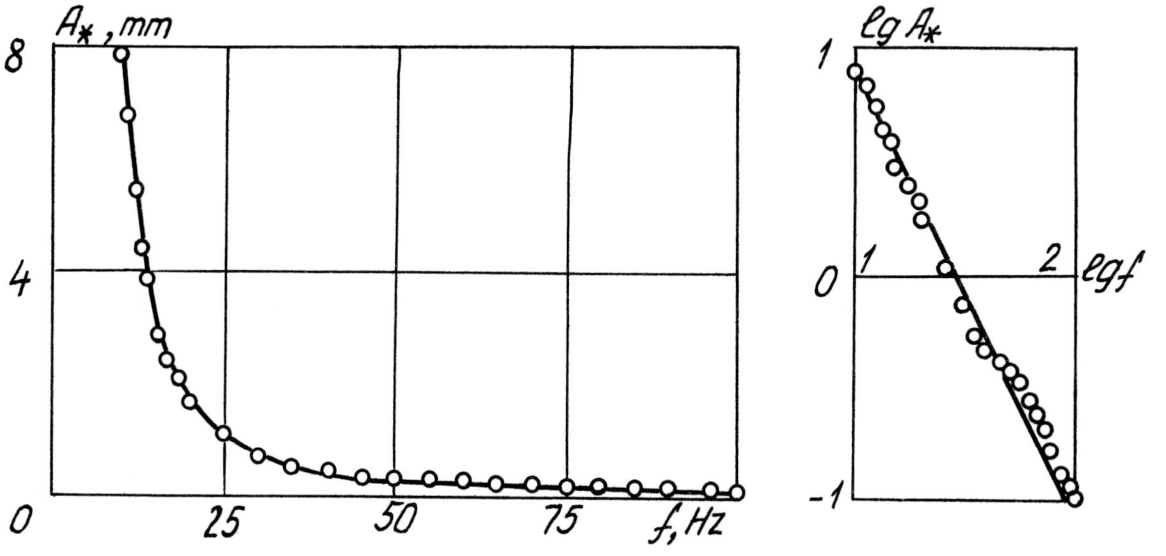

Figure 10. Critical values of amplitudes and frequences needed to restore the integrity of the layer

the leveling effect, we measured the critical values A_* of amplitudes and frequencies f of vibrations, which restored the integrity of an initially discontinuous layer. It can be seen from the left graph in Figure 10, that at high frequencies the weakening of the thermocapillary convection becomes appreciable at very low amplitudes. On the right graph this critical curve is shown in the logarithmic coordinates. All the experimental points fall well on one straight line corresponding to the constant vibration acceleration: $A_* f^2 = const = 794$. This experimental result, obtained at various values of the frequency and amplitude of vibrations, give evidence that the restoration of the layer unity is determined exactly by vibration acceleration. The Marangoni numbers $Ma = (\partial\sigma/\partial T)\ \Delta T\ h/\eta\chi$ in the experiment reached values of 10^4 and did not depend on the presence of vibrations. For the case of vibrations we can determine number $\widetilde{Ma}$ not through the real thermal difference, but through the thermal difference which produces analogous deformations in the absence of vibrations. As it is shown in Figure 11, this Marangoni number, characterizing the magnitude of the thermocapillary deformation, with the growth of vibrational acceleration from zero to value of $3g_0$ decreases several times.

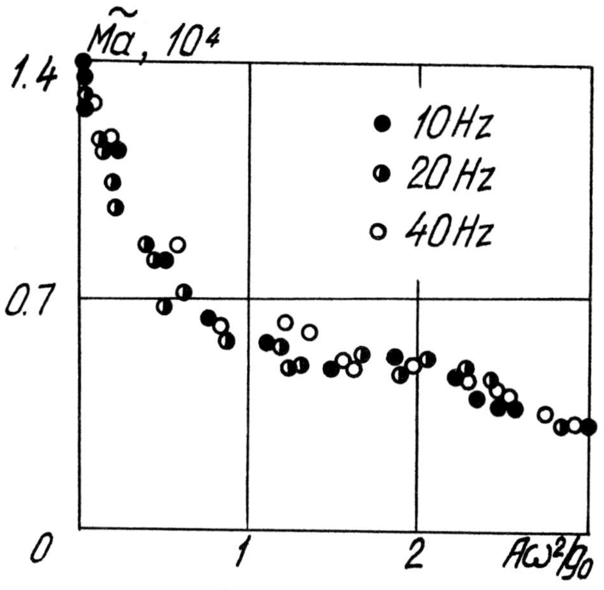

Figure 11. Action of vibrations on the deformation of liquid layer

REFERENCES

1. A.L.Zuev and A.F.Pshenichnikov, *Prikl. Mekh. i Tekh. Phiz.* (Applied Mechanics and Technical Physics), 3 (1987) 90-95
2. Yu.K.Bratukhin, V.A.Briskman, A.L. Zuev and A.F.Pshenichnikov, *Gidromekhanika i teplo-massoobmen pri poluchenii materialov* (Hydromechanics and Heat / Mass Transfer in Material Processing), Nauka, Moscow 1990 273-281
3. J.C.Loulerque, *J. Thin Solid Films*, 82 (1981) 61-71

MARANGONI INSTABILITY DUE TO EVAPORATION

A.G.Belonogov, Y.A.Buyevich, V.M.Kiseev, N.A.Korolyeva.

Ural State University, Lenin Street, 51, Sverdlovsk, USSR.

ABSTRACT

Experimental study of process of evaporation from wetting meniscus surface in glass capillary is carried out. Currents near evaporating surface observed in experiment are explained by thermal and concentrational convection. To support this explanation stability linear analysis of gas - liquid system with free evaporating surface is suggested.

Keywords: *Marangoni Convection, Liquid Condensational Films, Wetting Mechanism of Capillary Walls, Surface Tension Coefficient, Substances Concentration on the Evaporation Surface.*

1. INTRODUCTION

Under conditions of weightlessness or close to that, Marangoni convection acquires a predominant role in the processes of heat and mass transfer. One of the mechanisms behind the appearance of Marangoni convection may be connected with concentration and temperature change in the course of rapid evaporation from the free surface. This results in appearance of convective currents at the surface, which intensities all the processes of transfer and liquid evaporation.

Thermal and concentrational capillary convection as the intensifying factor is of interest for the cases of non-condensable gas transfer (e.g. CO_2) through liquid condensational films to absorbing surfaces for calculations of material corrosion resistance in liquids, liquid condensation in porous media and fillings, as well as for processes of evaporational cooling arrangement, for heat pipe type systems.

2. EXPERIMENT

The processes of thermal and concentrational convection were simulated in experiments with liquid evaporation from meniscus surface in a glass capillary (ref.1). In the process of orderly convective current in the vicinity of the meniscus was observed. The plant diagram is shown in figure 1.

Glass capillary with inner diameter of cylindrical part of 0.2 mm was tapered at one end with connection to atmosphere. The other end of horizontal capillary was connected to a glass tube of 2.2 mm inner diameter, positioned 10 mm below the capillary. From the tapered end the capillary could be heated with electric heater without blocking the exit for the steam produced into the atmosphere (the heater is not shown in figure 1).

During filling the glass tube with heat carrier the capillary filled itself forming a meniscus in the tapered section.

Electric heater caused evaporation of the heat carrier and the meniscus was moving into narrower part of the tapered end. Visual observation with help of particles suspended in liquid showed the picture of velocity field in the area of the evaporating meniscus. This picture is shown in figure 1 a,b,c. The surface current was directed towards the center of meniscus, i.e. against the integral hydrodynamic heat

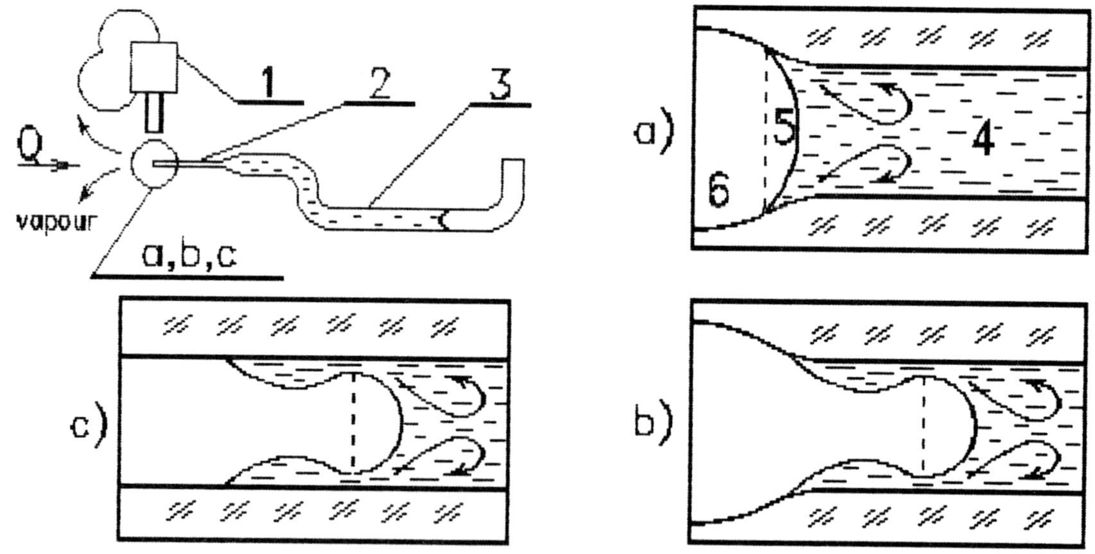

Figure.1. Diagram of experimental plant and evaporation meniscus picture made from photographs (a,b,c). 1 - movie camera, 2 - capillary, 3 - glass tube with divisions, 4 - liquid, 5 - meniscus shadow, 6 - film area

carrier mass stream.

As is known, self filling of capillaries with liquid is accompanied by development of eddy currents under the progressing wetting meniscus from meniscus center to the wetting film area with a return current near the capillary wall (ref.2). In our experiments a reverse current was observed.

It is to be noted, that evaporation under conditions of capillary liquid transfer changes the wetting mechanism of capillary walls overheated in relation to saturated temperature.

Presence of a film covering capillary walls before the meniscus cannot be explained by the surface force theory (ref.2), as its thickness amounted to 1 ... 10 μm for water, acetone and penthane, which is at least by an order bigger than maximum radius of surface forces´action.

For this reason it deems natural to link the actually observed current with the dependence of surface tension coefficient on evaporation surface temperature as well as concentration near it. As simple estimates demonstrate, the second factor may be quite substantial even with very small initial concentrations in connection with rapid evaporation and hampering diffusion exchange of the subsurface layer with the rest of the liquid.

Thus the forced convective currents and film formation at meniscus may be naturally linked to the increase of non-volatile substances concentration on the evaporation surface and decrease of volatile substance concentration as well as appearance of the corresponding surface tension gradient.

3. LINEAR STABILITY ANALYSIS

As a simplest model let us consider the problem of free surface instability in a gas-liquid system, consisting of two half-spaces $z > -h$ and $z < -h$, filled respectively with gas and liquid (areas 1 and 2) and separated with a flat boundary $z=-h$. Evaporation causes movement of the interface surface with velocity $U = dh/dt$. In the liquid phase there is heat transfer in direction perpendicular to the phase separation surface.

In the system of coordinates, moving together with the separating boundary $z' = z + Ut$, the undisturbed temperature fields are described by relationships (further on the apostrophe is omitted) $T_j^o = T_b^o - \sigma z_j$,

$j = 1,2$, this being based on conditions of heat current continuity $\lambda_1\beta_1 = \lambda_2\beta_2$. Here β_j, λ_j are temperature gradients and heat conductivity coefficients of different phases respectively, T_b^o is the undisturbed surface temperature, which we assume to be equal to boiling temperature for the sake of simplicity.

In the situation, where the liquid contains non-volatile admixture, evaporation from the free surface leads to admixture concentration growth at this surface. This is equivalent to supply of C_sU admixture molecule stream through this surface. During stationary evaporation of the solution from the free surface, admixture concentration distribution C changes continuously in time t. This distribution has the form of

$$C^o = C_\infty[1+(1+U^2t/D+UZ/D)\exp(UZ/D)],$$

where C_∞ - admixture concentration at infinite distance from the boundary, D - diffusion coefficient. Assuming that the disturbance time scale is by far smaller than the scale of concentration undisturbed field change (equal to $D/U^2 \sim 10^{-5}$ from (ref.2), the latter can be investigated as quasistationary.

At small disturbances T_b, C_s of surface temperature T_b^o and concentration C_s^o, the surface tension coefficient can be approximated by the linear function $\sigma = \sigma^o - a_1T - a_2C$, for parameter $a_1 > 0$ and $a_2 > 0$. The latter relationship is true for surface active substances.

The disturbed temperature and admixture concentration on the surface are connected by a linear relationship $T_b = mC_s$ ($m > 0$).

For the problem considered the Squire theorem is true, i.e. it is sufficient to investigate stability only in relations to flat disturbances. It can be easily shown, that the heat relaxation time is much smaller than the characteristic time of stationary distribution establishment of admixture concentration, for that reason we will consider temperature fields to be quasistationary.

Linearized equations for current function Ψ_j, convective heat conductivity for temperature disturbance T and convective diffusion for concentration disturbance C , as well as boundary for those, possess in the moving system of coordinates a form, analogous to the system in (ref.3).

$$\left\{\nu_j\Delta - \frac{\partial}{\partial t} - U^o\frac{\partial}{\partial Z}\right\}\Delta\Psi_j = 0,$$

$$\Delta = \frac{\partial^2}{\partial^2 X} + \frac{\partial^2}{\partial Y^2}, \quad j = 1,2$$

$$\Psi_1 \to 0,\ Z \to \infty; \quad \Psi_2 \to 0, \quad Z \to -\infty;$$

$$\frac{\partial \Psi_j}{\partial X} = 0, \qquad \frac{\partial \Psi_1}{\partial Z} = \frac{\partial \Psi_2}{\partial Z} \ . \qquad (1)$$

$$\left(\frac{\partial^2}{\partial Z^2} - \frac{\partial^2}{\partial X^2}\right)\left(\mu_2 \Psi_2 - \mu_1 \Psi_1\right) = \frac{\partial \sigma}{\partial X}, \quad Z=0 \ ;$$

$$æ_j \Delta T_j = \beta_j \frac{\partial \Psi_j}{\partial X}, \qquad T_1 = T_2 = T_b$$

$$\lambda \frac{\partial T_1}{\partial Z} = \lambda \frac{\partial T_2}{\partial Z}, \quad Z=0 \ . \quad (2)$$

$$\frac{\partial C}{\partial t} + U^{o} \frac{\partial C}{\partial U} - D\Delta C = \left(-U + \frac{\partial \Psi_2}{\partial X}\right)\frac{dC}{dZ},$$

$$C \to 0,\ Z \to -\infty; \ D\frac{\partial C}{\partial Z} = CU^{o} + C^{o}U, \ Z=0. \ (3)$$

Here $\mu_j = \rho_j \nu_j$ and ν_j are dynamic and kinematic viscosities, the last condition in (1) reflects the balance of tangential tensions on the interface surface. For simplicity reasons the surface tension coefficient is assumed to be fairly big in order to assume this surface to be flat also in the disturbed state.

We will limit ourselves to analysis of monotonous instability. Solution of problems (1)-(3) we find in the form of $\exp(\alpha t - ikx)$ with the actual positive wave number k and amplitudes , depending on z . The instability in question is correlated to the respective neutral instability curve, defined by the relationship of $\alpha = \alpha(k) = 0$. The system appears to be instable in relation to sufficiently long waves (see figure 2) which imposes a lower limitation on meniscus size (and respectively on pore size) where appearance of secondary convective currents and correspondingly intensification of exchange is possible.

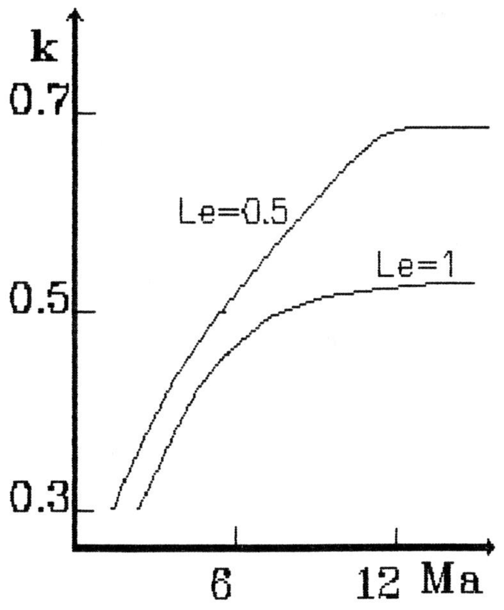

Figure 2. Neutral stability curves, instability areas are situated under the curves, $Ma = aql/(\lambda_2 \mu_2 æ_2)$, $Le = D/æ$, l - characteristic length scale, q - heat flux, $a = a_1 + a_2/m$

4. CONCLUSION

Thus, generation of subsurface currents at evaporating meniscuses in capillary-porous materials can be really explained from the Marangoni effect.

REFERENCES

1. A.G.Beionogov, V.M.Kiseev, Y.A.Buyevich. *Heat pipes: Theory and practik. Part 1.* ITMO AN BSSR, Minsk. (in Russion)
2. N.V.Chuyraev. *Fiziko-himija proccesov massoperenosa v poristyh telah.* Moscow, Chemistry Press (in Russion)
3. Y.A.Buyevich. *Injenerno-fiziches-kii J. 1985. T.48, N 2, pp.230 - 239.* (in Russion).

CONCENTRATION-DEPENDENT OSCILLATORY AND STATIONARY CONVECTION IN ONE - OR TWO - LAYERED SYSTEMS

Zh. Kozhoukharova, A. Nepomnyashchy*, I. Simanovskii**, S. Slavchev
Institute of Mechanics and Biomechanics, Sofia, Bulgaria.
* Technion, Haifa, Israel.
** Perm State Pedagogical Institute. USSR.

ABSTRACT

The influence of surfactants on the stability of one- or two-layer systems with a deformable free surface (interface) is investigated. The deformation of the interface can lead to the new type of oscillatory instability.

Keywords: convection, stability, surfactant, interface, deformation.

1. INTRODUCTION

When a gaseous surface-active solute absorbs or desorbs on a free surface of a liquid layer placed at a horizontal rigid wall, convective motions can take place. This instability of the fluid is caused by surface tension forces due to changes in surface concentration of the surfactant. It is shown theoretically that concentration-dependent stationary convection exists nevertheless the free surface is considered undeformable or deformable [1-7]. When deflections of the free surface are taken into consideration, stationary convection can appear in two different modes with a finite or infinite wavelength [6,7]. Brian and Smith [8] have proved that oscillatory convection can not be present if the surface is flat. But some experiments show that oscillatory motions are possible in binary systems and they can lead to travelling or standing waves [9]. Moreover, in analogous considerations of thermocapillary instability in a liquid layer heated from below both stationary and oscillatory motions have been established when the free surface is deformable [10-13]. In two-layer systems the surface-active agents lead to the appearance of a new type of oscillations [14-16].

In the present report, based on the linear stability analysis, the conditions for presence of stationary and oscillatory convection in one- or two-layer systems with a deformable free surface (interface) during absorption/desorption of a surface active solute are discussed.

2. ONE-LAYERED SYSTEM

Consider a physical absorption/desorption of a surface active solute on the free surface of an infinite liquid layer placed at a horizontal rigid wall. Restricting our analysis to dilute solutions, the fluid viscosity η and density ρ are assumed constant. The physical quantities that vary are the solute concentration C, the surface excess concentration Γ, and the surface tension depending linearly on C. In the equilibrium state the liquid is at rest and the concentration gradient $(-\beta)$ across the layer is constant being positive at absorption (e.g. $\beta<0$) and negative at desorption $(\beta>0)$.

Performing the linear stability analysis, we obtain the following set of equations for the evolution of the z-dependent amplitudes of the disturbances - the z-component velocity W and the concentration C:

$$[Sc^{-1}\omega-(D^2-l^2)](D^2-l^2)W=0, \qquad (1)$$
$$[\omega-(D^2-l^2)]C=W$$

with the boundary at the free surface (z = 1)

$$W-\omega Z=0,\ -\omega\gamma Cr Sc^{-1} l^2 W=Cr[Sc^{-1}\omega-$$
$$-(D^2-3l^2)]DW+(Bo+l^2)l^2 Z, \quad (2)$$
$$(D^2+l^2)W+Ma\,l^2(C-Z)+\omega\gamma Sc^{-1}DW=0,$$
$$DC-NaDW+(L+l^2S+\tilde{\delta}\omega)(C-Z)=0,$$

and at a rigid wall (z = 0)

$$W=DW=DC=0 \quad (3)$$

("insulating" wall)

Here $D=d/dz$, $l^2=l_x^2+l_y^2$ is the square of the modulus of the two-dimensional wave number (l_x, l_y), $\omega=\omega_r+i\omega_i$ the time constant, and Z the deviation of the surface. In the equations and the boundary conditions the following dimensionless parameters also appear: $Ma=\alpha\beta d^2/\rho\nu D_L$ the Marangoni number, $Na=\Gamma_0/\beta d^2$ the adsorption number, $Cr=\rho\nu D_L/\sigma_0 d$ the Crispation number, $Sc=\nu/D_L$ the Schmidt number, $Bo=\rho g d^2/\sigma_0$ the Bond number, $L=Hk_G d/D_L$ the Biot number, $S=D_S\delta/D_L d$, $\tilde{\delta}=\delta/d$, $\gamma=\Gamma_0/\rho d$, $\alpha=-d\sigma/dT$, $\delta=\alpha/RT$ "Gibbs depth", d the layer thickness, D_S the surface diffusivity, D_L the bulk diffusivity, k_G the gas-liquid transfer coefficient, H Henry's law constant, R the gas constant, T the temperature, g the gravity, σ_0, Γ_0 are reference values of the surface tension and the concentration, respectively.

The deformability of the fluid surface is characterized by the Crispation number and the adsorption number accounts for the surface convection of the solute. When Gibbs adsorption is assumed, e.g. $\Gamma=\delta C_i$ (C_i the bulk concentration), Ma and N_a have the same sign [2].

The eigenvalue problem (1-3) has been solved analytically for $\omega=0$ (stationary convection) [6,7] and for $\omega_r=0$, $\omega_i\neq 0$ (overstability) [13]. The solubility condition is of the form

$$Ma=F(l,\omega,Na,Cr,Bo,L,Sc,S,\gamma,\tilde{\delta}) \quad (4)$$

where F is a real-valued function of the parameters in parentheses. In particular, Brian and Smith's results [8] for a flat liquid surface ($Cr=0$) are obtained explicitely.

In the case of the oscillatory convection Ma is a complex function for arbitrary values of the enumerated parameters, e.g. $Ma=Ma_r+iMa_i$. But, the solution of the problem is only reasonable for real values of Ma; a numerical search has been conducted to find the values of ω_i for which $Ma_i=0$. Here, for simplicity, S, γ and $\tilde{\delta}$ are assumed zero.

In Figure 1 the imaginary part of Ma is plotted against its real one for $l=0.2$, $Na=0.01$, $Cr=0.2\cdot10^{-3}$, $Bo=0.1$, $L=1$ and different values of the Schmidt number (curve 1-0.1, 2-10, 3-100). For Sc = 0.1 the Marangoni number has a real value (Ma = 1292) at $\omega=0$ only, while for larger Sc there are two positive values of Ma, one at $\omega_i=0$ and another at non-zero ω_i (Ma=1068, ω_i =0.85 at Sc=10 and Ma=1067, ω_i =0.86 at Sc=100). The point ω_i= 0 corresponds to a stationary mode while a non-zero time constant to an oscillatory one. The computational analysis shows that Marangoni number increases with N_a till the adsorbtion number reaches some "saturation" value N_a^* depending on the other parameters.

Neutral curves for the above mentioned values of N , L and Bo, Sc= = 10, and two values of Cr equal to 0.001 (curves 1 and 2) and 0.01 (curves 3 and 4) are shown in Figure 2. Curves 2 and 4 correspond to the oscillatory convection while 1 and 3 to the stationary one. In the later case the Marangoni number doesn't depend on $Sc, S, \gamma, \tilde{\delta}$.

The surface deformability has a

destabilising effect as the critical Marangoni number becomes smaller with increasing the Crispation number. It is seen that at larger Cr the instability sets in as an overstability rather than as a stationary convection.

3. TWO-LAYERED SYSTEM

The space between two horizontal solid plates $y = a_1$ and $y = -a_2$ with different temperatures (the full temperature drop is θ) is filled up by two immiscible viscous fluids. The densities, coefficients of dynamic and kinematic viscosities, heat diffusivities and heat conductivities are equal to $\rho_m, \eta_m, \nu_m, \varkappa_m, \chi_m$; m=1 for the upper layer, m=2 for the lower layer. The concentration of surfactant af the inferface is small enough, so it's molecules form an "interface gas". The coefficient of surface tension is the linear function of temperature and concentration: $\sigma = \sigma_0 - \alpha T - \alpha_s \Gamma$ ($\sigma_0, \alpha, \alpha_s$ are constants). At the mechanical equilibrium state the concentration of surfactant is constant: $\Gamma = \Gamma_0$.

The transition equation for disturbances of concentration Γ_d after linearization can be written:

$$\frac{\partial \Gamma_d}{\partial t} + \Gamma_0 \frac{\partial v_x}{\partial x} = D_0 \frac{\partial^2 \Gamma_d}{\partial x^2}$$

where v_x is the horizontal component of fluid's velocity at the interface, D_0 is the coefficient of surface diffusion.

The boundary value problem can be written in the view:

$$(\lambda + i\omega) D\psi_m = -c_m D^2 \psi_m,$$

$$-(\lambda + i\omega) T_m - ikA_m\psi_m = d_m P^{-1} DT_m$$

$$y = 1: \psi_1 = \psi_1' = T_1 = 0,$$

$$y = -a: \psi_2 = \psi_2' = T_2 = 0$$

$$\psi_1''' - \eta^{-1}\psi_2''' + [(\lambda + i\omega)(1 - \rho^{-1}) - 3k^2(1 - \eta^{-1})]\psi_1' + ik[Ga(\rho^{-1} - 1) + Wk^2]h = 0; \quad \eta(\psi_1'' + k^2\psi_1) - (\psi_2'' + k^2\psi_2) - ikMr[T_1 - s(1 + \varkappa a)^{-1} h] + k^2 B(\lambda - D_s k^2)^{-1} \psi_1' = 0$$

$$\psi_1' = \psi_2', \quad \psi_1 = \psi_2 = -i(\lambda + i\omega)k^{-1} h,$$

$$T_1 - T_2 = s(1 - \varkappa)(1 + \varkappa a)^{-1} h,$$

$$\varkappa T_1' - T_2' = 0$$

Here ψ is the stream function, k is the wave number, $\lambda + i\omega$ is the complex decrement, h is the deflection of interface, $c_1 = d_1 = 1$, $c_2 = \nu^{-1}$; $d_2 = \chi^{-1}$; $\rho = \rho_1/\rho_2$, $\eta = \eta_1/\eta_2$, $\nu = \nu_1/\nu_2$, $\varkappa = \varkappa_1/\varkappa_2$, $\chi = \chi_1/\chi_2$, $a = a_2/a_1$, $Ga = g a_1^3/\nu_1^2$ is Galilee number, $P = \nu_1/\chi_1$, is Prandtl number, $W = \sigma a_1/\eta_1\nu_1$, $Mr = \alpha\theta a_1/\eta_2\nu_1$ is parameter analogous to Marangoni number, $B = \alpha_s \Gamma_0 a_1/\eta_2\nu_1$, $D_s = D_0/\nu_1$; $s = 1$ for heating from below, $s = -1$ for heating from above.

Let's consider the model system: $\nu = 0.5$; $\varkappa = \chi = a = P = W = 1$; $\rho = 0.999$.

Without deformations ($h = 0$) and surfactant ($B = 0$) the oscillations are the only possible mechanism of instability in this system. For $B = 0$, $h \neq 0$, $s = 1$ the oscillations keep for the values of the wave number, larger than critical one; in the longwave region the monotonous disturbances are the most dangerous. The neutral curves are shown in Figure 3. ($Ga = 10^7$; $B = 0$ (lines 1, 2; .1 (3, 4); 5 (5)).

At heating from above ($s = -1$) the deformation of the interface leads to the new type of oscillatory instability [17]. The neutral curve has the form of the "sack" (Figure 4a: $Ga = 10^4$; $B = 0$ (line 1), $B = 5$ (2), $B = 15$ (3)); in the region of $k < k_1(B)$ two values of $s \cdot Mr$ correspond to any value of wave number. The dependence of ω on k with the same numeration of lines is shown in Figure 4 b).

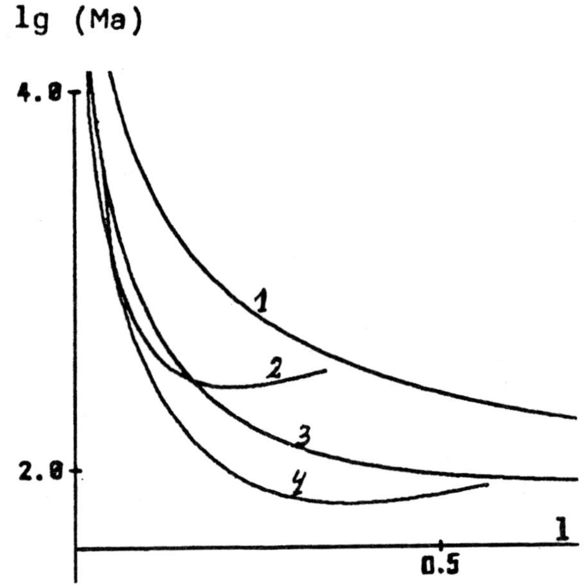

Figure 1: The dependence of $Ma(\omega)$.

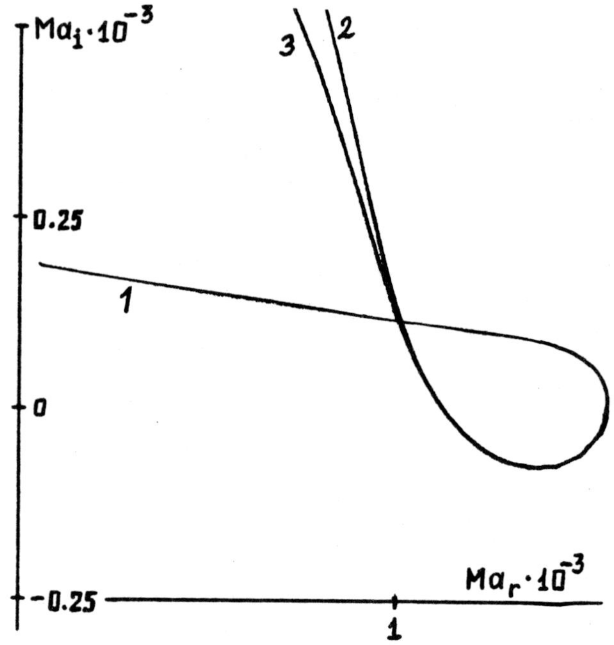

Figure 2: Neutral stability curves.

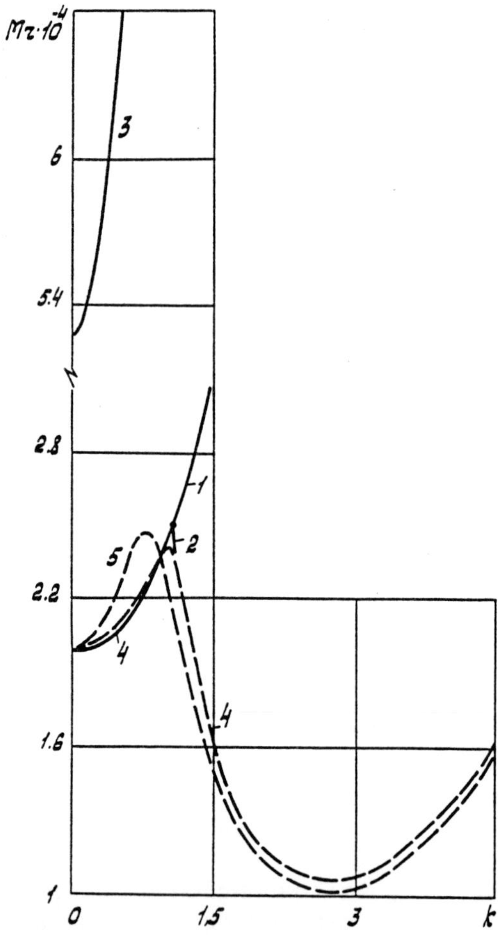

Figure 3: Neutral curves (s = 1).

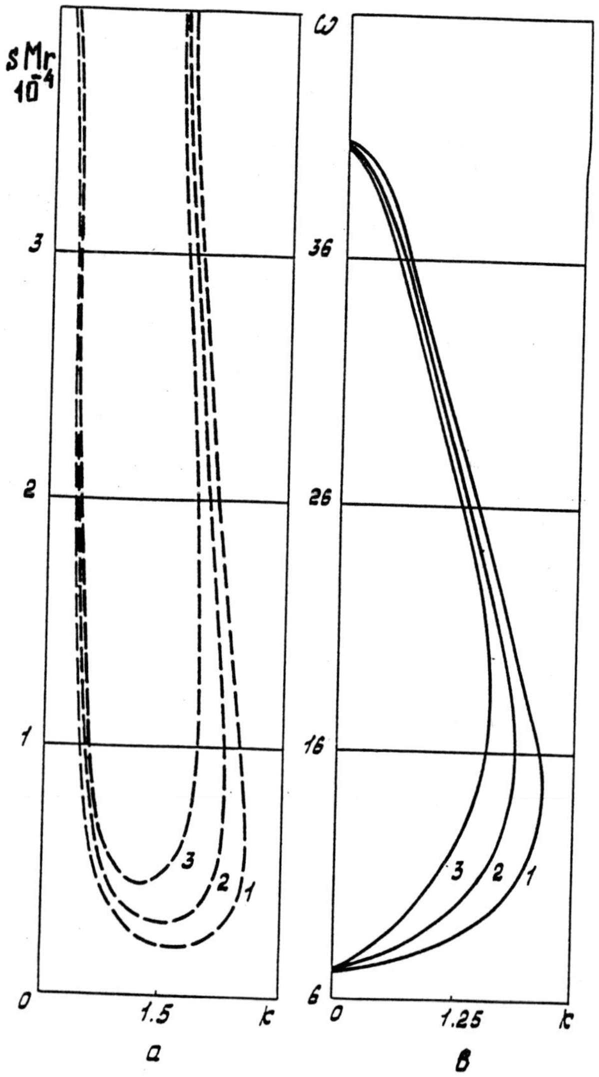

Figure 4: Neutral curves (a) and dependence of ω on k (b); s = − 1.

4. CONCLUSIONS

i. The influence of surface - active agents and the deflection of the interface on the monotonous and oscillatory neutral curves may be considerable.

ii. The deformation of the interface leads to the new type of oscillatory instability.

REFERENCES

1. C.V. Sternling , L.E. Sriven , A.I.Ch.E. Journal 5 (1959) 514.
2. P.L.T. Brian , A. I. Ch. E. Journal 17 (1971) 756.
3. M. Hennenberg, P.M. Bish, M. Vignes-Adler, and A. Sanfeld , J. Colloid Interface Sci. 74 (1980) 495.
4. X.- L. Chu , and L.- Y. Chen , Commun. Theor. Phus. (Beijing). 6 (1986) 237; 8 (1987) 167.
5. X. - L., Chu and M. G. Velarde, j. Colloid Interface Sci. 127 (1989) 205.
6. Zh. Kozhoukharova and S. Slavchev, Engng.Trans. Warsaw 37 (1989) 403.
7. S.G. Slavchhev and Zh. D. Kozhoukharova , "Proceedings of the Sixth National Congress of Theoretical and Applied Mechanics. Varna. Sept. 1989". Publishing House of the Bulg. Akad. Sci., Sofia. 3 (1990) 346.
8. P. L. T. Brian and K.A. Smith , A.I.Ch.E. Journal 18 (1972) 231.
9. H. Linde , M.Z. Kunkel, Warme - und Stoffubertragung, 2 (1969) 60.
10. M. Takashima, J. Phys. Japan, 50 (1981) 2745.
11. G. Gouesbet, J. Maquet, C. Roze , R. Darrigo , Phys. Fluids. A 2. (1990) 903..
12. C. Perez-Garcia, and G. Carneiro, Phys. Fluids A 3, 292 (1991).
13. Zh. Kozhoukharova , Theor. Appl. Mech. (1991) (in press).
14. A. A. Nepomnyashchy, I. B. Simanovskii, Izv. Akad. Nauk. SSSR Mech. Zhidk. i Gaza 2 (1986) 3.
15. A. A. Nepomnyashchy, I. B. Simanovskii, Izv. Akad. Nauk SSSR Mech. Zhidk. i Gaza 2 (1988) 187.
16. A. A. Nepomnyashchy, I. B. Simanovskii, Dokl. Akad. Nauk. SSSR V. 306. 2 (1989) 310.
17. A. A. Nepomnyashchy I. B. Simanovskii, Izv. Akad. Nauk. SSSR Mech. Zhidk.i Gaza 4 (1991) 11.

THERMOCAPILLARY MOTION OF A TWO LAYERED LIQUID WITH NONLINEAR DEPENDENCE OF THE SURFACE TENSION ON THE TEMPERATURE.

Shevtsova V.M., Indeikina A.E. and Ryazantsev Yu. S.

Institute for Problems in Mechanics
117526 Moscow pr. Vernadskogo, 101

ABSTRACT

The paper considers the problem of thermocapillary motion in two layered system, due to the existence of temperature gradient. These problems admit self-similar solution (in the generalized sense) within the framework of the Navier-Stokes equation. At the approximation of small Marangoni number it was obtained that in the case $\alpha = 2$ the lower liquid is motionless.

Keywords: Two layered system, Marangoni number

Let us consider thermocapillary motion in the system consisting of two layers of viscous incompressible liquids of thickness H_1 and H_2 respectively. It is assumed that the fluids are immiscible and the density in the upper layer is less than in the lower one ($\rho_1 > \rho_2$). The physical domain investigated in this problem is shown schematically in figure 1.

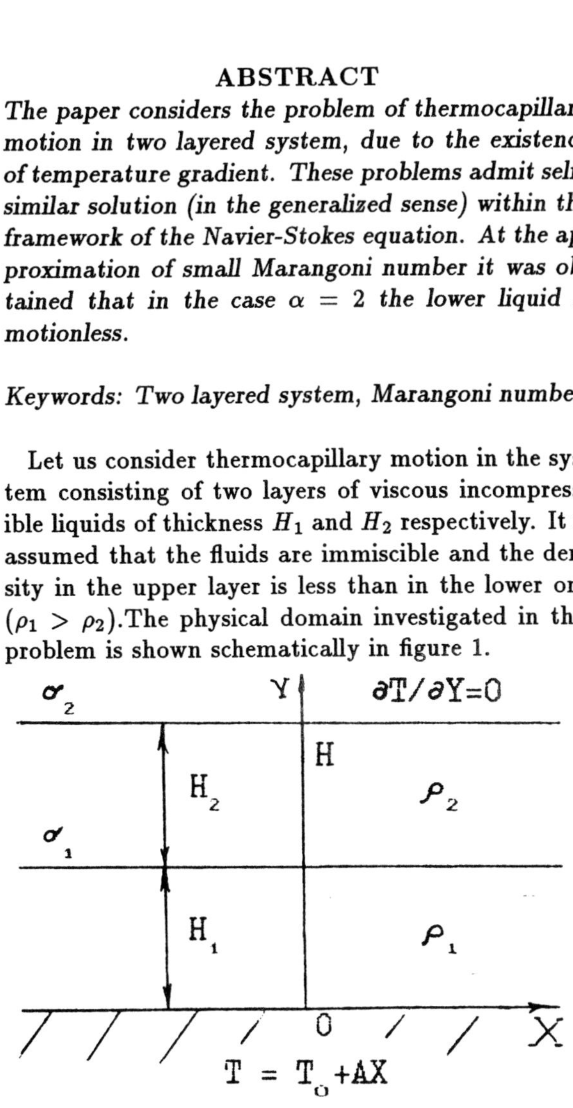

Figure 1: Geometry of the system

We shall assume that a constant linear temperature distribution is maintained at the bottom of the system, and that the free surface of the upper layer is heat–insulated. A tangential thermocapillary force acts on the interface liquid– liquid due to the non–uniform temperature distribution, on which, by assumption, dependence of the coefficient of the sur face tension on the temperature is the following (Refs.1,2)

$$\sigma_1 = \sigma_{01} + \frac{1}{2}\alpha_1(T_1 - T_0)^2, \sigma_{01} = const, \alpha_1 = const$$

Here T_0 is the temperature corresponding to the extremal value of the coefficient of surface tension.

We shall find the distribution of temperature and velocities in each layers in the next cases:

1. the upper boundary of the system is a solid surface
2. the upper boundary is a free surface, where the coefficient of surface tension depends on nonlinear law as

$$\sigma_2 = \sigma_{02} + \frac{1}{2}\alpha_2(T_2 - T_0)^2, \sigma_{01} = const, \alpha_1 = const$$

Owing to the heat conductivity of the liquids the temperature distribution on the boundary liquid–liquid and on the free surface will be nonuniform, which leads to the emergence of tangential thermocapillary stresses and causes motion in each layers. We shall consider the steady regime of this type of motion in which the tangential thermocapillary stress on the surface of each layer is balanced by the acting of viscous forces.

With the standard simplifying assumptions the mathematical formulation of the problem includes the Navier-Stokes , heat conduction and the continuity equations

$$U_i\frac{\partial U_i}{\partial X_i} + V_i\frac{\partial U_i}{\partial Y_i} = -\frac{1}{\rho_i}\frac{\partial P_i}{\partial X_i} + \nu_i\nabla^2 U_i,$$

$$U_i\frac{\partial V_i}{\partial X_i} + V_i\frac{\partial V_i}{\partial Y_i} = -\frac{1}{\rho_i}\frac{\partial P_i}{\partial X_i} + \nu_i\nabla^2 V_i - g, \quad (1)$$

$$U_i\frac{\partial T_i}{\partial X_i} + V_i\frac{\partial T_i}{\partial Y_i} = \chi\nabla^2 U_i, \quad \frac{\partial U_i}{\partial X_i} + \frac{\partial V_i}{\partial Y_i} = 0$$

Here and below the subscript $i = 1$ and $i = 2$ corresponds to lower and upper layers respectively.

The boundary conditions at the bottom ($Y = 0$) are nonslip ones and the linear temperature distribution is maintained

$Y = 0,\ U_1 = V_1 = 0,\ T_1 = T_0 + AX,\ A = const$

On the interface between fluids $Y = H_1$ the following conditions are imposed.

Continuity of velocity and temperature

$U_1 = U_2, \quad T_1 = T_2.$

Impermeability condition $V_1 = V_2$.

Continuity of the heat flux

$$k_1 \frac{\partial T_1}{\partial Y} = k_2 \frac{\partial T_2}{\partial Y}$$

The balance of viscous and thermocapillary forces:

$$\eta_1 \frac{\partial U_1}{\partial Y} = \eta_2 \frac{\partial U_2}{\partial Y} + \frac{\partial \sigma_1}{\partial T_1} \frac{\partial T_1}{\partial X}$$

At the upper surface $Y = H$ we have:

1. In the case of solid surface (enclosure problem) boundary conditions are nonslip ones $U_2 = V_2 = 0$ and the surface is thermally insulated
2. In the case of free surface we have the balance of viscous and thermocapillary forces

$$\eta_2 \frac{\partial U_2}{\partial Y} = \frac{\partial \sigma_2}{\partial T_2} \frac{\partial T_2}{\partial X}$$

the surface is undeformable $V_2 = 0$ and heat-insulated $\partial T_2 / \partial Y = 0$. It will be shown that the assumption of plane surfaces is valid for the heavy liquids or under the condition of high thermocapillary pressure σ_1 and σ_2. Let us introduce dimensionless variables and parameters: $x = X/H$, $y = Y/H$, $h_1 = H_1/H$, $h_2 = H_2/H$, $h = h_2/h_1$, H is the thickness of two layered system, $Pr_i = \nu_i/\chi_i$ is the Prandtl number. We shall seek a self-similar solution of the problem in the form

$$U_i = \frac{\nu_i}{H} x \Psi_i'(y), \quad V_i = -\frac{\nu_i}{H} \Psi_i(y),$$

$$T_i = T_0 + aHx\Theta_i(y), \qquad (2)$$

$$P_i = P_{0i} - \frac{1}{2} \frac{\rho_i \nu_i^2}{h^2} (\lambda_i x^2 + f_i(y)) - \rho_i g H y$$

where $P_{01} = P(0,0) = constant$, and $P_{02} = P(0,1) = constant$ are the pressures at the lower and upper boundaries respectively.

The self-similar solution in a one layer of liquid was considered before in Refs.3,4.

In order to determine the new unknown functions $\Psi_i(y)$, $\Theta_i(y)$, $f_i(y)$ and the constant λ_i, which are eigenvalues of the problem, we obtain from (1)-(2) the following two-point boundary-value problem for the nonlinear system of ordinary differential equations:

$$\Psi_i''' + \Psi_i \Psi_i'' - \Psi_i'^2 + \lambda_i = 0, \quad f_i' = 2\Psi_i'' + 2\Psi_i \Psi_i',$$

$$\Theta_i'' - Pr_i(\Psi_i' \Theta_i - \Psi_i \Theta_i') = 0. \qquad (3)$$

the boundary conditions at the bottom are the next: $y = 0$, $\Psi_i(0) = \Psi_1(0) = 0$, $\Theta_1(0) = 1$

The upper surface are assumed to be thermally insulated $y = 0$, $\Theta_2(1) = 0$, and

1. in the case of solid surface $\Psi_2(1) = \Psi_2'(1) = 0$.
2. in the case of free surface $\Psi_2 = 0$, $\Psi_2'' = m_2 \Theta_2^2(y)$, here $m_2 = \alpha_2 A^2 H^3 / \eta_2 \nu_2$ - is the Marangoni number for upper layer. At the interface $y = h_1$ we have

$$\Psi_1' = \nu \Psi_2', \ \Psi_1 = \Psi_2 = 0, \ \Theta_1 = \Theta_2,$$

$$\Theta_1' = k\Theta_2', \ \Psi_1'' - \eta\nu\Psi_2'' = m_1 \Theta_1^2$$

Here $\nu = \nu_2/\nu_1$, $\eta = \eta_2/\eta_1$, $k = k_2/k_1$ and $m_1 = \alpha_1 A^2 H^3 / \eta_1 \nu_1$ is the Marangoni number for the lower layer.

In order to clarify the properties of the thermocapillary flow being considered, we obtain an approximate analytical solution of the problem in each layer for the low values of the Marangoni numbers, assuming that the Prandtl numbers are of the order of unity.

In the case of open system there are two Marangoni numbers and it is possible to use any Marangoni number for the expansion. For generalizing we shall choose interface Marangoni number m_1 as expansion parameter assuming that the both value m_1 and m_2 are of the same order of magnitude.

When $m_1 = m_2 = 0$ the problem has the solution

$$\Psi_i = 0, \quad f_i = 0, \quad \lambda_i = 0, \quad \Theta_i = 1. \qquad (4)$$

which corresponds to a liquid at rest with a uniform temperature distribution along the depth.

For $m_i < 1$ we shall construct the solution by the method of small perturbations in the form

$$\Psi_i = m_1 \Psi_i^{(1)} + m_1^2 \Psi_i^{(2)} + \ldots, \quad f_i = m_1 f_i^{(1)} + \ldots,$$

$$\lambda_i = m_1 \lambda_i^{(1)} + \ldots, \quad \Theta_i = 1 + m_1 \Theta_i^{(1)} + \ldots \qquad (5)$$

After substituting (5) into (3) neglecting terms of second order in m_1, we obtain the following linear boundary value problem

$$\Psi_i''' + \lambda_i = 0, \quad f_i' = 2\Psi_i'', \quad Pr_i \Psi_i' = \Theta_i'', \qquad (6)$$

To avoid the mixture of indexes here and later the upper index "one" determing the first term of expansion is omitted.

The boundary conditions are the next:

$$\Psi_1(0) = \Psi_1'(0) = 0, \ f_1(0) = f_2(1) = 0,$$

$$\Theta_1(0) = \Theta_2'(1) = 0. \qquad (7)$$

1.$\Psi_2(1) = \Psi_2'(1) = 0$.

2.$\Psi_2(1) = 0, \ \Psi_2''(1) = \mu, \ \mu = m_2/m_1$

At the interface $y = h_1$

$$\Psi_1(h_1) = \Psi_2(h_1) = 0,\ \Psi_1'(h_1) = \nu\Psi_2'(h_1),$$

$$\Theta_1(h_1) = \Theta_2(h_1),\ \ \Theta_1'(h_1) = k\Theta_2'(h_1),$$

$$\Psi_1'' - \nu\eta\Psi_2'' = 1.$$

Solving the problem (6) with condition (7) we shall obtain eigenvalues λ_1 and , respectively, first terms in expansion (5).

1. The case of solid upper boundary (enclosure)

The solution of (4)-(7) with accuracy to terms $O(m_1^2)$ for the temperature, velocity and pressure has form

$$\lambda_1 = -\frac{3}{2}\frac{h}{h_1(\eta + h)},\quad \lambda_2 = -\frac{3}{2}\frac{h}{\nu h_2(\eta + h)},$$

$$V_1 = \frac{\nu_1 m_1 \lambda_1}{6H} y^2(y - h_1),$$

$$V_2 = \frac{\nu_2 m_1 \lambda_2}{6H}(y-1)^2(y - h_1),$$

$$U_1 = \frac{x\nu_1 m_1 \lambda_1}{6H} y(3y - 2h_1),$$

$$U_2 = \frac{x\nu_2 m_1 \lambda_2}{6H}(y-1)(3y - 2h_1 - 1), \qquad (8)$$

$$P_1 = P_{01} - \rho_1 g y H + \frac{m_1 \rho_1 \nu_1^2 \lambda_1}{6H^2}[3(y^2 - x^2) - 2h_1 y],$$

$$P_2 = P_{02} - \rho_2 g y H + \frac{m_1 \rho_2 \nu_2^2 \lambda_2}{6H^2}[3(y^2 - x^2) - (2h_1 + 4)y + 2h_1 + 1],$$

$$T_1 = T_0 + AHx[1 - \frac{m_1 Pr_1 \lambda_1}{72} y^3(3y - 4h_1)],$$

$$T_2 = T_0 + AHx\{1 - \frac{m_1}{72}[Pr_2\lambda_2(y - h_1)^2 (3(y-h_1)^2 - 8h_2(y - h_1) + 6_2^2) - Pr_1\lambda_1 h_1^4]\}$$

This case corresponds to the motion of two layer liquid in the gap between plates with heated from below.

Fig.2 shows stream lines and the profile of the longitudinal component of the flow velocity in the region $x > 0$ for the case of the same thickness of layers$h_1 = h_2 = 0.5$. Due to the fact that the boundary conditions for the both layers are the same in the case when $m_1 = 0$ the temperature distribution is uniform eqs.(4). We shall obtain in the first order of accuracy that the picture of flow is symmetrical in respect to the line of interface $y = 0.5$. Substituting values λ_i and m_i into (8) we shall find estimation for the velocities

$$V_i(y) \approx 1/(\eta_1 h_2 + \eta_2 h_1),\ U_i \approx 1/(\eta_1 h_2 + \eta_2 h_1)$$

As follows from these expressions, the variations of viscosity influence only on the scale of velocity and doesn't break the symmetry of flow.

At this approximation the deviation of temperature from initial profile doesn't influence on the velocity field, the dependence upon χ will appear in the next order to term m_1

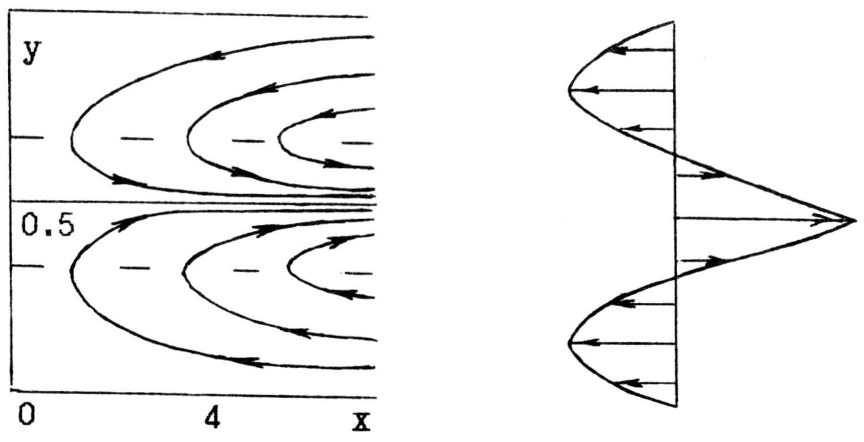

Figure 2: Streamlines and velocity for enclosure problem.

2. The upper boundary is a free surface

At this case the solution of (6)-(7) can be written in the form

$$\lambda_1 = \frac{3h(\alpha-2)}{h(3\eta+4h)}, \quad \lambda_2 = -\frac{3(\eta+\alpha(\eta+2h))}{\nu h\eta(3\eta+4h)},$$

$$V_1 = \frac{\nu_1 m_1 \lambda_1}{6H} y^2(y-h_1),$$

$$V_2 = \frac{\nu_2 m_1 \lambda_2}{6H}(y-1)^2(y-h_1)(y-a),$$

$$U_1 = \frac{x\nu_1 m_1 \lambda_1}{6H} y(3y-2h_1),$$

$$U_2 = \frac{x\nu_2 m_1 \lambda_2}{2H}(y-b_+)(y-b_-), \qquad (9)$$

$$P_1 = P_{01} - \rho_1 gyH + \frac{m_1\rho_1\nu_1^2\lambda_1}{6H^2}[3(y^2-x^2)-2h_1 y],$$

$$P_2 = P_{02} - \rho_2 gyH + \frac{m_1\rho_2\nu_2^2\lambda_2}{6H^2}[3(y^2-x^2) - 2(a+h_1+1)y + 2(a+h_1-0.5],$$

$$T_1 = T_0 + AHx[1 - \frac{m_1 Pr_1\lambda_1}{72} y^3(3y-4h_1)],$$

$$T_2 = T_0 + AHx\{1 - \frac{m_1}{72}[Pr_2\lambda_2(y-h_1)^2 (3(y-h_1)^2 + 4(y-h_1)(h_1-h_2-a) + 6h_2(a-h_1)) - Pr_1\lambda_1 h_1^4]\}$$

here

$$a = \frac{\eta(1+h_2)+\alpha(\eta(h_1-h_2)+2h_2)}{\eta+\alpha(\eta+2h)},$$

$$b = \frac{1+a+h_1}{3} \pm \frac{1}{3}\sqrt{1+a^2+h_1^2-a-h_1-ah_1}$$

. In this problem the normal stresses at the interface and at the free surface do not keep constant value. This situation has to lead to the distortion of both surfaces. Let us assume that the distortion vanishes due to the large value of g. At this case, as follows from (8) and (9) the pressure is basically hydrostatic. As in the previous case the distribution of velocity in the first order of accuracy to m_1 doesn't depend upon the changing of temperature profile and the thermal characteristics χ and Pr respectivly.

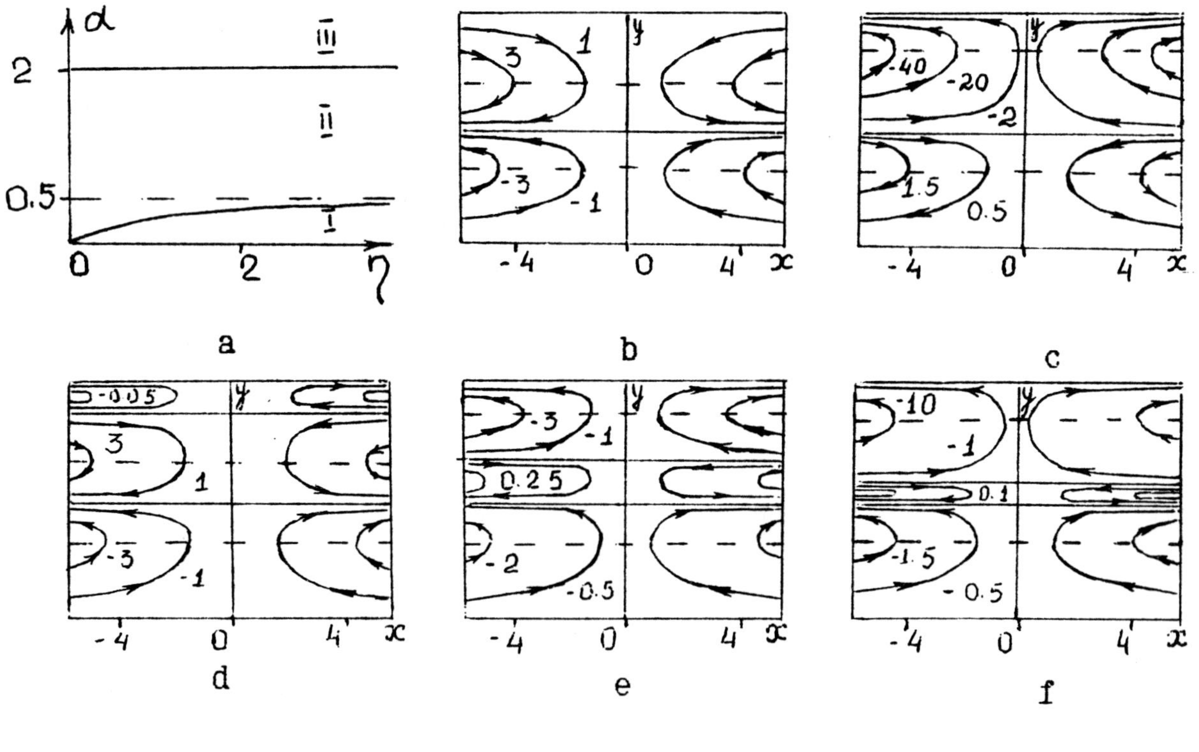

Figure 3: Stream lines for the different regimes.

As follows from expressions for the velocities U_i and V_i in eqs.(9) there are a few different regimes of steady motion in a two layered system depending on the values of parameters α, η and the relative thickness h.

1 regime:

$$\alpha < \alpha^* = \frac{\eta}{2\eta + h},$$

$$a > 0,\ b_+ > 1,\ h_1 < b_- < 1,\ V_1 < 0,\ V_2 > 0.$$

2 regime:

$$\alpha^* < \alpha < 2,\ h_1 < a < 1,\ h_1 < b_+ < 1$$

$V_1 > 0$, V_2 – change the sign

3 regime:

$\alpha > 2,\ a < h_1,\ b_- < h_1,\ h_1 < b_+ < 1,$
$V_1 < 0,\ V_2 > 0.$

Figure 3 shows stream lines for different regimes when $\eta = 1/3$ and equal thickness of layers $h_1 = h_2 = 0.5$. The values of stream function, normalized by the quantity $-10^2\nu_1 m_1$ are plotted near the stream lines.

In the case of small values of α , corresponding to the first regime $\alpha < \alpha^*$ (figure 3–b,$\alpha = 0.1$) the surface tension force is substantially larger than that at the free surface. The intensity of motion in both layers is of the same order and the direction of vortexes and the flow picture is similar to the case of enclosure system.

In the second regime $\alpha^* < \alpha < 2$ there is a competition of motions formed by interface Marangoni force and free surface one (figure 3–d,e,f). For rather small value α , close to α^*, one more vortex (figure 3–d) with the opposite direction of circulation appears near the surface. This results to decreasing of the intensity of motion in a lower liquid. Under the further increasing of a (figure 3-e,f) the size and intensity of vortex near the free surface are enlarged, while middle vortex is compressed with decreasing of intensity. As the result the intensity of motion in the lower layer goes down.

There is a characteristic value of parameter α which is $\alpha = 2$. At this case as it follows from (9) the liquid 1 becomes motionless, and there is one vortex motion at the upper layer, and the structure of flow doesn't depend upon the ratio of layer thicknesses.

It is necessary to take into account for the analysis of solution of eqs.(9) that the ratio of thickness upper layer to that of lower one $h = h_2/h_1$ has to be order of m_1 or larger. At the third regime $\alpha > 2$ (figure 3–c) the liquid motion is controlled by the free surface Marangoni force. The direction of vortex motion in liquid 1 changes to the opposite in comparison with the first regime but the flow intensity in the upper layer is much more stronger that at the previous cases.

Figure 3–a shows the boundaries of different regimes in dependence of parameters α and η. At any regimes the direction of flow changes at a depth equal $y = 2h_1/3$ in a liquid 1. The influence of viscosities ratio $\eta = \eta_2/\eta_1$ on liquid motion is weaker that the influence of parameter $\alpha = \alpha_2/\alpha_1$. If $\alpha < 0.5$ the variation of η replaces the boundary between vortexes and the plane of turning of liquid but it can not changes the regime of motion . For the $0 < \alpha < 0.5$ it is observed transition from first to second regime when $\eta > 2\alpha h(1 - 2\alpha)$.

Figure 4 shows the distribution of longitudional velocity with height for the equal thickness of layers (figure 4–a) and for different thicknesses. It is seen that the structure of flow doesn't change for different h_1 . The intensity of motion slightly changes in liquid 1,but in the upper layer in regimes 2 and 3 (curves 2–4) it decreases with increasing the height of lower layer.

Figure 5–a shows deviation of temperature from the initial profile $\Theta = 12[T_i - (T_0 + AHx)]/(m_1 Pr_1 AHx)$, $i = 1, 2$.

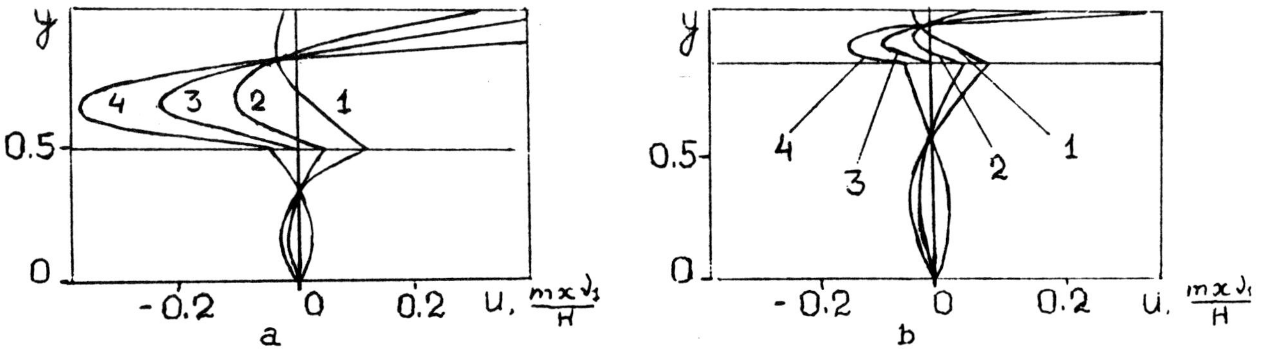

Figure 4: Disribution of velocity.

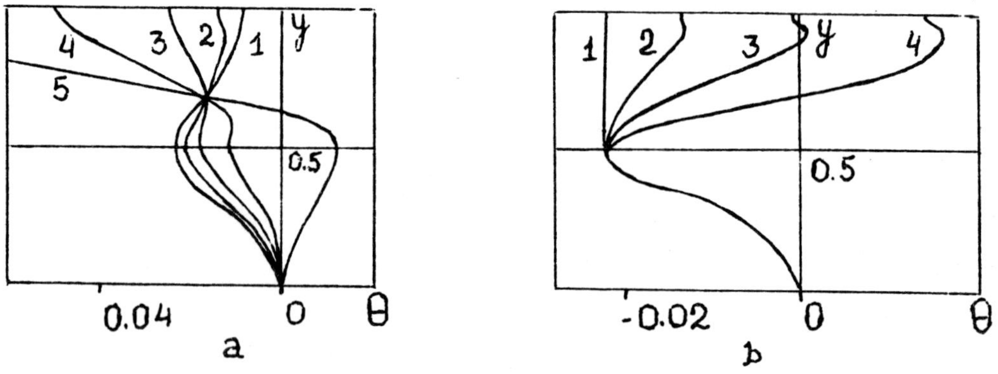

Figure 5: Disribution of temperature deviation.

As follows from comparising figures 5-a and 3 that extremumes of $\Theta(y)$ corresponds to the boundaries between vortexes, that is plane where intensity of motion achieves maximum.

Let us consider in detail the changing of temperature profile in 1 and 2 regimes with height (curves 1–4). As we maintain constant temperature gradient at the bottom than the temperature is enlarged with increasing of distance from the center $x > 0$. The liquid particles from the region near the bottom are involved by vortex and moved into the colder central part where they lose heat and achieve the interface with the temperature lower than initial . In the upper layer the circulation is directed at the opposite side. Liquid is moved into the hot region, where it is heated and goes back to interface (or to boundary of vortexes in the 1 regime) with temperature heigher than initial. This value of temperaure gives local maximum of temperature field.

In the second regime when there are three vortexes the heat transfer near the free surface has the same mechanism as in the lower layer. In the third regime, when the motion is controlled by Marangoni force at the free surface the liquid is heated on the wayto the interface and cooled at the back way.

Figure 5–b shows the distribution of temperature with height to the different values $\chi = \chi_2/\chi_1$ .Curves 1–4 corresponds $\chi = \infty$, 2, 0.825, 0.5. In the upper liquid there is a uniform temperature distribution when $\chi \to \infty$(curve 1) and the magnitude of deviation from the initial profile $(T_0 + AHx)$ is defined by the interface temperature. For the smaller χ the influence of heat conductivity is decreased in comparison with the convective heat transfer.

REFERENCES

1. R. Vochten and G.Petre, *Colloid Interface Sci.* 42 (1973) 320
2. Yu.S. Ryazantsev , V.M. Shevtsova and Yu.V. Valtsiferov, *Journal of Engng. Physics* 57 (1989), No.1, p.p.746-751.
3. Yu.S. Gupalo and Yu.S. Ryazantsev, *Fluid Dynamics* 22 (1987), No.5, p.p.752-757.
4. J.F. Brady and A. Acrivos, *J.Fluid Mech.* 112 (1981) 127

COMPARISON OF THE FLOW STRUCTURE FOR A HIGH INTENSIVE THERMOCAPILLARY CONVECTION, BUOYANCY CONVECTION AND THEIR INTERACTION

K.G.Dubovik
Institute for Problems in Mechanics, Academy of Sciences of the USSR
prospect Vernadskogo 101, 117526 Moscow, USSR

ABSTRACT

The flow and temperature field structures in a region with a free boundary for buoyancy convection, thermocapillary one and their interaction are under consideration. Flow patterns are obtained for the transition from the normal gravity to a theoretical wheitlessness condition. The hypothesis for the dependency of the thermocapillary convection on the history of the flow is presented.

Keywords: Thermocapillary convection, buoyancy convection, numerical modelling.

1. INTRODUCTION

Numerical investigation of different types of convection in regions of simple form was carried out for a several tens of years, which is due to importance of a such work for better understanding of the flow and heat and mass transfer, caused by convection. From a wide number of publications on the problem we shall point out only (Ref.1,2) which were the first to investigate buoyancy convection in the side heating square cavity with adiabatic and linear temperature distribution on the horizontal boundaries. The paper (Ref.3) was one of the first to consider thermocapillary convection in the square cavity for moderate and high Marangoni numbers.

Experiment (Ref.4) was carried out in the square cavity for the terrestrial conditions, when interaction of thermocapillary convection and buoyancy one should be taken into account. Thermocapillary flows in the long rectangular layers were investigated for different Prandtl numbers in experiment and numerically (Ref.5,6,7,8).

Numerical experiment brings a beautiful possibility for the analyses of the flow not only for the interaction of the convective mechanisms, but in separate manner too, the latter is the main purpose of the present paper.

2. FORMULATION OF THE PROBLEM

The flow in the square cavity of the height H is under consideration. The lateral boundaries are maintained at constant but different temperatures $T_1=T_0+\Delta T$ and $T_2=T_0+\Delta T$, where $2\Delta T$ is the characteristic temperature difference. The top and bottom boundaries are adiabatic. All the fluid properties are constant with the exception of the surface tension σ and density ρ, which are considered to be

$$\sigma=\sigma_0+\delta(T_0-T);\quad \rho=\rho_0+\beta(T_0-T) \qquad (1)$$

where σ_0, ρ_0, β, δ - are positive constants for the liquid properties at temperature T_0. The free surface is flat.

For the flow description we shall

use the Navier-Stokes and heat transfer equations written in vortisity ω, stream function ψ and temperature θ variables

$$\omega_t+(u\omega)_y+(v\omega)_y=(\omega_{xx}+\omega_{yy})/Re+Gr\theta_x/Re^2 \quad (2)$$

$$\psi_{xx}+\psi_{yy}=-\omega \quad (3)$$

$$\theta_t+(u\theta)_x+(v\theta)_y=(\theta_{xx}+\theta_{yy})/Ma \quad (4)$$

where $u=\psi_y$; $v=-\psi_x$; $\omega=-u_y+v_x$

All the variables are dimensionless, the terms H, $\delta\Delta T/(\rho\nu)$ and $H\rho\nu/(\delta\Delta T)$ being used as the scales of the length, velocity and time. The non dimensional temperature θ was introduced as

$$\theta=(T-T_o)/\Delta T$$

The boundary conditions were

on rigid boundaries

$$u=v=0 \quad (5)$$

on the free surface

$$\omega=0 \text{ or } \theta_x \quad (6\text{ a,b})$$

and for eq. (4)

$$0<y<1 \quad x=0: \theta=1;$$

$$x=1: \theta=-1; \quad (7)$$

$$0<x<1, \quad y=0;\ 1: \quad \theta_y=0$$

The system of eq. (2)-(4) includes parameters, i.e. Reynolds $Re=\delta\Delta TH/(\rho\nu^2)$, Grashoff $Gr=g\beta\Delta TH^3/(\rho\nu^2)$, Prandtl $Pr=\nu/\alpha$ and Marangoni $Ma=Re\ Pr$ numbers, with ν - kinematic viscosity, α - temperature conduction, g - gravity acceleration.

The boundary condition (6a) was used for the case of gravitational convection without thermocapillary effect. When thermocapillary convection was taken into account, the (6b) condition was used.

Eqs. (2)-(4) were solved numerically with the programm package of finite differences (Ref.9). The eq. (3) for stream function was solved by iterations with optimal parameters (Ref.10). The boundary condition (5) was realized with the help of method after V.I.Poleshaev & V.L.Grjasnov well described in (Ref.11).

Non uniform 97x81 grid was used, being condensed to the lateral boundaries and the free surface.

The dimensionless parameters in the computations presented were $Re=1000$, $Pr=17$, $Gr=3.5\ 10^4$, $Ma=1.7\ 10^4$, which are close to the data of (Ref.4).

The initial data for the buoyancy convection were

$$u=v=0, \quad \theta=(1-2x)$$

After reaching the steady flow structure the thermocapillary effect was included into computations. At the final stage of computations the flow development from the normal gravity to the sudden zero gravity flow was considered. This process could be of interest as an example of transient processes in a drop tower or on a space vehicle after an engine switching off.

3. DISCUSSION OF RESULTS

Fig.1 shows the flow and temperature fields for the buoyancy convection, interaction of the buoyancy and thermocapillary convection and the thermocapillary convection. The first case is not of high interest, analogous results were obtained by various of authors. The only difference of the results presented from the usual problem of buoyancy convection is a slight distortion of the central symmetry of the flow, which is due to slip conditions on the top boundary. The flow structure is stable.

In the second case (Fig.1b), there is a pronounced distortion of the central symmetry of the flow due to the influence of the thermocapillary convection. Stream lines are attracted to the contact points of the free surface and lateral boundaries. The internal flow has a good resolving structure. The changes in the temperature field are of maximum near the free surface and, surprisingly, in the bottom region of the flow. The isotherms behavior in the bulk was of minimum change.

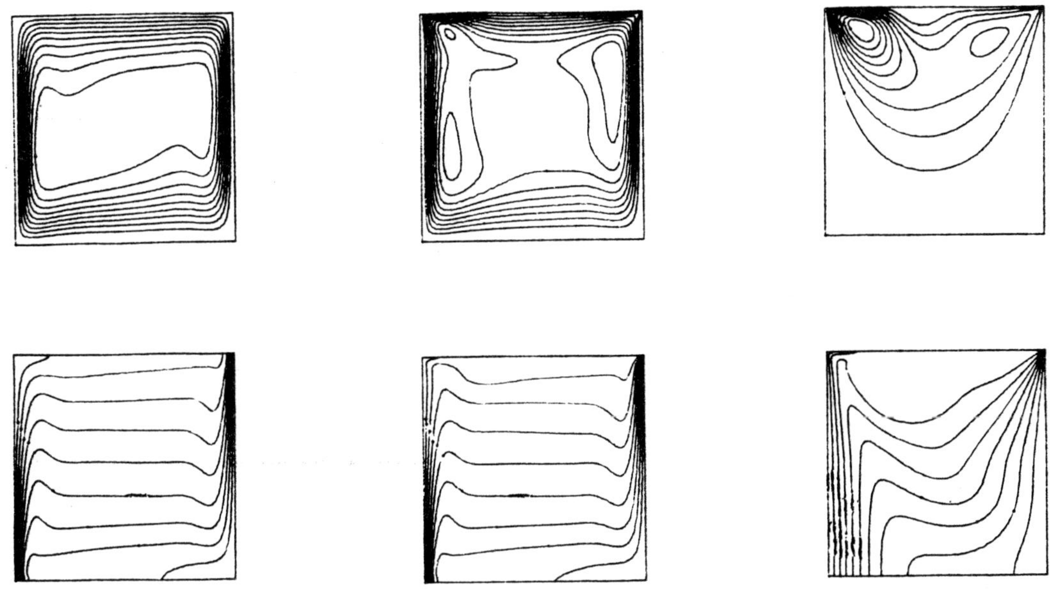

Figure 1. Stream lines (a) and isotherms (b) for buoyancy convection with free surface (left), $Gr=3.5x10^4$; interaction of the buoyancy convection and thermocapillary one (middle), $Gr=3.5x10^4$, $Ma=1.7x10^3$; and thermocapillary convection only (right), $Ma=1.7x10^3$.

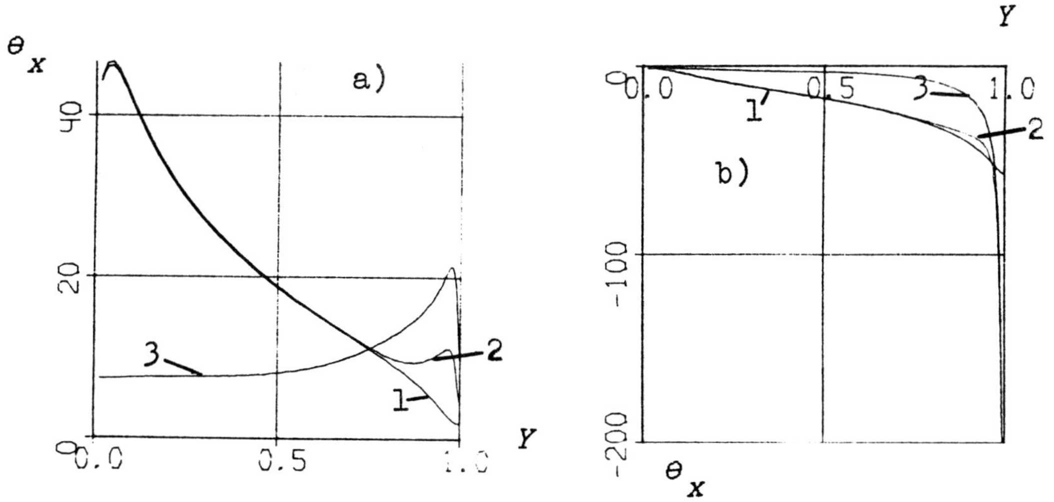

Figure 2. Heat flux distribution on the hot (a) and cold (b) boundaries of the layer.
Curve 1 - buoyancy convection only, Gr=3.5x10
Curve 2 - interaction of buoyancy and thermocapillary convection. $Gr=3.5x10^4$, $Ma=1.7x10^3$.
Curve 3 - thermocapillary convection only. $Ma=1.7x10^3$.

The increase in the heat flux over the lateral boundaries near the free surface is characteristic of the thermocapillary convection and is well seen from the isotherms.

Another important distinct form the buoyancy convection is slow oscillation of the inner flow structure. More detail analyses of this item is being planed in an other publication.

The thermocapillary flow (Fig.1c) presents three vortexes near the free surface, one of them being of the opposite rotation. The liquid near the free surface mixed well and the temperature difference in this part of the flow is less then $0.1\Delta T$. Outside the thermal boundary layer there is a lengthy region of the reverse temperature gradient. The flow is subjected to slow oscillations.

The heat flux distribution on the lateral boundaries (Fig.2) is, perhaps, the most important and sensitive characteristic of the problem. There are pronounced qualitative changes in it associated with the flow regimes. For the case of buoyancy convection it has a maximum near the bottom of the layer and monotonically falls along the hotter wall. For the case of interaction of the buoyancy and thermocapillary convection there is an additional maximum near the free surface, associated with the influence of the thermocapillary convection. At the same moment, the maximum in the bottom region slightly decreases.

For the only thermocapillary case the heat flux was uniform on the 1/2 of the layer height, with pronounced maximum near the free surface. On the cold boundary the heat flux increases monotonically for all regimes and the main change was in it´s increase at the contact point of the free surface and cold wall when the thermocapillary convection was taken into account. As it was mentioned above, the transient flow from the interaction of the buoyancy and thermocapillary convection to thermocapillary only may be realized in a drop tower experiment or in a sounding rocket flight. Fig.3 presents the flow and temperature fields of the transient regime; the numbers under the pictures are dimensionless time.

Immediately after setting gravity to zero, the intensive vortex forms near the hotter wall, the intensity of the flow in the bulk of the liquid falls down. Then the vortex is stretched and divided into two. The second takes the central place and then moves to the cold boundary. As it moves, the third counter rotating vortex forms near the free boundary, which plays the conciliatory role. The tear-off vortex intensity decreases with the intensity of the conciliatory one. But at this moment the hot drop of the liquid starts from the hot boundary, which leads to arising of the vortex paire again. In further computations this structure does not change.

The transient flow depends on the previous flow regime. But one can assume, that the final state of the thermocapillary flow can be of such dependence too. This hypothesis, certainly, needs for more detail an experimental and theoretical investigation. Ability of the different flow regimes for the thermocapillary convection as a result of the flow history was mentioned in (Ref.11).

4. CONCLUSION

In the present paper we considered flow and temperature field structures in a region with a free boundary for buoyancy convection, thermocapillary one and their interaction. Flow patterns have been obtained for the transition from the normal gravity to a theoretical wheitlessness condition. The hypothesis for the dependency of the thermocapillary convection on the history of the flow has been presented.

References

1. Oniyanov V.A., Tarunin E.L. *Proc.of the Perm´s State University. Hydrodyn.* 1970. N 216. Iss. 2. P. 163-175. (in Russian).

2. Vahl Devis G. De. *Int.J. for numerical methods in fluids.* 1983. V.3. N 3. P.249-264.

Figure 3. Flow structure (a) and temperature field (b) development for a sudden transition from the normal ($Gr=3.5x10^4$, $Ma=1.7x10^3$) to zero gravity ($Gr=0$, $Ma=1.7x10^3$). Numbers under the pictures - non dimensional time.

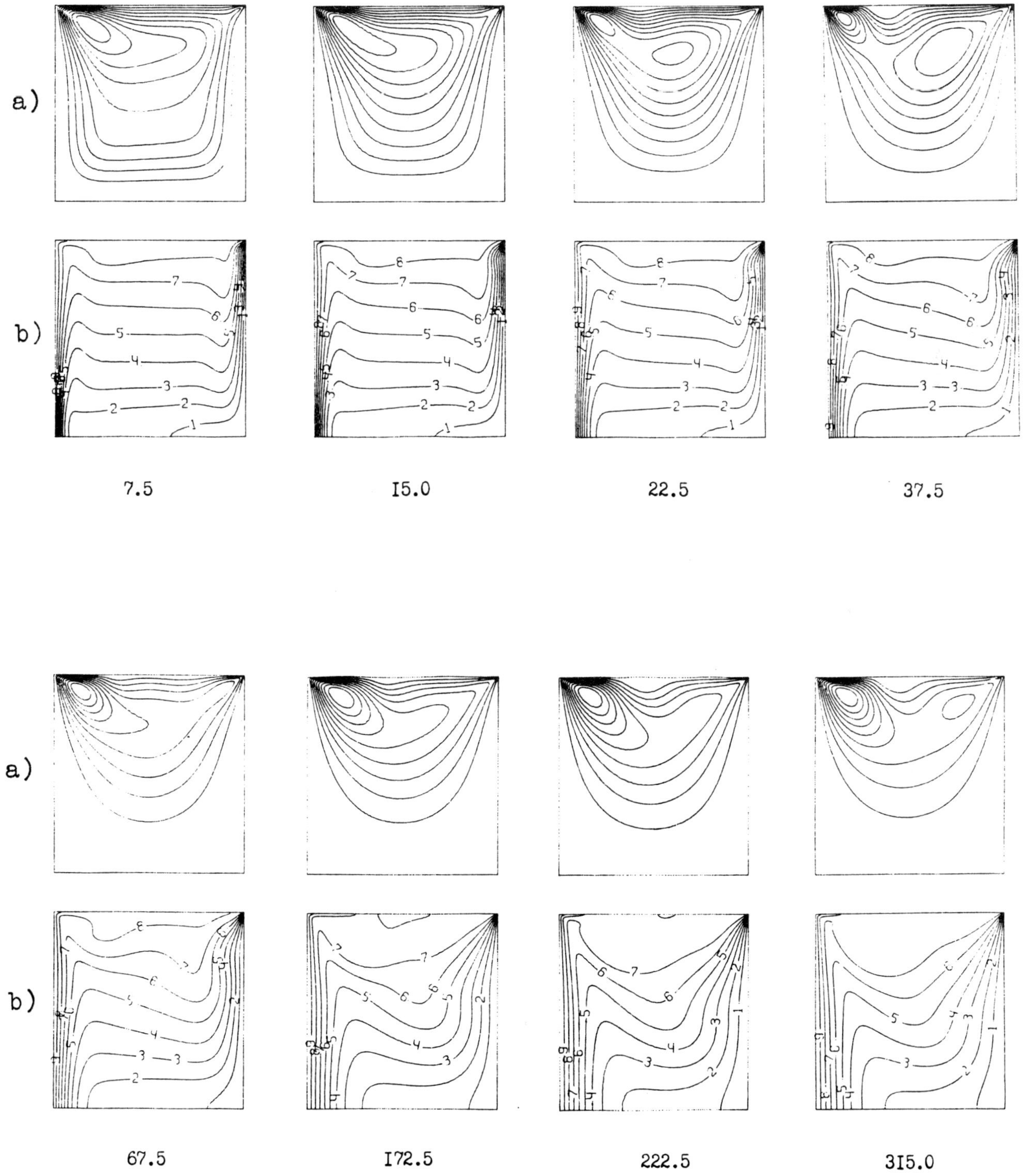

3. Zebib A., Homsy G.M., Meiburg E. *Phys. Fluids.* v.28, N 12, 1985, p.3467-3476.

4. Schwabe D., Metzger *J. Crystal Growth.* v.97. 1989. p.23-33.

5. Kirdiyashkin A.G. *Hydrodynamics and Transport Phenonema in Wieghtlessness.* Sverdlovsk. Ural. Sci Center Acad. Sci. of the USSR. 1983. P.126-135. (in Russian).

6. Berdnikov B.S., Zabrodin A.G., Markov B.A. *Structure of forsed and thermogravitational flows.* Novosobirsk. Sibir. Department of the Acad. of Sci. of the USSR. 1983. P. 147-163. (in Russian).

7. Hadid H.B., Roux B., Laure P., Tison P., Camel D., Favier J.J. *Adv.Space Res.* v.8. 1988. p.293-304.

8. Laure P., Roux B., Hadid H.B. *Phys. Fluids.A2(4).* 1990. p.516-524.

9. Poleshev V.I., Bune A.V., Dubovik K.G. et all. *Mathematical modelling of convective heat transfer on the base of Navier-Stokes equations.* Moscow. Nauka Publisher. 1987. 272p. (in Russian).

10. Fedorenko R.P. *Uspehi Matematicheskih Nauk.* 1973. Iss. 2. P.121-182. (in Russian).

11. Poleshev V.I., Verezub N.A., Dubovik K.G. et all. *Convective processes in weightlessness.* Moscow. Nauka Publisher. 1991. 240p. (in Russian).

AXISYMMETRIC THERMOCAPILLARY FLOWS IN CYLINDER AND CYLINDRICAL LAYER

V.K. Andreev and O. V. Admaev
Computing Centre SB USSR Academy Krasnoyarsk 660036

ABSTRACT

The deformation of cylinder and cylindrical layer, exposed to thermocapillary forces is studied under the conditions close to weightlessness. The dependence of the surface tension on the temperature has a form $\sigma = \sigma_0 + \kappa(\theta - \theta_0)$, or $\sigma = \sigma_0 + \kappa_1(\theta - \theta_0)^2$. In the first case the temperature is chosen to be distributed by the square law along the surface, in the second one – by the linear law. The corresponding axisymmetric solutions describe thermocapillary motions nearby the critical points, at which the outer temperature has maximum or minimum. With that, there occur both stationary and non-stationary flows.

Keywords: *Thermocapillary Convection, Weightlessness, Surface Tension, Marangoni Number, Cylindrical Layer*

1. NONSTATIONARY FLOWS

We consider the flow of a liquid infinite cylinder with a free surface with axisymmetric heating and absence of gravity (Ref.1). The temperature on its boundary is assumed to have maximum or minimum at the point $z = 0$. Then, in the vicinity of this point the outer temperature can be approximated by parabolic law:

$$\theta_{ou}(z,t) = a_{ou}(t)z^2 + b_{ou}(t).$$

Assume that the surface tension

$$\sigma(\theta) = \sigma_0 + \kappa(\theta - \theta_0), \kappa = \text{ const } < 0.$$

As it was shown in (Ref.2) the equations of viscous heat conducting fluid motion assume the solutions of the form:

$$\begin{aligned} u &= u(r,t), v = 0, w = w(r,t)z, \\ p &= p(r,t), \theta = a(r,t)z^2 + b(r,t), \end{aligned}$$

where u, v, w are the velocity components, p is the pressure.

We introduce the following dimensionless variables and functions:

$$\begin{aligned} t &= \tau h_0^2/\nu, \\ r &= h_0 g(\tau) y, \\ w &= W(y,\tau)\nu/h_0^2, \\ u &= U(y,\tau)\nu/h_0, \\ a &= A(y,\tau)T/h_0^2, \\ b &= B(y,\tau)T. \end{aligned}$$

Here T is a characteristic temperature, $g(\tau)$ is a dimensionless cylinder radius at the point $z = 0$ and subjected to be determined in the course of the problem solution, y is Lagrangian coordinate, h_0 is the initial radius of a liquid cylinder.

To determine the cylinder radius $g(\tau)$, the problem is reduced to the system of three integral differential equations with respect to the unknown functions A, W, g with dimensionless parameters: $M = Th_0\kappa/(\rho\nu^2)$ is the Marangoni number, Pr – the Prandtl number, $Bi = \beta h_0/\chi$ is the Biot number, χ, β are thermal conductivity and thermal surface conductance.

The solution of the boundary problem obtained has been determined by Galerkin technique. As base functions we used the shifted Jacobi polynomials $R_m^{(0,1)}(y)$. The approximate solution has been sought in the form

$$\begin{aligned} W^n(y,\tau) &= \sum_{m=0}^{n} W^m(\tau) R_m^{(0,1)}(y), \\ A^n(y,\tau) &= \sum_{m=0}^{n} A^m(\tau) R_m^{(0,1)}(y). \end{aligned}$$

With that, the function $g(\tau)$ is always defined only by the first expansion member $W^0(\tau)$:

$$\frac{dg}{d\tau} = -\frac{1}{2} W^0(\tau) g.$$

The other thermocapillary flow of the cylinder can be obtained, if we assume, that the outer temperature is linearly dependent on z:

$$\theta_{ou} = a_{ou}(t)z + b_{ou}(t).$$

With that σ is a square function:

$$\sigma(\theta)=\sigma_0+\kappa_1(\theta-\theta_0)^2.$$

In this case the temperature distribution can be considered to be linear:

$$\theta=a(\tau,t)z+b(\tau,t).$$

Calculations for both problems have been fulfilled by Runge-Kutta technique of the fourth order with $M=10, \Pr=0.2, Bi=2$. The outer temperature has been taken periodically depending on τ, $a_{o\infty}=-0.05\sin 5\tau, b_{o\infty}=0$.

The curve I corresponds to parabolic law of temperature distribution on z, (Figure 1). The curve II shows the calculation of the cylindrical radius with the linear temperature dependence on z. Here a monotonic increase of the radius occurs due to the temperature coefficient of the surface tension being always negative. The curve III corresponds to $\kappa_1>0$ and the fluid always flows from the point $z=0$, which results in the decrease of the cylindrical jet radius.

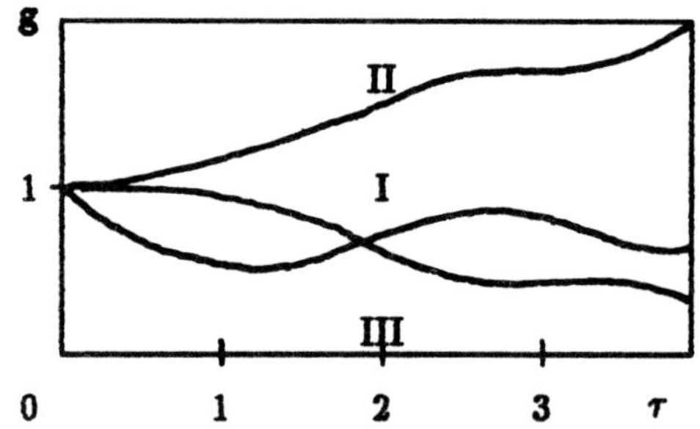

Figure 1:
The variation of cylindrical radius for
$M=10$, $\Pr=0.2$, $Bi=2$

2. STATIONARY FLOWS

Consider a stationary thermocapillary motion in a cylindrical layer. Assume, that on the inner solid body of the prescribed radius r_0 a constant linear distribution of the temperature along z – axis is maintained, on the free surface $r=h_0$ the heat flux is absent. The length of the inner body and of the cylindrical layer are large enough; due to this we can neglect the boundary effects.

On the solid surface the condition of impenetrability is satisfied, on the free surface - the conditions of impenetrability, balance of tangential and viscous stresses. We also assume, that the free surface is undeformable (the value σ_0 is large enough). The reference point in the cylindrical coordinate system is on the solid surface, at the point, where the temperature value θ_0 is achieved.

The axisymmetric self-similar solution has been sought in the form (Ref.3), (Ref.4).

$$\begin{aligned}
u &= -\nu\psi(R)/(h_0R),\\
w &= \eta\nu\psi'(R)/(h_0R),\\
\theta &= \theta_0+Ah_0\eta a(R),\\
p &= p_0-\frac{1}{2}\rho(\nu/h_0)^2[f(R)+\lambda\eta^2],\ A=\text{const},\\
\sigma &= \sigma_0+\kappa_1(\theta-\theta)^2,\ \kappa_1=\text{const}>0,
\end{aligned}$$

where u, w are components of the velocity vector on the axis r and z, p is the pressure, $\psi\eta\nu/(Rh_0)$ is a stream function, ν – kinematic viscosity, κ_1– surface tension coefficient, ρ – density, θ_0, P_0 – temperature and pressure at the reference point on the solid surface, corresponding to the extreme value of the surface tension, λ is unknown constant, which must be determined in the course of solution, $R=r/h_0$, $\eta=z/h_0$ are dimensionless variables.

The equations of motion and energy lead to the boundary problem for the system of non-linear ordinary differential equations

$$f=\psi^2/R^2+2\psi'/R,$$

$$\psi'''=(\psi''-\psi''\psi+\psi'^2)/R+(\psi\psi'-\psi')/R^2-\lambda R,$$

$$a''=[\Pr(\psi'a-a'\psi)-a']/R,$$

$$\begin{aligned}
R &= 1:\psi=0,\psi''-\psi'=Ma^2,a'=0,\\
R &= d_0=r_0/h_0:\psi=\psi'=0,a=1,
\end{aligned}$$

which has been solved numerically. Here

$$M=\kappa_1A^2h_0^3/(\rho\nu^2).$$

The results of numerical integration are presented in Figure 2 for Prandtl number equal to 1. Parameter λ is represented as a function of Marangoni number.

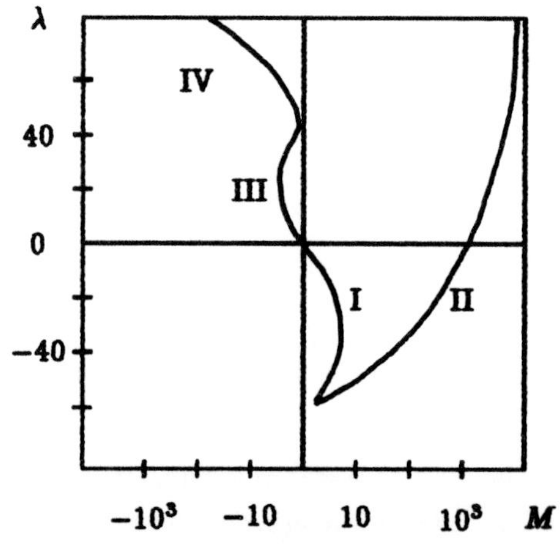

Figure 2:
The dependence of λ on M for $\Pr=1$.

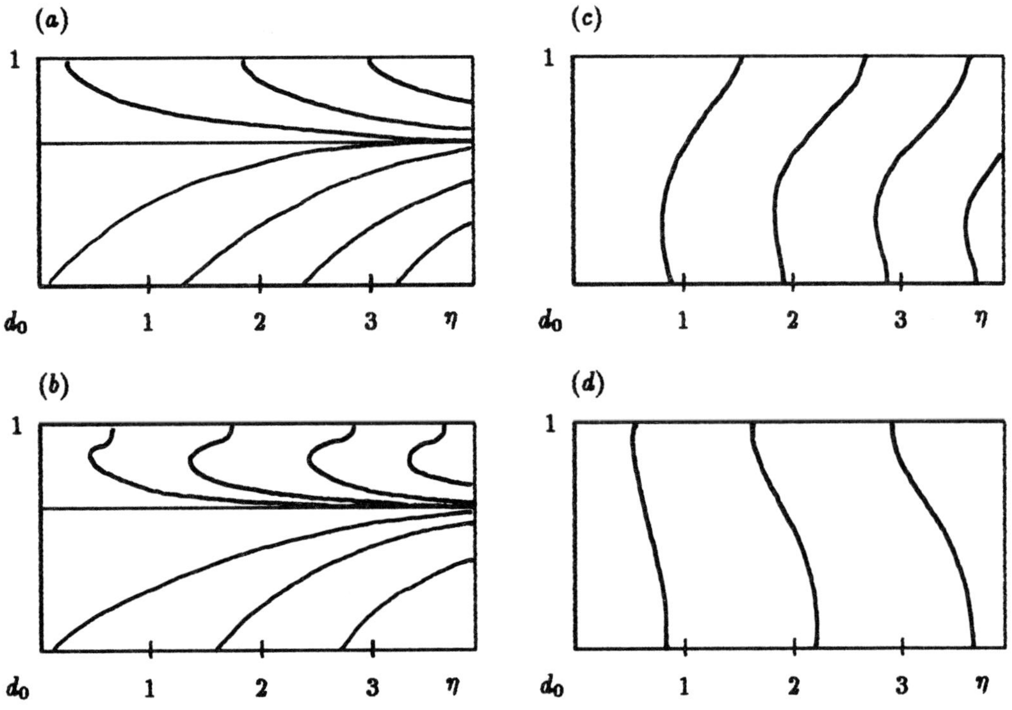

Figure 3 :

The temperature distributions inside the layer for Pr = 1. The figures 3(a), 3(b), 3(c), 3(d) correspond to the groups of solutions IV, II, I, III accordingly.

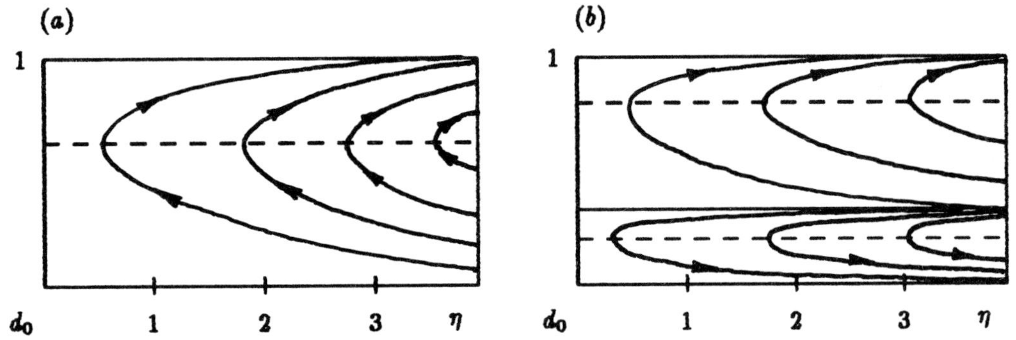

Figure 4 :

(a) Streamlines for $M = 100$, $\lambda = 10$, Pr = 1.
(b) Streamlines for $M = 2.2 \cdot 10^4$, $\lambda = 10^3$, Pr = 0.005.

We can distinguish four groups of solutions, according to the temperature distribution inside the layer. Isotherms for every group are represented in Figure 3.

While solving a dynamic problem, two different situations arise. The first one is realized when $M > 0$, i.e. when temperature coefficient of the surface tension is more than zero, and, besides, the sign of the Marangoni number doesn't depend on heating or cooling of the fluid, and the second one – when $M > 0(\kappa_1 < 0)$. In the first situation (groups I, II) the fluid on the free surface flows away from the centre $\eta = 0$. In the second one (groups III, IV) – towards the centre. It occurs due to the fact, that the fluid on a free surface flows in the direction of the greater value of the surface tension coefficient. With $\lambda < 658$ the flow always changes the direction on the depth, which is equal to 2/3 of the cylindrical layer thickness for different values of Marangoni number, but for the solutions of the group II with large $M(M = 7 \cdot 10^5)$ the turn point "moves" to the free surface, Figure 4a. With $\lambda > 658$ and $M > 0$ the second vortex occurs, Figure 4b. With the increase of Marangoni number the secondary vortex domain increases, and with the further increase of Marangoni number the intensity of the lower vortex begins to prevail the intensity of the upper one.

Under the similar conditions we consider the motion of the fluid with a linear dependence of the surface tension coefficient on the temperature with $\mathrm{Pr} = 4.1$. The dependence of λ on M is represented in Figure 5.

In the case under consideration six groups of solutions arise, and the transition from one group to another one can be characterized by the qualitative change of the solution as moving along the curve, illustrated in Figure 5.

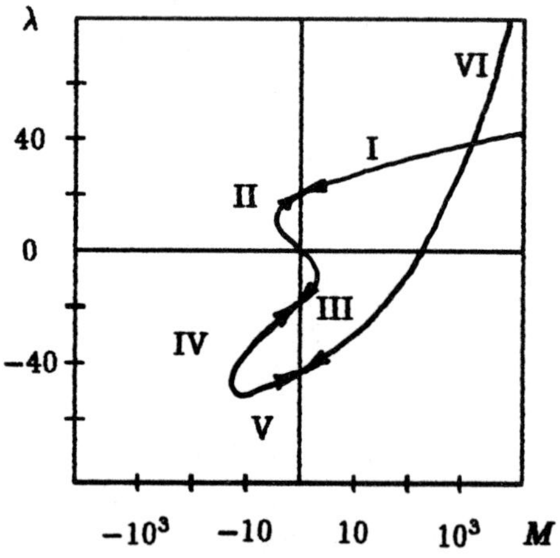

Figure 5:
The dependence of λ on M for $\mathrm{Pr} = 4.1$.

In particular, it occurs in going from the group V to the group VI, when the number M tends to zero on the right and on the left. But with the numerical solution of the dynamic boundary problem with $M = 0$ the unique solution $\lambda \equiv 0$, $\psi(R) \equiv 0$ has been obtained. Therefore, the given transition occurs due to the change of the temperature field with $\lambda \Rightarrow \lambda_* + 0$ and $\lambda \Rightarrow \lambda_* - 0$.

The similar changes of the temperature field occur also while going from group I to the group II, from II to III, from III to IV.

In conclusion we note, that with the fixed values of Marangoni and Prandtl numbers, several different motion regimes are possible. Thus, for example, with $\mathrm{Pr} = 4.1$ and $M = 0.3$ four solutions are possible, Figure 5.

3. REFERENCES

1. V.K. Andreev and O.V. Admaev, *Modelling in Mechanics.* 4 (21) (1990) 73.

2. V.K. Andreev and V.V. Pukhnachev, *Numerical Methods of the Continuum Mechanics.* 14 (1983) 3.

3. Yu.P. Gupalo and Yu.S. Ryasantsev, *Mech.of Fluid and Gas.* 5 (1988) 132.

4. J.F. Brady and A. Acrivos, *J.Fluid Mech.* 112 (1981) 127.

STABILITY OF BUOYANCY STATIONARY CONVECTIVE FLOWS UNDER ACTION OF ELECTROMAGNETIC, THERMOCAPILLARY OR VIBRATIONAL FORCES

A.Yu.Gelfgat and B.J.Martuzans

The University of Latvia, Research Institute of Mathematics and Computer Science, 29 Rainis boul., Riga, LATVIA, 226250

ABSTRACT

The main aim of the investigation is the numerical analysis of stability of stationary buoyant convective flows under action of some additional force which may occur significant in microgravity or may be used for laboratory modeling of microgravity on the Earth. The other object of interest is the possible application of the considered additional forces to the control of convective stability. The investigation is carried out for three different problems: 1 - the convection in a square cavity heated from below and vibrating in the vertical direction with high frequency and small amplitude; 2 - the buoyant-thermocapillary convection in a laterally heated square cavity with free upper surface and 3 - buoyant convection of an electrically conductive fluid in the homogeneous (vertical or horizontal) magnetic field. All the computations were carried out for low Prandtl number fluid with Pr=0.02. The spectral Galerkin method with the trial functions defined in the whole region of the flow was used for numerical solution

Keywords: free convection, stability, simulation of microgravity.

1. THE EFFECT OF HIGH-FREQUENCY VIBRATIONAL FORCE

The effect of high-frequency vibrational force was studied for the natural convection in square cavity heated from below with rigid perfectly conducting walls. The cavity may oscillate in the vertical direction with the amplitude b and the circle frequency ω, b and ω are supposed to satisfy the conditions $b << l/\beta\Delta T$ and $\omega l^2/\nu << 1$, what means high-frequency vibrations with small amplitude. Such a supposition allows to use averaged Boussinesq equations and to take into consideration only "slow" component of the flow (see Ref.1). The influence of the vibrational force is characterized by the vibrational Rayleigh number $\mathbf{Ra_v} = (\beta b\omega\Delta T l)/2\nu\chi$. Here β is the thermal expansion coefficient, ΔT is the temperature difference between hot and cold walls, ν is the kinematic viscosity, χ is the thermal diffusivity, l is the characteristic length.

The map of stability was plotted in the coordinates $\mathbf{Ra_g}$ and $\alpha = \mathbf{Ra_v}/\mathbf{Ra_g}$ ($\mathbf{Ra_g} = g\beta\Delta T/\nu\chi$ - gravitational Rayleigh number) and is shown in Figure 1 . The solid line in the lower part of the figure corresponds to the convective stability of quasi-quiescent non-uniformly heated fluid. Below the solid line there is no averaged convective motion and above the line such a motion appears. With the increasing of α the critical value of $\mathbf{Ra_g}$ grows as it is well-known for the case of infinite horizontal fluid layer (see Ref.1). On the other hand in confined cavity a salient point on the stability curve appears (at $\alpha = 1.6$). Two smooth parts of the stability curve correspond to two different patterns of secondary flows. When $\alpha < 1.6$ the instability leads to the central symmetric secondary convective flow, but when $\alpha > 1.6$ the secondary flow is symmetric with respect to the horizontal axes and has one-above-another vortex structure. The oscillatory instability of these two different motions was studied separately.

The dashed line in Figure 1 is the stability curve for central symmetric flows. Central symmetric flows are stable between solid and dashed lines, and unstable outside of this region. Axial symmetric flows are stable between solid and dash-and-dot lines. The oscillatory instability of central symmetric flows onsets outside the region bounded by dashed

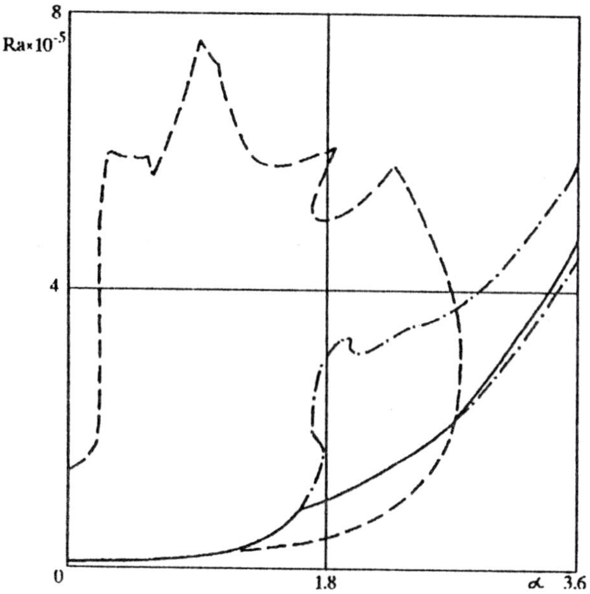

Figure 1. The map of stability for the stationary buoyant- vibrational convective flows. Pr = 0.02 .

and solid line above dash-and-dot line, but below the dash-and-dot line the instability is monotonic and corresponds to the change of the spatial symmetry of the flow. Similarly, axial symmetric flows are oscillatory unstable above dash-and-dot line outside the region bounded by dashed and solid lines, and monotonically unstable inside this region above and left from the dot-and-dash line. Rather strong vibrational force causes the hysteresis phenomenon (see Figure 1): stationary flows remain stable below the marginal stability curve of the quiescent fluid. So, the high-frequency vibrational forces have a strong influence both on the stability and the spatial structure of stationary convective flows. This influence must be taken into account when the magnitudes of vibrational and buoyant forces are comparable, and, perhaps, may be used for the control of the convective stability and/or of the spatial structure of the convective flows.

2. THE EFFECT OF THERMOCAPILLARY FORCE

The effect of thermocapillary force was studied for the stationary buoyant-thermocapillary convection in laterally heated square cavity with free upper boundary (for details see Ref.2). The cavity with conducting vertical and adiabatic horizontal boundaries was considered. The diagram of stability in coordinates **Gr-Ma** is shown in Figure 2 . Here $\mathbf{Gr} = \mathbf{Ra_g/Pr}$ - the Grashof number, $\mathbf{Ma} = (\partial\sigma/\partial T)\Delta Tl/\rho\nu^2$ - the Marangoni number, where σ - the surface tension coefficient, ρ - the fluid density.

Positive and negative values of **Ma** mean respectively the decrease and the increase of the surface tension with the temperature. The stationary buoyant-thermocapillary flows are stable in the shaded region and oscillatory unstable outside this region. Obtained critical value of the Marangoni number in zero-gravity (**Gr** = 0) is **Ma** $= \pm 1.1 \cdot 10^5$. As it is seen from Figure 2 the coupled action of thermocapillary and buoyancy forces does not necessary cause the destabilization of the flow. It is possible to increase **Gr** and **Ma** in

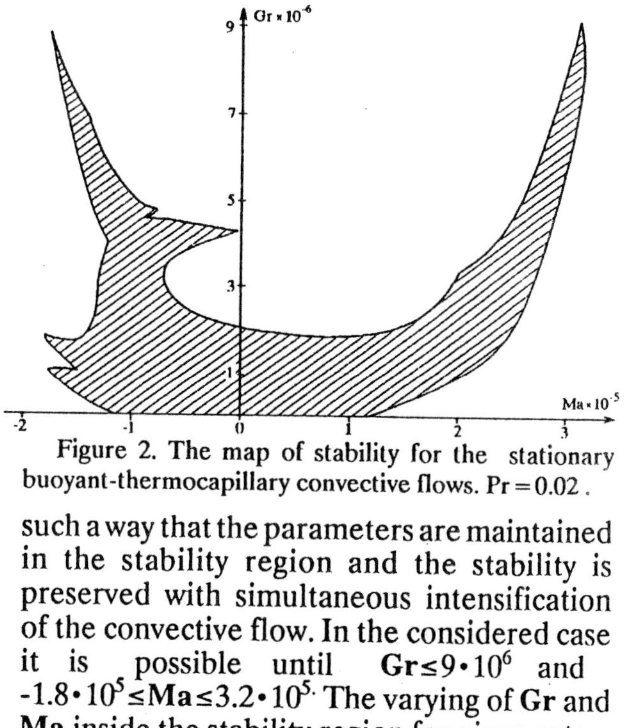

Figure 2. The map of stability for the stationary buoyant-thermocapillary convective flows. Pr = 0.02 .

such a way that the parameters are maintained in the stability region and the stability is preserved with simultaneous intensification of the convective flow. In the considered case it is possible until $\mathbf{Gr} \leq 9 \cdot 10^6$ and $-1.8 \cdot 10^5 \leq \mathbf{Ma} \leq 3.2 \cdot 10^5$. The varying of **Gr** and **Ma** inside the stability region for given external conditions and given material is possible only with the varying of the characteristic length. So, the requirement of stable stationary coupled buoyant-thermocapillary convection leads to the problem of proper choose of the size of technological device.

3. THE EFFECT OF MAGNETIC FIELD

The effect of externally imposed magnetic field on the stability of the stationary buoyancy convective flows was studied for the convection in a laterally heated square cavity (for details see Ref.3). Rigid adiabatic horizontal boundaries and rigid conducting vertical boundaries of the cavity were considered. The magnetic field was supposed to be constant and homogeneous. The influence of the flow on the magnetic field was neglected. The electromagnetic force is characterized by the Hartman number $\mathbf{Ha} = Bl(\gamma/\rho\nu)^{1/2}$, where B is the induction of the magnetic field, γ is the electric conductivity. Figure 3 shows the diagram of stability for both vertical and horizontal directions of the magnetic field. The suppression of the convective flow by the electromagnetic force causes rapid stabilization of the flow with the growth of the Hartman number. The direction of the magnetic field as it is seen from the Figure has no significant influence on the critical value of **Gr**.

To obtain a qualitative estimate of the possibility to use the magnetic field for the laboratory modeling of reduced gravity the pulsative and averaged components of the flow and the trajectories of liquid particles (all obtained from the straight-forward solution) were studied. The investigations show that the

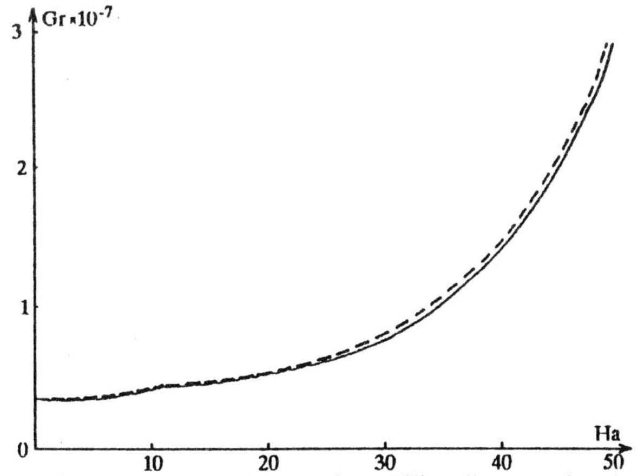

Figure 3. The oscillatory instability diagram for the stationary buoyancy convection in the laterally heated square cavity in the externally imposed magnetic field. Pr = 0.02. Solid line - vertical magnetic field. Dashed line - horizontal magnetic field.

averaged component of the flow and trajectories of liquid particles (see Figure 4) change very similarly when the Grashof number decreases or the Hartman number increases. Basing on these comparisons the action of the magnetic field may be considered as the analog of reduced gravity. On the other hand

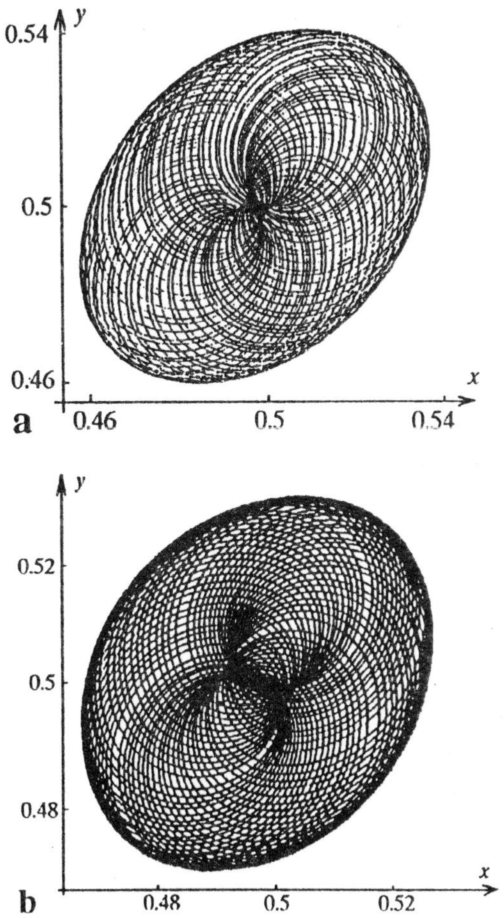

Figure 4. The trajectory of a liquid particle starting at the point $x = y = 0.5$. $Pr = 0.02$
a - $Gr = 6 \cdot 10^6$, $Ha = 0$; b - $Gr = 10^7$, $Ha = 30$.

further calculations with increasing **Pr** show that the changes in the pulsative component of the convective flow show some analogies between increasing Hartman and Prandtl numbers.

1. Gershuni G.Z. and Zhukhovitsky E.M. Fluid Mechanics - Soviet Research, 15, 63(1986).
2. Gelfgat A.Yu. and Martuzans B.J. Fluid Dynamics, 25, 169(1990).
3. Gelfgat A.Yu. Magnetohydrodynamics, 3, 324(1988).

MARANGONI OSCILLATION ONSET IN SIMULATED FLOATING ZONE IN MICROGRAVITY

Prof. R. Monti(*); Dr. R. Fortezza(^)

(*) Univ. of Naples
P.le Tecchio,80; I-80125 Naples Ph. +817682356 - FAX +81632044
(^) MARS Centre
Via Diocleziano, 328; I-80125 Naples Ph. +815272756 - FAX +817252750

ABSTRACT

Several experiments were performed, on board sounding rockets, aimed to the measurement of the thermo-fluid-dynamic conditions in which the convective Marangoni flow in liquid bridge configuration changes from an axi-symmetric pattern to an oscillatory regime. So far, the attempts to find a "critical parameter" able to predict the transition between the axial symmetric motion and the oscillatory one were unsuccessful. The experiments have been designed to obtain, in the limited micro-g time available, the maximum number of transitions to oscillations, measuring the relevant parameters. During the last flights the Telescience operative mode has been successfully implemented in order to maximize the scientific return. Several oscillation onsets have been obtained on liquids with different viscosities and at different temperature ramps. The experimental results are discussed and compared with numerical simulations.

Keyword: Microgravity, Floating-zone, Marangoni Flows, Marangoni Instabilities

1. INTRODUCTION

The aim of the performed experiments was the study of the thermo-fluid-dynamic behaviour of a silicone oil liquid bridge, simulating a floating zone, formed between two copper disks. The liquid was stimulated by imposing a temperature difference between the two disks. The temperatures of the two disks followed different linear ramps, selected by the P.I. during the experiment runs.

The effect of different temperature ramps on the onset of the oscillation was studied: it was necessary to increase the thermal difference across the liquid bridge several times using low temperature ramps; this implied problems due to the limited time available for the experiment. The use of telescience operation, with a direct interaction between the PI and the experimental module, was the only possible solution.

The thermo-fluid-dynamic fields inside the liquid bridge was compared with a numerical computation of the unsteady conditions obtained during the experiment. The temperatures and the velocity fields are examined and compared to identify the commonalities existing among the different experimental conditions at the very onset of the oscillations.

The microgravity fluid-dynamic experiments were performed on Spacelab (SL-1 and D1)[ref 1-3] and on the Texus 9, 14b, 23 and MAXUS 1 sounding rocket [ref 3-10]. The last two ones have been characterized by the introduction, for the first time, of the real operational mode in Telescience. The experimental module was fully controlled by the PI directly from MARS Centre. Unfortunately due to a failure of the rocket system, the microgravity condition was not reached during the MAXUS 1 flight and the scientific results were missed. However, in spite of the tumbling attitude of the rocket, the telescience link was successfully tested and the video/data/audio communication was correctly established between MARS and the launch base located at Esrange (S).

The experiments are performed in the framework of the space research conducted at MARS Centre and at University of Naples.The results obtained are being used for the "Critical Marangoni Flow" experiment definition to be flown on the D-2 Mission and carried out in the Advanced Fluid Physics Module (AFPM).

2. EXPERIMENTAL SET-UP

The experiments performed on board sounding rockets have been conceived following a step-wise evolutive approach. The selected facility was the TEM 06-4 that has been updated for each

flight, including new features in the links and improving of the diagnostics.
The liquid bridge was formed by silicone oil. The liquid was partially contained between two copper disks (1.8 cm diameter). The kinematic viscosity selected was 5 Cs for the first two flights and 2 Cs for the last ones.
A less viscous oil was selected for the recent flights to obtain higher values of Marangoni velocity (V_M) at a given temperature difference ΔT among the two disks. The value is given by: $V_M=\sigma_T \Delta T/\rho \nu$ where σ_T is the temperature derivative of surface tension, ρ is the density and ν is the kinematic viscosity. V_M represents a measurement of the driving force for the convective flow generated at the surface by the unbalance of the surface tension. Therefore the transitions have been reached at lower temperature differences, reducing also the time required for each run. Additional new features concern the thermal control system. Heating of the upper disk was performed by a Joule heater on Texus 9, and by a Peltier element on the other flights. During the last flights the thermal sub-system was changed and based on two Peltier elements controlling, simultaneously, the temperature ramps of the two disks in order to achieve symmetric temperature profiles with respect to the initial temperature. Each disk was instrumented by heat flow-meters, and the temperature inside the liquid measured using a mobile thermocouple comb controlled from ground. The illumination system consisted in a He-Ne laser, producing a light sheet to illuminate tracer particles in the meridian plane of the bridge.The visualization system included a cine-camera and a telecamera This duplication has been motivated to ensure image recording also in case of TV link failure. In the last flight the liquid motion was visualized using tracers illuminated by two different laser light cuts. The first laser beam was oriented orthogonally to the main optical path of the first CCD. The images detect the tracers motion contained in the cylinder meridian plane of the liquid bridge. An additional laser beam, generated using a laser-diode with different wave-length, was used to illuminate a plane parallel to the cylindrical liquid bridge generatrix, directed in the central part of the liquid. A second CCD is used to observe the illuminated liquid surface line and to analyze the motion of the tracers on the surface where the unbalanced surface tension is acting. The two CCDs optics were adequately filtered to separate the two images.

3. EXPERIMENT CONDUCT

The experiments were performed using similar operative procedures. After automatic formation of the liquid bridge (+70s from lift-off), the first temperature ramp was applied to the disks. Successively the thermocouple comb was inserted in the zone and located close to the free surface. The position of the free surface has been detected by the increase of the thermocouple signals during the insertion of the comb in the liquid bridge. Seven oscillation onsets were reached during the flights according to the following table:

Flight	Temp ramp [°C/s]	$\Delta T_{critical}$ [°C]	Oil [Cs]	Tele-science
TX9	1	50	5	no
TX14b	0.4	28.5	5	Yes
TX14b	0.5	33	5	(Kiruna)
TX23	-0.5 +0.5	28	2	+
TX23	+0.5 -0.5	28	2	Yes
TX23	+0.25 -0.25	27	2	(Naples)
TX23	+0.1 -0.1	24	2	+

After the oscillation onset, the temperature difference along the disks was kept constant for few seconds, then the investigator activated the cooling temperature ramps to damp the oscillation and start a new run. During the Texus 23 the same ramp was experimented twice using two opposite thermal gradients. The results show that the critical ΔT is not affected by the residual acceleration ($5x10^{-6} \div 1x10^{-5}$) or by the geometrical difference of the disks.
After a sustained oscillation phase, during which the ΔT was kept constant, a new oscillation decay was set. When the temperature across the liquid bridge reached a value ranging between $10 \div 15$ [°C] the oscillations were damped and a new run started with a different ramp.
From the analysis of the table presented, it is clear that the use of telescience increased dramatically the efficiency of the experiments in terms of scientific results.

4. EXPERIMENTAL RESULTS

A. Velocity field analysis

Velocity field has been obtained from the analysis of tracer motion recorded during the experiment. Typical flow fields and velocity vectors, corrected from the optical distortion caused by the cylindrical lens effect of the liquid bridge, are illustrated in Fig.1. The velocity fields in axial symmetrical configuration are in agreement with the those obtained by a numerical simulation based on Navier-Stokes equations solved using a finite difference method (the results are reported in []). Adiabatic conditions are assumed at the liquid-air interface. Assigned temperature ramps are assumed on both disks. Due to the ramp control the real steady state is never reached.

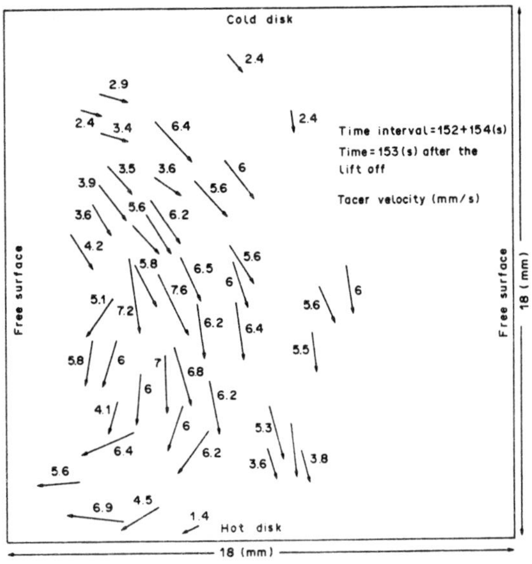

Fig. 1 - Flow Field and Velocity Vectors obtained from the analysis of the tracers motion

B. Temperature data analysis

During the runs, the boundary conditions have been monitored in real-time by measuring the temperatures of the two disks contacting the liquid. Four other thermocouples were used to measure the temperature inside the liquid.

From the data analysis it seems that the oscillations affect the temperature of the cold disk and that of the liquid nearest to it. This observation gives an indication for the monitoring of the oscillations in future experiments. The maximum value of the conditional Peclet number is of the order of 10^5, therefore a thermal boundary layer is formed on the cold and on the hot disks and normal to the surface. The time variation of Pe_m for the Texus 23 results is plotted in fig 2. A comparison of the temperature and velocity values measured and predicted by the numerical simulation is given in tab. 1.

D. Heat flows

Two thin foil heat flow-meters have been used to measure the heat flows transferred from the hot disk to the liquid and from the liquid to the cold disk The aim was to determine the relation among the transition and heat transfer rates.

Plots of the non-dimensional heat flow referred to the time and to Peclet Number (TX 23) are shown in Figs. 3 and 4.

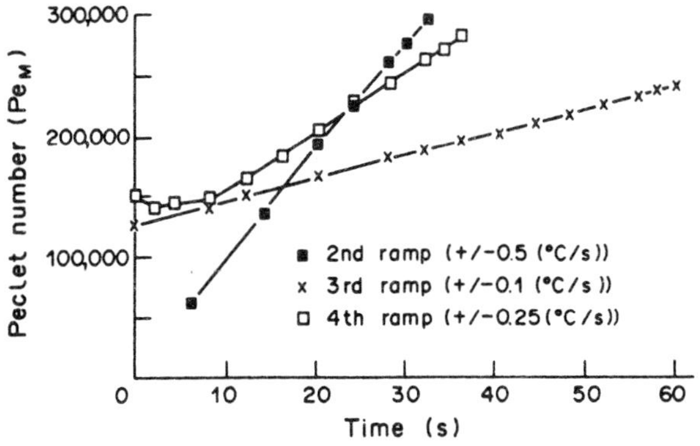

Fig. 2 - Peclet (Pe_m) number vs time for different temperature ramps

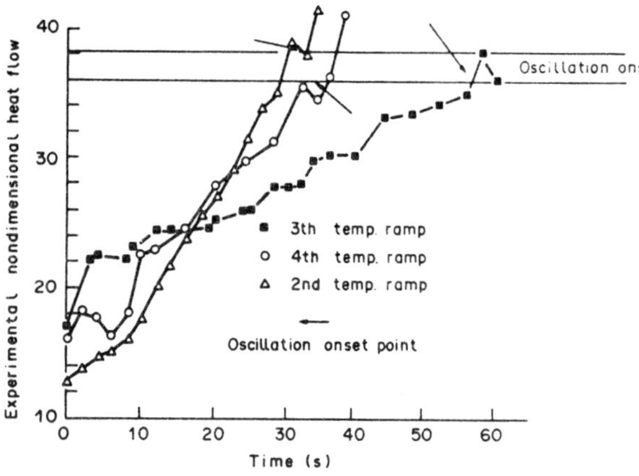

Fig. 3 - Non-dimensional heat flow vs time

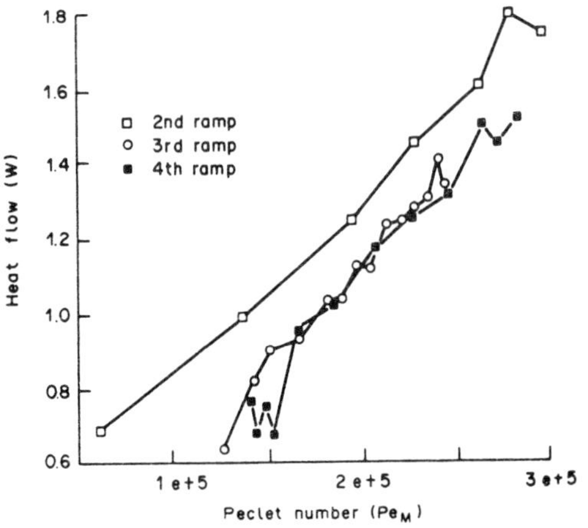

Fig. 4 - Experimental heat flow vs Peclet number

The non-dimensional heat flow is computed referring to the experimental value measured on the upper disk and to the purely diffusive value evaluated at the onset conditions for each ramp:

$Q_{ND} = Q_{up} / Q_{ref}$

$Q_{ref} = K_{oil} \Delta T_{crit} A/L$

where Q_{ND} is the non-dimensional heat flow, Q_{up} is the heat flow measured in the thin foil upper flow-meter, Q_{ref} is the reference heat flow, K_{oil} is the Silicon Oil thermal conductivity, A and L are respectively the cross area and the length of the liquid bridge and ΔT_{crit} is the temperature difference across the liquid bridge measured at the oscillation onset. The time indicated in the figures is referred to the beginning of each run.

An evaluation on the unsteadiness of the phenomena, from a thermal point of view, can be made by evaluating the power through the disk necessary to reach a constant temperature ramp and the net heat flow measured through the disk (Q_{ex}; experimental data). The first value can be derived under the assumption that the average liquid temperature follows 1/2 of the ramp imposed on the disk:

$$U = Q_a/Q_{ex} = \frac{\frac{r}{2}\frac{L}{2}\pi R^2 \rho c}{Q_{ex}}$$

where r is the temperature ramp set, L is the zone length, R is the disc radius, ρ and c respectively are the density and the heat capacity of the liquid.If the value of U<<1 then the thermal field is steady; vice versa, when U>>1 it is unsteady The values for U obtained from the experiments range between 0.5 and 2; it means that the heat flow transferred to the liquid is comparable with the energy required to heat the liquid with the assigned ramp. This result indicates that the phenomenon has to be considered as unsteady from a thermal point of view, because the fluid store a large amount of energy and is not able to transfer, rapidly, the heat by Marangoni convection mechanism.

The data collected during these flights show that the critical Marangoni number (P_{em}) values were generally higher than the ones obtained on ground. In Fig. 5 the critical Marangoni numbers obtained in the Teletexus experiment and in other investigations performed in micro-g and in 1-g conditions are shown. The large dispersion of the points suggests that the Pem number is not the characteristic number for the transition to the oscillatory regime and this may support the new proposed criteria to describe the oscillation onset [12,13] based on the dynamic Weber number.

E. Order of Magnitude Analysis (OMA).

In order to evaluate the influence of different factors on the experiment results the Order of Magnitude Analysis can be used. Following this approach it is possible:

a) to study the influence of unsteadiness on the process.In particular it is necessary to evaluate the ratio between the characteristic time of the boundary condition changes and those related to the thermo-fluid-dynamic phenomena.

b) to evaluate the accuracy of the numerical simulation results.

c) to evaluate the adequate grid spacing.

The value obtained allows the identification of the flow regime and the evaluation of the order of magnitude of the maximum velocity.

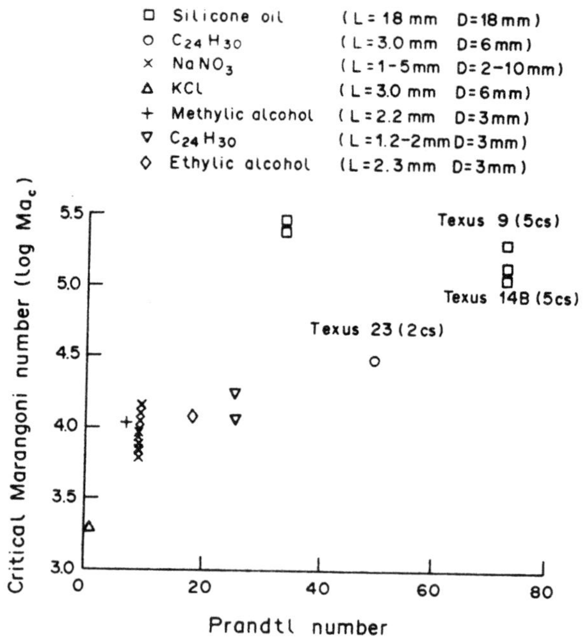

Fig. 5 - Critical Marangoni Number at the onset of oscillatory thermocapillary flow

It is important to point out that this approach is useful to characterize the process "a priori" on the basis of the geometric and thermo-fluid-dynamic parameters. The OMA (Napolitano method) theory can be deeply investigate in ref. [14,15]. In this context the method is applied and the results of the theoretical analysis are compared with the experimental data.

The first step, in general, is to evaluate the parameters that drive the Marangoni convection. Starting from the non-dimensional form of the system of balance equations of momentum, energy and species that describe the phenomena, it is possible to evaluate which is the parameter that drives the process and that has to be considered as a reference.

The non-dimensional weights of the balance equations terms are obtained by comparing the dimensional groups with the diffusive dimensional term in each equation. The ratios of these terms forms the characteristic numbers for our study: Prandtl number (Pr) and Schmidt number (Sc).

In the Marangoni convection experiment, the species equation is not considered (silicone oil is the only species present). Since the Prandtl is relatively high ($Pr=30\div50$), the OMA theory can be based on the assumption that the diffusive term used for the non-dimensionalization is the thermal diffusion. Under this assumption the energy balance equation describes the driving phenomena. The comparison between the experimental and OMA results should confirm the validity of the theory. If confirmed, the OMA theory represents an useful tool to predict "a priori" the experimental behaviour and can be used as a guide for the definition of the future researches.

In these experiments a reference parameter is the time t_r that has to be obtained from the boundary conditions. The experiment is characterized by a gradual variation in the temperature boundary condition. The temperature ramp is assigned, therefore the reference time is obtained from:

$$t_r = \frac{\Delta T}{\text{Temperature ramp}}$$

where ΔT is measured between the two disks and represents the maximum value reached during a run. The value could be modified "a posteriori" using the ΔT measured at the oscillation onset. The order of magnitude for the seven t_r related to the oscillation onsets are respectively:

$$t_{r1} = 50\ [s];\ t_{r2} = 70\ [s];\ t_{r3} = 70\ [s]$$
$$t_{r4\text{-}5} = 30\ [s];\ t_{r6} = 50\ [s];\ t_{r7} = 50\ [s]$$

The characteristic times, considered thermo-fluid-dynamic parameters of the problems, belong to three different groups. The first group is obtained using a time related to gravity convection given by:

$$t_g = \left(\frac{L}{\beta_t\,\Delta T\,g_r}\right)^{1/2}$$

where g_r is the g-level, β_t is the thermal expansion coefficient and L is the reference length (the liquid-bridge length). In the present experiment, considering that on-board the rocket the g-level achieved is 10^{-4}, the values of t_g obtained in the experiments are in the following range:

$$t_{g1\div7} = 5000 \div 6500\ [s]$$

The second group is related to Marangoni convection and is obtained from:

$$t_M = \frac{L}{V_M} = \frac{L\,\mu}{\sigma_T\,\Delta T}$$

where V_M is the Marangoni velocity, μ is the dynamic viscosity and σ_T is the surface tension gradient.This time is based on the Marangoni velocity evaluated at the critical conditions. The real convective time could be of a different order of magnitude and has to be obtained using a reference velocity obtained from OMA.

The related values are:

$$t_{M1} = 0{,}025\ [s];\ \ t_{M2} = 0.045\ [s];$$
$$t_{M3} = 0.039\ [s];\ t_{M4\text{-}5} = 0.018\ [s];$$
$$t_{M6} = 0.020\ [s];\ \ t_{M7} = 0.019\ [s];$$

The last group is formed by the diffusive terms. Since the energy balance equation contains the terms that drive the phenomena, the diffusive time is indicated as:

$$t_D = \frac{L^2}{\alpha}$$

In this equation α is the thermal diffusivity of silicone oil. The value obtained from the experiments is:

$$t_D = 5143\ s$$

Comparing t_r con t_g t_M e t_D we have the following two possibilities:

$t_r < \min(t_g,t_M,t_D)$	unsteady
$t_r > \min(t_g,t_M,t_D)$	quasi-steady.

From the analysis of the data obtained the second inequality is verified.

Referring to the quasi-steady conditions there are two different possibilities:

$t_D < (t_r,\ t_M)$	quasi-steady diffusive
$t_M < (t_r,\ t_D)$	quasi-steady with Marangoni effect

In the experiment performed the value of t_M satisfies the last inequality. It is important to consider that due to the very low value of t_M, the phenomenon has to be considered quasi-steady also for high values of the temperature ramp. This result implies that all experimental runs, performed using different temperature ramps, can be considered similar and compared among them.

The successive step in the non-dimensional analysis consists in the evaluation of the fluid reference velocity and of the scale factors of the phenomena. The theory reported in ref.[14] and [15] allows to combine in the same expression the velocity and the scale factor. In the present case it is possible to identify two possible velocities: one related to the thermo-capillary convection (Marangoni velocity) and the other to the buoyancy forces. The Marangoni velocity is given by:

$$V_M = \frac{\sigma_T \Delta T}{\mu} l$$

where l is the non-dimensional scale factor.
The buoyancy velocity is evaluated using:

$$V_g = \frac{g \beta_t \Delta T L^2}{\mu} l^2$$

The reference velocity, included in the definition of Marangoni and Grashof numbers, is selected between the above characteristic values according to:

$$V_r = \max (V_M l, V_g l^2) \qquad (A)$$

This approach ensures that the reference non-dimensional number be of the order of 1:

$$\max(Ma\, l, Gr\, l^2) = O(1)$$

To evaluate the scale factor other conditions are needed.
It is necessary to verify that:

$$St_D l^2 \leq O(1) \qquad (B)$$

and:

$$Pe_M l^2 \leq O(1) \qquad (C)$$

with the following condition:

$$l \leq O(1)$$

The first parameter (where the number of Sthroual appears) is equivalent to the ratio:

$$\frac{t_D}{t_r} l^2$$

and measures the effects of diffusion respect to unsteadiness.The second parameter (containing the Peclet Number) is equivalent to the ratio:

$$\frac{t_D}{t_{con}} l^2$$

and compares the effect of diffusion respect to the convection.
The scale factor shall keep these ratios less than one and can be used to determine the region where the diffusive phenomena are of the same order of magnitude as the convective or unsteady (Stokes regime if l = 1; boundary layer l << 1).
From (A) we obtain:

$$V_r = V_M l$$

and:

$$t_{conv} = \frac{L}{V_r} = \frac{L}{V_M l} = \frac{t_M}{l}$$

the ratio $\frac{t_D}{t_{con}} l^2$ becomes:

$$\frac{t_D}{t_M} l^3 \qquad (D)$$

Combining (B) e la (D) it is possible obtain a new non-dimensional parameter where l (unknown) does not appear:

$$\frac{(St_D l^2)^3}{(Pe_M l^3)^2} = \frac{St_D^3}{Pe_M^2} = \frac{t_D^3 t_M^2}{t_r^3 t_D^2} = \frac{t_D t_M^2}{t_r^3} = F$$

The value of F can be determined on the basis of the planned ramp. The values obtained are:

$F_1 = 2.7\ 10^{-5}$; $F_2 = 2.8\ 10^{-5}$;
$F_3 = 2.7\ 10^{-5}$; $F_{4-5} = 7.8\ 10^{-5}$;
$F_6 = 1.3\ 10^{-6}$; $F_4 = 1.2\ 10^{-5}$

It is important to determine the scale factor l.
If $F \leq O(1)$ then $St_D l^2 \leq O(1)$ and the (B) is satisfied; the condition $Pe_M l^3 = O(1)$ implies the verification of (C) :

if $Pe_M \leq O(1)$ then $l = 1 \Rightarrow$ case 1

if $Pe_M > O(1)$ then $l = Pe_M^{-1/3} \Rightarrow$ case 2

The case 1 means that the convective and diffusive phenomena are of the same order of magnitude in the entire field.
In the case 2 the two effects are comparable only in a limited volume (boundary layer).
From the analysis of the value of F our experiment belongs to the second case
In a more general case also the other hypotheses have to be evaluated: if $F > O(1)$ the dominant phenomena is the unsteadiness implying that the inequality $Pe_M l^3 \leq O(1)$ is satisfied (in that case we impose $St_D l^2 = O(1)$):
If $St_D l^2 \leq O(1)$ than $l = 1$. It means that the variation in the boundary condition is balanced by diffusive phenomena (the fluid is adapting to the changes);
If $St_D l^2 > O(1)$ than $l = St_D^{-1/2}$; The fast changes in the boundary conditions ($t_r << t_D$ e $t_r << (t_D t_M^2)^{1/3}$) affect a limited zone in the fluid field.
The Peclet numbers related to the seven ramps are:

$Pe_{M1} = 200000$; $Pe_{M2} = 120000$;
$Pe_{M3} = 140000$; $Pe_{M4-5} = 280000$
$Pe_{M6} = 240000$; $Pe_{M7} = 270000$

therefore:

$l_1 = 0.017$; $l_2 = 0.020$; $l_3 = 0.019$
$l_{4-5} = 0.015$; $l_6 = 0.016$; $l_7 = 0.015$

that implies the existence of a boundary layers.
The values of the reference velocities $V_r = V_M l$ are:

$$V_{r1} = V_{M1}\, l_1 = 12 \text{ [mm/s]}$$
$$V_{r2} = V_{M2}\, l_2 = 8 \text{ [mm/s]}$$
$$V_{r3} = V_{M3}\, l_3 = 9 \text{[mm/s]}$$
$$V_{r4-5} = V_{M4-5}\, l_{4-5} = 15 \text{[mm/s]}$$
$$V_{r6} = V_{M6}\, l_6 = 13 \text{ [mm/s]}$$
$$V_{r7} = V_{M7}\, l_7 = 14 \text{ [mm/s]}$$

From the analysis of the experimental data it is possible to observe that, close to the oscillation onset, the liquid velocity is of the same order

(velocities are ranging between 8 and 10 [mm/s]). Another point to consider is that the maximum speed is achieved on the free surface where the driving effect is active, but, due to the cylindrical-lens effect, this zone cannot be observed. Therefore the velocities obtained from the OMA analysis are in a a good agreement with experimental velocities that demonstrate the validity of the approach.
Using the reference velocity it is now possible to evaluate the convective time obtained from:

$$t_C = \frac{L}{V_r} = \frac{L\,\mu}{\sigma_T\,\Delta T\,l}$$

The t_C obtained for the experiments is in the range 1÷3 that differs of one order of magnitude from the reference time related to the gradual variation in the temperature boundary condition, measured above. This difference is valid at the critical conditions, meanwhile in the first phase of the run, when ΔT is less than 10 [°C], t_r and t_C are comparable. This result implies a dependence of the oscillation onset from the imposed temperature ramp that, predicted by OMA, has been found in the experimental results and require further theoretical and experimental analysis (see fig. 6).

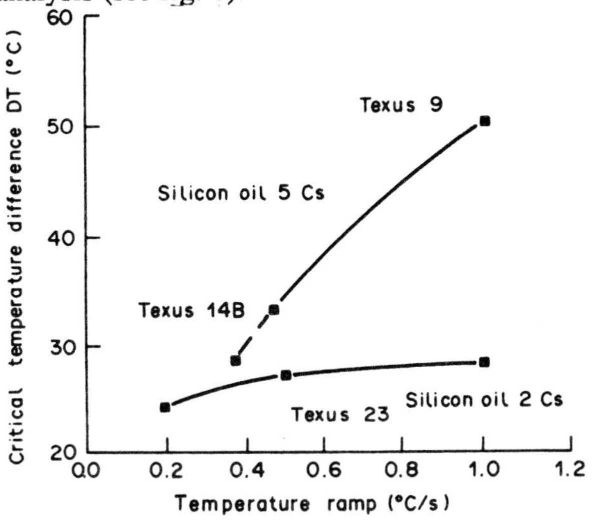

Fig. 6 - Critical temperature difference vs temperature ramp imposed.

The reference velocities can be used to evaluate the Reynolds numbers (V_rL/ν) reached at the oscillatory onset:

$Re_{r1} = 43$ $Re_{r2} = 30$ $Re_{r3} = 33$
$Re_{r4\text{-}5} = 135$ $Re_{r6} = 122$ $Re_{r7} = 132$

These values are used to define the envelope of the thermo-fluid-dynamic regimes simulated during the experiment in the Pem-Gr normogram illustrated in fig.7. In this figure it is possible observe that the flow regimes established in the onset experiments are located in the area where Marangoni and boundary layers co-exist.
Referring to the thickness of the thermal boundary layer L_M from ref.[12] [15]:

$$L_M = L\,Pe_M^{-(1/3)} = L\,1$$

from the ramps we obtain:

$$L_{M1\text{-}7} = 0{,}20 \div 0{,}30 \text{ mm};$$

Such a small value requires the generation of a really fine grid close to the boundary for the numerical simulation An adequate grid step for the simulation has to be 0,02÷ 0,04[mm].

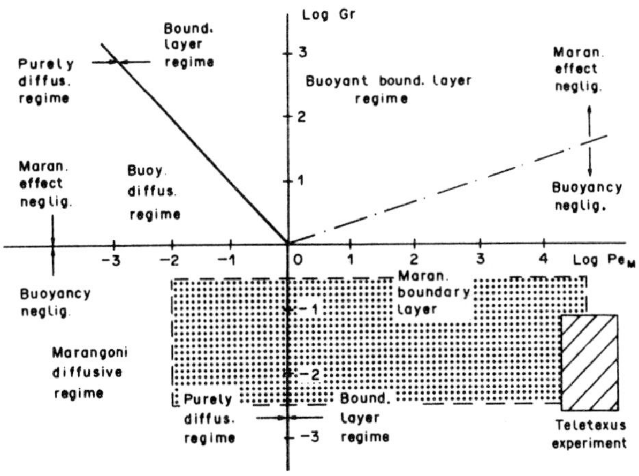

Fig. 7 - Oscillatory Marangoni Flow Experiments regimes in the plane Pem - Gr

5. CONCLUSIONS

The large amount of data obtained during the flights requires a particular effort in the numerical correlation for velocity and temperature fields and for the integrated heat flows through the end disks during the unsteady process induced by Marangoni flow. The OMA applied to this phenomena, has to be considered as a valid tool to predict the behaviour of the fluid and to define the characteristic parameters for the future experiments. The high Peclet (or Marangoni) numbers obtained and the relatively high Reynolds numbers attained in the liquid bridge is responsible for the presence of complex thermal and kinetic boundary layers on the solid disks and at the surface (Marangoni boundary layer) that together with the corner regions require very fine grids in the numerical simulation. Furthermore the unsteadiness of the process also suggests small time steps in the equation integration. Future plans of activities include the computer simulation of the experiment performed by other investigators and other microgravity experimentations to identify the onset conditions.

AKNOWLEDGMENTS

The authors wish to thank Dr. G. Desiderio, for the support given in the numerical computation, and the staff of Texus Program for the highly appreciated collaboration.
The research was supported by the Italian Space Agency (ASI) and by ESA.

REFERENCES

[1] L.G. NAPOLITANO, R. MONTI, G. RUSSO: "Experimental Study of Thermal Marangoni Flow in Silicon Oil Floating Zones"; *34th IAF Congress*, Budapest, September 1983 published on *Acta Astronautica **Vol II n° 7/8**,* pp 369-378, 1984.

[2] L.G. NAPOLITANO, R. MONTI: "Surface Driven Flow: recent Theoretical and Experimental Results"; *6th European Symposium on Material Sciences under Microgravity*; Bordeaux; 2-5 December 1986.

[3] L.G. NAPOLITANO, R. MONTI, G. RUSSO, C. GOLIA: "Comparison between D-1 spaceborn experiment and Numerical/ground experimental work on Marangoni Flow"; *6th European Symposium on Material Sciences under Microgravity*; Bordeaux; 2-5 December 1986

[4] R. MONTI, L.G. NAPOLITANO, G. MANNARA: "Texus flight results on convective flows and heat transfer in simulated floating-zones"; *Proceedings of the 5th Symposium on Material Sciences under Microgravity* - Schloss Elmau, November 1988.

[5] R. MONTI, G. MANNARA:: "Texus experiment on the convective heat transfer induced by Marangoni flows"; *Acta Astronautica* - ***Vol. 12, N° 7/8,*** 1986.

[6] R. MONTI, R. FORTEZZA, G. MANNARA: "Results of the TEXUS 14-B Flights Experiments on a Floating Zone-First approach towards Telesciences in Fluid Science", *XXXVIII IAF Congress - Brighton*, 10-17 October 1987.

[7] R. MONTI, R. FORTEZZA: "Experimental Results on Microgravitational Fluid-dynamics During Texus Program"; *IX AIDAA Congress* - Palermo, 26-28 October 1987.

[8] R. MONTI, R. FORTEZZA: "TELETEXUS Experiment: preliminary experience for the Columbus Programme"; *Columbus V - Symp. on Space Station Utilization* ; Capri, 3-7 July 1989 published on Space Technology ***Vol. 10 N° 1/2***, 1990

[9] R. MONTI, R. FORTEZZA: "Oscillatory Marangoni Flow in a Floating Zone: Design of a Telescience experiment for Texus 23"; *VII European Symposium on Material and Fluid Sciences in Microgravity*; Oxford, 10-15 September 1989 published on ESA SP-295

[10] R. MONTI, R. FORTEZZA: "TELETE-XUS": The Technical and Operational aspects of a Microgravity Experiment in Telescience"; *XXXX Congresso IAF*; Torremolinos, October 7-13, 1989

[11] R. MONTI, R. FORTEZZA: "The Scientific Results of the experiment on oscillatori Marangoni flow performed during Texus 23 " *Microgravity Quaterly, **Vol.1, No3*** - 1991.

[12] R.MONTI: "On the onset of Oscillatory regimes in Marangoni Flows"; *Transaction Note, Acta Astronautica **Vol 15 N°8** pp.557-560,* 1987.

[13] R.MONTI, R.FORTEZZA: "Thermo-capillary Driven Flow in Floating Zone"; *XXVII COSPAR Congress*; The Hague, 25 June - 6 July 1990

[14] L. G. NAPOLITANO: "Order of magnitude analysis of unsteady Marangoni and buoyancy free convection."; *XXXV IAF Congress* - Losanna, October 1984.

[15] L. G. NAPOLITANO: "Surface and buoyancy driven free convection."; *Acta Astronautica **Vol. 9** (1982).*

Table 1 - Comparison between numerical and experimental results

r [mm]	z [mm]	Velocity Experimental [mm/s]	Velocity Numerical [mm/s]		r [mm]	z [mm]	Temperature [°C] Experimental [°C]	Temperature [°C] Numerical [°C]
0.35	0.39	2.4	2.6		0.23	0.2	29.0	29.6
0.47	0.32	2.9	2.8		0.23	0.6	29.5	31.1
0.45	0.84	5.6	5.4		0.23	1.2	31.5	31.3
0.58	0.97	7.6	6.8		0.23	1.6	29.9	29.2
0.84	0.97	5.8	5.4					
0.58	1.37	6.0	5.9					
0.34	1.47	6.4	6.1					
0.16	1.68	5.6	5.0					

Hydrodynamics and Hydrostatics of Non-Uniform Media in Microgravity

SOME PROBLEMS OF STABILITY OF ZERO-g LIQUID BRIDGES

L.A. Slobozhanin

Institute for Low Temperature Physics and Engineering.
Ukrainian Academy of Sciences, 47, Lenin Avenue, Kharkov, 310164, Ukraine

ABSTRACT

Some problems of stability are considered for a weightless axisymmetric liquid bridge between two coaxial equal-radius cylindrical rods. Five types of resting bridges are investigated which are characterized by their particular situation of each of two wetting lines on the edge, on the flat end face or on the lateral surface of the rods. The stability of a rotating bridge is analyzed in a problem on simulation of the process of purification of materials and single crystals growing by the floating zone method.

The final results are represented as boundaries of the stability region in the space of the parameters characterizing the equilibrium of liquid bridge.

Key words: *weightlessness, liquid bridge, floating zone, stability, edge of solid.*

1. INTRODUCTION

In space technology there arises the problem of preservation of a weightless liquid bridge between two coaxial cylindrical rods with its axial symmetry of shape maintained. It will be assumed below that the rods have the same radius ξ_0 and their end faces are flat and perpendicular to their axis.

Introduce a cylindrical system of coordinates (ξ, θ, ζ) with the origin at the centre of one of the wetting lines and the axis ζ directed towards the liquid. The length τ of the arc will be measured along the profile (the axial cross-section θ = *const*) of the free surface from the initial point $\tau = 0$ for which $\zeta = 0$ to the end point $\tau = \tau_1$. Denote by v the volume of the liguid, by $\beta(\tau)$ the angle of incli nation of the tangent line at points of the profile, and by h the distanse between the wetting lines γ_0 and γ_1 (Figure 1).

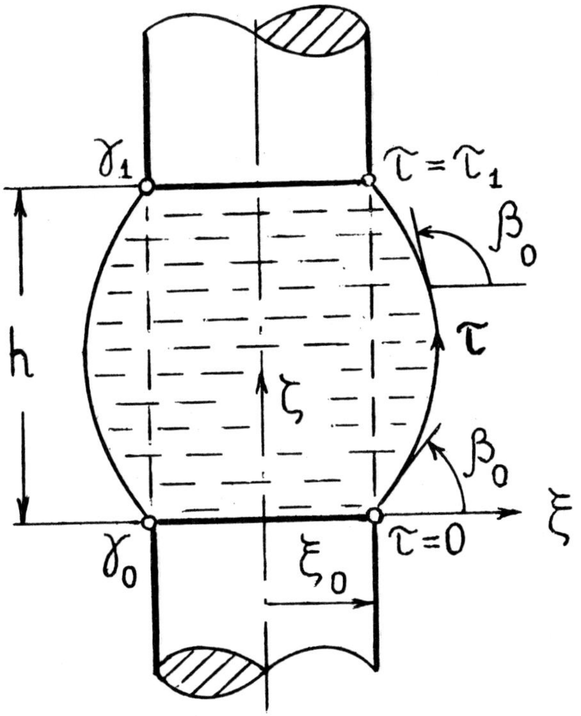

Figure 1: Geometry and coordinate system for the basic type of the liquid bridges with wetting lines on the edges of the rods.

2. STABILITY OF THE LIQUID BRIDGE WITH THE WETTING LINES ON THE EDGES OF THE RODS

Let us first consider the problem of stability of an axisymmetric bridge for the case where the lines γ_0 and γ_1 coincide with the edges of the rods (Figure 1).

We shall first discuss the perturbations upon which both the wetting lines remain fixed. Stability under axisymmetric perturbations of this type was investigated in Ref. 1 and that under arbitrary perturbations in Ref. 2 (see also Ref. 3). It was concluded that in this problem only "convex" ($\xi(\tau) > \xi_0$ for $0 < \tau < \tau_1$) or "concave" ($\xi(\tau) < \xi_0$ for $0 < \tau < \tau_1$) surfaces may be stable. These are symmetric with respect to the equatorial plane $\zeta = h/2$. Figure 2 shows the boundary of the region of stable equilibrium states constructed by calculation in the coordinates $\eta = h/\xi_0$ and $V = v / \pi\xi_0^2 h$. The stable region is unbounded, viz. along the non-intercrossing branches Am and Fn, $V \longrightarrow \infty$ as $\eta \longrightarrow \infty$.

It is seen that for the prescribed values of ξ_0 and h there exist a range of v values, where the equilibrium is stable. The convex ($V > 1$) equilibrium surface corresponds to the maximum v, and the concave ($V < 1$) one for $\eta < 2\pi$ and the convex one for $\eta > 2\pi$ correspond to the minimum v. The only exception are ξ_0 and h for which $0.722 < \eta < 0.809$ (between the abscissae of the points C and B). For these ξ_0 and h, there are two ranges of stable volume values. In the second range, all equilibrium surfaces are concave.

The convex profiles of the limiting surfaces rep resented by the branch of Am, and the concave profiles of the limit surfaces corresponding to the segment ABC are bounded by points with horizontal tangents. Such surfaces are critical under nonaxisymmetric perturbations. Up to

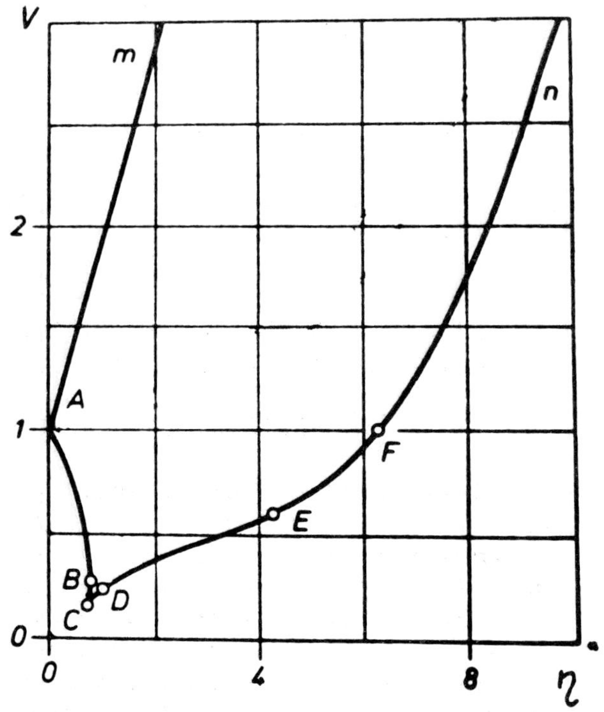

Figure 2: Boundary of the region of stable equilibrium of liquid bridges under perturbations for which the wetting lines remain on the rod enges.

a small multiplier the normal component of these perturbations is $N = \zeta'(\tau)\cos\theta$. Loss of stability under nonaxisymmetric perturbations can result either in breakage of the liquid zone, or in its transformation into a stable but nonaxisymmetric equilibrium state. To solve this alternative, one is to study into the problem of bifurcation of critical axisymmetric shapes.

The surfaces corresponding to the points of $CDEFn$ are critical with respect to axisymmetric perturbations. The profiles corresponding to the segment CDE do not contain any characteristic points. Hazardous here are axisymmetric perturbations symmetric with respect to the equatorial plane. It is characteristic of the profiles of the surfaces corresponding to the branch EFn ($\eta > 4.260$) that $\beta(0) = \beta(\tau_1) = \pi/2$. Here hazard is presented by perturbations of the form $N = \xi'(\tau)$. The points D ($\eta = 0.944$, $V = 0.240$) and F ($\eta =$

2π, $V = 1$) are determined by the critical surfaces of the catenoid and the cylinder.

Note that loss of stability under axisymmetric perturbations leads to breakage into equal parts when the segment *CDE* is crossed and into unequal parts when the segment *EFn* is crossed. The dynamics of breakage of the bridge, its volume becoming smaller than the minimum, is discussed in Refs. 4 and 5. Note that the lower boundary of stability of the liquid zone between discs of different radii was constructed in Refs. 6,7.

Are the perturbations, under which the contact lines remain unmoved, the most hazardous ? It is known (Refs. 3, 8) that the necessary condition of stability of equilibrium of the liquid in the case, where the contact line γ runs along the edge of the solid, consists in the inequalities

$$\psi \geq \alpha, \qquad \psi_o \geq \alpha_o \tag{1}$$

Here ψ and ψ_o are the dihedral angles formed by the liquid and the gas at γ, α and α_o the respective boundary angles at the contact with the smooth solid surface ($\alpha + \alpha_o = \pi$). If both the inequalities (1) are rigorous, then indeed the most hazardous are perturbations under which the line γ remains fixed. If the dihedral angle formed by one of the media is smaller than the corresponding boundary angle, then equilibrium is unstable under perturbations displacing γ towards this medium.

Fulfillment of the conditions (1) depends on the shape of the solid and the angle α. For a bridge whose free surface rests on the edges of the discs, the above constructed boundary is the boundary of stability under arbitrary perturbations in the case of $\alpha = 0$ (experimentally corroborated in Refs. 9 and 10), but for $\alpha \neq 0$ the lower boundary should be reconstructed (experiments of Ref. 11), as will be specified below.

If the bridge rests on the edges of the cylindrical rods, then, in view of the evident relations

$$\psi_o = \pi/2 + \beta_o, \qquad \psi = \pi - \beta_o$$

(β_o is the value of the angle β at the point $\tau = 0$), inequalities (1) are transformed into the forms

$$\pi/2 - \alpha \leq \beta_o \leq \pi - \alpha \tag{2}$$

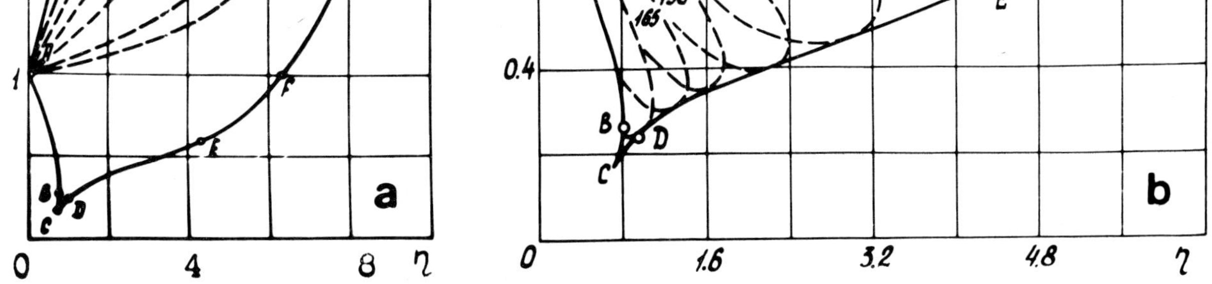

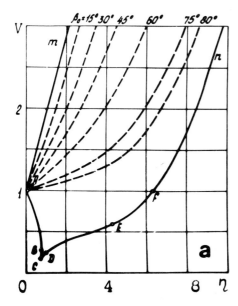

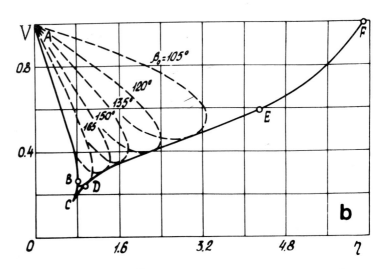

Figure 3: Lines corresponding to the states with β_o = *const* in the regions $V > 1$ (*a*) and $V < 1$ (*b*). These lines enable finding the region of stability for the bridges shown on Figure 1 under arbitrary perturbations

In the plane (η, V), the lines representing the states for which β_o = $const$, $0<\beta_o<\pi/2$, lie in the region $V > 1$, never intersect with the earlier constructed boundary of the stable region, and along these, $V \longrightarrow \infty$ as $\eta \longrightarrow \infty$ (Figure 3*a*). The lines corresponding to the states with β_o = const , $\pi/2 < \beta_o < \pi$, lie below the line $V = 1$ and are tangent to the branch *CDEFn* in the segment *CDE* (Figure 3*b*). For the states corresponding to the segment *AF* and the branch *EFn* , $\beta_o = \pi/2$. Since along the boundary *Am*, $\beta_o = 0$, and along the segment *ABC* , $\beta_o = \pi$, then the conditions (2) will be fulfilled, for $\alpha < \pi/2$, if the branch *Am* is replaced by the line $\beta_o = \pi/2-\alpha$ and the segment *ABC* by the segment of the line $\beta_o = \pi - \alpha$ bounded by the point *A* and the point of tangency. In the case of $\alpha \geq \pi/2$, only the lower boundary is replaced by the line $\beta_o = \pi - \alpha$. Such reconstruction enables finding the sought-for boundary of the region of stability under arbitrary perturbations.

3. OTHER TYPES OF AXISYMMETRIC BRIDGES

What can be the result of a disturbance of stability of the axisymmetric equilibrium state of the liquid with the wetting lines at the edges of the rods ? We have mentioned above a possibility of breakage of the liquid zone or of its transition to the stable nonaxisymmetric state. But if the conditions (2) are not fulfilled, the following new axisymmetric states of the zone can be distinguished with consideration for perturbations displacing the contact line to the smooth part of the solid surface, in which:

A) each of the wetting lines is situated on the smooth flat surface of the end face of the respective rod (Figure 4*a*);

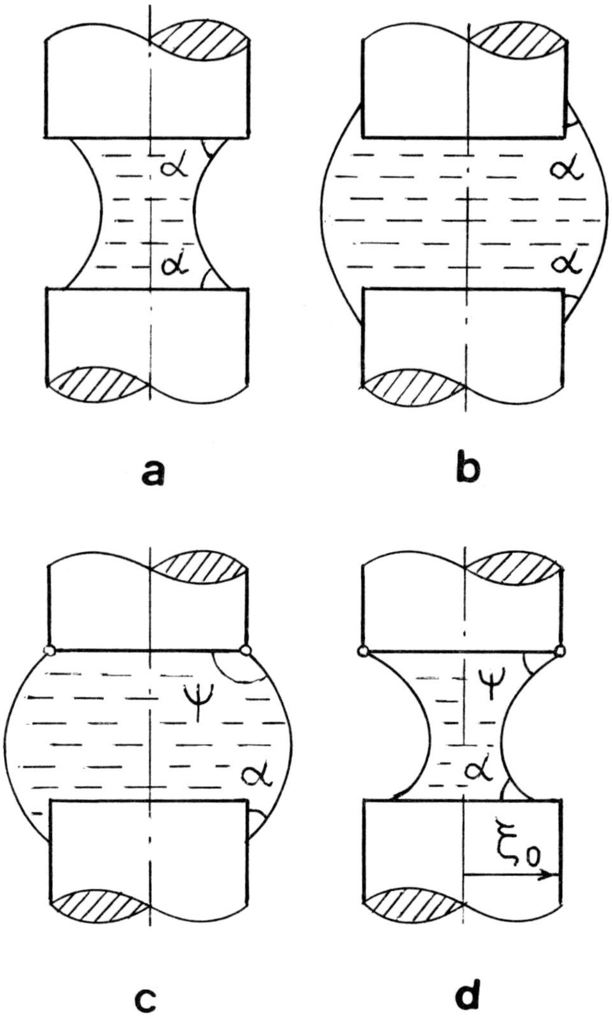

Figure 4: Sketch of other types of bridges between two cylindrical rods.

B) each of the wetting lines is situated on the lateral face of the respective cylindrical rod (Figure 4*b*);

C) one of the wetting lines is situated along the edge of the rod, and the other one on the lateral face of the other rod (Figure 4*c*);

D) one of the wetting lines is situated along the edge of the rod, and the other one on the smooth flat end face of the other rod (Figure 4*d*).

Some of these states have been observed experimentally (Ref. 12). Below we present the results of the investigation of their stability.

In the case A, the characteristic parameters of equilibrium are the angle α and the dimensionless volume v / h^3 .

Let us disregard perturbations which can displace the bridge as a whole in the direction parallel to the planes of the ends. Then, the region of stability lies above the boundary represented in Figure 5 (Refs. 2,3).

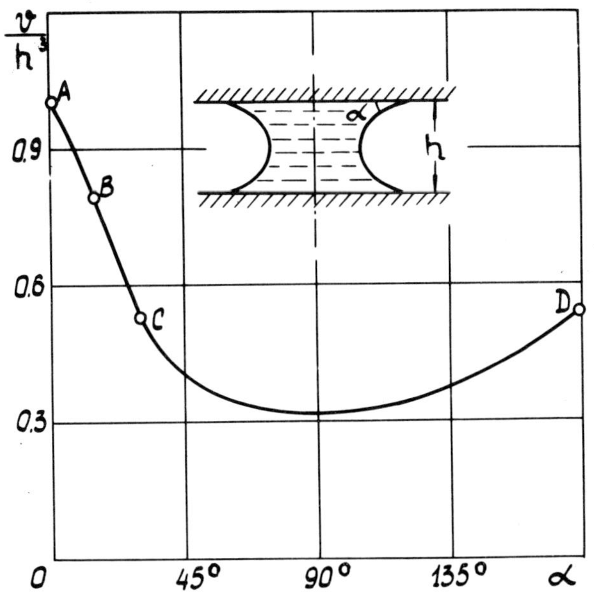

Figure 5: Boundary of the stability region for type *A* liquid bridges.

The profiles of stable surfaces do not contain inflection points. Loss of stability is caused by axisymmetric perturbations. At the segment *ABC* such perturbations are symmetric with respect to the equatorial plane (the point *B* corresponds to the critical catenoid). At the section *CD* they are antisymmetric and are given by $N = \xi'(\tau)$, and in the profiles of the respective critical surfaces the terminal points are inflection points. Hence, for $\alpha < \alpha_c$ ($\alpha_c = 31^0$) the bridge is divided into two equal parts, and for $\alpha > \alpha_c$, into unequal parts. For $\alpha = 0, 90^0$ and 180^0 the critical values of the parameter v/h^3 are 1, $1/\pi$ and $\pi/6$, respectively.

In the case B, the equilibrium states are characterized by the angle α and the dimensionless volume v_1/ξ_0^3 (v_1 is the volume of the region between the free and cylindrical surfaces). Let us disregard perturbations under which the free surface is displaced as a whole along the symmetry axis. Then, the stability region is bounded from below by the curve *b* shown in Figure 6 borrowed from Ref. 2. This boundary

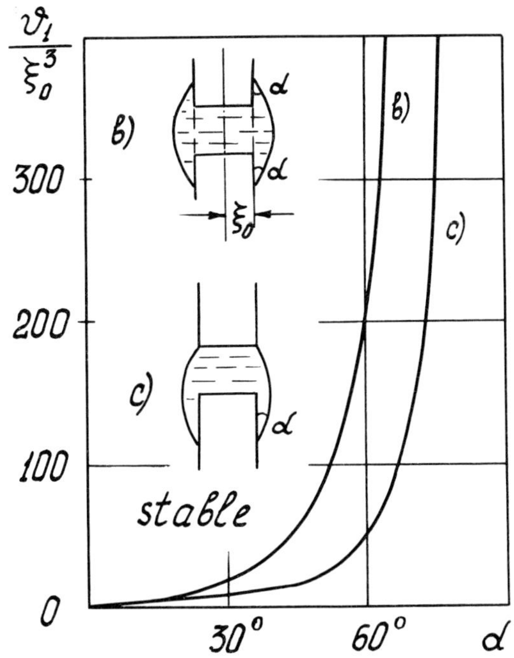

Figure 6: Dependences of the critical values of the dimensionless volume v_1/ξ_0^3 on the angle α for *B* and *C* types of liquid bridges (curves *b* and *c*, respectively).

asymptotically tends to the straight line $\alpha=90^0$. The profiles of the stable surfaces contain the inflection points within them. For the critical states, the terminal points of the profiles coincide with the inflection points, and stability is lost owing to nonaxisymmetric perturbations $N=\zeta'(\tau) \cos\theta$, which must cause the liquid, if in small quantity, to become asymmetric.

In the case C, the equilibrium parameters are the same as in the case *B*. The stability region is bounded by the curve *c* (Figure 6). It also bounds the stability region from below and asymptotically tends to the straight line $\alpha = 90^0$. Stability is also lost under nonaxisymmetric perturbations, but their structure is more complicated than in the case *B*, and the critical profiles do not have any characteristic features. Generally, the condition of the wetting line being fixed on the edge of the solid is stabilizing factor, and therefore the stability range is in this case wider than in B. Figure 7 shows the dependences of the elongation η in the critical state

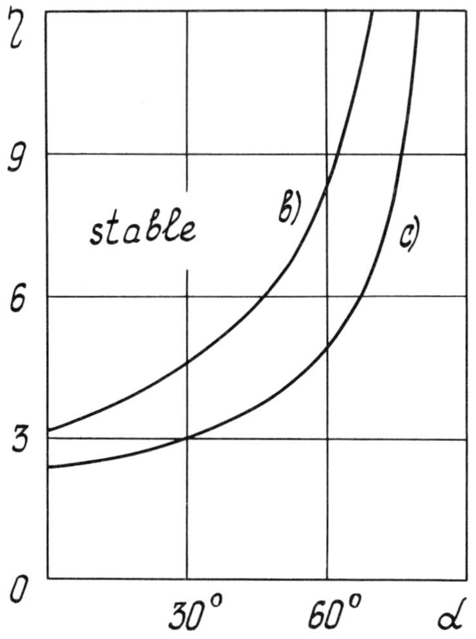

Figure 7: Dependences of the critical values of the elongation η on angle α for B and C types of liquid bridges (curves b and c, respectively).

on α for B and C. For stable states the elongation is larger than critical. It is worthwhile to compare the critical elongation with that calculated from the distance between the ends of the rods.

In the case D, the equilibrium system is specified by the three parameters: η, V and α. First, we shall assume the end face to be unbounded, i.e. not a circle of a radius ξ_0 but an unbounded plane. The boundaries of stability regions in the coordinates η and V for $\alpha = 0$, 30^0, 60^0, 90^0, 135^0 and 180^0 are shown by solid lines in Figure 8. The dashed line shows the boundary of the stability region represented also in Figure 2.

As earlier, all stability regions

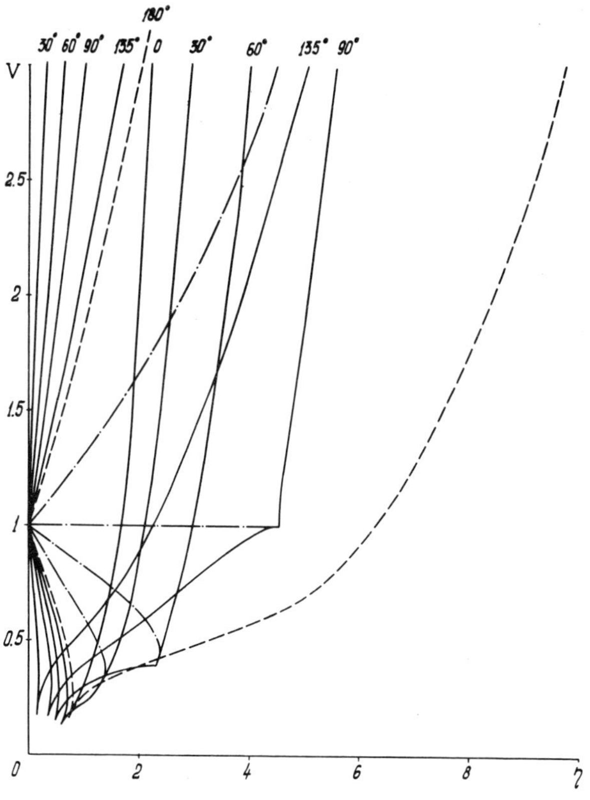

Figure 8: Stability regions of type D liquid bridges for some α values.

are unbounded. For $\alpha\neq 0$ and $\alpha\neq 180$ the following is typical. The upper branch of the boundary corresponds to equilibrium states for which the tangent line to the profile at the end point τ_1 lying on the edge of the rod is horizontal, so that $\beta(\tau_1)=180^0$. The left part of the lower boundary, up to the inflection point, corresponds to the states, for which the tangent to the profile at the end point, is also horizontal, but here $\beta(\tau_1)=0$. At these two parts, hazardous are nonaxisymmetric perturbations $N=\zeta'(\tau)\cdot\cos\theta$. At the right part of the lower boundary, near the inflection point, hazardous are axisymmetric perturbations. Note that in contrast to all the above problems, the free surfaces in this problem have not the equatorial plane of symmetry.

For $\alpha=0$, there is no upper boundary, and the left part of the lower boundary coincides with the respective part of the dashed line. On the contrary, when $\alpha=180^0$, all the stability region is bounded from above by the upper branch of the dashed line.

In the case D, area of the liquid bridge contact with the end of the rod cannot have a radius larger than ξ_0. This condition requires construction of curves representing states with the radius of contact area equal to ξ_0. Such curves shown for every α by dash-dot line are boundaries of real regions of stability of the bridge with the wetting line situated on the end face of the rod of the radius ξ_0. For $\alpha=0$, this dash-dot line coincides with the left part of the lower boundary, and for $\alpha=180^0$, with the whole boundary. But, while for $\alpha=0$, this means that there are no stable states satisfying the said restriction, then for $\alpha=180^0$, all the stable states satisfy this restriction.

4. STABILITY OF A RESTING AND A ROTATING WEIGHTLESS BRIDGE DURING ZONE MELTING

Let us come back to the problem of stability of a liquid bridge with fixed wetting lines situated on the edges of two coaxial cylindrical rods. Consider it as applied to purification of materials and growing of single crystals by the floating zone method. In zone melting, the angle β_0 is fixed at one of the ends which is the solidification front. The value $\beta_0=\pi/2 - \mu$, where μ is the so-called growth angle. The magnitude of μ is close to zero, and for real semiconducting materials (Si, Ge) is 10^0-15^0.

It follows from the results of section 2 that a resting weightless liquid zone at $\mu=0$ is stable only if it is cylindrical and $\eta < 2\pi$, and at $0 < \mu < 90^0$ it is stable if it is convex (stable for any η).

Let now the weightless zone of a density ρ and cylindrical rods rotate with a constant angular velocity ω around the axis of the rods. It can be shown (Ref. 13) that in the case as well, for $\mu=0$, only a cylindrical zone can be stable. The stability criteria for a rotating liquid cylinder are

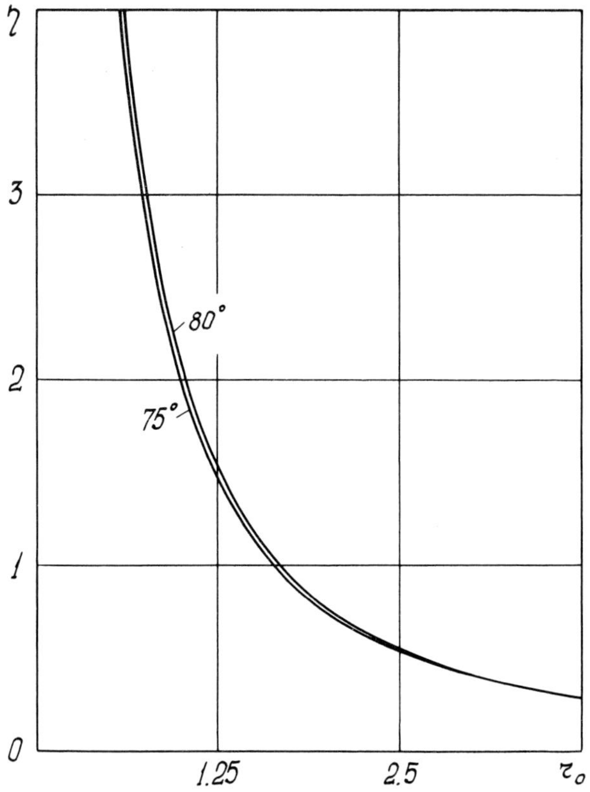

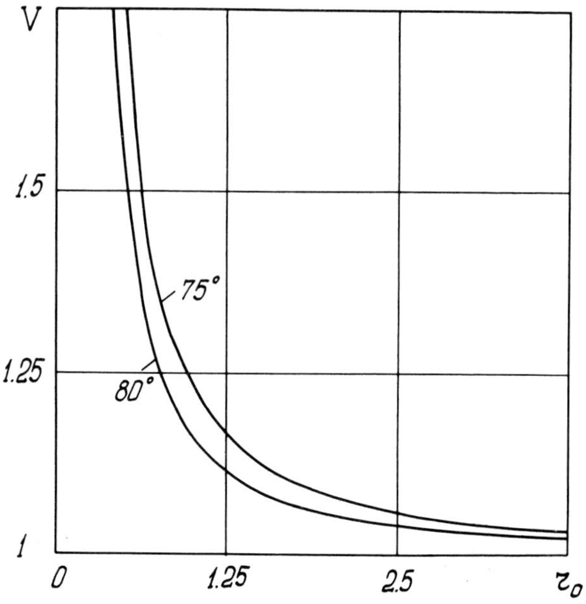

Figures 9 and 10: Dependences of the critical values of η and V on r_0 for a rotating liquid zone with $\beta_0=75^0$ and 80^0

known (Refs. 3, 14).

Now we are to determine the stability conditions for nonzero values of μ close to real. Stability of the whole family of equilibrium states of a rotating bridge is characterized by the three parameters: η, V and the dimensionless quantity $r_o \equiv \xi_o(\rho\omega^2/2\sigma)^{1/3}$ which expresses the ratio of the centrifugal forces to the surface tension forces (σ is the surface tension coefficient). In Refs. 15 and 16, with fixed values of one of the parameters, the curves of the neutral stability are constructed for the two other parameters.

Because we are interested not in the whole family of the equilibrium states, but only in those which are characterized by the prescribed value of β_o, we can find the whole relationship between the critical values of the above three parameters. We made studies for $\beta_o = 75^o$ and 80° which correspond to real growth angles in the interval $0.05 \leq r_o \leq 7$. Figures 9 and 10 show the critical values of η and V against r_o. The critical surfaces of the zone are always convex, are symmetric with respect to the equatorial plane and lose stability under nonaxisymmetric perturbations. As $r_o \longrightarrow 0$, the critical values of η and V tend to infinity, in agreement with the conclusions on stability of the weightless resting zone.

5. ACKNOWLEDGEMEHTS

The author would like to thank Drs. M.A.Svechkareva and E.P. Levchenko for their help in numerical calculations.

6. REFERENCES

1. R.D. Gillette and D.C. Dyson, *Chem. Eng. J.* 2 (1971) 44.
2. L.A. Slobozhanin in *Hydromechanics and Heat- and Mass Transfer under Zero-Gravity*, Nauka, Moscow (1982), p. 9 (in Russian).
3. A.D. Myshkis, V.G. Babskii, N.D. Kopachevskii, L.A. Slobozhanin and A.D. Tyuptsov, *Low-Gravity Fluid Mechanics*, Springer-Verlag, Berlin etc. (1987).
4. J. Meseguer, *J. Fluid Mech.* 130 (1983) 123.
5. J. Meseguer, A. Sanz and D. Rivas in *Proc. 4 th Europ. Symp. Mater. Sci. under Microgravity*, ESA SP - 191, Madrid (1983), p. 261.
6. I. Martinez, ibid., p. 267.
7. I. Martinez and J.M. Perales, *J. Crystal Growth* 78 (1986) 369.
8. L.A.Slobozhanin and A.D.Tyuptsov, *Izv. Akad. Nauk SSSR, Mekh. Zhid. Gaza* No 1 (1974) 3 (in Russian).
9. M.P. Elagin, A.P. Lebedev and A.V.Shmelev in *Hydromechanics and Heat- and Mass Transfer under Zero- Gravity*, Nauka, Moscow (1982) p. 24 (in Russian).
10. I.V. Barmin, N.A. Bezdenezhnykh, V.A. Briskman et al., *Izv. Akad. Nauk SSSR, Ser. Fiz.* 49 (1985) 698 (in Russian).
11. A. Sanz and I. Martinez, *J. Coll. Interface Sci.* 93 (1983) 235.
12. V.A. Briskman and N.A. Bezdenezhnykh, *Laboratory Simulation of Stability of the Liquid Zones under Stationary and Vibrational Loading*, publ. by Inst. Continuous Media Mech. UB of Acad. Sci. USSR, Perm (1989) (in Russian).
13. L.A. Sloboshanin, *Izv. Akad. Nauk SSSR, Ser. Fiz.* 49 (1985) 652 (in Russian).
14. L.A. Slobozhanin in *Mathematical Physics and Functional Analysis*, 2, publ. by Inst. Low Temp. Phys. and Engng. Ukr. Acad. Sci., Kharkov (1971), p. 169 (in Russian).
15. R.A. Brown and L.E. Scriven, *Phil. Trans. R. Soc. London* Ser. A 297 (1980) 51.
16. I.V.Barmin, B.E.Vershinin, I.G. Levitina and A.S. Senchenkov, *Izv. Akad. Nauk SSSR, Ser. Fiz.* 49 (1985) 661 (in Russian).

CAPILLARY SURFACES IN EXOTIC CONTAINERS

Paul Concus

Lawrence Berkeley Laboratory and Department of Mathematics
University of California, Berkeley, CA 94720, U.S.A.

Robert Finn

Department of Mathematics
Stanford University, Stanford, CA 94305, U.S.A.

ABSTRACT

A survey is presented of results to date for capillary surfaces in "exotic" containers. These containers have the property that each one admits a continuum of distinct equilibrium free surfaces, all bounding with the container walls the same volume of fluid, making the same contact angle at the triple interface curve, and having identical mechanical energies. The containers can be so designed that they are themselves axially symmetric but that the fluid configurations of minimizing energy cannot be axially symmetric.

Key words: *capillary surfaces, free surfaces, microgravity, symmetry breaking, space experiments*

The free surface of a liquid that partly fills a container under the action of surface and gravitational forces may assume, in general, one of several possible equilibrium configurations. An example for which only one configuration is possible is a vertical homogeneous cylindrical container of general cross-section, with gravity either absent or directed downward into the liquid; if the boundary of the free surface lies entirely on the cylindrical walls, then the surface is determined uniquely by its contact angle and the liquid volume.[1,2] Examples of other containers can be given for which there exist two or more distinct equilibrium configurations. Our interest here is in certain container shapes having the striking property that there is an entire continuum of such configurations.

Axially symmetric containers having this property can be constructed for any prescribed contact angle and Bond number, in such a way that all configurations in the family have identical volumes and yield identical mechanical energies, and so that no two of the free surfaces in the continuum are congruent to each other.[3,4,5] The procedure requires integration of a system of differential equations, in which the coefficients are determined as solutions of an auxiliary (nonlinear) differential equation system. Nevertheless, the global existence can be proved rigorously. It is shown in Refs. 4 and 6 that the families are all unstable, in the sense that the horizontal flat surface, which all families contain, can be deformed locally into a surface of smaller mechanical energy. It is possible to construct such "exotic" containers in such a way that no symmetric configuration can be energy minimizing.

The configurations in Figure 1 show seven surface meridians of the continuum (dashed curves) in the particular case of contact angle 60°, at varying Bond numbers; the solid curves are container meridians. The curves, which depict scaled height z/a as a function of scaled radius r/a, have been normalized by choosing a so that the containers have maximum radius unity. The Bond numbers $B = g\Delta\rho\ell^2/\sigma$ used for the numerical integration, which are indicated in Figure 1, were based—for com-

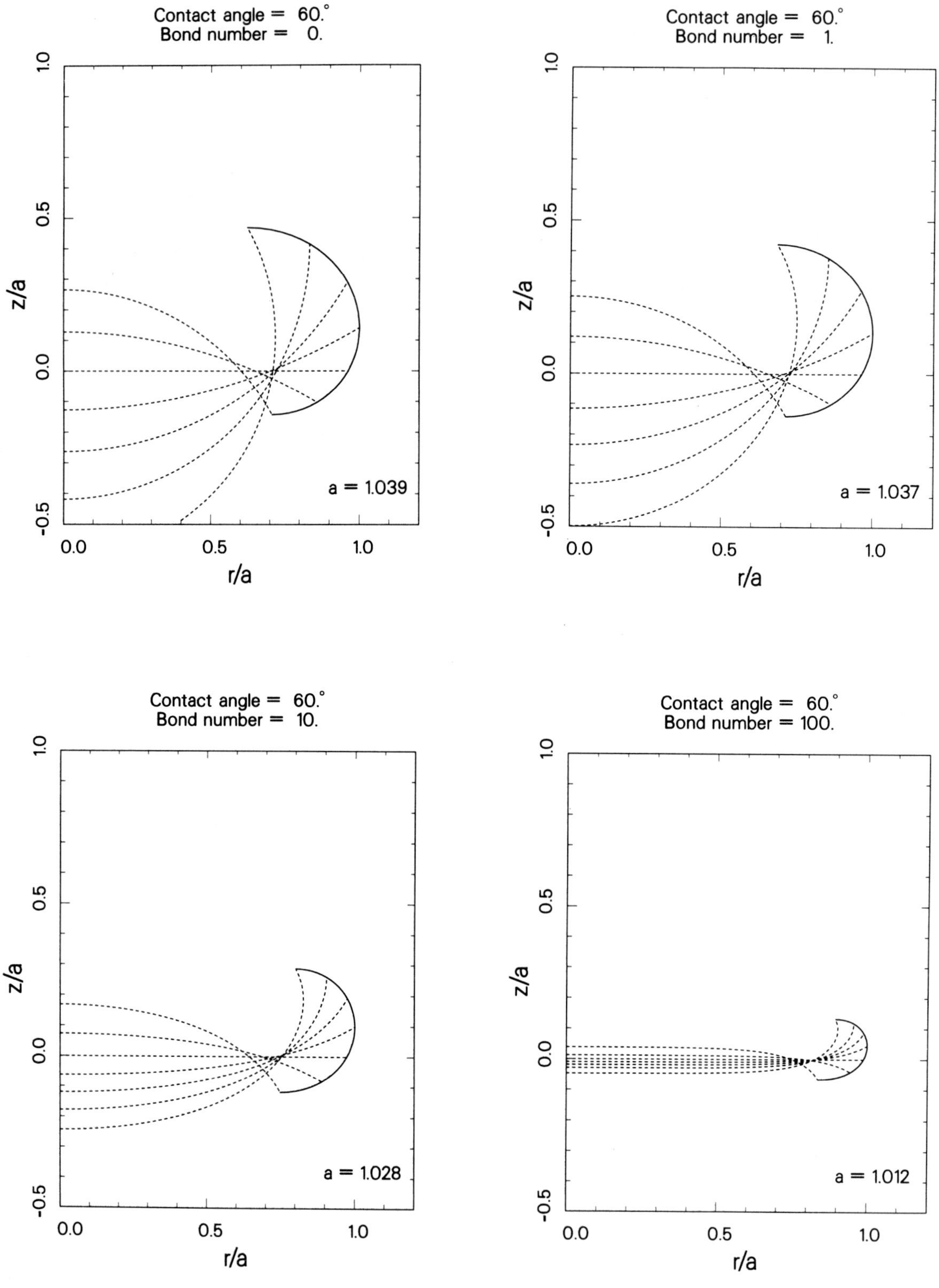

Figure 1. Meridian of container (solid curve) for contact angle 60° and several Bond numbers showing meridians of some of the symmetric equilibrium solution surfaces (dashed curves), all having the same contact angle and energy, and enclosing the same volume of liquid.

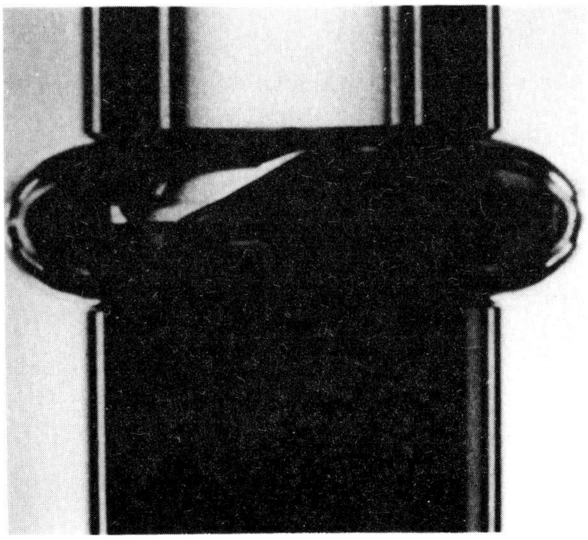

Figures 2a-b. Drop-tower experiment: 2a. (upper) Flat interface initial configuration; 2b. (lower) Non-symmetric terminal configuration.

putational reasons—on characteristic length ℓ equal to the radius of the flat interface[5] (g is the acceleration due to gravity, $\Delta\rho$ is the density difference across the free surface, and σ is the interfacial surface tension). Bond numbers based instead on maximum-radius characteristic length are Ba^2, where a is the indicated scaling factor.

Further particular cases are calculated and displayed in Ref. 5. It would be difficult to observe the exotic-container phenomenon in the earth's gravitational field, as the dimensions of the apparatus would have to be so small as to preclude accurate measurements. However, if gravity is absent the equations describing the containers are scale invariant, and experiments can be designed of any size, subject only to the usual laboratory constraints. Even in a microgravity environment that is not strictly zero-gravity, adequately large length scales for accurate observation and measurement are possible. An extended space experiment has been planned and is scheduled to be carried out on the NASA United States Microgravity Laboratory flight (USML-1) in 1992.

Preliminary drop tower experiments have been conducted by M. Weislogel on the five-second drop tower at NASA Lewis Research Center in Cleveland, Ohio.[7] An exotic container was fabricated corresponding to a contact angle of 80° (the contact angle for the materials used—acrylic plastic and water) and $B = 0$. The container, initially vertical, was filled with a volume of liquid corresponding to a horizontal planar equilibrium free surface meeting the container wall with contact angle 80°, as shown in Figure 2a. This is an equilibrium solution for any gravity level, but for zero g the interface is unstable in this container according to the mathematical theory, as are all members of the symmetric equilibrium family corresponding to the ones depicted in Figure 1. A resulting configuration after a five-second period of free fall subsequent to release of the container in the drop tower is depicted in Figure 2b. The liquid appeared essentially to have settled down to an equilibrium configuration by that time. The free surface exhibits a shape that is obviously non-symmetric, in accordance with the mathematical theory.

Independently, numerical studies were carried out in collaboration with M. Callahan. These studies are based on the Surface Evolver program developed by K. Brakke,[8] which deforms a given surface in such a way as to decrease (if that is possible) a discrete approximation to the total energy. Three local minima are suggested by the numerical investigation, of which the one with least energy looks remarkably like the surface in Figure 2b observed by Weislogel. It meets the container wall at the join between the bulge and the upper cylindrical wall for a good portion of the circumference, drops rapidly to the join with the lower cylindrical wall, and meets the lower join for the remainder of the circumference. The other locally energy-minimizing solutions are similar in nature, but have two or three excursions from top to bottom of the bulge (and back), instead of just one.[9]

Detailed derivations and further discussion can be found in the references.

This work was supported in part by the National Aeronautics and Space Administration under Grant NAG3-1143, by the National Science Foundation under Grant DMS89-02831, and by the Mathematical Sciences Subprogram of the Office of Energy Research, U. S. Department of Energy, under Contract Number DE-AC03-76SF00098.

REFERENCES

1. R. Finn, *Equilibrium Capillary Surfaces* (Springer-Verlag, New York 1986), Chapter 5. Russian edition: (Mir Publishers, Moscow 1988).
2. T. I. Vogel, Uniqueness for certain surfaces of prescribed mean curvature, *Pacific J. Math.* **134** (1988) 197–207.
3. R. Gulliver and S. Hildebrandt, Boundary configurations spanning continua of minimal surfaces, *Manuscr. Math.* **54** (1986) 323–347.
4. R. Finn, Nonuniqueness and uniqueness of capillary surfaces, *Manuscr. Math.* **61** (1988) 347–372.
5. P. Concus and R. Finn, Exotic containers for capillary surfaces, *J. Fluid Mech.* **224** (1991) 383–394; Corrigenda, *J. Fluid Mech.* (1991), to appear.
6. P. Concus and R. Finn, Instability of certain capillary surfaces, *Manuscr. Math.* **63** (1989) 209–213.
7. P. Concus, R. Finn, and M. Weislogel, Drop-tower experiments for capillary surfaces in an exotic container, *AIAA J.* (1991), to appear.
8. K. A. Brakke, *Surface Evolver Program*, Geometry Supercomputer Project, University of Minnesota, Minneapolis, MN, 1990.
9. M. Callahan, P. Concus, and R. Finn, Energy minimizing capillary surfaces for exotic containers, in *Computing Optimal Geometries* (with accompanying videotape), J. E. Taylor, ed., AMS Selected Lectures in Mathematics (Amer. Math. Soc., Providence, RI, 1991).

STABILITY AND VIBRATIONAL BEHAVIOR OF CYLINDRICAL LIQUID LAYERS IN MICROGRAVITY

H.F.Bauer

Institute of Spacetechnology, University of the German Armed Forces Munich, 8014 Neubiberg
Federal Republic of Germany

ABSTRACT

Natural damped and undamped frequencies as well as the response of an annular liquid layer are determined.

Keywords: Liquid bridge, Dynamics

1. INTRODUCTION

Adverse vibrations affect the quality of liquid experiments and manufactured products in microgravity. For that reason certain frequency ranges should be avoided during experiments. Therefore the natural and natural damped frequencies of an annular liquid layer around a center core are determined. In addition the natural frequencies deviate under the action of an axial microgravity field and exhibit also different response due to axial excitation if the Bond number is not vanishing.

2. Natural frequencies and damped natural frequencies of an annular liquid layer with anchored edges

2.1. Natural Frequencies

Treating the liquid as incompressible and frictionless one is able to determine the natural frequencies by solving the Laplace equation with vanishing normal velocities at the walls r=b and the upper and lower discs at $z=\pm h/2$ (Figure 1). In addition it is assumed that the liquid is anchored (stuck-edge) at the boundary of the disc (Ref.1). The liquid layer becomes unstable for $h \geq 2\pi a$ and exhibits larger natural frequencies in comparison with a liquid layer between two parallel plates, where a freely slipping edge is considered. The numerical results for the first three axisymmetric frequencies are presented in Figure 2 for a diameter ratio k= b/a =0.5. The natural frequencies decrease with increasing length h of the liquid layer.

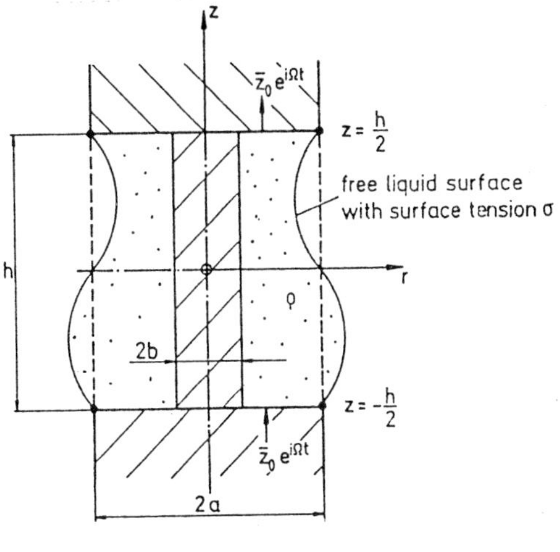

Figure 1: Geometry of System

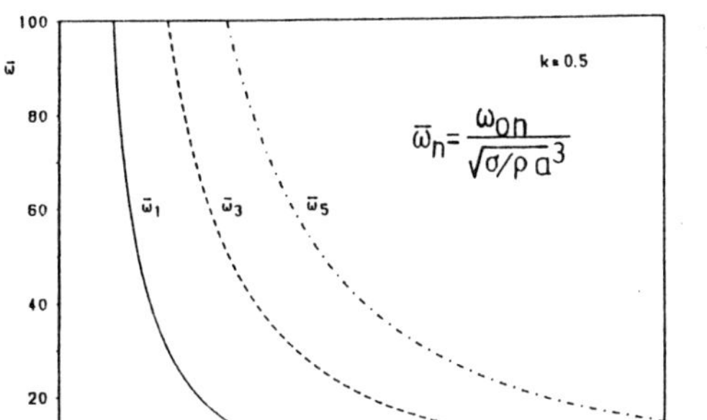

Figure 2: Natural frequencies for anchored edges

2.2. Response of frictionless liquid layer to axial excitations

Exciting the liquid layer axially with an harmonic function $z_o \exp(i\Omega t)$, where z_o is the excitation amplitude and Ω the forcing frequency, we obtain after solving the Laplace equation with the inhomogeneous disc boundary condition the response of the liquid layer (Ref.2). The response exhibits only resonances at the odd axisymmetric natural frequencies. The basic equations that have been solved are the Laplace equation

$$\Delta \Phi = 0 \tag{1}$$

with the boundary conditions

$$\frac{\partial\Phi}{\partial z} = \bar{z}_o i\Omega e^{i\Omega t} \quad \text{at } z=\pm\frac{h}{2},$$

$$\frac{\partial\Phi}{\partial r} = 0 \quad \text{at } r=b \tag{2}$$

and the free surface equations consisting of the kinematic and dynamic condition at r=a

$$\frac{\partial\Phi}{\partial r} = \frac{\partial\zeta}{\partial t}, \quad \frac{\partial\Phi}{\partial t} - \frac{\sigma}{\rho a^2}\left[\zeta + a^2\frac{\partial^2\zeta}{\partial z^2}\right] = \frac{\sigma}{a} \tag{3}$$

where ζ is the free surface displacement. In addition the anchored edge condition and volume preserving condition are given by

$$\zeta=0 \text{ at } z=\pm\frac{h}{2} \text{ and } \int_{-h/2}^{+h/2} \zeta dt=0. \tag{4}$$

The result in form of the response of the free surface is shown in Figure 3 for h/a =2 and various diameter ratios k at the location z= h/4. At the odd resonances the response exhibits a singularity.

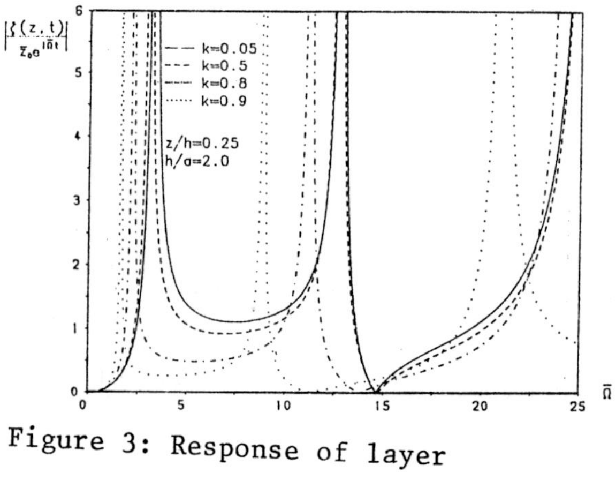

Figure 3: Response of layer

2.3. Damped natural frequencies

Treating the viscous and incompressible liquid layer for small velocities and surface displacements the Stokes equation

$$\frac{\partial\vec{v}}{\partial t} + \frac{1}{\rho}\,\text{grad}\, p + \nu\, \text{curl curl}\, \vec{v} = 0 \tag{5}$$

has to be solved with the continuity equation div $\vec{v}=0$ and the appropriate boundary conditions of

$$u=w=0 \text{ at } r=b, \quad w=0 \text{ at } z=\pm\frac{h}{2} \tag{6}$$

vanishing shear stress at the free surface, i.e.

$$\tau_{rz} = \eta\left[\frac{\partial w}{\partial r} + \frac{\partial u}{\partial z}\right] = 0 \tag{7}$$

as well as free surface condition at r=a

$$u=\frac{\partial\zeta}{\partial t}, \quad p-2\eta\frac{\partial u}{\partial r} + \frac{\sigma}{a^2}\left[\zeta + a^2\frac{\partial^2\zeta}{\partial z^2}\right] = \frac{\sigma}{a}. \tag{8}$$

The adherence condition at the disc in radial direction has been replaced by the weaker anchored edge condition $\zeta=0$ at $z=\pm h/2$ (Ref.3). The fundamental damped natural frequency is presented in Figures 4 and 5 for the surface tension parameter $\sigma^* \equiv \sigma a/\rho\nu^2=5$, 100 for a diameter ratio k=0.5. It may be seen, that for low surface tension parameter the layer performs only in the range of $0.09<h/2\pi a<0.2$ a decaying oscillation, while for small heights of the liquid layer $h<0.57a$ and for large heights $1.26a<h<6.28a$ only an aperiodic motion is possible.

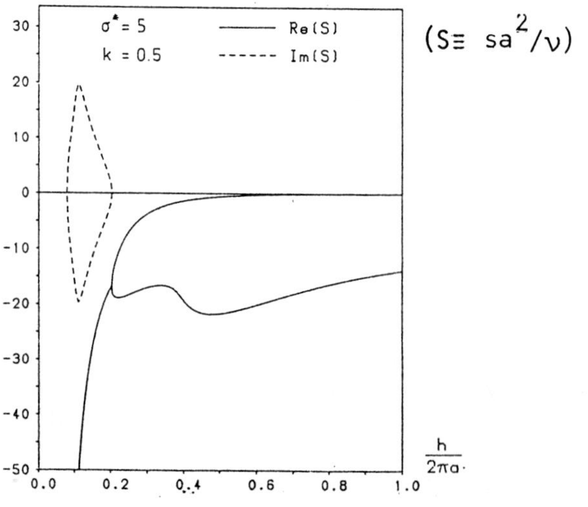

Figure 4: Damped natural frequency

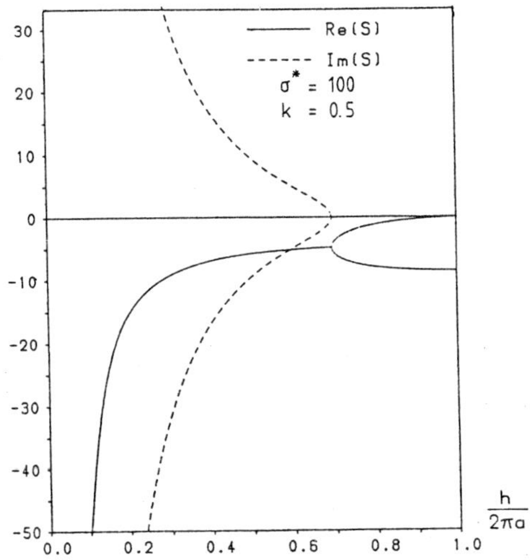

Figure 5: Damped natural frequency

In addition we notice that for increasing length of the layer in the region 0.57a<h<1.26a the oscillation frequency decreases with decreasing decay magnitude. For larger surface tension parameter $\sigma^*=100$ the oscillatory range is enlarged to a length of the layer $h\approx 4.4a$. The critical value of σ^*, where the motion of the layer changes its character from oscillatory damped motion to a decaying motion is presented for various diameter ratios, k=0, 0.5 and 0.7, in the motion identification chart of Figure 6.

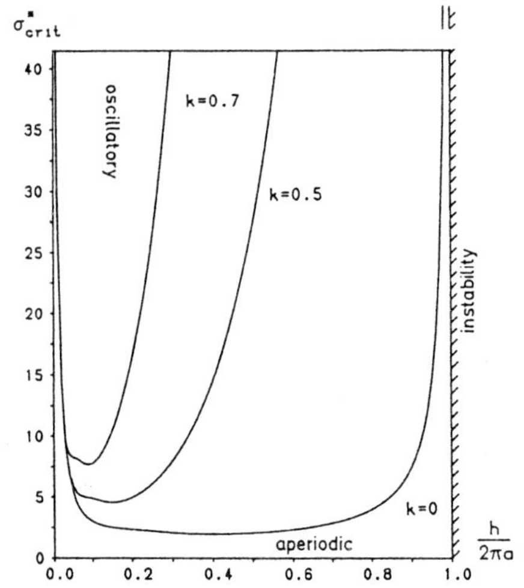

Figure 6: Motion identification chart

First of all it may be noticed that the layer is unstable for $h\geq 2\pi a$, and that with the increase of the diameter ratio k, i.e. the decrease of the thickness of the layer the oscillatory region becomes smaller, such that for certain layer heights h/a only an aperiodic behavior is possible for axisymmetric motion.

2.4. Response of viscous liquid layer to axial excitation

For an harmonic excitation $z_o \exp(i\Omega t)$ the right-hand side of the Stokes equation (5) will exhibit the value $-\Omega^2 z_o \exp(i\Omega t)\vec{k}$ where $\vec{k}$ is the axial unit-vector. The response of the free liquid surface displacement $|\zeta/z_o|$ is presented in Figure 7 for a layer height ratio h/a=4 and a surface tension parameter $\sigma^*=1000$. It is shown versus the reduced frequency $\bar{\Omega}\equiv\Omega a^2/\nu$ and for various diameter ratio k=0.01, 0.5 and 0.7. We notice that the resonance peak decreases with decreasing thickness of the layer and that according to the motion identification chart the resonance peaks disappear, since no oscillatory motion is anymore possible (Ref.3).

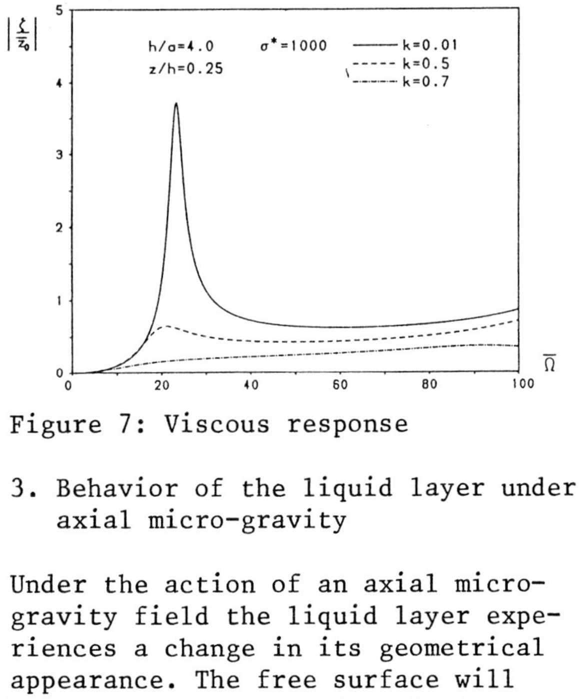

Figure 7: Viscous response

3. Behavior of the liquid layer under axial micro-gravity

Under the action of an axial microgravity field the liquid layer experiences a change in its geometrical appearance. The free surface will bulge out and may be described as of the shape of an amphora. The equilibrium free surface deflection deviation from the circular cylindrical form may be expressed as (Figure 8) (Ref.4)

$$r_o(z)=Bo\cdot r_1(z)+Bo^2\cdot r_2(z) \quad (9)$$

where the Bond-number is $Bo=\rho g a^2/\sigma$ and $r_1(z)$, $r_2(z)$ may be expressed analytically from the solution of the Laplace-Young equation.

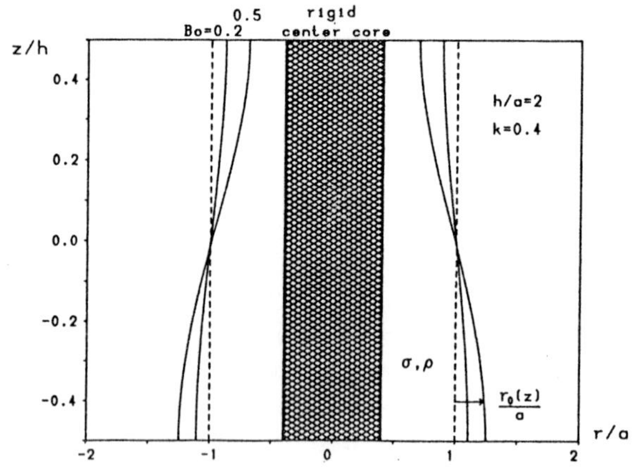

Figure 8: Equilibrium position

The solution of the problem for frictionless and incompressible liquid may be found from the Laplace equation $\Delta\Phi=0$, with the boundary condition $\partial\Phi/\partial r=0$ at $r=b$ and the free surface conditions at $r=a+r_o(z)+\zeta(z,t)$, i.e. the kinematic condition in linearized form in Φ and ζ:

$$\frac{\partial\zeta}{\partial t}=\frac{\partial\Phi}{\partial r}+r_o(z)\frac{\partial^2\Phi}{\partial r^2}-\frac{\partial\Phi}{\partial z}\frac{dr_o}{dz}+ + \frac{1}{2}\frac{\partial^3\Phi}{\partial r^3}r_o^2-r_o\frac{dr_o}{dz}\frac{\partial^2\Phi}{\partial r\partial z} \quad \text{at } r=a \tag{10}$$

and the dynamic condition

$$\frac{\partial\Phi}{\partial t}+r_o\frac{\partial^2\Phi}{\partial t\partial r}+\frac{1}{2}r_o^2\frac{\partial^3\Phi}{\partial t\partial r^2}-\frac{\sigma}{\rho a^2}[r_o+\zeta+ + a^2\frac{\partial^2\zeta}{\partial z^2}+a^2\frac{d^2r_o}{dz^2}]+\frac{\sigma}{\rho a^3}[r_o^2+2r_o\zeta - -\frac{a^2}{2}[(\frac{dr_o}{dz})^2+2\frac{\partial\zeta}{\partial z}\frac{dr_o}{dz})]-\frac{\sigma}{\rho a^4}\{r_o^3+3r_o^2\zeta - -\frac{a^2}{2}(r_o(\frac{dr_o}{dz})^2+2r_o\frac{dr_o}{dz}\frac{\partial\zeta}{\partial z}+\zeta(\frac{dr_o}{dz})^2) - -\frac{3a^4}{2}[(\frac{dr_o}{dz})^2(\frac{d^2r_o}{dz^2}+\frac{\partial^2\zeta}{\partial z^2})+2\frac{dr_o}{dz}\cdot \cdot\frac{d^2r_o}{dz^2}\frac{\partial\zeta}{\partial z}]\}=-gz \quad \text{at } r=a. \tag{11}$$

3.1. Free oscillations

For the free oscillations the boundary condition at the top and bottom wall is $\partial\Phi/\partial z=0$ at $z=\pm h/2$. The results of the analysis are given in Figure 9 and 10 for the axisymmetric fundamental vibration mode m=0, n=1 a height ratio h/a =2 and various diameter ratios k=0.01, 0.5, 0.8, 0.9 and 0.95. It may be seen, that the natural frequency decreases in comparison with the pure cylindrical layer with increasing Bond number. It also exhibits a stronger decrease for thinner layers.

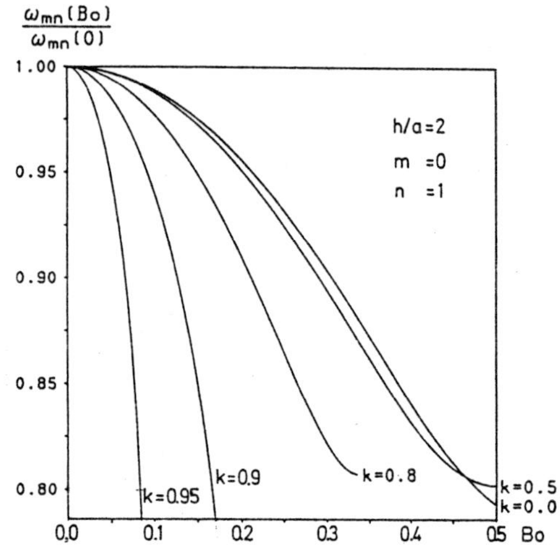

Figure 9: Change of natural frequencies

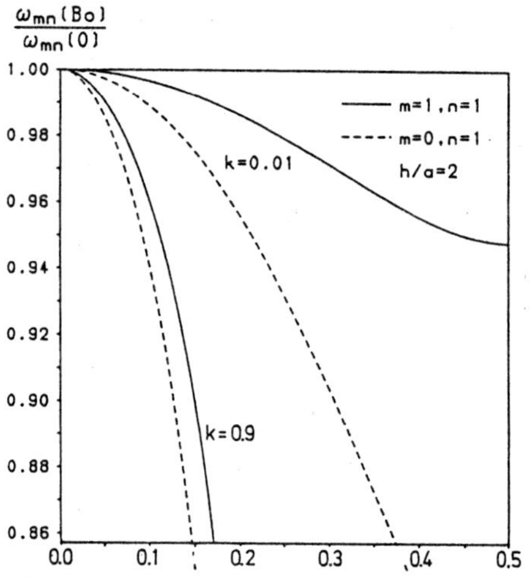

Figure 10: Change of natural frequencies

For example a layer of thickness 0.1a shows for an axial Bond number Bo=0.17 a 26% decrease of the natural frequency. In Figure 10 the natural frequency of the amphora-type geometry is shown for the modes m=1, n=1 and the axisymmetric mode m=0, n=1. It may be noticed that the axisymmetric mode (---) exhibits the largest deviation in the natural frequency.

3.2. Axial response to harmonic excitation

For a harmonic axial excitation $\bar{z}_o \exp(i\Omega t)$ the above wall boundary condition has to be changed to $\partial\Phi/\partial z = \bar{z}_o \exp(i\Omega t)$ at $z = \pm h/2$. The results for the response of the free liquid surface are presented in Figures 11 und 12 for the height ratio h/a=2 and various diameter ratios k=0.01, 0.5, 0.8 and 0.9 at the location z=0.2h. Figure 11 exhibits the response for Bo=0,

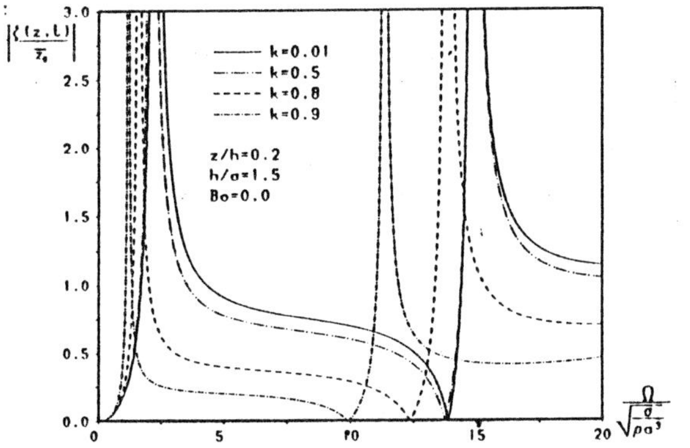

Figure 11: Response for Bo=0

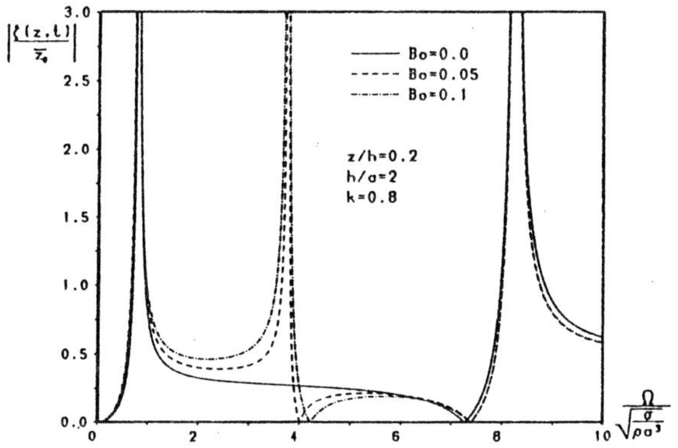

Figure 12: Response of Bo $\neq$ 0

where we notice the shifting of the resonance locations to lower frequency values for decreasing thickness of the layer, while Figure 12 shows for a diameter ratio k=0.8 the response of the free liquid surface at z=0.2h and h/a =2 for various Bond number Bo=0, 0.05 and 0.1. First of all we notice that response for nonvanishing Bond numbers exhibits for all natural axisymmetric frequencies a resonance peak, while for Bo=0, i.e. the circular cylindrical geometry only odd resonances appear. In addition the response magnitude is between first and second resonance larger for larger Bond numbers.

4. REFERENCES

1. H.F.Bauer, Natural axisymmetric frequencies of cylindrical liquid column with anchored free surface. JSME International Journal, Series III, Vol.34, Nr.4 (1991)

2. H.F.Bauer, Response of a nonviscous liquid column and layer with anchored liquid surface to axial excitation. JSME International Journal Series II, Vol.34, No.4 (1991), 474-481

3. H.F.Bauer, Response of an anchored viscous liquid layer to axial excitation in zero gravity. ZAMM (to appear)

4. H.F.Bauer and Kammerer,L., Natural frequencies and axial response of a liquid layer under the influence of steady microgravity field. Forsch.Ing.Wes. (to appear)

IMPULSIVE MOTION OF VISCOUS AXISYMMETRIC LIQUID BRIDGES*

J. Meseguer[1), J.M. Perales[1) and N.A. Bezdenejnykh[2)

1) Lamf-μg, Laboratorio de Aerodinámica, E.T.S.I. Aeronáuticos, Universidad Politécnica, 28040 Madrid, Spain

2) Institut Mekhaniki Sploshnykh Sred, Permski Nauchnyi Centr (Uralskoe Otdelenie AN SSSR), 614061 Perm, USSR.

ABSTRACT

This paper deals with the dynamics of isothermal, axisymmetric, viscous liquid columns held by capillary forces between two circular, concentric, solid disks. The transient response of the bridge to an excitation consisting of a small change in the value of the acceleration acting along its axis has been considered. To compare existing theoretical results with experimental ones a number of experiments have been performed; in these experiments, a small liquid bridge with its axis placed vertically was suddenly displaced downwards for a small distance and the response of the interface recorded, the agreement between theoretical and experimental results (damping time constant) being good enough.

Keywords: *Liquid bridge, Microgravity, g-jitter.*

1. INTRODUCTION

The liquid bridge configuration considered here consists, as sketched in Figure 1, of an isothermal, axisymmetric mass of liquid held by surface tension forces between two parallel, coaxial, circular solid disks. Such a fluid configuration can be uniquely identified by the following set of dimensionless parameters: the ratio of the radius of the smaller disk, R_1, to the radius of the larger one R_2, $K=R_1/R_2$; the slenderness, $\Lambda=L/2R_o$, where $R_o=(R_1+R_2)/2$ and L is the distance between both disks; the dimensionless volume of liquid, $V=V/R_o^3$, V being the physical volume; the Bond number, $B=\rho g R_o^2/\sigma$; and the capillary number, $C=\nu(\rho/\sigma R_o)^{1/2}$, where ρ is the liquid density, g the axial acceleration, σ the surface tension and ν the kinematic viscosity.

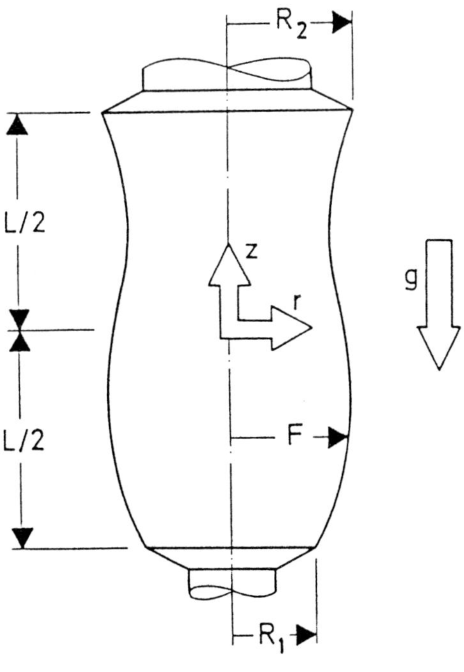

Figure 1. Geometry and coordinate system for the liquid bridge problem.

* Dedicated to the memory of Prof. I. da Riva

Most of the published studies related to the liquid bridge response to forcing perturbations are concerned with harmonic perturbations (Refs. 1 to 5). Non-harmonic perturbations have been considered in Ref. 6, where a one-dimensional inviscid slice model was used to analyze the liquid injection or removal in the liquid bridge and in Ref. 7, where the same problem here considered is analytically solved by using a one-dimensional Cosserat model, including viscosity effects, although in that case the analysis is restricted to cylindrical liquid bridges ($V=V_c=2\pi\Lambda$) between equal disks (K=1) and in gravitationless conditions (B=0) (in this case the interface at rest is a cylinder).

A linear theoretical model of the dynamic response of viscous axisymmetric liquid bridges due to small changes in the value of the axial acceleration acting on the liquid has been developed (Ref. 8) by using the Cosserat's one-dimensional model for continuum media. This model has been used for capillary jet problems and for liquid bridge problems, and has proved to give satisfactory results, when compared with 3-D models results, provided that the slenderness of the liquid bridge is large enough ($\Lambda > 1$).

In order to check the above theoretical predictions, the transient response of liquid bridges subjected to a small perturbation in the axial acceleration has been measured. Experiments have been carried out by working with very small liquid bridges ($R_o=0.5 \times 10^{-3}$ m) and using distilled water as working fluid. The liquid bridge, initially at rest, is suddenly displaced vertically a distance of the order of R_o and the subsequent evolution recorded, so that the damping time constant of the evolution until the final equilibrium state is measured. A number of experiments have been performed, varying either the volume of liquid or the distance between disks, experimental results showing the same trends as theoretical predictions.

2. THEORETICAL RESULTS

Before pursuing further it would be convenient to point out some characteristics of the liquid bridge response, stated in Ref. 7 for cylindrical liquid bridges between equal disks (obviously the behaviour is qualitatively the same no matter the values of V, K, and B are, provided the liquid bridge configuration is stable). In the following we will denote as $h_n=\gamma_n+i\omega_n$ the roots of the secular equation deduced by the analysis of the problem in the Laplace domain. For a given liquid bridge configuration, that is, once Λ, K, B and V are fixed, all the roots are imaginary (γ_n=0) if C=0, and consequently the liquid bridge response will be oscillatory, without damping, as one could expect from an inviscid movement, and all the oscillation modes will be present in the liquid bridge response. As C increases, the value of γ_n becomes more and more negative, whereas the absolute value of ω_n decreases. That means that all oscillation modes are damped, their oscillation frequencies being smaller as C grows. For each oscillation mode there is a critical value $C=C^*_n$ beyond which the associated movement is only damped, without oscillation (ω_n=0).

Concerning the importance of the different oscillation modes in the liquid bridge dynamics it must be pointed out that the only significative oscillation mode is the first one by two reasons. The first one concerns the amplitude of the different oscillation modes, much larger in the first oscillation mode than in the following ones and the second is that, in the case of damped oscillations, the absolute value of γ_n increases as the index of the oscillation mode, n, grows; therefore, all oscillation modes are damped much quicker than the first one (oscillation modes different from the first one are only important for short times when all of them appear to fulfill the initial conditions).

The same qualitative behaviour can be observed when a variation of the Bond number, the disk diameter ratio or the volume is considered. Even more, for a given value of C the resonance frequency, ω_1, decreases and the absolute value of damping factor, γ_1, increases as the liquid bridge configuration approaches the corresponding stability limit; close enough of the stability limit there is a region in which ω_1=0 and the absolute value of γ_1 decreases. The damping factor becomes zero at the stability limit and it is greater than zero for unstable liquid bridge configurations (Refs. 7-9)

The dependence on the dimensionless parameter of viscosity, C, of the real and imaginary parts of the roots corresponding the first oscillation mode, γ_1 and ω_1, are shown in Figure 2, the main characteristic to be pointed out from this plot being that the damping factor γ_1 varies almost linearly with C no matter the values of the remaining liquid bridge parameters are. This linear dependence of γ_1 on C is of great help when theoretical results are compared with experimental ones because it allows the comparison of both types of results independently of the value of surface tension, which is difficult to measure.

3. EXPERIMENTAL RESULTS

The experimental set-up, as sketched in Figure 3 (not to scale) consists of a liquid-bridge rig (although in Figure 3 only the liquid bridge is shown (1)), background illumination (2), a microscope (3) to which a CCD

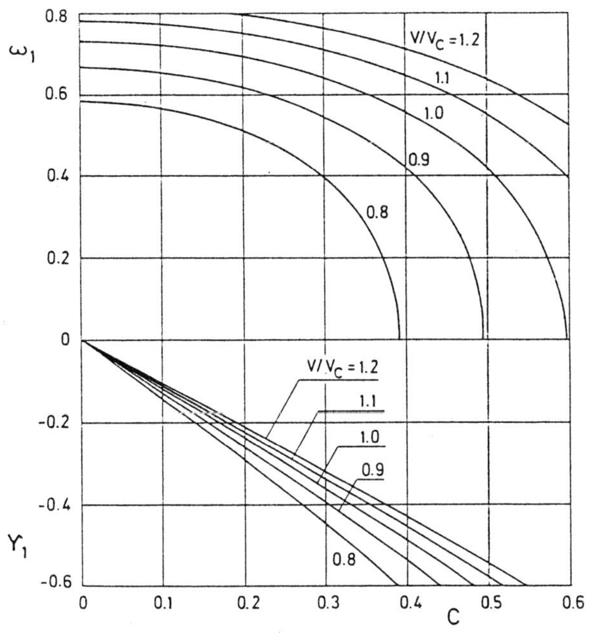

Figure 2. Variation with the dimensionless viscosity parameter, C, of the resonance frequency, ω_1, and damping factor, γ_1, corresponding to the first oscillation mode. Numbers indicate the value of V/V_c. These results correspond to liquid bridges with Λ=2, K=1 and B=0 (from Ref. 8).

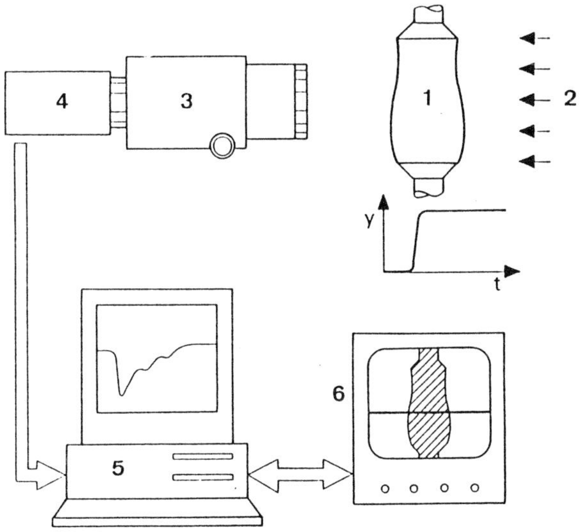

Figure 3. Experimental set-up. 1) Liquid bridge; 2) illumination; 3) microscope; 4) CCD camera; 5) computer; 6) TV monitor.

camera (4) is connected, a desktop computer (including data acquisition and image processing systems (5)) and a TV monitor (6). The liquid-bridge rig (LB rig) consists of a closed test chamber having two opposite lateral sides made of glass to allow visualization. The liquid bridge is formed inside the test chamber between two coaxial solid disks made of Perspex (10^{-3} m in diameter) their common axis being vertical during experiments. The bottom disk is fixed to the rig frame, whereas the upper one can be displaced upwards and downwards as well as rotated. During experiments a small volume of water was poured at the bottom of the test chamber to keep the air inside the test chamber saturated, thus avoiding the evaporation of water (which was used as working liquid) from the liquid bridge once formed, which could cause the variation with time of the liquid bridge volume. The bottom disk incorporates the filling duct (which ends at a hole, less than 0.5×10^{-3} m in diameter, at the centre of the bottom disk). The liquid bridge is formed by injecting some amount of water from an external syringe; once a small drop of liquid is formed on the bottom disk, the upper one is displaced downwards until it is contacted and wetted by the water drop, and a short liquid bridge spanning between both disks results. From now on the distance between disks is slowly increased, injecting water at the same time, until the desired slenderness and volume of liquid are reached.

The whole LB rig is mounted on an electric shaker which is driven by an electric-signal generator. The shaker was adjusted in such a way that its axis moves upwards a distance of 0.3×10^{-3} m, as sketched in the insert in Figure 3 and in Figure 4, when required.

For illumination, to avoid the excessive heating of the LB rig, a cold source of light was used: a lamp placed some distance apart of the rig, the light being conducted through an optic–fibre cable. Using the appropriate background illumination the liquid bridge appears in the TV screen as a black shadow with a well defined contour on a white background. With the capabilities of the image processing system used it is possible to get either the whole contour, which is used to calculate the slenderness of the liquid bridge as well as its volume (it takes about 1 s to determine Λ and V), or to get the time evolution, at a rate of 25 frames/second, of up to three different sections of the liquid bridge.

The experimental procedure was as follows: once the liquid bridge is formed and the slenderness and volume determined as indicated above, the second choice for the image processor (variation with time of the liquid bridge contour at some prefixed section) is selected and a starting pulse is sent to the shaker (averaged nominal profiles of the displacement, velocity and acceleration experienced by the whole liquid bridge are shown in Figure 4). Because of the impulsive acceleration acting on the liquid bridge its interface distorts and a damped oscillation with the corresponding natural frequency starts. Once the liquid bridge is at rest again, and the

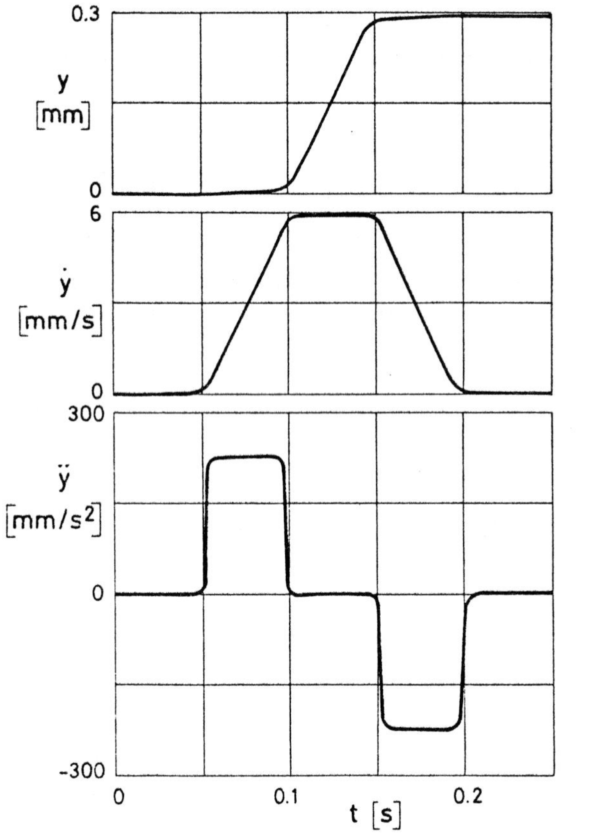

Figure 4. Typical variation with time, *t*, of the axial displacement, *y*, velocity, $\dot{y}$, and acceleration, $\ddot{y}$, of the LB rig during experiments.

shaker axis returned to its initial position, some amount of liquid is injected (or removed) into the liquid bridge to get a new volume of liquid (provided the liquid bridge is still formed, otherwise it would be necessary forming a new liquid bridge as explained above) and the next experiment can start.

Unfortunately, this experimental set-up is only appropriate to measure damping time constant but not natural frequencies, the reason being that the scanning rate in normal video signal is 25 frames/second whereas natural frequencies of the tested configuration are higher than 25 Hz (Ref. 1). Typical evolutions of different liquid bridge configurations (the variation with time, *t*, of the radius of the liquid bridge, $f=F(t)$ - $F(0)$, at the measurement section) are shown in Figure 5. Note that in these plots the final value of f is not zero; this is because of the perturbation imposed (a quick axial displacement of the liquid bridge): the liquid bridge moves in respect to the scanning line in the TV screen so that the time variation of the interface radius is measured at a different section of the initial one. For each experiment a time-exponential function has been fitted to experimental data and the damping factor obtained.

A problem arising when comparing experimental and theoretical results is that surface tension appears in most of the parameters needed for data reduction, namely: characteristic time, $t_c=(\rho R_o^3/\sigma)^{1/2}$; viscosity parameter, $C=\nu(\rho/\sigma R_o)^{1/2}$ and Bond number, $B=\rho g R_o^2/\sigma$. Surface tension of water can vary almost an order of magnitude

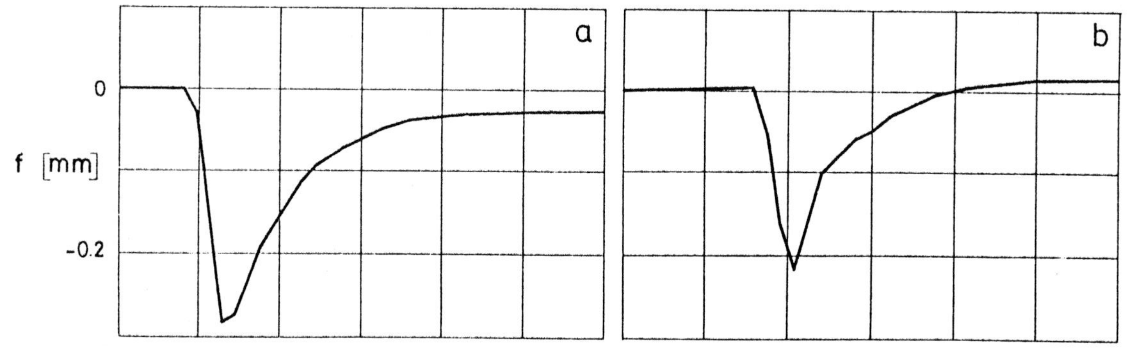

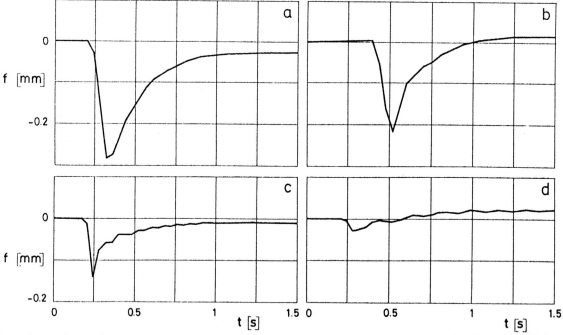

Figure 5. Variation with time, *t*, of the radius of the liquid bridge interface at the measurement section, $f=F(t)$ - $F(0)$. a) Λ = 2.04, V/V_c = 0.88; b) Λ = 2.04, V/V_c = 1.10; c) Λ = 2.49, V/V_c = 0.98; d) Λ = 2.49, V/V_c = 1.35, where $V_c=2\pi\Lambda$.

depending on the degree of surface contamination, assuming a variation from 0.06 $N.m^{-1}$ to, let say, 0.02 $N.m^{-1}$ (in other experiments using water as working fluid performed with the same set-up the value σ=0.06 $N.m^{-1}$ was measured, Ref. 10) and taking into account the nominal values of the remaining physical parameters of the experiments (R_o=5 x 10^{-4} m, ρ=10^3 $kg.m^{-3}$, ν=1.08 x 10^{-6} $m^2.s^{-1}$, g=9.81 $m.s^{-2}$), one gets that Bond number can range between 0.04 and 0.12, that viscosity parameter would be between 0.006 and 0.010 and that the value of characteristic time is between 1.4 x 10^{-3} s and 2.5 x 10^{-3} s.

The damping factor depends on Λ, K, B, V and C. Once the four first parameters are fixed, the dependence on C is linear provided C is small enough (as shown in Figure 2). Therefore, the dimensionless damping factor can be expressed as γ_1=mC where m=$m(\Lambda,K,B,V)$, so that the physical damping factor, $\bar{\gamma}_1$ will be

$$\bar{\gamma}_1 = \frac{\gamma_1}{t_c} = \frac{mC}{t_c} = \frac{m\nu}{R_o^2} = 4.3m(\Lambda,K,B,V) \quad \left[s^{-1}\right]$$

In the experiments performed both disks were equal in diameter (K=1) and the slenderness and the volume of liquid were measured with high accuracy as explained above. Nevertheless, there is some uncertainty concerning the value of Bond number due to the uncertainty in the value of the surface tension. The different experiments performed are summarized in Figure 6. In this plot each one of the experiments is represented by a circle in the Λ–V stability diagram (K=1). Also in this plot the stability-limit curves corresponding to B=0, 0.05 and 0.1 are shown. Obviously, since all the liquid bridge configurations tested were stable, this plot gives an upper limit for the value of Bond number which corresponds to a high value of surface tension (σ ~ 0.05 $N.m^{-1}$).

Experimental results (damping factor versus reduced liquid bridge volume, V/V_c, where V_c=$2\pi\Lambda$) are compared with theoretical ones in Figure 7, for liquid bridges with Λ=2.04, and in Figure 8 (for the case Λ=2.49). In these plots theoretical curves corresponding to different values of Bond number have been represented. Note that experimental data shows the same trends as theoretical predictions, although they are too scattered.

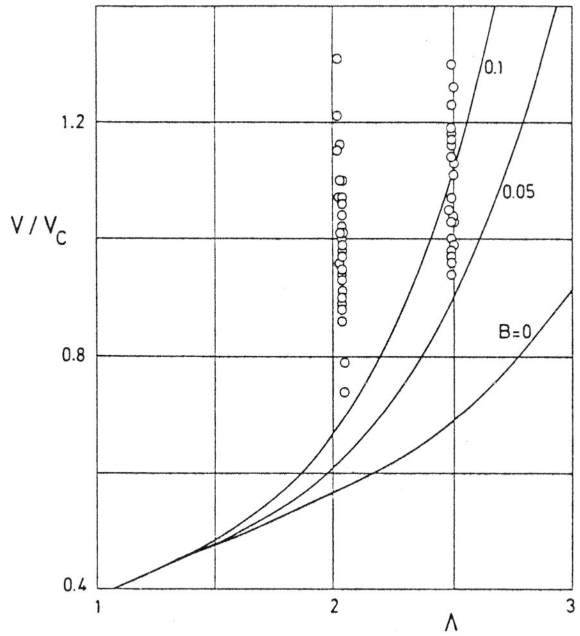

Figure 6. Minimum volume stability limit for different Bond numbers (0, 0.05 and 0.1). Circles show conditions where damping time constant has been measured.

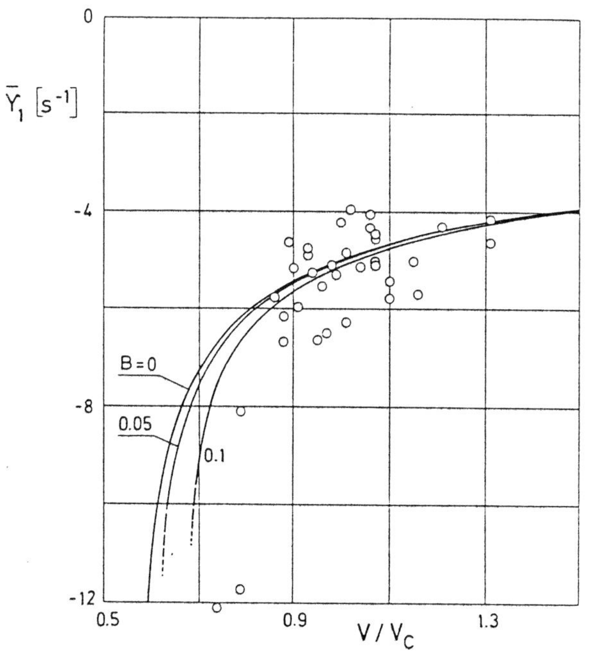

Figure 7. Damping factor, $\bar{\gamma}_1$, as a function of reduced volume, V/V_c for liquid bridges with slenderness Λ = 2.04. Circles indicate experimental results whereas solid lines show the theoretical ones corresponding to different values of Bond number.

Acknowledgements

This work has been performed during a stay of Dr. Bezdenejnykh at Lamf-µg (Madrid) sponsored by the European Space Agency (ESA) and the Universidad Politécnica de Madrid. This work has been supported also by the Comisión Interministerial de Ciencia y Tecnología (CICYT), Project No. ESP88-0359.

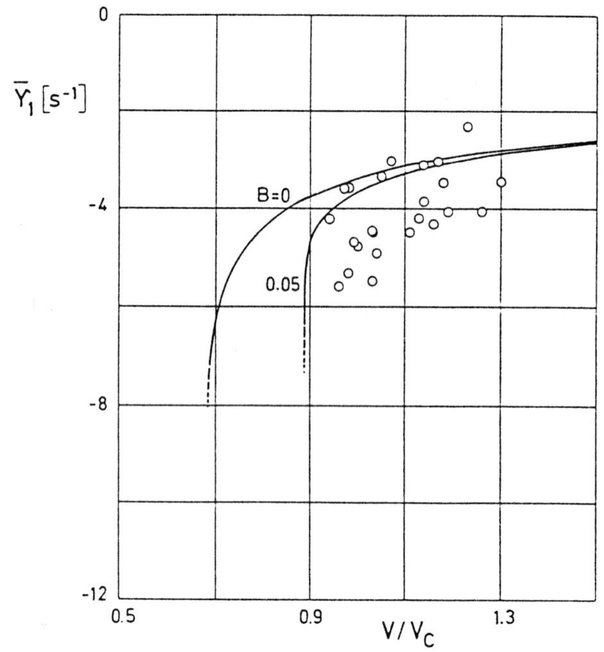

Figure 8. Damping factor, $\bar{\gamma}_1$, as a function of reduced volume, V/V_c for liquid bridges with slenderness $\Lambda = 2.49$. Circles indicate experimental results whereas solid lines show the theoretical ones corresponding to different values of Bond number.

REFERENCES

1.- J. Meseguer, Axisymmetric Long Liquid Bridges in a Time-Dependent Microgravity Field, *Appl. Microgravity Technol.* **1**, 136 (1988).

2.- J. Meseguer, A. Sanz and J.M. Perales, Axisymmetric Long Liquid Bridges Stability and Resonances, *Appl. Microgravity Technol.* **2**, 186 (1990).

3.- Y. Zhang and J.I.D. Alexander, Sensitivity of Liquid Bridges Subject to Axial Residual Acceleration, *Phys. Fluids A* **2**, 1966 (1990).

4.- J.M. Perales, Dinámica de Puentes Líquidos, *Tesis Doctoral*, Universidad Politécnica de Madrid (1990).

5.- J.A. Nicolás, Frequency Response of Axisymmetric Liquid Bridges to an Oscillatory Microgravity Field, *Microgravity Sci. Technol.*, in press.

6.- J. Meseguer and A. Sanz, One-Dimensional Linear Analysis of the Liquid Injection or Removal in a Liquid Bridge, *Acta Astronautica* **15**, 573 (1987).

7.- J. Meseguer and J.M. Perales, A Linear Analysis of g-Jitter Effects on Viscous Cylindrical Liquid Bridges, *Phys. Fluids A*, in press.

8.- J. Meseguer, J.M. Perales and N.A. Bezdenejnykh, A Theoretical Approach to Impulsive motion of Viscous Liquid Bridges, *Microgravity Quarterly*, submitted.

9.- J. Meseguer, The Breaking of Axisymmetric Slender Liquid Bridges, *J. Fluid Mech.* **130**, 123 (1983).

10.- N.A. Bezdenejnykh and J. Meseguer, Stability Limits of Minimum Volume and Breaking of Axisymmetric Liquid Bridges between Unequal Disks, *Microgravity Sci. Technol.*, in press.

DYNAMIC STABILIZATION OF THE CAPILLARY INSTABILITY OF A CYLINDRICAL LIQUID ZONE

D.V.Lyubimov, A.A.Cherepanov

Perm State University, 15, Bukirev Str., 614600, Perm, USSR

ABSTRACT

It is known, that a cylindrical coumn of a liquid in the absence of external forces is unstable due to the development of the Rayliegh instability. The dynamic mechanism of the liquid cylinder stabilization has been investigated. It is shown that vibrations of the circular polarization may suppress the development of Rayliegh instability, if the cylindrical column is surrounded by a liquid layer of different density. The conditions of the liquid zone stability under the influence of vibrations have also been determined.

Keywords: *liquid zone, instability, vibration.*

Let us consider a vessel containing immiscible incompressible liquids and accomplishing vibro-translation motion with the circular polarization, so that every point of the vessel moves in conformity with the law:

$$\vec{R}(t)=\vec{R}_0+\vec{i}a\cos\omega t+\vec{j}a\sin\omega t \qquad (1)$$

where $\vec{R}_0$ - vessel point coordinates in own reference frame; orthogonal units $\vec{i},\vec{j}$ and the amplitude of the vibrations do not depend on t. In the reference frame, connected with the vessel the equations liquids motion look as:

$$\rho\left(\frac{\partial\vec{v}}{\partial t}+\vec{v}\cdot\nabla\vec{v}\right)=-\nabla p+\eta\Delta\vec{v}+\omega^2\rho(\vec{i}\cos\omega t+\vec{j}\sin\omega t), \qquad (2)$$

$$\operatorname{div}\vec{v}=0.$$

It must be noted that although every point of the vessel moves along a circular trajectory, there is no place for Coriolis forces in (2) since the chosen reference frame does not change in its orientation.

It is necessary to complete the equations (2) with the boundary conditions. On the rigid wall no slip condition must be fulfilled, while at the interface, described by the equation, $F(\vec{r},t)=0$ - the conditions of velocity continuity, stress balance and kinematic condition:

$$[\vec{v}]=0,$$
$$[\sigma]\cdot\vec{n}-[p]\vec{n}=\alpha\vec{n}\operatorname{div}\vec{n}, \qquad (3)$$
$$\frac{\partial F}{\partial t}+\vec{v}\cdot\nabla F=0.$$

Here σ - viscous tension tensor, α surface tension coefficient, $\vec{n}=\nabla F/|\nabla F|$ - vector, normal to the interface, the jumps of the variable at the interface are marked by square brackets.

When the frequency of the vibrations is large enough ($\omega \gg \nu/L^2$, where ν - kinematic viscosity, L - characteristic size) and the amplitude of vibrations is small ($a \ll L\bar{\rho}/\delta\rho$ where ρ is the average density and $\delta\rho$ - is the deference of densities) it is possible to separate the motion into averaged and oscillating parts and to derive the corresponding equations and boundary conditions. The method of averaging is as described in (Ref.1), so we shall give only the results.

To an approximation of small amendments the field of the velocity may be presented in the form:

$$\vec{v} = a\omega(\vec{V}_s \sin \omega t - \vec{V}_c \cos \omega t) + \vec{u} \tag{4}$$

where average velocity $\vec{u}$ and the amplitudes of the oscillating velocities are "slow" functions of time, that is the characteristic times of their changing are large compared with the period of vibrations. The fields of $\vec{u}$, $\vec{V}_s$, $\vec{V}_c$ can be found from the equations:

$$\frac{\partial \vec{u}}{\partial t} + \vec{u}\cdot\nabla\vec{u} = -\frac{1}{\rho}\nabla\Pi + \nu\Delta\vec{u}$$

$$\text{rot } \vec{V}_c = 0, \quad \text{rot } \vec{V}_s = 0,$$

$$\text{div } \vec{u} = 0, \text{ div } \vec{V}_c = 0, \text{ div } \vec{V}_s = 0, \tag{5}$$

$$\Pi = p + \frac{\rho b^2}{4}(\vec{V}_s^2 + \vec{V}_c^2)$$

where $b = a\omega$.

The boundary conditions on the rigid wall are:

$$\vec{u} = 0, \quad V_{cn} = 0, \quad V_{sn} = 0. \tag{6}$$

For the amplitudes of the velocity oscillations we lay down impermeability conditions and not no-slip ones because in our approximation we have to neglect thin viscous skin-layer.

On the interface we have conditions:

$$[\rho\vec{W}_s]\times\vec{n} = 0, \quad [\rho\vec{W}_c]\times\vec{n} = 0,$$

$$[\vec{u}] = 0, \quad [\vec{W}_s]\cdot\vec{n} = 0, \quad [\vec{W}_c]\cdot\vec{n} = 0,$$

$$\frac{1}{4}b^2\left(\left[\vec{W}_s^{(1)}\right]\cdot\left[\vec{W}_s^{(2)}\right] + \left[\vec{W}_c^{(1)}\right]\cdot\left[\vec{W}_c^{(2)}\right]\right)[\rho]\vec{n} - \tag{7}$$

$$[\sigma]\cdot\vec{n} + [\Pi]\vec{n} + \alpha\vec{n}\,\text{div }\vec{n} = 0$$

The designations have being used here are:

$$\vec{W}_s = \vec{V}_s + \vec{j}, \quad \vec{W}_c = \vec{V}_c + \vec{i}.$$

Viscous tension tensor σ has been calculated according to the averaged field of the velocity. Upper indices in brackets are used to distinguish the liquids.

We used the obtained equations and boundary conditions for the problem of the cylindrical liquid zone stability.

Let us consider a cylindrical liquid column of the radius R_1, surrounded by the liquid layer of another density. This system is placed into a rigid cylindrical vessel of the radius R_2, coaxial with the internal liquid zone. In the absence of gravity and other external forces such a state with the cylindrical interface is equilibrated. It is known, that this equilibrium is unstable with respect to axisimmetrical perturbations if the length of the liquid cylinder is large enough (Rayleigh instability). If the density of the external liquid is more than that of the internal one than the development of the instability may be prevented by rotating of the system about its axis. In the opposite case of more dense internal liquid the rotation leads to additional destabilization due to the Rayleigh-Taylor instability in the field of centrifugal forces being added to the capillary instability. In (Ref.2) the stability of the described system has been considered for the case of linear high frequency vibrations along the axis of a symmetry. It has been shown that in this case one more instability mechanism arises being connected with the Kelvin-Helmholz instability.

We have studied the stability of this system under the influence of circularly polarized oscillations (1), when the unit vectors $\vec{i}, \vec{j}$ are perpendicular to the axis of the cylinders.

Let us name quasiequilibrium such a state, where averaged motion is absent though pulsating components of the velocity are not equal to zero.

The problem of the equilibrium can be obtained from (5,7) if we put down $\vec{u}=0$ and $\partial/\partial t=0$. The fields $\vec{W}$ in the both liquids satisfy to the equations:

$$\mathrm{rot}\,\vec{W}=0,\quad \mathrm{div}\,\vec{W}=0 \qquad (8)$$

with the boundary conditions on the rigid wall

$$W_{sr}^{(2)}=\sin\beta,\quad W_{cr}^{(2)}=\cos\beta \qquad (9)$$

and on the interface $r=1+\zeta(z,\beta)$:

$$W_{sn}^{(1)}=W_{sn}^{(2)},\quad W_{cn}^{(1)}=W_{cn}^{(2)}$$

$$W_{s\tau}^{(1)}=\rho W_{s\tau}^{(2)},\quad W_{c\tau}^{(1)}=\rho W_{c\tau}^{(2)}, \qquad (10)$$

$$B(\rho-1)(\vec{W}_s^{(1)}\cdot\vec{W}_s^{(2)}+\vec{W}_c^{(1)}\cdot\vec{W}_c^{(2)})+\mathrm{div}\,\vec{n}=0$$

Here the quantities related to the internal liquid are marked by the indices 1, and to the external one - by the indices 2. We use the cylindrical reference system Z,r,β; the value R is chosen as a unit for length; $\rho=\rho_2/\rho_1$ is the relative density of the external liquid, $B=b^2 r_1 R_1/4\alpha$ is the vibrational parameter.

The problem (8-10) allows the solution with the cylindrical interface $\zeta=0$ and nonzero components of fields $\vec{W}$:

$$W_{sr}^{(1)}=(P+1)\lambda\sin\beta,\quad W_{sr}^{(2)}=(1+\frac{P}{r^2})\lambda\sin\beta,$$

$$W_{cr}^{(1)}=(P+1)\lambda\cos\beta,\quad W_{cr}^{(2)}=(1+\frac{P}{r^2})\lambda\cos\beta,$$

$$W_{s\tau}^{(1)}=(P+1)\lambda\cos\beta,\quad W_{s\tau}^{(2)}=(1+\frac{P}{r^2})\lambda\cos\beta,$$

$$W_{c\tau}^{(1)}=-(P+1)\lambda\sin\beta,\quad W_{c\tau}^{(2)}=-(1-\frac{P}{r^2})\lambda\sin\beta,$$

(11)

Conventional signs introduced in (11) are

$$P=\frac{\rho-1}{\rho+1},\quad R=\frac{R_2}{R_1},\quad \lambda=\frac{R^2}{P+R^2}$$

Let us consider small perturbations of the solutions (11). We introduce the potentials ϕ, ψ for perturbations of the fields $\vec{W}_s$, $\vec{W}_c$, linearize the problem (8-10) in the vicinity of the solution (11), carry the boundary conditions to the unperturbed interface, and as a result - obtain ϕ and ψ satisfying to the Laplace equation. The conditions on the external boundary $r=R$ are:

$$\phi_r^{(2)}=\psi_r^{(2)}=0, \qquad (12)$$

on the interface:

$$\phi_r^{(1)}-\phi_r^{(2)}=-2P\lambda(\zeta\sin\beta)_\beta,$$

$$\psi_r^{(1)}-\psi_r^{(2)}=2P\lambda(\zeta\cos\beta)_\beta,$$

$$\phi^{(1)}-\rho\phi^{(2)}=2\rho P\lambda(\zeta\cos\beta), \qquad (13)$$

$$\psi^{(1)}-\rho\psi^{(2)}=2\rho P\lambda(\zeta\sin\beta),$$

$$2B\rho P\lambda(\cos\beta(\phi_r^{(1)}+\frac{1}{r}\psi_\beta^{(1)})+$$

$$\sin\beta(\psi_r^{(1)}-\frac{1}{r}\phi_\beta^{(1)}))+\zeta+\zeta_{\beta\beta}+\zeta_{zz}=0$$

where low indices stand differentiation on the corresponding variable.

The conditions of the problem (12-13) nontrivial solutions existence define the bifurcational surface in the parameters space being simultaneously the surface of neutral stability.

The equation of this surface for the periodical on β and z solutions with the azimutual number m and wave number k is

$$2B\rho P^2\lambda^2(k\rho I_m(Q_{m+1}+Q_{m-1}) - (\rho-1)(m+1)I_{m+1}Q_{m+1} + (\rho-1)(m-1)I_{m-1}Q_{m-1}) - -1+k^2+m^2=0, \quad (m\geq 1) \tag{14}$$

where $I_m(k)$ and $K_m(k)$ are modified Bessel functions and

$$Q_{m\pm1}=\frac{I_m+\vartheta_{m\pm1}K_m}{(\rho-1)I'_{m\pm1}(I_{m\pm1}-\vartheta_{m\pm1}K_{m\pm1})-\vartheta_{m\pm1}/k}$$

$$\vartheta_m=\frac{I'_m(kR)}{K'_m(kR)}$$

For axisimmetric perturbation with m = 0 we have instead of (14):

$$4B\rho^2P^2\lambda^2\frac{(I_0+\vartheta_1K_0)(\rho kI_0-(\rho-1)I_1)}{Q_1} - 1+k^2=0 \tag{15}$$

Neutral curves of the liquid zone stability are qualitatively descraibed in Fig.1 for three different values of the vessel and the interface radii ratio R. Instability regoins are shaded.In the case of B = 0 the cylindrical interface is unstable in respect of longwave perturbations with $k<1$. The vibrations supress longwave perturbations and on the other hand lead to the appearance of a new type of instability - the wave relief generation.

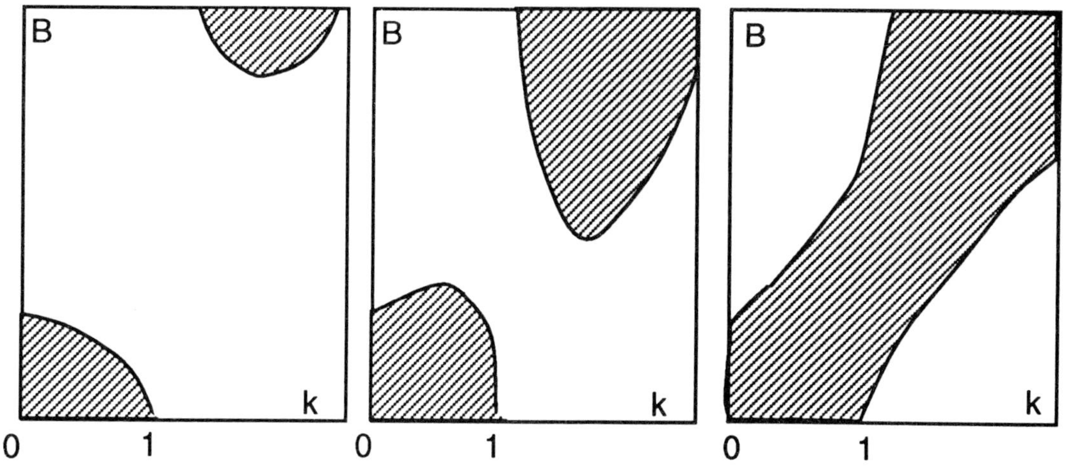

Fig.1 Neutral curves for the different values of radii ratio R: a) $R=R_a$, b) $R=R_b$, c) $R=R_c$; $R_a < R_b < R_c$

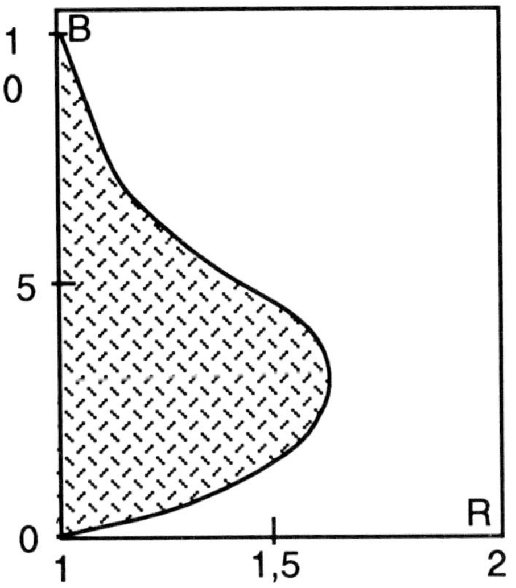

Fig.2 Stability diagram.

For small radii ratio there exists the interval of the vibrational parameter B values, where longwave perturbations supressed and shortwave are not generated. This interval becomes narrower with increasing R and at R = R* instability areas overlap and the stabilization of a cylindrical interface becomes impossible. Critical radii ratio R* has fond to be 1,578 . Stability diagram in the parameters plane B - R is qualitatively descraibed in the Fig.2. The stability region is shaded. It must be noted that the shape of this region depends on the densities ratio, but the value of R* does not change with the varying of ρ. The calculations show, that nonaxisymmetric perturbations ($m \neq 0$) are less critical and do not influence on the general stability behavior.

REFERENCES

1. D.V.Lyubimov and A.A.Cherepanov, *Izv. AN SSSR, Mekhanika zhidkostei I Gazov* 6(1986)8. (in Russian)

2. V.A.Briskman, D.V.Lyubimov and A.A.Cherepanov, *The numerical and experimental modelling of hydrodynamical phenomena at the weightlessness,* Sverdlovsk, UrO AN SSSR, (1988)18. (in Russian)

EQUILIBRIUM SHAPES AND STABILITY OF LIQUID BRIDGES SUBJECTED TO A.C. ELECTRIC FIELDS

A. Castellanos, H. González and A. Ramos.
Dpto. Electrónica y Electromagnetismo, Universidad de Sevilla, Spain.

Abstract

We have performed theoretical, numerical, and experimental studies of liquid bridges under the influence of an a.c. electric field.

The stability boundaries of cylindrical bridges as a function of the bridge slenderness, the ratio of permittivities, electrical Bond number and Weber number have been determined by means of a statical analysis based on normal mode decomposition. Experimental results on this geometrical configuration have been obtained using a non-rotating Plateau tank facility. Experimental and theoretical results are in good agreement. Finite elements have been implemented to obtain the equilibrium shapes and bifurcation points for axisymmetric non-cylindrical bridges with arbitrary volume subjected to an electric field.

The Lyapunov-Schmidt method has been used to determine the effect of residual gravity upon the stability of the perfectly cylindrical liquid bridge. Results obtained by means of the finite element method have confirmed the previous analysis and extended their results a finite distance from the bifurcation points.

Keywords: *Electrohydrodynamics, Liquid Bridge, Microgravity.*

1 Introduction

The study of the behaviour of liquid menisci under effects such as gravity, rotation, and electromagnetic fields is a difficult electro-fluid mechanical problem of great interest for its potential technological application to several industrial processes.

A typical example is the floating zone melting technique which has been widely employed to produce single crystals of high quality and to purify materials with high melting points. Renewed interest on these problems has arisen from the possibility of using these techniques in a microgravity environment. Experiments on crystal growth have been already performed during Skylab, Salyut and Spacelab missions. In particular, the knowledge of the mechanical behaviour of the molten zone is of interest in order to obtain pure crystals of high quality (Hurle et al., 1987).

In this paper, we present both theoretical and experimental results on the equilibrium shapes and stability of floating liquid zones (liquid bridges) under the influence of an intense a.c. electric field in microgravity environments.

2 Formulation of the problem

Let us consider an isothermal dielectric incompressible liquid of density ρ_i and electrical permittivity ϵ_i anchored on two sharp-edged metallic rings of radius R and heigth h $(h \ll R)$ welded on two opposite plane-parallel electrodes, a distance L apart and subjected to a potential difference ϕ_0. This liquid bridge is surrounded by another fluid of density ρ_o, electrical permittivity ϵ_o and both are assumed to be immiscible, with σ the interfacial surface tension (see figure 1 for a schematic drawing). In the absence of gravity the outer medium could be the vacuum itself. The liquid bridge is assumed to be axisymmetric with free contact angles between the liquid and the electrodes, and rotating about the symmetry axis with angular velocity Ω. Both liquids are assumed to be homogeneous with no free charges, therefore the electric forces that appear are of dielectric origin. Because the liquids are incompressible no electrostriction effect is considered.

In a reference system that rotates with the same angular velocity Ω, static solutions are governed in the volume by the Laplace equation and on the interface by the augmented Young-Laplace equation,

$$\nabla^2 \phi = 0 \tag{1}$$

$$\Delta \Pi - Bz + \frac{W}{2} f^2 + \frac{\chi}{2} \Delta \left[\epsilon E^2\right] - \frac{\chi}{|\nabla F|^2} \Delta \left[\epsilon (\nabla F \cdot \mathbf{E})^2\right] + \nabla \cdot \mathbf{n} = 0 \tag{2}$$

where ϕ, Π and f are the electric potential, the total pressure and the interface radial coordinate, $\mathbf{E} = -\nabla\phi$ is the electric field , $\mathbf{n} = \nabla F/|\nabla F|$ with $F \equiv r - f(z) = 0$ is the normal to the interface (pointing into the surrounding medium), and the

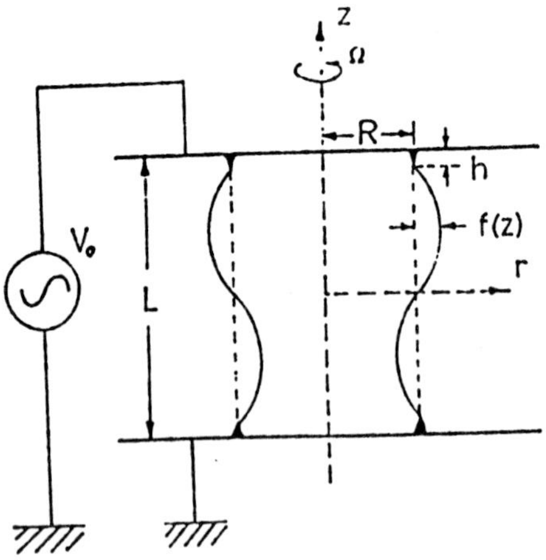

Figure 1: Liquid bridge configuration.

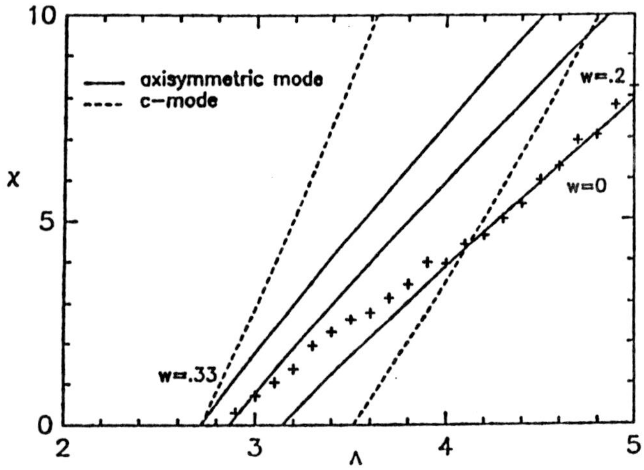

Figure 2: Bifurcation curves in the $\chi - \Lambda$ plane for $\beta = 0.55$ and three values of the Weber number: (a) $W = 0$; (b) $W = 0.2$; (c) $W = 0.33$. Solid lines correspond to bifurcation to an axisymmetric shape, and dashed lines to a c-mode (m=1). Experimental points (+) correspond to $W = 0$.

symbol Δ represents the difference in a quantity as we cross the interface, $\Delta X = X_o - X_i$. Equations (1) and (2) are already dimensionless because length is measured in units of R, the electric field in units of $E_\infty = \phi_0/L$, and ϵ in units of the inner permittivity. The parameters are: $\Lambda = L/2R$ the slenderness; $\chi = \epsilon_i E_\infty^2 R/\sigma$, the electrical Bond number; $\beta = \epsilon_o/\epsilon_i$, the nondimensional permittivity of the outer medium; $B = gR^2 \Delta \rho/\sigma$ the gravitational Bond number; and $W = \Delta\rho\Omega^2 R^3/\sigma$ the Weber number.

The total pressure jump $\Delta\Pi$ is determined by constraining the liquid bridge volume V to be fixed

$$\frac{1}{2}\int_0^{2\pi} d\theta \int_{-\Lambda}^{\Lambda} dz\, f^2 = \frac{V}{R^3} = 2\pi\Lambda\tau \qquad (3)$$

where $\tau \equiv V/\pi R^2 L$ is the ratio between the real volume and that of the cylinder with the same R and L.

The associated boundary conditions are

$$f(\theta, \pm\Lambda) = 1, \qquad (4)$$

$$\phi(r,\theta,\Lambda) = 2\Lambda, \quad \phi(r,\theta,-\Lambda) = 0, \qquad (5)$$

$$\frac{\partial\phi}{\partial r} = 0 \quad \text{en } r = 0, \quad \lim_{r\to\infty}\phi = z + \Lambda \qquad (6)$$

$$\Delta\phi = 0, \quad \nabla F \cdot \Delta(\epsilon \mathbf{E}) = 0 \qquad (7)$$

Equations (1)–(3) with conditions (4)–(7) define a boundary problem whose solution determines the interface position as a function of the space coordinates and time.

3 Liquid bridges in the absence of gravity

3.1 Cylindrical solution stability

The cylinder is the only analytical solution known in closed form, that verifies the equilibrium equations for all values of Λ, χ, β and W. The linear stability of this solution is examined by means of a static modal analysis by writting $f(\theta, z) = \sum_m f_m(z)e^{im\theta}$ and $\phi(r,\theta,z) = \sum_m \phi_m(r,z)e^{im\theta}$. This leads to an independent set of equations for each mode. The method of solution is based on a Fourier expansion of all the functions depending on z, as shown in González et al. (1989) We then look for the bifurcation points in the parameter space $\{\Lambda, \beta, \chi, W\}$, i.e., those for which a nontrivial solution of the linearized equations exists. It is found that the bifurcation points form an infinite family of nested surfaces in the four-dimensional parameter space.

In figure 2 the stability map for three selected values of the Weber number W and $\beta = 0.55$ is presented. The value of $\beta = 0.55$ was selected in order to compare with the experimental data obtained from a liquid bridge formed with silicone oil, surrounded by isodense ricinus oil, in a nonrotating ($W = 0$) Plateau tank facility. The experimental data points (+) serve to verify the theoretical analysis both qualitatively and quantitatively, for the case of $\beta = 0.55$ and $W = 0$ (González et al., 1989).

A dynamical study for $W = 0$ (González et al., 1988) shows that the stable region is bounded by the surface with the minimum value of Λ, in accord with the experiments. Continuity arguments lead to the same statement for all W. Now two different ways of destabilization are possible: the axisymmetric mode $m = 0$ (solid lines), and the c-mode ($m = 1$), with dashed lines (González and Castellanos, 1990). The strongly stabilizing effect of the applied electric fields, as well as the destabilizing role played by the rotation is readily apparent from this figure.

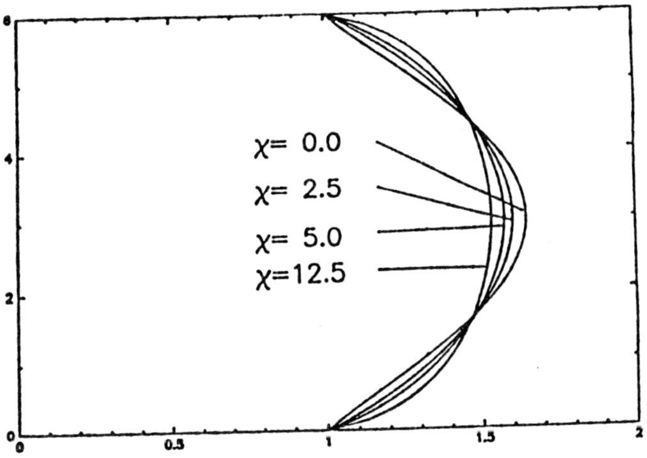

Figure 3: Equilibrium shape interface for liquid bridges with $\tau = 2.0$, $\beta = 0.5$, $\Lambda = 3.0$ and $\chi = 0.0, 2.5, 5.0$ and 6.25.

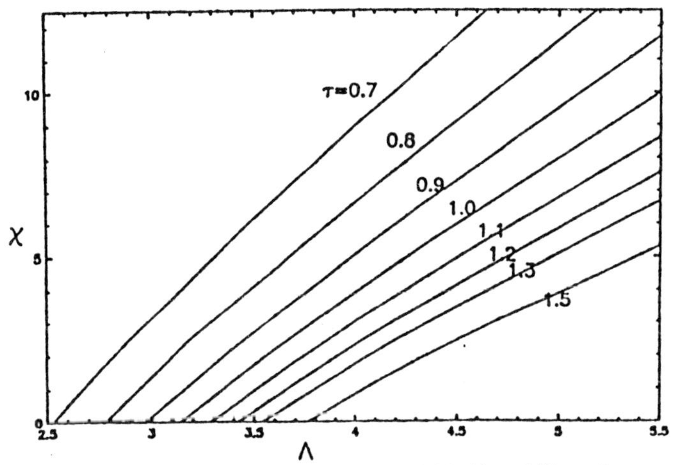

Figure 4: Graph in χ–Λ plane of liquid bridges bifurcation curves for a set of different τ values when β is fixed to 0.55.

3.2 Axisymmetric shapes with arbitrary volume

Here we will restrict ourselves to axisymmetric deformations. Since in this case the equilibrium and Laplace equations cannot be solved in analytical form, numerical methods have to be used (Ramos and Castellanos, 1991). The equilibrium equation (2) was discretized using a centered finite difference scheme of second order for z-derivatives, the potential was obtained using linear finite elements, and finally the volume constraint was discretized using also linear finite elements for $f(z)$. This leads to a system of nonlinear equations whose solution was obtained by the Newton-Raphson method.

Equilibrium shapes. We present in figure 3 the equilibrium shapes corresponding to liquid bridges with $\beta = 0.5$, $\tau = 2.0$ and $\Lambda = 3.0$ subjected to different electric field intensities χ from $\chi = 0.01$ to $\chi = 12.5$. It may be easily shown from this figure that the main effect of the field is to align the interface with the applied electric field except for a small region near the fixed contact lines. The same behaviour was obtained for different permittivity ratios and volumes.

Bifurcation points. To determine the set of bifurcating points we started from a known stable equilibrium shape and increased a chosen parameter until the jacobian of the Newton method went to zero. This method was used to determine the bifurcation curves in the $\chi - \Lambda$ plane for a fixed relative permittivity, $\beta = 0.55$ and a set of different values of the parameter τ. These curves are plotted in figure 4. The value of β was chosen to compare with experimental data to be obtained in the Plateau tank facility reported in González et al. (1989). Notice that the curve for $\tau = 1$ is the same as the one given in figure 2. When $\chi = 0$, the results of minimum volume are coincident with those given by Martínez and Perales (1986). Due to the relationship between bifurcation and neutral stability, these curves delimitate the region of stability.

4 The effect of axial residual gravity

Gravity will be considered to be small enough to be treated as and imperfection of the previous theoretical model. Hence the Lyapunov-Schmidt method, extensively used in the absence of electric fields, is suitably generalized to be applicable to our problem (see Myshkis et al. 1987 for a theoretical description of the method).

Define the vector $\mathbf{x}$ and the inner product of two vectors $\mathbf{x_1}$, $\mathbf{x_2}$ as

$$\mathbf{x} = (\phi^{in}(r,\theta,z), \phi^{ex}(r,\theta,z), f(\theta,z), \Delta\Pi) \quad (8)$$

$$\langle \mathbf{x}_1, \mathbf{x}_2 \rangle \equiv \chi_0 \int_{V_{in}} dV \phi_1^{in}\phi_2^{in} + \beta\chi_0 \int_{V_{ex}} dV \phi_1^{ex}\phi_2^{ex} + \int_{S_l} dS f_1 f_2 + \Delta\Pi_1 \,\Delta\,\Pi_2 \quad (9)$$

Then the linear part $\mathcal{A}$ of the nonlinear operator $\mathcal{F}$ corresponding to eq. (1)–(3) and boundary conditions (4)–(7) is self-adjoint. Denoting by $\mathbf{v}$, the nontrivial solution to the homogeneous equation $\mathcal{A}\mathbf{v} = 0$, which is the eigenvector corresponding to the critical eingevalue (bifurcation point), we write:

$$\mathbf{x} = \varepsilon\mathbf{v} + \mathbf{x}_\perp, \quad \text{with} \quad \mathbf{x}_\perp \equiv (\varphi^{in}, \varphi^{ex}, u, \pi) \quad (10)$$

$$\mathbf{x}_\perp = \sum_{i+j+k>0} \bar{\chi}^i \varepsilon^j B^k (\mathbf{x}_\perp)_{ijk} \quad (11)$$

where $\bar{\chi} = \chi - \chi_0$ and B is the Bond number. Substituting series expansion (11) into the operator equation $\mathcal{F}(\chi, B, \mathbf{x}_\perp) + \psi\mathbf{v} = 0$, with the scalar

ψ also expanded in series of the same parameters $\psi = \sum_{i+j+k>0} \bar{\chi}^i \varepsilon^j B^k \psi_{ijk}$, we end up with the following bifurcation equation

$$\varepsilon(\psi_{110}(\Lambda, W)\bar{\chi} + \psi_{030}(\Lambda, W)\varepsilon^2) + \psi_{001}(\Lambda, W)B + \ldots = 0 \quad (12)$$

The coefficients are determined integrating the hierarchy of non-homogeneous linear operator equations by the method of lines, discretizing in z and solving analytically for the potential in r.

For $B = 0$ and $W = 0$ we obtain a subcritical pitchfork bifurcation for the cylindrical brigde (notice that the stable region is for values of $\chi > \chi_0$). For $B \neq 0$, i.e. when an axial residual gravity is present, the pitchfork bifurcation for cylindrical bridges subjected to electrical fields has changed to a rupturing of the bifurcation point into two limit points, as shown in figure 5. The lower part of the upper branch corresponding to the values $\chi_c \leq \chi < \infty$ is the physical observable branch.

The bifurcation diagram has been also obtained by means of a method of subdomains, tessellated into a set of curvilinear quadrilateral elements (with nine nodes), bordered by fixed spines parallel to the r-axis. Galerkin/finite element weighted residuals of the augmented Young-Laplace equation for the surface and of the Laplace equation for each volume are formed. Together with the constraint of fixed volume a final set of nonlinear equations on the unknowns is obtained, which is solved by Newton-Raphson method. To continue along any particular solution at the bifurcation points, the eigenvector of the Jacobian are first computed and then continuation proceeds on the desired direction. In anticipation of turning points that may appear in the system a continuation method based on the arc length parameter was implemented. The tangent vector to the branch solution, normalized to one, was added as another equation. The augmented Jacobian remains now regular at the possible turning points.

The numerical results obtained by this method are also presented in figure 5. There is a satisfactory agreement with those obtained previously by the Lyapunov-Schmidt method. Experiments to compare with the obtained theoretical results are now under way.

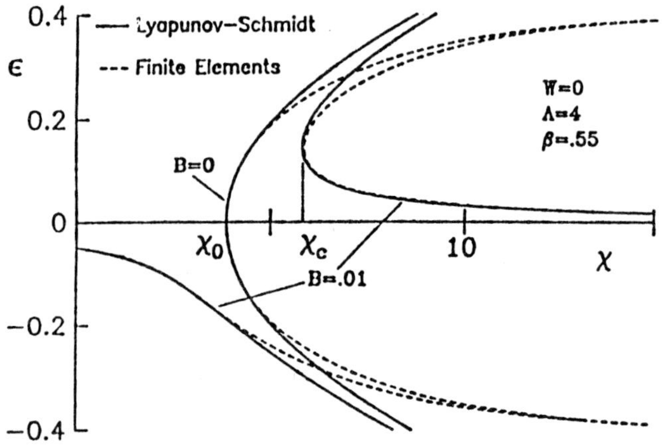

Figure 5: A typical bifurcation diagram for a liquid bridge subjected to an axial electric field and residual gravity ($B = 0$ and $B = 0.01$). Solid lines are local results obtained by the Lyapunov-Schmidt method and dashed lines are numerical results computed with a Galerkin/finite element method (see text).

Acknowledgement

We acknowledge the financial support provided by the Consejería de Educación de la Junta de Andalucía. We also wish to acknowledge fruitful discussions with Prof. A. Barrero.

REFERENCES

1. Hurle, D.T.J., Muller. G., Nitsche, R. Crystal growth from the melt. In *Fluid Sciences and Materials Science in Space*, cap. X. Springer, 1987.

2. González, H., McCluskey F. M. J., Castellanos, A., Barrero, A. 1989. Stabilization of dielectric liquid bridges by electric fields in the absence of gravity. *J. Fluid Mech.*, **206**, pp. 545–561.

3. González, H., McCluskey, F. M. J., Castellanos, A. and Gañán, A. 1988. Small oscillations of liquid bridges subjected to a.c. fields, in *Synergetics, Order and Chaos*. Ed.: Velarde, M.G., World Scientific, Singapoore.

4. González, H. and Castellanos, A. 1990 The effect of an axial electric field on the stability of rotating dielectric cylindrical liquid bridge. *Phys. Fluids A* **2** (11), pp. 2069-2071.

5. Ramos, A. and Castellanos, A. 1991 Shapes and stability of liquid bridges subjected to a.c. electric fields. *J. Electrostat.* (to appear).

6. Martínez, I., Perales, J. M. 1986 Liquid bridge stability data. *J. Crystal Growth*, **78**, pp. 369-378.

7. Myshkis, A.D., Babskii, V.G., Kopachevskii, N.D., Slobozhanin, L.A., Tyuptsov, A.D. 1987. *Low-gravity fluid mechanics*. Springer Verlag. Berlín. Heidelberg.

Numerical Model of Liquid Bridge Forming Under Microgravity Considering Wettability

Hiroshi SAKUTA, Yasushi FUKUZAWA, Masaaki OKADA
Yo KOJIMA, Daisuke MARUYAMA

Nagaoka University of Technology,
Department of Mechanical Engineering.
1603-1 Kamitomioka Nagaoka, 940-21 Japan.

ABSTRACT

Estimation of physical properties concerning microgravity, mainly the contact angle between liquid and wall of container, wettability and surface tension of liquid, is difficult due to various experimental restrictions. A technique to compare experimental data and results of numerical analysis and to estimate these properties is presented. Experimental results are taken as the initial conditions for numerical analysis of physical properties. In numerical procedures, the solution is optimized in the surface shape, and then converged to a proper value. The procedures have been implemented in a workstation.

In this paper, the results in zero gravity are reported. In some cases, the shape of outer surface of liquid bridge and contact angles are obtained through the analysis system proposed.

Keywords: *Contact Angle, Wettability, Surface Tension*

1. THEORETICAL BACKGROUND

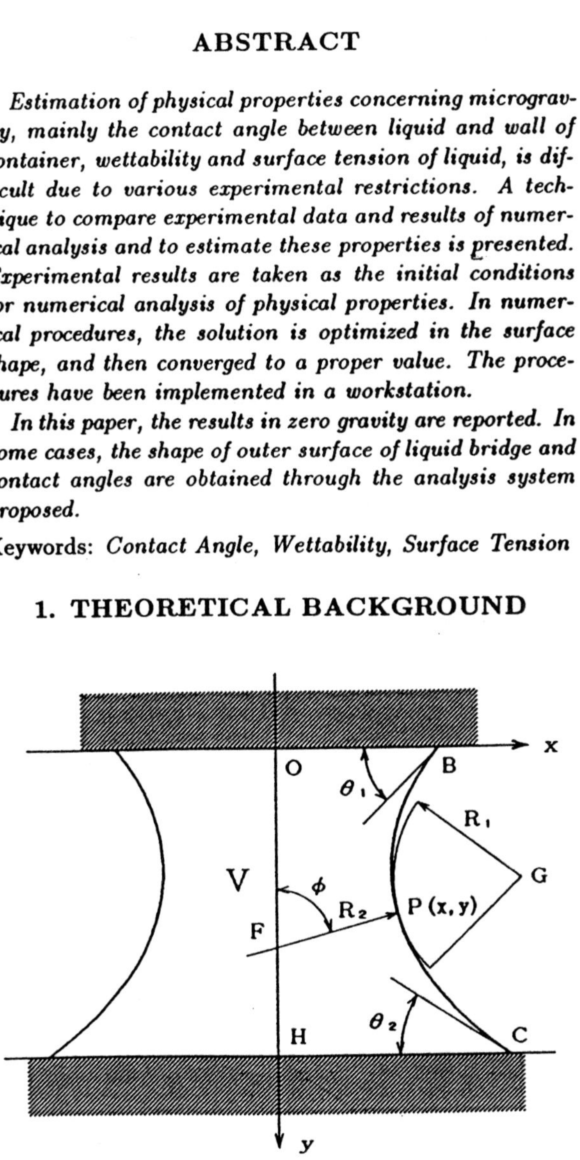

Fig.1 Liquid bridge model.

The Laplace equation of capillarity is a statement of mechanical equilibrium between two homogeneous fluid separated by an interface. It relates the pressure difference across a curved interface to the surface tension and the curvature of the interface:

$$\Delta p = \gamma \left(\frac{1}{R_1} + \frac{1}{R_2} \right) \tag{1}$$

where γ is the surface tension, R_1 and R_2 represent the two pricipal radii of curvature, and Δp is the pressure difference across the surface. For an axisymmetrical fluid surface in the absence of external forces, other than gravity, the Laplace equation can be written (ref. to Fig.1 for the definition of the geometrical variables x, y, R_1 and R_2) in the next form:

$$\frac{\Delta p_0 + \rho g y}{\gamma} = \frac{1}{R_1} + \frac{1}{R_2} = \kappa_1 + \kappa_2 \tag{2}$$

where ρ is the liquid density, g is the gravity, Δp_0 is the pressure difference at the top (point 'O' in Fig.1).

Through the Laplace equation, we know that at every point on the surface the difference in pressure due to the local radius of curvature and the surface tension must balance the hydrostatic pressure due to the liquid column held above the bulk liquid level.

Now, we consider an axisymmetric surface, and define this surface as follows:

$$p(x, y) = (x \cos\phi, x \sin\phi, y), \text{ for } 0 \leq \phi \leq 2\pi \tag{3}$$

Then, the mean curvature $H = \frac{1}{2}(\kappa_1 + \kappa_2)$ is given by the differential geometry, as follows (from Ref.[2]):

$$2H = \frac{1}{x\sqrt{1 + \left(\frac{dx}{dy}\right)^2}} - \frac{\frac{d^2x}{dy^2}}{\left\{1 + \left(\frac{dx}{dy}\right)^2\right\}^{\frac{3}{2}}} \tag{4}$$

Therefore, Eq.2 is substituted and arranged. Finally, the differential equation is

$$\frac{d^2x}{dy^2} = \frac{1}{x}\left\{1 + \left(\frac{dx}{dy}\right)^2\right\} - \frac{p_0 + \rho g y}{\gamma}\left\{1 + \left(\frac{dx}{dy}\right)^2\right\}^{\frac{3}{2}} \tag{5}$$

2. CALCULATION METHOD

The solution for Eq.5 is only available in an analytical form in a few cases and thus should be solved in general by numerical means. The proper boundary conditions for the above differential equations are two contact angles and the liquid volume. conservation, that is:

1) CONTACT ANGLES:

$$\begin{cases} y = 0 \ : \dfrac{dx}{dy} = -\cot\theta_1 \\ y = H : \dfrac{dx}{dy} = \cot\theta_2 \end{cases} \tag{6}$$

2) LIQUID VOLUME:

$$\pi\int_0^H x^2 dy = V(\text{const.}) \tag{7}$$

Where θ_1 and θ_2 are , respectively, the contact angles at upper wall and lower walls. H is the distance between both walls. and V is the volume of liquid bridge that must be constant.

This liquid bridge form is obtained as the solution of a boundary value problem of nonlinear second-order differential equation. To solve this problem, we replace it by an initial value problem, and define the following system of equations:

$$\begin{cases} \cot\theta_2 = \theta(x_0, p_0) = \left.\dfrac{dx}{dy}\right|_{y=H} \\ V = V(x_0, p_0) = \pi\displaystyle\int_0^H x^2 dy \end{cases} \tag{8}$$

where x_0 and p_0 are values at $y = 0$, (the initial values.)

The numerical methods used are Runge-Kutta-Gill method for initial value problem and Newton-Raphson method for simultaneous equations. The calculations were programmed in C compiler and carried out on a Sun386i (Sun micro systems) workstation.

3. RESULTS

The working liquid is glycerin and materials of wall are Nickel and/or Teflon whose physical properties are measured by a series of experiments. The parameters (bridge height H, volume V and gravity g) are assumed to be independent of time. The physical properties and parameters of calculation are:

Table 1. Simulation parameters.

Surface tension γ	0.0634	[N/m]
Liquid density ρ	1264	[kg/m^3]
Liquid volume V	4.0	[cm^3]
Bridge height H	10.0	[mm]
Contact angle (Nickel)	30.0*	[deg]
Contact angle (Teflon)	60.0*	[deg]

* : from Experiment.

The results of liquid bridge simulation for some combination of wall materials are shown in Table 2. In each sketch outer shape of liquid bridge is represented.

Table 2. Results of calculation.

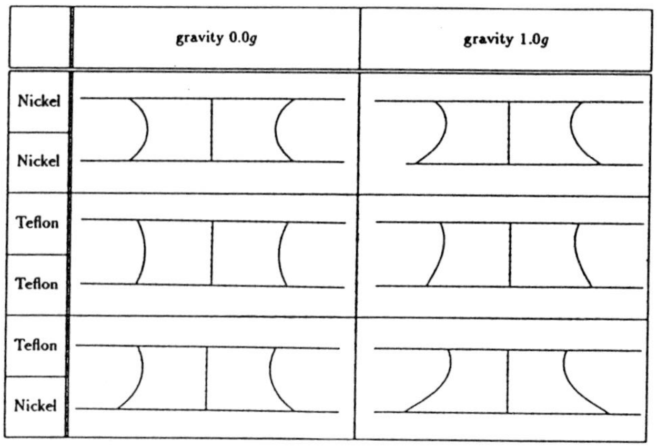

The experimental results are shown in Table 3 (from Ref.[6]).

Table 3. Results of experiment.

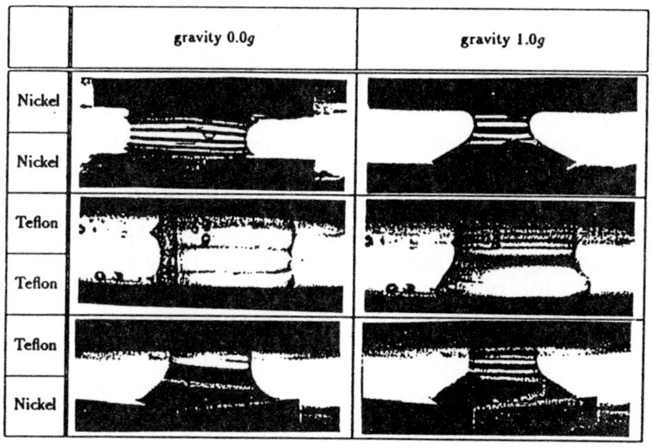

4. DISCUSSION

Numerical results can be compared with experiments under $1.0g$ and $0.0g$. Each photograph on Table 3 corresponds with Table 2. Generally, the wettability of Nickel is larger than that of Teflon. Then, for every condition, numrical results agree with experimental ones. This demonstrates the applicability of the procedure.

A new technique for calculating the liquid bridge shape considering liquid-solid wettability has been presented. The parameters for calculation are contact angles at upper/lower walls and the volume of liquid bridge.

In order to determine physical properties (i.e. in Table 1), this theory can be applied the problem. The outer shape of liquid bridge is compared with the experimental results.

5. CONCLUSION

The work presented above gives following results:

1. The liquid bridge shape can be computed by this theory with the proposed calculation method.

2. The parameters of liquid-solid wettability are defined by the contact angles as well as the pressure distribution at the wall.
3. The relationship of gravity and wettability is influenced by the pressure distribution.

6. REFERENCE

[1] K. Harada and Y.Muramatsu,
"Measurements of the Surface Tension and Contact Angle of Molten Metals by Contour Curve Fitting Method", J.Japan Inst. Metals, Vol.52, No.1(1988), pp.43-49.

[2] S. Kobayashi,
"Differential geometry of a curves and surfaces", Selection of Basic Mathematics No.17, Syoukabou.

[3] S. Ono,
"Surface tension", Physics One Point No.9, Kyouritsu Syuppan.

[4] R.A.Brown, F.M.Orr, JR., and L.E.Scriven,
"Static Drop on an Inclined Plate: Analysis by the Finite Element Method", Journal of Colloid and Interface Science, Vol.73, No.1, Jan. 1980.

[5] Perales J.M.Perales,J.Mesegure and I.Martinez,
"Journal of Crystal Growth", Vol.110(1991),pp.855-591.

[6] H. Kimura, M. Shimizu, Y. Fukuzawa, M. Okada, K Yamaguchi and S. Ishikura,
"Evaluation of wettability on the solid-liquid interface under microgravity by liquid bridge method", in this symposium.

EVALUATION OF WETTABILITY ON THE SOLID-LIQUID INTERFACE UNDER MICROGRAVITY BY LIQUID-BRIDGE METHOD

Y.Fukuzawa*, H.Kimura**, M.Shimizu**, M.Okada*,
K.Yamaguchi*, A.Kofuji* and S.Ishikura***

*:Nagaoka University of Technology Department of Mechanical Engineering. 1603-1 Kamitomioka Nagaoka,940-21,Japan.
**:National Space Development Agency of Japan. Shiba Ryoushin Bldg. 2-5-6, Shiba Minato-ku, Tokyo, 105.Japan.
***:Japan Space Utilization Promotion Center.

ABSTRACT

On the space material and biological experiments came out to be a failure due to neglecting the wetting phenomena at the liquid-solid interface. In order to examine the effects of gravitational conditions for the wettability between the liquid of interest and the wall of the container, the liquid bridge method was carried out at the ground (1g) and below the 1/100 g at a parabolic flight. The wetting behaviours were observed by the liquid bridge shapes and the contact angles, which were estimated after observing the video recording systems. The statical reduced wetting behaviours were evaluated and on the parabolic flight the gravitational effects were detected from 0.1g to 2.5g. The variation of the materials combination affected significantly the gravitational effects for the wetting behaviours. The liquid bridge shapes were analyzed by the hydro-mechanical model considering the wetting behaviour and gravitational effects.

Keywords: Liquid bridge, parabolic flight, microgravity, wettability, interface

1. INTRODUCTION

It is well known that some material experiments had led to fail due to neglecting the effects of wettability between the molten material (liquid state) and the container (solid state) in space. (Ref.1) Thus, it is required to estimate the effects of the gravity on the wetting behaviour at the liquid-solid interface. Many attempts had been tried to obtain precisely the value of the static contact angle at the interface on the ground (≒ 1g) in the last 2 decades (Refs.2,3,4,5) but few experimental data exist on wettability concerning the hysteresis contact angle and the microgravity effects. In this report, we propose a new experimental method to evaluate the wettability at the liquid-solid interface under microgravity, and obtain some experimental data by using an aircraft in parabolic flight.

2. EXPERIMENTAL PROCEDURE

The materials used are a teflon and nickel plate as the solid state material, and commercial glycerin as the liquid. The wetting behavior was observed by the liquid-bridge method as shown in Figure 1. This method was used in hydrodynamics area without considering the wetting effects between the solid-liquid interface.(Ref.5) In this system, the compressed glycerin liquid is pull out together with the parallel solid plate at the beginning of the microgravity environment, and after 6s of pulling, the elongated liquid-bridge is held statically untill the end of the parabolic flight. On a standard cace, the microgravity condition continued about 20s like in Figure 2, and after that the gravity changed from μ g to 2.5g and returned to 1g finally. The experimental conditions are shown in Table 1. The liquid bridge shapes were measured with CCD video recording system and by

Table 1. Experimental condition

Tensile speed	$(\times 10^{-3}$m/s)	0.58
Moving time	(s)	5.9
Test temperature	(K)	296~298
Air humidity	(%)	22.8~23.5
Atmospheric pressure	(kPa)	94.2~95.2
G value	(G :1G = 9.8 N)	0.0~2.2
Average μG time	(s)	17

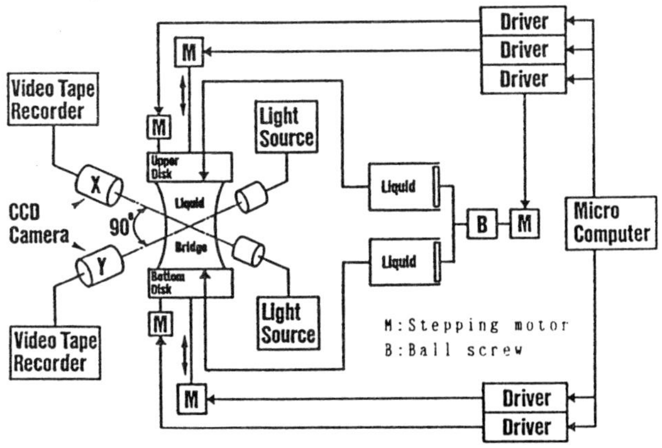

Figure 1:Block diagram of experimental system

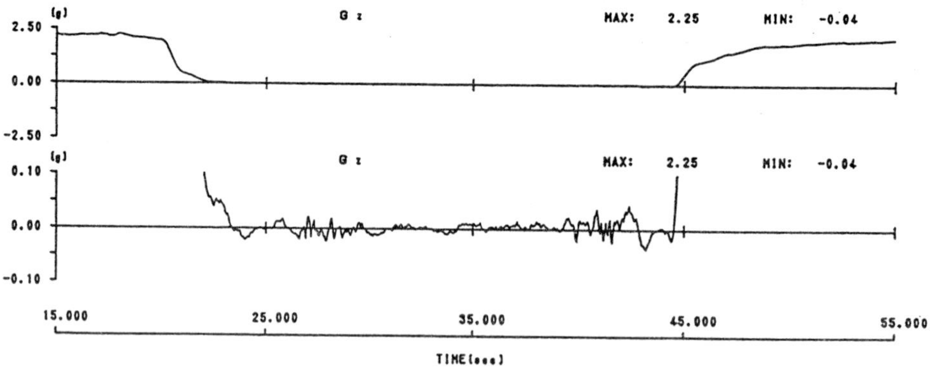

Figure 2:Typical pattern of μ g change at Z direction

observing the shapes the image-processing system, and the contact angle were determined. The gravitational effects on the wetting behaviour were observed and considered.

3. RESULTS AND DISCUSSION

The wetting behaviors were detected from the liquid-bridge shapes under 1g and μ g with the following material combinations;

(a) Teflon(upper)-Teflon(lower),
(b) Nickel(upper)-Nickel(lower),
(c) Teflon(upper)-Nickel(lower).

The typical liquid-bridge shapes at 1g and μ g are shown in Figure 3 and Figure 4. As the liquid volume is held constant during the disk motion, the area of solid-liquid interface become smaller than in the original position. It is well known that there is a lot of difference between the advancing and the receding contact angle. So special attention should be paid to determine the wettability in these cases. We have been able to measure the dynamic and static receding contact angles. The following results were obtained from the photographs.

(1): Teflon had a large contact angle near 80° to the glycerin at 1g.
(2): Nickel had a small contact angle about 40° to the glycerin at 1g.
(3): Under 1g, the liquid-bridge shape was broken in case of teflon/nickel.
(4): The aspect ratio changed between upper and lower disk under 1g, but turned to almost the same value under μ g.
(5): Some gravitaional effects were detected on the liquid-bridge shape and contact angle.
(6): On the teflon/nickel under μ g,nickel side contact angle is smaller than in the nickel/ nickel case.

On the parabolic flight experiment, the gravitational condition was changed from μ g to 2.5g continuously, so we are able to detect the gravitational effects on the contact angle. These dependence is represented in Figure 5 (a),(b) and (c). The experiments were carried out with the poor and good wettability plate of teflon and nickel. The following phenomena were observed:

On the combination (a), teflon/teflon, the contact angles became almost the same value under microgravity than on ground, but the lower side contact angles (θ_{TL}) became lager than the upper side and increased to the constant value rapidly with increasing value of gravity. On the other hand, the upper side contact angles (θ_{TU}) remained constant independent of the gravity conditions.

On the combination (b), nickel/nickel, below the 1g conditions the contact angles remained as under microgravity condition, over the 1g conditions the lower side values (θ_{NL}) decreased but the upper side values (θ_{NU}) increased slightly.

On the combination (c), teflon/nickel, the upper side contact angle (θ'_{TU}) changes continually and had a large value than the θ_{TU}, the lower side values (θ'_{NL}) however became smaller than θ_{NL} at the microgravity condition. Below the 1g conditions, the θ'_{TU} values decreased, but the θ_{NL} increased with gravitaty. Over the 1g conditions, θ'_{TU} and θ'_{NL} stay constant values. In this case, the (θ'_{TU})+(θ'_{NL}) value kept almost constant from 110° to 120° on each gravity conditions. The result indicate that the liquid bridge shapes were controlled by the wettability on the solid-liquid interface and the hydromechanical factors. So it will be solved by the hydromechanical model considering the wetting behaviour and the gravitational effects (Ref.6).

From the facts described above, we may concluded that the gravitational effects to the wetting behaviors became a large at the poor wetting combination than the good one and depended on the gravitational direction and magnitude.

4. CONCLUSIONS

The gravitational effects on the wettability were observed by the liquid-bridge method using a parabolic flight method. The following conclusions were obtained.

(1) On account of gravity, the variation of the contact angle can be divided according to types of the side (lower plate side):
(a)increases with the poor wetting combination.
(b)decreases with the good wetting combination.
(2) On the hanging drop side(upper plate side), the precise dependency on gravity was not observed.
(3) On the Teflon(poor wettability) and Nickel (good wettability) mixture case,the wettability depended on the hydromechanical effects and interfacial tension at the solid-liquid interface.

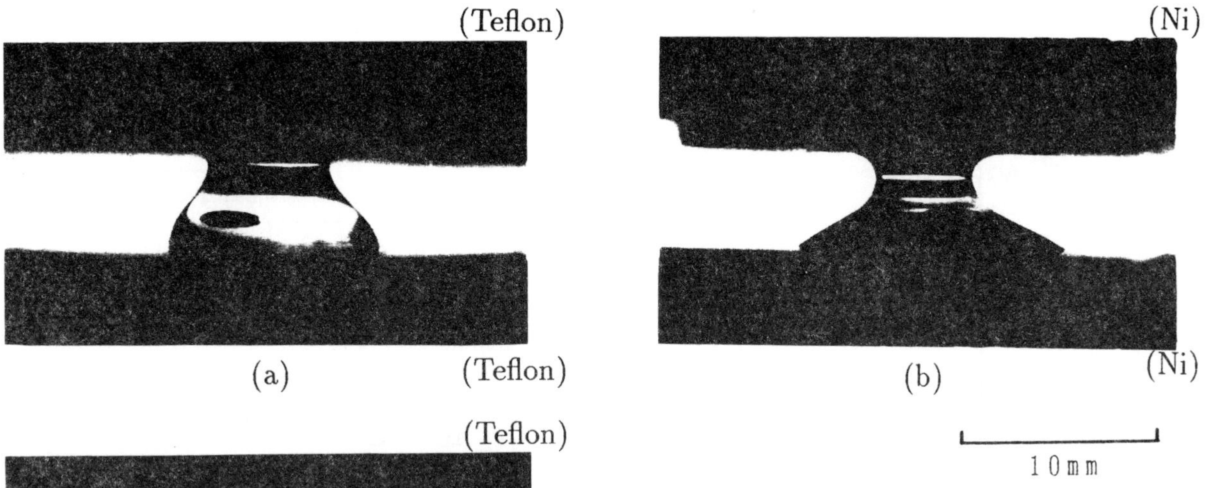

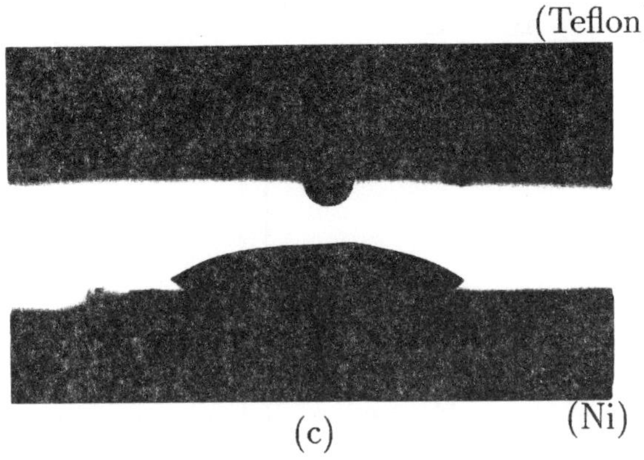

Figure 3:Liquid-bridge shape at 1g

(a) Teflon(upper)/Teflon(lower)
(b) Nickel(upper)/Nickel(lower)
(c) Teflon(upper)/Nickel(lower)

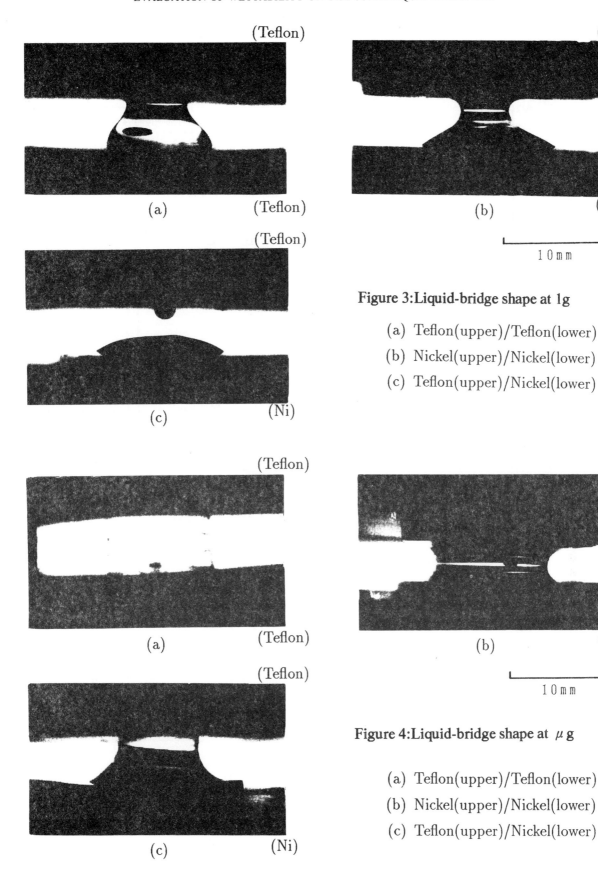

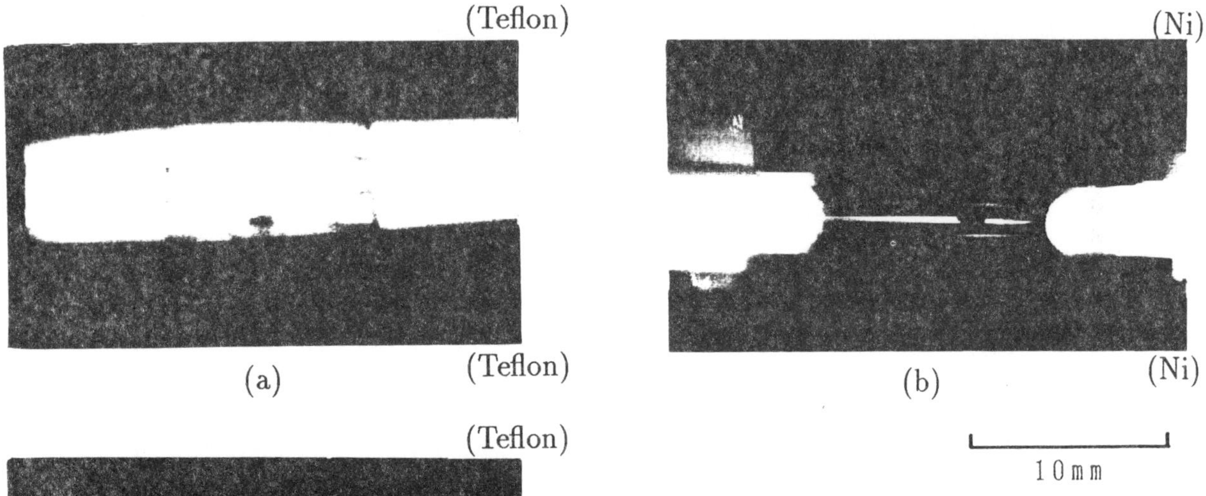

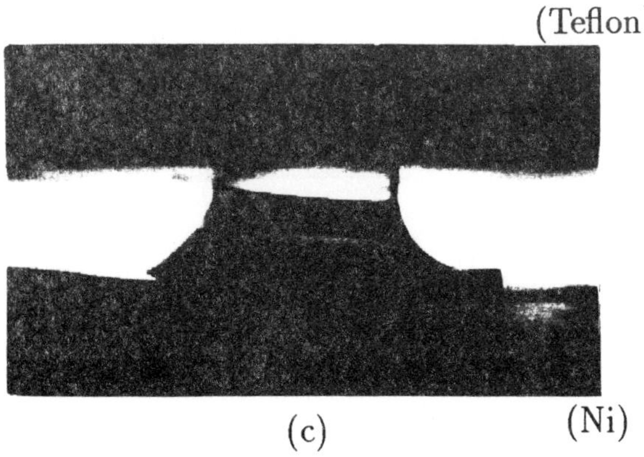

Figure 4:Liquid-bridge shape at μg

(a) Teflon(upper)/Teflon(lower)
(b) Nickel(upper)/Nickel(lower)
(c) Teflon(upper)/Nickel(lower)

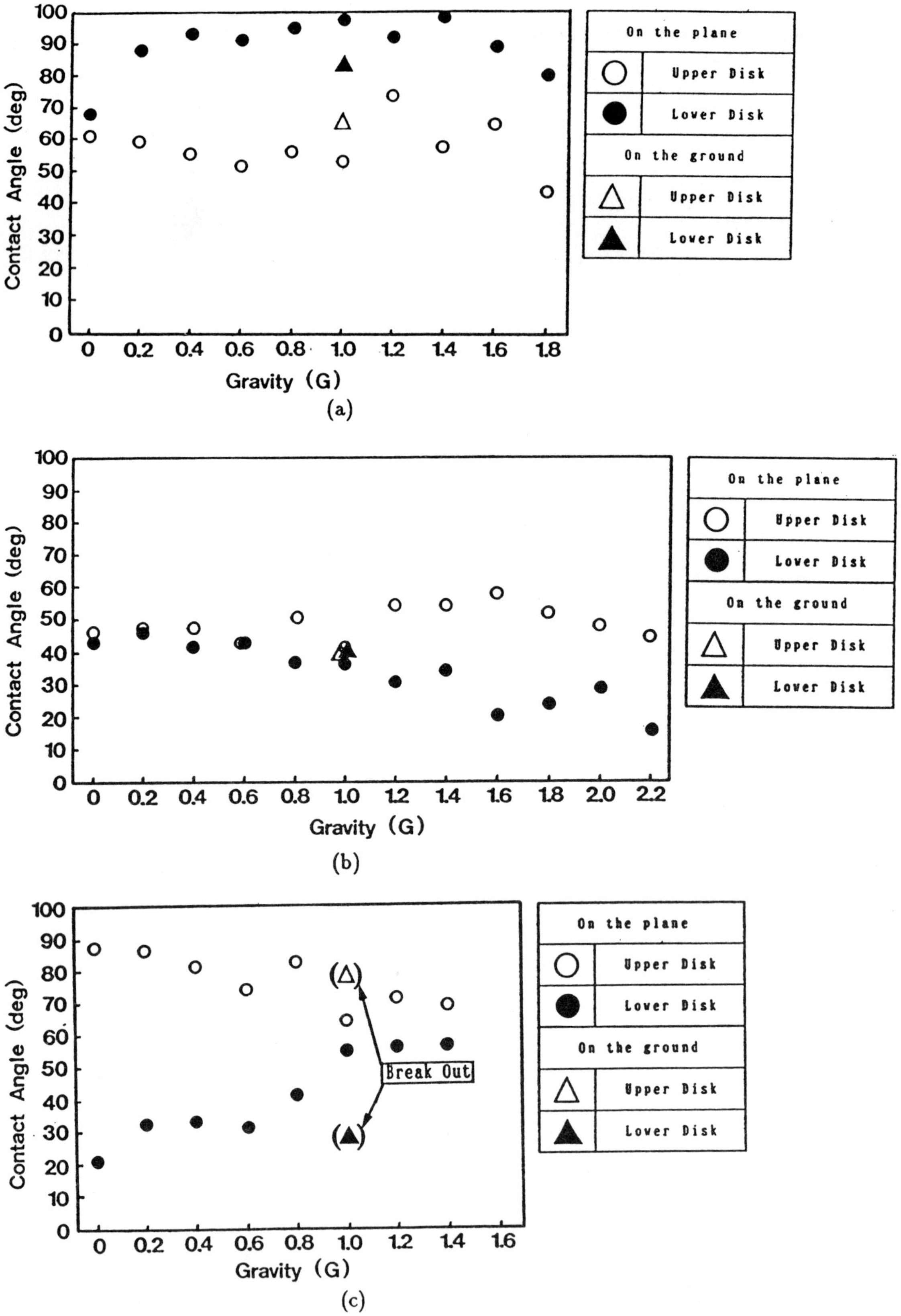

Figure 5 : Dependence with gravity of contact angle
(a) Teflon (upper)/Teflon(lower),(b) Nickel(upper)/Nickel(lower),(c) Teflon(upper)/Nickel(lower)

5. REFERENCES

1. H.C.de.Groh III and H.B.Probst: AIAA, 89-0304 Jan, (1989), 1
2. A.Sanz and I.Martinez: J.Colloid Int.Sci., 93,(1983), 235
3. D.Rivas and J.Meseguer: J.Fluid Mech., 138, (1984), 417
4. A.Sanz: J.Fluid Mech.,156, (1985),101
5. J.Meseguer and A.Sanz: J.Fluid mech., 153, (1985), 83
6. H.Sakuta, Y.Fukuzawa, M.Okada, Y.Kojima and D.Maruyama: in this symposium.

THE SIMULATION OF HYDRODYNAMIC PROCESS IN DRAINING TANK UNDER MICROGRAVITY CONDITIONS.

E.L.Kalyazin

Moscow Aviation Institute, 125871, Moscow, USSR.

ABSTRACT

Problems of liquid flows without gas bubbles from fuel tanks in microgravity are considered. The complex of mathematical and physical methods is proposed to simulate the drain process with taking into account surface tension. Asymptotic and finite-difference methods are used to solve nonlinear equations of ideal liquid motion in the simple geometry tanks. Results of calculation are compared with experimental data that are obtained in the drop tower, the flying laboratory .

Keywords: *Draining Tank, Microgravity, Mathematical Model, Experiment*

The draining of the liquid which fill partly the tanks of space vehicles and space rocket stations takes place in the fields of weak forces or under microgravity conditions. The free surface of liquid distorts strongly owing to the predominant effect of the surface tension forces. The fluid and gas phases can produce nonlinear motion under different outside distributions. The free surface of liquid distorts over outlet draining because of the variable velocity profile. A great deformation of the free surface and the following burst of the gas phase into the main breaks the regime of fuel supply. These phenomena are particularly impermissible by liquid propellant supply into the jet engine.

Therefore the investigation of the processes in draining tanks by change of the acceleration or the expenditure regime are necessary to solve solution the practical problems of designing the gas-liquid systems for the present and perspective space vehicles.

Investigate the nonlinear motion of liquid into draining tank under microgravity conditions, using the methods of the mathematical and physical simulation. Consider the tanks of the simple geometrical forms, that are filled partly by the ideal incompressible liquid. The motion of a liquid is nonwhirlwind. The apparent acceleration a, which directs along the axis of the tank, defines the external field of mass forces.

Connect with the tank the cylindrical coordinate system OxrO (Figure 1). Mark the free surface of fluid - $f(x,r.\Theta,t)$, the wetting surface of the tank - $f_1(x,r,\Theta)$, the surface of the outlet - $f_3(x,r,\Theta)$, the wetting contour - Γ.

Record the system of the partial differential equations

for the definition of the nondimensional fluid velocity potential $\varphi\,(x,r,\Theta,t)$ and the free surface form $x=\zeta\,(r,\Theta,t)$

$$\varphi_{xx}+\varphi_{rr}+\frac{1}{r}\varphi_r+\frac{1}{r^2}\varphi_{\theta\theta}=0 \qquad in\ \tau; \qquad (1)$$

$$\varphi_x=\varphi_{rr}+\frac{1}{r^2}\zeta_\theta\varphi_\theta+\zeta_t \qquad on\ f; \qquad (2)$$

$$\varphi_t+\frac{1}{2}(\nabla\varphi)^2-r_0\varphi_x+\frac{1}{F_r}\zeta-\frac{1}{B_0F_r}(\zeta_{rr}+\frac{1}{r}\zeta_r+\frac{1}{r^2}\zeta_{\theta\theta}+\frac{1}{r^2}\zeta_\theta^2\zeta_{rr}-\frac{2}{r^2}\zeta_{r\theta}\zeta_r\zeta_\theta+\frac{2}{r^3}\zeta_\theta^2\zeta_r+\frac{1}{r}\zeta_r^3+\frac{1}{r^2}\zeta_r^2\zeta_{\theta\theta})(1+\zeta_r^2+\frac{1}{r^2}\zeta_\theta^2)^{-3/2}=0 \qquad on\ f; \qquad (3)$$

$$\varphi_{n1}=0 \qquad on\ f_1; \qquad (4)$$

$$ctg\vartheta=\frac{\bar{n}\cdot\bar{n}_1}{\bar{n}\times\bar{n}_1} \qquad on\ \Gamma; \qquad (5)$$

$$\varphi_x=\begin{cases}0, & when\ t=0^-\\ r_0^2-1, & when\ t=0^+\end{cases} \qquad on\ f_3. \qquad (6)$$

Here r_0-nondimensional radius of the outlet; $Fr=V^2a^{-1}R^{-1}$- Froud Number; $Bo=\rho aR^2\sigma^{-1}$-Bond Number; $We=\rho V^2R\sigma^{-1}$ - Weber Number; R- nondimensional radius of the tank; V-typical liquid velocity (draining velocity); ρ and σ - fluid density and the coefficient of surface tension; ϑ - angle of wetting; $\bar{n}$, $\bar{n}_1$ - vectors of normals to free and wetting surfaces.

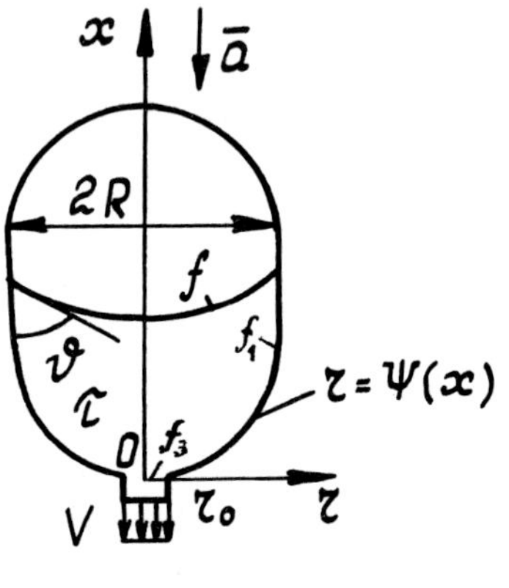

Figure 1: The scheme of tank.

It is possible to use two methods to solve this problem depending on Bond Numbers. The first method is used to solve draining liquid problem under any conditions , especially if the surface tension forces exceed the gravity forces, and the equilibrium fluid free surface is near-by spherical one.

The solution is sought by the finite-difference method. The axissymetrical case is considered. Record the equation of side tank surface in form $r=\psi(x)$. The condition

$$\varphi_r=0 \text{ by } r=0 \qquad (7)$$

supplement the system and equation (5) will take the form:

$$(\zeta_r+\psi_x)(1-\zeta_r\psi_x)^{-1}=ctg\vartheta \qquad on\ \Gamma. \qquad (8)$$

The Laplace equation (1) with boundary conditions (2)-(4) is solved for the knots of calculated net in the fluid volume τ for every moment of time t_i. The net is connected with the motionless boundary of the volume so as its knots will be always on this boundary. Using Green theorem and the simple transformations, write down the partial differential analogy of Laplace equation for the element of volume $\Delta\tau$ in form

$$A_{k,\ell}\varphi_{k+1,\ell}+\bar{A}_{k,\ell}\varphi_{k,\ell+1}+C_{k,\ell}\varphi_{k-1,\ell}+\bar{C}_{k,\ell}\varphi_{k,\ell-1}+B_{k,\ell}\varphi_{k,\ell}=F_{k,\ell} \qquad (9)$$

Where $A_{k,l}$;$\bar{A}_{k,l}$;$C_{k,l}$;$\bar{C}_{k,l}$ - the geometrical characteristics of the areas S_1,S_2,S_3,S_4, which limit the elementary volume (Figure 2); $B_{k,l}=-A_{k,l}-\bar{A}_{k,l}-C_{k,l}-\bar{C}_{k,l}$; $F_{k,l}$ - the item considering the boundary conditions on the free surface and on the surface of the outlet into the knots of a net near these boundaries. The formulas for $A_{k,l}$;$\bar{A}_{k,l}$;$C_{k,l}$; $\bar{C}_{k,l}$; $B_{k,l}$;$F_{k,l}$ are different and depend on the location of the knot relatively the boundaries of the considering field. The potential of fluid velocity φ is found by the method of the successive upper relaxation (Ref.1) for the moment of time t_i into volume τ . Then the form of the free surface and the velocity potential on it is found by integrating the equations (2), (3) for time t_{i+1}. The finite-different analogies of the equations (2) and (3) are integrated by the evident scheme. In order to increase the precision of the calculation of the derivatives the free surface form and the velocity potential on it is approximated by power-functions by the spline-method (Ref.2). The typical regimes of the tank draining were revealed after numerous calculations which were fulfilled for the spherical tank with radius of the outlet $r_0=0.02$ - 0.1 and the initial levels filling $h < 0.5$. These regimes are conveniently distinguished, using the conception of the critical height - that level of the fluid into the tank, by which the gas phase bursts in the expended main during the given moment.

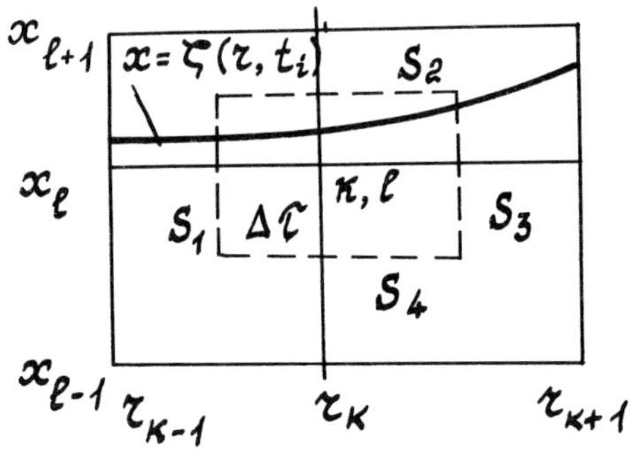

Figure 2: The knot of calculating net.

The first regime - oscillating - is observed by the great initial levels of the tank filling and the small fluid expenditures. This regime is characterized by the continuing cycles of the tank draining without the gas phase burst, a small magnitudes of the critical height and the small rest of fluid which it is impossible to take from the tank.

The free surface is distorted quickly in the second regime which is observed when initial levels are small. It reaches the outlet quickly, after that the gas-liquid mixture begins to go into the main instead of the fluid stream. The fluid volume, which remains in tank in this moment, may be considerable and the cycle of the feeding supply may be short. The calculated profiles of the free surface are shown on Figures 3a,b for two regimes of the draining spherical tank with the radius of the outlet $r_0=0.08$: with the monotone distortion of free surface and with its oscillation . The detailed results of this calculations are given in article (Ref.3). In the second method the asymptotical series are used to solution draining problems if Bo>100. The equilibrium fluid free surface is near to the plan, which is normal to the acceleration a. The unknown velocity potential and the unknown form of the fluid free surface will write down in the form of the asymptotical power series on the parameter ε :

$$\varphi(\varepsilon, x, r, \theta, t) = \sum_{j=0}^{N} \varepsilon^j \varphi^{[j]}, \qquad (10)$$
$$\zeta(\varepsilon, r, \theta, t) = \sum_{j=0}^{N} \varepsilon^j \zeta^{[j]}.$$

Using the well-known method, instead of equations (1) -(8) will receive j systems of the equations . Every system corresponds to j approximate solution: when j=0 we have the solution of the linear system , when j=1 - the first approximate solution of this system and so on.

The equations, which describe the first, the second and the following approximations, have insignificant distinctions from equations for j=O: the functions, which appear in the right parts of conditions (2),(3) when j = 1, 2, 3,..., depend on solutions of the (j-1) approximation. The condition of fluid draining (6) is taken into account only once, when j=0.

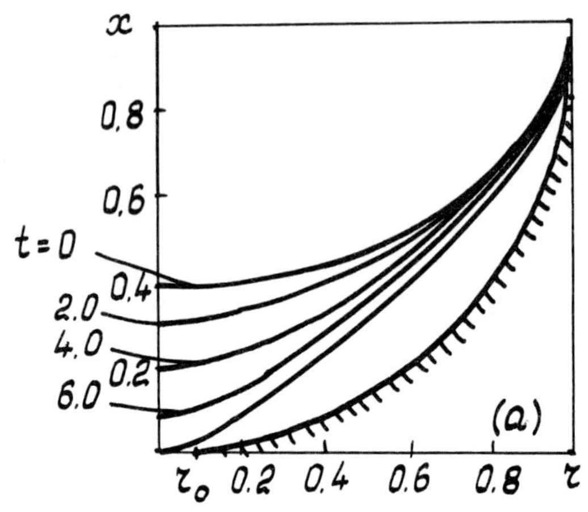

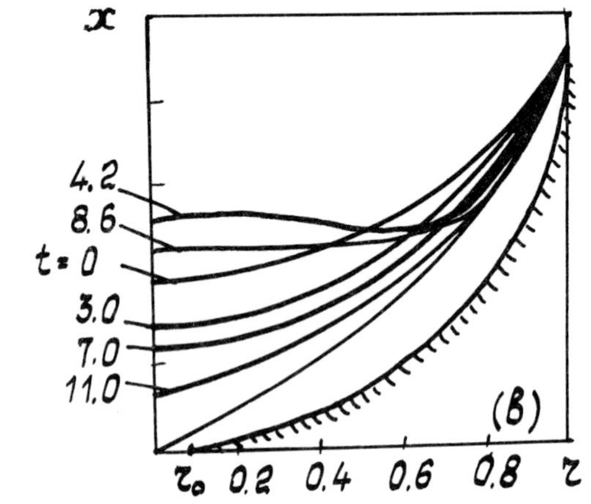

Figure 3: The calculating profiles of the liquid free surface in 0-gravity.

a - We=50, Bo=0, r_0=0.08, ϑ=30°;
b - We=33, Bo=0, r_0=0.08, ϑ=30°

The problem is solved by number integration methods. The calculation results, which are received to cylindrical tank in case Fr = 390, Bo = 270, r_o = 0,05 taking into account 3 approximations, show that when the fluid level is enough high h > 0,3 , the first and the second approximations give small addition to the linear problem solution. Calculations with using the asymptotical method confirm the results, which are received by finite-difference method. Also two draining regimes are appeared : oscillating regime and regime of monotonous distorting of the free surface. Surface tension forces stabilize the fluid free surface. This influence reduces the critical height on 15 - 20 % in field Bo = 100 - 200. Figure 4 shows the dependence of nondimensional height h from Froud Numbers for various Bond Numbers.

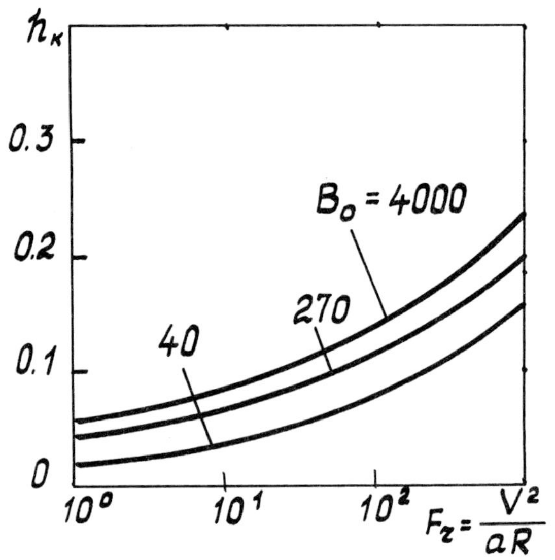

Figure 4: The dependence of nondimensional height h from Froud Numbers.

The results of calculations on the asymptotical method are compared with experimental data, which are obtained on the drop tower. Experimental critical heights to cylindrical tank in field of Bond Numbers 1000 > Bo > 100 are approximated by formula

$$h_K = 0{,}82\ th(1{,}2\, Fr_*^{\,0{,}35}) \,,$$

where modification Froud Number Fr_* calculated by the average velocity of fluid level decrease in the tank $\dot{h}$ (Figure 5).

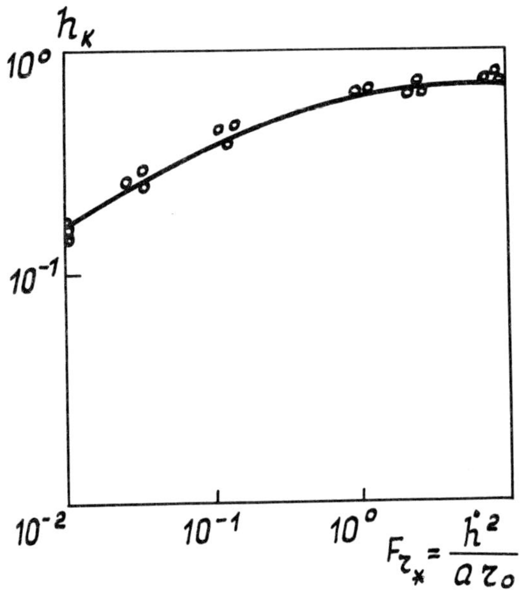

Figure 5: The dependence of nondimensional height h from modification Froud Number.

Therefore, the finite-difference and asymptotical methods of calculations allow to consider the nonlinear effect because of the combined action of the gravity and surface tension forces and give the results near-by the experimental ones. It is necessary to provide small amplitudes of free surface oscillations into draining tanks in order that the rests of liquid were small. The choice of rational regimes fluid outflow, geometry of outlet and acceleration of external force field allows significantly to reduce the useless rest of fluid in tanks.

REFERENCES

1. D.M. Young, Iterative methods for solving partial difference equations of elliptic type. *Trans. Amer. Soc.*, 76,92, (1954).

2. C.H. Reinsch, Smoothing by Spline functions. *Numerics Mathematic*, 10, half 3, (1967).

3. V.M. Amahin, E.L. Kalyazin, A.V. Voronkov, The numerical calculation of draining tank processes in the weak forces field, *The questions of Aerohydromechanics of flying Vehicles,* Collected Work of Moscow Aviation Institute, (USSR) Numb.323, (1975) p.110.

DROP TOWER INVESTIGATION OF CAPILLARY INDUCED FLUID MOTION IN SURFACE TENSION SATELLITE TANK MODELS

M. Dreyer, A. Delgado, H.J. Rath, G. Netter[1] and H.D. Bruhn[1]
Center of Applied Space Technology and Microgravity
University of Bremen, W-2800 Bremen, Federal Republic of Germany
[1] *ERNO-Raumfahrttechnik GmbH, Hünefeldstr. 1-5, W-2800 Bremen*

ABSTRACT

The fluid transport in capillary vanes under microgravity has been investigated experimentally in the Drop Tower Bremen. The rise of the fluid meniscus was observed with a videocamera and depicted dimensionless versus time. Using the Ohnesorg number $\eta/\sqrt{\sigma\rho a}$ as scaling parameter yields a common relation for two different test liquids. A prediction of meniscus rises for other fluids is therefore possible.

Key Words Surface tension, capillary vanes, drop tower, satellite tanks, Ohnesorg number

INTRODUCTION

The primary design objective for a surface tension tank is to cover the tank outlet with propellant whenever outflow is demanded. This ensures that only the propellant, and none of the pressurisation gas, will leave the tank until it is nearly depleted. Surface tension propellant management systems consist of devices to establish propellant flow under low acceleration levels to the outlet from wherever the bulk liquid is located in the tank and to position sufficient propellant to support all engine operations when the acceleration on the tank is not settling liquid over the outlet (Refs. 1,2).

The transport of liquids in a special vane has been investigated in the Drop Tower Bremen (free fall time 4.74 seconds). Other types of capillaries are considered elsewhere (Ref. 3). The vanes are open capillaries with a cross section partly bounded by solid walls. Under microgravity the fluid rises into the capillaries perpendicular to the previous flat fluid surface. The drop tower operation (step from 1g to zero-g) is a good simulation for a part of a satellite mission profile. After an acceleration due to thruster firing the capillary vanes have to be filled under microgravity conditions until the next firing.

EXPERIMENTS

All experiments were performed in the Drop Tower Bremen. The system Drop Tower Bremen consists of the 110 m high drop tube with an internal diameter of 3.5 m and the 11 m high deceleration chamber at its foot. The drop tube and the deceleration chamber form a vacuum system with a volume of 1.700 m^3 and can be evacuated with the help of a system of 18 pumps with a nominal pumping capacity of 32.000 m^3/h (about 9.000 l/s) down to a residual pressure of 1 Pa in about 1.5 hours. This yields to a free fall time of 4.74 seconds and a residual acceleration on the drop capsule of about 10^{-5} times the earth gravity (9.81 m/s^2).

The experiments have been carried out in a specially-constructed capsule, which is designed to be both pressurized and shock-proof. Before free fall, the experiment capsule is held by an electropneumatical tensioning element fixed at the release platform. To decelerate the capsule, a specially constructed tank filled with fine-graded polystyrol is used. The system guarantees a deceleration below 30 g. The experiment tank is made from plexiglass and contains the capillary vanes and the propellant management device (PMD).The vanes are folded plates and mounted with a defined distance (a) from the outer tank wall. Figure 1 shows the tank model with capillary vanes and the PMD. The filling height is depicted as H.

Under microgravity the fluid rises into the gap between the vanes and the tank wall due to capillary action. The rise of the meniscus was observed with a ccd-video-camera. The observation area was illuminated with a diffuse bright field light from the back side for better visibility of the fluid meniscii. The time was measured by a

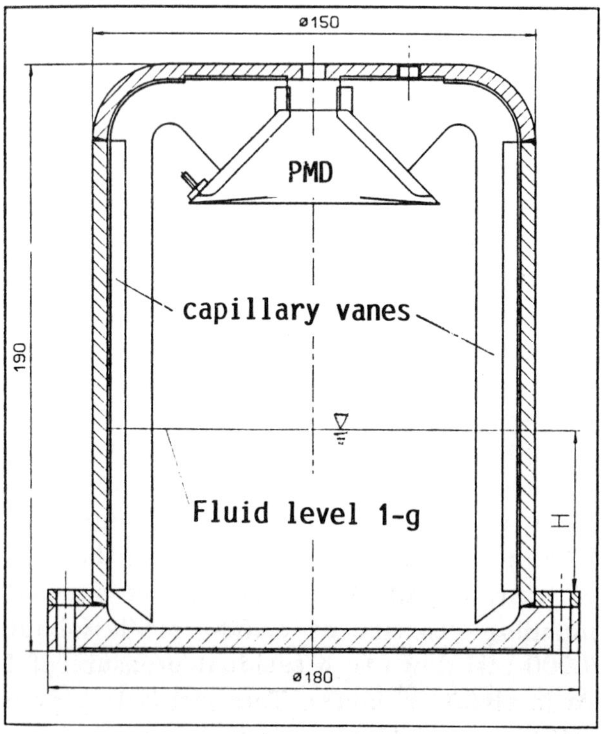

Figure 1: Satellite tank model

clock with an accuracy of 1 millisecond. The highest point of the meniscus was selected as reference point for the evaluation of the videotapes. The meniscus height h [mm] was measured from the original flat surface under normal gravity up to the reference point.

A sequence of videoprints for an experiment with FC-77 (filling height H=40 mm) is shown in figure 2. The meniscus height h is depicted in the videoprints. A second meniscus is formed at the inner corner of the vane, this gives an amount of fluid but it is not considered in this investigation.

The liquids used in this investigation are Dow Corning 200 Silicon Fluid 0,65 cSt and FC-77. A small amount of color was added for better visibility of the fluid meniscii. The properties of the test fluids are at 20^0 Celsius (Table 1).

TABLE 1 Fluid properties

Fluid	*ρ [g/cm³]*	*σ [mN/m]*	*η [cP]*
SF 0.65	0.761	15.9	0.5
FC-77	1.78	15	1.424

Density is denoted as ρ, surface tension and viscosity as σ and η, respectively. Concerning their properties the fluids are selected to simulate as well as possible the behaviour of real propellants like MMH and N_2O_4.

RESULTS

Figure 3 shows the results of the meniscus rise in original coordinates. Because the surface tension for both liquids has about the same value, the differences of the meniscus heights are due to the different viscosities. The viscosity of FC-77 is about three times higher than that of silicon fluid 0.65 cSt, thus the friction in the capillary gap is higher and the meniscus velocity lower. The density affects only the inertia of the fluid motion in the beginning. The liquid rise

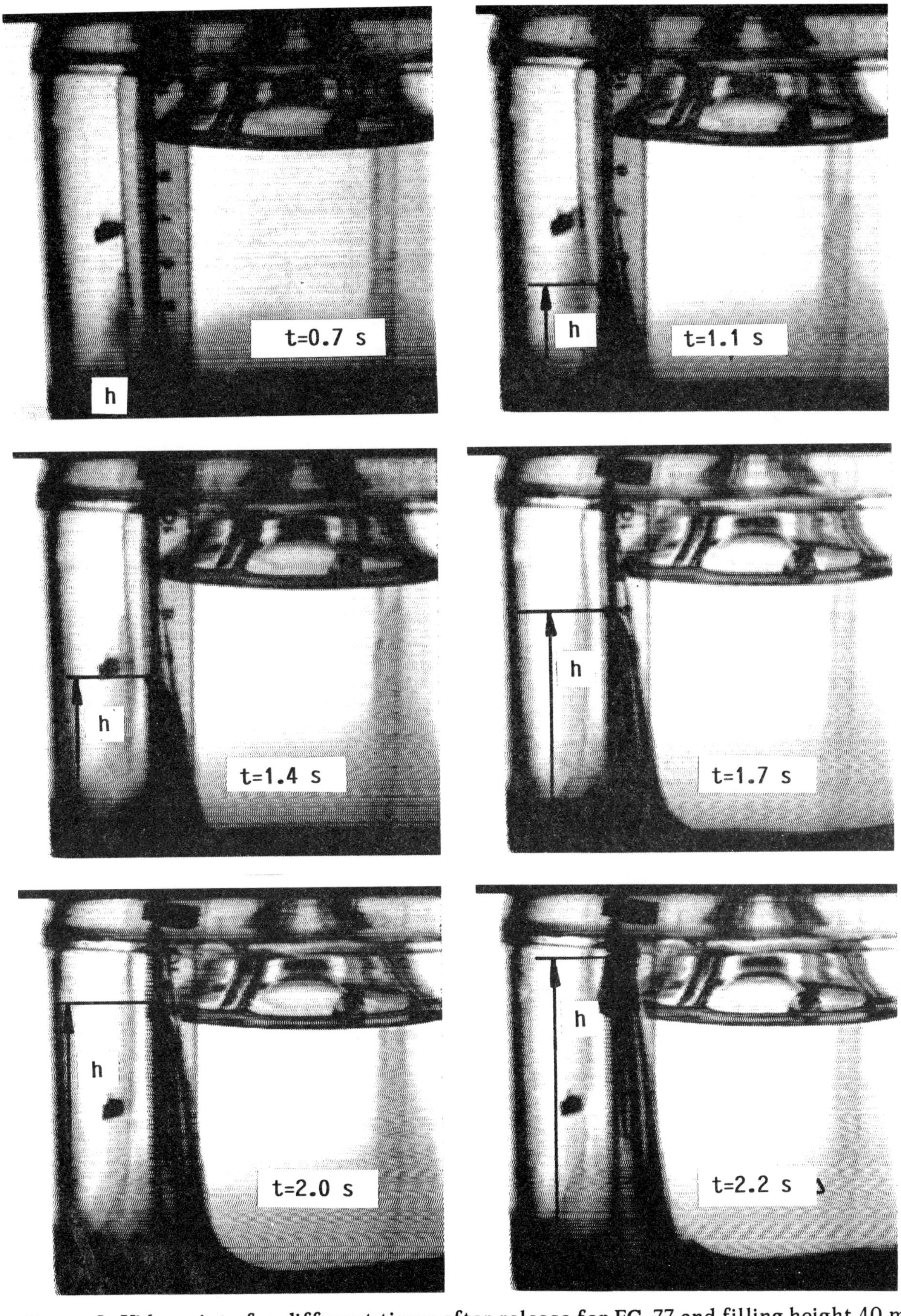

Figure 2: Videoprints for different times after release for FC-77 and filling height 40 mm

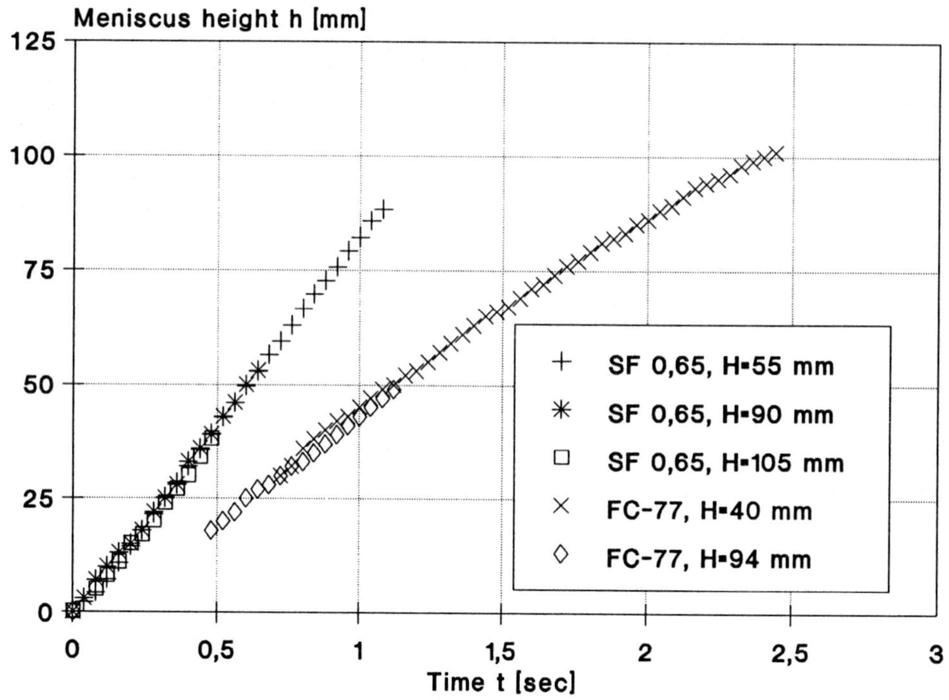

Figure 3: Meniscus heights h versus time for different filling heigths H

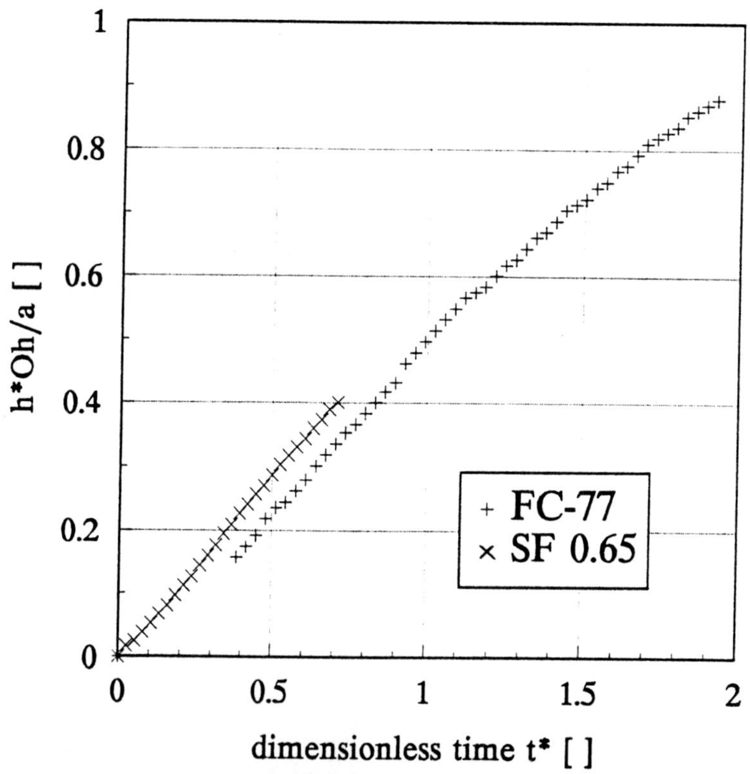

Figure 4: Dimensionless representation of the experimental data

does not depend on the filling height H.

A dimensional analysis was performed to generalize the experimental data. It was assumed that a relationship of the meniscus height yields to the form $h=f(\sigma,\rho,\eta,a,t)$, where (a) means the gap width. One finds that $h/a=F(Oh, t^*)$, where Oh is the Ohnesorg number $\eta/\sqrt{\sigma\rho a}$ and t^* a dimensionless time $\eta t/\rho a^2$. Depicting the data graphically in the form $h \cdot Oh/a = G(\eta t/\rho a^2)$ (with a=1 mm in this investigation) yields the graphs in figure 4. The Ohnesorg numbers are for a=1mm: SF 0.65, Oh=0.004546 and FC-77, Oh=0.008715.

Other dimensionless numbers are considerable, like the Reynolds number $Re=\rho va/\eta$, the Weber number $We=\rho v^2 a/\sigma$ and the Capillary number $Ca=\eta v/\sigma$, where v denotes a characteristic velocity. But in the unstationary meniscus rise a characteristic velocity is not known a piori; the velocity of the meniscus rise decreases from the nearby constant value in the first second to an unknown velocity for longer times. The surface tension force remains constant during the whole wetting process, while the resistance due to viscous friction depends on the length and the velocity of the liquid column and thus increases with time. Thus the Ohnesorg number, yielding the relations $Ca/Re=Ca^2/We=Oh^2$, is considered as a possible scaling parameter in unstationary capillary wetting processes. The slope of both curves in the range 0.4-0.7 of the x-axis is the same, the delay of the FC-77 meniscus in the beginning is considered to depend on the inertia of the liquid. The density of FC-77 is more a twice the density of SF 0,65. Further experiments are planned to investigate the beginning of the wetting process and to observe longer meniscii rises. The curves in figure 4 are a first approach to predict the meniscus height versus time for other fluids, for example propellants like MMH and N_2O_4.

REFERENCES

1. Otto, E.W., *Chemical Engineering Progress Symposium Series,* Vol. 62, No. 61, pp. 158-177, (1977)

2. Netter, G., Beig, H.-G., *Proceedings of Symposium:' Fluid dynamics and space',* VKI, Rhode-Saint-Genese, 25/26 June 1986 (ESA SP-265, August 1986)

3. Dreyer, M., Delgado, A., Rath, H.J., *to be published on IUTAM Symposium on microgravity fluid flows, to be held in Bremen, Germany, September 1991*

SPINNING FLUID DYNAMICS UNDER THE CONDITIONS OF LOW GRAVITY. EXPERIMENT.

M. I. Galace, A. S. Makarova, N. E. Boitsun

Dniepropetrovsk State University,
72, Gagarin Av., Dniepropetrovsk 320625,
USSR

ABSTRACT

Drop-tower experimental investigations of a rotating liquid dynamics in cylinder with flat, spherical and concave bottom are carried out. The reorientation process of a rotating liquid under the action of longitudinal load factor differential from one to zero is modelled. The influence of the vessel angular rotating rate, physical properties of the liquid, filling level of the vessel is investigated. Analysis is done of the influence of the liquid spin-up rate in the vessel on its dynamics when passing to zero-g conditions. It has been noted that a short-time stop of vessel rotation up to discontinuous lowering of the load factor is the most efficient method to limit the rotating liquid displacement to the side walls of the vessel.

Keywords: *vessel rotation; liquid dynamics; zero-gravity; drop-tower; liquid spin-up; emptying bottom.*

1. INTRODUCTION

Investigation of the dynamics of spinning fluid is of practical interest for designing the fuel tanks of the rotating space vehicles. In modern practice there is a number of publications devoted to the definition of the equilibrium forms of spinning fluid (Refs. 1,2), to the calculation of its spinning-up under the conditions of normal gravity (Ref. 3), to the study of reorientation as the angular velocity changes under the conditions of zero-gravity state. However the dynamics of spinning fluid under the conditions of longitudinal load factor drop is studied insufficiently. The experimental works (Ref. 4) being available are devoted to the particular tasks for mastering the specified designs of tanks and these works do not propose general recommendations. The present work describes the experimental works on reorientation of the spinning fluid at step wise reduction of the value of longitudinal load factor from 1 to 0.

2. EXPERIMENTAL INVESTIGATION RESULTS

The experimental research was carried out on the test bed of zero-gravity state without guides having 15 m height. The time of container free fall with the model made 1.35 s. Figure 1 shows the general view of the container. Filming of fluid dynamics was realized dy two cine cameras. Through the side wall of the model the level of fluid raising was fixed, and from below the process of bottom emptying was filmed. The frequency of filming made 96 exposures per second. As a model vessel was used a cylinder of acrylic plastic of 12 sm diameter with replaceable bottoms: spherical concave, spherical convex and flat. The diameter of the spherical part of the bottoms was equal to 16 sm. The experiments were conducted as follows. The container with the model filled with the fluid was lifted to the specified height. After the full dampening of the fluid in the model a command was given for spinning. Bringing the fluid spinning to speed was performed for a specified time and then the cameras were brought into use and the throw down was realized; the rotating model free falling took place for 1.35 s. In some experiments 2 seconds prior the throwing the model rotating was stopped. Table 1 presents the values of main parameters that were changed in course of experiments. In two last experiments ethyl alcohol was used as working fluid. All other experiments were realized with octane.

Some results of investigations are shown in figures 2-4. The curves of the figure 2 are plotted on the base of the data of experiments 1, 2 and 8. They illustrate the process of fluid level raising during the free falling at the side walls of the cylindrical vessel with flat bottom. The curve 1 corresponds to static conditions (spinning are absent), the curve 2 is plotted for the fluid being spinned-up before the throw up to the rotation of a solid body with angular velocity ω =32 rpm (experiment 8), the curve 3 is plotted as per the data of the experiment 2, i.e. for the fluid spinning before the throw as a solid body with the angular velocity of 72 rpm. As it is seen from figure 2 fluid spinning exerts considerable influence on the process of reorientation. For example for ω =32 rpm the height of fluid raise is increased three times as compared with the static conditions. And it should be noted that the spinning with angular velocity of 32 rpm under the conditions of normal gravity visually does not provoke noticeable curvature of the flat free surface.

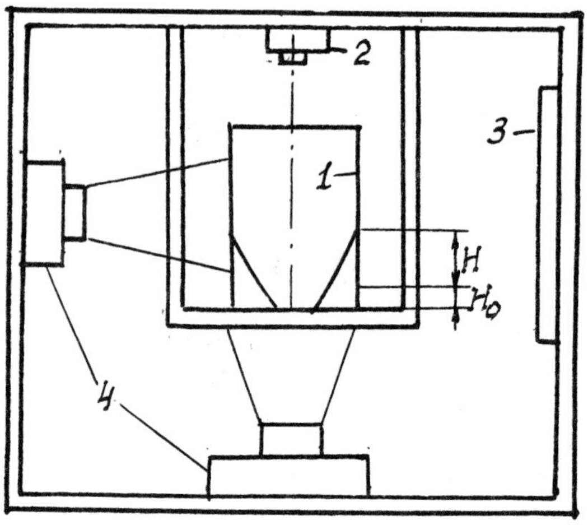

Figure 1: General view of container. 1 - vessel; 2 - vessel rotating facility; 3 - lighting; 4 - cine camera.

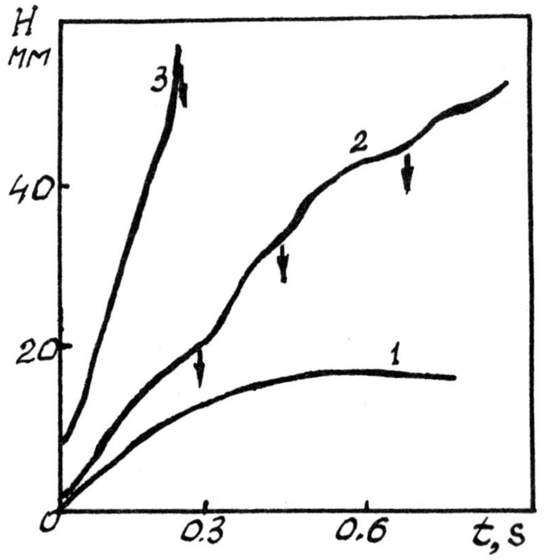

Figure 2: Level raise of the liquid on the side wall of a flat-bottom cylinder following the transition from normal gravity to zero-g. 1 - stationary liquid; 2 - fully spinned-up liquid, ω =32 rpm; 3 - fully spinned-up liquid, ω =72 rpm.

TABLE 1 Programm of experiments

No.	*Bottom*	*Volume of filling with fluid (ml)*	*Angular velocity of spinning (rpm)*	*Time of spinning before throw (s)*	*Spinning stop for 2 s before throw*
1	Flat	160	0	0	No
2	Flat	160	72	60	No
3	Flat	160	72	60	Yes
6	Flat	160	72	10	No
8	Flat	160	32	60	No
9	Flat	160	32	60	Yes
11	Flat	160	32	10	No
14	Flat	280	0	0	No
15	Flat	280	32	10	No
16	Flat	280	32	120	No
21	Convex	160	0	0	No
22	Convex	160	32	60	No
23	Convex	160	32	10	No
25	Convex	160	32	60	Yes
31	Concave	75	0	0	No
32	Concave	75	32	60	No
33	Concave	75	32	60	Yes
37	Convex	50	32	60	No
38	Convex	50	32	60	No
39	Convex	50	32	60	Yes

The reorientation process of the spinning fluid as the load factor is relived differs from the reorientation of the initially unmovable fluid not only in quantity but also in quality. If in the initial state the fluid was unmovable load factor relieving provokes first of all the raise of the thin layer of the fluid on the side surface at practically flat level in the center. Then starting from 0.35 s at nearly unchangeable level at the side walls the widening and deepening of fluid meniscus became the dominating process. On the background of meniscus in its central part, axially symmetric oscillations of the free surface take place that do not affect great mass of the fluid.

Somewhat other processes take place during reorientation of fully spinning-up fluid. In this case there are axially symmetric oscillations of the surface in the central part of the vessel. The intensive raise of fluid as per tank walls rapidly leads to the bottom emptying. At ω =72 rpm the maximum area of bottom emptying makes 70% and is reached after 0.33 s of falling. But the process of reorientation is not yet over. On the side walls reverse waves (ripples) are formed periodically and flow down. The time of their appear is marked on the curves of figure 2 by the aid of arrows oriented downwards. These waves result in decrease of the thickness of fluid on the side wall and provoke flowing of thin layers of fluid on the empty bottom. The thin layers are accumulated gradually and a front of fluid is formed on the bottom and its ring is contracted. For example at ω =32 rpm (curve 2, figure 2) from 0.6 s up to 0.78 s the area of empty bottom is reduced from 45% to 10% and at t =0.8s the bottom is fully covered with fluid. Then it is emptied once more. At ω =72 rpm the analogous process takes place but the fluid layer being flowed on the bottom is in this case much thinner (1-2 mm). Figure 3 shows the data of reorientation of poorly spinned-up fluid being in the state of spinning braking. The curve 1 illustrates the raise of the level of unspinned fluid, the curve 2 is plotted for fully spinned-up fluid (experiment 3). The curve 3 is plotted according to the data of experiment 9 in which the rotation of the vessel with fully spinned-up fluid was stopped for 2 seconds before the throw. The curve 4 presents the raise of fluid level in the cylinder rotating for 10 seconds before the throw (experiment 11). The curves 1, 2 have been examined earlier in the figure 2 and are presented here for comparison. The curves 3, 4 in figure 3 are plotted for small periods of time. It is associated with the fact that in experiments 9 and 11 together with the raise of the level on the walls in course of time an intensive reverse wave is formed. The curves 3 and 4 are interrupted as the reverse wave directed down wise becomes prevailing (through the contact line "creeps along" upwards), but it is impossible to fix the level of the fluid flowing along the preliminary wetted wall. The reorientation of the fluid in the experiment 9 (curve 3) was realized in the following manner. As the fluid layer adjusting the walls of the tank and its bottom before the throw was braked the raise of the level along the side wall took place almost as in statics. However in the center of the vessel the process was developed as for the fully spinned-up fluid: the bottom was empty by 45% over 0.33 seconds and to the moment - 0.75 second in was once more covered with fluid. Entirely in an other manner the process has been developed in the experiment 11 (curve 4). As in this

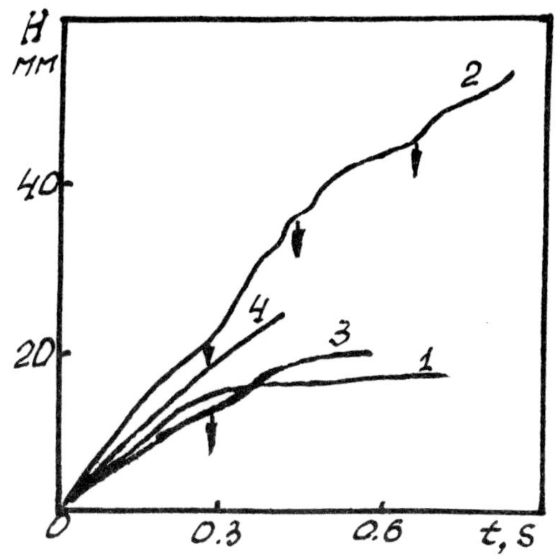

Figure 3: Level raise of the liquid on the side wall of a flat-bottom cylinder following the transition from normal gravity to zero-g. 1 - stationary liquid; 2 - fully spinned-up liquid, ω =32 rpm; 3 - fully spinned-up liquid at ω =32 rpm and stopped for 2 s before the fall; 4 - spinned-up liquid for 10 s, ω =32 rpm.

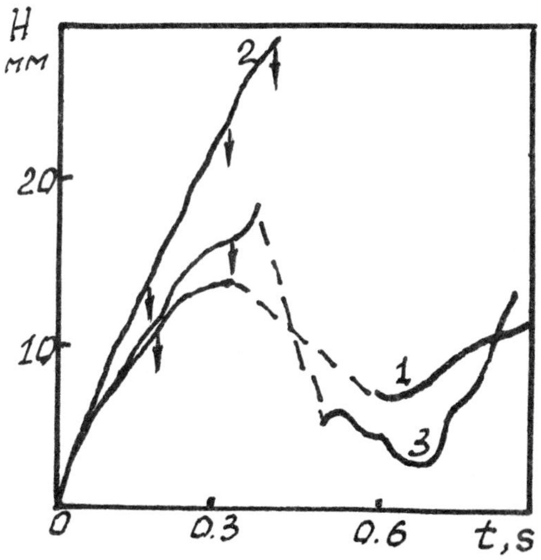

Figure 4: Level raise of the liquid on the side wall of a concave-bottom cylinder following the transition from normal gravity to zero-g. 1 - stationary liquid; 2 - fully spinned-up liquid, ω =32 rpm; 3 - fully spinned-up liquid at ω =32 rpm and stopped for 2 s before the fall.

oase the wall adjusting layers were involved in spinning and fluid mass in the center was unspinned, an intensive raise of the level took place along the side wall and in the center there were humps and craters i.e. there were axially symmetric oscillations typical for unmovable fluid. But 0.8 s after the throw start the bottom in the experiment 11 was empty. The results of analogous experiments conducted for the cylinder with a concave spherical bottom are presented in figure 4. The curve 1 is plotted as per data of experiment 31 i.e. for static conditions. The curve 2 illustrates the raise of spinned fluid along the walls of cylindrical tank with concave bottom (experiment 32), the curve 3 is plotted for the conditions of experiment 33. The main trends of fluid behavior are analogous to that being described for cylindrical tank with a flat bottom. But it should be noted that for a specified geometry of the bottom in course of oscillations (experiments 31, 33 curves 1, 3) much more fluid takes part than in the case of flat bottom. For example to the moment of time 0.55 - 0.6 s (curves 1, 3) the fluid surface had once more a flat shape that is not typical for a flat bottom.

3. CONCLUSION

The present work describes the nonstationary hydrodynamical processes taking place in a rotating cylindrical vessel after passing from the conditions of normal gravity to the free fall. The differences are noted in the development of processes of reorientation in the volumes of fluid spinning as a solid body, poorly spinned-up and being in course of spinning braking. Relations between the fluid level on the side walls of the models and time are plotted. The obtained data permit to recommend the optimal modes to relieve the longitudinal load factor in the rotation areas for the different forms of the fuel tanks.

4. REFERENCES

1. R.J. Hung and F.W. Leslie, *J. Spacecraft and Rockets* 25 (1988).
2. R.J. Hung, Y.D. Tsao, B.B. Hong and F.W. Leslie, *AIAA Paper* 88-045.
3. J.F. Homicz and N. Gerber, *AIAA Paper* 86-1121.
4. J.E. Anderson, D.A. Fester and D.W. Dugan, *AIAA Paper* 76-598.

THE MOTION OF SOLID BODY IN A LIQUID UNDER THE INFLUENCE OF A VIBRATIONAL FIELD

D.V. Lyubimov, A.A. Cherepanov and T.P. Lyubimova [1)]

Perm State University, 15, Bukireva Str., Perm, 614600, USSR

1)Institute of Continuous Media Mechanics, UB Acad. Sci.USSR, 1, Akad. Korolyov Street, Perm, 614061, USSR

ABSTRACT

Solid bodies in a vessel containing a liquid and subjected to the high frequency vibrations can display an unusual behavior: the bodies of a more large density than that of a surrounding liquid can float, while light bodies can sink under certain conditions. It is shown that the vibrational field leads to the appearance of an average force acting to the bodies. The conditions of the bodies floating under the influence of vibrations have been determined.

Keywords: *vibrations, average force, floating.*

1. INTRODUCTION

The vibrations of the container with the heterogeneous liquid lead, generally speaking, to the motion of this liquid with respect to the walls of a container. Such a motion can be observed, in particular, if the solid particles are included into a homogeneous liquid of density different from that of the particles. The interaction of these bodies with the flow leads to the appearance of the force acting to the bodies, from the liquid. In the present work the behavior of solid bodies inside the liquid under the influence of high frequency vibrations of small amplitude has being examined. It is possible in this case to separate the motion of the liquid and the bodies into quick oscillating and slow average components; that permits to calculate the average force acting to the solid bodies from the oscillating flow.

2. EQUATIONS OF SOLID BODY MOTION IN AN OSCILLATING LIQUID

Let us consider the vessel containing the liquid of density ρ and performing the translational harmonic vibrations of the frequency ω and amplitude a. There is the solid body of the volume Ω and the density ρ_s inside the liquid. At the reference frame connected with the vessel the motions of the liquid and solid body are described by Navier-Stokes and Newton equations with the adding of inertia forces:

$$\frac{\partial \vec{v}}{\partial t} + \vec{v}\cdot\nabla\vec{v} = -\frac{1}{\rho}\nabla p + \nu\Delta\vec{v} + g + b\omega\vec{k}\,\sin\omega t \qquad (1)$$

$$\mathrm{div}\,\vec{v} = 0 \qquad (2)$$

$$m\frac{\partial^2 \vec{R}}{\partial t^2} = \vec{f} + mb\omega\vec{k}\,\sin\omega t + \oint_F \sigma\cdot d\vec{S} \qquad (3)$$

Here $b = a\omega$ - the amplitude of the vibrations velocity, $\vec{k}$ - unit vector along the axis of vibrations, $m = \rho_s \Omega$ - mass of the body, $\vec{R}$ - radius-vector of its centre of inertia, σ - stress tensor, $\vec{g}$ - density of an external mass force (it might be, in particular, the static gravitational force $\vec{f} = m\vec{g}$). The integration in (3) is over the surface F of the body, the other designations are the same as usually accepted ones. We consider below such a symmetry of the body, that allows to limit ourselves by the translational motion of the body, that is why we do not analyze the equations describing its rotation.

It is necessary to complete the equations (1-3) with no-slip conditions on the walls of the container and on the surface of the body.

Let us separate the variables in (1-3) into quickly oscillating and slow (on a scale of ω^{-1}) components. Procedure of the averaging will be effective if we ignore the terms $\nu\Delta\vec{v}$ and $\vec{v}\cdot\nabla\vec{v}$ in comparison with $d\vec{v}/dt$ in oscillating part of the equation (1). It might be done if the parameters of vibrations satisfy to the conditions:

$$a \ll L, \quad \nu \ll \omega L^2 \tag{4}$$

where L is the characteristic hydrodynamic lengh (it may be, in particular, the size of the body). In conformity with the above we can write down the variables (1-3) in the form:

$$\vec{v} = b\vec{V}\,cos\omega t + \vec{u},$$

$$p = \rho\omega\beta P\,sin\omega t + p_0 \tag{5}$$

$$\vec{R} = \vec{R}_0 + \frac{\mu b}{\omega}\vec{k}\,sin\omega t$$

where the fields $\vec{V}$, $\vec{u}$, P, ρ_0, $\vec{R}_0$ are slow functions of time, i.e. they do not vary significantly during the time of the order of ω^{-1} . Substituting (5) into (1-3) and collecting the terms proportional to ω^{-1} we obtain:

$$\nabla P = \vec{V} + \vec{k}, \quad div\,\vec{V} = 0,$$

$$(\mu+1)\vec{k} = \frac{\rho}{m}\oint P\,d\vec{S} \tag{6}$$

In the next order which corresponds to the averaging over the time, large comparatively with ω^{-1}, we obtain the equations, describing slow averaged motions of the liquid and of the solid body:

$$\frac{\partial\vec{u}}{\partial t} + \vec{u}\cdot\nabla\vec{u} = -\frac{1}{\rho}\nabla p + \nu\Delta\vec{u} + \vec{g},$$

$$div\,\vec{u} = 0 \tag{7}$$

$$m\frac{d^2\vec{R}}{dt^2} = \vec{f} + \oint_F \sigma\cdot d\vec{S} + \frac{1}{4}\rho b^2 \oint_F [\nabla P + (\mu+1)\vec{k}]^2\,d\vec{S} \tag{8}$$

where tensor σ is calculated on the field of averaged velocity $\vec{u}$ and pressure p, lower indices "0" of R_0 and p_0 are left.

Let us discuss now the boundary conditions. Since viscous term was ignored in the equation of an oscillating component of the velocity, for $\vec{V}$ on the wall of the vessel and on the surface of the solid body we have to ignore the viscous skin-layer and put down impermeability condition instead no-slip one. Hence taking into account (6), we can write:

$$\left[\frac{\partial P}{\partial n}\right]_S = \vec{k}_n, \quad \left[\frac{\partial P}{\partial n}\right]_F = (\mu+1)\,k_n \qquad (9)$$

Here $\vec{n}$ is the vector normal to the solid surface, S stands the surface of the container wall. For the average velocity $\vec{u}$ no-slip condition should be kept.

The equations (6-8) with the corresponding boundary conditions form the closed system for the determination of the averaged motion of the solid body in the oscillating liquid.

As it follows from (8) the vibrations of the vessel lead to the appearance of the specific "vibrational" force acting to the body. This force is described by the last term of (8).

We change the variables by introducing the function Ψ:

$$P = (\mu+1)\vec{k}\cdot\vec{r} + \mu\psi \qquad (10)$$

($\vec{r}$ - radius-vector of an arbitrary point of the liquid).

As it follows from (6) P and, hence, Ψ are harmonic functions:

$$\Delta\Psi = 0 \qquad (11)$$

and from (9-10) we have:

$$\left[\frac{\partial \psi}{\partial n}\right]_S = k_n, \quad \left[\frac{\partial \psi}{\partial n}\right]_F = 0 \qquad (12)$$

The equation of solid body motion (8) takes form:

$$m\frac{d^2\vec{R}}{dt^2} = \vec{f} + \oint_F \sigma\cdot d\vec{S} + \frac{1}{4}\rho b^2\mu^2\oint_F (\nabla\psi)^2 d\vec{S} \qquad (13)$$

where μ, as it follows from (6) is

$$\mu = \frac{m_e}{\rho\oint_F \psi\vec{k}\cdot d\vec{S} - m_e}, \quad m_e = (\rho_s - \rho)\Omega \qquad (14)$$

3. INTERACTION OF SOLID BODY WITH THE PLANE WALL OF THE VESSEL

We illustrate the effect of vibrations on the example of solid sphere interaction with the plane wall of the vessel, when the vibrations axis direction is normal to the wall. The radius of sphere is R, its centre is situated at the distance h from the wall. Let us superpose the plane $z=0$ of Cartesian system with the wall of the vessel and direct z axis to cross the centre of the sphere. We untroduce dimensionless variables choosing sphere radius R as a unit for length and for function Ψ. It is convenient to consider the problem in bispherical coordinate system with the coordinates τ, σ, ϕ being connected with the Cartesian ones by the formulas:

$$x = \frac{C\,\sin\sigma\,\cos\phi}{ch\tau - \cos\sigma},$$

$$y = \frac{C\,\sin\sigma\,\sin\phi}{ch\tau - \cos\sigma},$$

$$z = \frac{C\,sh\tau}{ch\tau - \cos\sigma}$$

In bispherical system the plane of the wall coincides with the coordinate surface $\tau=0$ and the surface of the sphere corresponds to $\tau=\tau_1$. For our choice of units we have $C = sh\tau_1$ and the distance between the centre of the sphere and the wall is $h=ch\tau$. The symmetry of the problem allows to put down $d/d\phi=0$.In this case the Laplace equation (11) for the function Ψ looks as:

$$\frac{\partial}{\partial \tau}\left(\frac{\sin \sigma}{ch\tau - cos\sigma}\frac{\partial \psi}{\partial \tau}\right) + \frac{\partial}{\partial \sigma}\left(\frac{\sin \sigma}{ch\tau - cos\sigma}\frac{\partial \psi}{\partial \sigma}\right) = 0 \qquad (15)$$

while the boundary conditions with the taking into account $\vec{k} = \nabla z$ may be put down in the form:

$$\left[\frac{\partial \psi}{\partial \tau}\right]_{\tau_1} = 0, \quad \left[\frac{\partial \psi}{\partial \tau}\right]_0 = -\frac{sh\,\tau_1}{1 - cos\,\sigma} \qquad (16)$$

The solution of the equation (15) satisfying to the conditions (16) is:

$$\psi = -2\sqrt{2}\, sh\,\tau_1 \sqrt{ch\,\tau_1 - cos\,\sigma} \times \sum_{k=0}^{\infty}\left(\sum_{j=k+1}^{\infty} U_j\right) ch\frac{2k+1}{2}\tau\, P_k(cos\,\sigma) \qquad (17)$$

where

$$U_j = \frac{1}{2\,sh\frac{2j+1}{2}\tau_1}\left(e^{-\frac{2j+1}{2}\tau_1} - \frac{(j-1)\,sh\,\tau_1}{2\,sh\frac{2j+1}{2}\tau_1}\right)$$

and P_k - Legendre polynomial of k-th power.

Substitution of the function (17) into (13-14) allows to define the value of the "vibrational" force acting to the sphere. Numerical tabulation of the series shows that this force is directed towards the wall and grows monotonously from zero at large value of h to the infinity when the sphere approaches to the wall ($h \to 0$). The existence of this force may lead to unusual effects. Let us consider, for example, the interaction of the body with the lid of the vessel in the gravitational field. If the density of the sphere is more than that of the liquid, then in the absence of vibrations the sphere will go down, sink. In the presence of vibrations , since the vibrational attraction towards the wall infinitely increases with the approach to the wall, always would be found such a distance between the body and the lid of the vessel, where all acting forces (gravity, buoyancy and vibrations) would balance. This equilibrium is unstable. If the distance up to the lid is less than that of the equilibrium then heavy body would rise to the surface. Quite a similar argument shows that the light sphere would sink if it was placed near the bottom of the vessel. The same result was obtained in (Ref.1) by another method.

4. INTERACTION OF TWO BODIES IN AN OSCILLATING LIQUID

Let us consider two long circular cylinders both of radius R with parallel axes, being placed inside the vessel with a liquid oscillating in the direction normal to the axes of the cylinders. We direct the axis X of Cartesian reference frame along the line connecting the centres of cylinders and the axis Z - parallel to the axes of cylinders. The origin of the coordinate system is placed in the middle of the stretch, connecting the centres of cylinders. The axis of oscillations is situated in the plane X0Z and forms the angle α with the Z- axis. We assume that the size of the vessel is large compared with R and the distance between the cylinders, so that it is possible to consider the medium as infinite. The influence of the walls would display in the fact, that at large distance from the cylinders the function Ψ tends asymptotically to:

$$\psi_0 = \vec{k}\cdot\vec{r} \qquad (18)$$

This value is equal to that of a vessel filled with the homogeneous liquid without any particles. The vector $\vec{k}$ has only two nonzero components:

$$k_x = cos\,a, \quad k_y = sin\,a \qquad (19)$$

and we may examine the problem in two-dimensional formulation. It is convenient to introduce bicylindrical reference frame:

$$x = \frac{C\, sh\, \sigma}{ch\tau - cos\sigma}, \quad y = \frac{C\, sin\, \sigma}{ch\tau - cos\sigma}, \quad z=z \tag{20}$$

If we choose the radius of the cylinders R as a unit of length, we obtain $C = sin\ \tau_1$ for the constant C in (20) and $h = (ch\ \tau_1)/2$ for the distance between the cylinders axes ($\tau = \pm\tau_1$ correspond to the surfaces of the cylinders). This transformation conformally transfers the plane XOY into the plane τ,σ, so that the function Ψ is the solution of Laplace equation in the coordinates τ,σ. It satisfies to the boundary conditions $\partial\psi/\partial\tau = 0$ at $\tau = \pm\tau_1$ and transfers to (18) at large distance from the cylinders:

$$\psi=2sh\,\tau_1\, cos\,\alpha \sum_{k=0}^{\infty}\left(e^{-k\tau} + \frac{e^{-k\tau_1}}{ch\, k\tau_1}\, sh\, k\tau\right) cos\, k\sigma \tag{21}$$

$$\psi=2sh\,\tau_1\, sin\,\alpha \sum_{k=0}^{\infty}\left(e^{-k\tau} + \frac{e^{-k\tau_1}}{sh\, k\tau_1}\, ch\, k\tau\right) sin\, k\sigma$$

The solution (21) is written for the area $\tau > 0$. When τ is negative it is necessary to change the sign of τ in the exponents. The substitution of (21) into (13-14) allows to calculate the force of the cylinders interaction. Because of the unwieldy of the full formula we limit ourselves by the case of large distance between the cylinders (compared with R). As it follows from (14), $\mu=1$ at $h \gg R$ and we have for "vibrational force" F per unit of length

$$F_x = -F_0\, cos\, 2\alpha, \quad F_y = -F_0\, sin\, 2\alpha,$$

$$F_0 = \frac{\pi\rho b^2 r^4}{4h^3} \tag{22}$$

So, the cylinders repel if the axis of vibrations is directed along the line connecting their centres ($\alpha = 0$) and attract if these directions are orthogonal ($\alpha = \pi/2$). For any other values of α the vibrational force are is not directed along the line connecting the centres of the cylinders.

5. CONCLUSION

Our results show that there is the force acting to solid bodies in an oscillating liquid, that is caused by the interaction of the bodies with the flow. The method developed in the present paper allows to calculate the intensity of this force comparatively simply.

6. REFERENCES

1. B.A. Lugovtsov and V.L. Sennitsky, *DAN SSSR* 289(1986)314. (in Russian)

THE RISING BUBBLE AND DROP HYDROMECHANICS UNDER THE FAINT GRAVITATION

P. K. Volkov
Institute of Thermophysics, 630090, Novosibirsk,
Siberian Division of Academy of Sciences of the USSR

ABSTRACT

The maps of flow regimes for the problem of stationary streamlining of a bubble or a drop by liquid which moves in infinity with constant velocity have been constructed on the basis of numerical solution of the complete Navier - Stokes equations. The suggested choice of the parameters to be determined makes the streamlining problem to be analogous to the bubble rising problem. The parameters, in which each medium is shown by a straight line, whose inclination is determined by the Morton number, are used as coordinates for the maps of flow regimes. This allows the interpretation of the results for the conditions of a faint gravitation.

Keywords: *Rising Bubble, Drop, Wake, Map regime's.*

1. INTRODUCTION

During the recent hydrodynamics investigations under decreasing gravity conditions as in a space satellite the governing equations to be taken exclude the terms of gravity because of its negligible magnitude by 10^3 -10^6 times less than on earth. So, for isothermal case the static problem, namely the equilibrium of liquid shapes and their stability need to be solved (Refs.1,2).

2. DIMENSIONALITY ANALYSIS AND MAPS OF FLOW REGIMES

The motion of bubbles, drops - particles in a liquid has one common peculiarity, i.e. the rising (submerging) velocity u is the determined value depending on the shape and structure of the flow (even in the case of solid sphere, i.e. a sphere sinks with a cerfain velocity, but the problem of a flow past it can be solved for different velocities of incident flow). This imposes certain restrictions in the solution of mathematical problem in a system of coordinates related to the particle, in which it rests, but liquid impinges it having the velocity far away which is equal by the value of rising (submerging) velocity.

After the dimensionless procedure is commonly used the sphere radius a and velocity u the main four dimensionless complexes appear in the problem: $Re=u2a/\nu$, $We=\rho u^2 2a/\sigma$, $Fr=u^2/ga$, $Pd=(p_k-p_\infty)2a/\sigma$, which are named the Reynolds, Weber, Froude numbers, respectively. The Pd number is the dimensionless pressure. According to the theory only two numbers for example Re and We are independent, so the other two Fr and Pd should be searched for simultaneously with the streamline functions (Ref.3). Now the Fr number will be searched as the main feature parameter serving as the rising velocity.

After solving the flow streamlining problem for a bubble with the constant Re and We numbers, it is easy to know what kind of liquid parameters is needed to a bubble with radius a to rise with velocity u. While undertaking this process with different values of Re and We numbers the explicit information could be obtained. The most suitable way to summarizes this information is the regime map drawing using special scales. The inner scales $\delta_\nu=(\nu^2/g)^{1/3}$, $\delta_\sigma=(\sigma/\rho g)^{1/2}$ where the last one is in the Laplace number (Ref.4). Using these inner scales one could obtain another two independent dimensionless numbers: $R_\sigma=a/\delta_\sigma$, $R_\nu=a/\delta_\nu$ (Ref.5), as the most suitable coordinates for regime map drawing, where every medium is shown by a straight line according to Morton's number $M=\rho^3\nu^4 g/\sigma^6=(R_\sigma/R_\nu)^6$. Having plotted the isolines of the Froude number in the coordinates R_σ and R_ν and noted the regions with characteristic types of flows one can obtain the data on bubble rising for any liquid having calculated the value of M for it. Since the coordinates R_σ, R_ν do not incorporate the rising velocity u, the Fr number is the only number which provides the complete information on rising velocity. This parameters are linked with the following relationship: $R_\sigma=(We/2Fr)^{1/2}$, $R_\nu=(Re^2/4\ Fr)^{1/3}$.

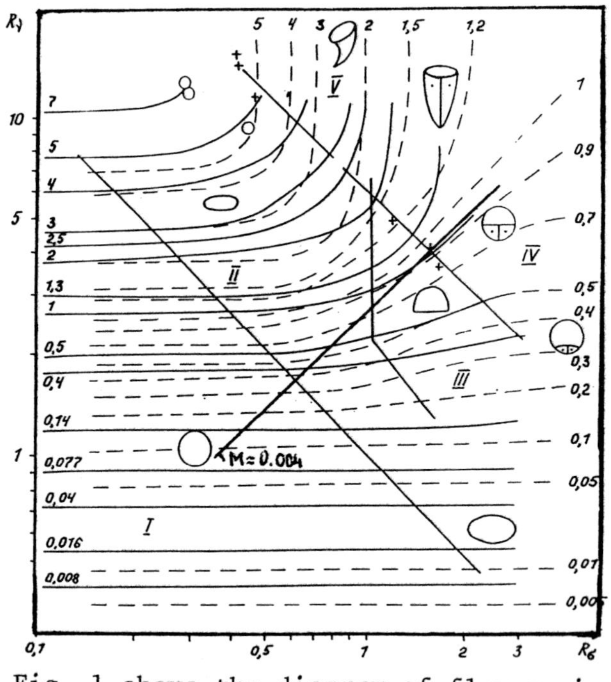

Fig. 1 shows the diagram of flow regimes for bubble rising in an infinite volume of liquid. Here solid lines are the constant Froude number isolines, obtained during the calculations; dashed lines reflects the experimental results of many authors (Ref.6). Here the distinct regions are numerated as: 1 - spherical bubbles, II - ellipsoidal ones, III - spherical caps with the plane bottom, IV - bubbles with closed vortex trace, V - bubbles with protracted trace. The upper part of the map is associated with swirling ascending bubbles. The crosses are here local maxima of rising velocity function with respect to the bubbles radii, circles - the same ones for water and methyl alcohol solution (Ref.7). An inclined line corresponds to liquid with M= 0.004, which (by classification (Ref.8)) divides the media into two types: with M > 0.004, where the dependence of rising velocity on the bubble size grows monotonically, and with M < 0.004, where this dependence has the local maximum.

In this kind of the representation of results no problem appears to estimate the gravity influence on rising process. If the gravity acceleration g for example decreases by 100 times it will lead to corresponding scaling for M, δ_σ and R_σ. The last one is by 10 time decreased and the whole picture shifted to the left on a map to much less distorted bubbles. The only parameter represented for this process is the M number, so the faint gravity is also included into the map.

It should be taken into account that great influence of admixtures appears only if $R_\sigma < 1$. Opposite to it, when $R_\sigma > 1$, there are no admixtures influence on bubble rising (Ref.5).

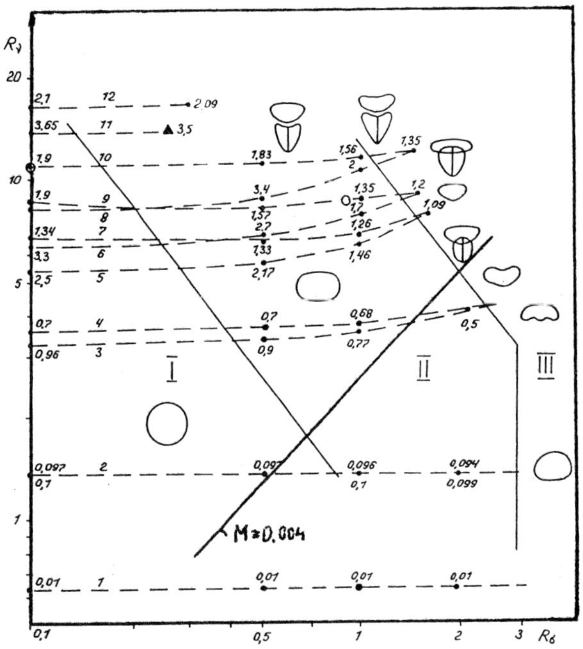

Figure 2 represents the total scheme of rising drops. There the relative density is equal to 0,1 which is suitable for example for freon with vapour. Lines 1,2 ..., 12 correspond to calculations in terms of We with constant values of Reynolds numbers inside Re_1 and outside Re_2 of the drop (Ref.9): 1 - Re_1=0.4, Re_2=0.4 (is equal Re_1=60, Re_2=0.4; 2 - Re_1=0.4, Re_2=4 (is equal Re_1=60, Re_2=4); 3 - Re_1=60, Re_2=12; 4 - Re_1=0.4, Re_2=12; 5- - Re_1=60, Re_2=40; 6 - Re_1=60, Re_2=60; 7 - Re_1=0.4, Re_2=40; 8 - Re_1=60, Re_2= 100; 9 - Re_1=0.4, Re_2=60; 10 - Re_1=0.4, Re_2=100; 11 - Re_1=20, Re_2=200; 12 - Re_1=0.4, Re_2=200.

The regions here mean: I - spherical drop, II - ellipsoidal drop, III - a toroidal wake is behind the drop or the boundary in the rear is wavy. Here the scales are: $R_\sigma = a/(\sigma/\rho_2 g)^{1/2}$, $R_\nu = a/(\nu_2^2/g)^{1/3}$, so the inclination angle for everv medium is 45°. The motion inside the drop affects the rising velocity and the structure of external flow. But after the toroidal vortex appears, the rising velocity bacomes much less influenced by the media parameters. Figures near the points on the lines are Fr values obtained while colving the problem.

To obtain the result under a faint gravity condition it is needed to calculate the Morton number values for interested liquid and observe them according to the regime map in figure 2.

Analogously the similar data could be searched for rising bubbles inside the vertical pipes (Ref.10,11) in a faint gravity.

3. CONCLUSIONS

The constructed maps of the flow regimes allow one to compare the results on rising of bubbles and drops of exact size in different liquids using simple comparison of data at the corresponding points, evaluate the effect of the droplet medium on rising. The change in external conditions will affect the value of Morton number and, hence, one can estimate qualitative and quantitative changes during bubble and drop rising.

REFERENCES

1. A.D.Mischkis, Gidromechanika v nevesomosti (Moskva, Nauka, 1976).
2. E.L.Kalyasin and A.G.Mednov. Gidromechanika i teplo-massoobmen v nevesomosti (Moskva, Nauka, 1982).
3. C.I.Christov and P.K. Volkov. J. Fluid Mech., 158 (1985).
4. S.S.Kutateladze. Analis podobiy v teplofisike (Novosibirsk, Nauka, 1982).
5. P.K.Volkov and E.A.Chinnov, Heat Transfer: Soviet Research, 15(5), (1983).
6. E.A.Chinnov, Sovremennye Problemy Teplofiziki (Novosibirsk, 1984).
7. W.L.Haberman and R.K.Morton. Proc. Amer. Soc. Civil Engns., 49(387), (1954).
8. D.Bhaga and M.B.Weber. J.Fluid Mech. 105 (1981).
9. P.K.Volkov. Modelirovanie v mechanike, 4(21),(Novosibirsk, 1990), 5.
10. P.K.Volkov. J. App.Mec.Tec.Fys., 6 (1989).
11. P.K.Volkov. J.App. Mec.Tec.Fys., 4 (1991).

Figure 1: Map of flow regimes for bubble rising.

Figure 2: Map of flow regimes for drop rising.

STUDY OF GAS BUBBLE BREAKUP IN LIQUID INDUCED BY IMPULSIVE ACCELERATIONS

W.M. Mironov, F.M. Starikov

The Scientific Research Institute
of Thermal Processes
8, Onezhskaya, Moscow, 125438 USSR

ABSTRACT

A drop tower was designed for conducting experiments on falling capsule filled with water or water-glycerine solutions. First, the capsule was falling freely (at a state of weightlessness) and then in the middle of falling a long- or short-term accelerations were imposed on the capsule by mechanical means. Gas bubbles of various sizes were introduced in the capsule and they became spherical in shape before the action of gravity. Depending on the bubble size, levels of gravity and its duration, the bubble could be destructed or might remain uncollapsed. The region of bubble destruction in terms of dimensionless parameters was found. Physically, the pattern of collapse includes formation of cumulative splash, conversion of the spherical bubble in toroidal one and then destruction of it in isolated spherical bubbles.

Keywords: fuel tank, collapse, gas bubble, liquid, gravity, weightlessness.

When a spacecraft is moving in space under low gravity disturbances arise from multiple startup of main engines. For the settling of fuel in the vicinity of tank outlet, station-keeping engines are turned on for a short time. As a result, gas ullage will experience an impulsive acceleration and can turn into a multitude of smaller bubbles, distributed throughout the whole volume of fuel. So, a breakup of the gas ullage will influence the trouble-free engine starting because of possible collection of gas bubbles at the tank outlet.

We consider a gas bubble placed in the liquid under the action of impulsive acceleration. In zero-g conditions the bubble takes a spherical shape under the action of surface tension forces and is at rest in the liquid. When acceleration begins the initial equilibrium is upset. There will be distribution of hydrostatic pressure along the vector of acceleration and also on the bubble surface. The movement of liquid in the tank will result in a change in the bubble shape.

If gravity grows slowly, a floating bubble will have time to acquire equilibrium shape and quasistable velocity. At some critical level of acceleration the bubble will be unstable and collapses into smaller bubbles.

Nonstationary changing of bubble shape due to a sudden or rapid application of acceleration may also cause its collapse. Such a collapse may take place both: in cases when time of acceleration action is large (larger than is needed for bubble destruction) and when time is smaller than needed for destruction.

SIMULATION OF BUBBLE COLLAPSE

A bubble collapse phenomenon is governed by: initial size of bubble d, the physical properties of liquid, density ρ, viscosity μ and surface tension coefficient σ, and also by the level of acceleration, ng (where g is the gravity on Earth) and by time interval of its action, τ. This set of dimensional parameters yields the following dimensionless Parameters: a Bond number $B_0 = \frac{\rho n g d^2}{\sigma}$; modified Weber number $We = \frac{\rho d (n g \tau)^2}{\sigma}$ and Galileigh number

$$Ga = \frac{n g \rho^2 d^3}{\mu^2}.$$

Consider two extreme cases of collapse resulting from suddenly applied acceleration, illustrated by Fig. 1. In the first case (a) the time of action is smaller than the time for bubble breakup. As the experimental results have shown, the breakup will depend upon the product of acceleration and the time interval it acted, $ng\tau$. The Dimensionless parameter governing breakup when viscosity influence is neglected, is the Weber number (We).

In the second extreme case (Fig. 1,b), when the duration of acceleration exceeds the time for breakup, the impulse acquired, that is $ng\tau$, will not readilly define the breakup. Under given conditions the breakup process is defined by the value of acceleration, ng. It is necessary to take a Bond number (Bo) as a governing dimensionless parameter, when viscosity is also neglected.

If we consider the breakup process as dependent both on Bo and We numbers, it is possible to obtain a generalized relationship describing the breakup of bubbles in the liquid under acceleration; however then, the influence of viscosity is supposed negligible. If the last condition is not satisfied, it is necessary to take such relationships for a given range of Ga number.

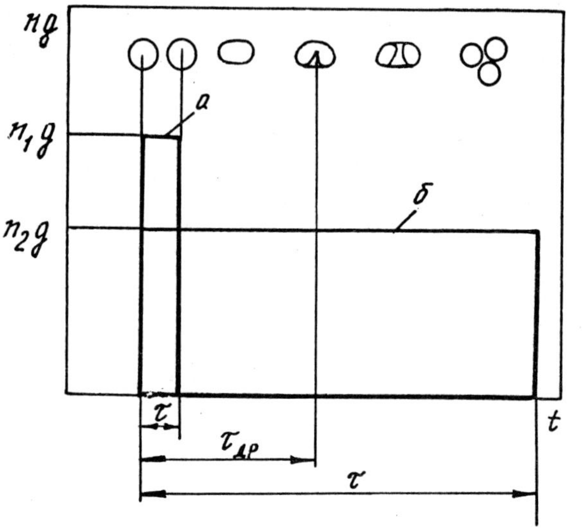

Figure 1. Scheme of Deceleration Action and Pattern of Bubble Collapse: a) Time of deceleration action is smaller than time needed for collapse; b) Time of deceleration action is greater than time needed for collapse

DESCRIPTION OF EXPERIMENTAL EQUIPMENT

The experimental study of bubble breakup when weightlessness is followed by acceleration has been conducted on the drop tower shown in Fig. 2. A rigid post (3) has a rest (2) in its upper end. A bracket (7) contains the drop mechanism (1) to which a suspension 8 is connected. The suspension consists of two lobes (upper and lower), tightened one to another by pins. The lower lobe suspension is connected to a replaceable loaded bar (4) the other end of which holds a drop capsule, consisting of a vessel (5) with model liquid, equipped with a system (6) injecting bubbles in the lower part of the vessel. The vessel has two transparent windows to take photos of bubbles in the liquid. A receiving tank (9) in the bottom part of the equipment serves for capturing the drop capsule.

Experiments were conducted in the following manner: before the beginning of the experiment the whole system was placed in the position shown in Fig. 2a. In this position air was pumped in a separate volume under a pressure of $6 \cdot 10^5$ N/m^2. After that bubbles were injected in the lower part of the vessel filled with liquid. The size of injected bubbles was determined by sizes and shapes of replaceable orifices, through which gas was supplied.

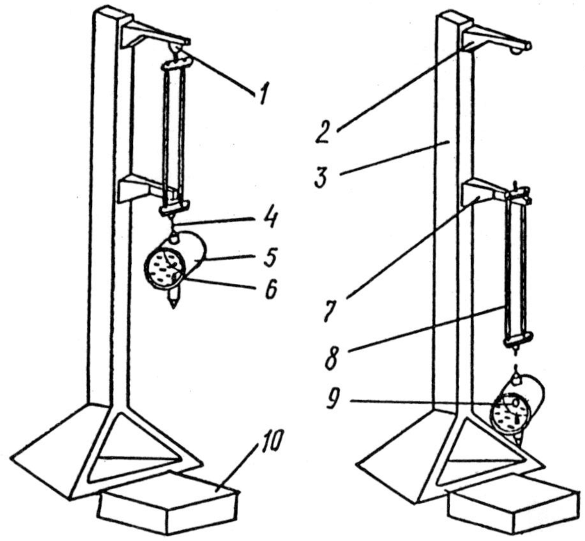

Figure 2. Experimental Equipment БН-I: a) Initial Position; b) After Action of Deceleration

Furthermore, by using the drop mechanism the suspension with the vessel, fixed in the upper position, was released. The whole system (suspension, loaded bar and vessel) was beginning to fall freely. This time the liquid and the bubbles were in a state of weightlessness. By the action of surface tension forces the bubbles acquire spherical shape. The length of free fall, before the beginning of braking deceleration, was approximately one meter. After that, the upper lobe of the suspension struck the rest. At this moment the drop capsule began to experience deceleration, caused by forces stretching and destructing the loaded bar. The value of deceleration was determined by mechanical properties of the bar and its geometry. Time of deceleration action depended on the length of the loaded bar and on the velocity of the drop capsule just before the collision. On deceleration, the drop capsule was being braked, and after bar destruction the drop capsule continued free fall. The height of free fall before the capsule hit the receiving tank amounted nearly to 0.5 m. In Fig. 2b the drop equipment is shown in free fall after braking.

A spark discharge technique was used for taking photographs of bubbles before and after braking. The first frame was taken when the upper lobe of suspension touched the rest, the second frame was taken some time after the bar destruction and end of braking. The first frame, made just before impact, gave an initial diameter of the bubble d. The second frame registered the fact of bubble collapsing or non-collapsing, and also the changing of the bubble shape during collapsing.

The value of deceleration realized in experiments, was calculated as ratio between the force necessary for destruction of bar and mass of the drop capsule. The force of the bar destruction was determined on the laboratory test machine "UM-UP" and also by calculations based on given mechanical properties of the material and sizes of the bar. In the similar manner and by measuring the bar length after the test, an elongation value, Δl, caused by destruction, was determined.

The time interval τ during which deceleration was acting, has been calculated from the velocity of the fall at the moment of deceleration start and from elongation of the loaded bar. This interval was also measured by an oscillograph fed by position signals, from the beginning of deceleration untill after the destruction of loaded bars, when deceleration ceased. An agreement of results, obtained by both methods, was observed. The loaded bars were made of various alloys of steel and of copper. Their diameter ranged from $1.5 \cdot 10^{-3}$ to $4.5 \cdot 10^{-3}$ m and the length from $20 \cdot 10^{-3}$ m to $80 \cdot 10^{-3}$ m. The length of copper bars reached a value of $2.5 \cdot 10^{-1}$ m. These bars permitted to change the decelerationg impulse, $ng\tau$, applied to the drop capsule, from $2 \cdot 10^{-2}$ m/s up to $100 \cdot 10^{-2}$ m/s.

The time of deceleration action was changing from $0.5 \cdot 10^{-3}$ s to $20 \cdot 10^{-3}$ s. The diameter of bubbles was in the range of (3-12) $\cdot 10^{-3}$ m. With the use of distilled water in experiments there was possibility for variation of We number from $1.5 \cdot 10^{-2}$ up to $2 \cdot 10^{2}$. The variation range of Bond number was from Bo = 5 to Bo = $3 \cdot 10^{2}$.

Evaluation of experimental accuracy has shown that errors in determining Bo and We numbers could reach the 30 percent value.

EXPERIMENTAL RESULTS

A consequence of bubble collapse in its various phases is shown in Fig. 1a and 1b. In the lower part of the bubble a comulative blowup of liquid piercing the bubble is observed. As a result, the bubble acquires a toroidal shape. The latter is collapsing into individual bubbles, the size and number of which depend on a level of deceleration and on the time interval in which it is acting.

When a bubble is acted upon by gravity during a very short time interval, the bubble shape will remain practically spherical. But when the level of deceleration is sufficiently large when the value of the product of $ng\tau$ is sufficiently large), so that We > We_{cr}, the bubble will collapse. The purpose of the main part of experiments was to define boundaries of a dimensionless parameters region inside which a collapse will occur. The results of experiments are presented in Fig. 3. Clear signs correspond to experiments, in which collapse took place, blacked signs denote experiments when collapse was not observed. It is not difficult to separate a region in. Fig. 3, inside which collapse is

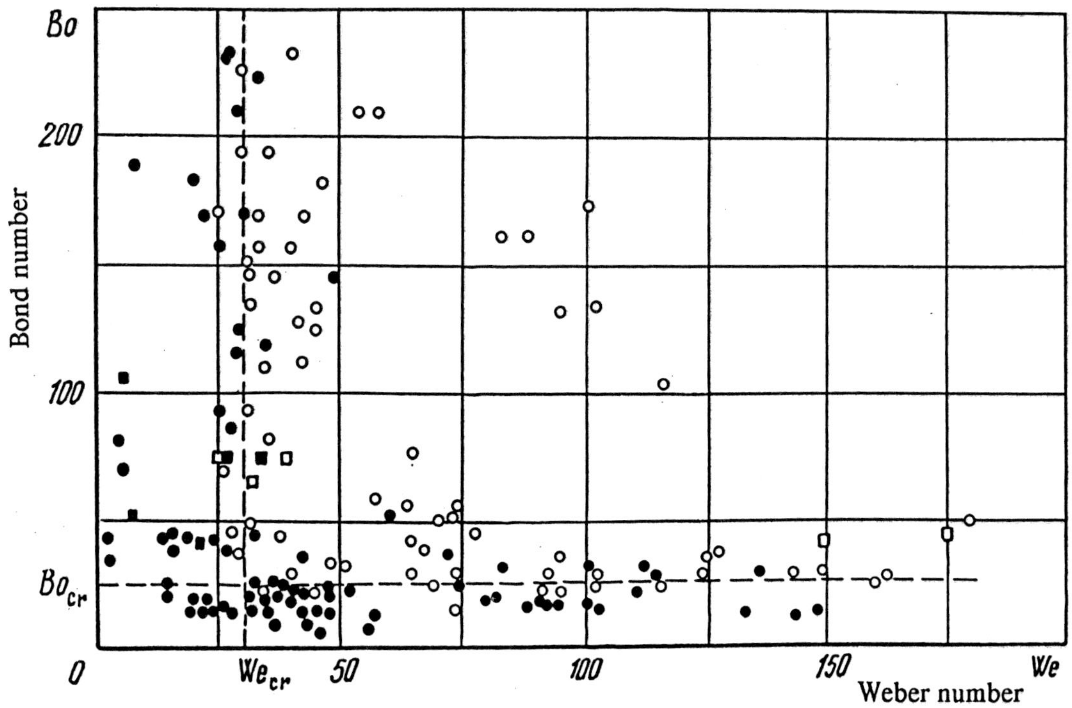

Figure 3. Experimental Results on Bubble Collapse

occuring. This region is restricted to a We number $We = \frac{\rho d(ng\tau)^2}{\sigma}$ by a value of $We_{cr} \approx 30$ and to a Bond number, $Bo = \frac{\rho ng\,d^2}{\sigma}$, of $Bo_{cr} \approx 25$.

Thus, for cases when both dimensionless parameters exceed their critical values the initially unbroken gas bubble in the liquid will collapse. But when any of these parameters is smaller than its critical value, the collapse will not occur.

It is seen from the figure that the region of collapse contains also experimental points, corresponding to non-collapsing bubbles. It can be explained by the probabilistic nature of the bubble collapse after becoming toroidal-shaped. That is, an unstable toroidal-shaped bubble being close to the collapse boundary, can again spring back into a single spherical bubble, or it may collapse into several bubbles.

Since all experiments were conducted with distilled water at room temperature, the data obtained correspond to a single value of viscosity and are valid for a strictly defined region of Ga number. As for the tests conducted, the variation of Ga number was in the interval of Ga = 10^7-10^9. When the values of Ga number are large, the influence of viscosity on bubble collapse should be small, because in this case friction forces are small enough as to be compared with gravity forces. In other words, beginning with some values, a similarity will be observed on Ga number. This fact permits to extend the region of using experimental results on large values of Ga number.

To reveal the possible influence of the Ga number, a set of tests was conducted in which a 59 wt. % glycerin-water solution was used. The viscosity of this solution is 10 times greater than that of water, thus the Ga number was lowered down to a 10^5 value.

The experimental points corresponding to this set of experiments are plotted in the same Fig. 3 by clear (collapsed) and blacked (non-collapsed) squares. These results confirm the absence of a strong influence of viscosity on the collapse mechanism in this region. Since for a full-scale spacecraft the values of Ga number are, as a rule, higher, it is possible, having in mind similarity on Ga number, to use the results presented in Fig. 3 for prediction of bubble collapse.

FLOW OF GAS SLUGS UNDER MICROGRAVITY CONDITIONS

Hsueh-Chia Chang

Department of Chemical Engineering
University of Notre Dame
Notre Dame, IN 46556

ABSTRACT

We study the microgravity flow of gas slugs in a liquid-containing circular conduit. Under neutrally buoyant and slow slug speed conditions, the slugs squeeze the liquid into a thin annular film which enlarge into thick static regions between two neighboring slugs. Using a matched asymptotic expansion, we are able to estimate the slug shape and the pressure drop necessary to drive a train of slugs at constant speed. It is shown that the pressure drop for a slug train is not determined by the length of the train but by the number of slugs in the train. Surprisingly, the pressure drop across one member slug in a train is independent of the compression between slugs and is identical to that of an isolated slug. A correlation for the apparent viscosity of slug flow under neutrally buoyant conditions is then derived in closed form from first principles.

Keywords: *Multi-phase Flow, Microgravity, Matchet Asymptotic Analysis, Laplace-Young Equation, Apparent Viscosity.*

1. INTRODUCTION

For the same reasons that two-phase flow is of paramount importance in heat and mass transfer processes and in pipelines on earth, it will also become a major concern in the design of efficient space stations. However, in the absence of gravity, two-phase flow becomes extremely unstable due to interfacial forces and its flow characteristics are quite different from those on earth. Consider, for example, the core-annular flow configuration in a circular pipe with gas flow in the middle and liquid flow in the outer film. Due to the absence of gravity, the annulus is axisymmetric and the configuration is unstable to an axisymmetric "sausage" instability driven by the azimuthal curvature [1]. As a result, the gas phase is quickly dispersed into slugs as necking occurs along the pipe. These slugs are approximately the same length as the wavelength of the most unstable disturbance which is approximately the circumference of the original gas-liquid annular interface. Compression ensues and the slugs are compressed into a train as shown in figure 1. The slugs are flattened against one another so that adjacent slugs are

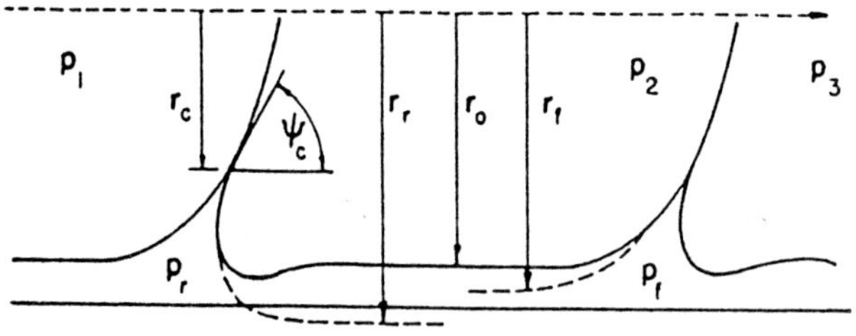

Figure 1: Schematic of a bubble train. The contact angle ψ_c and contact radius r_c for the lamellas are shown. Limiting radial positions r_f and r_c are the extrapolation of the static lamellas and r_o is the flat-film cylinder radius.

separated only by a thin lamella similar to those surrounding soap bubbles. Although the liquid within the lamella will drain out eventually without the stabilizing effects of surfactants, such coalescence occurs very slowly and the slug train travels essentially as a unit. We are interested in the pressure drop necessary to drive such a slug train to travel at a certain velocity, viz. the apparent viscosity of the two-phase flow.

2. ANALYSIS

We shall consider a slug train consisting of arbitrably long air slugs which are separated by identical lamellae. This implies that the degree of compression at each lamella is identical although the slugs can be of different length. Because of the assumption of identical lamellae, the dimensionless gas pressure p scaled by σ/R within each slug must be related by

$$p_1 - p_2 = p_2 - p_3 = \dots = p_i - p_{i+1} \quad (1)$$

However, the pressure difference across each lamella is related by the contact angle ψ_c and contact radius r_c by a simple force balance

$$(p_1 - p_2)\,\pi\, r_c^2 = 2(2\pi\, r_c \cos\psi_c) \quad (2)$$

However, due again to the periodicity of the lamellae, the pressures p_r and p_f in the "static" regions immediately below the front and back lamellae are related by

$$p_1 - p_r = p_2 - p_f \quad (3)$$

Hence, combining (2) and (3), one obtains

$$\cos\psi_c = \frac{r_c}{4}\,\Delta p \quad (4)$$

where $\Delta p = p_r - p_f$ is the total pressure drop in the liquid across a member slug of the train. This is the quantity we seek.

Consider the thin annular film around the slug. Assuming a sufficiently thin film such that the azimuthal curvature is negligible, one can use Bretherton's lubrication analysis [2,3] to derive an equation for the film region,

$$\frac{d^3\eta}{d\xi^3} = \frac{3(\eta - 1)}{\eta^3} \quad (5)$$

where $\eta = h/h_\infty$ is the normalized film height and $\xi = \zeta h_\infty/Ca^{1/3}$ is the scaled axial coordinate. The quantity h_∞ is the dimensionless thickness at the middle of the infinitely long slug scaled with respect to the pipe radius, $Ca = \mu U/\sigma << 1$ is the capillary number which is also the dimensionless slug velocity and $\zeta = \xi$ - Ca t is the coordinate moving with the velocity of the bubble. The usual lubrication scaling arguments are in play here. Without gravity, the liquid flow in the film is driven by the pressure gradient which is related to the gradient in the axial curvature. Hence, a balance between the viscous and pressure term in the streamwise momentum balance yields that the ratio between the normal length scale and the tangential length scale is $Ca^{1/3}$. At the two ends, the curvature of the lamellae is of O(1) to leading order which implies that the axial curvature in the film h_{xx} must also be of order unity to allow proper matching. As a result, the film thickness h_∞ is of $O(Ca^{2/3})$ and it also leads to the scaled variable ξ.

Equation (5) can be converted into a dynamical system by using the variables

$$\underline{x} = \begin{pmatrix} x_1 \\ x_2 \\ x_3 \end{pmatrix} = \begin{pmatrix} \eta - 1 \\ \eta_\xi \\ \eta_{\xi\xi} \end{pmatrix} \tag{6}$$

such that

$$\underline{x}_\xi = \begin{pmatrix} 0\,1\,0 \\ 0\,0\,1 \\ 3\,0\,0 \end{pmatrix} \underline{x} + \begin{pmatrix} 0 \\ 0 \\ \dfrac{-3x_1^2(3x_1 + 3 + x_1^2)}{(1 + x_1)^3} \end{pmatrix} \tag{7}$$

The linear Jacobian has eigenvalues $3^{1/3}$, $-\frac{3^{1/3}}{2}(1 \pm i\sqrt{3})$. The fixed point $\underline{x} = \underline{0}$ corresponds to the middle of the film region where η_ξ and $\eta_{\xi\xi}$ vanish. Trajectories of (7) from a small neighborhood of the origin in the positive ξ direction quickly approach the real unstable eigenvector and η blows up exponentially from one in the positive direction. Nonlinear effects become important as the trajectory departs from the origin and the unstable eigenvector evolves into a one-dimensional unstable manifold. As η becomes large, it is obvious from (5) that $\eta_{\xi\xi\xi}$ vanishes and the unstable manifold becomes tangent to the plane $x_3 = 1.338$. The limiting behavior of η as $\xi \to +\infty$ is then

$$\eta \sim 1.338\,(\xi - \xi_f)^2 + \eta_{min}^f \tag{8}$$

where $\eta_{min}^f = 2.780$. This result was first obtained by Bretherton who also noted that the number ξ_f is unimportant. Behavior in the $-\xi$ direction has yet to be fully explored. Trajectories in this direction evolve on the two-dimensional stable eigen space and the subsequent two-dimensional unstable manifold at larger distance. Since the corresponding stable eigenvalues are complex, the back interface exhibits characteristic undulations with wavelength of order $Ca^{1/3}$. However, since the amplitude increases by a factor of $\exp(2\pi/\sqrt{3}) \sim 40$ per wavelength, the interface amplifies rapidly to approach the static region and one typically sees only one or two humps (see figure 1). Because the stable eigen space is two-dimensional, there is a one-parameter family of trajectories emanating from the origin which remain on the stable manifold. This family is parameterized by the initial direction which is determined by the relative contribution of the two unstable eigenvectors. Each trajectory leads to a different limiting behavior of η as ξ approaches $-\infty$. However, since the front and back lamellae are identical, their curvatures must also be the same to leading order and hence so should the limiting $\eta_{\xi\xi}$ of (5) in both directions. We hence choose a particular initial direction which yields the asymptotic behavior at $\xi \to -\infty$

$$\eta \sim 1.338\,(\xi - \xi_b)^2 + \eta_{min}^b \tag{9}$$

where $\eta_{min}^b = -0.732$ and ξ_b is distinct from ξ_f but is also inconsequential.

Both limiting profiles (8) and (9) must now be matched into the static lamellae at the ends. In terms of the unscaled axial coordinate ζ and the radial coordinate r, the two interfaces are governed by the nodoidal solution of the Laplace-Young equation given by the elliptic integral

$$\zeta - \zeta_i = - \int_{r_i}^{r} \frac{\alpha_i (r^2 - \beta_i^2)\, dr}{[r^2 - \alpha_i^2 (r^2 - \beta_i^2)^2]^{1/2}} \tag{10}$$

where subscript i denotes b or f and the boundary conditions that the lamellae have zero contact angles ($\psi = 0$) at $\xi_i = \zeta_i h_\infty / Ca^{1/3}$ where $r = r_i$ have been imposed (see figure). The parameters α_i and β_i are specified by the boundary conditions,

$$\psi = \psi_c \quad \text{at } r = r_c \quad \text{at the front}$$
$$\psi = \psi_c - \pi \quad \text{at } r = r_c \quad \text{at the back} \tag{11}$$

where ψ is the angle the interface makes with the horizontal. The parameters α_i and β_i are then

$$\alpha_i = (r_i \pm r_c \cos \psi_c)/(r_i^2 - r_c^2) \equiv \frac{1}{2R_i} \tag{12a}$$

$$\beta_i^2 = r_i^2 - r_i/\alpha_i \tag{12b}$$

The parameter α_i also defines the front and back radii of curvature R_f and R_b.

As the lamellae approach the film region, the film thickness $1 - r_i = c_i$ is of $O(Ca^{2/3})$ and the elliptic integral can be expanded with proper scaling. After considerable algebra, one obtains

$$h(\zeta) = 1 - r \sim \frac{1}{2} \left[\frac{(1+r_c^2)}{(1-r_c^2)}\right] (\zeta - \zeta_i)^2 - c_i \tag{13}$$

which is the limiting behavior of the lamellae near the film. We hence match (8) and (9) with (13) to get

$$h_\infty (r_c) = [(1-r_c^2)/(1+r_c^2)]\ 1.338\ Ca^{2/3}$$
$$= [(1-r_c^2)/(1+r_c^2)]\ h_\infty (0) \tag{14a}$$

$$c_i = - \eta_{min}^{i}\, h_\infty (r_c) \tag{14b}$$

and the liquid pressure drop

$$\Delta p = \frac{1}{R_f} - \frac{1}{R_b} = 2\, (\alpha_f - \alpha_b) \tag{15}$$

Simplifying α_f and α_b of (12a) by expanding in $1-r_i = c_i$ and substituting (4) and (14b), one obtains

$$\Delta p \sim 2\, (c_b - c_f)\, [(1+r_c^2)/(1-r_c^2)] + O(Ca^{4/3})$$

$$\sim 2(\eta_{min}^{f} - \eta_{min}^{b})\, h_\infty (0) \sim 9.40\ Ca^{2/3} \tag{16}$$

We hence obtain the surprising result that Δp is independent of the contact radius r_c to leading order and is identical to that of an isolated bubble. The film thickness, however, decreases with increasing r_c (compression) as is evident from (14a).

3. CONCLUSION

Our analysis shows that the pressure drop across a member slug of a train is identical to that of an isolated slug regardless of the degree of compression. This simplifies the modeling of the apparent viscosity of slug flow. Consider a sequence of separated slug trains all travelling at a speed U. Then the apparent viscosity defined by a Hagen-Poiseuille relation is

$$\mu_A = (R^2/8U)\, (\Delta\hat{p}_t/L) \tag{17}$$

where L is the conduit length and $\Delta\hat{p}_t$ is the total dimensional pressure drop across the entire length of the conduit. This pressure drop can then separated into two contributions

$$\Delta\hat{p}_t = \Delta\hat{p}_l + \Delta\hat{p}_s \tag{18}$$

where $\Delta\hat{p}_l$ is the dimensional pressure drop occurring in the liquid segments between uncompressed slugs and $\Delta\hat{p}_s$ comes from the slug trains. Hence, from (16)

$$\Delta\hat{p}_s = n\, 9.40\, Ca^{2/3}\, \sigma/R \tag{19}$$

where n is the number of slugs and

$$\Delta\hat{p}_l = \frac{8\mu\, U}{R^2}\, l \tag{20}$$

where l is the total length of liquid segments. Hence, combining (17) to (20), we obtain

$$\mu_A = \mu\, (\frac{l}{L} + n\, 9.40\, Ca^{-1/3}\frac{R}{8L}) \tag{21}$$

We again note that the contribution from the air slugs is independent of the total slug length (L-l) but is proportional to the number of slugs n and inversely proportional to the one-third power of the dimensionless speed Ca.

REFERENCES

1. P. S. Hammond, *J. Fluid Mech.*, 137 (1983) 363.

2. F. P. Bretherton, *J. Fluid Mech.*, 10 (1961) 166.

3. J. Ratulowski and H.-C. Chang, *J. Fluid Mech.*, 210 (1990) 303.

Mixtures, Thermodynamics and Physical Properties

INSTABILITY WITH SORET EFFECT : THE GLOBALIZED FORMULATION, ITS PHYSICAL INTERPRETATION, AND SPECIFIC SOLUTIONS FOR LONG CYLINDERS

D. Henry - G. Hardin** - B. Roux*- R. Sani***

Laboratoire de Mécanique des Fluides
et d'Acoustique-Ecole Centrale de Lyon
69131 ECULLY-FRANCE

**Institut de Mécanique des Fluides*
1 rue Honnorat
13003 MARSEILLE-FRANCE

***N.I.S.T.*
325 Broadway-BOULDER
COLORADO 80303-3328-U.S.A.

**** Department of Chemical Engineering*
University of Colorado-BOULDER
COLORADO 80309-0432-U.S.A.

ABSTRACT

The linear stability of a mixture submitted to a Soret separation by applying a temperature gradient is studied in the case of a vertical cylinder. A physical analysis of the contribution of the Soret effect to instability is presented, that allows a clear development of the equations leading to the globalized formulation. Some specific solutions are found for long cylinders.

Keywords : *Instability, Soret effect, Cylinder, Globalized Formulation.*

INTRODUCTION

Thermal diffusion in a mixture (often called the Soret effect) is the diffusion of the constituents under the action of a temperature gradient, leading to their separation. This effect is counterbalanced by ordinary diffusion, and the diffusive stationary state corresponds to a separation called "the Soret separation" which is characterized by the Soret coefficient St=D'/D.

Although being generally a secondary effect, the Soret effect can become important in experiments concerned with mixtures submitted to temperature differences : it usually leads to some separation of the constituents, and can be the cause of undesirable inhomogeneities or perturbations in physical or chemical processes. These behaviours, combined with the still fundamental interest related to such cross effects, have stimulated specific measurements of this Soret effect.

In fact, these measurements can often be perturbated by buoyancy driven convection, which has motivated flight experiments aboard spacecrafts in order to minimize these perturbations (Refs.1,2) and numerical studies (stability or simulation) to understand these phenomena (Refs.3,4,5,6).

In the case of the linear stability study of vertical situations (vertical temperature gradient), the equations with Soret effect can be written at steady state under a globalized formulation with only two parameters : a generalized Rayleigh number $\tilde{R}$ and a generalized Soret parameter ψ. The objectives of this presentation will be first to show a step by step derivation of the stability equations allowing to see the role of these parameters and their physical signification : this will give a better insight in the physical interpretation of the stability curves. Then, in the case of long cylinders, simple specific solutions can be found that compare well with the numerical solution of the stability problem.

HYPOTHESIS AND MATHEMATICAL DESCRIPTION OF THE PROBLEM

We consider a binary newtonian fluid mixture confined in a vertical cylinder with an aspect ratio A=L/$\underline{R}$ (L length, $\underline{R}$ radius)(figure 1). We use the Boussinesq approximation : the physical properties are assumed to be constant except for the density $\bar{\rho}$, which is assumed to be a linear function of the temperature $\bar{T}$ and of the mass fraction $\bar{X}$ (of the denser component) in the buoyancy term :

$$\bar{\rho}=\bar{\rho}_0[1-\alpha(\bar{T}-\bar{T}_0)+\beta(\bar{X}-\bar{X}_0)] \qquad (1)$$

where α and β are respectively the thermal and solutal expansion coefficients, and an overbar denotes a dimensional quantity. We use the following phenomenological equations relating the heat and mass fluxes ($\overline{J}_Q$ and $\overline{J}_X$) to the thermal and solutal gradients :

$$\overline{J}_Q = -\lambda\, \nabla\overline{T} - \overline{\rho}\, D_f\, \nabla\overline{X} \qquad (2)$$

$$\overline{J}_X = -\overline{\rho} D' \overline{X}(1-\overline{X})\nabla\overline{T} - \overline{\rho} D \nabla\overline{X} \qquad (3)$$

where λ is the thermal conductivity, D is the mass diffusivity, and D' and D_f are the phenomenological transport coefficients for the Soret and Dufour effects, respectively.

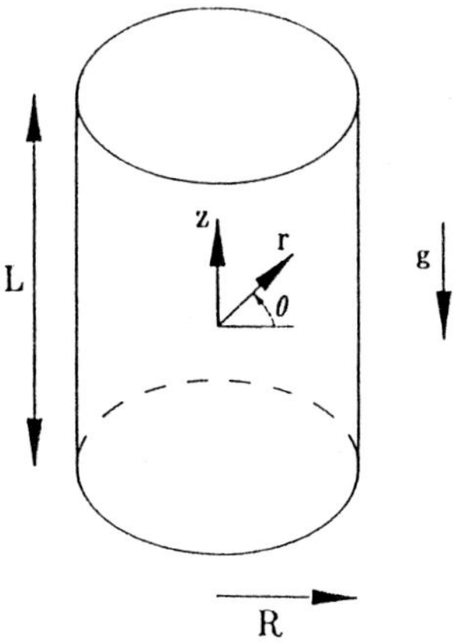

Figure 1 : Problem geometry and co-ordinate system.

The contribution of the temperature gradient to the flux of matter (Soret effect) is the leading phenomenon in our study : it creates the separation of the constituents. The contribution of the mass-fraction gradient to the flux of heat (Dufour effect), which is typically small, particularly for liquids, will be neglected in this study.

In a vertical cylinder subjected to a temperature difference between its parallel faces, the linear stability study of the Soret diffusive stationary state (no-flow base state) leads to the following set of dimensionless perturbation equations :

$$\nabla . V = 0 \qquad (4)$$

$$\partial V/\partial t = -\nabla p + Gr(T+SX)e_z + \nabla^2 V \qquad (5)$$

$$Pr\partial T/\partial t = Pr V z + \nabla^2 T \qquad (6)$$

$$Sc\partial X/\partial t = Sc V z - \nabla^2 T + \nabla^2 X. \qquad (7)$$

Here T is nondimensionalized by $\Delta\overline{T}/A = (\overline{T}_B - \overline{T}_T)/A$ (where $\overline{T}_B$ and $\overline{T}_T$ are the imposed temperature at the bottom and top ends of the cylinder, respectively), X by $(\overline{X}_B-\overline{X}_T)_{Sor}/A = (-D'\overline{X}_0(1-\overline{X}_0)\,\Delta\overline{T})/(DA)$ (where $(\overline{X}_B-\overline{X}_T)_{Sor}$ is the "Soret separation", i.e. the concentration difference between the bottom and top ends due to the Soret effect as a result of the applied temperature difference), r by $\underline{R}$, z by L, V by $v/\underline{R}$, t by $\underline{R}^2/v$ and ∇ by $1/\underline{R}^2$. The dimensionless parameters are the Grashof number $Gr = \alpha g \Delta T \underline{R}^3/(Av^2)$ ($Gr \geq 0$ when the heating is from the bottom, and $Gr \leq 0$ when the heating is from the top), the Prandtl number $Pr = v/\kappa$, the Schmidt number $Sc = v/D$, and the separation or Soret parameter $S = D'\overline{X}_0(1-\overline{X}_0)\beta/(D\alpha)$ giving the density change due to the Soret separation compared to that due to the applied temperature difference. The associated boundary conditions are T=0 at z=0 and z=1; T=0 at r=1 (conducting sides) or $\partial T/\partial r = 0$ at r=1 (adiabatic sides); V=0 and $(J_X)_n = \partial T/\partial n + \partial X/\partial n = 0$ on all the boundaries (no mass flux through the walls).

PHYSICAL ANALYSIS OF THE CONTRIBUTION OF THE SORET EFFECT TO INSTABILITY

For a better understanding of the stability results, as the perturbations of the linearly stratified fields of the temperature and mass fraction evolve to deformations of these fields, we study first the influence of the thermal boundary conditions and of the values of Pr and Sc on the deformations of the temperature and mass fraction fields within a convective situation with Soret effect.

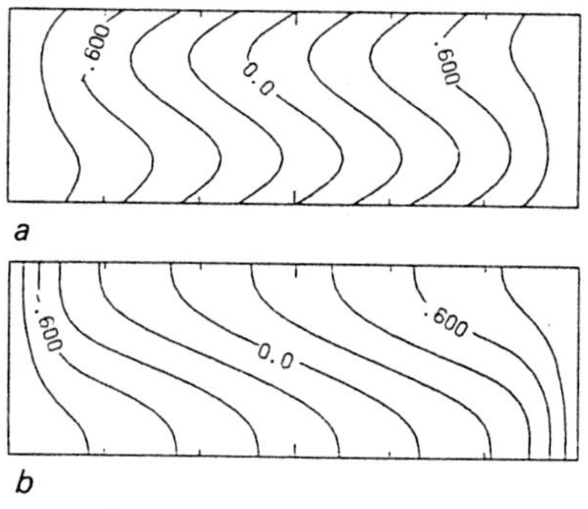

Figure 2 : Iso-values of the temperature in the V plane for Gr=278, Pr=0.6 and S=0. a) Conducting thermal boundary conditions; b) Adiabatic thermal boundary conditions.

Deformations induced by convection

We made some simulations in a long horizontal cylinder (A=6), at Pr=0.6 (which is representative of molten salts) and Gr=278 (a value strong enough to obtain deformations of the isotherms). The parameter S has been chosen equal to zero, a case where the mass fraction has no influence on the velocity. We considered different thermal boundary conditions (conducting (a) and insulating (b)), and different Sc values. The pictures presented here correspond to contours of T or X in the vertical symmetry plane (plane of main circulation, denoted the "V plane").

The deformations of the isotherms in the V plane with the two different thermal boundary conditions are presented in figure 2. The two cases look different : in the conducting case, the values of T are imposed at the boundary (presence of a lateral thermal flux) and we obtain deformations with an S shape; in the insulating case, the zero-flux condition does not constrain the sidewall temperature profile, resulting in more elongated T profiles with stronger deformations along the axis, although the convection is weaker in this case. The mass-fraction profiles corresponding to these situations have been obtained for different values of Sc.

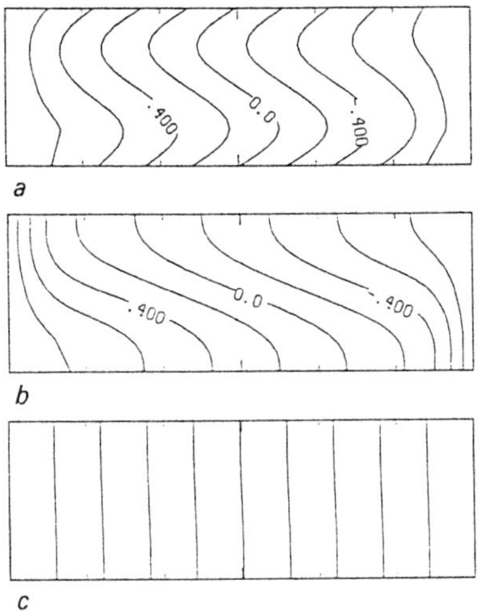

Figure 3 : Iso-values of the mass-fraction in the V plane for Gr=278, Pr=0.6, S=0 and Sc=0.006. a) Conducting thermal boundary conditions, Xbot=0.995; b) Adiabatic thermal boundary conditions, Xbot=0.987; c) Fictitious case with no deformation of T, Xbot=1.

For small values of Sc (Sc=0.006, i.e., Sc=0.01 Pr), the mass fraction profiles in both cases (a) and (b) result in both a good separation (the normalized separation $Xbot=(\Delta Xm)/(\Delta X)_{Sor}$, where ΔXm is the difference between the mean value of X at the bottom and at the top, is near 1) and a strong deformation (figure 3). In fact, the mass fraction profiles can be exactly superimposed on their respective (dimensionless) temperature profiles. This is confirmed by a fictitious case with no deformation of T (denoted as case (c) in the figures), where a quite similar X profile with no deformation is obtained. In fact, for small values of Sc, the convective transport term in the mass fraction equation becomes negligible compared to the diffusion terms (thermal diffusion and ordinary diffusion). In this case, we have a separation profile of the mass fraction directed only by the thermal diffusion effect. This mass fraction profile is generated by the temperature profile and reproduces identically its properties, i.e. the differences between its extreme values (leading to a good separation) and its deformations (for appropriately non-dimensionalized fields). This behaviour can be formally established with the equations (see Ref.7).

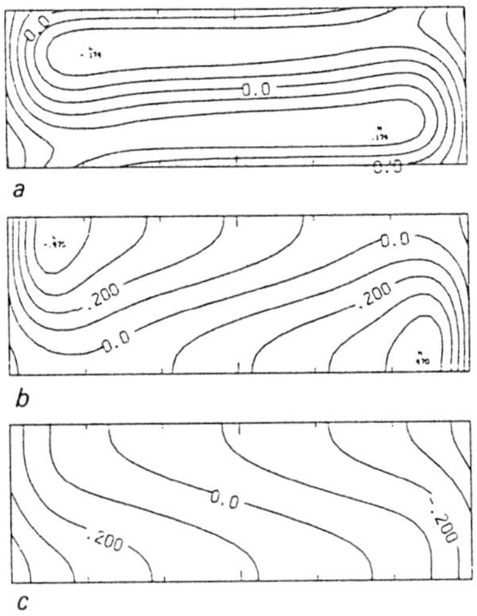

Figure 4 : Iso-values of the mass-fraction in the V plane as in figure 3 but with Sc=0.6. a) Conducting thermal boundary conditions, Xbot=0.208; b) Adiabatic thermal boundary conditions, Xbot=0.036; c) Fictitious case with no deformation of T, Xbot=0.345.

For larger Sc values (Sc=0.6, i.e., Sc=Pr, see figure 4), we obtain more tilted mass fraction profiles, for which the deformation comes not only from the repercussion of the deformation of the isotherms by the Soret effect, but also from the convective transport of the mass fraction which is no longer negligible. The case (c) gives a deformation caused almost solely by the convective transport term.

As a main result, we can say that in similar conditions insulated walls lead to stronger deformations of the isotherms than conducting walls. Moreover, by thermal diffusion, these deformations of the isotherms have a nearly perfect corresponding repercussion on the mass fraction at steady state, and can give concentration deformations even for small Sc where the convective transport term is negligible. Stronger values of

Sc can add to this deformation a contribution of the convective transport term.

Phenomenology of the instability

The instability is connected to the behaviour of the perturbations of the individual variables rather than simply on the overall density gradient. A look at the stability phenomenon by considering the convective state that occurs just beyond the instability threshold is then more appropriate and gives the following indications.

First, because the base-states under consideration are vertical, the temperature and mass-fraction fields act on the motion only if they are deformed so as to cause a horizontal stratification of the T and X fields in addition to the vertical stratification present in the initial state of mechanical equilibrium. This reflects the fact that in the perturbation equations, the variables that determine the stability threshold are the perturbations to the initial fields of T and X.

If T acts on the motion X will act too, thanks to the Soret effect. In fact, a not-too-small value of Sc will lead to a deformation of X that is stronger than that of T because of the additional deformation of X due to the convective-transport term in the species-mass balance. But, the real contribution of X to the motion compared to that of T will depend on the value of S.

Finally, because at steady state there is an automatic repercussion of the deformations of T on X due to the Soret effect, and an additional deformation of X by convection, it is useful to contrast the total influence on the motion of the deformation of T along with its repercussion on X against the influence of the "self" deformation of X due to the convective transport term in the X equation. This approach can also be developed from a careful consideration of the perturbation equations.

ANALYSIS OF THE PERTURBATION EQUATIONS

Didactic development of the equations

We consider the initial system of perturbation equations (4-7), bearing in mind the fact that local perturbations evolve to global deformations of the originally linearly stratified T and X fields.

The zero-mass-flux condition on the boundaries suggests the following transformations : the mass fraction X is replaced by a new dimensionless variable Xc such that X=T+Xc (the condition is then $(J_X)_n=\partial Xc/\partial n=0$). If the convective transport term is null, equation (7), at steady state and with the associated boundary conditions, gives the solution X=T (see Ref.7). This change of variables corresponds then to what has been observed previously, i.e., that the deformations of the temperature lead at steady state to identical deformations of the mass fraction, to which is added a "self" deformation of the mass fraction due to the convective transport term, Xc. We obtain then (we will not repeat further the continuity equation (4)) :

$$\partial V/\partial t=-\nabla p+Gr((1+S)T + S\ Xc\ )e_z+\nabla^2 V \quad (8)$$

$$Pr\partial T/\partial t=PrVz+\nabla^2 T \quad (9)$$

$$Sc(\partial T/\partial t+\partial Xc/\partial t)=ScVz+\nabla^2 Xc. \quad (10)$$

In this form equations (8-10) at steady state give similar expressions for the perturbations of the temperature, T and the "self" perturbation of the mass fraction, Xc. It is then interesting to define new perturbations T' and Xc', which are of the same order of magnitude and satisfy identical stationary equations, by the following change: T=Pr T' and Xc=Sc Xc'. We have then:

$$\partial V/\partial t=-\nabla p+Gr(Pr(1+S)T'+Sc\ S\ Xc')e_z+\nabla^2 V \quad (11)$$

$$Pr\partial T'/\partial t=Vz+\nabla^2 T' \quad (12)$$

$$Pr\partial T'/\partial t+Sc\partial Xc'/\partial t=Vz+\nabla^2 Xc' \quad (13)$$

Equation (11) indicates that the buoyancy forces which generate the motion are proportional to the perturbations of the temperature T' with the factor GrPr(1+S) and to the perturbations of the mass fraction Xc' with the factor GrScS. It must be pointed out that the term containing the perturbations of the temperature, T' corresponds to the perturbations of the temperature but also to its repercussion on the mass fraction field by the Soret effect, whereas the term containing the perturbations of the mass fraction, Xc' corresponds only to the "self" perturbation of the mass fraction due to the convective transport term. T' and Xc' being now perturbations of the same order of magnitude, the relative importance of the two terms will directly depend on the factors GrPr(1+S) and GrScS, and will be correctly represented by the parameter $\psi=ScS/(Pr(1+S))$ which was first proposed by Schechter et al. (Ref.8) and Gutkowicz-Krusin et al. (Ref.9).

From the above discussion we see that if $\psi\rightarrow 0$, the main influence comes from the perturbation of the temperature, whereas if $\psi\rightarrow\infty$, the main influence comes from the "self" perturbation of the mass fraction. Equation (11) can then be written as a function of this parameter ψ and of the Rayleigh number $\tilde{R}=R(1+S)=GrPr(1+S)$, which leads to :

$\partial V/\partial t=-\nabla p+\tilde{R}(T'+\psi\ Xc')e_z+\nabla^2 V,$ (14)

and then to the global system (12-14).

Comprehension of the role of ψ will help in the correct interpretation of the stability curves (Ref.7). But as we have seen, ψ has a physical meaning only for stationary situations. (For the numerical solution of (12-14) see Ref.3.)

Simple analytic solutions in some particular cases

At steady state, simple analytic solutions can be found from the system (12-14), particularly for long cylinders.

-In the pure thermal case (S=0 so that $\psi=0$ and $\tilde{R}=R$), the system (12-14) at steady state gives:

$-\nabla p+R\ T'e_z+\nabla^2 V=0$ (15)

$Vz+\nabla^2 T'=0$ (16)

We know that at steady state, for a given aspect ratio A, there is a positive value R_{co} such that $R_{cr}{}^{st}=R_{co}$ for conducting lateral thermal boundary conditions, and a positive value R_{ad} such that $R_{cr}{}^{st}=R_{ad}$ for adiabatic lateral thermal boundary conditions.(For A=6, $R_{co}\approx 240$ and $R_{ad}\approx 80$.)

-Considering a mixture with Soret effect, but such that $\psi\approx 0$ (i.e., the disturbance arises directly or indirectly from the temperature field), we obtain the following system at steady state:

$-\nabla p+\tilde{R}\ T'e_z+\nabla^2 V=0$ (17)

$Vz+\nabla^2 T'=0$ (18)

$Vz+\nabla^2 Xc'=0$ (19)

We see that the perturbation Xc' has no influence on the stationary threshold, which is only determined by the two first equations (17-18). These two equations are the same as those obtained in the pure thermal case (15-16) with R replaced by $\tilde{R}$, and have the same boundary conditions. We have then:

$\tilde{R}_{cr}=R_{cr}(1+S)=R_{co}$ (20)

and $\tilde{R}_{cr}=R_{cr}(1+S)=R_{ad}$ (21)

for conducting and adiabatic boundary conditions, respectively. This case corresponds, for example, to mixtures with very small Sc/Pr ratios.

-In the asymptotic case $\psi\rightarrow\infty$ (so that instability arises from perturbations in Xc') and at steady state, the system can be written :

$-\nabla p+\tilde{R}\ \psi\ Xc'e_z+\nabla^2 V=0$ (22)

$Vz+\nabla^2 T'=0$ (23)

$Vz+\nabla^2 Xc'=0$ (24)

In this case, T' has no influence on the stationary threshold which is determined by the equations (22,24). These equations are still identical to the system (15-16), if we consider that T' is replaced by Xc' and R by $\tilde{R}\psi$. However, since the variables are different, the boundary conditions also differ. The lateral boundary conditions on Xc' are conditions of zero normal derivative and can correspond to the adiabatic lateral boundary conditions for T'. A difference is only observed on the conditions imposed at the extremities of the cylinder, which are in one case $\partial Xc'/\partial z=0$ and in the other case T'=0. In fact, the motion obtained in such long cavities in low convective situations is a one roll flow with a nearly uniform (i.e., not varying with z) velocity profile over most of the height of the cavity and a decrease of this velocity in the end parts which are recirculating zones (Refs.4,5). Just beyond the instability threshold, the deformation of the iso-mass fractions is generally small, but especially near the ends of the cavities where the velocity is small (Ref.5). The value of the mass fraction is then pratically constant at the ends and not very different from the equilibrium value before perturbation, almost corresponding to Xc'=0.

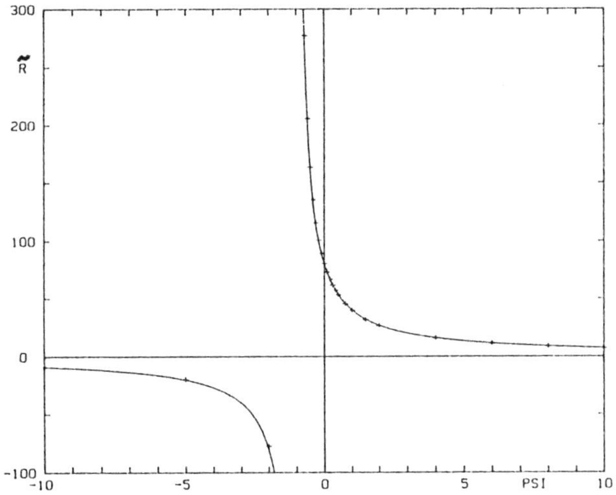

Figure 5 : Stationary stability curves $\tilde{R}_{cr}{}^{st}$ as a function of ψ; A=6 : +, numerical results obtained in the adiabatic case; solid line, curve $\tilde{R}=R_{ad}/(1+\psi)$.

For $\psi\rightarrow\infty$ the system is then pratically equivalent to the one for the pure thermal case with adiabatic boundary conditions. By similitude with the relation $R_{cr}=R_{ad}$, we have here:

$\tilde{R}_{cr} \approx R_{ad}/\psi$ (25)

The results presented in the adiabatic and conducting case in the figures 5 and 6 show that the asymptotic relation (25) is closely approximated for $|\psi| \geq 10$ in the case with A=6. On the contrary, for flat cells corresponding to small A this relation is no longer correct (Ref.9).

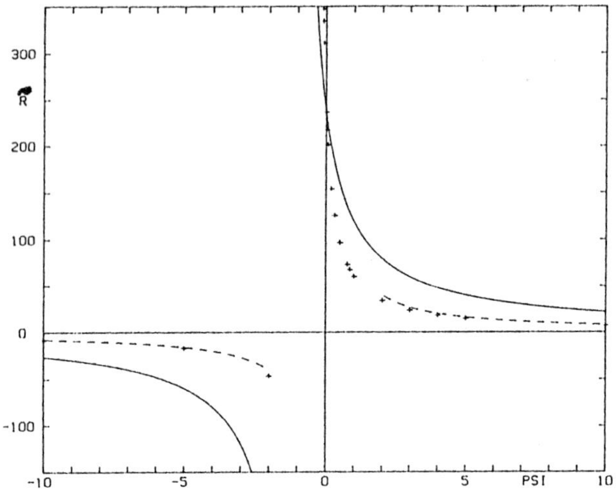

Figure 6 : Stationary stability curves $\tilde{R}_{cr}^{st}$ as a function of ψ; A=6 : +, numerical results obtained in the conducting case; solid line, curve $\tilde{R}=R_{co}/(1+\psi)$; dotted line, asymptotic curve ($|\psi|\rightarrow\infty$) $\tilde{R}=R_{ad}/\psi$.

-In the case of adiabatic lateral boundaries, the similarity of the boundary conditions on Xc' and T' mentioned above lead us to look for an analytical expression of the instability threshold $\tilde{R}$ valid on a large domain of ψ. The general system (12-14) can be transformed at steady state by adding ψ times (13) to (12). We obtain :

$$-\nabla p+\tilde{R}(1+\psi)[(T'+\psi Xc')/(1+\psi)]e_z+\nabla^2 V=0 \quad (26)$$

$$Vz+\nabla^2[(T'+\psi Xc')/(1+\psi)]=0. \quad (27)$$

This system is still identical to the system (15-16) with $(T'+\psi Xc')/(1+\psi)$ as a new variable replacing T' and R replaced by $\tilde{R}(1+\psi)$. Following the remarks made above for large A (i.e., tall cylinders), the boundary conditions on the new variable can be considered as nearly identical to those of T', i.e. a zero normal derivative on the lateral boundaries and a zero value at the ends. We then obtain a similar solution:

$\tilde{R}_{cr} \approx R_{ad}/(1+\psi)$. (28)

This relation agrees well with the asymptotic behaviours mentioned above, namely $\tilde{R}_{cr}=R_{ad}$ for $\psi\rightarrow 0$ and $\tilde{R}_{cr}=R_{ad}/\psi$ for $\psi\rightarrow\infty$. Figure 5 shows that this relation is in perfect agreement with the numerical results for A=6, except possibly in the vicinity of ψ=-1. The analogous relation in the conducting case has no relevance (figure 6): the conducting conditions on T' are too different from the zero normal derivative conditions on Xc'.

CONCLUSION

In the vertical case and at the marginal steady state, the convective motions in a mixture with Soret effect are generated by two forces, one due to the perturbations of the temperature and to their repercussion on the mass fraction by Soret effect (that we can call the "total" force due to T and proportional to GrPr(1+S)), and the other due to the "self" perturbations of the mass fraction (that we can call the "self" force due to X and proportional to GrScS). The total force due to T is in fact represented by the generalized Rayleigh number $\tilde{R}$, whereas the ratio of the "self" force due to X to the "total" force due to T is represented by the generalized Soret parameter ψ. That explains why at steady state the globalized formulation only depends on these two parameters and can then express the stability thresholds as general curves giving $\tilde{R}$ as a function of ψ.

In the more general case including Dufour effect, it is still possible by a similar analysis on the perturbations of the T and X fields that takes into account multiple repercussions to understand the role of the two parameters $\tilde{R}_D$ and ψ_D present in the globalized formulation (Ref.9).

The globalized formulations are then useful because they allow a reduction of the number of the governing parameters and as a consequence a diminution of the calculations, but also because they give indications on the real physics of the problem.

REFERENCES

1. J. Bert, I. Moussa and J. Dupuy 1987 Space thermal diffusion experiment in a molten AgI-KI mixture. In *Proc. Sixth European Symposium on Material Sciences under Microgravity* (**ESA SP-256**), pp. 471-475. ESA Publ. Division c/o ESTEC, Noordwijk, The Netherlands.

2. J.P. Praizey 1987 Thermomigration dans les alliages métalliques liquides. In *Proc. Sixth European Symposium on Material Sciences under Microgravity* (**ESA SP-256**), pp. 501-

508. ESA Publ. Division c/o ESTEC, Noordwijk, The Netherlands.

3. G.R.Hardin, R.L. Sani, D. Henry and B. Roux 1990 Buoyancy driven instability in a vertical cylinder: Binary fluid with Soret effect. Part 1. General theory and stationary stability results. *Intl J. Num. Meth. Fluids* **10**, 79-117.

4. D. Henry and B. Roux 1986 Three dimensional numerical study of convection in a cylindrical thermal diffusion cell : its influence on the separation of constituents. *Phys. Fluids* **29**, 3562-3572.

5. D. Henry and B. Roux 1987 Three-dimensional numerical study of convection in a cylindrical thermal diffusion cell : inclination effect. *Phys. Fluids* **30**, 1656-1666.

6. D. Henry and B. Roux 1988 Soret separation in a quasi-vertical cylinder. *J. Fluid Mech.* **195**, 175-200.

7. D. Henry, B. Roux, G.R. Hardin and R.L. Sani 1991 The role of the Soret effect in buoyancy-driven instability in tall vertical cylinders. Submitted to *J. Fluid. Mech.*

8. R.S. Schechter, I. Prigogine and J.R. Hamm 1972 Thermal diffusion and convective stability. *Phys. Fluids* **15**, 379-386.

9. D. Gutkowicz-Krusin, M.A. Collins and J. Ross 1979 Rayleigh-Bénard instability in nonreactive binary fluids. I- Theory. II-Results. *Phys. fluids* **22**, 1443-1460.

THERMOPHYSICAL PROPERTIES OF UNDERCOOLED MELTS AND THEIR MEASUREMENT IN MICROGRAVITY

Iván Egry and Berndt Feuerbacher

Institute of Space Simulation, DLR, Cologne, Germany

ABSTRACT

The knowledge of the thermophysical properties of melts is of great scientific and technological importance. However, the available data on metallic melts are scarce and partially incompatible, due to the inherent experimental difficulties; this is particularly true for the undercooled regime.

In the present paper we want to show how the method of electromagnetic levitation in combination with microgravity conditions can be used to derive reliable data on thermophysical properties. In detail, we will discuss the non-contact measurement of

- *enthalpy*
- *specific heat*
- *density*
- *thermal expansion*
- *surface tension*
- *viscosity*
- *thermal conductivity*
- *electrical conductivity*

Keywords: Thermophysical Properties, Undercooling, Containerless Processing, Levitation, Microgravity.

1. INTRODUCTION

Thermophysical properties of undercooled melts are essentially unknown, due to the experimental difficulties in accessing this metastable state [1, 2].

The knowledge of such properties is however urgently needed to increase our understanding of the liquid-solid phase transition, in particular its hydrodynamics, kinetics and thermodynamics [3].

The specific heat c_p is the key to the thermodynamics of the system. For example, the entropy can be obtained from the specific heat by integrating over temperature:

$$S(T_1) = S(T_0) + \int_{T_0}^{T_1} dT \frac{c_p(T)}{T} \qquad (1)$$

On the other hand, the specific heat plays also a central role in describing heat and mass transfer in fluids. This becomes evident in considering the dimensionless numbers that govern hydrodynamics, and in particular convective phenomena. For example, the Péclet number *Pe*, describing the ratio between convective and conductive heat transfer, is given by:

$$Pe = \frac{v L \rho c_p}{\lambda} \qquad (2)$$

Here, v is the flow velocity, L is a characteristic length, ρ is the density and λ is the thermal conductivity. In addition to the specific heat c_p, the Péclet number also contains the density and the thermal conductivity. Since these parameters usually appear as a group, it is convenient to define the thermal diffusivity κ, in analogy to the (mass) diffusion coefficient:

$$\kappa = \frac{\lambda}{\rho c_p} \qquad (3)$$

Spatial variations of the surface tension can lead to convective instabilities. The onset of these instabilities is controlled by the Marangoni number *Ma*

$$Ma = \frac{\nabla T L^2}{\kappa \eta} \frac{\partial \gamma}{\partial T} \qquad (4)$$

where γ is the surface tension, η the viscosity and T is the temperature.

As a final example, let us mention the Rayleigh number *Ra*

$$Ra = \frac{\rho \beta \nabla T L^4 g}{\kappa \eta} \qquad (5)$$

which controls buoyancy driven convection, driven by the gravity force ρg and thermal expansion β.

In addition to hydrodynamics, viscosity and specific heat also influence the solidification from the undercooled melt. The nucleation rate is given by [4]:

$$I \sim \frac{1}{\eta} \exp\left(-\frac{16\pi\tilde{\gamma}^3}{3\, kT\, \Delta G^2}\right) \qquad (6)$$

where $\tilde{\gamma}$ is the liquid-solid interfacial tension and ΔG is the difference in free energy between liquid and solid. ΔG can be calculated from c_p.

2. LEVITATION AND MICROGRAVITY

2.1 Levitation

Experiments in the undercooled regime can only be carried out if nucleation is suppressed or at least considerably reduced. For electrically conducting melts, electromagnetic levitation provides containerless heating and positioning; this helps to eliminate heterogeneous nucleation sites.

The principle of levitation melting can be formulated following Okress *et al* [5] as follows: An inhomogeneous, alternating electromagnetic field has two effects on a conducting, diamagnetic body; Firstly, it induces eddy currents within the material, which, due to ohmic losses, eventually heat up the sample (inductive heating) and secondly, it exerts a force on the body pushing it towards regions of lower field strength (Lorentz force). The latter effect can be used to compensate the gravitational force acting on the body.

2.2 Microgravity

On earth, strong magnetic fields are needed to compensate the 1-g gravitational force. The microgravity environment offers the unique possibility to minimize the magnetic positioning fields. This has a number of consequences:

- Inhomogeneous forces, such as the magnetic field, acting on a liquid sample inevitably deform its shape. Under microgravity, there is practically no deformation of the sample, the spherical shape is maintained.
- There is no turbulent flow in the melt, and flows which could exert centrifugal forces on the surface are greatly reduced.
- No additional damping of oscillations due to magnetic fields occurs in the sample.
- The heat induced in the sample by the positioning field is reduced and, therefore, no gas stream is required for cooling the sample and undercooling becomes possible in ultra high vacuum.
- Due to the strong electromagnetic fields needed on earth, terrestrial levitation experiments are essentially restricted to refractory metals and good conductors. In space, processing and undercooling of metals with a low melting point, such as many glass forming alloys becomes possible. There is also the potential to levitate semiconductors like Si or Ge.

The spherical shape of the sample is essential for measurements of surface tension, viscosity, density, thermal expansion and electrical conductivity. For the measurement of viscosity it is also crucial that no additional damping mechanisms exist. UHV conditions are mandatory for the measurement of the specific heat.

The first electromagnetic levitation facility operating in microgravity will be the German TEMPUS facility, built by Dornier under DARA contract. It is scheduled to fly on the IML-2 mission in 1994. Its potential for measuring thermophysical properties has been described in [6].

3. THERMOPHYSICAL PROPERTIES

3.1 Enthalpy

The enthalpies of liquid metals can be measured by combining levitation and drop calorimetry. This method has been used by Frohberg and coworkers [7] to determine the enthalpy as well as the heat and entropy of fusion for some refractory metals. In addition, this method also allows to measure the heat of mixing by alloying two liquid metals *in situ* during levitation and detecting the related change in temperature.

3.2 Specific Heat

As mentioned in the introduction, the specific heat is the central parameter which controls thermodynamics, hydrodynamics, nucleation and solidification of undercooled melts. Again, a non-contact method is needed for its measurement as a function of temperature in the liquid, undercooled regime. Fecht and Johnson [8] have proposed to use a modulation technique. By superposing a sinusoidally varying heat input P_ω onto the stationary heat input P_0, a temperature variation is induced in the sample, which can be related to the specific heat.

The heat balance can be written as:

$$P_{in} = P_0 + P_\omega, \qquad P_\omega = P_{\omega 0}\cos^2(\omega t/2) \qquad (7)$$

where P_{in} is the heat input to the sample. Under vacuum conditions, the heat loss is entirely due to radiation and is given by:

$$P_{rad} = A\,\varepsilon\,(T^4 - T_0^4) \qquad (8)$$

where ε is the total emissivity. The temperature of the sample will follow the oscillatory behaviour of the heating power, subject to some internal and external relaxation mechanisms. Fecht and Jonson showed that the temperature variation is given by:

$$(\Delta T)_\omega = \frac{P_{\omega 0}}{2\omega c_p}\left(1 + \left(\frac{1}{\omega\tau_1}\right)^2 + (\omega\tau_2)^2\right)^{-1/2} \qquad (9)$$

Here τ_1 is an external relaxation time and τ_2 is the internal relaxation time. The latter can be expressed as

$$\tau_2 = \frac{3c_p}{4\pi R\lambda} \qquad (10)$$

R being the radius of a spherical sample.

By choosing the appropriate frequency, one can satisfy the following inequality

$$\tau_2 \ll 1/\omega \ll \tau_1 \qquad (11)$$

In this case, one obtains the specific heat from:

$$c_p = \frac{P_{\omega 0}}{2\omega(\Delta T)_\omega} \qquad (12)$$

Incidentally, when c_p is known, one can choose a different frequency and determine the thermal conductivity from τ_2.

3.3 Density and Thermal Expansion

The first measurements of density and thermal expansion using electromagnetic levitation have been made by Shiraishi and Ward in the sixties [9]. They have employed high speed photography to determine the shape of levitated liquid drops. Assuming a spherical shape of the drop, i.e. discarding photographs showing aspherical drops, they determined the radius and hence the volume V of the drop. Since the mass of the drop can be measured by weighing it before and after the levitation process, the density ρ can be calculated from

$$\rho = M/V \qquad (13)$$

Repeating the measurement at different temperatures, also the thermal expansion $\beta = 1/V\,(\partial V/\partial T)$ can be derived.

3.4 Surface Tension and Viscosity

Surface tension and viscosity can be measured by exciting oscillations of a freely floating drop. The angular frequency ω of the oscillations is related to the surface tension γ, while the damping $1/\tau$ yields the viscosity η.
The radius of a viscous sphere undergoes oscillations of the form

$$R_n \sim R\cos(\omega_n t)\, e^{-t/\tau_n}\, P_n(\cos\vartheta) \qquad (14)$$

where n is a label for the different modes and P_n is a Legendre polynomial.
Frequency ω_n and damping τ_n are given by [10]:

$$\omega_n^2 = \frac{4\pi}{3}\, n(n-1)(n+2)\frac{\gamma}{M} \qquad (15)$$

$$1/\tau_n = (n-1)(2n+1)\,\frac{\eta}{\rho R^2} \qquad (16)$$

where R is the radius, ρ the density and M the mass of the sphere. The oscillations can be detected as variations in the shape of the sample.

Employing the modified Stokes-Einstein relation [1]

$$D = \frac{kT}{\alpha\, v_a^{1/3}\, \eta} \qquad (17)$$

where v_a is the atomic volume and α is a number between 5 and 6, also the diffusion coefficient D can be derived from this type of experiments.
Unfortunately, the above analysis is valid for spherical samples only. On earth, levitated samples are always deformed and are, at best, axially symmetric. The oscillation spectrum of aspherical drops is much more complicated; the single peaks corresponding to a particular value of n are split and shifted with respect to (15), (16) which make a quantitative evaluation of the oscillation spectra obtained on earth more difficult, but not impossible [11]. A comparison of terrestrial measurements of the surface tension using this technique is given in [12]. For viscosity measurements, levitation has not yet been used so far. The concept of a microgravity experiment is discussed in [13].

3.5 Electrical Conductivity

In metals, the electrical as well as the thermal conductivity can be attributed almost entirely to free conduction electrons. During the past three decades, theoretical considerations on electronic transport properties of liquid metals, and particularly their electrical conductivities, have made great progress [3]. However, relatively few experimental studies have been carried out on the electrical properties of liquid metals and alloys [14, 15].
Electromagnetic levitation offers the possibility to determine the electrical conductivity by a non-contact, inductive measurement. The levitation coil and the sample inside the coil are part of a high frequency oscillatory circuit, to which power is supplied by a generator. A change in the sample's conductivity modifies the total impedance $Z(\sigma)$ of this circuit. Provided the metallic droplet has a simple geometry and its position relative to the coil is fixed, the function $Z(\sigma)$ can theoretically be determined and the specific electrical conductivity can be derived from measurements of the impedance.
A change in the impedance can be easily measured:

- There is a change in the power loss P, which for fixed voltage U is given by

$$P = U^2\, Re\left\{\frac{1}{Z(\sigma)}\right\}. \qquad (18)$$

- The phase difference φ between coil current I and voltage U changes

$$tg\varphi = \frac{Im\{Z(\sigma)\}}{Re\{Z(\sigma)\}}. \qquad (19)$$

- The resonance frequency ω of the oscillatory circuit is shifted

$$\omega = \frac{1}{Im\{Z(\sigma)\}\, C}, \qquad (20)$$

  where C is the capacity of the circuit.

On earth, this inductive method has already been applied to metallic melts enclosed in a cylindrical crucible [14, 15]. Apart from the contamination problems, which would also adversely affect undercooling experiments, a crucible can only be used for materials having a low melting point. The levitation technique could, in principle, extend this method to all metals and into the undercooled regime.
Based on the same conduction mechanism, the electrical and thermal conductivity σ and λ are closely related via the Wiedemann-Franz law [3]

$$\frac{\lambda}{\sigma T} = L$$

$$L = 1/3\left(\frac{\pi k}{e}\right)^2 = 2.45\cdot 10^{-8}\left[\frac{W\,\Omega}{K^2}\right] \qquad (21)$$

where the L is the Lorentz number. This theoretical relation is well established for the majority of metals, i.e. metals in which the electronic transport is carried by free electrons. Hence, we can use it to determine the thermal conductivity from measurements of the electrical conductivity that can be performed with higher precision and without being influenced by any convective transport. Alternatively, by combining independent measurements of thermal and electrical conductivities on the same sample, the validity of (21) can be checked and an experimental value for L be derived.

4. CONCLUSION

We have shown how electromagnetic levitation technique can be used to measure most of the relevant thermophysical properties of metallic melts, both above and below their melting points. Some of these methods have been already used successfully on earth; the microgravity environment offers either substantial improvement in precision or is an indispensable prerequisite for carrying out the experiment at all.

Results from microgravity experiments are expected from the first flight of the TEMPUS facility in 1994.

REFERENCES

[1] T. Iida, and R.I.L. Guthrie, *The Physical Properties of Liquid Metals* Clarendon Press, Oxford, 1988

[2] A. Nagashima *Int. J. of Thermophysics*, 11 (1990), 417

[3] M. Shimoji, *Liquid Metals* Academic Press, London, 1977

[4] D. Turnbull *Contemp. Phys.* 10 (1969) 473

[5] E.C. Okress, D.M. Wroughton, G. Comenetz, P.H. Brace, J.C.R. Kelly *J.Appl.Phys.*, 23, (1952),545

[6] G. Lohöfer, P. Neuhaus, I. Egry, *to appear in: High Temperature - High Pressure, (1991)*

[7] G. Betz, M.G. Frohberg *High Temperatures-High Pressures* 12, (1980), 169

[8] H.J. Fecht, W.L. Johnson *Rev.Sci.Instr.* 62, (1991), 1299

[9] S. Y. Shiraishi, R.G. Ward *Canadian Metallurgical Quarterly*, 3 (1964), 117

[10] W.H. Reid *Q. Appl. Math.* 18 (1960), 86

[11] S. Sauerland, K. Eckler, I. Egry *to appear in Journal of Materials Science Letters*

[12] B.J. Keene, K.C. Mills, A. Kasama, A. McLean, W.A. Miller *Metallurgical Transactions* 17B, (1986), 159

[13] I. Egry, B. Feuerbacher, G. Lohöfer, P. Neuhaus *Proceedings of the 7th European Symposium on Materials Sciences under Microgravity*, ESA-SP295, (1989), 257

[14] B. Delley, H.U. Künzi, H.J. Güntherodt *J. Phys. E* 13 (1980), 661-664.

[15] M. Lambeck *Z. Metallkunde* 75 (1984), 806-808.

PHASE SEPARATION IN FLUIDS WITHOUT GRAVITY

P. Guenoun, Y. Jayalakshmi, F. Perrot, B. Khalil and D. Beysens
Service de Physique de l'Etat Condensé
Département de Recherche sur l'Etat Condensé, les Atomes et les Molécules
CE-Saclay, 91191 Gif-sur-Yvette Cedex, France

Y. Garrabos, B. Le Neindre
LIMHP (CNRS), Université Paris Nord
Avenue J.B. Clément, 93430 Villetaneuse, France

ABSTRACT

Phase separation in fluids is a phenomenon where gravity can play a major rôle because of the density difference of the fluid phases. Consequently, it is an experimental challenge to suppress the influence of gravity in order to reveal the internal thermodynamical mechanisms which govern the phase separation. We show here that gravity effects can be nearly suppressed in a partially deuterated mixture. For a mixture at critical concentration, the spinodal decomposition mechanism has been studied over large time scales. The suppression of gravity effects has thus been checked and new phenomena are shown up as the influence of walls or that of concentration gradients. Off-critical mixtures are also of interest but, in this case, the matching of densities is not precise enough to provide unambiguous results on Earth. Pure fluids, though fundamental, have been less studied due to the complexities of their specific thermodynamic behavior in the critical region. We present results obtained in microgravity which show that pure fluids quantitatively behave as binary mixtures in spite of their numerous thermodynamic specificities.

Keywords: *Phase Separation, Spinodal Decomposition, Microgravity, Growth Laws, Self-similarity*

1. INTRODUCTION

Under some circumstances, one thermodynamically stable phase can be driven in a state which is not stable. This for instance occurs when temperature is lowered and the entropic effects are reduced such as two more ordered phases appear (Ref.1). Note that an increase in temperature can play the same role when less trivial interactions (e.g. hydrogen bonding (Ref.2), Van der Waals interactions (Ref.3), depletion forces (Ref.4)) are present. Phase separation (PS) is the process by which both phases emerge and grow to become two macroscopic phases. Figure 1 resumes the above depicted situation by means of a phase diagram including a critical point. Most experiments of PS described in this paper are performed by making a thermal quench (labelled by (2) in Figure 1). The vicinity of the critical point is essential for two main reasons. First, the critical slowing down provides acceptable experimental time scales. Second, the scaling behaviors close to the critical point

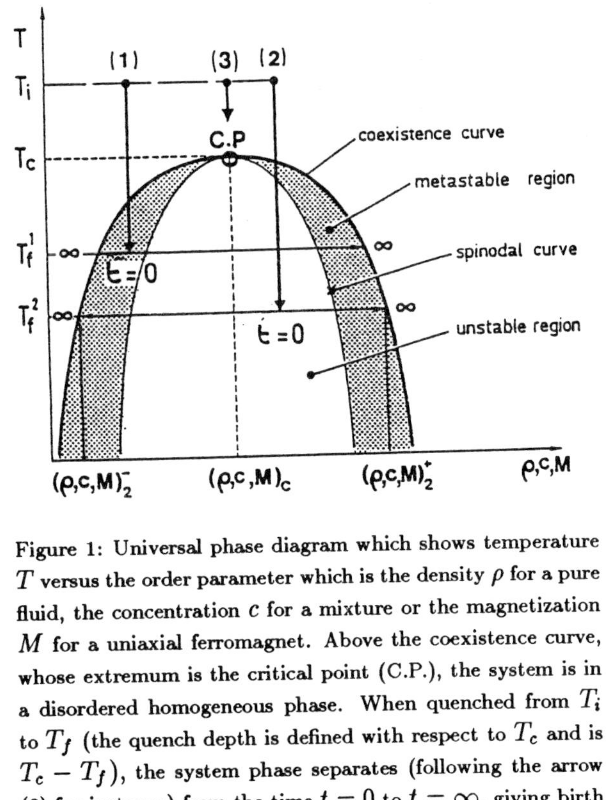

Figure 1: Universal phase diagram which shows temperature T versus the order parameter which is the density ρ for a pure fluid, the concentration c for a mixture or the magnetization M for a uniaxial ferromagnet. Above the coexistence curve, whose extremum is the critical point (C.P.), the system is in a disordered homogeneous phase. When quenched from T_i to T_f (the quench depth is defined with respect to T_c and is $T_c - T_f$), the system phase separates (following the arrow (2) for instance) from the time $t = 0$ to $t = \infty$, giving birth to the equilibrium phases of order parameter ρ_2^- and ρ_2^+ (or (c_2^-, c_2^+) or (M_2^-, M_2^+)).

imply the existence of two natural units of length and time. The first unit is the correlation length of the fluctuations of the order parameter, ξ at a given temperature. The second unit is the typical lifetime of the fluctuations τ at a given temperature. In the case of fluid phases, gravity plays a major role in the growth since the denser (lighter) phase moves downwards (upwards). This should prevent large spatial scale patterns of PS to be observed because of sedimentation and convective flows. This paper shows, in the first part, that the action of gravity can be made negligible when studying binary liquid mixtures. A second part takes benefit of this to examine the bulk growth mechanisms. Third, we investigate the surface growth laws. A fourth part extends some of the previous results to the case of pure fluids.

(a)

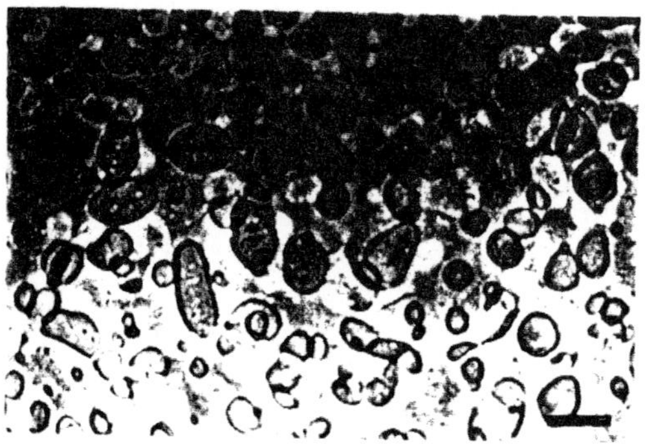

(b)

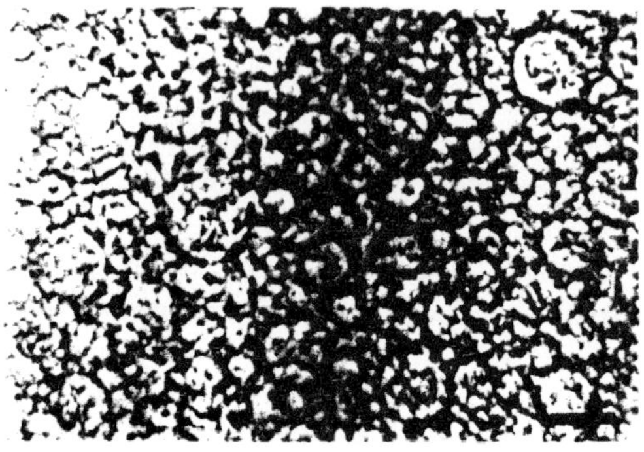

Figure 2: Two pictures of phase separation in a critical mixture (a) and in a slightly off-critical mixture (b) ($c = c_c + 1\%$). (a) is a pattern of interconnected tubes revealing two bicontinuous phases. (b) is a pattern of dense droplets where one of the phases is no longer bicontinuous. The droplet phase is now the minority phase. The bar is in both cases 1 mm.

2. DENSITY-MATCHED BINARY MIXTURES

The order parameter for binary fluid mixtures is the difference in concentration $c - c_c$ where c_c is the critical concentration. Some binary mixtures are made of components whose densities are very close such as cyclohexane ($\rho_c = 0.77$ g.cm^{-3} at 25°C) and methanol ($\rho_m = 0.79$ g.cm^{-3} at 25°C). In this latter case, a very fine matching of densities can be achieved by partial deuteration of the cyclohexane since the fully deuterated cyclohexane has a density larger than ρ_m ($\rho_{dc} = 0.89$ g.cm^{-3}). A careful study has shown (Ref.5) that such a ternary mixture exhibits the same phase diagram as a binary mixture. Following a thermal quench from the homogeneous region into the two-phase region (Figure 1), this mixture phase separates and shows large scale spatial structures (Figure 2). These structures exhibit a characteristic size $L_m(t)$ which grows with time meaning that the two phases separate at the expense of each other. Moreover, in the case $c = c_c$, the patterns are self-similar with time as shown in Figure 3. A statistical analysis has also been carried out by means of numerical Fourier transform of the pictures and has quantitatively confirmed these results. A microgravity experiment has also demonstrated that the residual influence of gravity was negligible (Texus 13 (Ref.6)). This has been tested by using the self-similarity of the patterns (see (Ref.6) for details) and the growth laws (see below) as criteria.

3. LARGE SCALE GROWTH MECHANISMS

As soon as a binary mixture is quenched into the two-phase region, the fluctuations in concentration will rule the PS behavior of the mixture. For simplicity, we present only a mean-field description of this behavior (Figure 1). In a first case, the fluctuations can be unstable, at least in a definite range of wavevectors k. For $k \longrightarrow 0$ (large scale) the conservation of mass prevents the divergence of the fluctuations whilst for $k \longrightarrow \infty$ (small scale) it costs too much energy to get diverging fluctuations. As a result, there exists a wavevector k_m which corresponds to the maximum growth rate of the fluctuations. Non-linearities then occur and make k_m decrease with time. The length $L_m = \frac{2\pi}{k_m}$ naturally appears as the characteristic length of the pattern in the direct space as seen in Figure 3. This process is called "spinodal decomposition" (path (2) in Figure 1) and we focus on it in the following because the concentrations of the mixtures that we have studied are close to the critical concentration. The second case is called "nucleation" and means that the mixture is quenched towards a point where the fluctuations are only metastable (path (1) in Figure 1). These fluctuations need to reach a critical size in order to develop and grow. It is generally believed that this process is associated to concentration gradients in the fluid which drive the growth. On the

a

b

c

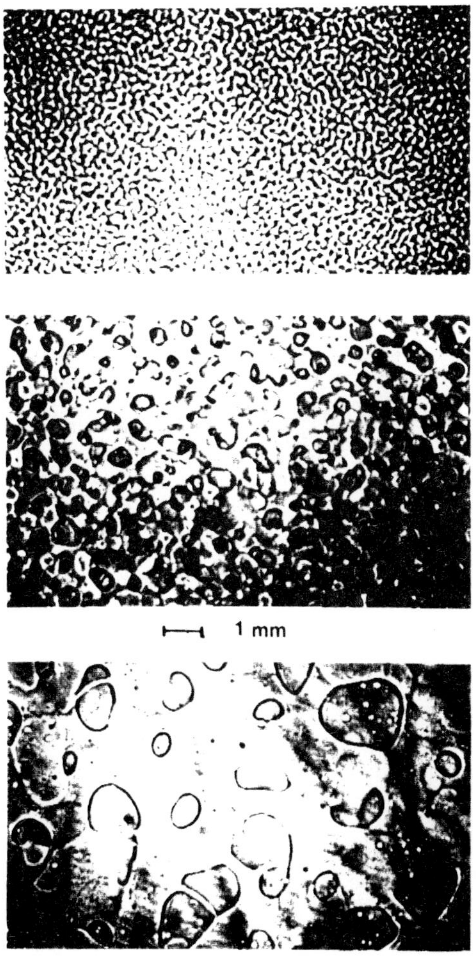

Figure 3: Three pictures of phase separation in a critical mixture $(T_c - T_f = 5 \text{ mK})$ at times $t = 40$ s (a), $t = 3$ min (b) and $t = 9$ min (c). Self-similarity of the growth merely expresses that (c) can be regarded as a blow-up of (b) whereas (b) can be regarded as a blow-up of (a), taking into account the loss in optical resolution at small scale in (a).

contrary, spinodal decomposition makes the concentrations reach their equilibrium values in a time far shorter (or order τ) than the typical time of observation of large scale patterns. Light scattering studies show (Ref.7,8) that the beginning of the growth is roughly described by a power law $L_m \sim t^{1/3}$. For critical mixtures whose two phases are of equal volume an hydrodynamic growth then takes place because of the interconnected pattern (see Figure 2). This has been explained by Siggia (Ref.9) in terms of capillary instabilities and leads to $L_m \sim t$. An alternative explanation has been proposed (Ref.10) in terms of coalescence of the phases which would be nearly close-packed. Our study of the large scale interconnected patterns has checked the previous growth law over several decades as shown in Figure 4 (Ref.8). When the mixture is not critical and the symmetry of the

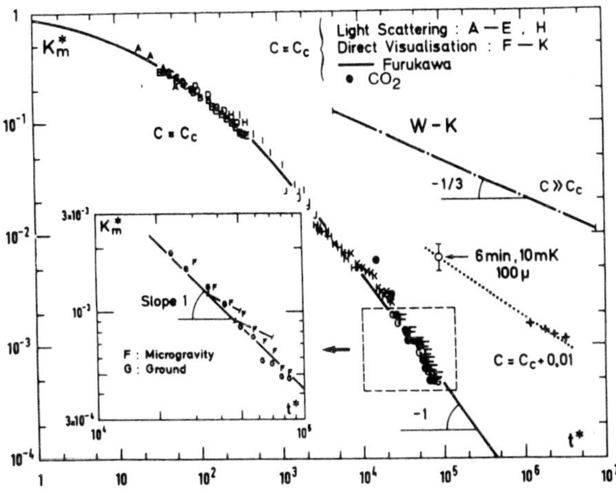

Figure 4: Master curve of the growth during spinodal decomposition for $c = c_c$. The quantities plotted are $k_m^* = \frac{2\pi}{L_m}\xi$ and $t^* = \frac{t}{\tau}$. This enables the data of different experiments (various quench depths) to collapse on a single curve. The solid line is a heuristic fit by Furukawa. The late stage growth in t (slope -1 in the above representation) is well established. The insert shows that the same experiment performed either on Earth or in microgravity yields similar results, proving that gravity effects are actually removed. The data corresponding to the phase separation in a pure fluid ($\bullet$) are in agreement with the previous results.

Results about off-critical mixtures are more ambiguous. The data + + + show results for $c - c_c = 10^{-2}$, extrapolated (...) at a 6 min duration time in order to match microgravity results ($\circ$). The line $\cdot - \cdot - \cdot$ corresponds to the Wong and Knobler data (Ref.7) in samples far from criticality ($c - c_c > 4.10^{-2}$).

volume of the phases is no more respected (see Figure 2) such a hydrodynamic growth is not observed and the growth is far slower. Residual gravity effects can be relevant and unambiguous results are delicate to obtain without the help of microgravity experiments (see Figure 4). Recent experiments have been carried out with a mixture exhibiting a controlled concentration gradient. The central region corresponding to a narrow extension around c_c is growing rapidly whereas the surrounding zones are seen to expand on a much slower time scale (Ref.11) (Figure 5).

4. PHASE SEPARATION CLOSE TO A WALL

Large scale patterns of spinodal decomposition are modified close to a wall. This is due to the fact that one phase should wet completely the wall, close to a critical point, and thus excludes the other phase from it (Ref.12). One observes that the wetting phase which grows along the wall is rippled and connected to the bulk (Figure 6). In contrast, the non-wetting domains form a parallel layer along the wall and grow

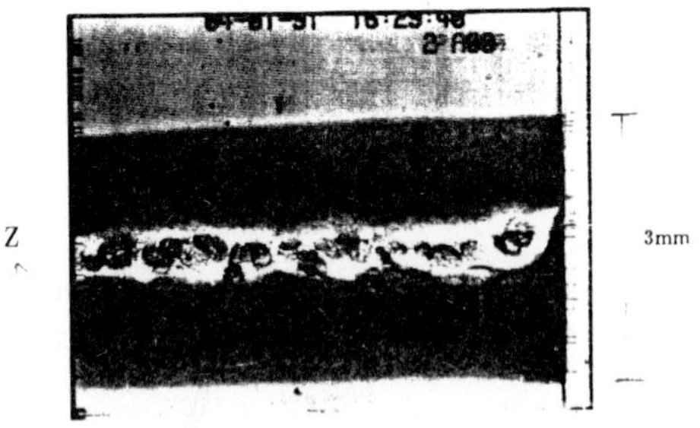

Figure 5: Phase separation under a concentration gradient. The gradient is in the z direction. The brighter zone corresponds to concentrations around c_c and shows fast growth. The darker zone corresponds to slow growth.

in an anisotropic way (Figure 7). The length $L_m(t)$ is no more sufficient to describe the growth and one must introduce the wavelength λ of the ripples and the thickness e of the wetting phase. They behave as $\lambda \sim e \sim t^{0.6}$ (see Figure 8 and (Ref.13) for details).

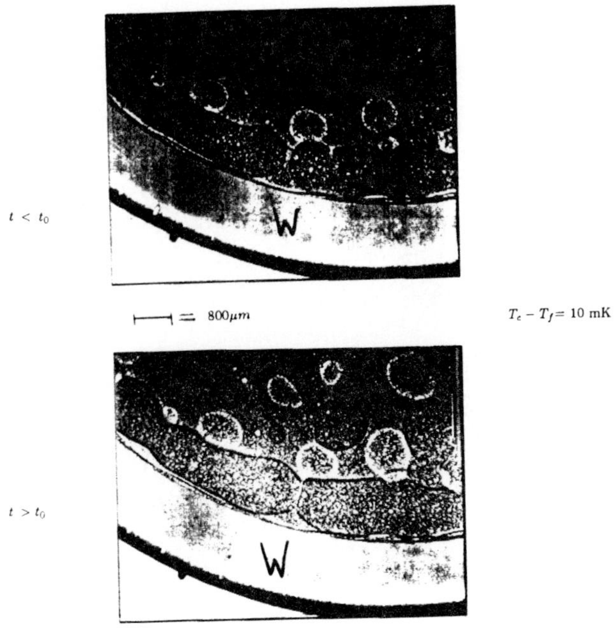

Figure 7: Non-wetting domains coalescing in a surface parallel to the wall W. These events create anisotropic domains of the non-wetting phase.

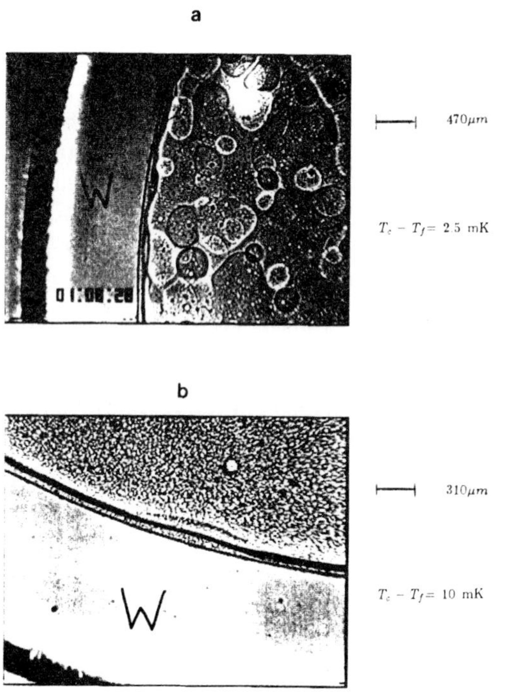

Figure 6: Close to the wall (W), one observes the rippling of the wetting layer, (a) in the plane of the picture and (b) in the plane perpendicular to the picture. In the bulk, the usual pattern is recovered.

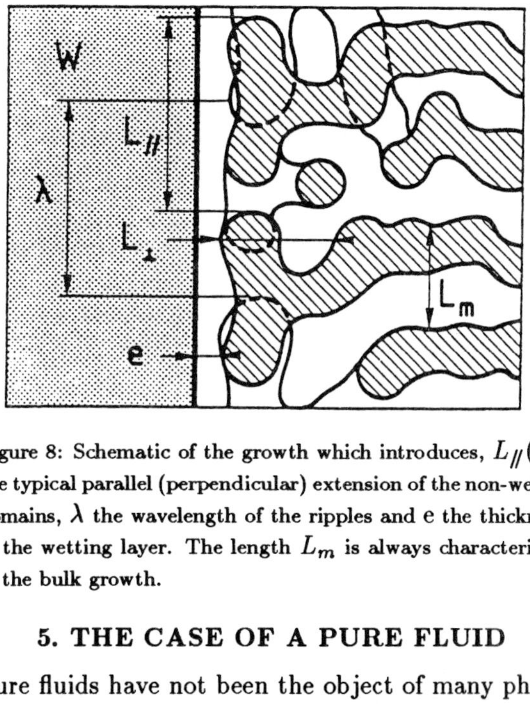

Figure 8: Schematic of the growth which introduces, $L_{//}(\perp)$ the typical parallel (perpendicular) extension of the non-wetting domains, λ the wavelength of the ripples and e the thickness of the wetting layer. The length L_m is always characteristic of the bulk growth.

5. THE CASE OF A PURE FLUID

Pure fluids have not been the object of many phase separation experiments (Ref.14). This is due to the time scale of evolution which is nearly two orders of magnitude faster than for binary mixtures and to the severe anomalies of the thermodynamic coefficients which prevent to get an homogeneous pure fluid on Earth. For instance, the isothermal compressibility K_T diverges as one approaches the critical point, making the fluid to be stratified under its own weight. Microgravity appears thus mandatory in order to make a valuable experiment. Another problem, however, appears very serious in absence of gravity: the thermal diffusivity D_T goes to zero close to the critical point. Since convections are suppressed in microgravity, one does not understand which mechanism can provide the quick thermal equilibration necessary to perform the thermal quench. Fortunately, it was recently understood (Ref.15) that the high compressibility of the fluid can generate acoustic waves which provide a fast and homogeneous thermal equilibration (the piston effect). This has been recently checked during a micrograv-ity experiment (see the other communication in the same proceedings by Guenoun et al. and the Refs. therein). A spinodal decomposition experiment has been able to be carried out on a sample of critical CO_2 and has shown the same interconnected structures as for a mixture (see Figure 9). The hydrodynamic growth $L_m \sim t$ has also been recovered indicating that no anomalies due to the high compressibility of the fluid are present (see Figure 4) (Ref.16).

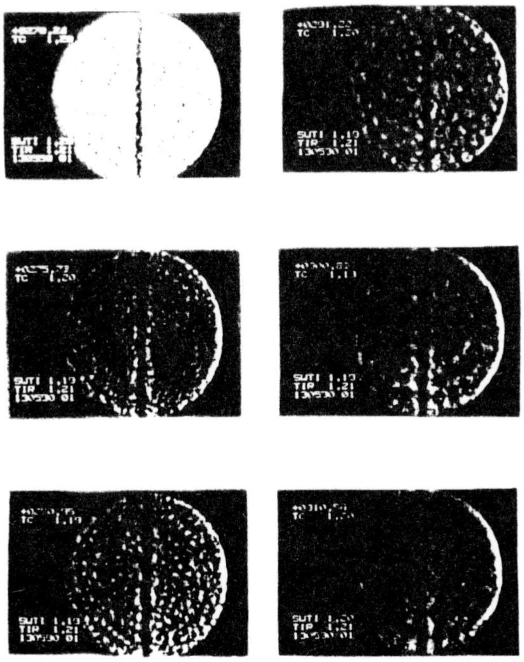

Figure 9: Pictures of spinodal decomposition in a critical sample of CO_2. Time in seconds is written on the top line (the beginning of the quench is $t = 250$ s and the quench depth is 1 mK). The bottom line is the data and the other lines are related to temperature measurements. The diameter of the sample is 1 cm. The vertical thread in the middle of the cell is dedicated to thermal measurements.

6. ACKNOWLEDGEMENTS

We thank CNES and ESA for their financial support.

7. REFERENCES

1. See e.g. L. Landau, E. Lifshitz, *"Statistical Physics"* (Pergamon Press).
2. J.C. Wheeler, *J. Chem. Phys.* **62** (1975) 433.
3. B. Vincent, *Colloïds and surfaces* **24** (1987) 269.
4. S. Sanyal, N. Easwar, S. Ramaswamy and A.K. Sood submitted to *Phys. Rev. Lett.*.
5. C. Houessou, P. Guenoun, R. Gastaud, F. Perrot and D. Beysens, *Phys. Rev. A* **32** (1985) 1818.
6. D. Beysens, P. Guenoun and F. Perrot, *Phys. Rev. A* **38** (1988) 4173.
7. Y.C. Chou and W.I. Goldburg, *Phys. Rev. A* **23** (1981) 858.
   N.C. Wong and C.M. Knobler, *Phys. Rev. A* **24** (1981) 3205.
8. P. Guenoun, R. Gastaud, F. Perrot and D. Beysens, *Phys. Rev. A* **36** (1987) 4976.
9. E.D. Siggia, *Phys. Rev. A* **20** (1979) 595.
10. P. Guenoun, *unpublished.*
11. Y. Jayalakshmi, B. Khalil and D. Beysens, *unpublished results.*
12. J.W. Cahn, *J. Chem. Phys.* **66** (1977) 3667.
13. P. Guenoun, D. Beysens and M. Robert, *Phys. Rev. Lett.* **65** (1990) 2406.
14. K. Nitsche and J. Straub, *Proceed. 6th European Symposium on Materials Sciences under Microgravity Conditions,* 1986, ESA SP-256, p.109.
15. B. Zappoli, D. Bailly, Y. Garrabos, B. Le Neindre, P. Guenoun and D. Beysens, *Phys. Rev. A* **41** (1990) 2264.
    A. Onuki, H. Hao and R.A. Ferrell, *Phys. Rev. A* **41** (1990) 2256.
    H. Boukari, J.N. Shaumeyer, M.E. Briggs and R.W. Gammon, *Phys. Rev. A* **41** (1990) 2260.
16. Y. Garrabos, B. Khalil, B. Le Neindre, P. Guenoun and D. Beysens, *to be published.*

MEASUREMENT OF THERMAL CONDUCTIVITY USING TRANSIENT HOT WIRE METHOD UNDER MICROGRAVITY

Shin NAKAMURA and Taketoshi HIBIYA

Space Technology Corporation
Kudan-kita 1-4-5, Chiyoda-ku, Tokyo 102, Japan

Fumio YAMAMOTO and Takao YOKOTA[1]

NEC Corporation
Functional Devices Research Laboratories
Miyazaki 4-1-1, Miyamae-ku, Kanagawa 213, Japan

ABSTRACT

The thermal conductivity of molten InSb was measured on board a TEXUS-24 sounding rocket by the transient hot wire method using a thermal conductivity measurement facility (TCMF). The effects of convection were confirmed to be sufficiently suppressed in a microgravity environment, while measurements were affected by convection on the ground. The thermal conductivity of molten InSb was 15.8 (W m^{-1} K^{-1}) and 18.2 (W m^{-1} K^{-1}) at 830 K and 890 K, respectively.

The thermal conductivity of mercury was measured on board a Mitsubishi MU-300 aircraft in 10^{-2}, 10^{-1} and 1 G environments. We determined a critical Rayleigh number below which the effects of wire-heating induced convection on thermal conductivity measurement using TCMF could be suppressed. From this Rayleigh number we concluded that the effects of convection caused by wire heating were negligible in the TEXUS-24 experiment.

Key Words: *Transient Hot Wire Method, Thermal Conductivity, Molten InSb, Microgravity.*

1. INTRODUCTION

A microgravity (μG) environment in which buoyancy convection is suppressed is considered to be an ideal environment to measure thermal conductivity. Several thermal conductivity measurements and the apparatus for μG environments[1-3] have been proposed. However, to date, only one experiment has been carried out on board the MASER 1 Swedish sounding rocket[4], where the thermal conductivity of ethanol was measured by the transient hot wire method at 307K and 327K. Further, the suppression of convection was confirmed in the μG environment of the rocket.

Compared to the thermal conductivities of organic liquids, those of high temperature melts such as liquid metals and molten semiconductors are seldom measured, although they are required for crystal growth processes using these melts. In the melts, buoyancy convection readily occurs, and the suppression of convection is difficult to achieve on the ground. We think that the thermal conductivity of high temperature melts can be accurately measured in a μG environment, reducing the effect of convections.

[1] present address: NEC Corporation , Space Development Division, Ikebe-cho 4035, Midori-ku, Yokohama 226, Japan

We have developed a method [5,6] and apparatus [7] to measure the thermal conductivity of high temperature melts, aimed at attaining a measurement technique which is suitable for a μG environment. The method is based on the transient hot wire method. In order to apply this method to the melts which are corrosive and electrically conductive, a ceramic probe and apparatus for the TEXUS sounding rocket have been developed [8]. With this apparatus (Thermal Conductivity Measurement Facility; TCMF), the thermal conductivity of molten InSb was measured on board the TEXUS-24 sounding rocket in December 1989 [9].

We also performed a thermal conductivity measurement on mercury in μG environments achieved by parabolic flights of the Mitsubishi MU-300 aircraft. The objective behind this measurement was to determine the conditions required to measure thermal conductivity without the effects of convection.

2. EXPERIMENT

Measurement principle

As described previously, measurements were performed with ceramic probes which have been developed for corrosive, electrically conductive, and high temperature melts [5,6]. The measurement technique using these probes is basically the same as the conventional transient hot wire technique. However, the probes were different from conventional ones which utilize only thin metallic wire. Instead, a printed wire on a thick ceramic substrate was used as a sensor to measure temperature increase ΔT when a constant electric power is input to the wire. In order to avoid any leakage of current through electrically conductive melts, the printed wire was coated with a thin ceramic layer.

For this kind of probe structure, Takegoshi et al. have given a working equation, based on the assumption that the wire is an ideal line heat source sandwiched between two materials [10]. The materials in the present experiment are a liquid specimen and the substrate of the probe, which respectively have thermal conductivities λ_L and λ_S. The sum of thermal conductivities $\lambda_L + \lambda_S$ is expressed as follows,

$$\lambda_L + \lambda_S = \frac{Q}{2\pi} \Big/ \frac{d(\Delta T(r,t))}{d(\ln(t))} \quad , \qquad (1)$$

where Q is heat input power per unit length of wire and t is time. λ_L is determined by subtracting λ_S from the sum of thermal conductivities.

Measurement cell structure and facility

Figure 1 shows the structure of measurement cell [8]. The ceramic probe and the sample to be measured (high temperature melt) are placed in a carbon crucible. A platinum wire was printed on the flat surface of a substrate made of alumina ceramic. A wire of 70 mm in length, 15 μm in thickness and 100 μm in width was coated with a 60 μm alumina layer to prevent any leakage of current through the electrically conductive melt. The structure of the ceramic probe is described in two previous papers [6,8]. The crucible containing the melt and probe was contained in a nickel cartridge. The nickel cartridge was set vertical to the ground inside a furnace within the measurement facility.

Two facilities were integrated for the present experiments. One was a flight module in which the thermal conductivity of molten InSb was measured on board the TEXUS-24, and the other was a laboratory module for terrestrial experiments. The laboratory module was functionally the same as the flight module. In the aircraft experiment, the laboratory module was used to measure the thermal conductivity of mercury at room temperature.

Figure 2 shows the flight module for the thermal conductivity measurement facility (TCMF) in the TEXUS experiment [8]. The flight module consists of three parts; upper, middle, and lower parts. Two furnaces (A and B) are maintained in the upper part, and the electronic system in the middle part. The lower part contains the batteries. The furnace to maintain the sample has a large ϕ40 mm x 70 mm isothermal area and is capable of a maximum temperature of 1300K. In the isothermal area, a small temperature gradient is shown along the center line from the top to bottom; the temperature at the top part of a nickel cartridge was designed to be 5 K higher than that at the bottom part.

The electronic system for measurement consists of two parts; constant current supply and digital voltmeter[8]. When a constant current is supplied to a wire in the probe, the increase in resistance of the wire is measured with the digital voltmeter. Temperature increase ΔT is calculated with the temperature coefficient of resistance of the platinum wire.

3. RESULTS AND DISCUSSION

TEXUS flight experiment

The measurement of thermal conductivity of molten InSb was performed on board the TEXUS-24 sounding rocket [9]. Solid InSb samples in the furnaces of the TCMF were melted before launching, since a 6 minute microgravity period was believed to be too short to homogenize temperature distribution. Furnace A had temperatures of 843K on the upper cartridge surface and 838K on the lower surface, while furnace B recorded temperatures of 903K on the upper surface and 898K on the lower surface. The temperature difference between the upper and lower surfaces of molten InSb in the crucible was thought to be less than 2 K. In both furnaces, the temperature of the upper surface was higher so that the occurrence of convection before and during launching could be suppressed.

Figure 3 shows apparent thermal conductivities calculated by Eq(1) as a function of ln(t) for furnace A. The A_1 to A_7 are 1G reference measurements on the ground 500 seconds before the launching. The measurements of A_9 to A_{13} were performed at an interval of 60 seconds in a μG environment of about 10^{-4} G. The applied current was 1.5 A which corresponds to 87.6 (Wm^{-1}). The apparent thermal conductivities $\lambda_L + \lambda_S$ in the 1G reference measurements reached higher values during later measurements, suggesting that convection may be accelerated by successive heating of the wire.

On the other hand, the calculated apparent conductivities were mostly constant in a μG environment. This suggested that the

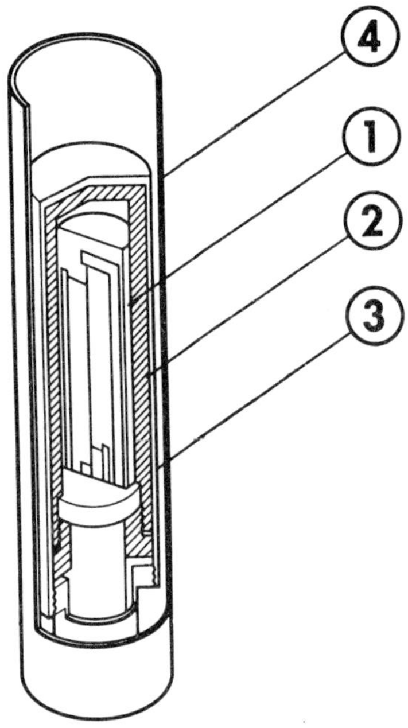

Figure 1 Measurement cell structure
1: Ceramic probe, 2: Carbon crucible
3: Nickel cartridge 4: Inner tube of furnace

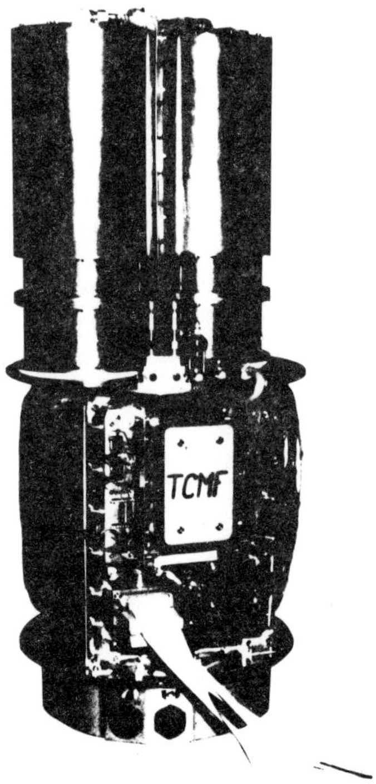

Figure 2 Flight module of thermal conductivity measurement facility (TCMF) for TEXUS sounding rocket. This module is 900 mm in height and 400 mm in diameter.

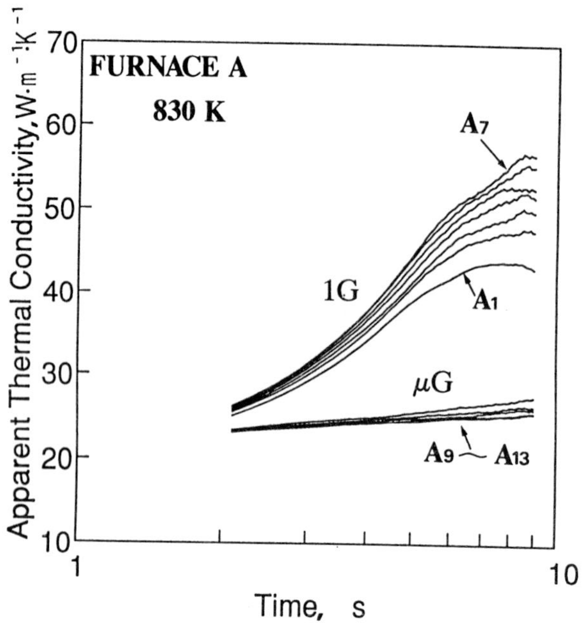

Figure 3 Apparent thermal conductivity as a function of logarithmic measurement time ln(t) for furnace A in TEXUS-24; A_1 to A_7 were measured as reference measurements on the ground, A_9 to A_{13} in μG environment.

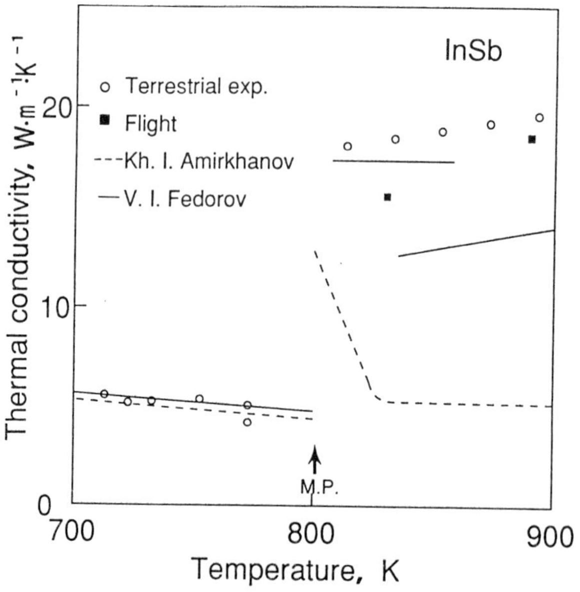

Figure 4 Thermal conductivity of InSb Amirkhanov[11], Fedorov[12], Flight[9], Terrestrial[7]

contribution of convection to apparent thermal conductivity was to a great extent reduced under a μG environment.

Figure 4 shows the thermal conductivity of InSb for both molten and solid states. The thermal conductivities of InSb were determined to be 15.8 (W m^{-1} K^{-1}) at 830K and 18.2 (W m^{-1} K^{-1}) at 890K, respectively.

As indicated in Figure 3, the difference between the measurements on the ground (1G) and under a μG environment is caused by buoyancy convection on the ground. In the present measurement cell, two kinds of convection need to be considered; one is caused by the inhomogeneous temperature distribution within the melt, and the other by wire heating during measurement.

Although the temperature of upper part in furnace was set to be higher, convection might have occurred due to the above mentioned inhomogeneous temperature distribution and the high acceleration during launching. If this convection had occurred, measured thermal conductivities might have been higher than the real values in the early stages of the μG environment, and the thermal conductivities would have decreased during the period of the μG. However, there is no remarkable change of the measurements from A_9 to A_{13} in the μG environment. This indicates that convection is considered to be weak or to be suppressed after entering the μG environment.

MU-300 parabolic flight experiment

In order to quantitatively investigate the effects of the above mentioned convection due to wire heating on measured thermal conductivity values, thermal conductivity measurements should be carried out under various gravitational acceleration levels on which the magnitude of convection depends, as follows,

$$Ra = \frac{g\beta\Delta TL^3}{\alpha\upsilon} , \qquad (2)$$

where α is thermal diffusivity, β is volumetric expansion coefficient, υ is kinematic viscosity and g is gravitational acceleration. L is the characteristic length of the system. Although the magnitude of convection depends not only on gravitational acceleration but also on input heat power and the depth of measurement cell and thermophysical properties, only gravitational acceleration can readily be changed in orders of magnitude. Using the parabolic flight of an aircraft, gravitational acceleration of 10^{-2} and 10^{-1} G can be realized.

The thermal conductivity of mercury was measured at room temperature with the laboratory model of the TCMF at μG levels of 10^{-2} and 10^{-1} G using the parabolic flight of a Mitsubishi MU-300 jet aircraft. Gravity levels were varied from 10^{-2} to 10^{-1} by changing the pitch angle of the aircraft. The g-jitter ranged from $\pm1\text{x}10^{-2}$ to $\pm2\text{x}10^{-2}$ G for both cases of gravity. Since the gravity level varied during the entry phase of the parabolic flights, every measurement was commenced five seconds after the microgravity was attained. Input powers to the wire were 139 (W m^{-1}) 315 (W m^{-1}), 559 (W m^{-1}) and 874 (W m^{-1}). These correspond to the ΔT of 4.7K, 14.8K 29.5K and 48.8K, respectively.

Let us discuss the effects of convection suppression using parameter ζ. In order to express the ratio of total heat transfer to heat transferred by conduction, a Nusselt number is often used. However, it was difficult to define the number for this experiment. Therefore we introduced ζ which is similar to a Nusselt number. This is the ratio of apparent thermal conductivity containing the contribution of convection to the sum of the thermal conductivities for mercury and the substrate as follows,

$$\zeta = \frac{\lambda_L+\lambda_S+\lambda_C}{\lambda_L+\lambda_S} \quad . \tag{3}$$

λ_C is the contribution of heat transferred by convection and is defined as an increase in apparent thermal conductivity value due to convection as shown in Figure 3. This ζ expresses the ratio of heat transfer to heat conduction for this measurement. $\zeta = 1$ when there is no contribution by convection to the apparent thermal conductivity; $\lambda_C = 0$. $\lambda_L+\lambda_S+\lambda_C$ was determined from apparent conductivity at 9 s. $\lambda_L+\lambda_S$ was the sum of

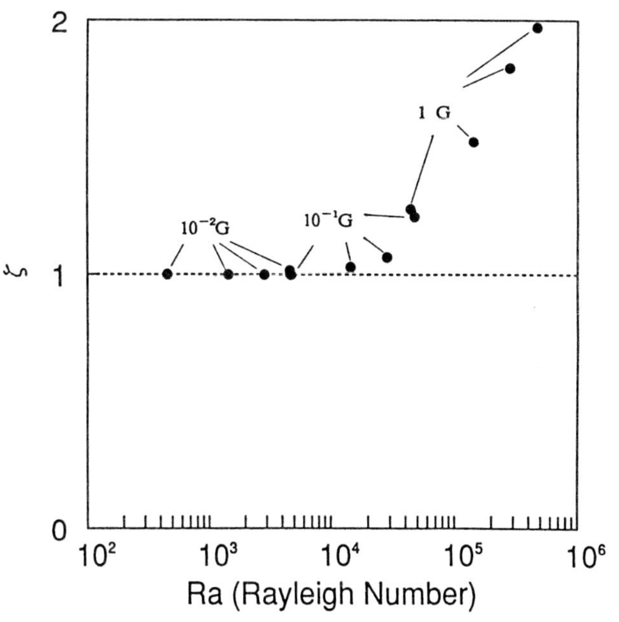

Figure 5 ζ as a function of Rayleigh number.

thermal conductivities of the two materials. Figure 5 shows ζ as a function of Ra. Below 10^3 of Ra, ζ is 1. This suggests that the contribution of convection to the measured thermal conductivity seems to be negligible when $Ra<10^3$ for the system. The Ra for the experiment on the TEXUS 24 sounding rocket was calculated to be less than an order of 10^1. As a result, we confirmed that the wire-heating induced convection had a negligible effect on the thermal conductivities measured on the TEXUS sounding rocket.

4. CONCLUSION

The method of measurement to determine thermal conductivity in a μG environment was established by using the transient hot wire method. We consider that this method is suitable for μG experiments which are normally of short duration.

The thermal conductivity of molten InSb was determined by this method in the experiment on board the TEXUS sounding rocket. The other experiment in the MU-300 jet aircraft indicated that the wire-heating induced convection had negligible effect on the thermal conductivities measured in the

TEXUS rocket.

5. REFERENCES

1 H. Coenen, *Tech. Mitt. Krupp Forsh. Ber.*, 37, 83 (1979)

2 J. C. Perron, *Proc. 6th European symp. on Material Sciences under Microgravity conditions* (ESA, SP-256, 1987), p.509.

3 K. Wanders, H. Steinbichler, G.P.Görler, *Proc. 4th European Symp. on Material Sciences under Microgravity Conditions* (ESA, SP-191, 1987), p.403.

4 S. Aalto, S. Andersson, M. Eklof, S. Engstrom, K. Hedvoll, U. Hogman, N. O Jansson. B. Svensson, *Thesis, Charmers University of Technology*, Sweden, 1986, 1987

5 S. Nakamura, T. Hibiya and F. Yamamoto, *Rev. Sci. Instrum.*, 59, 997 (1988)

6 S. Nakamura, T. Hibiya and F. Yamamoto, *Rev. Sci. Instrum.*, 59, 2600 (1988)

7 S. Nakamura, T. Hibiya and F. Yamamoto, *Int. J. Thermophys.*, 9, 933 (1988)

8 F. Yamamoto, S. Nakamura, T. Hibiya, T. Yokota, D. Grothe, H. Harms and P. Kyr, *Proc. CSME Mechanical Engineering Forum*, (Toronto, 1990) p.1.

9 S. Nakamura, T. Hibiya and F. Yamamoto, *Int.J. Thermophysics in press*

10 E. Takegoshi, S. Imura, Y. Hirasawa and T. Takenaka, *Bull. JSME*, 25, 395 (1982)

11 Kh. I. Amirkhanov, Ya. B. Magomedov, *Soviet Phys. Solid State*, 8, 241 (1966)

12 V. I. Fedorov, V.I. Machuev, *High Temp.*, 8, 419 (1970)

NUMERICAL SIMULATIONS OF A MICROGRAVITY EXPERIMENT: THE M.I.T.E.: MEASUREMENT OF INTERFACE TENSION EXPERIMENT

Amalia Ercoli Finzi*; Alessandra Maulino**
** Aerospace Dept. Politecnico di Milano - Via Golgi 40, 20133 MILANO*
*** CISI ITALIA S.p.A. - Piazza della Repubblica 32, 20124 MILANO*

ABSTRACT

The thermal design and numerical simulations of a microgravity experiment are carried out. The experiment is the MITE, a facility to measure interface tension between immiscible liquids in microgravity, aboard the sounding rocket MASER IV (ESA/SSC). On the preliminary design of the experiment module, based on a merely passive thermal control, a finite element model is created upon which the thermal analyses are conducted using two finite element codes: NASTRAN and SAMCEF. The aim of the analyses is to verify the thermal behaviour of the experiment during pre-launch, launch and microgravity phases, when it is subjected to stringent thermal requirements. The use of two finite element codes is due not only to the need for redundancy imposed by the great accuracy and reliability required by the experiment, but also to the interest in comparing performances and capabilities of the two codes in the solution of a complex thermal problem, involving radiation exchange and view factors calculations. Numerical and accuracy problems related to the use of the two codes are pointed out. On the basis of the results obtained from each analysis the thermal design is step by step enhanced and the final configuration is reached.
The flight of the experiment (March, 29, 1990) confirmed the thermal behaviour as good, as well as the accuracy and reliability of the numerical analyses.

Keywords: Finite Element Simulations, Thermal Analysis, Interface Tension.

1.INTRODUCTION

The thermal design of a microgravity experiment involves several problems due mainly to the kind of environment in which the experiment is to operate and to the requirements imposed by the experiment to be successfully performed.

The thermal requirements of a microgravity experiment are generally connected to the necessity for the payload not to reach too high or too low temperatures. Most of microgravity experiment, for example those involving fluids and chemico-physical phenomena, require also stable temperatures and small thermal gradients inside. Such requirements can be very restrictive and aim not only to the survival of the experiment devices but also to the preservation of the physical principles upon which the experiment is based.

On the other hand the space thermal environment does not have its known and stable characteristics: in fact the thermal surroundings of a space experiment can be very different when the experiment works, for example, on an orbiting platform, in a shuttle cargo bay, or in a sounding rocket module. In all these cases the time during which the experiment is in contact with a certain environment can vary from a period of few minutes to many days or even to some years.

For these reasons the design of the thermal control of a microgravity experiment must rely on a very accurate prevision of the operative conditions, of the solicitations and above all of the experiment response, in order to guarantee its correct behaviour.

At this aim the thermal numerical analysis becomes extremely important: a numerical model must be developed representing the system and its environment, which could lead to an accurate prevision and allow to easily follow all the changes apported to the design in the various project phases.

The numerical analysis of the MITE experiment has been carried out using the two computer codes MSC/NASTRAN and SAMCEF, both being general purpose programs based on the finite element method, which can be considered a general approximation technique for the solution of physical systems described by field equations in continuous media.

2.THE MITE: MEASUREMENT OF INTERFACE TENSION EXPERIMENT

The experiment has been ideated by CNR/ICFAM (Consiglio Nazionale delle Ricerche / Istituto di Chimica-Fisica Applicata ai Materiali) and its object is the measurement of interface tension between immiscible liquids exploiting microgravity conditions.

The experimental technique is the following (see figure 1): a bubble of liquid A is obtained at the tip of a nozzle in a cell containing liquid B, moving at constant speed the piston of a precision syringe by means of an electrical motor. The two liquids are in communication with the two inputs of a pressure transducer which measures the pressure difference at the interface. The interface tension σ is then evaluated by the Laplace law, which in absence of gravity has the form: $\Delta P = 2\sigma/R$, where the bubble radius R is calculated knowing the liquid volume displaced by the syringe.

The experiment has been carried out aboard the sounding rocket MASER IV (ESA/SSC, march 29, 1990) which provided about 7 minutes of microgravity.

The structural design of the experiment has been carried out by CISE and its preliminary layout is shown in figure 2.

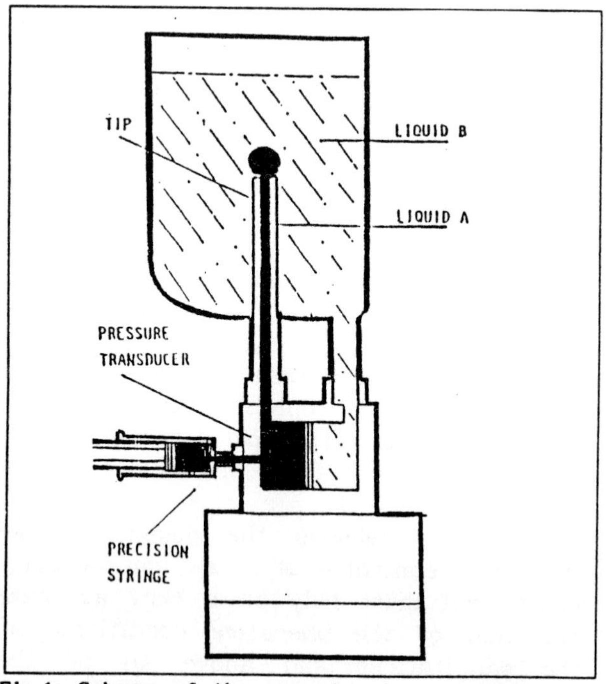

Fig.1: Scheme of the experiment.

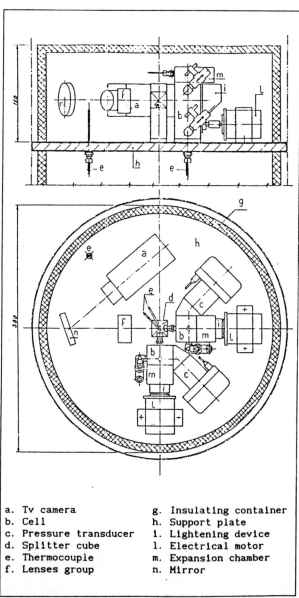

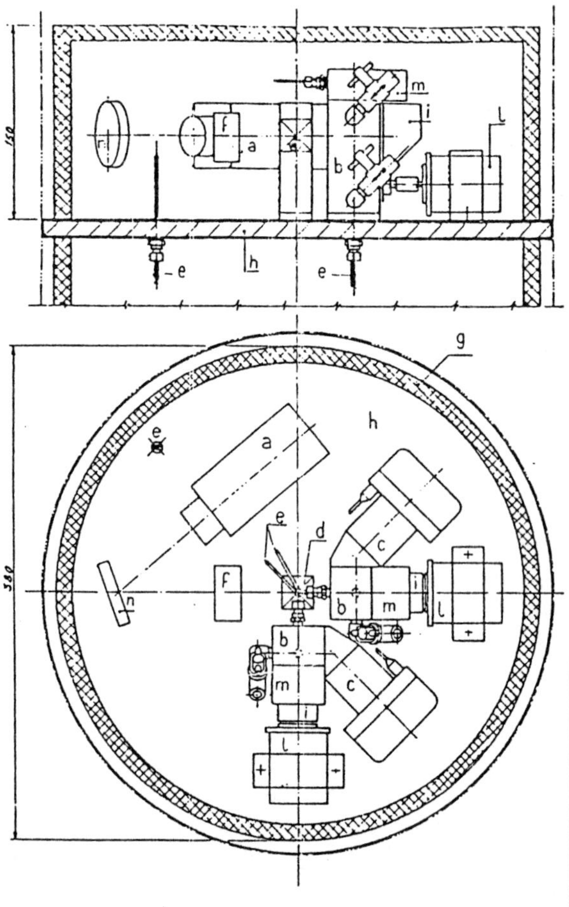

a. Tv camera
b. Cell
c. Pressure transducer
d. Splitter cube
e. Thermocouple
f. Lenses group
g. Insulating container
h. Support plate
i. Lightening device
l. Electrical motor
m. Expansion chamber
n. Mirror

Fig.2: MITE structural design

2.1 Thermal requirements and solicitations

The experiment requires a constant temperature with a tolerance of ± 0.5°C and a maximum temperature gradient of 0.1°C/cm in the measurement cell during both launch and coasting phases, when the external temperature is quickly arising and the internal electronics are heating sources.

The thermal solicitations acting on the experiment have not yet been defined in the first phase of the design, therefore the following hypotheses have been made:

Maser wall $\begin{cases} T = 20°C & t < 75s \\ T = 150°C & t > 75s \end{cases}$

Neighbouring modules $\{ T = 20 + 0.06t\ °C$

Tv camera	6 W
Electrical motors	2 W × 2
Electronic subsystem	60 W

Since a specific temperature is not needed but a temperature stability is required for a relatively short time (400s), the thermal control can be based on a merely passive concept.

Figure 3 shows the preiliminary configuration of the experiment module and of the component of the thermal control system.

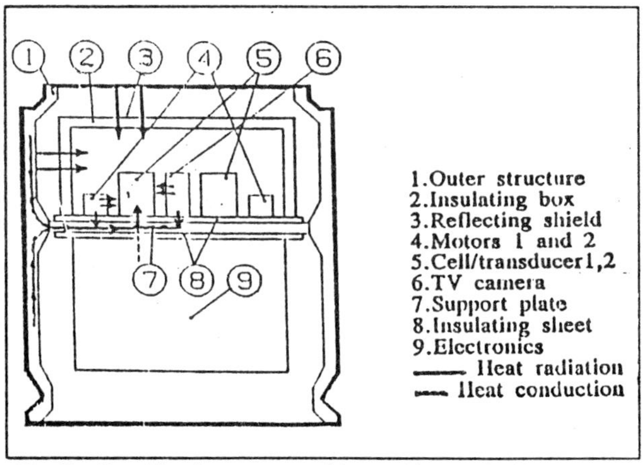

Fig.3: Preliminary configuration of the experiment module.

3.THERMAL ANALYSIS

3.1 Preliminary analysis

On the basis of the preliminary layout of the whole module (by CISE and OFFICINE GALILEO) a finite element model has been developed upon which the preliminary analyses have been performed taking in account only the effect of conduction; the radiative exchange has been taken in account in a second phase, even if its contribution to the thermal behaviour of the experiment cells has been estimated as minimal.

Each analysis consists of a linear steady state analysis simulating the pre-launch phase and a linear transient response analysis simulating launch and microgravity phases (0 ÷ 400 s).

The finite element model consists of 1046 nodal points connected by 730 tridimensional isoparametric conduction elements.

The detail of the discretization is quite different in the various parts of the experiment and the mesh is thus quite inhomogeneous. This fact is due to the need for having a good detail in certain zones of the experiment (like the measurement cells) and to the parallel impossibility to keep the same detail for all the components, which would lead to extremely high analysis costs.

This solution therefore introduces some accuracy problems in the transient response analysis.

The results obtained by the first analysis show actually some inaccuracies: they show the presence of oscillating temperatures in the cells and in other parts of the model, although heat is continuously apported to them; such oscillations have the same order of magnitude as the maximum allowed temperature variation.

This effect is due to a relatively coarse mesh in presence of high solicitations and thus to high thermal gradients.

It has been found in fact that if the size of the space discretization near highly solicited walls is greater than the penetration depth, that is the "thermal conduction boundary layer", the internal nodes near the wall show in the first time instants after the application of the solicitation a lower temperature than the initial one, even if heat is apported to the wall.

This penetration depth is given by:

$\delta(t) = K \sqrt{\beta_w \, t}$ where:

β_w = thermal diffusivity of the wall: $\lambda/\rho c$

K = numerical factor (2÷4)

t = time elapsed since appearance of the solicitation.

and represents the zone in which the whole temperature variation is confined at time t.

It is thus necessary to choose a spatial discretization in such a way that going through the discretized medium in an orthogonal direction with respect to the wall, many layers of elements are encountered within the above zone.

Such problems have been solved with the use of conductive non-isoparametric elements, which are internally composed of 10 tetrahedra and provide a kind of mesh refinement effect.

Figure 4 shows the temperature distribution in the cell axis at various time steps. As it can be seen the thermal requirements are met only in a small zone near the top, while near the bottom of the cell there is a temperature variation of 5 °C and a thermal gradient of 3 °C/cm.

Some changes in the design are needed to guarantee the full respect of the requirements.

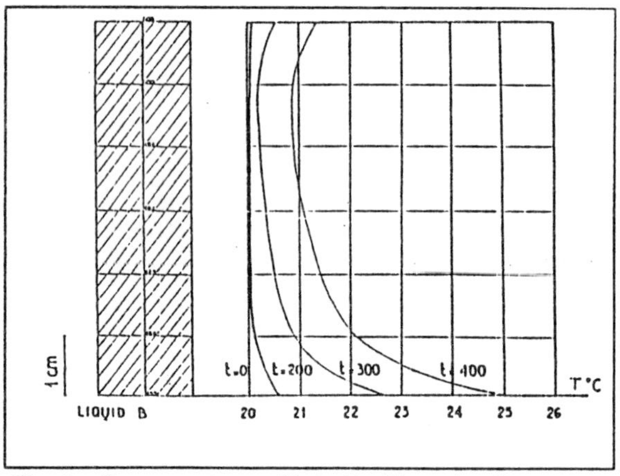

Fig.4: Temperature distribution in the cell axis at various time steps.

3.2 Radiation analysis

In order to evaluate the effect of radiative heat exchange the non-linear analyses have been carried out. The second aim of these analyses is the comparison between performances and capabilities of the two codes in the solution of the radiation problem.

The two codes are in fact based upon different modellization methods for the radiative exchange simulation and use different techniques for the solution of the non-linear system.

The two modellization methods for the radiative exchange are the "emissivity element method", used by NASTRAN and the "radiosity node method", used by SAMCEF. The characteristics of the two methods are:

- Emissivity element method:

- Each radiating surface element is represented by a boundary element with given emissivity and absorptivity.
- The problem is expressed in terms of nodal temperatures.

- Radiosity node method:

- The primary unknowns are the surface radiosities.
- Each radiating area is represented by a boundary element connected to a dummy node associated to the radiosity and with an unknown temperature.
- All the radiosity nodes are connected to one another by link elements related to the view factors.

It can be pointed out that the second method involves the introduction of a great number of elements and unknowns, with the consequent loss in the matrix sparsity, so that the problem becomes heavier and heavier from a calculation point of view.

The results obtained in the experiment cells confirm the negligibility of the radiation effect, whose contribution is of the order of 1 ÷ 2% of the total temperature increase, that is some hundredths of degree. In these zones the two programs provide the same results.

On the contrary, in the points where the effect of radiation is more relevant, like the insulating container, a great difference in the results of the two codes has been detected (see figure 5). In these points the NASTRAN analysis provides very inaccurate results, which are even physically absurd.

The reason of such a difference in the behaviour of the two codes in presence of high non-linearity can be found in the solution technique used by each program.

The two codes actually exploit the same

solution algorithm but in NASTRAN the solution is conducted without any updating of the iteration matrix (which contains the linearized terms of radiation); the residuum of the iteration is not updated and can even diverge giving place to very inaccurate results.

SAMCEF on the other hand performs the solution with an updating strategy of the iteration matrix, which can be planned by the user on the basis of the non-linearity level of the problem. In this case the attempt to perform the analysis without any updating of the matrix, which means emulating the NASTRAN solution, caused the interruption of the calculation for lacking of convergence thus demonstrating the inadequacy of the NASTRAN technique.

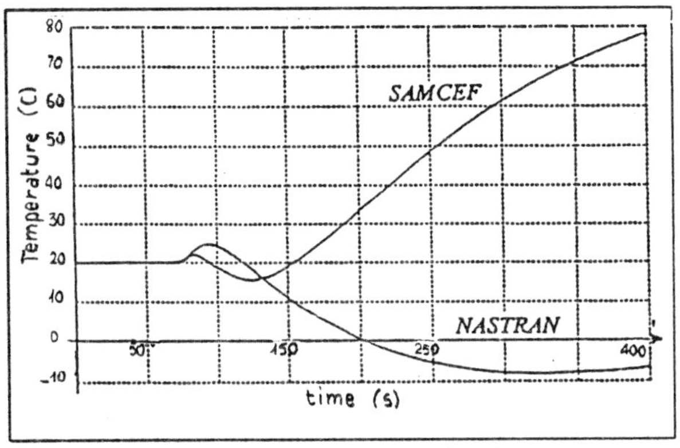

Fig.5: Results obtained with the two codes in non linear analysis.

3.3 The final design and analyses

On the basis of the results obtained from the preliminary calculations the thermal design of the experiment has been changed. The main changes in the design are:

- Thermostating chamber: introduced in order to impose an adequate temperature to the experiment during the pre-launch phase, when the rocket is externally heated with air at 10÷25 °C and the functional tests of all the modules are performed. Such chamber consists of a hollow wall inside which a fluid is flowing which is thermally controlled by an external unit of 560 W.
- Increase of the insulating sheet between the experiment and the support plate; introduction of an insulating tablet under the cells.
- Reduction of power and size of the electrical motors.
- Introduction of a 70 W transmitter in the electronic subsystem with consequent removal of the insulation between the plate and the subsystem in order to allow the dissipation of the generated heat.
- More accurate modellization of the cells: also the pyrex windows for TV monitoring, the nozzles and the injection channels have been modelled.

The final analyses include steady state analyses for the pre-launch phase under different boundary conditions, among which a "cold" and a "hot"case.

The steady state analyses confirmed the adequacy of the thermostating chamber. The transient analyses have been carried out for the cold and hot cases starting from the thermal distribution calculated in the steady state analyses.

The results of these final analyses are shown for the hot case in figure 6, which shows that the thermal requirements are fully satisfied.

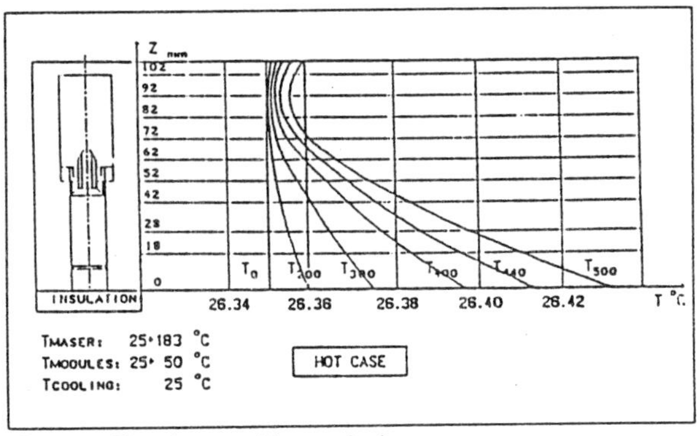

Fig.6: Final results - hot case

4.THE FLIGHT

The data recorded during the flight show that the thermal behaviour of the experiment fully met the requirements on temperature and temperature gradient.

Temperature data concerning one experiment cell are represented in figure 7, where the two curves concern two thermoresistances placed inside the cell near its bottom and its top.

As it can be seen from this plot the maximum increase in the experiment cell was about 0.45 °C, where a maximum of 1.0 °C was allowed.

The plot also shows an offset in the temperatures of the two thermoresistences due probably to a difference in calibration; the same offset was detected during thermal ground tests.

The maximum temperature gradient along the cell axis, neglecting the above offset, was about 0.03°C/cm where a maximum of 0.01°C/cm was required.

The thermal behaviour of the experiment can thus be considered good, although the experiment was not carried out in the foreseen thermal environment.

On the basis of the previous analyses in fact it was found that an optimum pre-launch temperature would have been 25÷30°C but, due to an improper use of the thermostating unit (which was kept for many hours at about -20°C), the thermostating temperature could not rise above 17°C.

4.1 Correlation with numerical analysis

Since the flight thermal conditions were out of the foreseen range, a direct correlation between the results of the thermal simulations and flight data was not possible. For this reason a "post flight" numerical analysis has been carried out based on the flight boundary conditions, in order to verify the accuracy of the numerical simulations. The analysis has been carried out under the hypothesis of absence of radiative exchange, which contribution to the total temperature increase was found not to be relevant.

The results of this analysis are shown in figure 8, representing respectively the temperature growth of a point corresponding to the bottom thermoresistance inside cell 1. The solid lines represent the calculated temperatures while the dashed ones represent the recorded temperature, fitted and shifted to the calculated starting temperature.

As it can be seen, the difference between calculated and real temperature increase is about 0.1°C. This difference can be seen as the contribution of radiation, that has been neglected in the numerical analysis.

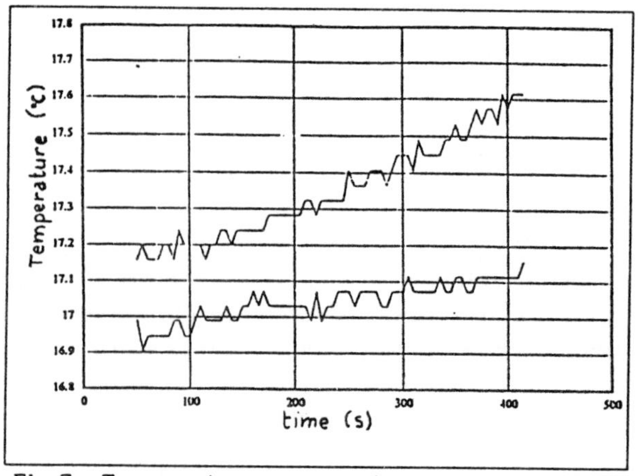

Fig.7: Temperatures recorded during the flight.

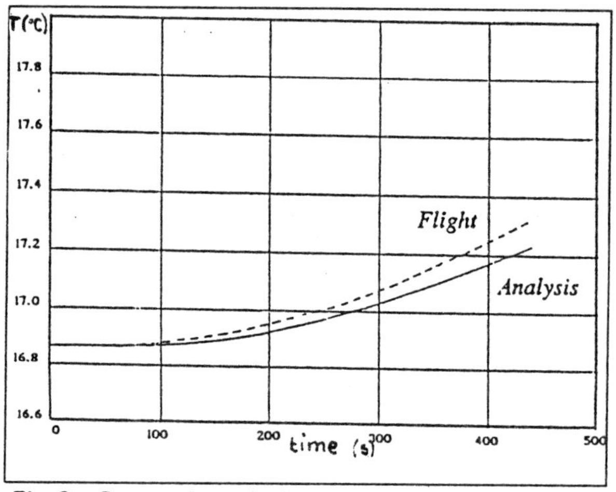

Fig.8: Comparison between recorded and calculated temperature.

5.REFERENCES

A. Maulino, *Il progetto termico di esperimenti spaziali: il M.I.T.E*; Apr.1990, Degree thesis - Aerospace Dept. Politecnico di Milano.

A.Ercoli Finzi, A.Maulino, F.Rossitto, *M.I.T.E - Maser IV experiment - Engineering problems in microgravity experiments: the MITE,* Paper presented at VII European Symposium on Materials and Fluid Sciences in Microgravity, Oxford, U , 10-15 September 1989, ESA SP-295 (January 1990).

HEAT TRANSPORT MECHANISMS IN SUPERCRITICAL FLUIDS: NUMERICAL SIMULATIONS AND EXPERIMENTS

P. Guenoun, B. Khalil, D. Beysens
Direction des Sciences de la Matière,
Service de Physique de l'Etat Condensé,
C.E. Saclay F-91191 Gif-sur-Yvette cedex

F. Kammoun, Y. Garrabos, B. Le Neindre
LIMHP (C.N.R.S), Université Paris Nord
av. J.B. Clement 93430 Villetaneuse, France

B. Zappoli
C.N.E.S, 18 av. E. Belin 31055 Toulouse cedex, France

ABSTRACT

Thermal transport, on Earth, in pure fluids close to the critical point cannot be dissociated from gravity. This is due to the fact that both the isothermal compressibility and the Rayleigh number diverge in the vicinity of the critical point. As a consequence, any thermal gradient can induce convections which in turn govern the thermal transport. However, it was recently suspected that the extreme compressibility of the critical fluid combined with its very low thermal diffusivity (critical slowing down) lead to another novel mechanism of heat transport. This latter mechanism was named "Piston effect" as it involves the thermal expansion of the thermal diffuse boundary layer appearing close to any hot spot in the fluid. Results of numerical simulations in zero gravity are presented which describe in detail the thermal equilibration in the fluid. Recent experimental evidences are discussed here thanks to the results provided by the Texus 25 flight (sounding rocket).

Keywords: *Supercritical fluid, Acoustic Heating, Numerical Simulation of Compressible Fluids, Heat Transfer*

1. INTRODUCTION

The transport of heat in pure fluids classically involves the mechanisms of convection, diffusion and radiation. In this communication, experimental evidences are given that a fourth mechanism, related to the compressibility of the fluid, is present and can be dominant. Recent numerical simulations of the entire set of Navier-Stokes equations (Ref.1) have shown that this fourth mechanism can be named the "Piston Effect" (PE). When heated near a wall, the thermal diffusive boundary layer of a compressible fluid expands very much. This expansion generates an acoustic wave which travels back and forth in the fluid, and acts as a piston. The thermal conversion of this wave, in turn, heats the bulk fluid. Note that this heating is spatially uniform as soon as a few acoustic times are elapsed. Pure thermodynamic calculations (Refs.2,3) are in qualitative agreement with these results. The precise determination of the equilibration behaviors is however still a matter of interest (Refs.2,4). The proximity of the critical point is particularly favorable in order to reveal PE since the compressibility diverges at this point. The above numerical studies (Refs.1,2,3) have been concerned with the critical region but one should keep in mind that PE can be relevant far from the critical point. However, PE is not easy to evidence under the Earh gravitationnal field because of the predominant influence of gravity-driven convections. Both Rayleigh and Grashof numbers can be very large close to the critical point because of the diverging compressibility. This means that even very small temperature changes in the fluid can generate gravity-driven flows, as discussed below. These flows are in turn responsible for the fast thermalisation of the fluid. Diffusion is a poorly efficient process since the thermal diffusion coefficient goes to zero as one approaches the critical point.

A first part resumes our numerical findings which evidence the PE. In the following part all details are given about the experimental set-up and the procedure which enables the thermal behavior of the fluid to be analyzed. Such a study under microgravity has been already undertaken in different configurations, leading to ambiguous but promising results (Ref.5). The principle of the experiment is the optical observation of a transparent cell filled with carbon dioxide (CO_2) at critical density ($\rho'_c = 467.8$ kg.m^{-3}). The cell is included in a high precision thermostat and follows a thermal cycle close to its critical temperature ($T'_c = 304.14°$ K). Local temperature is deduced from light transmission measurements through the cell.

The third part is concerned with the thermal equilibration process under Earth gravity and the convective flows which can be generated in the fluid. The main results are discussed in a fourth part, where we show that PE is the predominant mechanism of heat transport when a thermal quench is performed under microgravity. It is demonstrated that the average thermal equilibration of the fluid is much faster than a pure diffusive behavior. This is of practical importance since pure diffusion would lead to equilibration times of order one week at a distance of 1mK of T'_c (where the diffusion coefficient is $D_T \sim 2\ 10^{-11}$ m^{2}s^{-1}). On the other hand, it is shown that the thermal gradients which preexist prior to the quench are less affected by PE and go on relaxing by diffusion.

2. NUMERICAL MODELING

2.1. The governing equations

The numerical modeling is based on the solution of the Navier-Stokes equations written for a van der Waals gas contained within a 1-D slit-like container of length $x' = L$. For time $t' < 0$, the gas is at rest at a temperature T'_0 above the critical temperature (T'_c). The density ρ' is equal to the critical density ρ'_c. For $t' \geq 0$ the wall at $x' = 0$ is heated by 1 mK during 50 ms; the heat perturbation works on a time scale comparable to the heat diffusion time $t'_d = \frac{L^2}{D_T}$, a time scale much longer than the acoustic time $t'_a = L/c'_0$ (c'_0 is the sound velocity). One defines also a parameter $\epsilon = \mathcal{P}_r \frac{t'_a}{t'_d}$, $\epsilon \ll 1$, where $\mathcal{P}_r$ is the Prandtl number of the fluid. The non dimensional 1-D Navier-Stokes equations can be written as (1-3), where the subscripts t and x stand as derivatives versus time and space,

$$\rho_t + (\rho u)_x = 0 \tag{1}$$

$$\rho u_t + \rho u u_x = -\gamma^{-1} P_x + \frac{4}{3}\epsilon\, u_{xx} \tag{2}$$

$$\rho T_t + \rho u T_x = -(\gamma - 1)(P + a\rho^2)u_x + \epsilon\left\{\frac{\gamma}{\mathcal{P}_r} T_{xx} + \frac{4}{3}\gamma(\gamma - 1)(u_x)^2\right\} \tag{3}$$

Similarly the van der Waals equation of state writes,

$$P = \frac{\rho T}{1 - b\rho} - a\rho^2 \tag{4}$$

where $\rho = \frac{\rho'}{\rho'_c}$, $P = \frac{P'}{P'_0}$, $T = \frac{T'}{T'_c}$, $t = \frac{t'}{t'_a}$, $x = \frac{x'}{L}$, $P'_0 = \rho' R' T'_c$ and $u = \frac{u'}{c'_0}$. u' is the fluid velocity, P' the pressure, T' the temperature, γ is the ratio of specific heats at constant pressure and volume, $a = \frac{9}{8}$, $b = \frac{1}{3}$ and R the ideal gas constant.

The boundary conditions are

$$u = 0 \quad T = T(t) \quad \text{at } x = 0 \tag{5}$$

$$u = 0 \quad T = 1 \quad \text{at } x = 1 \tag{6}$$

and the initial conditions are

$$\rho = 1, \quad u = 0, \quad T = 1 + \tau, \quad P = -\frac{9}{8} + \frac{3}{2}(1 + \tau) \ \ at \ \ t = 0, \tag{7}$$

where $\tau = (T'_0/T'_c) - 1$. The wall at $x = 0$ is heated "slowly" that is, on time scale of order of the long diffusive time scale θ defined by

$$\theta = \frac{t'_a}{t'_d}\mathcal{P}_r t = \epsilon t. \tag{8}$$

2.2. The numerical method

The numerical method is based on a semi-implicit primary variable algorithm which solves the Helmholtz equation for the pressure correction. The algorithm called P.I.S.O. (Ref.6) for Pressure Implicit with Operator Splitting is able to solve low Mach number compressible flow problems.

2.3. The results of the numerical modeling

The numerical results have been obtained for carbon dioxide CO_2 considered as a van der Waals gas and for initial conditions such as $T'_0 - T'_c = 3.8$ K on the isochore $\rho' = \rho'_c$. As the wall is heated up, the expanding thermal boundary layer generates a velocity field (Figure 1) which then acts as a piston and causes a sound wave to propagate in the bulk. Then, as time goes by, the boundary layer becomes thicker and thicker but the adiabatic heating causes the temperature in the bulk to increase faster than by diffusion. Figure 2 shows the evolution of the temperature in the bulk for different times on a larger time scale. The flatness of the temperature profile in the bulk is a characteristic of this process. This is due to the adiabatic heating of the bulk by the acoustic field: energy is first transferred from thermal to kinetic within the expanding boundary layer, and then turned again into thermal energy within the bulk by the work of pressure forces during the slowing down of the fluid velocity.

3. EXPERIMENTS

The experimental cell (Figure 3) is illuminated by a white light parallel beam and imaged on a CCD video camera. A gold thread (0.3 mm diameter) is immersed in the sample. It is thermally isolated from

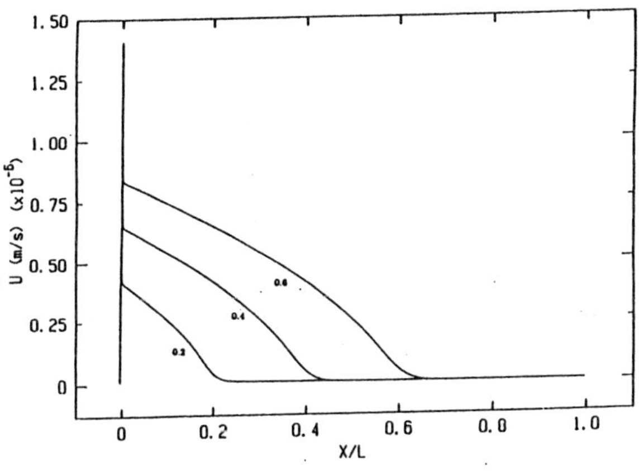

Figure 1: Velocity profile for the times $t = 0.2$, 0.4 and 0.6 in units of t_a.

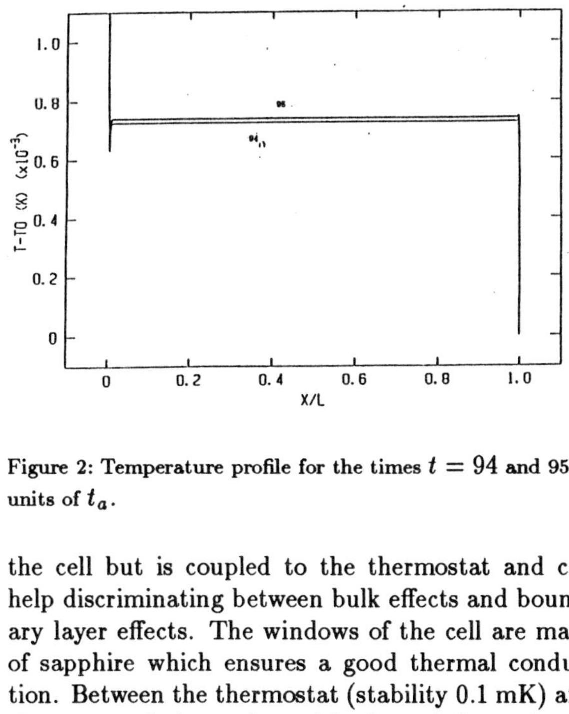

Figure 2: Temperature profile for the times $t = 94$ and 95 in units of t_a.

the cell but is coupled to the thermostat and can help discriminating between bulk effects and boundary layer effects. The windows of the cell are made of sapphire which ensures a good thermal conduction. Between the thermostat (stability 0.1 mK) and the camera is located a grid whose deformation of the shadow gives a measure of the density gradients within the cell through the measure of optical indices gradients (Ref.7). The whole set-up was loaded in the TEXUS 25 rocket (13/05/90) whose trajectory gives access to 6 minutes of reduced gravity (at least 10^{-4} times lower than the terrestrial gravity, randomly oriented). Before the launch the T_c' value is accurately determined and the temperature of the cell is set 2 mK above T_c'. An unavoidable waiting period follows and causes density gradients in the cell because of the very high compressibility of the fluid (see below). During the launch and before the beginning of the microgravity period, the rocket is spinned at a rate of 2 s^{-1} which generates flows in the cell. These flows stop a few seconds after the beginning of microgravity. A first thermal quench of 1 mK is performed, leading the system at 1 mK above T_c'. After 1 min 30 sec of equilibration, the fluid is again quenched from $T_c' + 1$ mK to $T_c' - 1$ mK, inducing a phase separation process of the fluid which lasts 1 min 30 sec. This phenomena is reported in a separate paper (see the other communication by P. Guenoun et al. at this conference). The last stage of the experiment consists in heating up the system back to the one phase region at $T_c' + 2$ mK (quench of 3 mK). We will focus here on the two extreme quenches.

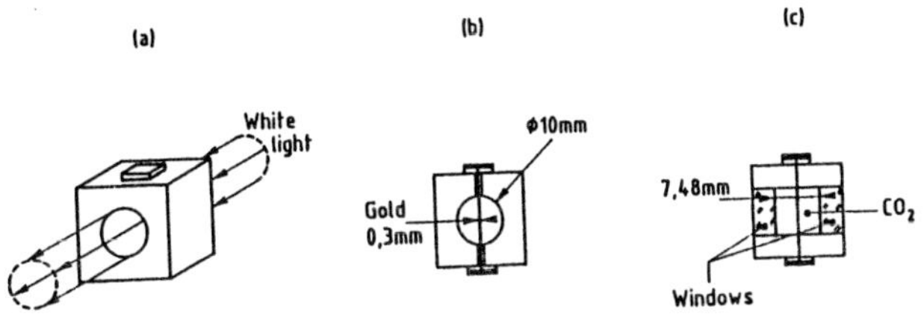

Figure 3: Sketch of the experimental cell. (a): general view; (b) front view; (c): side view.

The light transmission through the sample can be used as a local probe of the fluid temperature since the transmission is directly coupled to the scattering of light by the critical fluctuations. The transmission values are known to be directly related to the difference $T' - T_c'$(Ref.8). At $T' - T_c' = 2$ mK, one imposes the average of the transmission data to be equal to the known value coming from the above relation. This fixes the transmission data to be an absolute measure of the shift $T' - T_c'$ all along the following thermal evolution of the fluid. However, the recovering of the data is not straightforward since the gain of the CCD camera is automatically adjusted. An original procedure of recovering the gain value at each time through the measurement of the noise signal of each picture has been developpped. The gain is shown to be proportionnal to the variance of the noise (Ref.9). The noise signal is thus extracted from each image through the histogram of the intensity of the pixels. Moreover, it has been checked that the automatic adjustment of the gain is done such as the pixels of maximum illumination reach the maximum grey level of 255 (every pixel is coded over 8 bits).

The gain is inversely proportional to the illumination of these pixels. The maximum illumination has been chosen as a measure of the light transmission through the cell. Another choice can be made by computing an average intensity over the pixels and taking this value as a transmission measurement. This last procedure has the disadvantage that the averaging procedure is very sensitive to the shape of the histogram which, in this case, is not simple (e.g. gaussian). As the shapes evolve with time, this method provide more scattered results. However, this gives access to a measure of $\delta T'$, the temperature fluctuations in the cell. This measure is only qualitative but provides informations on the dynamics of $\delta T'$.

4. CONVECTIVE HEAT TRANSPORT

Experiments have been also carried out with the setup described above on Earth. They enable us to appreciate how important is the convective heat transport for a critical fluid. Such a task is very difficult to perform by theoretical means because of the geometry which is involved. For example, it is worth computing a Grashof number which gives a measure of the relative importance of viscous and inertial effects (Ref.10). This number writes as:

$$Gr = g\alpha \ominus L^3/\nu^2 \qquad (9)$$

where g is the gravitationnal acceleration, α is the isobaric coefficient of expansion, $\ominus$ is the temperature difference in the fluid, L is the spatial scale over which extends $\ominus$ and ν is the kinematic viscosity. As expressed by (9), the calculus of Gr is very sensitive to the value of L, a quantity which is difficult to estimate a priori. Fig.4 shows a picture of the cell during a thermal quench of 25 mK at a distance $T' - T'_c = 48$ mK (T' is the final temperature). Convective structures are clearly evidenced and a similar observation can be done a close as at $T' - T'_c = 3$ mK after a 5 mK quench. These structures typically appear and evolve during a dozen of seconds which is the time necessary for the thermostat to reach a constant temperature. The spatial scale of these structures is a very natural guess for L. A rough value is $L = 0.15$ mm for $T' - T'_c = 3$ mK. This makes possible an estimation of Gr at $T' = T'_c + 1$ mK. At this temperature one has $\alpha = 7.10^3$ K^{-1} and $\nu = 9.10^{-8}$ m^2 s^{-1} whilst $g = 10$ m s^{-2}and $\ominus$ is assumed to be 1 mK in order to simulate the quench performed in microgravity. This leads to $Gr = 1.1\ 10^5$ with $L = 0.15$ mm. The uncertainty over L is of order of a factor of 3 which means that Gr is exact within an order of magnitude. Such a large value of Gr suggests that experiments performed so close to the critical point are delicate to prevent from the influence of convection (Ref.11). On the other hand, a reduction of g by 4 orders of magnitude gives $Gr = 11$. This number and the random nature of the residual accelerations indicate that the convective flows are negligible.

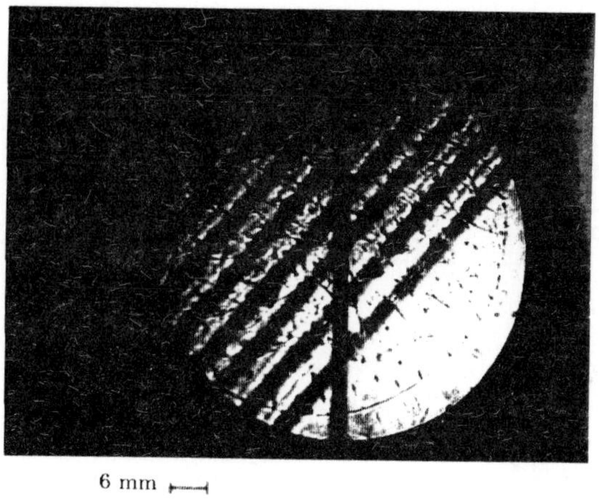

Figure 4: Convective cells observed during a quench of the fluid (see text). Note the gold thread and the shadow of the grid which measures the density gradients. Gravity is directed perpendicular to the picture.

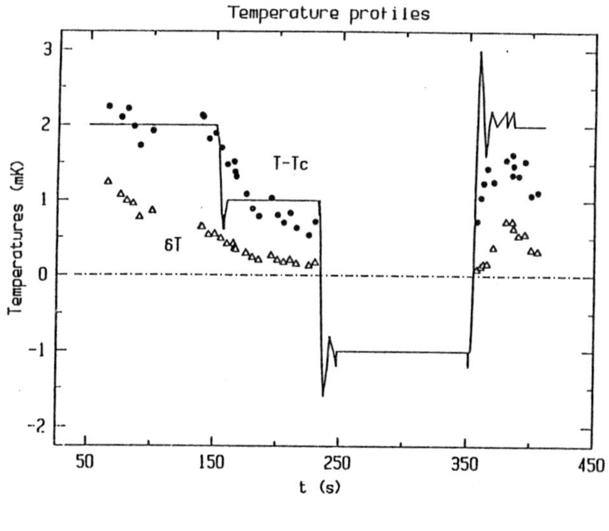

Figure 5: Temperature profile of the thermostat (continuous line). The symbols • show the average temperature within the fluid. The triangles Δ show the evolution of the fluctuations in temperature, $\delta T'$, but the scale in temperature is not correct for this quantity.

5. RESULTS

Figure 5 shows the main results associated to the thermal quenches which have been performed. The relaxation of the temperature deduced from the maximum illumination measurements is seen to occur on a time of order 30 s (first quench). This is far more rapid compared to the typical diffusion time which is of order one week at 1 mK far from T_c'. Moreover, the change in temperature is seen to be homogeneous in the fluid, as suggested by the numerical simulations of the PE. This is confirmed during the second quench where a very homogeneous phase separation is observed. The third quench is also associated with a rapid thermal equilibration, showing that PE is still efficient in a diphasic fluid. On the other hand, the qualitative evolution of $\delta T'$ during the two quenches under study is more comparable to a diffusion process. However, it is worth noticing that, during the third quench, $\delta T'$ is first increasing. This is because the departure point of the fluid is a diphasic state where isothermal differences in density exist. When quenched up, these differences are converted into temperature differences which relax diffusively.

6. ACKNOWLEDGEMENTS

We thank CNES and ESA for their financial supports.

7. REFERENCES

1. B. Zappoli, D. Bailly, Y. Garrabos, B. Le Neindre, P. Guenoun and D. Beysens, *Phys. Rev.* **A41,** (1990) 2264.
2. A. Onuki, H. Hao and R.A. Ferrell, *Phys. Rev.* **A41,** (1990) 2256.
   A. Onuki and R.A. Ferrell, *Physica* **A164,** (1990) 245.
3. H. Boukari, J.N. Shaumeyer, M.E. Briggs and R.W. Gammon, *Phys. Rev.* **A41,** (1990) 2260.
4. B. Zappoli (preprint, 1991).
5. K. Nitsche and J. Straub, Proceed. 6th European Symposium on Materials Sciences under Microgravity Conditions, 1986, ESA SP-256, p.109.
6. R.I. Issa, *J. Comput. Phys.* **62,** (1986) 40.
7. V. Gurfein, D. Beysens, Y. Garrabos and B. Le Neindre, (preprint, 1990).
8. J.H. Lunassek and D.S. Cannell, *Phys. Rev. Lett.* **27,** (1971) 841.
9. F. Kammoun, J.P. Astruc, D. Beysens, P. Hède and P. Guenoun (1991) to be published.
10. D.J. Tritton *"Physical Fluid Dynamics"* Oxford Publication (1988).
11. R.P. Behringer, A. Onuki and H. Meyer, *J. Low Temp. Phys.* **81,** (1990) 71.
    H. Boukari, M.E. Briggs, J.N. Shaumeyer and R.W. Gammon, *Phys. Rev. Lett.* **65,** (1990) 2654.

Apparatus and Physical Experiments in Space

EQUIPMENT FOR TECHNOLOGICAL EXPERIMENTS: PRESENT STATE AND DEVELOPMENT TENDENCIES

Igor V.Barmin and Alexander V.Egorov

SPLAV Technical Center, 9 Baikalskaya St., 107497, Moscow, Russia

ABSTRACT

The authors opinion is presented on the problem of developing production equipment for experiments to produce nonorganic and biologically active materials in space at the transfer to pilot industrial production. Advanced prototypes of equipment are considered. Requirements for the equipment being designed and space apparatus for space production are analyzed.

Keywords: *Technological Equipment, Technical data of Technological Units, System of requirements*

From 1969, when the first technological experiments in space were carried out on board the spacecraft (SC) Sojuz 6, thousands of experiments were conducted to study particularities of substance behaviour under microgravity, to produce metal, semiconductors, composite, optical, polymeric and biological materials under space conditions. The experiments were carried out using various flight vehicles: laboratory aeroplanes, high-altitude rockets, manned SV and permanent orbital stations, automatic satellites and also in drop towers. A great experience was gained in creating and operating on-board technological equipment. The analysis of the experimental results conducted in space, made possible a considerable advance in understanding the particularities of physical and chemical processes proceeding under microgravity, in development of technological processes with good prospects and at the same time showed that the quality of materials produced in space greatly depends on perfection of the on-board technological equipment and conditions for conducting experiments on board the spacecraft, first of all on the microacceleration level. Up to the present a great number of various models of technological equipment was created, from the simplest - for conducting demonstration experiments - to rather complex precision units. Table 1 lists various types of technological equipment. Of course, it is quite impossible to describe all varieties and types of equipment. Therefore by way of example we give characteristics of a number of units, operated earlier or used at the present time, prepared for work in the nearest future or being created (Table 2). Let us consider the tendencies in the development of technological equipment. In addition we`ll take into account, that the space technology and material science has reached such a development stage, that works will soon be carried out on experimental and industrial production of some types of materials. The equipment used must correspond to this purpose. Figure 1 shows a standard block diagram of the on--board complex of technological equipment (CTE), based on technological units and including a number of support systems. The energy used by the CTE can be generated by an on-board power-supply system (PSS) or by additional or fully autonomous energy sources of the CTE. The basis of the on-board PSS of the most part of the SC are solar batteries. For the PSS functioning on shadow orbit sections, the PSS must include buffer accumulator batteries. In the future, at a stage of

TABLE 1 Classification of Technological units (TU)

TU group	Classification sign	TU variety
	Purpose	For producing materials and products
	Type of energy for technological process	Electric Using energy Solar
All unit type	Recoverability	Recoverable Non-recoverable
	Automation degree	Automatic Semi-automatic (with operator participation)
	Type of working medium	Vacuum With gaseous atmosphere (reducing medium, inert atmosphere, special atmosphere) Liquid
For producing materials	Nature of materials produced	For producing non-organic materials For producing biological preparations
	Principle of technological process	For growing (heating and without heating) Electrophoretic
	Heating method	Electrothermal Exothermal Solar heating
Heating for growing	Temperature profile in working zone	Isothermic (thermostats) Gradient Universal (multizone)
	Technological process method	Volumetric solidification Directional crystallization Epitaxial Universal (using several method)
	Method of fixing sample in working zone	Crucible (ampoule) Without crucible With levitators
Electrothermal	Design of the heating assembly	Resistive Radiant heating with halogen lamps (mirror) Induction Arc Laser
Electroforetic	Realization method	With separation in free liquid medium With separation in liquid flow
	Method of electrophoresis	Isoelectric focusing Zone electrophoresis Isotachophoresis

developed production, when the PSS power consumption should be tens of kilowatts, it may be reasonable to create the PSS on the basis of reactor power-plants. It is possible to use solar concentrators to directly heat materials in furnace working zones, but this problem has not been realized because of basic difficulties. For maintaining the stationary operating conditions of the equipment, the development engineers of the CTE must solve a problem of dissipating heat from the technological units and control equipment and transferring it to radiating surfaces in the thermal conditions ensuring system (TCES) of the SC. For solving the problem of ensuring thermal conditions

Table 2 Technical data of on-board technological units

	Splav 02	IHF	MHF	GFQ	ELLI	Zona 03	Zona 04	Gallar	Crater B	AMF	Konstanta 02	AGHF	ZMF	Zona 08	Konstanta 04
Temp. Range [C]	500-1050	200-1600	200-2100	400-1500	400-1400	500-1400	480-1150	400-1300	400-1300	400-1200	<800	Max 1400	Max 1500	500-1400	600-1200
Gradient [C/cm]	up 100			110-150		10-30	20-50				<5		20-70	10-30	<2
Reloadable Samples	Yes 12	Yes	Yes	Yes	Yes	Yes 4	Yes 4-6	Yes 6	Yes 8	Yes 24	Yes 12			2	1
Vacuum Inert Gas	Yes No	Yes He	Yes Ar	Yes He/Ar	Yes Ar	Yes No	Yes No			Yes Ar	Yes	Yes Ar		Yes	Yes
Ampoule D Dim. [mm] L	20 130	40 100	20 130	10 180	30 60	30 360	30 230	33 546	56 700	20 150	41 150	22 140	28 140	40 200	85 400
Heating Rate		54 C/mm				6	No control				400 C/h			200-400 C/h	150 C/h
No.of Heat-ing Blocks	1		2		1	3x4	1x6 1x4			1	1		5	5x2	3
Heating Ele-ments/Bloc	Resist	Resist	Halogen Lamps	Resist	Halogen Lamps	Resist	Resist	Resist	Resist	Halogen Lamps	Resist	Resist	Resist	Resist	Resist
Magnetic Field	No	No	No	No	No	No	Yes Static/Rotat	No	No	No	No	No	No	Yes Rotat	No
Sample Trans Lenght [mm]	No		110		40	200	120			50	No			205	No
Sample Trans speed [mm/h]			0.06-600.0		0.06-600.0	0.1-45.0	0.2-15.0			0.06-600.0	No	0.06-600.0		0.1-10	No
Cooling Rate [C/h]	2.8; 5.6; 11.5; 22.5;			50 [C/s] Quench		8-617	Passive	12-600	12-600		22.5 C/h (800-700 C) 45 C/h else			1 - 60 C/h	5.0; 10; 25; 50; 150;
Data Dounlik	Yes	Yes	Yes	Yes	Yes	Yes	Yes	Yes	Yes	Yes	Yes			Yes	Yes
Exploitation Begining	1985	1983	1983			1990	1987	1989	1990		1991			1994	1994
Carrier	PHOTON	SL	SL	SL	SL	MIR	PHOTON	MIR	MIR	EURECA	PHOTON	SL	SL	NIKA	NIKA

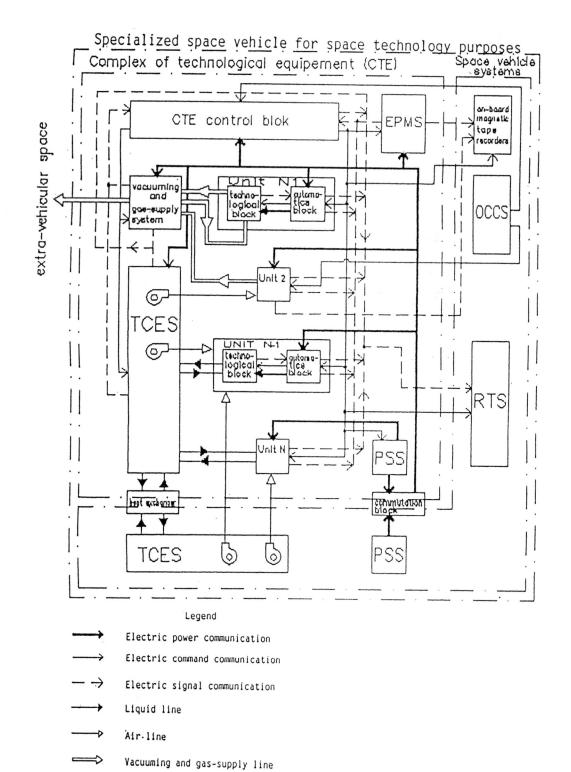

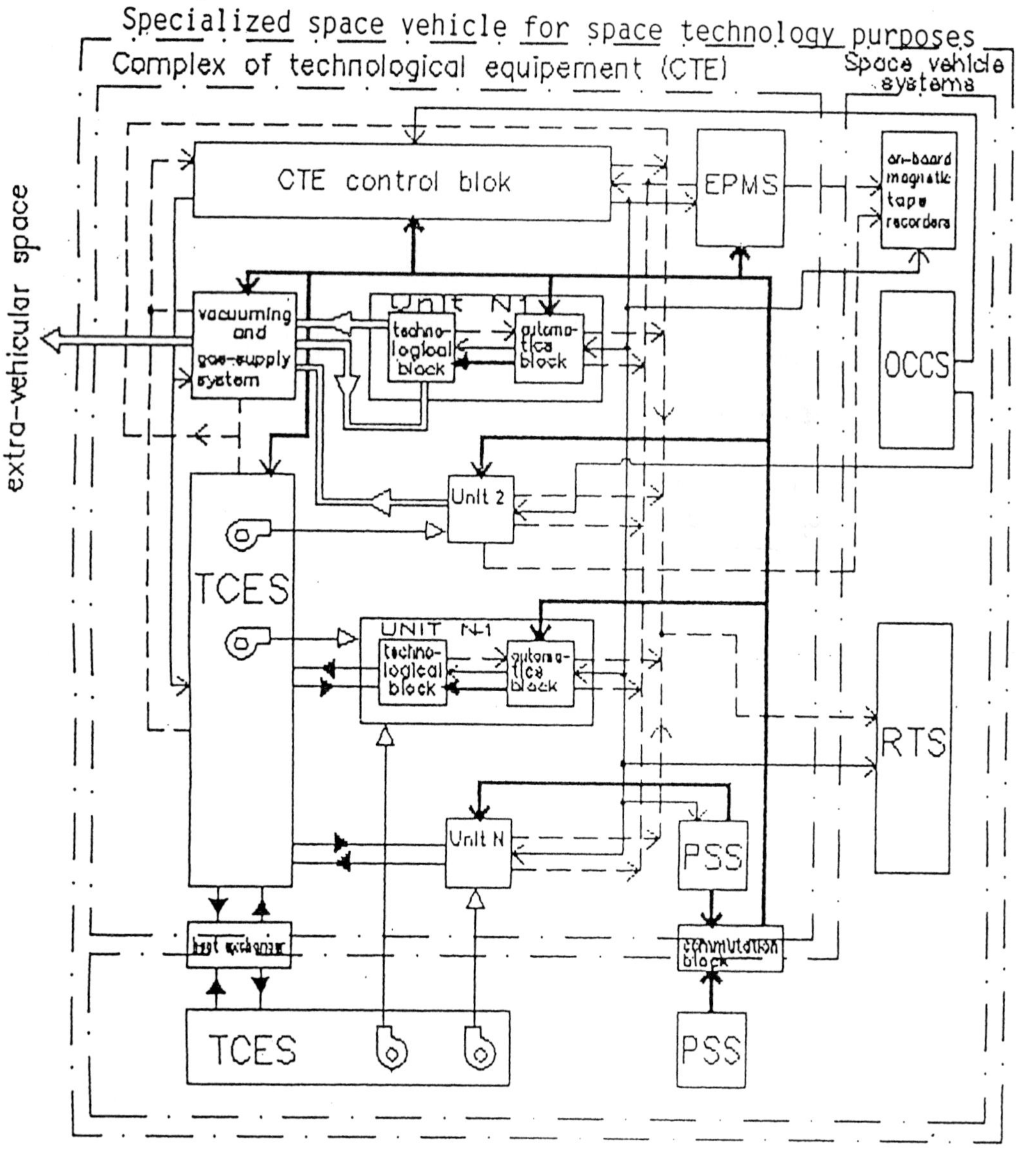

Legend

Electric power communication

Electric command communication

Electric signal communication

Liquid line

Air-line

Vacuuming and gas-supply line

Figure 1.

of specific equipment, it is necessary to take into account various factors, including: appearance, parameters, structural features, operating conditions of the equipment; appearance, parameters, structural features, operating conditions of the equipment; appearance, parameters, structural, features, operating conditions of the SC, the environment parameters such as temperature, pressure, composition and velocity of atmospheric gases circulation around the equipment. Much depends on correct configuration of the apparatus and its arrangement inside the SC. For a forced heat transfer it is possible to use gas-cooling of the SC atmosphere with the help of fans, on-board or autonomous, being part of the apparatus. For powerful (> 1 kilowatt) apparatus it is reasonable to use a liquid cooling circuit. Moreover, the main liquid circuit of the SC may be used or an autonomous liquid circuit may be provided for in the CTE. When placing technological units in unpressurized compartments or outside the SC, it is possible to ensure heat removal by directly radiating energy into space. Sometimes it is reasonable to equip the apparatus with autonomous cooling devices, some of which use the Peltier effect, or evaporation refrigerators. Much depends on a correct choice of heat insulation in the units. The most effective type of heat insulation is shield-vacuum heat insulation (SVHI). The use of vacuum allows also to ensure a long service life of the heater. For ensuring proper evacuation of the chambers of the units and for creating the required atmosphere in them, the CTE includes a vacuum and gas-supply system (VGSS). In particular case it is simply a vacuum system (VS), consisting of a set of pipes, valves, vacuum gauges, sensors and pressure relays. The use of extra-vehicular space allows on easy solution to the problem of evacuating the SVHI and the unit pressurized chambers, but since increasing the vacuum pipes diameter results in a great increase of pipe mass, and especially stop valves, a practically attainable vacuum (at d_n=20 - 50 mm) is of an order of 10^{-1} - 10^{-2} Pa. Therefore, it the technological process requires a deeper vacuum, it is necessary to include special vacuum pumps in the evacuation system or place the unit outside the SC, behind a special shield. The CTE control system is usually formed according to a hierarchic principle. The upper level in the system, when there is a possibility to interfere in the process of control from Earth, using a command radio line is a flight control center. The next level is an on-board complex control system(OCCS) of the SC, which, as a rule, is not directly connected with the process of the technological units working conditions control, but generates switch-on and switch-off commands to the CTE. The work of all the components of the CTE is coordinated by a complex control unit, which determines an order and operation sequence of all the unites and systems of the complex. Operation control of individual technological units is carried out, as a rule, with the help of individual automatic equipment units. The structural scheme of the control system may be not hierarchic, built on the basis of an on-board digital computer. A correct analyse of the experiments results is possible when trustworthy information on the parameters of the equipment operation and environment is available. With that end in view the units must be equipped with means for measuring the working conditions parameters. The environment parameters measuring system (EPMS) must ensure receiving information on a real pattern of microacceleration during the technological processes running, on a value of residual pressure in the evacuated chambers, on a pressure and temperature of gaseous atmosphere in the pressurized cabin of the SC, on a temperature and pressure of the liquid heat carrier in the cooling circuit. The measurement results must be directly transmitted to Earth with the help of a radio-telemetry system (RTS) of the SC and for then must be recorded on the on-board information recorders, such as tape recorders. The on-board technological equipment being created must satisfy a system of requirements, given in Figure 2. Let's considers some of them. The first group of requirements are those determined by special features of the technological processes. Since there is a great variety of such processes, there is a large number of requirements too. The most complex requirements to meet, with the desired accuracy, is the required law of changing the operation parameters according. The problem of using a microgravity factor determines a requirement for all drives and moving parts of the units not-to create considerable, especially dangerous low-frequency accelerations. The second group are general technical requirements. They include requirements for standardization and unification of designs, requirements, if necessary, to use cheap and non-scarce materials and furnishing parts during the equipment development. The general technical requirements also include a requirement

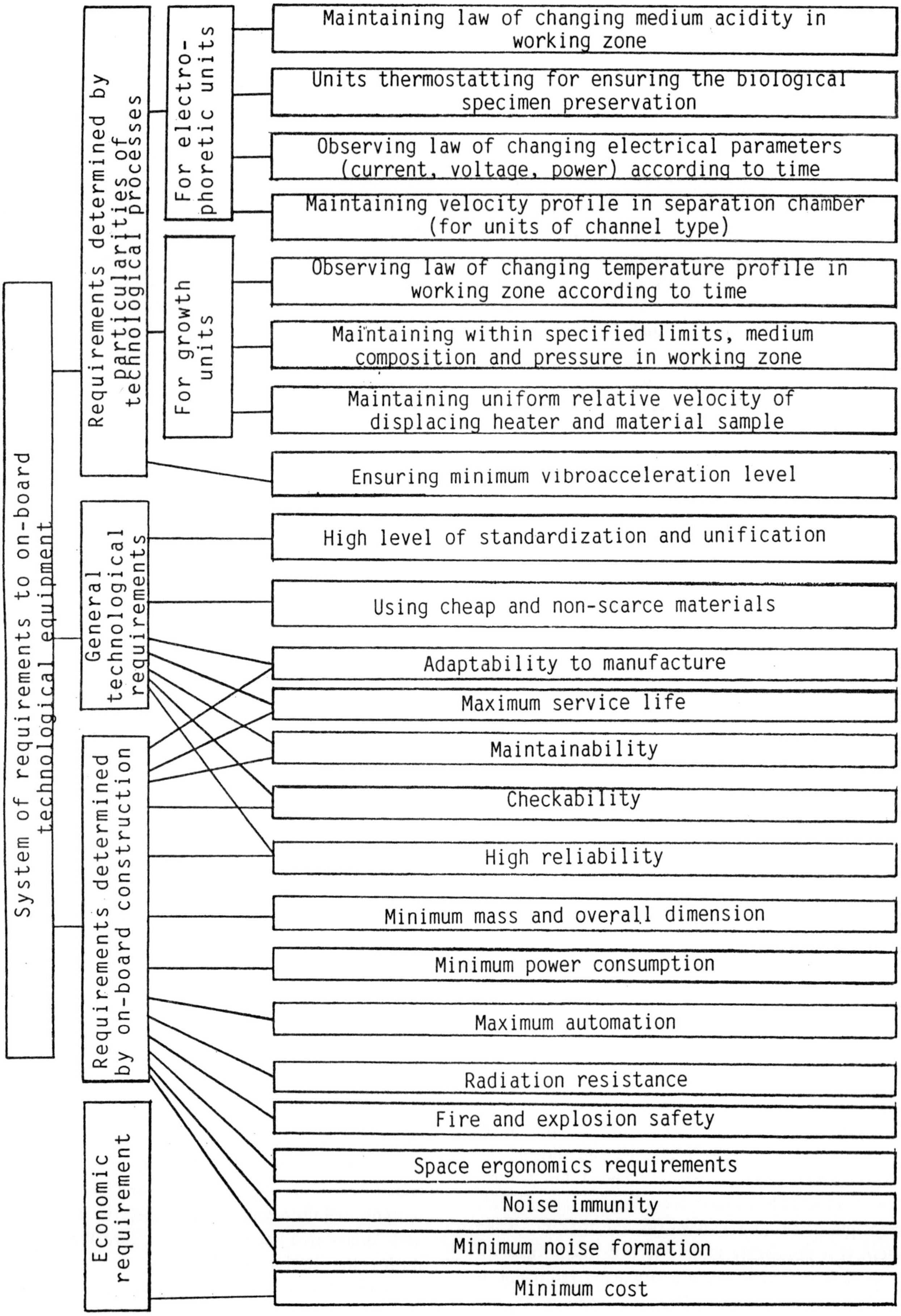
System of requirements to on-board technological equipment
Requirements determined by particularities of technological processes
For electro-phoretic units
For growth units
General technological requirements
Requirements determined by on-board construction
Economic requirement
Maintaining law of changing medium acidity in working zone
Units thermostatting for ensuring the biological specimen preservation
Observing law of changing electrical parameters (current, voltage, power) according to time
Maintaining velocity profile in separation chamber (for units of channel type)
Observing law of changing temperature profile in working zone according to time
Maintaining within specified limits, medium composition and pressure in working zone
Maintaining uniform relative velocity of displacing heater and material sample
Ensuring minimum vibroacceleration level
High level of standardization and unification
Using cheap and non-scarce materials
Adaptability to manufacture
Maximum service life
Maintainability
Checkability
High reliability
Minimum mass and overall dimension
Minimum power consumption
Maximum automation
Radiation resistance
Fire and explosion safety
Space ergonomics requirements
Noise immunity
Minimum noise formation
Minimum cost

Figure 2.

of adaptability to manufacture, a maintainability requirement and a checkability requirement, which have special features, connected with on-board construction of the equipment. The third group are requirements, determined by the on-board construction and also by particularities of the space equipment operation on an injection portion, during the orbital flight, during recovery from orbit and landing on Earth, and at the stage of delivery from the landing site to the place of post-flight research. It is necessary to specially mention the need of a specified high reliability during operation. This requirement must be satisfied even in the light of necessity of minimizing the mass and overall dimensions of the equipment being created. This is determined by a very high cost of launching a payload into orbit (thousands of dollars per 1 kg). A maximum service life requirement is determined by the same factor. A requirement of minimizing power consumption is no less important, as is described by both an energy gap on board the existing SC and by a difficulty of heat removal under space conditions. Ergonomic requirements are important, particularly in connection with ensuring the health of cosmonauts and flight safety. Finally, a composite economic requirement for equipment is a minimum cost of its creation and operation. It is justifiable to state a problem of optimization according to a minimum cost criterion. The engineers, developing the CTE, proceed from the required operation program and possibilities of the SC, which will be equipped with the CTE: the power of the PSS and TCES of the SC, the volume, given for placing the CTE equipment, the SC configuration, availability or absence of pressurized compartments. The aim is to use the SC possibilities at most. Of course, the best result may be obtained only when the technological equipment is a specially created technological SC. The problem of designing a specialized production spacecraft (SC) is a point on the agenda. Such SC must satisfy a number of special requirements. For example, their orbits must be high enough and close to a polar one to decrease overloads. The active lifetime must be long (from several months to several years). The SC must have a powerful power-generating system (> 10 kw) and a corresponding temperature-control system. It will be necessary to have an effective radio-telemetry system to get complete information on the technical processes under way and to have a possibility to interfere with their running, using a command radio line. The on-board systems must not create overloads exceeding threshold values, this limiting to same absent the technical processes which can be realized on-board the SC. Several design concepts of production SC can used. One of them provides for launching such a SC into orbit and for its return to Earth with the help of the recoverable space vehicle Space Shuttle or Buran, as envisaged for the EURECA mission. Admittedly, in my view, using such an expensive transportation system will make it difficult to get a pronounced economic effect. Another concept can be based on using cheaper, nonrecoverable launch vehicles. The SC can be launched into orbit using a onelaunch scheme. In this case it must contain a re-entry vehicle with production equipment and special recovery capsules for delivering the manufactured products. Admittedly in the last case the SC design is more complex owing to introduction of a system for transferring the products from the technological units into the recovery capsules. Examples of such spacecraft may be the SC "Nica" being created by "Foton" Design Bureau and SC "Technology", Space Biotechnological Complex and "Orbital Plant" being designed by "Salyut" Design Bureau. The SC can be also launched into orbit, using a two-launch scheme. In this case it is necessary to first launch into orbit a permanent module, equipped with powerful on-board systems of power supply, temperature control, telemetry, motion control. Then technological modules, containing a re-entry vehicle with production equipment, are sent to them for mating The re-entry vehicles may be recoverable and contain various changeable technological equipment.

REFERENCES

1. I.V. Barmin, E.I. Gorjunov, A.V. Egorov and others, *Equipment of space production* (Mashinostroenie, Moskva 1988).

2. I.V. Barmin, *Soviet space-processing programme and prospects for utilisation of Soviet achievements by foreign users* (Materials of European space conference, Paris 1991).

3. H.V.Walter and Ph. Willekens, *Material Sciences and Fluid Sciences Muli-User Facilities for Spacelab and EURECA* (Euopean Space Agency 1990).

4. *Microgravity Experiment Facilities, Instrumentation* (Dornier ,Deutsche Aerospace, march 1990).

A NEW PROGRAM FOR MICROGRAVITY EXPERIMENTATION BY MEANS OF SOUNDING ROCKETS AND TELESCIENCE

R. Monti

Univ. of Naples
Institute of Aerodynamics "U. Nobile"
P.le Tecchio, 80; I-80125 Naples Ph. +81616526 - FAX +815932044

ABSTRACT

A program for Sounding Rockets experimentation in microgravity is presented that has been conceived and implemented directly by the PI-team to best score the scientific goals and to optimize the scientific returns of the space operations. The paper shows how the experience made fits within a rational program aimed at implementing the Telescience as the operational mode for Columbus and for the microgravity platforms in general.

1. TELESCIENCE DEFINITION AND MOTIVATIONS

Telescience (TS) is a user concept that can be implemented in different disciplines, on different platforms, in support of different users located in appropriate control rooms.
The goal of Telescience is basically an experimentation procedure that sees a principal investigator (PI), confortably seated in his User Home Base (UHB) and autonomously carrying out his own experiment hosted in a facility accomodated on a space platform.
From the above definition one can recognize that Telescience poses much more severe requirements as compared with any other "productive" teleoperated system. The requirements refer to the coverage of up- and down-link transmission, to the time delays of the data and commands, and mostly, on the on-line data elaboration and display; these are dictated by the need of helping the PI on ground to take his decisions during the experiment conduct.
It is appropriate to recall and to underline the substantial difference that exists between what has been implemented, up to now, and what should be implemented in the future for microgravity Space experimentation. The partial interactive remote experimentation that already takes place in most Space-based Astronautical Telescopes (IUE, Exosat, Hubble, etc.), is much less demanding than Telescience for Microgravity in many respects, in particular in terms of Telepresence, Telediagnostics, Telecommand and Telemanipulation and of decision support equipment.
Typically on-line decisions are to be taken for complex experiments carried out in sophisticated multiuser facilities on the basis of many digital, TV and other field data (interferometry, thermography, etc.). These decisions, to be taken on the spot, are motivated by the fact that the Microgravity Environment is unique and the fluid behaviour is highly unpredicable. Therefore a User Center (or UHB) must provide Decision Support Ground Equipment (DGSE) that utilizes dedicated H/W, S/W and Data Banks to help quick decisions based on the outcome of the previous experiment steps. These requirements typically imply full (or almost full) coverage, small time delays in the transmission lines, high resolution field data and large

downlink bit rates. Furthermore one would like to be able to conduct experiments, hosted in the same facility, from different sites, in the perspective of Space Station utilization that imply long experimentation by many users from different geographical locations.

The need of Telescience is felt not only for long duration experiments and missions but also (and, in some cases, even more critically) for short duration missions (e.g. Sounding Rocket) because of the necessity to utilize the limited microgravity period at its best.

For these missions Telescience can help in converting typical look-and-see experiments, into quantitative, scientifically valid experiments. The first experiences made (Texus 23) helped identifying the requirements of the ground support equipment and of the experiment control logic. Even if implemented in a different scenario, the Telescience experience can be transferred to other similar sounding rockets or to other platforms experimentation. In fact all the basic ingredients for a fully implemented system have been realized: Telescience Work Station, up- and down-link from/to the platform, RX/TX equipment, management system for data and commands, and, mainly, ground decision support equipment, Figure 1.

In particular the direct involvement of the Principal Investigator in his experiment in short duration unmanned missions ensured the following relevant advantages:

1. a more rational and efficient conduct of the experiment,
2. possibility of on-line quick analyses and experiment assessment,
3. utilization of a number of complex ground facilities,
4. best utilization of the available microgravity time.

2. TRENDS IN TELESCIENCE PROGRAMS

Telescience is supposed to be the basic operative mode for both the Columbus Free Flyer and for the Columbus Attached Laboratory. The evolution towards the full Telescience calls for a gradual implementation of many aspects that depend on the platforms, Ground Infrastructures, the Telelinks and the experiment oriented Ground Support equipment. The trend is summarized in a somehow chronological order in Table 1. Different combinations can be realized according to the flight opportunities, to the experimenters needs and to the available telecommunication equipment.

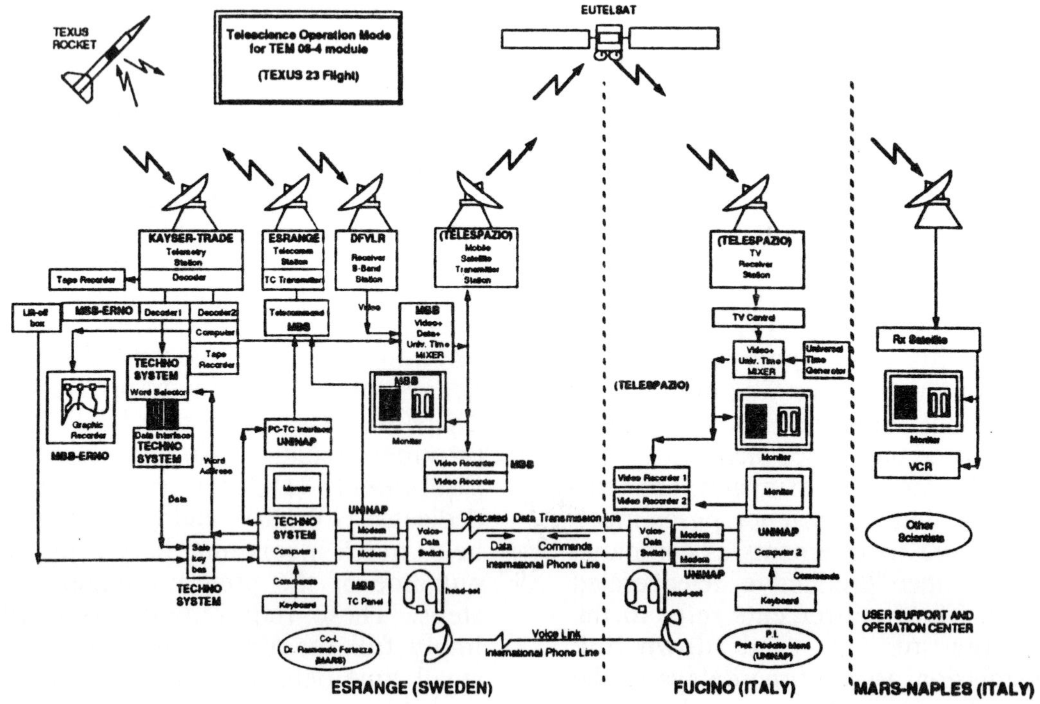

Fig. 1 - Telescience arrangement for the Texus 23 Experiment

Apart from the obvious impact on infrastructures, on the Facilities and on the PS skill and operations, it is important to understand how Telescience will impact on the PI himself.
In fact the deep involvement and the active participation of the PI during the experiment require specific experimental expertise, not otherwise needed if preprogramming the experiment or if relying on the Payload Specialist to perform it.
This may create difficulties to proposers of experiments that are not necessarily experimentalists and therefore an effort should be made to implement really user friendly MMI's. However the PI training will become a key issue for the Space experiment success. It must be considered that PI training "forces" the PI to spend time sitting at his Telescience Work Station to better define the achievable scientific objectives and to identify the best operational experimental procedures.
In conclusion it appears that in order to get the maximum benefit from Telescience also the PI role must evolve together with the evolving scenario.

As shown in Table 1 experience was made with Sounding Rockets that resulted in obtaining scientific results not otherwise obtainable and to get acquainted with the H/W, the S/W and similar operational problems to be encountered in long duration manned or unmanned missions. The two first successful experiences made with sounding rockets are illustrated in the next Section.

3. THE SOUNDING ROCKET EXPERIENCES

The first experience was made by the author (Co-I Ing. Fortezza) and his team with the Texus 23 mission that flew on a Sounding Rocket launched from ESRANGE in Kiruna on November 23, 1989. The purpose of this experiment was the identification of the conditions at which the Marangoni flow changes into an oscillatory regime in a liquid bridge (simulating a floating zone) established between two circular disks at different heating rates through the disks.

TABLE 1 - TRENDS IN THE TELESCIENCE PREPARATORY PROGRAM

	Yesterday	Today	Tomorrow	Future Columbus Scenario
On board Equipment	Existing single user Facility	Dedicated single user Facility	Multiuser Facilities	Dedicated Multiuser Facilities
Platform	Sounding Rockets	Extended duration Sounding Rockets	- Piggy Back Ariane - Small Satellites	Columbus Platforms (Attached Lab, Free Flyer)
Telecommands	Telephone + Commercial Satellites	Experimental Satellites	Dedicated Satellites	TDRS with IOC
Ground Infrastructure	Satellite Control Station	User Support Center	User Home base	- Disseminated User Home bases - Simultaneous Operations
Ground Support Equipment	Data/image display	Data and image elaboration and display	On-line computing for decision Support	- Data Base network - High Speed Computers for experimentation assistance - Teleconference

Measurements were taken on the thermo-fluidynamic flow field in a liquid (Silicon Oil) bridge between two coaxial disks (D=1.8 [cm]). The liquid motion was generated by the surface tension unbalance (Marangoni flow) induced by a temperature difference across the bridge. The flow was visualized by tracers; temperatures and heat fluxes across the bridge were measured and different temperature ramps were applied to the disks (the ramp was selected by ground commands) (Figure 2).

The PI at the Fucino Centre in Italy, received the TV images of the flow field inside the liquid bridge through TX-RX equipment and through a commercial TV satellite; data from the rocket, through the Ground console and through TX-RX modems were received and displayed by the Fucino Computer to the PI (See Figure 1). The PI on the basis of this visual and Digital information and with the help of an appropriate software, took decisions at a number of branch points, during the experiment.

More specifically, he was able to select the temperature time profile of the two disks (that hold the liquid bridge) on the basis of the onset (or of the decay) of the Marangoni oscillations that occurred at certain times during the run. The very same sequence of operations could have been performed by the co-PI sitting at the Kiruna Control Room.

Other control levels were implemented ensuring the experiment to be executed in all the foreseable contingencies (i.e. problems on the tele-link from/to rocket to ESRANGE, on the link from ESRANGE to Fucino for TV images or on the data/command link from/to Fucino to/from ESRANGE). An effort was made to facilitate the experiment conduct by realizing a user friendly Man Machine Interface (MMI) (As shown in Figure 2).

A high-performance Personal Computer has been connected to the data channel and a program has been developed to present these data in graphic and digital form in real time.

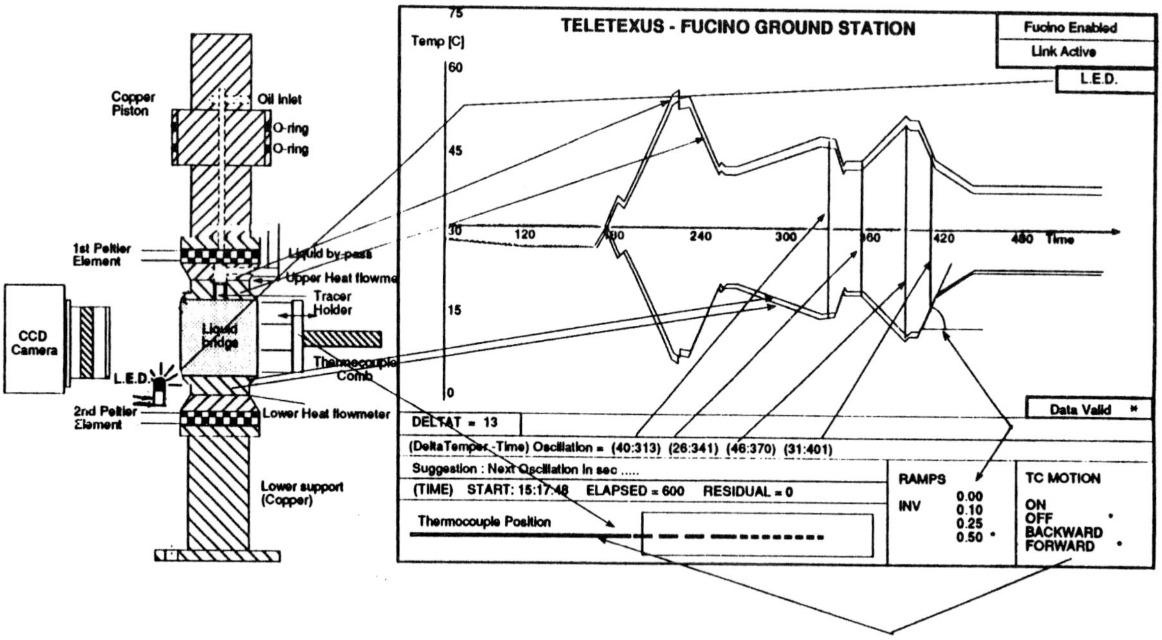

Fig. 2 - Functional Sketch of the TEM 06-04 Flight Cell and Typical temperature time profiles obtained during the remote control tests

The down-link data flow, transmitted from the rocket using PCM format, are received by the tracking station, and de-multiplexed using a word selector connected directly to the Control Work-Station (CWS) located in the Control Room in Kiruna. Using the same CWS the co-PI is able to send the commands to the module through an interface by the tele-command. The commands are activated using a single-key function for a quick interaction; they are filtered using a time-function matrix to prevent any wrong command.

The commands selected are shown on a screen window and are monitored as Facility Status (feedback link for the command activation is not available on the module). The reception of the commands to the module can be detected by observing the system response.

Another screen window (lower left) is used for the animation of the thermocouples comb motion (Figure 2).

The floating zone silhouette is constructed on the screen during the first phase of the flight run. The free surface is drawn by the computer at the position of the thermocouple tip at the instant of time the tip enters the liquid bridge (this location is detected by the thermocouple signal). This information is necessary for the correct positioning of the thermocouples just beneath the liquid surface where the Marangoni effects are dominant.

Other useful data for the experiment conduct are monitored in the central part of the screen (lift-off time, elapsed and residual time, current temperature difference across the liquid bridge). The upper part of the monitor is dedicated to the temperature plot during the experiment.

An essential feature of the software is the possibility to help the investigator during the run by suggestions based on the analysis of the data obtained during previous phases of the experiment.

The organization and the experiment procedure reproduce, in a small scale, the possible scenario for Telescience activities that will, in perspective, be carried out on permanent platforms from User Support and operation Centers.

The Teletexus experiment was 100% successful both in proving the benefit of telescience in these types of experimentation and in the achievement of scientific results. One can claim that quantitative results have been obtained at the place of qualitative experiments when performed by a similar equipment but in a preprogrammed mode.

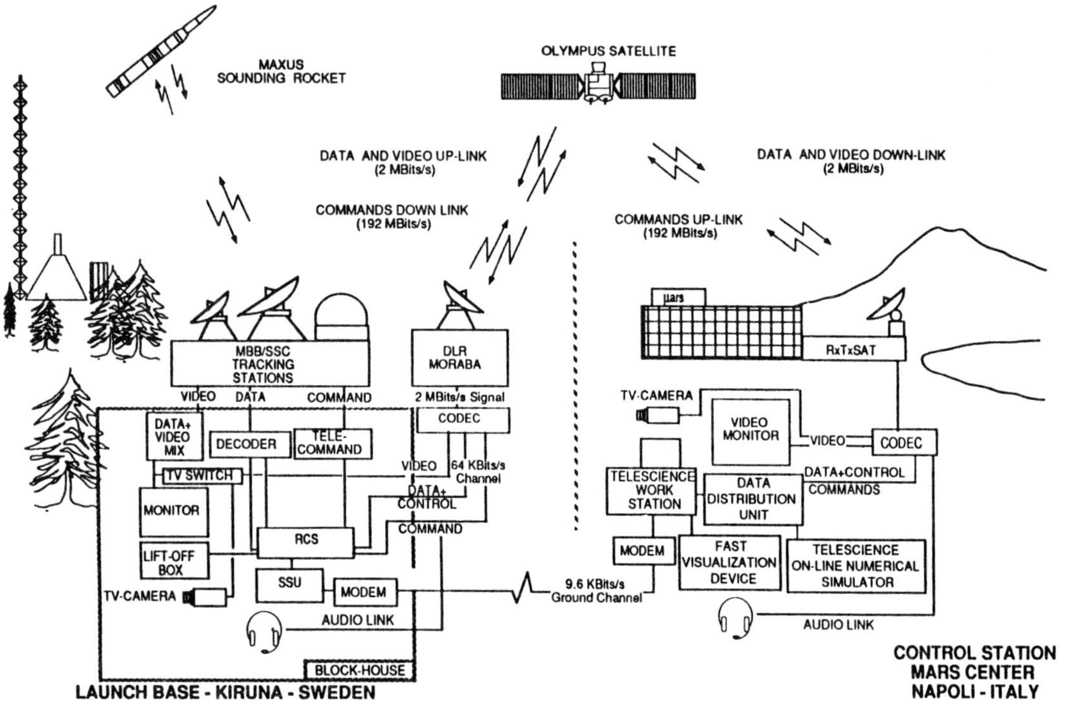

Fig. 3 - Telescience links for Maxus 1 Mission

The second step towards a fully implemented Telescience was made with the project Telemaxus that was aimed at testing a Telescience Service for ESA users that fly their experiments on Texus and/or Maxus. The system layout is shown in Figure 3. The possibility exists of two experiments being accomodated on the same Sounding Rocket driven from two separate Centers (in Europe). The main differences with the previous Texus 23 experiment can be grouped into two categories:

I - <u>Experiment cell</u>
- The liquid bridge has a larger diameter (2 [cm])
- It is possible to change the bridge length ($1.5 \leq L \leq 2.5$ [cm])
- It is possible to visualize the tracers velocity at the liquid-air interface

II - <u>Telescience system</u>
- The experimenter sits in his Laboratory (MARS Center in Naples)
- The data, TV and commands all follow the same route (Olympus satellite)
- The TV images are compressed together with voice and data into a low cost 2 Mbits/s channel
- The experimenter is able to check all the links from his Home basis
- The experimenter is able to utilize a powerful Ground Equipment for Telescience Decision Support (GETDS) (Figure 4)
- The flight data was elaborated in real time and presented to the PI for decision on the subsequent experiment step

The Telemaxus project was also 100% successful as far as the Telescience aspects were concerned. In fact during the entire mission data and commands were exchanged between the Facility and the TWS in Naples. Unfortunately something went wrong with the rocket propulsion system that made the Payload tumble so no microgravity conditions were reached. Consequently the liquid Bridge was not established. The PI however simulated the entire experimental conditions by setting the three temperature ramps, the three zone lengths and by moving the thermocouples forward and backward (Figure 5).
As result the choices made on the H/W and S/W proved to be the right ones. The same equipment (on board and on ground) will be used for the MAXUS reflight in 1992.

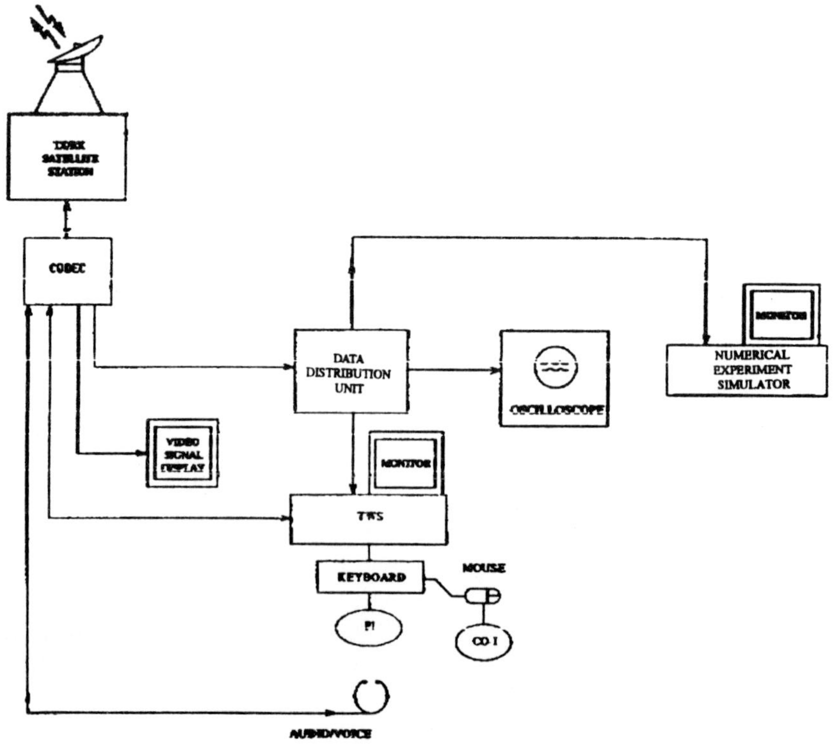

Fig. 4 - Ground Equipment for Telescience and Decision Support (GETDS)

4. CONCLUSIONS

The experience accumulated with Texus 23 and MAXUS 1 and the experiments performed show that the scientific return in microgravity short duration flight is strongly ameliorated using Telescience approach. In particular the execution of the experiment performed directly from the User Home Laboratory has offered the advantages of avoiding the interruption of the research activities caused by the long lasting presence at the Launch base, of utilizing the equipment that is available in the Lab to elaborate in real time the experimental data, of receiving the support of the entire research team available at home, and finally, of performing the experiment in a more comfortable environment. The success of the Telescience is however guaranteed only if the experimenter team is directly involved in all the preliminary phases of the operations. In fact only the experimenters team can take in due consideration the scientific/operational requirements and is able to reduce the risks related to the remote interactive conduct of the experiment by taking advantage of the experience acquired in the past.

The technical solutions found and the H/W and S/W implemented and developed represent an important experience available for any further telescience activity for sounding rocket missions as well as for orbiting platforms. Based on this achievements, a Telescience Service is now available for the European microgravity community.

5. REFERENCES

1. R. Monti, R. Fortezza, G. Capuano: *System and technologies utilized in the first Telescience Experiment during Texus 23 Mission*, 41st Congress of the International Astronautical Federation - Dresden, October 6-10, 1990

2. R. Monti, R. Fortezza: *The scientific results of the experiment on oscillatory Marangoni flow performed in Telescience on Texus 23* - Presented at the same Congress

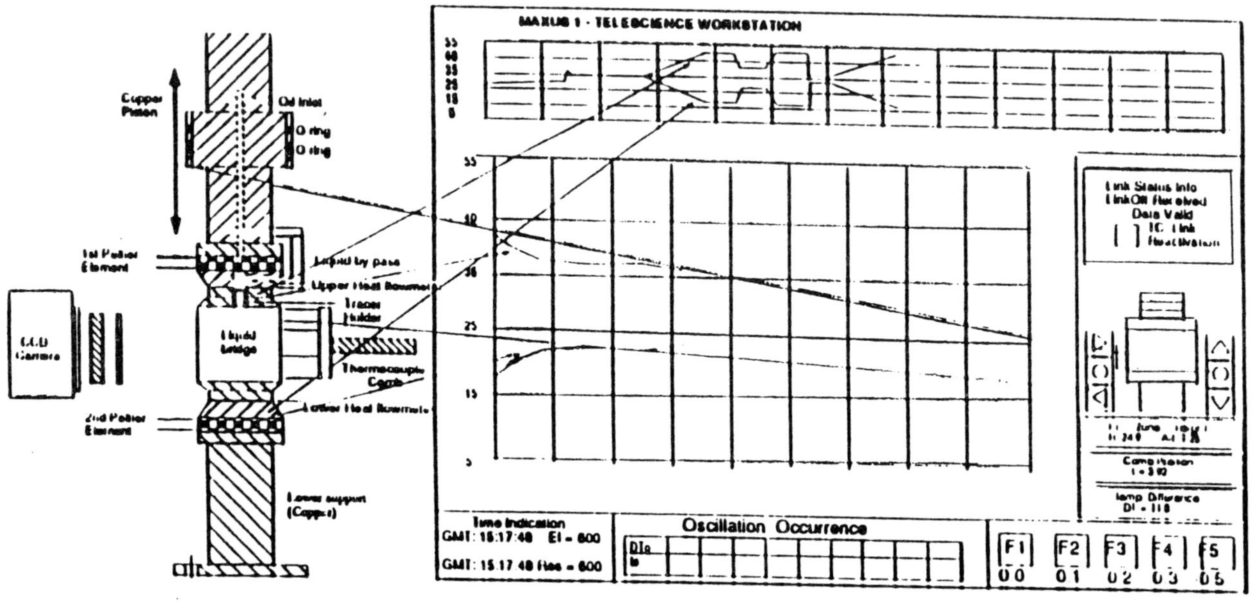

Fig. 5 - Functional Sketch of the TEM 06-04 Flight Cell and of the Control Screen for the MAXUS Mission

MOMO - MORPHOLOGICAL TRANSITION AND MODELL SUBSTANCES

K. Leonartz

ACCESS e. V., Intzestraße 5, D-5100 Aachen, Germany

ABSTRACT

Scientific goal of the MOMO project is the investigation of the convective stability of the solidification front in binary mixtures. The parameters are: concentration, temperature gradient, solidification velocity and gravity level. The stability model of Coriell, et al. (Ref. 1) will be prooved quantitatively. The interaction of the temperature and concentration field with the solidification front will be described. In order to enable in situ observation of the processes, transparent organic materials which solidify like metals will be used, e. g. Succinonitrile-Acetone. During the IML-2 Mission, scheduled for 1994, experiments will be carried out in the NIZEMI - Slow Rotating Centrifuge - and the BDPU - Bubble, Drop and Particle Unit.

Keywords: Model Substances, Succinonitrile-Acetone, Convection, Morphology, Stability, Centrifuge

1. INTRODUCTION

The morphology of the solidification front during directional solidification has been extensively investigated during more than three decades. Classical result is the morphological stability criterion G/v - temperature gradient versus solidification front velocity - by Tiller et al. (Ref.2) and the stability analysis by Mullins, Sekerka (Ref. 3, 4).

Nowadays there is still large interest in this matter, since directional solidification is an important processing technique for advanced engineering components. It is desirable to find the correlation between the microstructure and thus the properties of the material and the processing conditions.

Coriell, et al. have developed a numerical model which in addition to the classical model takes density changes and convection into account (Ref. 1, 5). The results of these calculations can be summarised in a stability diagram, see fig.1. While the G/v criterion, represented by the graph with negative slope, predicts that the solid/liquid interface is stable if the solidification parameters are on the left hand side of the graph, the theory of Coriell et al. predicts an unstable regime above the curved graph which is controlled by convection. Convection occurs under the force of the Earth gravity field if density gradients exist. Thus the influence of gravitation on the stability of solidification fronts is obvious.

A system is considered for which the temperature gradient alone would cause a negative density gradient and the solute gradient alone would cause a positive density gradient along the sample axis.

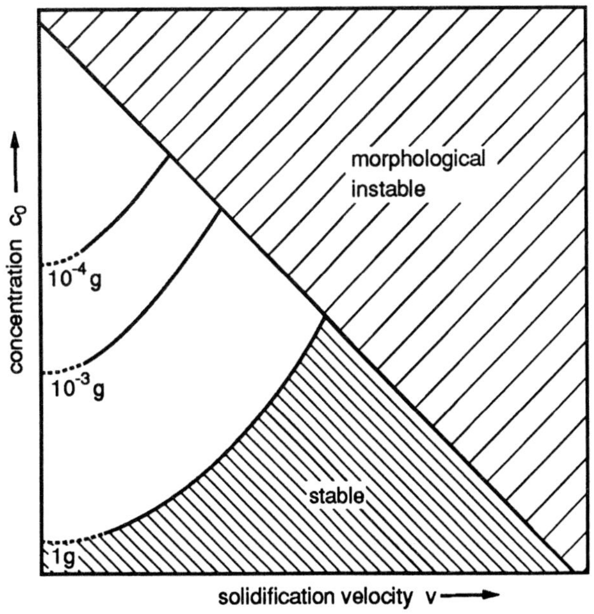

Figure 1: Sketch of a Stability diagram for a vertical directional solidification according to the theory of Coriell et al. (Ref. 1). The velocity is in the order of µm/s and the concentration in the order of 10^{-2} wt%.

2. SCIENTIFIC GOAL

Scientific goal of the MOMO project is the investigation of the convective stability of a planar solidification front during directional solidification of a binary mixture. They are in detail:

- a quantitative test of the stability model of Coriell et al., especially by variation of the imposed acceleration level;
- a morphological description of the planar-cellular transition in dependence of the parameters concentration c0, temperature gradient G, solidification velocity v and a variable gravity (acceleration) level g as driving force for instabilities;
- investigation of the interdependence between the temperature and concentration fields in the melt in front of the advancing solidification front and its shape. The correlation between the time dependent change in the temperature and concentration fields and the morphology of the solid/liquid interface will be investigated, (fig. 2).

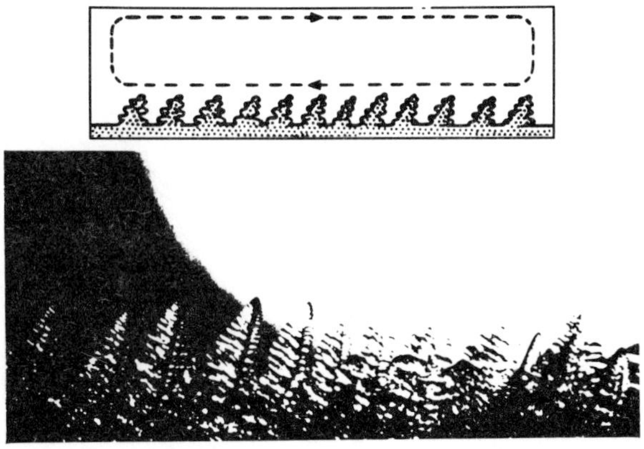

Figure 2: Objective of the MOMO project is the investigation of planar and cellular solidification patterns. Nevertheless a descriptive example is the influence of a change in the temperature and concentration fields due to convection on a dendritic solidification front. The solidification direction is tilted against gravity and temperature gradient direction (Ref.6).

3. SUBSTANCES

In order to achieve the scientific goals in situ observation of the solidification process is necessary. By using transparent organic materials which solidify like metals this prerequisite can be fulfilled. Analogy experiments for the solidification of metals with transparent substances have been used for some decades. As examples the work of Jackson and Hunt (Ref. 7) and Podolinsky (Ref. 8, 9) are mentioned. For the work here outlined the Succinonitrile-Acetone system will be used, for which a good database exits.

4. μ-g EXPERIMENTS

During the Space Shuttle mission IML-2 scheduled for 1994 μ-g experiments will be carried out in the NIZEMI-Slow Rotating Centrifuge -and the BDPU-Bubble Drop and Particle Unit- to supplement the 1-g research.

5. BDPU - BUBBLE, DROP AND PARTICLE UNIT

The BDPU (Ref. 10) is a facility for fluid dynamic experiments. It is scheduled for the IML-2 mission. The BDPU permits investigations on:

- bubble, drop and particle behaviour in fluids under the influence of temperature and concentration gradients and electric fields,
- formation and dynamics of solidification fronts,
- interaction between inclusions and solidification fronts,
- flow phenomena in general: e. g. Marangoni convection,
- electrolysis,
- crystal growth from solutions (nucleation, growth),
- crystal growth in transparent systems.

Basic concept

The BDPU consists of an experiment dedicated fluid cell which can be exchanged by the pay-load specialist, (fig.3). Investigations can be conducted by the following means: direct visualisation, two orthogonal directions, point diffraction interferometry, thermography, Schlieren optics. The following modes of illumination can be chosen: monochromatic (laser), diffuse background illumination, sheet illumination. The fluid can be homogenized by mechanical or acoustic stirring. Bubbles or drops can be generated by an injection system. Thermal gradients or electric fields can be applied as external stimuli.

Experiments are automatically controlled by a microprocessor; the payload specialist can in addition adjust the process and modify the time-temperature profile on the basis of the displayed data to optimize the process. Real time monitoring (passive) from the ground of experiment parameters and optical and infrared images and voice communication link is possible. The specifications are summarized in table 1.

The experiment series called MOMO-BDPU in the frame-work of the outlined project is prepared in cooperation with B. Billia and H. Jamgotchian, Laboratoire de Physique Cristalline, Marseille, France, S. R. Coriell, National Institute of Standards and Technology, Gaithersburg, USA, and S. Rex, ACCESS e. V., Aachen, Germany. Additional features of these space experiments are the development of cells during solidification and the cellular-dendritic transition.

TABLE 1 Technical and operational specifications

dimension of the BDPU	34 panel units (1 PU = 4.5 mm)
weight	approx. 180 kg
stowage requirements	60 kg / 100 litres
power requirements	400 W average
data transmission	2.1 Mb/s HRM for video 5 Kb/s via the Remote Data Acquisition Unit
observation system	1 CCD camera 1 thermo camera 1 cine camera interferometer with CCD camera field of view 6 cm x 4 cm
illumination	diffuse background illumination, sheet illumination or illumination by laser for interferometry
fluid cell (maximum)	7 x 7 x 5 cm^3 inside dimensions: 6 x 4 x 4 cm^3
injection, extraction system	syringes with positioning of the injection needle, monitoring of the injected volume
stirring	acoustic, mechanical
isothermal or gradient operation	between 0 °C and 120 °C maximum gradient across test cell: 40 K
sensors	temperature: thermistors (up to 10) and thermocamera; the electrical interface for pressure and electrical conductivity measurements is available
electric field to which the test cell can be subjected	28 VDC (2 A)

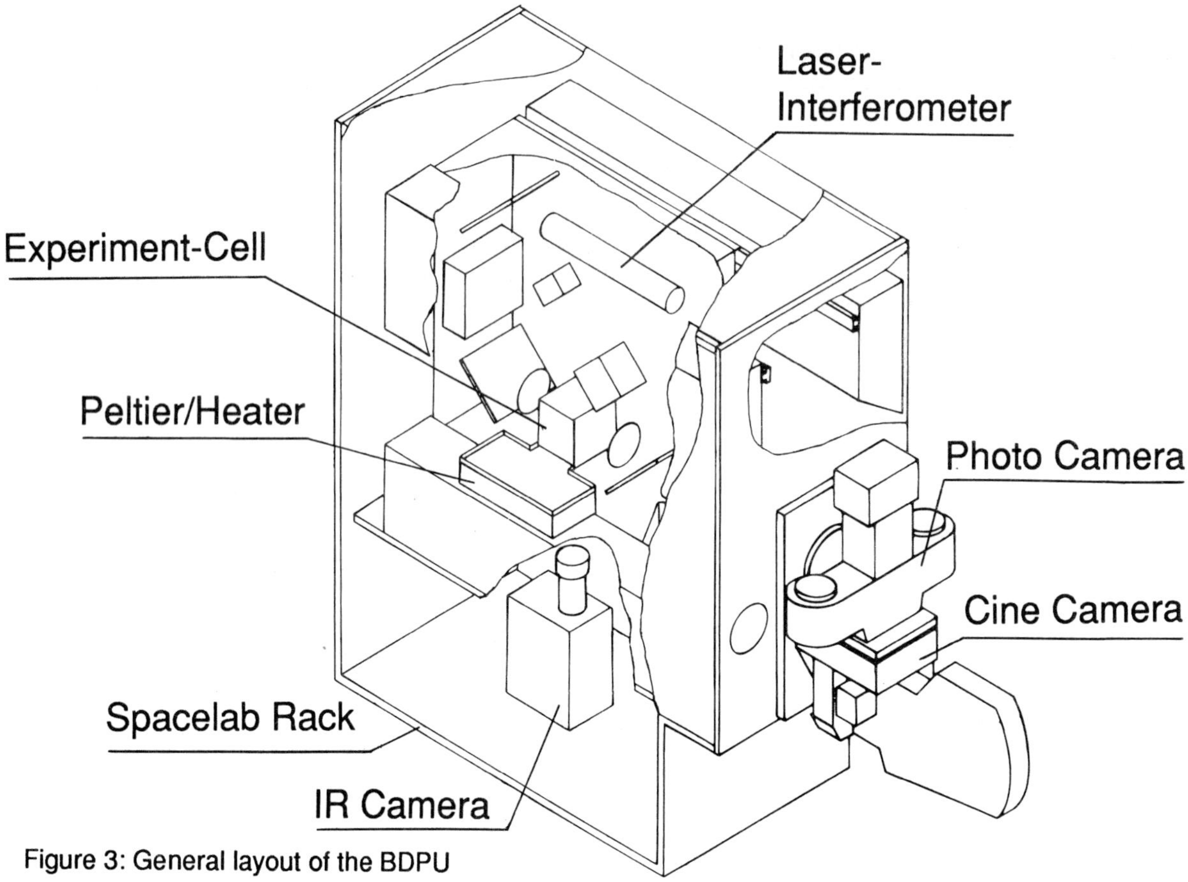

Figure 3: General layout of the BDPU

6. NIZEMI - SLOW ROTATING CENTRIFUGE

The NIZEMI (Ref. 11) is a Spacelab facility for optical investigations of small biological and non-biological specimens under variable accelerations from 10^{-3} up to 1.5 g. The facility is already available as a laboratory model, the flight model is scheduled for the IML-2 and E-1 missions.
NIZEMI will permit investigations on small biological systems like cells, plant seedings, protozoa, fungi and transparent non-biological systems like transparent model substances for metals, crystals and chemical systems.
While on Earth only accelerations of minimally 1 g can be obtained, in space the full range, beginning with µ-g can be investigated.

Basic Concept
NIZEMI consists of two functional parts: a microscope and a macro observation unit (fig. 4, table 2). The micro unit will have the full performance of a scientific microscope. Almost all functions are remote controlled. Normally only one unit should be used for investigations at one time. If the microscope is used, the stage of the macro unit will serve as an unbalance compensation. In case the macro observation is used, stage movement is not required. The prepared specimens can easily be positioned on the stage by a crew member. Visual observation is possible via onboard monitoring, onboard videorecording and optional video downlink. The cuvette temperature is automatically controlled. Within the standard interface, individual experiment containers can be customized.

The experiment series called MONI on NIZEMI in the frame-work of the outlined project are prepared in cooperation with S. R. Coriell, NIST, Gaithersburg, USA and S. Rex, ACCESS e. V., Aachen, Germany.

7. ACKNOWLEDGEMENT

I am thankful to ESA and DARA for providing the flight opportunities. To them and Dornier I am thankful for providing the facility descriptions and some pictures.

TABLE 2 Technical and operational specifications

Max. rotation speed	:	120 rpm	
Acceleration range	:	10^{-3} g - 1.5 g	
Experiment module (NEM)	:	(H) 612 x (W) 442 x (D) 612 mm^3	
Control module (NCM)	:	(H) 350 x (W) 442 x (D) 400 mm^3	
Data transfer	:	serial RAU interface, 28 words / 10 s, TV chanel	
Mass	:	115 kg	
Power consumption	:	260 W (peak) 142 W (mean)	
Sample temperature	:	18 °C - 37 °C for biological samples, app. 20 °C - 80 °C for model substances	
		Microscope	**Macroscope**
Field of scan	:	10 x 10 mm^2	
Field of view	:	depends on magnification	Ø 35 mm
Illumination	:	transmitted light	incident or transmitted light
Observation systems	:	- CCD-camera - Photo-camera (optional not on IML-2) - Zeiss standard objectives - different methods for contrast enhancement	- CCD-camera

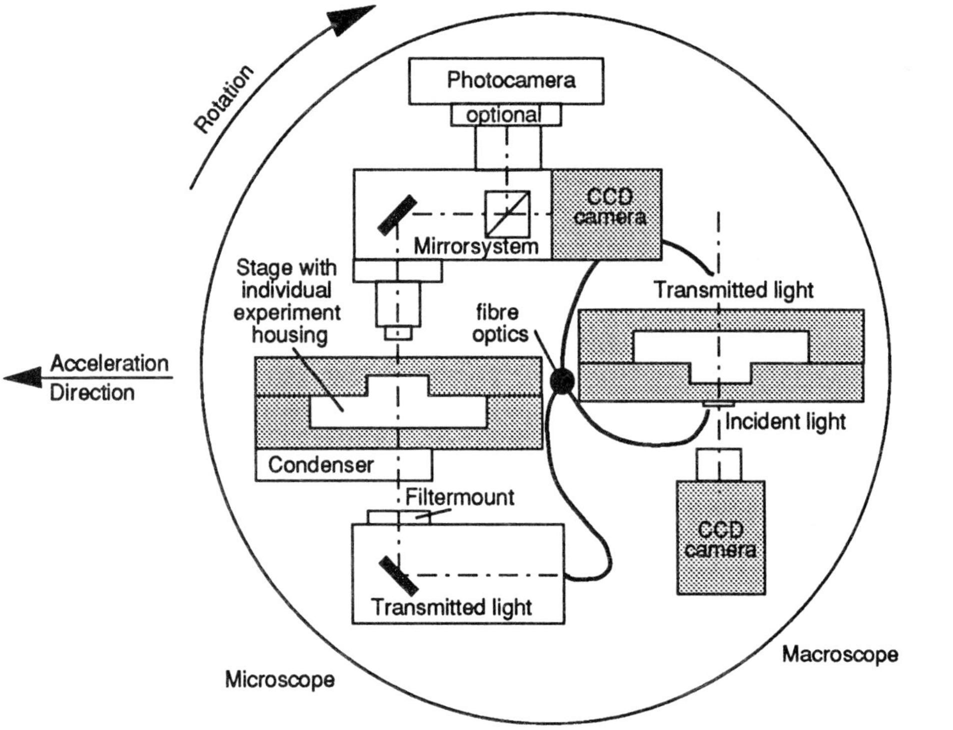

Figure 4: NIZEMI - top view of the principal set up - samples will be solidified against acceleration direction

8. REFERENCES

1. S.R. Coriell, M. R. Cordes, W. J. Boettinger and R. F. Sekerka, *J. Crystal Growth* 49 (1980) 13.

2. W. A. Tiller, K. A. Jackson, J. W. Rutter and B. Chalmers, *Acta Met.* 1 (1953) 428.

3. W.W. Mullins and R.F. Sekerka, *J.Applied Physis,* 35, (1964), 444.

4. W. Kurz and D. J. Fischer, Fundamentals of Solidification, *Trans Tech Publications* (1984).

5. R. J. Schäfer and S. R. Coriell, *Met. Trans. A 15 A* (1984) 2109.

6. A. Ecker, Thesis, Rheinisch-Westfälische Technische Hochschule Aachen (1985)58.

7. J. D. Hunt and K. A. Jackson, *Trans. Met. Soc. AIME,* 236 (1966) 1129.

8. V. V. Podolinsky, *J. Crystal Growth* 46 (1979) 511.

9. V. V. Podolinsky, Y. and N. Taran and V. G. Drykin, *J. Crystal Growth,* 96 (1989) 445.

10. H.U. Walter and Ph. Willeke, in: Spacelab and Eureca Materials Science and Fluid S*cience Experiment Facilities,* ESA (1990) 8.

11. DARA *Data sheet* - NIZEMI (1991).

SPACECRAFT FOR MATERIALS PRODUCTION

Dr. Anfisa E. Kazakova
The chief of designing sector of CSDB

Central Specialized Design Bureau & "Photon" Design Bureau
18, Pskovskaya st., Samara, 443 009, USSR

ABSTRACT

The paper presents information on the unmanned spacecraft which are designed for space materials and biopreparations manufacturing. The paper gives design data and technical characteristics of the spacecraft as well as environmental conditions inside the spacecraft during the experiments. The paper shows the spacecraft flight pattern and its operational features. The paper presents also information national economy oriented spacecraft which might carry out some additional experiments with the equipment to be installed both inside landing capsule and outside it in special scientific instrument containers mounted on the landing capsule surface (in this case experiments are carried out in outer space). The containers are to be delivered back to the ground together with the landing capsule. This information concerns the spacecraft being in service as well as the spacecraft being developed.

INTRODUCTION

The physical and chemical researches in hydrodynamics, heat and mass transfer, phase transition and other phenomena in various kinds of media under weightlessness conditions are providing a scientific and technical base for materials production in space for a number of sciences, industries and public health.

This is a fundamental direction of research, however it has a number of particular problems to be solved by manufactures of spacecraft. One of them is to minimize the microacceleration level.

The realization of a research programme in material science, biotechnology, technology development and their direct application to experimental and industrial production of materials under unique space conditions defined spacecraft particular requirements to solve these problems.

The requirements are as follows:

- providing enough space and weight reserve for scientific hardware;
- providing minimal g-loads in a technological equipment zone during the experiments;
- providing ample power supply;
- providing stable thermal conditions inside the technological equipment capsule under increased heat discharge during the experiments;
- providing the test equipment communication with the outer space vacuum;
- providing mission control and telemetry monitoring;
- providing specified inner conditions during the materials' delivery back to the Earth.

In addition, certain conditions during all preparations to experiments and after landing should be guaranteed.

Some information on spacecrafts for meeting the above requirements is presented below.

THE PHOTON SPACECRAFT

It was built (and has been in use since 1985) to verify theoretical assumptions concerning the manufacture of materials with improved properties, develop the structure of onboard technological equipment and study conditions of experiments on the first stage of space technology development.

The general view of the Photon spacecraft is shown in Figure 1.

It consists of the following modules:

- a descent vehicle (DV) which contains scientific hardware, a part of support equipment and a landing system;
- an instrument module containig control

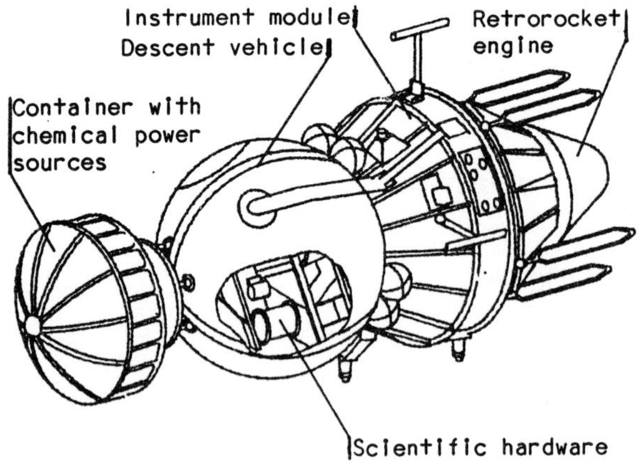

Figure 1: The general view of the Photon spacecraft

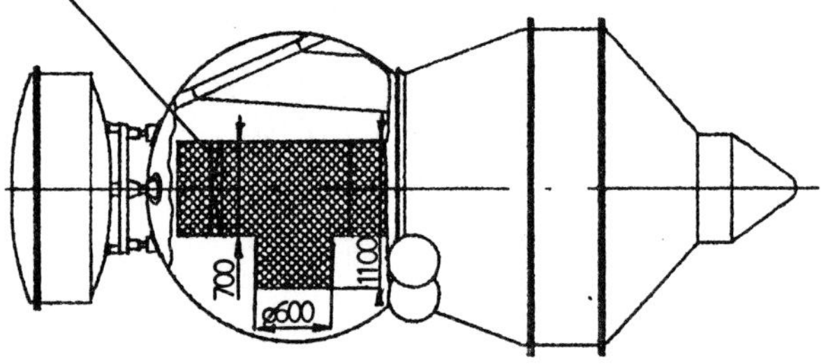

Figure 2: Photon's scientific payload zone

equipment, a telemetry system, attitude (orientation and stabilization) control instruments, a temperature control system and a power supply system. In the module's lower cone the retrorocket engine is installed;

- a container with chemical power sources.

Outside the spacecraft are mounted various aerials, sensors and cylinders with supply of compressed gas for the attitude (orientation and stabilization) control microjets.

The descent vehicle is a sphere with a diameter of 2.3 m.

The central part of DV has a space of 4.7 m^3 to accomodate recoverable scientific hardware up to 700 kg weight (Figure 3).

The vehicle's load bearing structure enables the scientific hardware to be installed onboard each particular spacecraft with no substantial modifications. The payload is positioned about the DV's center-of-gravity so as to minimize the g-load level. The

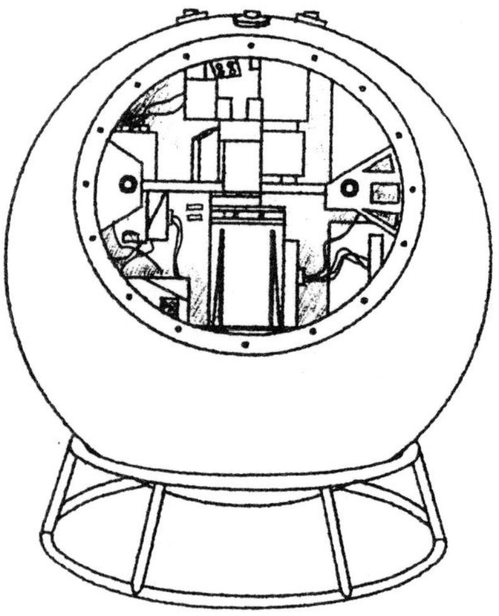

Figure 3: The Photon's descent vehicle with the scientific payload installed (the DV's cap is removed)

descent vehicle has an outer thermal protection providing the designed thermal conditions inside during descent when the temperature on the DV's surface rises up to 2200°C. The average dayly power consumption of the technological units is no less than 400 W. It is possible to increase the power consumtion up to 700 W for up to 1.5 hours. The power supply system uses chemical power sourses (27±4 V voltage). Maximal current is 25 A in stable mode and 45 A in pulse mode (for no longer than 200 msec). The source capacity enables 16 day's work of technological payload (under spacecraft average dayly power consumption of 400 W).

The inner DV's atmosphere temperature is kept during the normal operation within a range of 10-35°C when the scientific payload heat release ranges from 100 to 450 W. The temperature stability inside the descent vehicle under such widespread payload heat discharge is carried out by means of a temperature control system and by selecting certain lighting conditions in orbit when conducting the experiments, that's to say by selecting the spacecraft's launch time and launch date.

In order to minimize disturbances during experiments the Photon spacecraft operates all throughout its active lifetime period with the attitude control system off.

The attitude control is switched on before the beginning of the experiments in order to stop the spacecraft's rotation after its separation and must be turned off after one orbit.

At first the angular velocity is reduced to 0.03 deg/sec.

On the Photon spacecraft the technological hardware is linked to the outer space vacuum through an outlet in the descent vehicle's body and a vacuum system. Its valves uncover the port to the space vacuum once after launch.

The Photon spacecraft is usually launched from the Plesetsk launch complex. Its operational orbit altitude is chosen within 220-400 km with an inclination of 62.8°.

The spacecraft and its technological payload is controlled automatically under the commands of the onboard control unit and ground control stations. The information on the scientific hardware operation can be transmitted both directly or by using the intermediate data storage with its further playback during the communication contacts.

When the experiment is over the attitude control system is switched on again in order to point and stabilize the vehicle before re-entry. The samples of materials obtained are to be delivered in the descent vehicle on the landing site located at the Soviet Union territory.

The descent pattern is shown on the Figure 4.

The descent vehicle has a soft landing system which guarantees a vertical landing speed of about 3 m/s (10 m/s when using parachute only). That makes possible both DV's and research hardware recurrent use as well as carrying out experiments on biotechnology. If necessary, scientific payload can be returned right on the landing site.

THE RESURS-F SPACECRAFT

The Resurs-F spacecraft was originally intended for the Earth surface survey in the interests of natural resources exploration. But, having some reserve for extra payload, it is used for carrying out various experiments including material science and biotechnology.

The Resurs-F spacecraft consists of a descent vehicle identical to that of the Photon spacecraft, an instrument module, retrorocket and corrective engines (Figure 5 & 6).

The scientific hardware is installed inside the descent vehicle and in a special container mounted on its body outside. The container's cap is open in flight so the scientific hardware operates in outer space. Before re-entry the cap

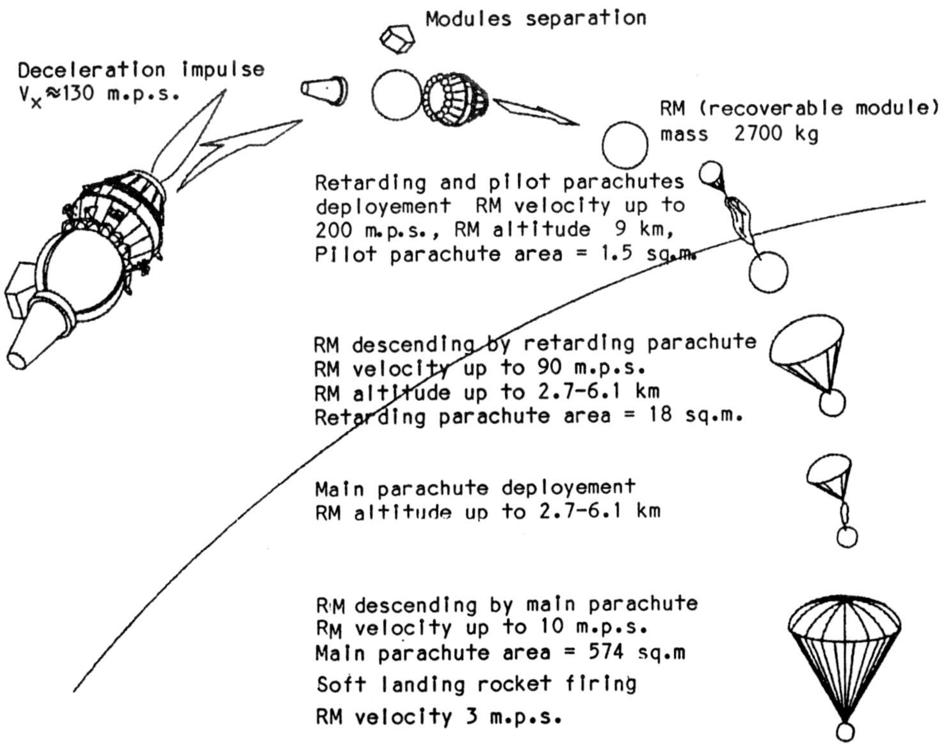

Figure 4: The descent pattern of the Resurs-F1 spacecraft

is closed providing hardware delivery to the Earth (Figure 7 & 8).

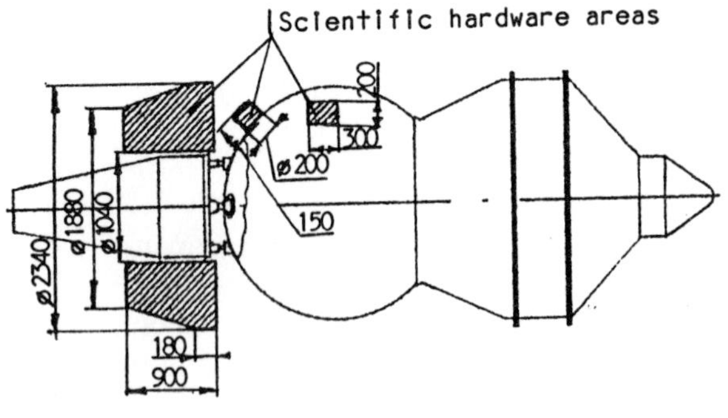

Figure 5: The scientific hardware areas of the RESURS-F1 spacecraft

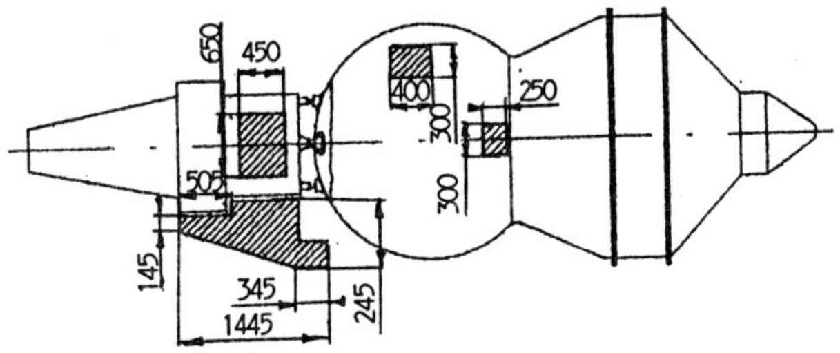

Figure 6: The scientific hardware areas of the RESURS-F2 spacecraft

Non-retrievable scientific payload can be installed outside the spacecraft in zones shown in Figures 5 & 6. The data obtained is to be transmitted through the telemetry channels.

Pre-launch Resurs-F preparations are performed at Plesetsk launch complex where all necessary operations including those related to additional scientific payload istallation are carried out.

At first Resurs-F is put into transfer orbit. Then in 1 or 2 days it transfers to operational orbit with altitude range 250-400 km. The spacecraft is oriented in an orbital co-ordinate system. The flight duration can be up to 23 days. Six days Resurs-F is orbiting in coasting flight with the attitude control off.

The spacecraft and its equipment run under control and telemetry monitoring carried out from ground stations located on the USSR territory.

After the mission program is completed the descent vehicle performs landing similar to

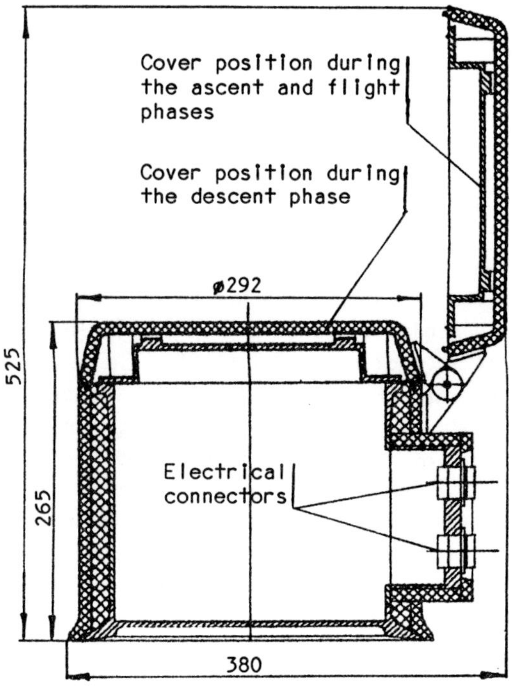

Figure 7: Scientific hardware capsule (High)

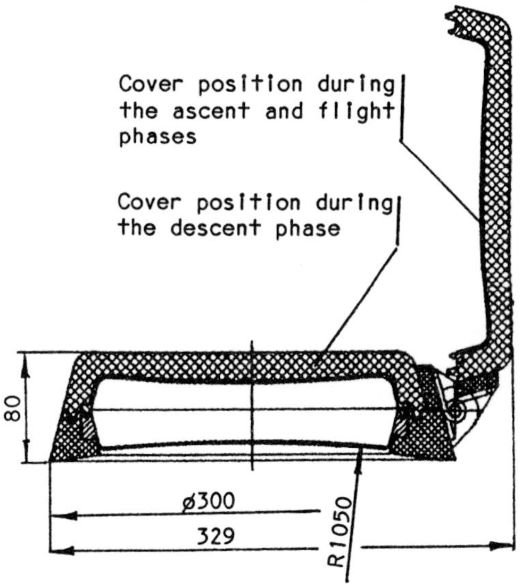

Figure 8: Scientific harware capsule (Low)

that of Photon. If necessary scientific payload and biological objects can be removed from descent vehicle right on the landing site.

At present have been launched seven Photon missions. The eight Resurs-F missions have been used for carrying out biotechnological experiments.

Certain test and research programmes have been completed on zero gravity physics and onboard technological processes learning.

However the data acquired are inadequate for unambiguous solving of problems of required properties of certain materials in space production management and for defInding correct experimental conditions.

The production of test - and - industrial materials in space requires a microgravity level reduction down to 10^{-4} - 10^{-6} g, substantial power supply increase, the increase of active lifetime, increase in amount of products obtained and their range extension and finally high automation of technological processes.

The new NIKA-T spacecraft being currently developed has a higher power capability and much longer active lifetime. For these reasons it is expected to be used for full-scale experiments on technological hardware construction, principles development, technological processes testing, space production automation system trials and test - and - industrial manufacture try-out.

The NIKA-T spacecraft.

The general view of NIKA-T spacecraft is shown in Figure 9.

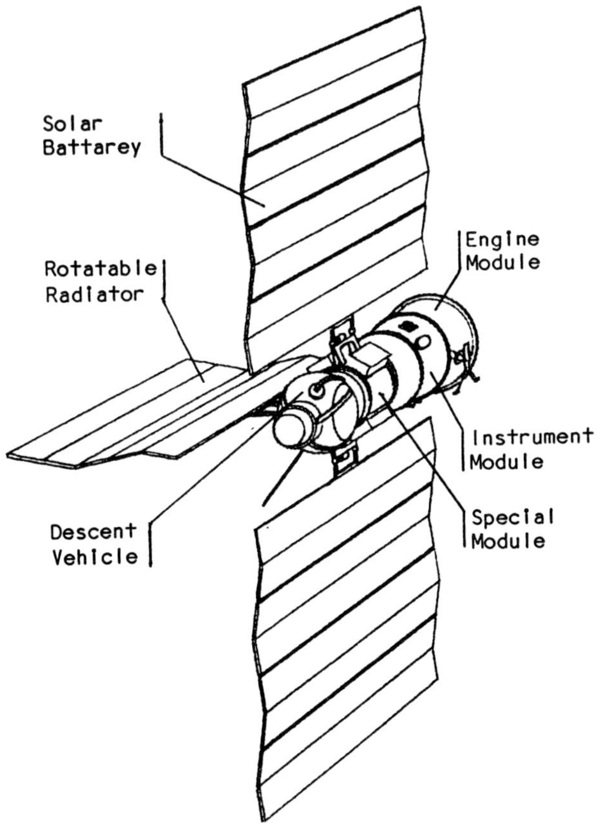

Figure 9: The general view of NIKA-T spacecraft

It comprises:

- a descent vehicle accomodating recoverable scientific payload and soft landing system;
- a special module which may contain non-recoverable scientific payload for researches of various kinds. The data from this payload are transmitted through telemetry channels;
- an instrument module containing onboard control panels and temperature control system hardware;
- an engine module containing engines and power supply system units.

Outside the spacecraft solar arrays with 72 m^2 total area are mounted, a deployable heat radiator with total radiating area 16.5 m^2 and chemical power sources covered with a special thermal insulation shrouds. There are several aerials and guidence system sensors installed on the instrument and on the engine modules outside.

A hung-out ring that holds the descent vehicle can serve as a base for additional payload to work at outer space. The spacecraft has ample reserve in power, control commands and telemetry for this extra payload which can be performed separably. For instance, it can be a small co-satellites.

The inner and outer payload accomodation zones are shown in Figure 10.

The basic NIKA-T's technical parameters are as follows:

- recoverable scientific payload mass — 1 200 kg;
- total payload — up to 2 000 kg;
- an average dayly power generation, including 4 500 W for technological equipment — 6 000 W;
- total lifetime — to 120 days.

The spacecraft is designed for sun-synchronous orbits with an altitude from 350 to 500 km.

When orbiting the spacecraft is oriented in an orbital co-ordinate system. The vibro-acceleration measurement works will be continued with NIKA-T spacecraft, which were originaly performed with Photon-7 was launched in September of 1991. The spacecraft is planned to be launched from Baikonur launch complex since 1994.

Like Photon and Resurs-F spacecraft NIKA-T will be able to carry foreign experimental scientific payloads.

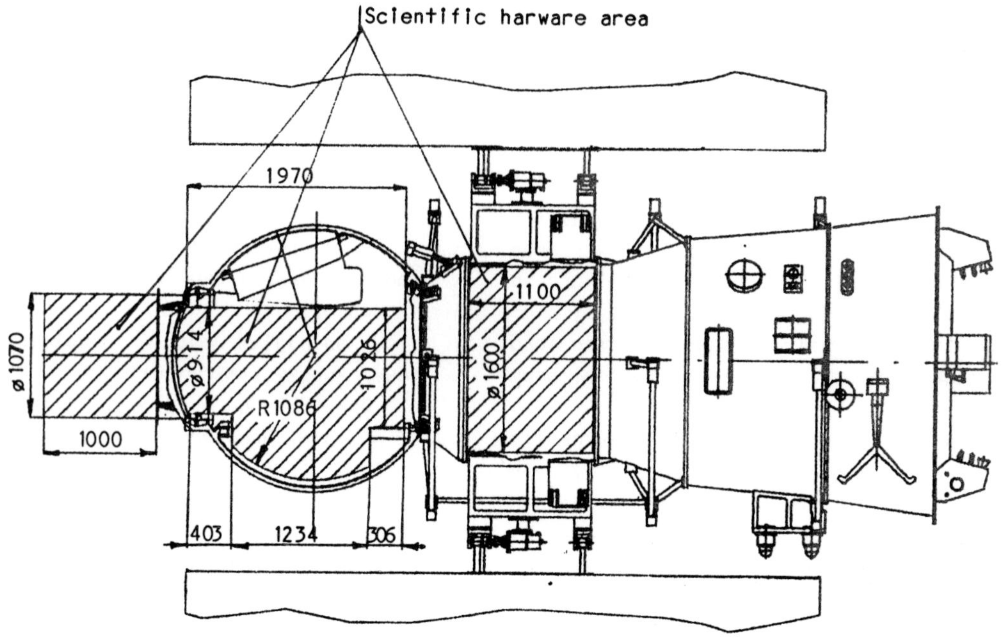

Figure 10: The scientific hardware areas of the NIKA-T spacecraft

REFERENCES

1. Avduevsky V.S., Uspensry G.V. The Industry of Space, Moscow, Machinery, 1989, 566 p.

2. Materials sciences in space. Edited by B. Feuerbacher, H. Hamacher, R. Naumann, Springer-Verlag Berlin, Heidelberg, 1986. Translation - 1989, Moscow, Mir, 479 p.

Phenomena Induced by Heat (Light) Sources

OPPORTUNITY OF INVESTIGATION OF FLUID INTERFACE INSTABILITY AND THERMOCAPILLARY PHENOMENA IN SPACE.

G.Gouesbet*andA.T.Sukhodolsky**

LESP, UA.CNRS.230, ROUEN INSA, BPO8, 76131, Mont-Saint-Aignan Cedex, France*.
USSR,117942 Moscow, Vavilov street,38, IOFAN**.

ABSTRACT

Thermal lens oscillations and associated phenomena have been systematically investigated on earth for more than ten years. It is here proposed that microgravity experiments would provide a better understanding of the involved instability mechanisms. A more general interest would concern microgravity processes in which local heating of interface occurs.

1. INTRODUCTION

There has been a long anchored tradition concerning Rayleigh-Bénard-Marangoni instabilities when a horizontal liquid layer is submitted to a vertical temperature gradient. Although the principle of exchange of stability holds when the layer is confined between two rigid walls, overstability transition leading to oscillatory motion is possible in the case of a liquid layer with a free surface. The case of a pure Marangoni mechanism has been studied by Takashima (Ref.1) . The case when both Marangoni and buoyancy effects simultaneously act together has been examined by Gouesbet and Maquet (Ref.2) and Gouesbet et al (Ref.3). The case when an additional shear is simultaneously present has also been investigated (Ref.4) . Generically, i.e. except for some special cases, overstability can only be observed when the rigid wall is cold (liquid layer suspended from a cold ceiling , or resting above a cold floor, i.e. with negative Marangoni numbers for standard liquids). In other words, overstability with heating of the free surface from the rigid wall is unlikely to occur . The above discussion concerns linear disturbance analysis. When nonlinear motions are considered, we may expect the appearance of free surface solitons when the layer is heated from above (Ref.5).

In contrast with the aforementioned cases, we now consider a new class of phenomena in which oscillatory behaviour and propagating waves appear when the layer is heated from below, namely thermal lens oscillations and associated hot-wire experiments.

2. THERMAL LENS OSCILLATIONS AND ASSOCIATED PHENOMENA.

Seemingly, the first observation of thermal lens oscillations has been reported in a short letter by Jakeman et al (Ref.6) but without any further investigation. The phenomenon has been independently fortuitously rediscovered by a Rouen team in 1982 (Ref.7) and systematically investigated since then. The reader may consult a review paper (Ref.8) in which many references up to 1990 are quoted. Basically, thermal lens oscillations may occur when a laser beam travels horizontally below the free surface of an absorbing liquid (so-called optical heart-beat 1, HB1) or vertically upwards (HB2), leading to oscillations of the beam outgoing of the liquid cell, associated with oscillatory convection in the bulk and propagating waves at the free surface (Fig.1).

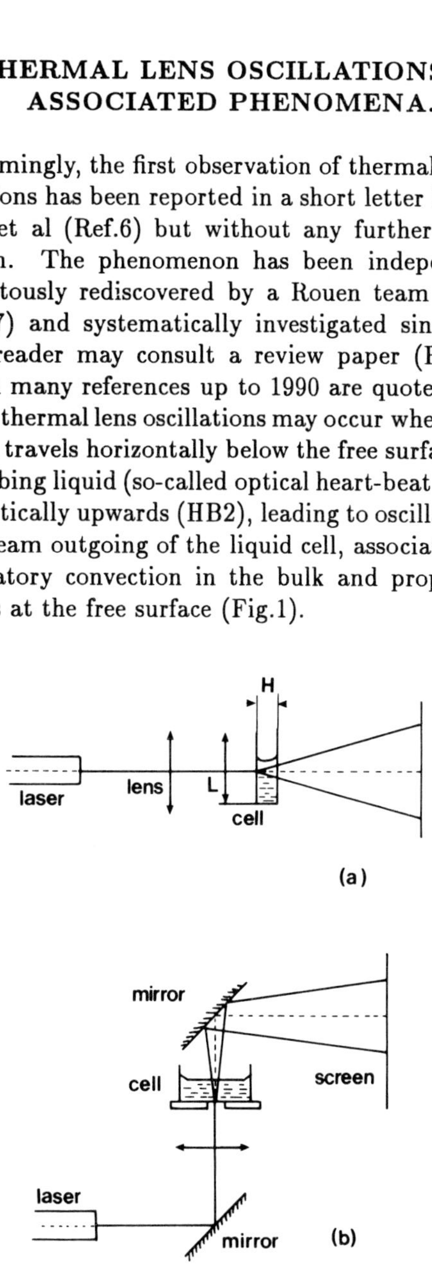

Experimental setups for the HB 1 (1 a) and HB 2 (1 b)

Figure 1:

HB1 may lead to various kinds of behaviour corresponding to typical keywords borrowed from the theory of nolinear dynamical systems: Hopf bifurcation, period-doubling cascades, quasi-periodicity, hysteresis associated with the coexistence of multiple attractors and chaotic attractors (Ref.9) .

The same phenomenon has also been later independently rediscovered by a Moscow team. Experimental observations basically agree with Rouen team results. Also, a new HB (HB3) has been designed (Fig.2) which, to some extent, combines HB1 and HB2 features. Associated phenomena concerning bubble dynamics (cavitator, photophoresis) have also been described. See Refs 10-12 from which earlier references may be found. HB1 has also been studied by another Russian team in Kiev, confirming the interest of these new instabilities (See for instance Refs 13-14).

Beside the interest in fundamental research concerning these particular hydrodynamical free surface instabilities, a more general interest potentially concerns any case of local heating near an interface. Applications to metrology like chemical species concentration measurements are also feasible by thermal lens oscillation spectroscopy (Ref.15).

Production of oscillatory convection with free surface wave propagation may also be obtained by hot-wire heating below an interface rather than by laser heating, leading to well controlled experiments,however exhibiting a more restricted range of behaviour up to now. In particular only the Hopf bi- furcation from steady to periodic dynamics has been observed by using a temperature-controlled hot-wire (Ref. 16,17). Such experiments are called HWE's (Hot-Wire Experiments) in contrast with HBE's (Heart-Beat Experiments).

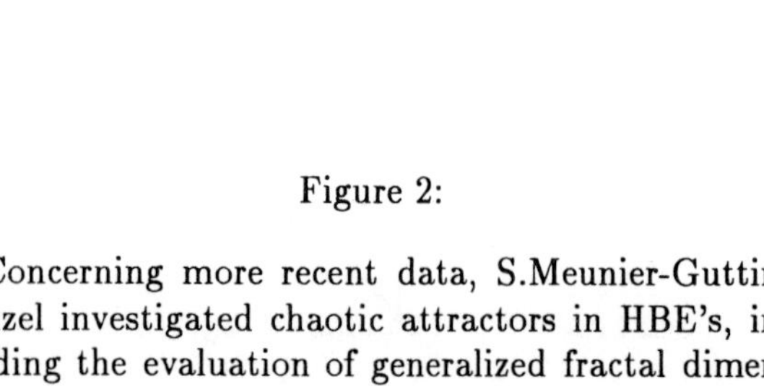

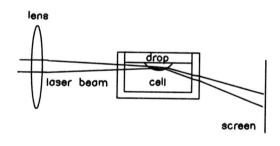

Figure 2:

Concerning more recent data, S.Meunier-Guttin-Cluzel investigated chaotic attractors in HBE's, including the evaluation of generalized fractal dimensions (Ref.18). C.Rozéinvestigated HWE's in the nonlinear domain beyond Hopf bifurcation with special emphasis on the study of the propagating waves produced by the instability (Ref.19).

Currently running (unconfirmed) HWE's with electric intensity control rather than with temperature control could indicate that HWE's might also be driven to chaos. Finally, a current study of wave propagation above a very long hot-wire (60 cm) might provide a good opportunity for experimental description of 1D-spatio-temporal chaos (Ref.20).

3. THEORETICAL UNDERSTANDING.

The best understanding available up to now is provided by two models, a so-called simple model designed in Rouen and a so-called circulator model designed in Moscow.

3.1. Rouen simple model.

This model identifies two characteristic time scales: t_1 for heat transfer from the heating source (HB1 or HWE's) to the free surface and t_2 for mass transfer by Marangoni disruption of hot blobs arriving at the surface. Clearly, if $t_1 \gg t_2$, the process takes a cyclic character. We therefore introduce a bifurcation parameter $R = \frac{t_1}{t_2}$ and state that bifurcation from steady to unsteady behaviour will occur when this parameter crosses a critical value $R_* = 0(1)$. Afterwards, we may also consider the free surface as being a mechanical oscillator with ξ the surface deformation, and $\frac{d\xi}{dt}$ the surface velocity. Notwithstanding nonlinear terms, the equation of motion of the oscillator may be written as

$$\frac{d^2\xi}{dt^2} - \mu\frac{d\xi}{dt} + \Omega^2\xi = 0 \quad (1)$$

$$\mu = \frac{(R-1)}{t_1} \quad (2)$$

$$\Omega = \frac{2\pi}{t_1 + t_2} \quad (3)$$

indeed indicating that a Hopf bifurcation occurs at $R = 1(\mu = 0)$ when the dissipation coefficient changes of sign and that the oscillator then exhibits a frequency $f = (t_1 + t_2)$.The model must be completed by evaluating expressions for times t_1 and t_2 and may be generalized to include more physical ingredients. Results of the model agree quantitatively well with experimental results using silicon oils and eventually allow the understanding of chaotic phenomena. Ref.21 provides details and earlier references on this modeling approach.

3.2 Moscow circulator

The spirit of the Rouen model is to consider that the actual differential equations governing the phenomena may be not essential. Any teacher in fluid mechanics for instance knows that some problems may be fully solved by dimensional analysis only, regardless of the value of constants that dimensional analysis cannot provide. In other words, it may happen that the solution of a set of equations is not imposed by the equations themselves but by more general and essential physical and/or mathematical constraints. Taking this point of view, first principles in the Rouen model are a qualitative analysis of the instability mechanisms, the use of dimensional analysis and of some concepts from nonlinear dynamics theory. Therefore ,this model starts with simplicity and must approach the reality by progressively adding more complex ingredients.

The Moscow model proceeds in the reverse way. Governing balance equations are the starting point as first principles. They form a set of Partial Derivative Equations(PDE's) which may be solved by using for instance finite volumes. However, the corresponding numerical problem is of excessive complexity (see Ref 22 for instance). Therefore, it is required ,starting from complexity to evolve toward simplicity. This is carried out by using only a limited number of cells specially selected for which heat and mass equation balances are written. These cells are chosen to be the most relevant ones for the description of the phenomena under study according to a prior qualitative understanding. Consider for instance the HB2-case (Fig 1b). One vertical column of cells accounts for the existence of the upward laser beam while one horizontal layer of cells below the free surface enables us to account for the Marangoni effect. When numerical integration of the equations are performed, oscillatory motions are produced which closely mimic the experimental ones. The fluid motion is essentially localized in a spatial cyclic volume, hence the name of circulator given to this model. See Ref 23 for details.

4. CONCLUSION.

This paper describes a new class of fascinating instability phenomena which attracts the interest of an increasing number of investigators. The present understanding is provided by two complementary models. Both models agree on the physical ingredients which are necessary, namely Marangoni effect as the mechanical engine for motion and heat transfer from the source to the surface as the fuel supply. Going to details, the reader would observe that heat transfer in the circulator results of a combination of heat diffusion and heat convection. Rouen model conversely pretends that heat transfer by conduction is dominant when the distance between heating source and surface is small, while, when this distance is big, convection is dominant. Therefore,the exact role of convection is still not perfectly understood. This issue of the role of convection is also found to be questionable by the Kiev 13-14 team (Ref. 13-14). Experiments in space would provide an acid test of the models and a better understanding of the involved phenomena. Beside a special interest for the described thermal lens oscillations and associated phenomena, we stress out that there would be a more general interest concerning hydrodynamical and surfacc instabilities produced by any kind of local heating. Space experiments could expectedly use solar radiation rather than laser sources to provide cheaper set-ups. Finally, we point out that such space experiments should preferably be prepared by preliminary parabolic flight experiments.

5. REFERENCES

1. Takashima M, J.Phys.Soc.Jpn. vol.50, no.8, 2751, 1981.
2. G.Gouesbet, J.Maquet. AIAA.J.Thermophysics and Heat Transfer, vol.3, no.1, 27,1989.
3. G.Gouesbet, J.Maquet, C.Rozé, R.Darrigo, Fluids A,2,6,903,1990.
4. C.Rozé,G.Gouesbet, J.Maquet, American Institute of Aeronautics and Astronautics, Washington, D.C.,1990.
5. M.Velarde, private communication.
6. E.Jakeman, E.R.Pike, J.M.Vaughan. R.R.E. Newsletter and Research Review, no.12,1973.
7. R.Anthore, P.Flament,G.Gouesbet,M.Rhazi, M.E.Weill, Appl. Optics, vol.21,no.1,2,1982.
8. G.Gouesbet, JSME Int.J., Series II, vol.32,3,301,1989.
9. G.Gouesbet and E.Lefort Physical Review A, vol.37,no.12,4903, 1988.
10. A.T.Sukhodolsky, Izvestija AN USSR, ser.physics vol 50, no 6,1095-1102,1986 (in Russian).
11. Rastopov S.F.,Sukhodolsky A.T., DAN USSR, vol 295,no 5, 1104-1107,1987 (in Russian)
12. Rastopov S.F., Sukhodolsky A.T., Physics Letters A,Vol 149, no5,229-232,1990.

13. V.Yu.Bazhenov, M.V.Vasnetsov, M.S.Soskin, V.B.Taranenko, Pisma v ZhETF vol 49,330,1989.

14. V.Yu.Bazhenov, M.V.Vasnetsov, M.S.Soskin, V.B.Taranenko, Appl.Phys. vol b49,485-489,1989.

15. Y.Enokida, M.Shiga, A.Suzuki, private communication.

16. M.E.Weill, M.Rhazi, G.Gouesbet. J. de Physique, vol.46, 1501, 1985.

17. G.Gouesbet, M.E.Weill,E.Lefort, AIAA J., vol.24, no.8, 1324, 1986.

18. S. Meunier-Guttin-Cluzel, Thèse de Doctorat, Rouen University, 1990.

19. C.Rozé, Thèse de Doctorat, Rouen University, 1989.

20. P.Bergé, M.Dubois private communication.

21. G.Gouesbet. Phys.Rev.A vol 42, no 10, 5928-5945,1990.

22. J.Maquet, G.Gouesbet, A.Berlemont, in Proceedings of the 5th International Conference on Numerical Methods for Thermal Problems, Montréal, Canada, 1987, edited by R.W.Lewis, K.Morgan and W.G.Habashi, Pineridge, Swansea, UK, 1987, vol 5, Pt1, pp 472-483.

23. S.A.Viznyuk, A.T.Sukhodolsky, Zh.Tekh.Fiz.,vol 58,1000,1988 (in Russian)

PHOTOINDUCED SOLUTOCAPILLARY CONVECTION: NEW CAPILLARY PHENOMENON

B.A. Bezuglyi

Tyumen State University, Tyumen, 625003, USSR

ABSTRACT

The results of the experimental studies of the solutocapillary convection induced and governed by heat effect of light in a thin liquid layer are reviewed. It is shown that the phenomenon is not associated with neither "photocondensation" (Refs.1,3) nor thermodiffusion (Refs.2,4), but caused by domination of the solute component of the surface tension gradient over the temperature one due to the solvent evaporation from the irradiated zone. A number of concomitant effects confirmed this mechanism has been demonstrated. Various applications are proposed.

Keywords: *Solutocapillary convection, surface tension driven flow, Marangoni effect, Capillary phenomena.*

1. INTRODUCTION

Photoinduced solutocapillary (SC) convection was discovered when the iodine melt in a glass ampoule heated above 300°C had been irradiated by an intensive light beam (Ref.1).As compared with the usual behavior of substance in a beam of light, Figure 1, the new phenomenon looks paradoxical. In the irradiated region there appear droplets of "photocondensate". They rush precipitately to the center of the beam and confluence to one drop, Figure 2, named "anomalous" owing to its uncommon behavior. 1. When the power of the beam is modulated the drop is "breathing" with reciprocal frequency changing its apparent diameter due to the changing of the surface curvature which demonstrates a change of its surface tension. 2. When the light beam is moved along the underlay surface the drop follows relentlessly after it striving to occupy a position in the focus of the beam, whereas the basic melt is evaporating and flowing out of the irradiated zone, Figure 3.

As both the set-up and the experimental methods were aimed at finding the so-called "photophase effect" (Refs.1,3) and a number of features of the discovered phenomenon corresponded to the conception of the hypothetical effect mentioned above, the first report gave an erroneous explanation based on the assumption of the key role of the photoexitations (Ref.1). It has been found later the concentrating of contamination in the light beam was essential in the mechanism of the phenomenon and this process was generated by the heat effect of light (Ref.2). These facts inspired the authors to consider thermodiffusion to be a mechanism of the new phenomenon. It is curious that this erroneous conclusion as to the cause of solute concentrating in the same systems was repeated by other researchers ten years later (Ref.4). The phenomenon proved to be deceptively simple even after its capillary caharacter was revealed (Refs.5-8). It is testified by the attempt to explain the solute concentrating only by thermocapillary (TC) mechanism (Ref.9).

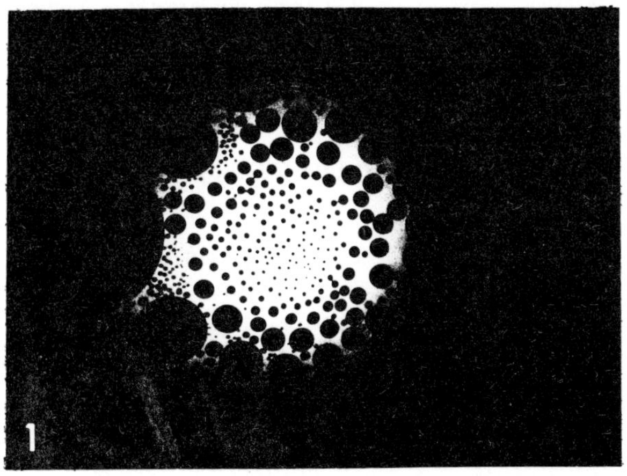

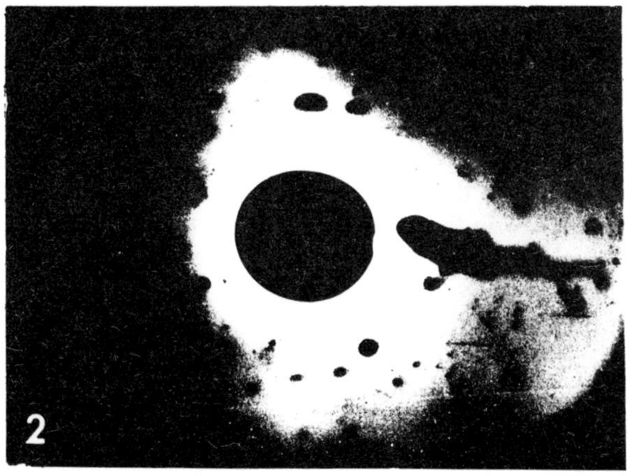

Figure 1: Usual evaporation of the melt iodine in an intensive light beam.

Figure 2: Creeping in and rolling down of the droplets to the center of the beam.

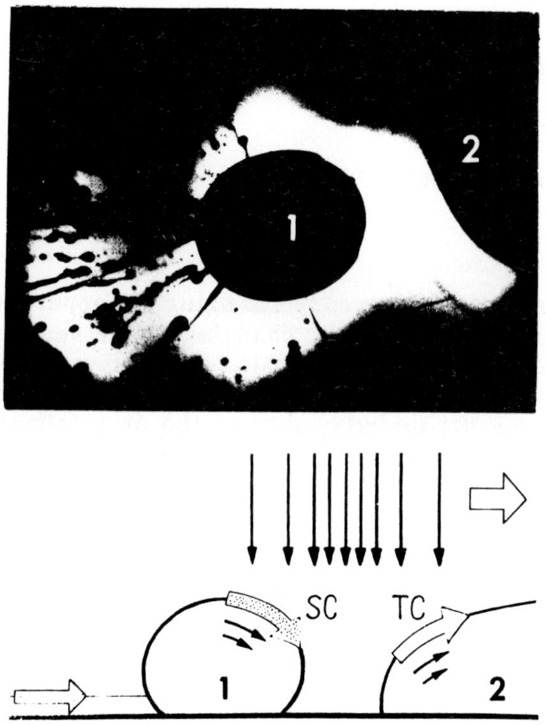

Figure 3: Two competing mechanisms, SC and TC, induced by the heat effect of light provoke different behavior of the anomalous drop 1 and the basic melt 2 in the beam of light.

2. BASIC STAGES OF THE PHENOMENON EVOLUTION

It has been established (Refs.5-8) that the photoinduced SC convection can be observed when a thin (down to a few nm) layer of the tensoactive solute (TAS) in a high volatile solvent is irradiated by a light beam if under the evaporation of the solvent the growth of its surface tension γ with the rising of TAS concentration X exceeds the lowering of γ due to heating and as a result there appears a field of $\nabla\gamma$ directed towards the zone of the maximum intensity of the beam, i.e.

$$\nabla\gamma=-(\partial\gamma/\partial T)\nabla T+(\partial\gamma/\partial X)\nabla X>0.$$

Thanks to viscosity the field of tangential stresses on liquid surface related to $\nabla\gamma$ induces a flow directed to the center of the beam. A subsequent rise of X in the liquid drawing to the beam lead to its phase separation and formation of a solitary drop. Figure 4a shows the basic stages of the phenomenon evolution from the moment of irradiating of the layer. For the viscous liquids as soon as the beam is switched-on a convexity is formed in the layer and then a thermocapillary depression is developed. Such phenomenon is described in (Ref.10), where it is explained by the "competition between capillary and gravitation forces". But in fact, as shown in Ref.8 the convexity is formed by thermal expansion, which is important until the velocity field generated by TC mechanism is settled.

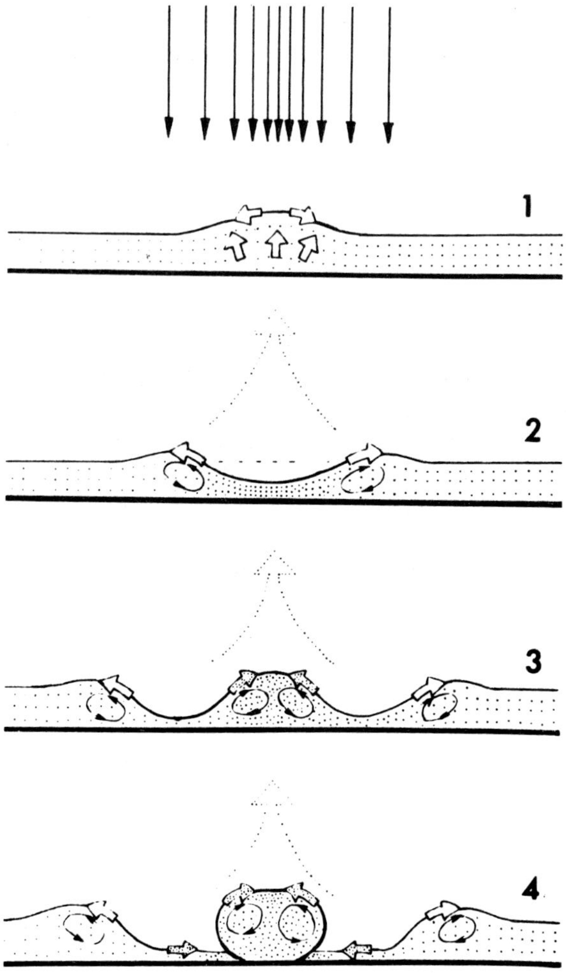

Figure 4a: Basic stages of the photoinduced SC convection evolution. 1.Rising the free surface due to the thermal expansion of liquid. 2. Photoinduced TC convection. Subjected to the action of TC forces the liquid is flowing out from heated zone resulting in the depression of the free surface. When the power of the beam is large and the layer is thin, the hollow can reach the bottom and owing to evaporation and/or spreading of the liquid and bed wetting of surface the rupture of the film can take place completing the capillary-convective process. 4. But if the condition of

$$(\partial\gamma/\partial T)\nabla T<(\partial\gamma/\partial X)\nabla X$$

is satisfied, then the stage of SC convection is sets in. 4. A further rise of X in the irradiated zone leads to an increase of γ so that liquid is separated from the underlay and anomalous drop is formed.

Photoinduced TC convection has been studied regularly since 1978 (Refs.8-11). Nevertheless, in case of TAS solution some features of its proceeding can lead us astray.

The 3d stage is characterized by the fact that when due to the switching of the SC mechanism in the TC depression a steady convexity arises. This phenomenon can be observed in rather thick layers of low-volatile liquids and with low-power beams, when due to low rate of evaporation the X is insufficient to increase γ so that to exceed the value of contact angle above $\pi/2$, that is necessary for liquid separation from underlay. It is such conditions that were described in Ref.12. However, the forming of convexity in the center of the TC depression is explained by a "hydrodynamics pressure of a stopped jet".

The 4th stage is the formation of a solitary drop with the above described uncommon features and it was observed in most systems except water solution of salts in which the SC effect is completed by formation of knoll with heightened solute concentration in it.

Increasing abruptly the power of sharp-focused beam it is possible to switch-on the TC mechanism repeatedly which is affecting now upon the body of the drop itself and causing its spreading. In the experiments with solution of 5% iodine in bromine the abrupt increase of the power of the beam give rise to the effect of "burning through" the drop, Figure 4b.

It should be noted specially stationary drop conditions when the solvent outflow by vaporization from the surface of the drop is equal to its SC inflow on the underlay, Figure 5.

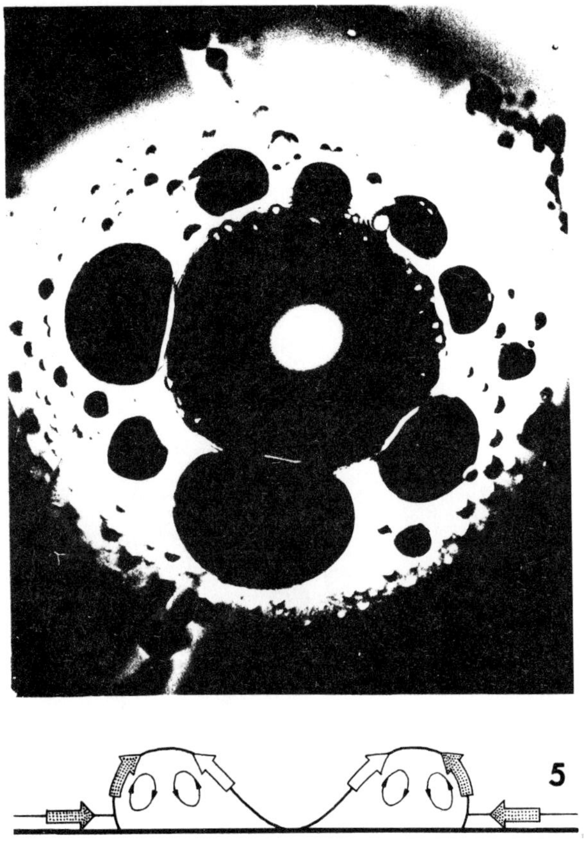

Figure 4b: The "burning through" of drop effect is related to the repeated switching of TC mechanism. It can be considered as fifth stage of the phenomenon evolution. Under the picture there is the scheme of flows in the drop.

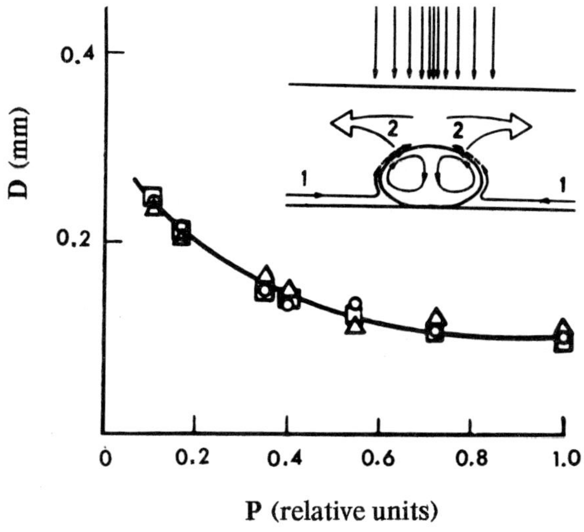

Figure 5: The diameter of the drop D (mm) as a function of power P (relative units) under its cyclic changing: o decreasing, □ increasing, Δ repeated decreasing of P. The inset: At stationary state the mass balance of liquid inflow 1 and vapor outflow 2 take place.

3. CONCOMITANT EFFECTS

By application the cell of Hele-Show type the basic features of the phenomenon were studied in the model systems (the solutions of iodine in acetone, bromine; the spirit solutions of the dyes,etc.), the methods of its investigations were devised , and a number of concomitant effects were found.

1. **Tearing of drop with a light beam when it crosses the interface in the direction of liquid-gas , Figure 6.** This effect can be observed only in those systems where the existence of the photoinduced SC convection is possible.

2. **Capture of the bubble with a light beam when it crosses the interface in the direction of gas-liquid.** We have paid attention to this effect wishing only to enrich the collection of paradoxes which had been commenced by G.Quinke, who described the "kicking" bubble paradox (Ref.13). The paradoxes of arising and of the sinking bubble are mentioned by Birkhoff (Ref.14). Our paradox of the "governing" bubble is related to TC mechanism and its nature is shown in Figure 7.

3. **Periodical drop generation at the interface.** If the beam is brought to the border of the solution at distance equal to 1-3 its diameters the drop is formed, in general, due to the surface flow from the side of the solution, Figure 8. As soon as the drop has reached a critical size it leaves the beam and a new drop arises at its place and it goes away too. The process is repeated with the period of 5-8 sec depending on the power of the beam, the distance from the border and the concentration of the solution.

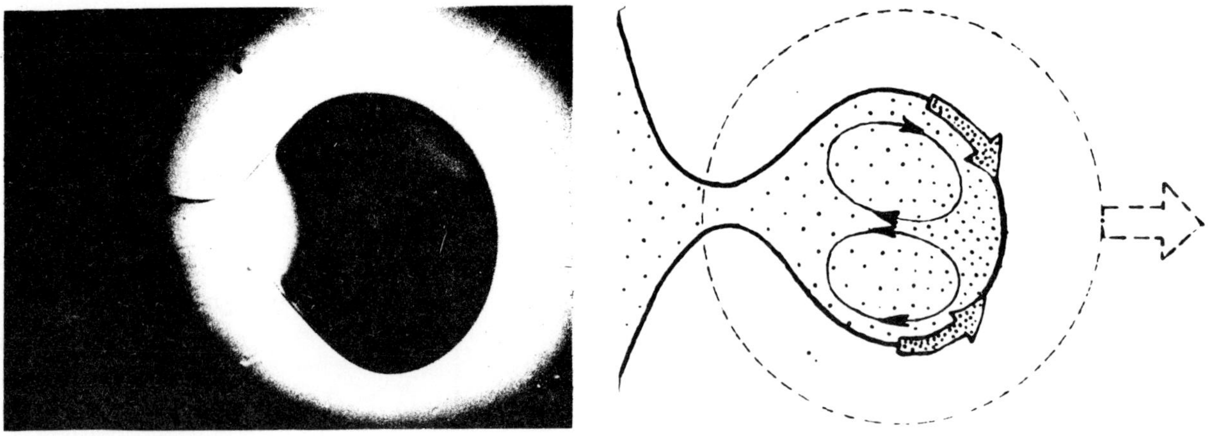

Figure 6: Tearing of the drop with a light beam.

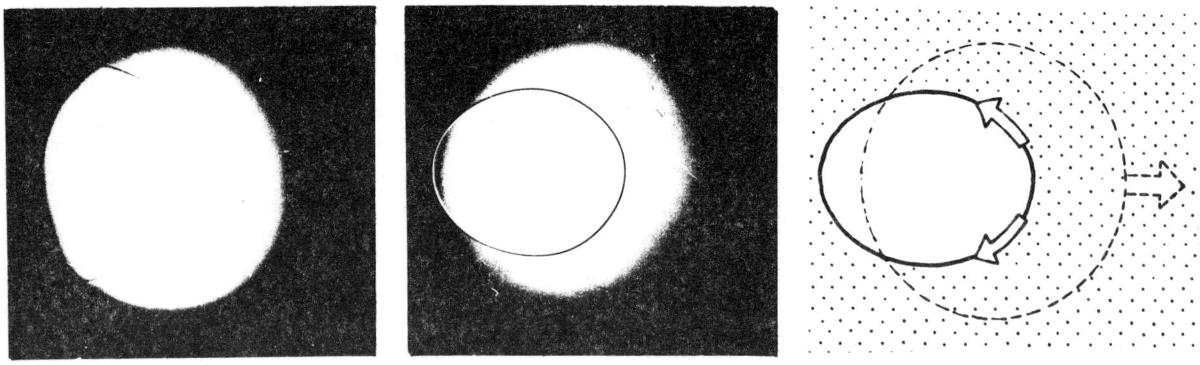

Figure 7: Capture of the bubble with a light beam.

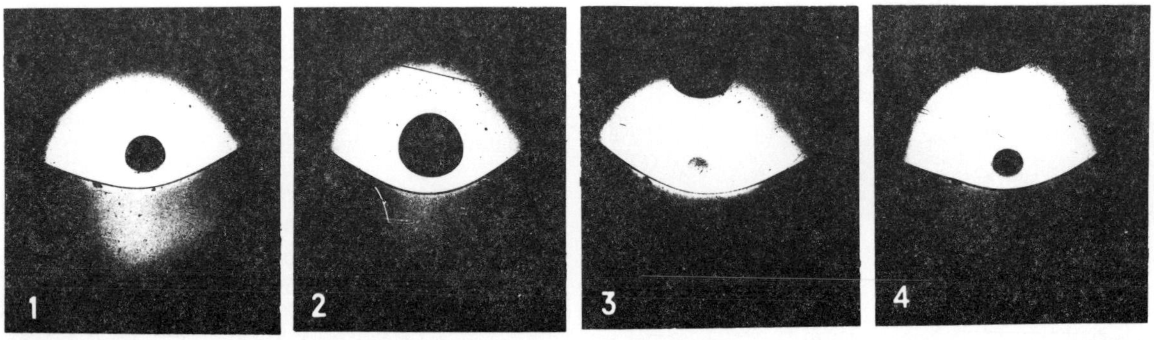

Figure 8: Periodical drop generation at the interface.

4. Star-shaped structure (Refs.5,8,16) is easily reproduced in the stationary drop when the power of the light beam is abruptly increased. We are struck here by the fast process of the separating of absorbing contamination in the body of the drop and the formation of the star-shaped structure due to power increase is proceeding so that the drop is cleared up at the center decreasing the dissipative losses of the light beam.

5. Anomalous drop as generator of the coherent structures. The drop brought to the border of the solution does not immerse, but periodically injects part of its liquid in the solution, Figure 9. A portion of the liquid injected through the border passes along the separating space ≈1 μm and in the mother solution forms a mushroom structure, known in the large scale hydrodynamics as buoyancy plume (Ref.17) and referred to the class of the coherent structures (Ref.18). After the ejection of a liquid portion the drop grows again to the former size and then process is repeated.

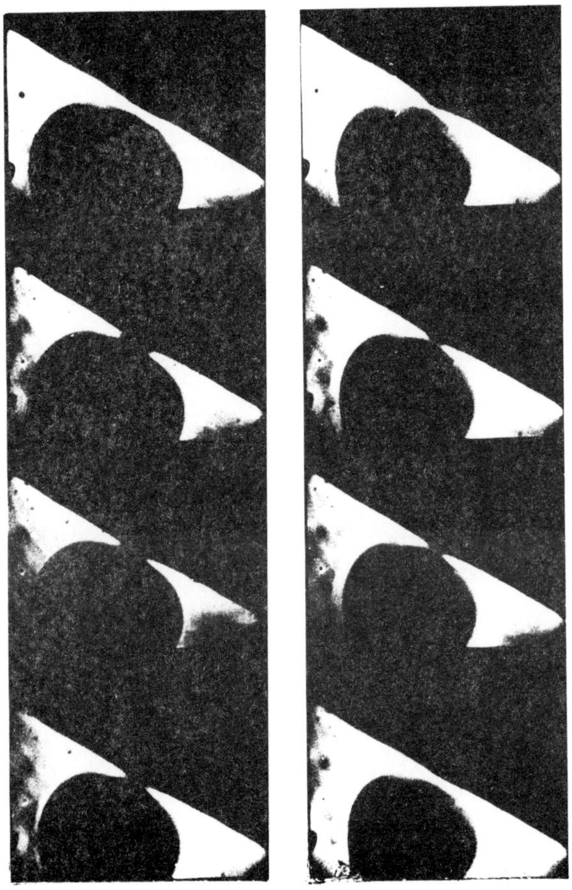

Figure 9: Periodical injection of the part of the anomalous drop liquid in the basic solution.

4. APPLICATIONS AND PROSPECTS

The high sensibility of liquid to shear stresses generated on its surface by heat effect of light allows to effective control mass transfer in the capillary systems (thin layers, menisci, drops, floating zones, etc.) without using any device, but arranging the capillary-convective flows alone. It is the liquid surface of capillary system that represents itself as the simplest device for transformation of the potential energy of the diffusive flow induced by a light beam in kinetic energy of the liquid flows.

Mastering SC mechanism for practical purposes has a number of advantages as compared with the TC mechanism. First of all, the relaxation time of the concentration disturbances is ($k/D \approx 100$) times greater. IT gives a more stable effect in time. Secondly, the created in liquid layer concentration disturbances are easily fixed (unlike the temperature ones) increasing the viscosity of the liquid in any way solvent evaporating, cooling to solidification, etc.). These advantages have been served as a basis for a number inventions.

1. Thermotensography. Differing from the well-known liquid layer systems (Refs.21-24) by an incomparably higher resolution ($\approx 600\ mm^{-1}$) the new method (Refs.21-24) first resolved the problem of fixing an image. Thanks to using concerted action of the adverse ∇T and ∇X the possibility of increasing its sensibility and the direct obtaining of a positive image has been demonstrated (Ref.8).

It is known that the viscous liquid layer subjected to photoinduced convection is deformed and forms a relief. By using this property a light sensitive composition for making a printing form has been proposed, and a number of thermo- and photosensitive systems have been invented (Refs.8,24).

2. The separation of substances in hermetic vessel, is considered in detaile in Ref.25.

3. The optical micro-metering divice can be designed on the basis of the periodical drop generation effect, Figure 9, if the drops leaving the beam are transferred by the governing beam to the receiving ampoules. Such device can be useful for operation with the radioactive or noxious matters.

5. CONCLUSIONS

The detailed experimental studies of the phenomenon prove the insufficiency of its explanation on the basis of the customary conceptions about the aqueous solutions of the surfactants and all the more insufficient for that purpose is TC mechanism alone.

Among the great number of the concomitant effects we presented the most interesting and easily reproducing ones. Although many of them can enrich the collection of curiouses, nevertheless, they give a splendid illustration of the actions of the TC and SC mechanisms both separately and in competition. Moreover, they also prove the possibility of effective operation of those mechanisms by the heat effect of light (Ref. 8).

6. ACNOWLEDGMENTS

The author would like to thank Prof. V.I. Polezhayev for his stimulating interest and invitation to the Symposium, Prof. L.A. Aslanov for his good advices and Prof. A.I. Ivandayev for the financial support.

The author is grateful to Mrs E.N. Kislovskaya and Mr. I.B. Zastenker for their assistance in preparation of the manuscript.

7. REFERENCES

1. B.A. Bezuglyi et al., *Pis'ma Zh. Eksp.Teor.Fiz. (USSR)* **22** (1975) 76.
2. B.A. Bezuglyi et al., *Pis'ma Zh. Tech. Fiz. (USSR)* **2** (1976) 832.
3. J.L. Katz et al., *J. Chem. Phys.* **75** (1981) 1459.
4. I.F. Bunkin et al., *Kvant. El. (USSR)* **12** (1985) 2391.
5. B.A. Bezuglyi & V.V. Nizovtsev, *Chimiya & Zhizn' (USSR)* No7 (1977) 33.
6. V.S. Mayorov, *DAN USSR* **237** (1977) 1073.
7. B.A. Bezuglyi & V.V. Nizovtsev, *Vestnik Mosc. State University. Ser.3*, **22** (1981) 37.
8. B.A. Bezuglyi, *PhD Thesis.* Moscow State University (1983).
9. A.T. Suchodol'skii, *Izv. AN. USSR, Ser. Fiz.* **50** (1986) 1095.
10. G. Da Costa, *J. Phys.* **17** (1982) 1053.
11. S. Slavchev, *Theor. & Appl. Mech. (Bulgaria)* **14** (1987) 54.
12. V.V. Nizovtsev, *Zh. PMTF (USSR)* No 1 (1989).
13. G. Quinke, *Ann. Phys.* **35** (1888) 593.
14. G. Birkhoff, *Hydrodynamics.* Princeton University, Princeton (1960).
15. J.T. Davies, *Turbulence Phenomena.* Academic Press, N.Y. (1972).
16. B.A.Bezuglyi, *Pis'ma Zh. Tech. Fiz.* **16** (1990) 55.
17. I.S. Turner, *Buoyancy Effects in Fluids*, Cambridge University (1973)
18. B.J. Cantwell, *Ann. Rev. Fluid Mech.* **13** (1981) 457.
19. A.S. Glushkov et al., *Pis'ma Zh. Tech. Fiz.* **5** (1979) 1223.
20. J.C. Loulerg et al., *J. Phys. D* **14** (1981) 1967.
21. B.A. Bezuglyi et al., *A.S. No 957155 (USSR) BI* No 33 (1982).
22. B.A. Bezuglyi et al., *Zh. Tech. Fiz. (USSR)* **52** (1982) 2415.
23. B.A. Bezuglyi, *Zh. Tech. Fiz. (USSR)* **53** (1983) 927.
24. B.A. Bezuglyi et al., *Avt. Svid. (USSR)* No: 1048943;1113774;1122137.
25. V.S. Mayorov et al., *Zh. Tech. Fiz. (USSR)* **48** (1978), 833, 2553.

Technical Application of Microgravity Fluid Mechanics

DROPLET COMBUSTION IN MICROGRAVITY

I. GÖKALP, Ch. CHAUVEAU and X. CHESNEAU

Centre National de la Recherche Scientifique
Laboratoire de Combustion et Systèmes Réactifs
45071 Orléans Cédex 2, France

ABSTRACT

New experimental results concerning the envelope flame of n-heptane droplets burning under reduced gravity are presented. In this work, the parabolic flights of an aircraft are used to create a reduced gravity environment of the order of 10^{-2} g. The experiments are performed in a droplet burning facility specially designed for this purpose. The burning of n-heptane droplets has been observed during parabolic flights with the Caravelle aircraft at the C.E.V. Brétigny, France. The results presented here constitute the first clear experimental evidence of the non-steady behaviour of droplet burning whereas the spherical symmetry is conserved, by showing essentially that the flame first moves away from the droplet, then back towards it, and that the ratio of flame to droplet diameter increases continuously during combustion. Preliminary results on the influence of pressure on droplet burning in microgravity are also presented.

1. INTRODUCTION

The burning of a liquid droplet is today largely accepted as one of the basic combustion problems which would benefit from the development of the domain of microgravity combustion (Refs. 1-2). Considerations of fundamental as well as applied nature guide this belief. In practical situations, such as rocket propulsion or diesel engines, very small droplets are sprayed in the combustion chamber. They are essentially transported by the gas phase velocity without decelerating, so that the droplet relative velocity is zero. This picture validates then the spherical symmetry hypothesis applied to the droplet and to the diffusion flame enveloping it and which facilitates significantly the handling of the conservation equations governing this chemically reacting, two-phase flow with phase change (Ref. 3).

The confrontation with experiments of the results deduced from theoretical and numerical studies which apply the spherical symmetry and other accompanying hypotheses to single droplet combustion pointed out that natural convection was the main source of departure from spherical symmetry of the burning droplet system in a stagnant environment. On the other hand, the presence of the natural convective effects not only disrupts the symmetry of the system, but introduces also additional coupled transport effects both inside and outside the droplet. The reduction of these asymmetrical unsteady buoyancy effects has been attempted by several workers. One method is to investigate small droplets similar to those found in practical situations, but the small physical size and reduced burning time limit the experimental resolution, both spatial and temporal, of the observed phenomena. Another method is to investigate suspended droplets burning in reduced pressure (Ref. 4). However, reduced pressure introduces modifications in the chemical kinetics which are difficult to assess. Reduced gravity appears then to be the most appropriate mean to achieve a buoyancy-free environment without introducing additional phenomena. It should also be added that droplet burning or vaporization near critical thermodynamic conditions introduces also the reduced gravity as the only mean to perform single droplet experiments because of the failure of the suspending filament technique due to the vanishing of the surface tension near critical conditions (Ref. 5)

In the work described in this paper, parabolic flights of an aircraft is used to create a reduced gravity environment of the order of 10^{-2} g. The most important feature of the parabolic flight possibilities we used is the operational reduced gravity times longer than 15 seconds which, together with the presence of an operator, allowed the observation of the entire droplet lifetime and the determination of the droplet and flame diameters for large fuel droplets up to 2 mm of initial diameter. The paper describes first the hardware and the diagnostics we used. In a second part, the latest results on the burning characteristics of suspended and free-floating n-heptane droplets are presented and discussed.

2. DESCRIPTION OF THE EXPERIMENTAL SET-UP.

2.1 The Droplet Burning Facility (DBF).

The experiments are performed in a specially designed facility : High Pressure Droplet Burning Facility (HP-DBF).

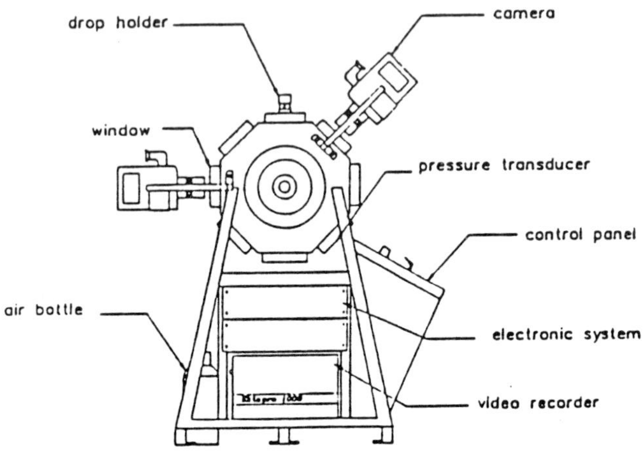

Figure 1. A schematic illustration of the HP-DBF.

The HP-DBF facility was designed to allow the investigation of the burning of suspended or free-floating droplets under variable pressure conditions, up to 120 atm. This apparatus is articulated around a cylindrical combustion chamber of 0.011 m^3 capacity, made of an Aluminium alloy (AU4G), and mounted on a support structure with a control panel. Two circular flanges are mounted on the two sidewalls of the octogonally shaped external faces of the chamber (Fig. 1). Ten identical openings are positioned on the chamber, two openings on the sidewalls for the droplet injection and ignition systems, and eight along the octagonal perimeter. The two injection systems are positioned face to face and are connected to an electrical pump, allowing the injection of a fuel amount of 0.5 microliters/min. A droplet is formed between the two injecting needles, which are retracted to let the droplet free float in the chamber during the reduced gravity period. Just before retracting the injection needles, the droplet is ignited by one pairs of electrodes, mounted around one injecting needle (Fig. 2). A tiltable (d=0.2mm) quartz fibre, mounted on the upper chamber wall, can be used and positioned between the two needles to hold the droplet in the case of suspended droplet experiments. The ignition spark is produced by a high tension (10 kV) solenoid under 24 V DC. In the case of high pressure droplet burning experiments, an improved version of the injection system and a heated coil for ignition are used. The sequence of injection and ignition operations is computer controlled, so that it is possible to observe the burning of two droplets per parabola. The HP-DBF has been used during our last four parabolic flight campaigns (NASA, August 1988 and CNES/CEV, June 1989, April 1990, October 1990).

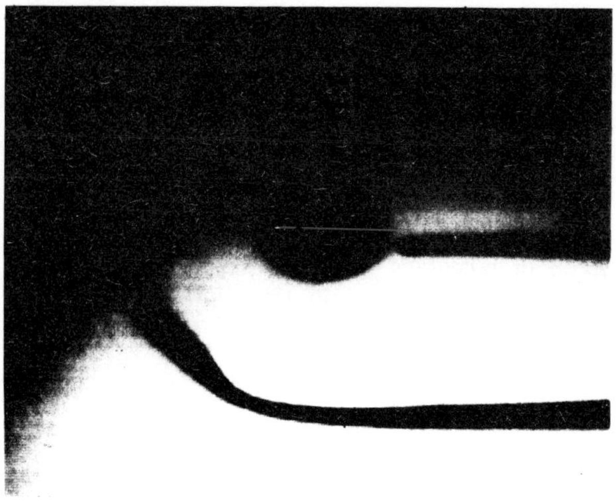

Figure 2. Droplet ignition just before the retraction of injection needles.

2.2. Diagnostics.

The principle diagnostic system we used is based on the visualization of the droplet burning. The recording equipment is composed of two systems. A high speed video camera, Kodak-Ektapro 1000, which can record up to 1000 full frames per second or up to 6000 partial pictures per second and allows the detailed analysis of the droplet burning. The Ektapro 1000 Motion Analyzer's live, real-time viewing makes possible to follow the investigated phenomena frame per frame. A standard video camera, Sony 8 mm PAL, is used to record the droplet and the flame in true colours, but only at 25 frames per second. A second camera has been added to the Kodak-Ektapro 1000 system, in order to be able to observe simultaneously the droplet and the flame with the appropriate magnification. The digitized images from the Ektapro system are transferred to a micro-computer, where the image analyses are performed. For the details of the image analysis, Ref. 6 should be consulted.

3. RESULTS AND DISCUSSION

The reduced gravity droplet burning program conducted at LCSR, Orléans is devoted to the determination of the burning characteristics of single component fuel droplets under various conditions. During the past parabolic flight campaigns, suspended and free-floating n-heptane droplets have been investigated. The results obtained during the first campaign have been presented in Refs. 7,8. This paper presents the results obtained during the last campaigns, with HP-DBF, for suspended and free-floating droplets under stagnant, normal and high pressure (up to 7 atm.) conditions.

3.1. Data reduction

The time evolution of the burning droplet dimensions and of the flame surrounding it is recorded with the high speed video camera. The digitized images of the droplet and the flame are used to extract their respective contours, from which the diameter of a sphere of equivalent surface area is determined. This new method of data reduction minimizes the average error in comparison with the previously used method (Ref. 7) where the diameter was defined based on the 45° line. The burning of several suspended and free-floating n-heptane droplets have been recorded and analyzed. The burning to completion is observed for suspended droplets. But in the case of free-floating runs, the burning to completion was not observed because of the residual gravity level which caused the drift of the droplet outside the viewing area.

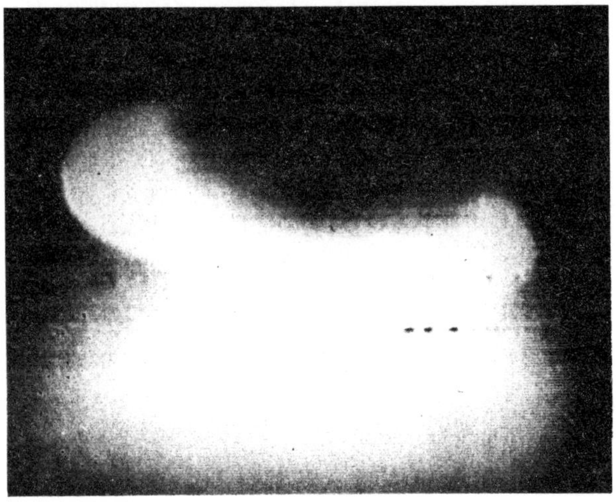

Figure 3. Picture of a free-floating burning droplet just after ignition.

The discussion of the burning droplet results are based on the time evolution of different parameters : the square of the droplet equivalent diameter, d_s^2, the flame equivalent diameter normalized by the initial droplet diameter, (d_f / d_{so}), and the flame-front standoff ratio, (d_f / d_s). The flame diameter is defined as the outer edge of the luminous zone. The presence of the suspending fibre influences essentially the sphericity of the droplet which is equal to 0.8 during the whole burning time. The sphericity of the flame is initially disturbed by the flow induced by the retraction of the needles, but recovers quickly a value of approximately 1 (Fig. 3).

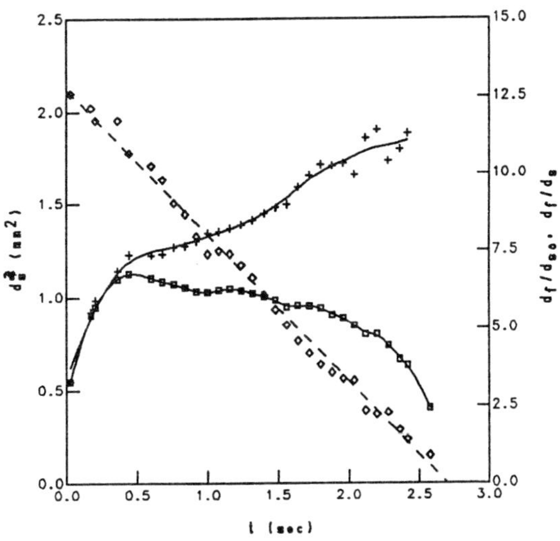

Figure 4. Evolution of the burning characteristics of a n- heptane droplet versus time.
d_{so} = *1.45 mm.* ◊ , d_s^2; □, d_f / d_{so} ; +, d_f / d_s.

3.2. Droplet burning experiments under normal pressure.

A typical suspended droplet burning run is presented on Fig. 4, where the initial droplet diameter is 1.45 mm. This figure shows that the surface regression rate of the droplet follows closely the d^2-law until its complete consumption, which lasts 2.7 s. A least-square fit gives an average slope of 0.79 mm^2/s for the surface regression rate. This value of the burning rate constant agrees quite well with the free-floating droplet results of Avedisian et al (Ref. 9) for an initial diameter of 0.5 mm. This is also a confirmation of the fact that the burning rate constant is independent of the initial droplet diameter. In addition, this value of the burning rate constant is very close to that given by the d^2-law, if the physical properties are calculated as recommended in Ref. 9. This prediction is also plotted on Fig. 4 by a broken line.

An important result of the present investigation is the possibility to determine precisely the time evolution of the flame diameter. This evolution seems to present three distinct phases (Fig. 4). During the first phase which lasts from the flame build-up to one-fifth of the burnout time, the flame diameter increases from three to seven times the initial droplet diameter, which is its maximum value. During this phase, in our spark-ignited experiment conducted in a cold environment, the fuel vapour concentration in the vicinity of the droplet is initially quite low but increases significantly after ignition. Consequently the flame lies in close proximity to the droplet during the beginning of this period. As the fuel vapour concentration increases, more fuel vapour can be used for combustion and the flame propagates outward. In the second phase, (d_f / d_{so}) decreases steadily and almost linearly down to 4.5. This phase lasts up to approximately four-fifth of the burnout time. During the last phase the decrease of (d_f / d_{so}) continues with an increasing rate until extinction. The rate of decrease of (d_f / d_{so}) during this phase is less than the rate of its increase during the flame build-up phase.

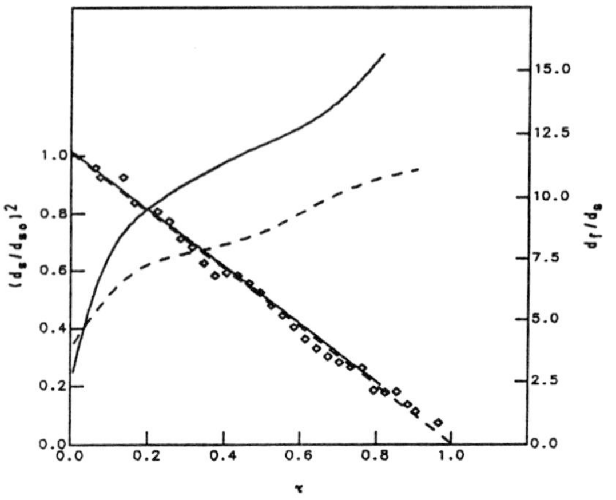

Figure 5. Comparison between our experimental results (--) and a numerical simulation (—) (Scherrer)

As a consequence of the previous evolutions, the flame-front standoff ratio (d_f / d_s) increases continuously to reach a maximum value around eleven just before extinction (Fig. 4). This result contradicts the well known prediction of the d^2-law concerning the constant behaviour of the flame-front standoff ratio, but is in accordance with the hypothesis of the fuel vapour accumulation (Ref. 4). The comparison between our experimental results and a numerical simulation based on a non-stationary model (Ref. 10) shows a good agreement for the droplet surface area regression rate and a satisfactory trend for the evolution of the stand-off ratio (Fig. 5)

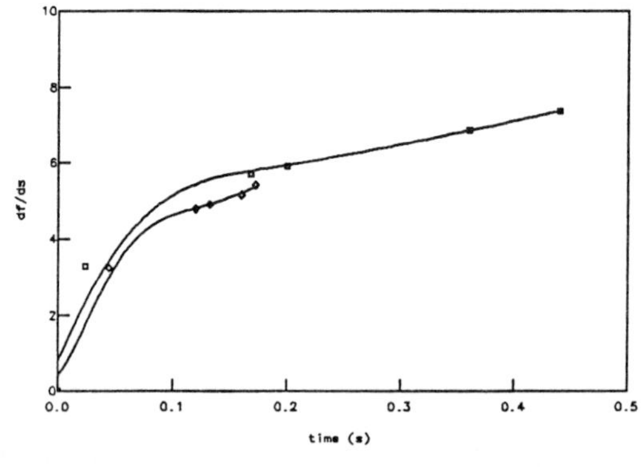

Figure 6. Comparison between the flame-front standoff ratio of a free-floating droplet (◊) and a fibre-mounted droplet (□).

As explained above, the burning of the free-floating n-heptane droplets is only observed for one-tenth of the burnout time, which corresponds to the flame build-up period. The similar behaviour of (d_f / d_s) during this phase for suspended and free-floating burning droplets is shown on Fig. 6.

3.3. Droplet burning experiments under high pressure

High pressure droplet burning experiments are an important part of our research program. The reduced gravity environment is an ideal situation to conduct basic studies on this topic. The general importance of high pressure droplet vaporization and burning is related to the occurrence of these processes in technological systems such as diesel engines, liquid rocket propulsion or gun systems using liquid propellants.

One important question related to these problems is to know if the liquid phase could reach its thermodynamical critical state during vaporization or burning. If critical conditions are approached, the quasi-steady gas phase hypothesis will not hold and in the presence of natural or forced convection, the "dense gas phase" will go through very complex geometrical shapes.

When the temperature of the droplet approaches the critical temperature of the liquid and if the ambient pressure is higher than the critical pressure, the surface tension and the latent heat of vaporization vanish; also the physical properties of the liquid phase, such as its density and transport coefficients, approach gas phase values. The system becomes totally non- stationary as the transport phenomena time scales and the time scale of the interface regression become comparable.

From the experimental perspective, Hall and Diedrichsen (Ref. 11) performed one of the first high pressure suspended droplet burning experiments. For hydrocarbon droplets, they observed the validity of the d^2 law up to 20 bars; the total burning time t_b was observed to decrease with $P^{-0.25}$ and the instantaneous flame position with $P^{-0.5}$. This means the increase of the heat transfer from the flame to the droplet and the increase of the droplet temperature. With the same technique, Goldsmith (Ref. 12) observed the burning of benzene droplets up to 5 bars. He showed that the burning rate was proportional to $P^{0.4}$.

Faeth et al. (Ref. 13) were the first to realize the importance of conducting high pressure droplet burning experiments in reduced gravity, in order to remove the enhanced natural convection effects, but also to be able to increase the operational pressure range, by preventing the droplet to fall off from its fibre because of vanishing surface tension at high pressures. They used a drop tower to create the reduced gravity environment. Their results for n-decane droplets showed a different pressure dependence of the burning time in subcritical and supercritical regimes : the combustion time decreased in the subcritical regime but increased with $P^{1/3}$ in the supercritical regime. The maximum flame diameter was found to vary with $P^{-1/3}$.

More recently, Kadota and Hiroyasu (Ref. 14) studied the pressure effect on the burning of n-heptane droplets, up to 1.5 times the critical pressure of the fuel. They conducted their experiments under normal gravity. Their results showed a three staged behaviour for the dependence of the burning time t_b, against the normalized pressure $P_r = P/P_{cr}$. For $P_r < 0.3$, t_b decreased with $P^{-0.2}$. For $0.3 < P_r < 1$, a sharper decrease of t_b with pressure was observed. Finally, for $P_r > 1$, they found also a decrease of t_b with P_r, but with a reduced rate. This emphasizes, when compared with the results by Faeth et al., the importance of the influence of gravity on high pressure droplet burning.

During the first phase of our experimental program, we investigated the influence of pressure on the burning rate of n-heptane droplets for a pressure range up to 0.7 MPa. Fig. (7) presents the comparison of the squared droplet diameter versus time for different pressures. The total burning time decreases with increasing pressure and consequently the average burning rate, K_m, which is calculated as the average slope of each of the previous curves, increases. This result confirms the above summarized earlier results for moderate pressure levels, $P_r < 0.3$.

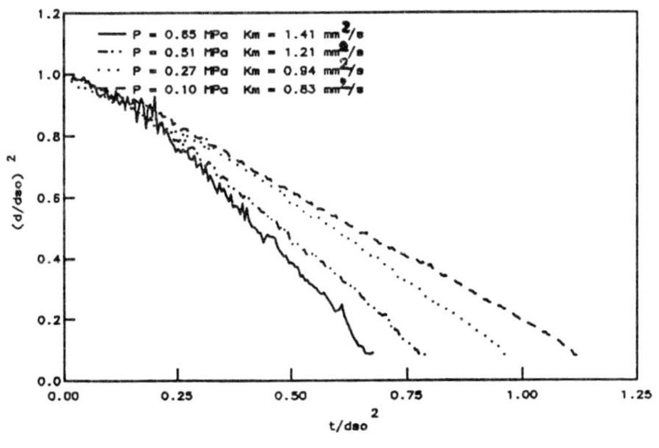

Figure 7. Variation of d_s^2/d_{so}^2 with time. Influence of pressure.

We evaluated the instantaneous burning rate from the time variation of the squared droplet diameter as $K(t) = -d(d^2)/dt$. Fig. 8 presents K(t) for the previous pressure values. For all pressures, the burning rate increases during a first period and approaches a quasi-stationnary value. For each pressure, the corresponding K_m value is also shown As expected, when the pressure increases, the non-stationary behaviour of the burning rate is accentuated.

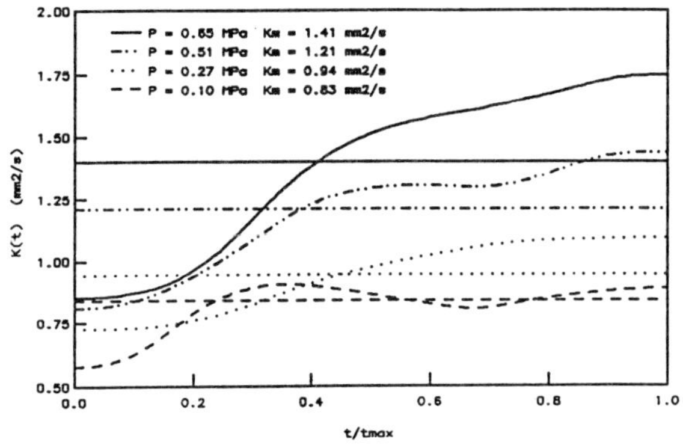

Figure 8. Instantaneous burning rate. Influence of pressure.

4. CONCLUDING REMARKS

Concerning the use of parabolic flights in burning droplet experiments, the results presented here show both their utility and their limits. The approximately 15 seconds of operational reduced gravity period allowed the simultaneous observation of the droplet and flame diameters for large fuel droplets, which, more easily than with small droplets, permit the differences between the buoyant and non-buoyant droplet combustion to be observed. The presence of an operator during the effective reduced gravity period is also an important factor in the conduction of the experiments. It should also be mentioned that the use of parabolic flights in reduced gravity combustion experiments may be considered as a first step to deal with the constraints of manned space experiments in this field. The limits of the parabolic flights are related to the relatively high level of the residual gravity. But this limit may be relaxed by allowing the free floating of the whole experimental set-up during the reduced gravity period of the parabolic flight.

However, the results presented here constitute the first clear experimental evidence of the non-steady behaviour of droplet burning whereas the spherical symmetry is conserved, by showing essentially that the flame first moves away from the droplet, then back towards it, and that the ratio of flame to droplet diameter increases continuously during combustion. In addition, preliminary results concerning the influence of pressure on the burning rate of fuel droplets are obtained; they are in good agreement with previous experimental results and theoretical predictions in the moderate pressure range. The high pressure aspect of our program will be deepened in our next parabolic flight campaigns and constitutes also a part of our future participation to the NASA program on Microgravity Combustion Science and to the ESA Columbus Precursor Flights.

Acknowledgments

This work is supported by the Microgravity Office of the European Space Agency and by the Centre National d'Etudes Spatiales (C.N.E.S.). We wish to acknowledge the efficient monitoring of these supports by H. WALTER (E.S.A.), O. MINSTER (E.S.A.), R. BONNEVILLE (C.N.E.S.) and B. ZAPPOLI (C.N.E.S.). Many thanks also to A. GONFALONE and V. PLETSER from the ESTEC/ESA staff at Noordwijk, Netherlands, for their enthusiastic support and assistance. It is also a pleasure to acknowledge the hospitality of the C.E.V. Brétigny.

References

1. Walter H.U. (Ed.) *Fluid Sciences and Materials Science in Space*, Chapter IX, Springer-Verlag, Berlin, 745 p., 1987

2. Sacksteder K. , *Combustion Science Research in the Space Station Freedom*, presented at the Modular Combustion Facility Assessment Workshop, NASA Lewis Research Center, May 17-18, 1989.

3. Williams F.A. , *Combustion Theory*, second edition, The Benjamin/Cummings, Menlo Park, California, 1985.

4. Law C.K., Chung S.H., and Srinivasan N., *Combust. and Flame*, **38** : 173-198, 1980

5. Chauveau C. and Gökalp I. Experiments on the burning of n- heptane droplets in reduced gravity. *Proceedings of 7th. European Symposium on Materials Sciences and Fluid Physics in Microgravity*. ESA-SP-295, pp. 467-472, 1990.

6. Chauveau C. Vaporisation et Combustion des Gouttes Isolées de n-Heptane. Etude en Microgravité et Influence de la Convection Forcée. *PhD. Thesis*, Université d'Orléans. 1990

7. Gökalp I., Chauveau C., Richard J.R., Kramer M., and Leuckel W. , *Twenty-Second Symposium (International) on Combustion*, The Combustion Institute, Pittsburgh, pp.2027-2035. 1988

8. Gökalp I, Chauveau C., Richard J.R., Kramer M. and Leuckel W. Droplet Vaporization and Combustion in Microgravity, *ESA SP-1113*, ESA Publications Division, p. 25, 1989.

9. Avedisian C.T., Yang J.C. and Wang C.H. 1988, *Proc. R. Soc. Lond.*, A **420**, 183-200, 1980

10. Scherrer D. ONERA RT 34/6112 EY, 1987.

11. Hall A.R. and Diederichsen J. *Fourth Symp. (Intern.) on Combustion*, Williams and Wilkins, p. 837, 1953 .

12. Goldsmith, M. Jet Propulsion, 26, p. 172, 1956

13. Faeth G.M., Dominicis D.P., Tulpinski J.F. and Olson D.R. *Twelfth Symposium (International) on Combustion*, The Combustion Institute, pp. 9-18, 1969

14. Kadota T. and Hiroyasu H. *Eighteenth Symposium (International) on Combustion*, The Combustion Institute, pp. 275- 282, 1981.

THE MECHANISMS OF TWO PHASE FLOWS SEPARATION AND LIQUID BOILING REALIZED IN LONG DURATION MICROGRAVITY ON THE BASIS OF SEMI-PENETRABILITY EFFECTS

V.N.Serebryakov

Science and Production Association "Energia", Kaliningrad, MD, USSR

ABSTRACT

The mechanism of separation of two-phase media by means of semi-penetrable (hydrophobic and hydrophilic) porous membranes has been investigated, and the characteristics of technological processes based thereof have been studied experimentally in long-duration microgravity. The effect of self-semilarity of the integral average characteristic of the two-phase flow in a channel with semi-permeable walls has been established that is of practical interest for apparatus designing. The possibility of controlling the maximum size of vapour bubbles, thermal characteristics and stability of the boiling process has been shown. In "pure" microgravity conditions a previously unknown mechanism of a highly efficient heat transfer into a stationary underheated liquid has been revealed, that is based on the phase transition "evaporation-condensation" and thermocapillary convection with a stationary form of vapour bubbles on the heater surface.

Keywords: *Two-phase Flows Separation, Boiling Crisis, Hydrophility (-phobity), Semi-penetrable Porous Membranes.*

Some investigation results which have defined the way and principles of implementation of technological processes running in two-phase gas-liquid media in microgravity are discussed in this report. The essence of the way consists in suction of phases through capillary-porous hydrophilic and hydrophobic membranes which posses selective phase penetrability (semi-penetrability), only for liquids or gases, respectively, provided that $0 < \Delta p_s < p_\sigma$. (Here $p_\sigma = 2\sigma \cos\Theta r^{-1}_{max}$ is a capillary pressure, σ is surface tension at the liquid-gas interface, Θ is a wetting angle, r_{max} is a maximum radius of the porous membrane capillaries, $\Delta p_s = k(r_{mid})U_s$ is a pressure differential created within the membrane to provide the required suction speed U_s of the wetting phase through the membrane, $k(r_{mid})$ is a porous structure penetrability coefficient which depends on the integral average radius of capillaries r_{mid} (Figure 1).

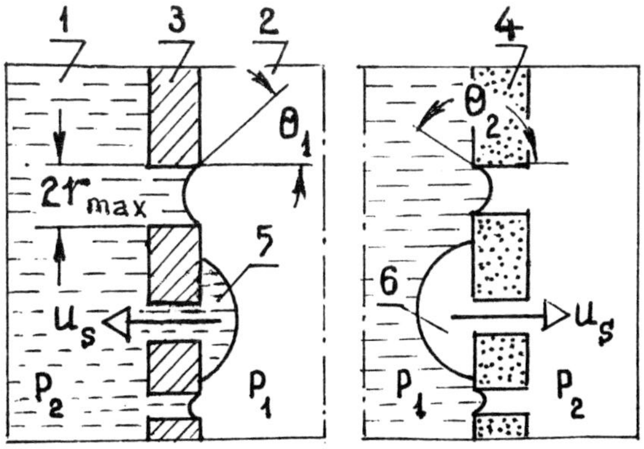

Figure 1: Phase separation by suction throw the semi-penetrable porous membranes (**1**- liquid, **2**- gas, **3**- hydrophilic membrane, **4**- hydrophobic membrane, **5**- liquid drop, **6**- gas bubble, U_s- suction speed, $p_1 - p_2 = \Delta p_s$ - a pressure differential, r_{max}- maximum radius of capillaries.

Previously, (Ref.1), it was shown through the analysis and experiment that gas-liquid phase separation occurred in the slots between the mentioned membranes in microgravity; in this case as suction through the porous membrane is being performed, any of the inclusions of the mixture (a drop, a bubble) attains some "critical" minimum volume, $v(r^*_0)$, when hydrodynamic stability of the gas-liquid interface is being lost (see Firure 2), and complete separation of inclusions from the non-wettable membrane or a break through the inclusion body takes place, (see Figure 3).

Figure 4 shows the results of the numerical analysis of the inclusion separation diameter, r^*_0 (r^*_0 on the non-wettable wall) as a function of the hydrophoby (Θ_1) and wetting ability (Θ_2) levels of the membrane surfaces, that was of practical interest for attaining maximum removal of phases from the slot between the membranes provided by the condition $r^*_0 \to 0$. The analysis was performed on a

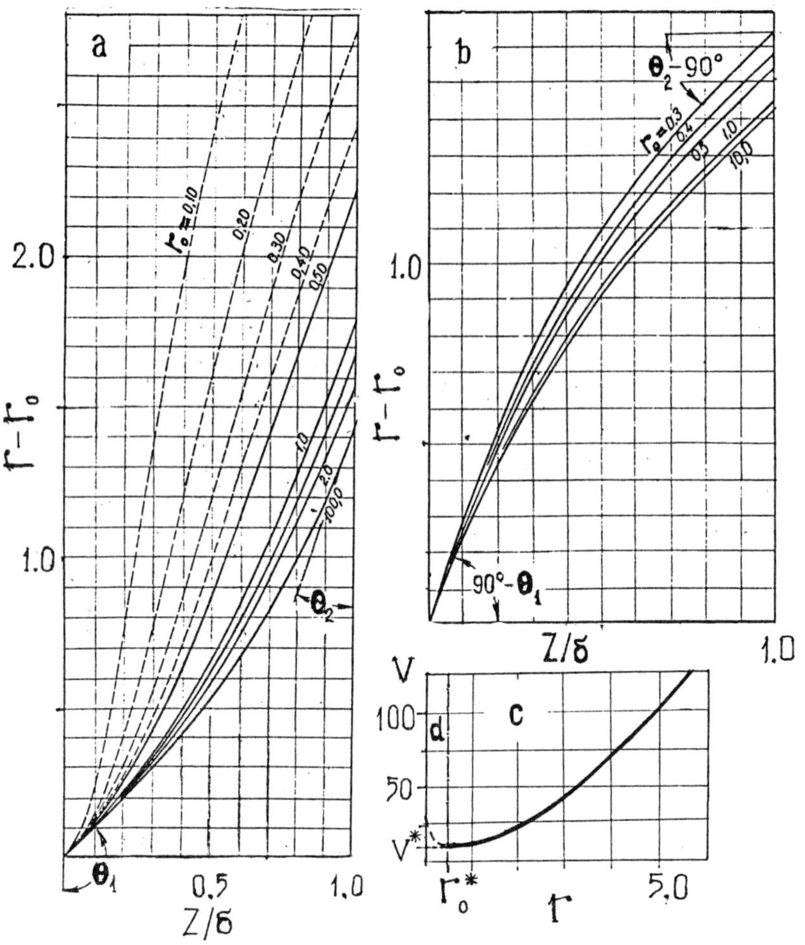

Figure 2: Drop (**a**) and bubble (**b**) surface form dynamics during the phase separation process, **c**- the drop contact radius $\mathbf{r_0}$ on the non-wettable wall and drop volume **v** relation, **d**- zone of instability, (————)- stable forms, (– – – –)- unstable forms, Θ_1=130°, Θ_2=20°, $\mathbf{r^*_0}$ and **v*** is a "critical" radius and volume.

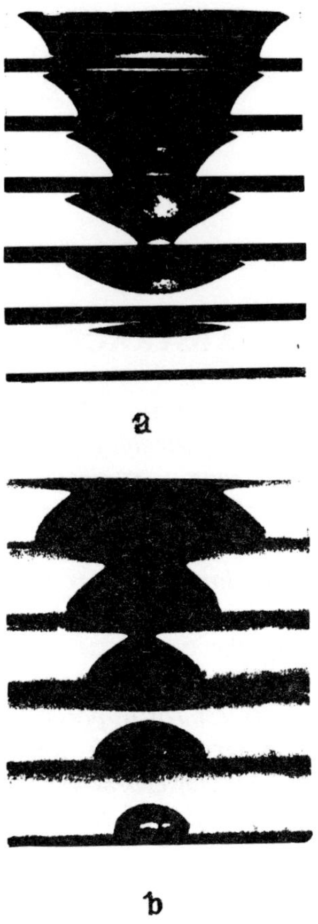

Figure 3: Separation of water drop **a** and air **b** bubble from the slot between hydrophilic (upper) and hydrophobic (lower) membranes (duration of process: **a**- 4s, **b**- 2s).

basis of variational methods through successive calculations of a series of equilibrium (minimal) surfaces with a decreasing volume to attaining a minimum volume **v*** (or radius $\mathbf{r^*_0}$) when a stable existence of the inclusion contacting with two walls is possible (according to the technique described in Ref.1). The calculation was based on the boundary problem solution with a search for eigenvalues of λ parameter characterizing the surface pressure at the liquid-gas interface: $rr'' - (1+r'^2)^{1/2} - (1+r'^2)^{3/2}\lambda = 0$ with the boundary conditions $r'(0)=tg(90°-\Theta_1)$, $r'(1)=tg(90°-\Theta_2)$ and a given current parameter - drop volume **v** or a radius on the hydrophobic wall $\mathbf{r_0=r(0)}$; $\mathbf{r=R/\delta}$ is a dimensionless cylindrical coordinate. Investigation throughout the possible hydrophoby (Θ_1=90°...178°) and wetting ability (Θ_2=5°...90°) ranges shows that the condition $\mathbf{r^*_0 \to 0}$ can be achieved only for convex surfaces ($\Theta_1+\Theta_2$>180°) which are characteristic for gas cavities in the slot. In case of convex-concave surfaces ($\Theta_1+\Theta_2$<180°) specific to a liquid drop, a hydrophoby increase above Θ_1=120°...130° makes no sense at all, and inthis case for readily liquid-wettable surfaces (Θ_2=5°...10°), a hydrophoby increase even results in some increase of the drop separation radius. (It should be kept in mind here that the wall wetted with liquid serves as a "hydrophobic" one for a gas cavity).

The method considered makes it possible to realize different technological processes in microgravity: mass exchange on bubbling of a liquid with a gas, liquid separation from two-phase flows, boiling and condensation (Ref.2), which hydrodynamic mechanism is apparent also from the mentioned above.

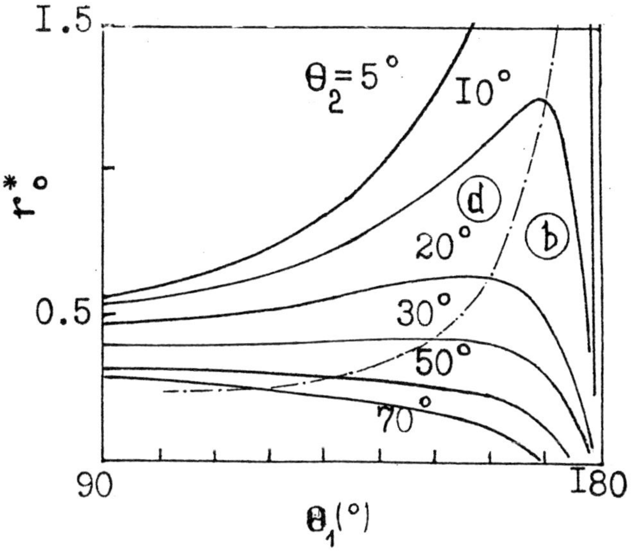

Figure 4: Drop ("**d**"-zone) and bubble ("**b**"-zone) separation as a function of hydrophoby, Θ_1, and wetting ability, Θ_2, levels.

For available usable two-phase mixtures with a small relative liquid flow rate (φ of about 1%) the separator design in Figure 5a is of the greatest practical interest. In this circuit the main point is to determine the path length L_s of the flow separation along the porous wall when the complete separation of the liquid is achieved, i.e. the content by volume of the liquid in the mixture in the given channel section is $\psi \to 0$ (see Figure 5a). Provided that liquid drops ("locks") in the separator slot channel are moving only contacting the slot walls and upon separating from impenetrable wall 3 as a consequence of suction through porous wall 2 the drop motion stops, the liquid mass balance condition in the channel with the width of **h** will be $h\delta \cdot dw = U_s h \psi(x)dx$ where $\psi(x)$ is a relative total portion of the channel section filled with the liquid; U_s is a liquid suction speed, **w** is average reduced liquid flow rate in this section, cm3/cm2s, related to the drop motion speed U_l by means of $\psi = w/U_l$. By integrating within $\int_{w_0}^{0} dw = U_s \delta^{-1} \int_0^{L_s} \psi(x)dx$ and introducing a dimensionless coordinate $X = x/L_s$ we receive $L_s/\delta = w_0/(\psi_m U_s)$, where w_0 is a reduced liquid flow rate at entry and $\psi_m = \int_0^1 \psi(X)dX$ is an integral average-by-length L_s relative value of the "wetted" surface of the porous membrane which determination is a critical point depending on hydrodynamic conditions in the channel. Considering the known asymptotic character of the dependence of the liquid portion ψ remained in the channel upon displacement by a gas

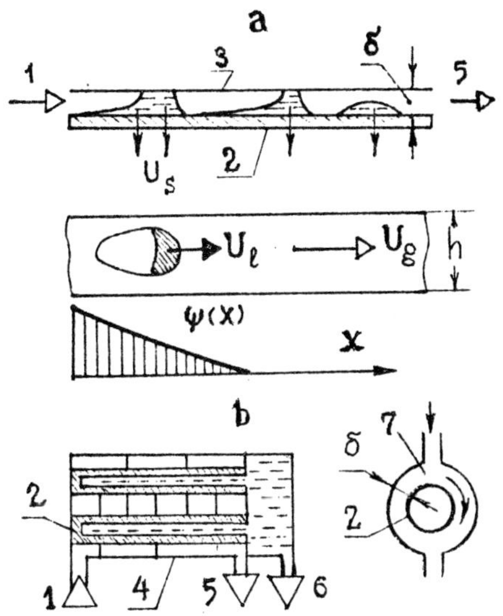

Figure 5: The channel circuit (**a**) and structure (**b**) of the separator for liquid (**1**- gas and liquid input, **2**- hydrophilic membrane, **3**- impenetrable wall, **4**- housing, **5**- gas output, **6**- liquid output, **7**- cross section of the separator channel.

on the Taylor number $\mu U_g/\sigma$, one can suppose that with $Te > 5 \cdot 10^{-2}$, the ψ_m value will be conservative relatively of the average speed of gas motion U_g in the channel. Figure 6 presents the

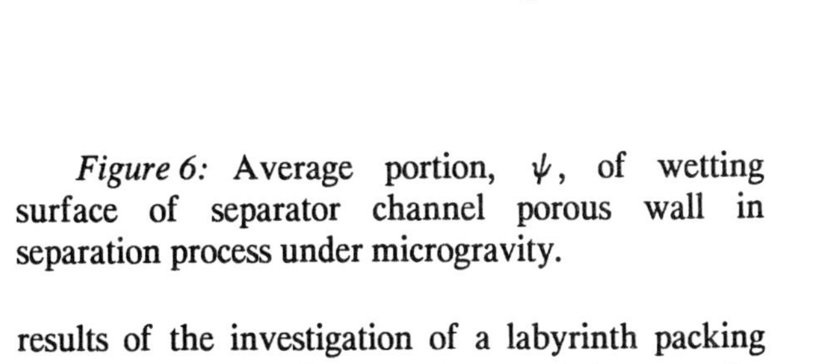

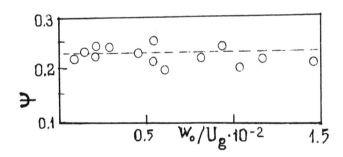

Figure 6: Average portion, ψ, of wetting surface of separator channel porous wall in separation process under microgravity.

results of the investigation of a labyrinth packing separator (see Figure 5b) carried out for $Te = (5...10) \cdot 10^{-2}$, $Re_{(gas)} = (6...12) \cdot 10^3$ and a relative water flow rate $\varphi = w_0/U_g$ from 0.5 to 1.5%. One can see that the $\psi(x)$ function retains similarity within the given range of hydrodynamic conditions, in this case the average portion of wetted surface ψ_m remains within 0.2...0.25, and the path length of separation can be expressedby the simple empirical relation mentioned above.

Figure 7: Separator of liquid with a transversal flow around porous tubes in a labyrinth channel under the microgravity condition (gas flow speed of 5 m/s).

Figure 7, present a typical water separation process from the air flow observed in short-term microgravity (40...45 s) during a Tu-104 flying laboratory flight, as well as during prolonged microgravity - on board a satellite with orbit altitudes from $3 \cdot 10^2$ to $3 \cdot 10^4$ km. It should be noted that with general convergence of the results vibrations and a residual g-load of about 0.002 change the liquid flow character in the first case -one can observe liquid break-up and bubble formation in the channel. In case of "pure" weightlessness uniform-periodical motion of the liquid "lock" along the sections of the separator is being observed with its gradual "spreading" over the channel walls.

Experimental studies with the use of the system "heater-hydrophobic membrane" show that the mechanism of nucleate boiling of a liquid in microgravity includes the following stages (see Figure 8):

- nucleation on the heating surface and growth in the spherical form till the moment of contact with the hydrophobic membrane (see State 1 in Figure 8a); in this case the bubble growth speed closely complies with the Fritz-Ende square law (see Figure 8b);
- bubble growth in a toroidal form till the moment of breaking the liquid film on the hydrophobic membrane (State 2);
- bubble suction with decreasing the "wetted" perimeter on the hydrophilic surface till the separating ($\mathbf{r} \rightarrow \mathbf{r}^*_0$) and subsequent suction into the hydrophobic membrane (see 3 and 4 in Figure 8a).

The results of experiments on water boiling in an evaporator with an electric heater show the mentioned dynamics of bubbles at the height of the

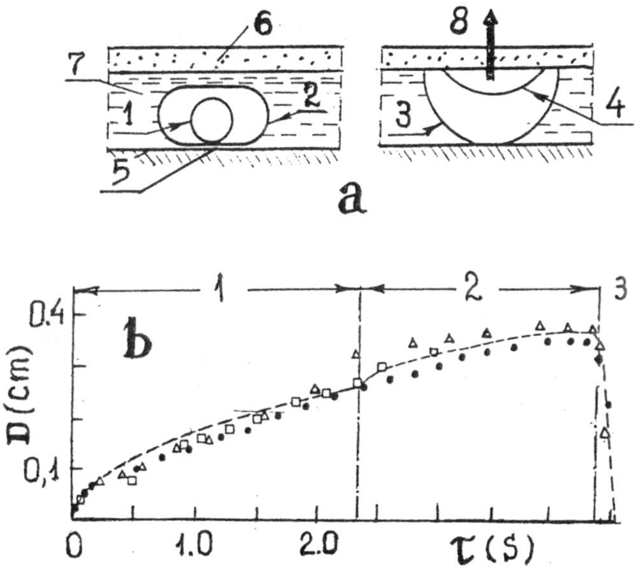

Figure 8: The mechanism of boiling on the hydrophobic membrane (**a**) and bubble growth dynamics at the height of slot σ **=0.3 cm (b)** (**1, 2**- bubble position before a break; **3**., **4**- at suction; **5**- heating surface; **6**- hydrophobic porous membranes; **7**- liquid; **8**- vapour suction.

slot between the porous membrane and impenetrable wall δ **= 3mm** (see Figure 8b and Figure 9).

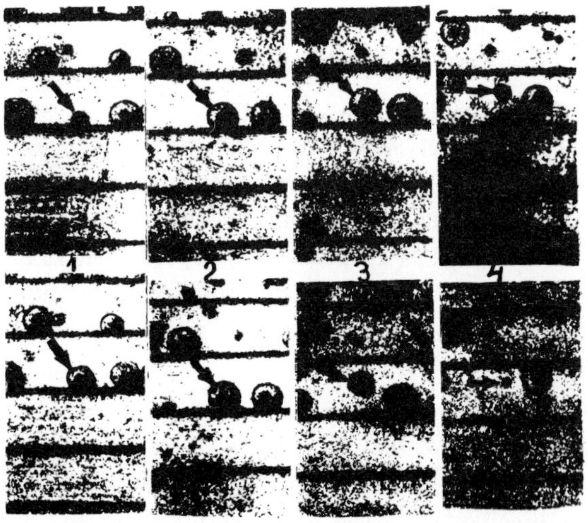

Figure 9: Bubble growth dynamics in the slot of 0.3 cm in experiment at microgravity.

Cinema frames in Figure 10 show the transition from the routine mechanism of boiling with bubble floating-up in the gravitational field to the mechanism of suction through the porous membrane when passing from **n = 1** to **n → 0.**

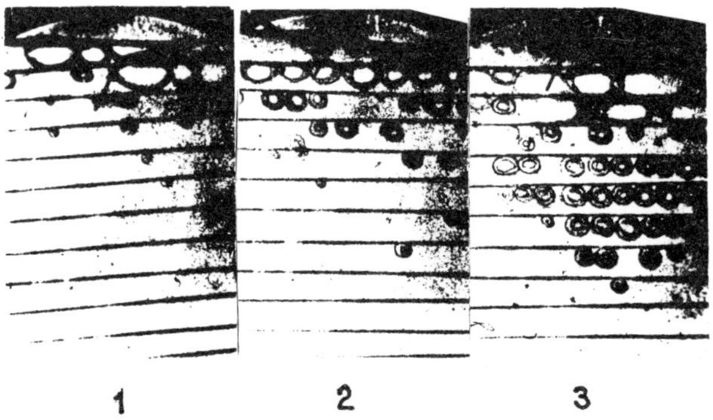

Figure 10: The passing from **n = 1** to **n → 0** process in liquid boiling.

A significant feature of this mechanism is a possibility to control the bubble separation diameter value $\mathbf{D_{max}}$ through the use of the values of the wetting angles, Θ_1, Θ_2, and the slot δ between the impenetrable and porous walls: $\mathbf{D_{max} \approx 2\,\delta\, r^*_0/(\cos\Theta_1+\cos\Theta_2)}$ and in turn by $\mathbf{D_{max}}$ to control the most essential thermal characteristics, stability and moment of boiling crisis.

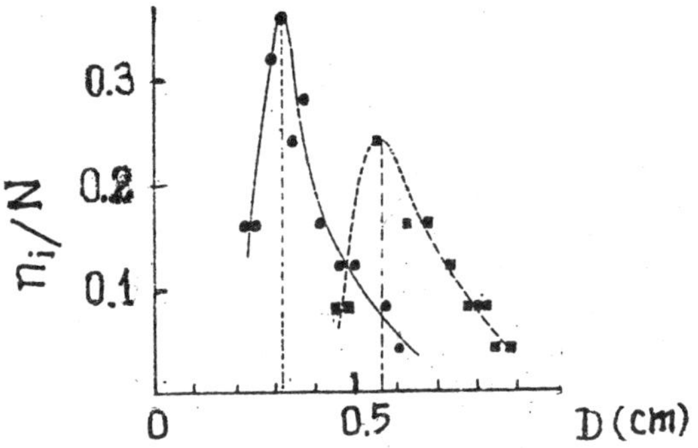

Figure 11: Bubble maximum size (diameter) distribution in the slots height of 0.28 cm (**1**) and 0.5 cm (**2**).

Figure 11 shows the measured distribution of the bubble diameter in the slots = **2.8** and **5.0mm**, and Figure 12 presents experimental dependences of heat transfer coefficients on the heat flow density, **q**.

The results confirm the above-mentioned possibility:

- at the slot height of δ = **2.8mm** the thermal characteristic reflects clearly defined nucleate boiling with signs of crisis at $\mathbf{q \approx 8 W/cm^2}$;

- at δ = **5.0mm** the coefficients are substantially lower (by 55 - 60%) and with flow densities $\mathbf{q \approx 4...6 W/cm^2}$ phenomena are observed which are typical of boiling crisis with coalescence (see cinema frames in Figures 13a, 13b). In this case, for both series of experiments the potential vapour suction

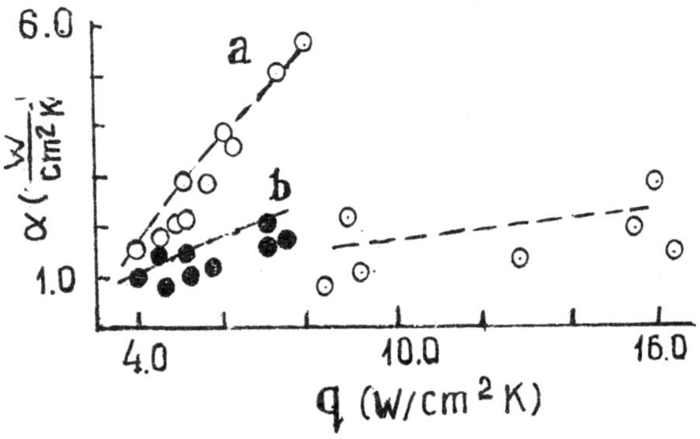

Figure 12: Experimental dependences of heat transfer coefficients, α, on the heat flux density, **q**, in the slots height of 0.28cm (**a**) and of 0.5cm (**b**).

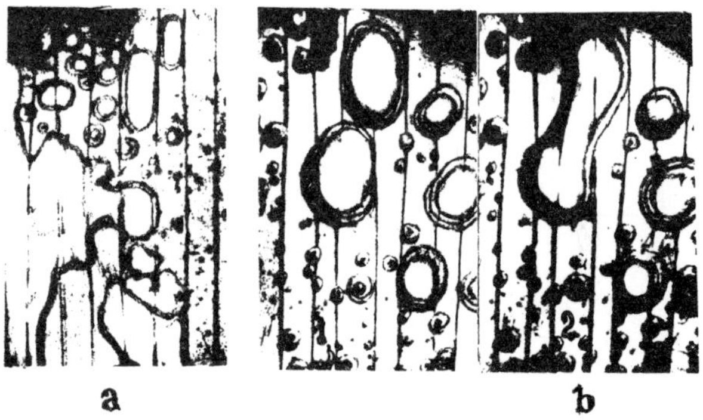

Figure 13: Boiling crisis at the slot height of 2.8mm **a**,$\mathbf{q=8W/cm^2}$,and 5.0mm **b** ,$\mathbf{q=4W/cm^2}$, under the microgravity.

speed provided by the hydrophobic membrane exceeds the vapour generation speed no less than by an order.

The most interesting fact revealed in the course of investigations in prolonged "pure" weightlessness is the existence of a stable mechanism of heat transfer with heat flows of high densities conforming to nucleate boiling but having no signs of discrete, periodic character of nucleate boiling. The temperature dynamics determined during the experiment (Figure 14) and cinema frames (Figure 15) illustrate the following features of the process of attaining a stationary regime by the evaporator with the constant heat flow density of $3 W/cm^2$.

The character of the temperature rise rate for the heater, $\mathbf{T_h}$, and the liquid being evaporated, $\mathbf{T_l}$, and their difference, $\mathbf{\Theta=T_h\text{-}T_l}$, comply with the heat transfer through heat conduction only in the initial period of about 1 min; upon reaching the regular thermal regime, still in underheated liquid, a sharp decrease of the thermal resistance and temperature

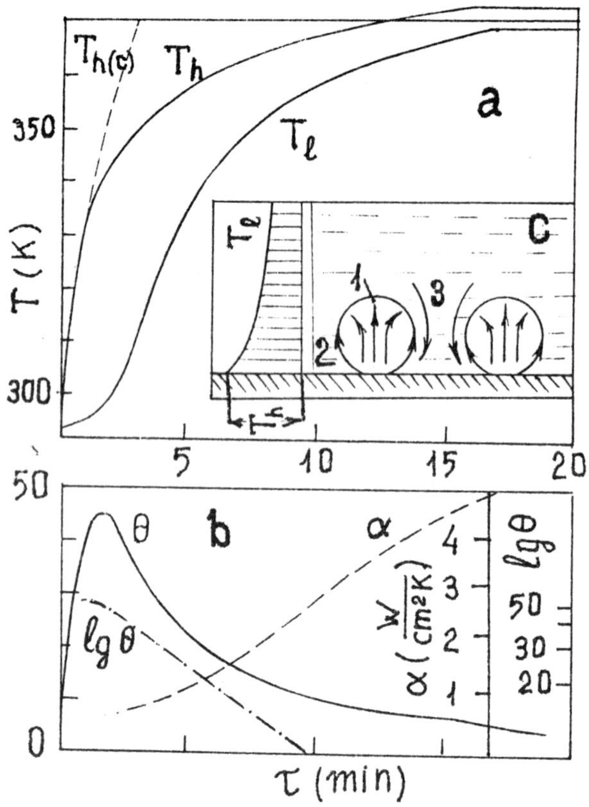

Figure 14: The liquid boiling in a long duration microgravity investigation results: **a**- the temperatures with time (**T_h**- heater temperature, **T_l**- liquid temperature, **$T_{h(c)}$**- heater temperature calculated by thermoconductive model); **b**- heat transfer characteristics throughout of the transition process; **c**- the mechanism revealed (**1**- vapour flow, **2**- thermocapillary convection, **3**- return liquid flow).

difference between the heater and liquid comes about. As may be seen from the cinema frames, this process is accompanied by formation of typical, stable with time and not mobile bubbles covering the surface of the wire electric heater. In this case, in the course of time the heat transfer intensity between the heater and liquid increases steadily (α up to $(3...4)\cdot 10^3 W/m^2K$), that is accompanied by the growth of the stationary bubble dimension. This effect can be explained by developing the two intensive heat transfer mechanisms such as "evaporation-condensation" within stationary vapour bubbles and the thermocapillary convection, which are provided by the action of the temperature gradient in the underheated liquid forming the bubble surface (see Figure 14c). It is significant for

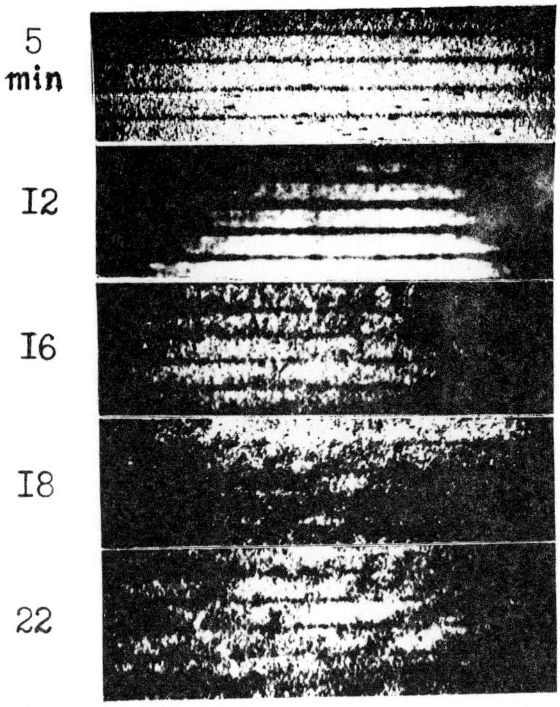

Figure 15: Cinema frames of process.

the mechanism observed that the rate of changing the temperature difference obeys the law of the regular thermal regime **$\ln(T_h-T_l)=K\tau$**, (**K - const.**) that points to the limiting role of heat conduction ofunderheated liquid film in the total heat transfer resistance (Figure 14b). It is evident also that temperature gradient along the heater surface between a neighboring vapour generation points exists as the measured average temperature of wire remains lower than boiling point at some time of the transition process. The normal boiling in this system is observed on attaining the saturation temperature by the liquid.

The results make more exact the popular conception that heater surface must be covered by an unstable vapour film with subsequent overheating of the surface in microgravity.

The observed mechanism of high intensive heat transfer during prolonged zero-g conditions is of interest for further theoretical and experimental study.

References

1. V.N.Serebryakov. Space Research (1966) v.4, iss.5.
2. V.N.Serebryakov. Designing Principles of Life-Support Systems for Spacecraft Crews. Moscow, Machinebuilding (1983) p.157.

A TWO-PHASE HEAT TRANSPORT LOOPS FOR LARGE SPACE PLATFORMS: THE TENDENCIES OF DEVELOPMENT, THE CONCEPT OF DESIGN, THE TECHNICAL AND SCIENTIFIC PROBLEMS.

A.A.Nikonov
SIC "Energy",Kaliningrad,Russia 141070
G.A.Gorbenko and V.N.Blinkov
Aviation Institute,Kharkov,Ukraine 310084

ABSTRACT

The increased thermal power levels and thermal control requirements on board of spacecraft stimulated the use of two - phase thermal control systems. The analysis of modern concepts of two - phase heat transport loops for large space platforms is presented. Some important scientific problems that are actual for any concept of two - phase loops are shown.

Keywords: Two - phase thermal control system, Boiling and condensation heat transfer in reduced gravity, Networks of heat exchangers, Mathematical models of two - phase loops.

Most of the technological experiments on space platforms require an exact thermal control. Growing thermal power levels and transport distances on board stimulated the use of the two-phase thermal control systems /TPS/ concept. Such systems are actively designed since the early 1980s for future platforms like Freedom /NASA/, Columbus /ESA/ and the Multipurpose Space Platforms MSP / SIC "Energy"/ /Refs. 3,10/. The data on two - phase flows hydrodynamics and heat and mass transfer are needed for qualitative TPS design.

The advantages of two - phase systems as compared to single - phase are:

- lower flow rates and mass of working fluid
- reduced pump power consumption
- higher rate of boiling and condensation heat transfer, which leads to reduced cold plates mass
- a virtually isothermal behaviour along the heat acquisition and rejection lines
- a possibility to develop a passive TPS with circulation due to the direct thermodynamic cycle.

According to estimates and calculations given in / Refs. 12,13 / the advantages of TPS manifest themselves evidently for a heat load of more than 10 KW. However, several problems must be considered:

- the dependence of hydrodynamics and heat transfer at low flow velocities upon gravity, which requires flight testing
- the necessity to develop the new types of heat exchangers / evaporators and condensers /

- the strong influence of heat load and vapour quality on pressure drops in lines and heat exchangers, that causes difficulties in the stabilization of flow rates through parallel legs
- the probability of steady and oscillatory instability
- the probability of cavitation in the pump and cavitative erosion.

There are significant benefits of the "thermal bus" concept to provide user's require - ments, long life time / up to 10 years /, reliability. According to the " thermal bus " concept, the primary elements of TPS are the central heat acquisition and transport loops, and radiator for heat rejection. The users interact with the system through heat exchangers - evaporators, which provide nearly isothermal behaviour of sinks. On the platforms mounted in space / Freedom / the individual thermal control systems access the central loop through mechanical contact heat exchangers. Three basic projects of TPS for Freedom Space Station are simultaneously being developed in the USA / Ref.4 /.

Grumman Aerospace Corporation works on the pumped two - phase systems which use plates responding to sensors that determine the amount of liquid remaining in the plates / Refs. 4,13 /. Lockheed Missiles and Space Corporation develops the pumped two - phase system where flow rates through evaporators and condensers are controlled by a porous material/Refs.4,7/. From our point of view the most attractive is the project developed by Boeing Aerospace and Sundstrand Corporation /Refs.4, 9,11 /. The key component of their concept is multifunctional rotary fluid management device /RFMD/ which provides:
- liquid supply to evaporators by the Pitot pumps
- vapor supply to condensers
- liquid/vapor separation after evaporators
- regeneration of heat
- control of liquid subcooling at the entrance to the cold plates.
- non-condensable gas venting
- control of liquid inventory in the loop /together with accumulator/.

The simple convective evaporating and condensing heat exchangers are intended to be installed. Cavitating venturis provide the flow control needed by the evaporators. The concept is characterized by:
- minimum power for a pump and need for active system control
- passive control during normal operation
- insensitivity to heat loads locations
- regeneration of heat
- liquid/vapor separation before vapor supply to condenser
- possibility for experimental evaluation of the concept on the ground
- unsensitivity to non-conden - sable gas blockade and easy venting.

Primary verification of the system concept was accomplished by ground and flight testing. All three projects provide temperature difference between acquisition and rejection lines of about ΔT = 3-5K. Presence of difference between source and sink temperature levels allows to organize the direct thermodynamic cycle, and to use its mechanical work for the pumping of working fluid. We call such a systems as "thermocirculating loops TCL" /Ref.1/, emphasizing the moving force of pumping is the transferred heat . TCL's do not consume an electrical power, they accept passive TPS design and they are characterized by high reliability and autonomy. Two-phase capillary pumped loop CPL by OAO Corporation is an example of TCL /Ref. 13/. CPL was tested in the USA and demonstrated the ability to transport 12 KW over 10 meters. The guaranteed pressure head of capillary pump is 20 KPa, thermal resistance is from 0,2 to 0,005 K/W /Ref.2/. The performance of CPL is influenced by gravity because of the low velocity level, which requires verification in flight experiments. The main applications of CPL are individual thermal control systems of small satellites and instrument modules. SIC "Energy" and Kharkov Aviation Institute develop, apart from pumped TPS, thermocirculating loops TCL on the basis of jet pumps /injectors and separation vapour-liquid pumps/ or turbo pumps. The admissible pressure drop in such a loops may exceed CPL pressure drop on one order of magnitude or even more. The pressure head ΔP and temperature difference between source and sink in TCL are given by the expression

$$\Delta P = \frac{\rho\, r\, x_{ev}\, \eta}{T_{ev}} \Delta T, \quad \text{where}$$

ρ - liquid density, r - latent heat of vaporization, x_{ev} and T_{ev} - evaporator exit quality and temperature, η - efficiency of energy transforming in jet pumps or in turbo pumps. The pressure head growth requires an increase of temperature difference. The TCL on the basis of injector with refrigerants R-113, R-114 and R-113 - antifreeze mixture as working fluids has been tested. The stable "selfcirculation" operation took place at 100 KPa loop pressure drop and 30-40 K source-sink temperature difference. According to our analysis the TCL with ammonia as a working fluid will have the pressure head 200...300 KPa when ΔT = 10K. TCL on the basis of turbo pump is the most effective option: with ammonia as a working fluid, such a TCL will have admissible pressure drop 200..300 KPa when ΔT=4-5K. The drawback of TCL with turbo-pump is the presence of moving parts. The described systems work stable at heat loads ranging within ± 30% of normal values. The auxiliary pump or reservoir are needed for startup. We consider the hybrid TCL with auxiliary mechanical pump as the most attractive concept. This concept combines the advantages and weakens the drawbacks of the " pumped " and " thermocirculated " concepts. The hybrid capillary mechanically pumped loop is described in /Ref.14/.

The authors participated in a testing program of TCL with auxiliary electrical - mechanical pump 7 /Figure 1/. The scheme is similar to the loop described in /Ref.5/. The vapor-liquid mixture enters the nozzle 1a. The vapor accelerates liquid droplets as it expands in the nozzle. The high velocity mixture enters the curvilinear separator 1b where the liquid separates from the vapor maintaining the liquid velocity. A major part of liquid through the capture slot 1c enters the diffuser 1d where it is decelerated and the dynamic pressure of liquid converted to a static discharge pressure. The rest of vapor goes to the exit 1f. The condensed liquid comes back through the secondary flow injection slot 1e. At normal operation the relation $P_3 > P_1 > P_2$ takes place. The working fluid returns from the diffusers exit 1d to the nozzle 1a via the hot leg evaporator 2. The vapor moves from the exit 1f through

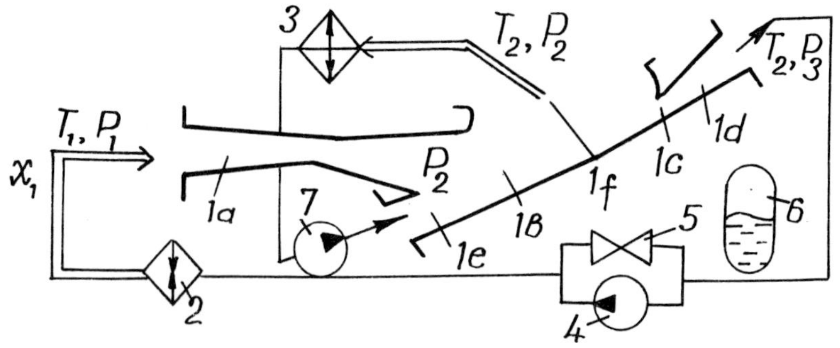

Figure 1: Selfcirculating loop with jet separation pump. —— liquid, ═══ vapor, 1a - nozzle, 1b - separator, 1c - capture slot, 1d - diffuser, 1e - injection slot, 1f - exit, 2 - evaporator, 3 - condenser, 4 - starting pump, 5 - valve, 6 - hydroaccumulator, 7 - auxiliary pump

the cold leg to the condenser 3 where it is condensed and subcooled. Condensate is pumped by the electrical-mechanical pump 7 to the secondary flow injection slot 1e. The experimental loop was able to overcome the hot leg pressure drop ΔP=200KPa when ΔT = 18K. The calculations show that using ammonia as a working fluid the loop will be able to reach ΔP = 400...600KPa when ΔT = 8...10K. It should be pointed out that two-phase jet pump in the described scheme performs almost the same functions as rotary fluid management device /RFMD/ in the Sundstrand-Boeing concept /pumping of near saturated liquid, liquid/vapor separation, flow rate and liquid inventory control/. However jet pump does not include moving parts and it is thus more reliable.

There are some scientific problems that must be solved for any option of TPS:

1. Reduced gravity pressure drop in straight and curved pipes, tees, valves, reduced - gravity boiling, evaporating and condensing heat transfer. The flight tests aboard the NASA - -JSC KC-135 /Refs.6,8/ with two-phase R-114 flows demonstrated the dependence of flow regime and friction / straight pipe L=1,85 m, D=15,8 mm / on gravitational acceleration ranging from near-zero to 1,8 g's. For qualities below 10% slug regime and for qualities above 15% annular regime were observed to exist that approximately correlates to Dukler flow regime map /Figure 2/.

The tests confirmed the prediction that two - phase pressure

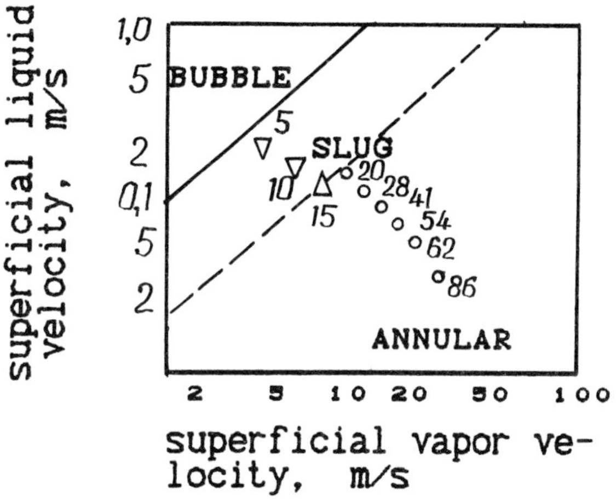

Figure 2: Dukler zero-gravity flow regime map.
Observed flow regime
○ - annular
Δ - slug/annular
∇ - slug
The number of data points are quality / % /.

drops in reduced gravity are greater than in one gravity due to prevailing of stratified regime for one gravity at low flow velocities. Reduced-gravity pressure drops were higher than one gravity pressure drops by almost 100% for pipes and by about 20% for pipes included fittings and valves. The authors analytically developed a flow regime map of boiling liquid / R-113 / in a straight heated channel for zero-gravity /Figure 3/.

The regime I is characterized by essential influence of Archimedes force on bubble departure from heated surface at ground conditions. In this region at zero-gravity conditions the developing of inverted annular flow with vapor film is expected to occur at zero-gravity conditions. The region II corresponds to the turbulent bubble /churn/ flow that is the result of film destruction due to Helmholtz instability.

Droplet /mist/ structure will occur in the region III. As for the heat transfer at zero-gravity, the methods of heat-transfer intensification represent the basic scientific problem for any concept of TPS /the use of porous and capillary materials, slots, swirl flows, ets/.

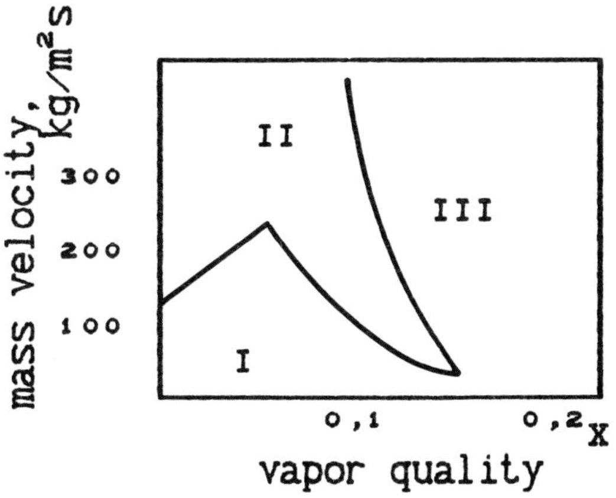

Figure 3: Predicted zero-gravity flow regime map in a heated channel
P_{in} = 0,1MPa, L = 1m, R-113, T_{in} = 313K, D = 0,01m

2. The networks of heat exchangers evaporators and condensers may include hundreds of series-parallel elements. The problem is to develop the algorithms and computer programs to predict the distribution of parameters: flow rates, qualities, pressure drops, heat transfer coefficients at any heat loads distribution. Such programs will help to analyze the effectiveness of passive flow rate requlators, cavitating venturis, capillary materials and throttles.

3. To develop the mathematical model for the prediction and analysis of steady TPS behaviour under variable heat loads and rejection. Such a model and computer program will be used at every stage of design and experimental work.

4. To develop the mathematical model for analysis of stability and transient behaviour: starting up, changes of heat load, oscillations and anomalies.

The authors presently work in the problems pointed out.

REFERENCES

1. M. A. Bespyatov, V. V. Bredikhin, G. A. Gorbenko , B. A. Goncharov , S. D. Frolov. Thermodynamic analysis of the thermocirculating loops with jet energy converter. Gas-thermodynamic of multi-phase flows in energy plants. Kharkov, KAI, pp. 85-93 /1987/.

2. Y. F. Maydanick, Y. G. Fershtater, V. G. Pastukhov. Contour heat pipes: design, analysis, technical computation. Termophysical institute, the Urals department of Soviet Academy of Sciences, Sverdlovsk, p. 52 /1989/.

3. V. J. Bilardo, W. Carlson. Space Station thermal control during on-orbit assembly. SAE Technical Paper Series 88-1070, p 16.

4. T. K. Brady. Spase Station thermal test bed status and plans. SAE Technical Paper Series 88-1068, p. 14.
5. D. J. Cerini. Circulation of liquids for MHD power generation. Electricity from MHD, Vol 3, p. 2019-2033 /1968/.
6. W. T. De Groff, C. S. Pietruszewski, R. S. Downing. Development status of a two - phase thermal management system for large space craft. AIAA 88-2703, p. 13.
7. A. Fox. Ground test unit system analysis for Space Station Freedom active thermal control system. SAE Technical Paper Series 89-9138 p. 5.
8. D. G. Hill, K. Hsu, R. Parish, I. Dominick. Reduced gravity and ground testing of a two - phase thermal management system for large spacecraft. SAE Technical Paper Series 88-1084, p. 12.
9. T. J. Kramer, D. L. Myron, M. P. Mc Hale. Two-phase ammonia thermal bus performance. AIAA 88-2701, p. 12.
10. P. Moller, H. Kreeb. Advanced thermal control technologies for European Space Station modules. SAE Technical Paper Series 85-1366, p. 7.
11. D. L. Myron, R. C. Parish. Development of a prototype two - phase thermal but system for space stations. AIAA 87-1628, p. 9
12. S. Ollendorf. Recent and planned developments at the Goddard Space Flight Center in thermal control technology. Proceeding of the International Symposium on Environmental and Thermal systems for Space Vehicles, Toulouse, France, p. 45, 4-7Oct. /1983/
13. J. G. Rankin, P. F. Marshall. Thermal management systems technology development for space systems applications. SAE Technical Paper Series 83-1097, p. 12
14. R. Schweickart, L. Ottenstein B. Cullimore, C. Egan, D. Wolf. Testing of a controller for a hybrid capillary pumped loop thermal control system. SAE Technical Paper Series 89-9476, p. 5
15. J. J. Tandler, V. J. Bilardo. An integrated model of the Space Station Freedom active thermal control system. AIAA 89-0319, p. 11.

INFLUENCE OF BIPHASE OF WORKING FLUID ON TRANSIENT PROCESSES IN LINES OF A POWER PLANT.

Y.M. Orlov.

Polytechnical Institute, 614600, Perm
Soviet Union.

ABSTRACT

The gas bubbles' or vapor appearance in the pressure lines of the aeroplane or space vehicle power plant increases the ranges of the pressure pulsation and amplitude of all the spectrum harmonics. This gives rise to increasing the dynamic loadings affecting the pump mechanizm elements and the lines walls.

1. INTRODUCTION

In the time of work of a power plant of an aircraft or of a space object the character of transient processes in the suction and pressure lines depends much on the state of the working fluid is a biphase medium. At the lack of pressure-tightness at the break of the suction line the fluid boils up and a biphase medium appears. The aim of the investigation is to determine conditions of the work of the power plant in the case of damage.

2. PRESSURE FLUCTUATION

An experimental investigation of a pressure fluctuation in pumping plant lines has been carried out using high frequencies pressure transducers and high frequencies electronic instrumentation. For pressure fluctuation measurement, piezo-electric pressure transducers were mounted in the pump inlet and outlet lines (Fig.1) of the piston pump, flush with the internal wall of the pipe in every 80 - 100 mm.

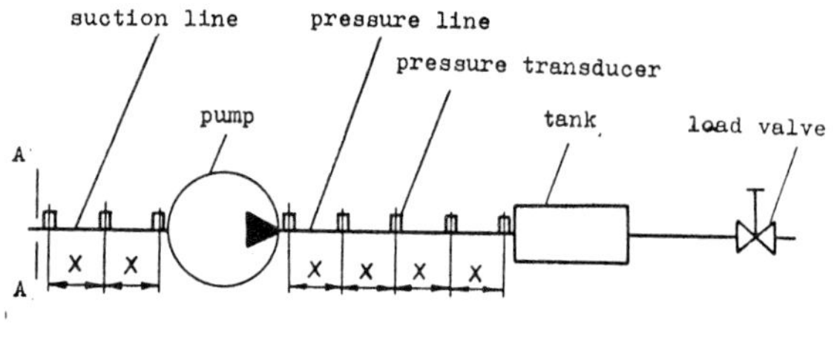

Figure 1: Schematic of the test setup.

Each transducer was connected to a cathode-ray oscillograph and a spectrum analyzer. Processes in the range of frequencies from = 20 Hz to = 10 kHz have been studied. Some gas: nitrogen or helium was injected into a section A - A of the suction line. The working fluid aviation fuel.

Gas content of the fluid in the suction line the pressure P_b = 0.2 MPa was:
0 - 10 % for nitrogen and
0 - 40 % for helium.

TABLE 1 Range pressure fluctuation

	P = 10 MPa	P = 8 MPa	P = 6 MPa	P = 4 MPa	P = 2 MPa
Pump Outlet					
Test N 1 Aviation fuel without gas MPa	2.2	1.95	1.35	0.65	0.2
Test N 2 Aviation fuel without gas MPa	2.15	1.8	1.35	0.8	0.35
Test N 3 Aviation fuel + nitrogen MPa	2.8	2.65	2.1	1.4	0.87
Test N 4 Aviation fuel + nitrogen MPa	3.0	2.6	2.1	1.45	0.7
Pump Inlet					
Test N 1 Aviation fuel without gas MPa	0.42	0.41	0.34	0.25	0.15
Test N 2 Aviation fuel without gas MPa	0.37	0.40	0.33	0.22	0.16
Test N 3 Aviation fuel + nitrogen MPa	0.07	0.06	0.03	0.22	0.01
Test N 4 Aviation fuel + nitrogen MPa	0.08	0.07	0.04	0.03	0.02

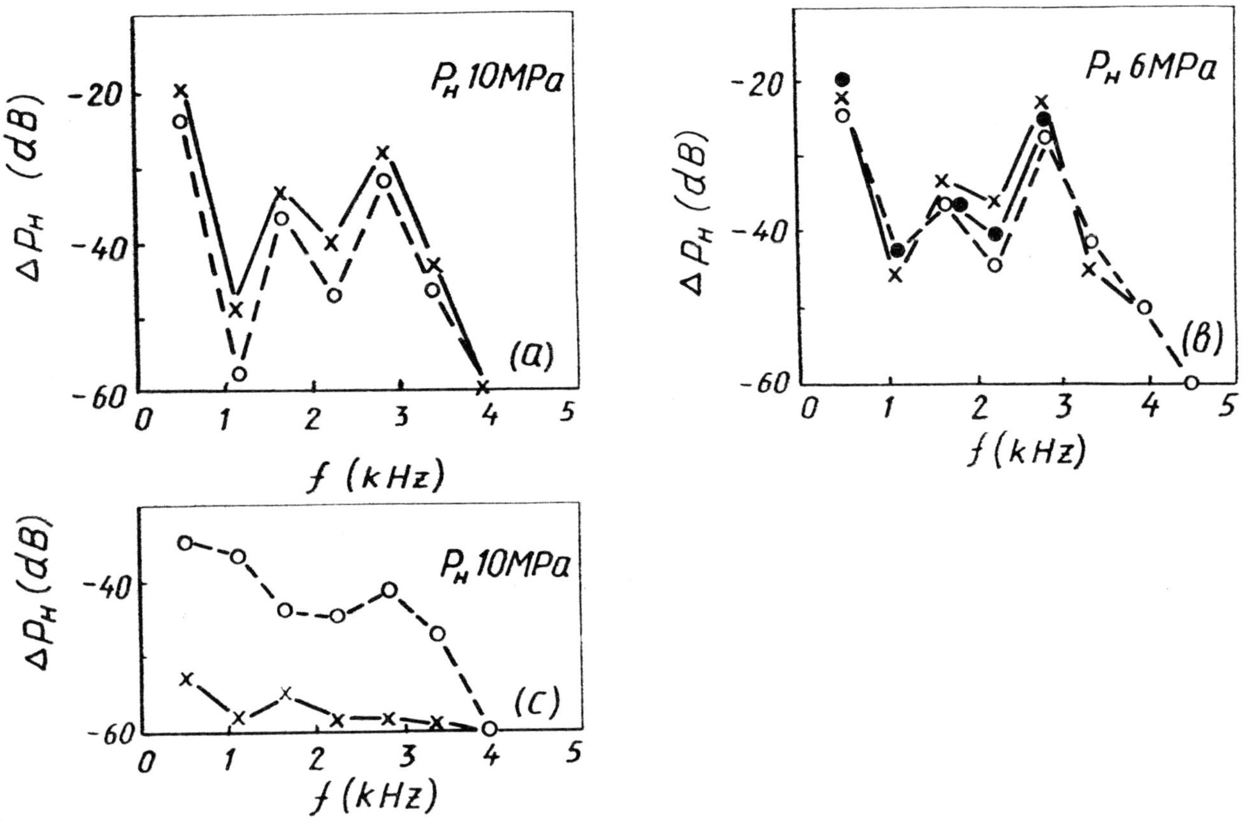

o aviation fuel without gas;

x aviation fuel + nitrogen;

• aviation fuel + helium.

Figure 2: Spectrum of piston pump generated pressure signal at the pump outlet (a,b) and the pump inlet (c). (speed, 4500 r/min; pump inlet P_b = 0.3 MPa).

The pump outlet pressure was changed from P = 2.0 MPa up to P = 10.0 MPa.
A biphase medium "fluid - gas" streamed along the suction line to the pump and further along the pressure line.
Experimental data showed (Table 1; Fig. 2) that the time of work of a test setup with a biphase medium the range of pressure fluctuations and the amplitude of all harmonics of spectrum on the pump outlet is increasing.

The maximum effect takes place in the time of work of a test setup on the aviation fuel with nitrogen.
Under the same conditions the range of pressure fluctuations and the amplitude of all harmonics of spectrum on the pump inlet is decreasing. Standing waves of range of a pressure fluctuations and of amplitudes for the first five harmonics of spectrum are observed along the length of the suction and pressure lines.

Parameters of standing wave in a pump outlet line increasing with the increase of the gas of the fluid (Fig. 3).

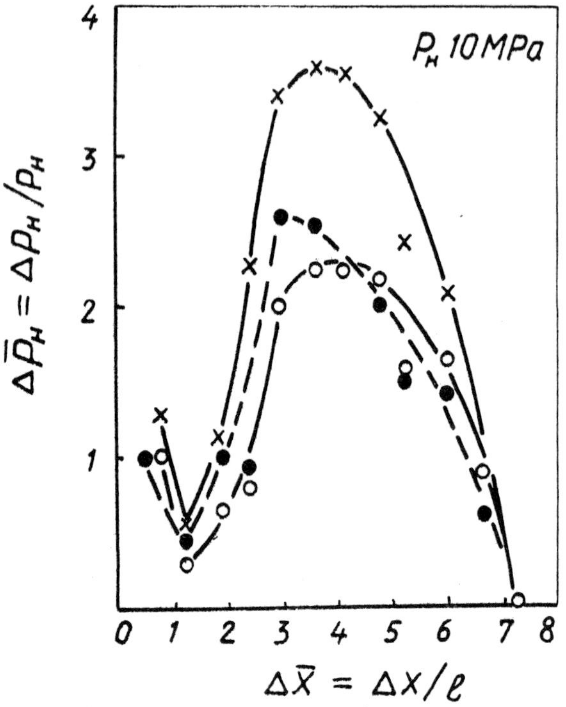

o aviation fuel without gas;
x aviation fuel + nitrogen;
• aviation fuel + helium.

Figure 3: Standing wave form of pressure pulsation at a piston pump pressure line.
(speed, 4500 r/min; pump inlet P_b = 0.3 MPa).

The increase of the maximum range of pressure fluctuation of a standing wave in a pump outlet line depends on the kind of gas in a biphase medium. When there are bubbles of nitrogen in the fluid the maximum range of a pressure fluctuation of a standing wave is two times bigger than the same meaning for the fluid without gas. For a suction line working under small pressures P_b 0.3 MPa and under small content of gas experimental data do not confirm existing theoretical propositions of Kogarko B.S. and Batchelor G.K. about possibility of a considerable decrease of the speed of the wave propagation in the biphase medium. Experimental values of speed of the wave propagation are similar to the value of speed of sound for fluid. So, the increase of dynamic loading actings on the elements of the pump mechanism and the walls of the pressure line is possible in the case of the damage of the suction line of a power plant of an aircraft or of space object.

TEMPERATURE FIELD MATHEMATICAL MODELING IN MULTILAYER SEMICONDUCTOR STRUCTURES BEING PRODUCED AND FUNCTIONING IN MICROGRAVITY

A.A. Melnikov, N.A. Kulchitsky and V.T. Khryapov
Scientific-Industrial Amalgamation "Orion", Moscow, USSR

ABSTRACT

This paper describes the model temperature field problems in multilayer semiconductor structures, by finite element method. This method is used to model temperature field problems in microphotodetectors operating in microgravity conditions.

NOMENCLATURE

T - thermal condactivity;
q - power of heat release sources;
Q - heat flow at the boundary S;
Ts - temperature of boundary S;
S - complete boundary of the calculated multilayer region V;
n - external normal vector to the boundary S;
T_i - nodal value of the desired function T;
x,y- rectangular coordinates;
F - functional.

1. INTRODUCTION

Fabrication technology of modern microelectronics including, photomicroelectronic devices requires the use of multilayer structures on Si-substrates. Temperature field modeling problems are impotant in such devices development. This opens the need for of investigating thermal processes in multilayer semiconductor structures (MSS) and, finally, of optimizing the device construction , that is produced and used in microgravity conditions.

In this connection the development of accurate numerucal simulation methods for thermal processes in MSS is needed. Numerical methods of mathematical modeling of thermal field in MSS are based on finite element techique, used for studying of hydromechanic and heat/mass exschange processes in semiconductor structure production.

2. MATHEMATICAL FORMULATION OF THERMAL FIELD PROBLEMS IN MSS

In the general case calculation of thermal fields in the MSS, produced and used in microgravity conditions, is reduced to:

$$\frac{\partial}{\partial x}\left(\lambda\frac{\partial T}{\partial x}\right)+\frac{\partial}{\partial y}\left(\lambda\frac{\partial T}{\partial y}\right)+q=0 \quad \text{in } V \qquad (1)$$

with boundary conditions, specified over different portions of the boundary in the examined region V

$$T = T_S \qquad \text{over } S_1 , \qquad (2)$$

$$\lambda\frac{\partial T}{\partial n}+Q=0 \qquad \text{over } S_2 , \qquad (3)$$

and with coupling conditions of the layerinterface, taken into account,

$$T_i = T_j , \qquad (4)$$

$$\lambda_i\frac{\partial T_i}{\partial n}=\lambda_i\frac{\partial T_j}{\partial n} \qquad (5)$$

3. FINITE ELEMENT SOLUTION OF THERMAL FIELD PROBLEMS IN MSS

We may say, that finite element solution of the thermal field problems in MSS (1) - (5) is equivalent to minimizing of the following functional

$$F=\iint_V \frac{1}{2}\left[\lambda\left(\frac{\partial T}{\partial x}\right)^2+\lambda\left(\frac{\partial T}{\partial y}\right)^2-2qT\right]dV+ \\ +\int_{S_2} QT\,dS \qquad (6)$$

To syplify the following calculations we introduce matrices

$$\{B\}=\left\{\frac{\partial T}{\partial x} \quad \frac{\partial T}{\partial y}\right\}' , \qquad (7)$$

$$[C]=\begin{bmatrix}\lambda & 0\\ 0 & \lambda\end{bmatrix}. \qquad (8)$$

Using (7), (8) and simplifying the equation, we can express the functional (6) as

$$F=\iint_V \frac{1}{2}\left[\{B\}'[C]\{B\}-2Tq\right]dV+ \int_{S_2} QT\,dS. \tag{9}$$

Assume, that calculated multilayer region V is divided into quadrangle superelements V_i^*, i=1,2...,N. The desired function T may be defined for each element $V^{(e)}$ as:

$$T^{(e)}=[N^{(e)}]\{T\} \tag{10}$$

Thus

$$\begin{bmatrix} \frac{\partial N_1^{(e)}}{\partial x} & \frac{\partial N_2^{(e)}}{\partial x} & \cdots & \frac{\partial N_S^{(e)}}{\partial x} \\ \frac{\partial N_1^{(e)}}{\partial y} & \frac{\partial N_2^{(e)}}{\partial y} & \cdots & \frac{\partial N_S^{(e)}}{\partial y} \end{bmatrix} \begin{Bmatrix} T_1 \\ T_2 \\ \vdots \\ T_S \end{Bmatrix} = [D^{(e)}]\{T\}. \tag{11}$$

The contribution of the finite element $V^{(e)}$ to the general functional value may be defined by the followig equations:

$$F^{(e)}=\iint_{V^{(e)}} \frac{1}{2}\{T\}'[D^{(e)}]'[C^{(e)}][D^{(e)}]\{T\}dV- \iint_{V^{(e)}} q[N^{(e)}]\{T\}dV+\int_{S_2} Q[N^{(e)}]\{T\}dS \tag{12}$$

From functional minimum:

$$\frac{\partial F}{\partial\{T\}}=\frac{\partial}{\partial\{T\}}\left(\sum_{e=1}^{E}F^{(e)}\right)=\sum_{e=1}^{E}\frac{\partial F^{(e)}}{\partial\{T\}}=0 \tag{13}$$

we have resulting equation system:

$$[\Lambda]\cdot\{T\}=\{f\}, \tag{14}$$

where

$$[\Lambda]=\sum_{e=1}^{E}[\Lambda^{(e)}],\ \{f\}=-\sum_{e=1}^{E}\{f^{(e)}\},$$

$$[\Lambda^{(e)}]=\iint_{V(e)}[D^{(e)}]'[C^{(e)}][D^{(e)}]dV,$$

$$[f^{(e)}]=-\iint_{V(e)} q[N^{(e)}]'dV+\int_{S_2} Q[N^{(e)}]'dS.$$

To take into account Dirixle boundary condition (2) the equation system (14) is simplified with the following algorithm. Assume, that numerical value of temperature of T_i is present, then bring all transpositions of matrix equation (14) down to the following : all coefficients of the i - th matrix row $[\Lambda]$, except diagonalones, are set equal to zero. Instead i-th element of the vector $\{f\}$ the product $\Lambda_{ii}T_i$ is substituted. All the other equations are transformed by the product substraction $\Lambda_{ij}T_i$ from f_i and by substitution Λ_{ij} =0, j = 1,2...,m , j ≠ i .

4. NUMERICAL REALIZATION OF THERMAL FIELD PROBLEMS IN MSS

Software package, intended for solving of thermal field boundary problems in MSS with arbitrary geometry, arbitrary arrangement of heat release sources, arbitrary thermal and physical characteristics of layer, was developed.

The mentioned software package includes data input module, initial data preparation module, module of autogeneration of finite element mesh in designed regions, module of visualization of finite element mesh, calculation module of desired values in nodes of finite-element mesh, module of data processing, analysis and data output, including module of data visualization with the help of thermal field distribution pictures.The software package uses the FORTRAN-4 programming language.

Holessky method is used for algebraic problem solution received as a result of initial boundary value problem discretization.

5. MODELING OF THERMAL FIELD IN MICROPHOTODETECTOR, FUNCTIONING IN MICROGRAVITY CONDITIONS

Let us consider thermal field problem hybrid photosensitiv MSS, block diagram of which is shown in figure 1.It consists of epitaxial photodiode CMT matrix and Si integrated circuit (IC), intended for signal processing. Epitaxial CMT layer is grown on a substrate, transparent in a suitable IR-region. Substrate has illuminated anti-reflection coating for reduce optical losses.

To connect photodiode with integrated Si circuit with the help of metal In contacts. All structure of hybrid micro- photodetector is glued to the heat sink, having the temperature of liquid nitrogen.

Thermal field problem in such microphotodetector, used in micrigravity conditions, is reduced to solving of partial differential equations (1) in layers V_i , i =1,2..., with boundary conditions

$$T = 77K \quad \text{over } S_1 , \qquad (15)$$

$$\lambda \frac{\partial T}{\partial n} + Q = 0 \quad \text{over } S_4 , \qquad (16)$$

$$\lambda \frac{\partial T}{\partial n} + Q = 0 \quad \text{over } S_5 , \qquad (17)$$

$$\frac{\partial T}{\partial n} = 0 \quad \text{over } S_2, S_3, S_6, \ldots S_K \qquad (18)$$

With the help of developed method and software package the temperature field of the hybrid multilayed MSS, shown in figure 1, was computed. Parameters of crystal and In interconnections, head release in integrated Si circuit and heat flow at boudaries S_4 , S_5 are taken into account.

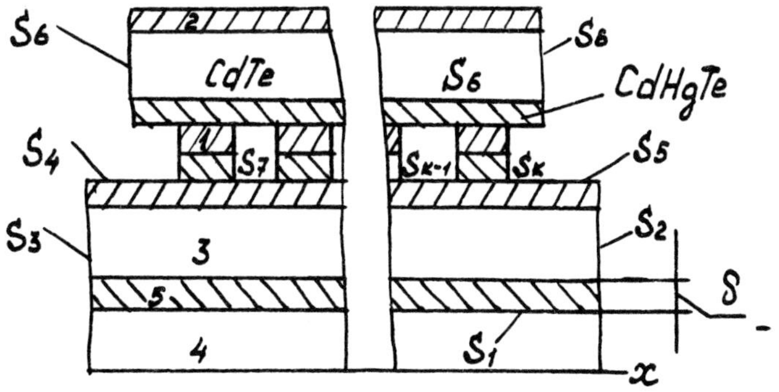

Figure 1: Computed block-scheme of the hybrid photosensitive multilayer semiconductor structure: 1-iterconnections; 2- anti-reflection coating; 3-Si integrated circuit; 4-heat sink; 5-glue layer

Temperature distribution in the working CMT layer versus heat release in Si crystal is shown in figure 2. Temperature distribution in the working CMT layer versus heat conduction of the glue is shown in figure 3. Temperature distribution in the working CMT layer versus heat input through integrated Si circuit interconnections with the housing leads is shown in figure 4. Total temperature dustribution in the working CMT layer is shown in figure 5.

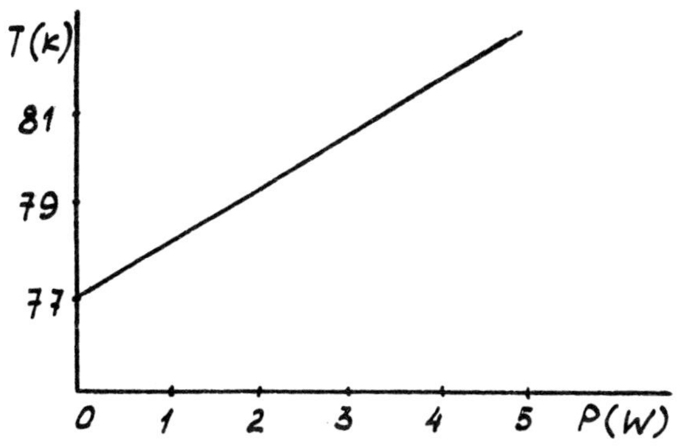

Figure 2: Temperature distribution in the working CMT layer versus heat release in IC with Q=0, δ =0.

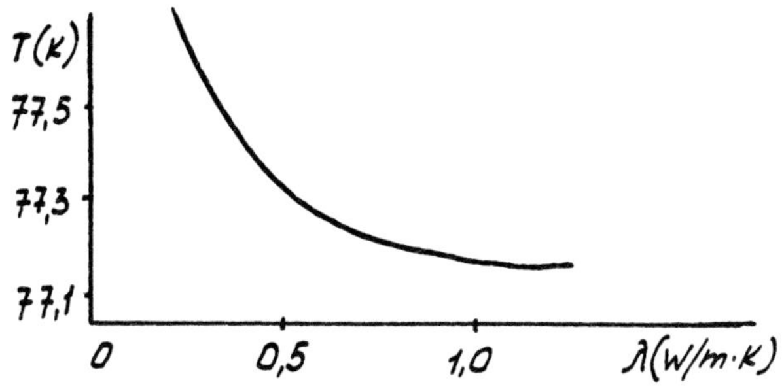

Figure 3: Temperature distribution in the working CMT layer versus heat condaction of the glue at P, in Si IC 0.5 W, Q=0.1 W/side, $\delta = 10^{-5}$ m

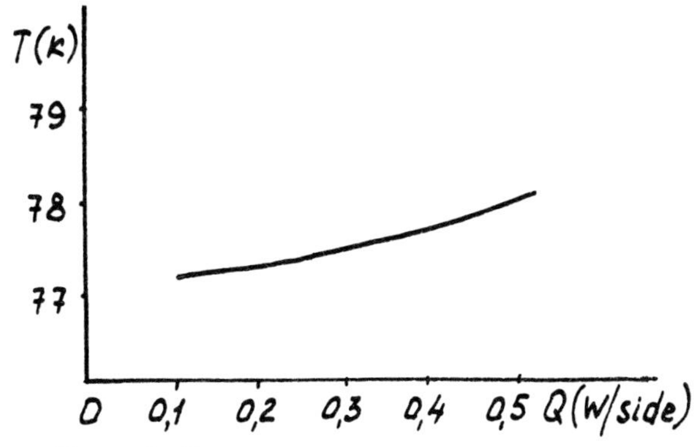

Figure 4: Temperature distribution in the working CMT layer versus heat input through Si IC interconnections with housing leads at P=0.5 W, $\delta = 10^{-5}$m, λ =0.2 W/m K

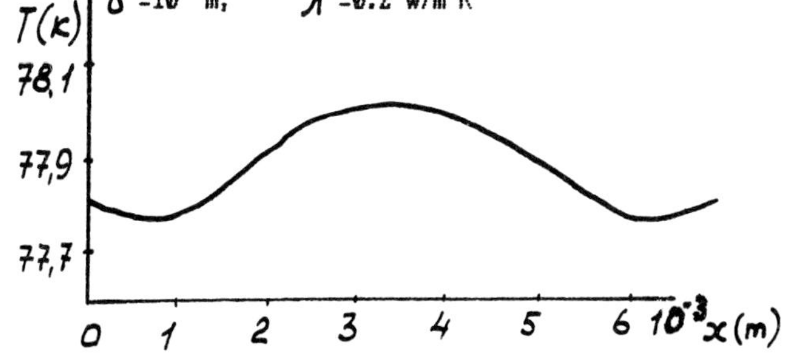

Figure 5: Total temperature distribution in the working CMT layer at P=0.5 W, $\delta = 10^{-5}$m, λ = 0.2 W/m K, Q= 0.1 W/ side

6. CONCLUSIONS

The usage of developed methods allows to refuse from expensiv and often impossible experiments, to define expediebcy of a device fabri cation technology on the basis of MSS, and to improve microphotoelectronic devices quality and reliability.

7. REFERENCES

1. O. Zienkiewicz , The Finite Element Method in Engineering Science. Moscow. World, 1975, 542.

2. O. Zienkiewicz and K. Morgan, Finite Elements and Approximation. Moscow, World, 1986, 318.

THE FLOW OF SUSPENSION IN TUBES UNDER MICROGRAVITY

B. Schwark-Werwach, A. Delgado, H.J. Rath

Center of Applied Space Technology and Microgravity (ZARM)
University of Bremen, Germany

ABSTRACT

The transport of solid particles with the aid of a fluid is important in many technical applications e.g. hydraulic transport of coal.

Under microgravity the influence of gravity and therefore sedimentation effects are eliminated. In this case it is possible to separate the effects of gravity from the interaction between solid particles, fluid and wall.

An intensive investigation program of suspension flow in tubes is being carried out in the drop tower Bremen. First the pressure drop for different flow rates and concentrations has been measured.

LIST OF SYMBOLS

Δp	pressure drop
l	measuring length
ρ	density
d_R	tube diameter
$\overline{c}$	averaged velocity
λ	friction factor
g	gravitational acceleration
g_o	gravitational acceleration on earth
η	viscosity
ν	kinematic viscosity
c_V	volume concentration
Re	Reynolds Number

subscripts:

G	general
F.Tr	Fluid
z	additional
sus	suspension

1. INTRODUCTION

In the Center of Applied Space Technology and Microgravity (ZARM) a project is being carried out, which investigates the flow behaviour of two-phase fluids under microgravity, especially pipe flow of suspensions.

Suspensions of particles in a fluid are common in a wide range of technical applications. Examples of the pipe flow range from the hydraulic transport of e.g. coal or sand to the blood in the veins.

The investigation under reduced gravity is fundamental for pipe flow applications on earth.

On the other hand, knowledge about hydraulic transport under reduced gravity is important for the design of facilities and plants in space.

If a fluid containing substances or phases of different density is exposed to a gravitational field sedimentation effects will occur. These effects complicate the pure flow and make the calculation of such two-phase flow difficult and inaccurate even today. As a result of this, the flow in vertical and horizontal pipes must be treated differently in theory.

Under microgravity, the influence of gravity and therefore sedimentation effects are eliminated, enabling the 'pure' study of such flows for the first time.

Suspensions showing small sedimentation effects can be treated by assuming that they are homogeneous fluids. The viscosity is then calculated from the pressure drop in the laminar region.

Measurements under microgravity can serve as a reference for determining the error connected to this assumption

When transporting a suspension with strong sedimentation (large heavy particles) in horizontal pipes the particles separate out at low flow velocities in the laminar region.

In vertical tubes the flow is defined by the sedimentation velocity. Under microgravity, a heterogeneous suspension (strong sedimentation) can be treated as a homogeneous suspension because sedimentation effects are absent. The viscosity and also a Reynolds number of the suspension can be determined.

A step-by-step analysis starting from the simple case (homogeneous fluid) building up to the more complex case (flow with sedimentation) is now possible.

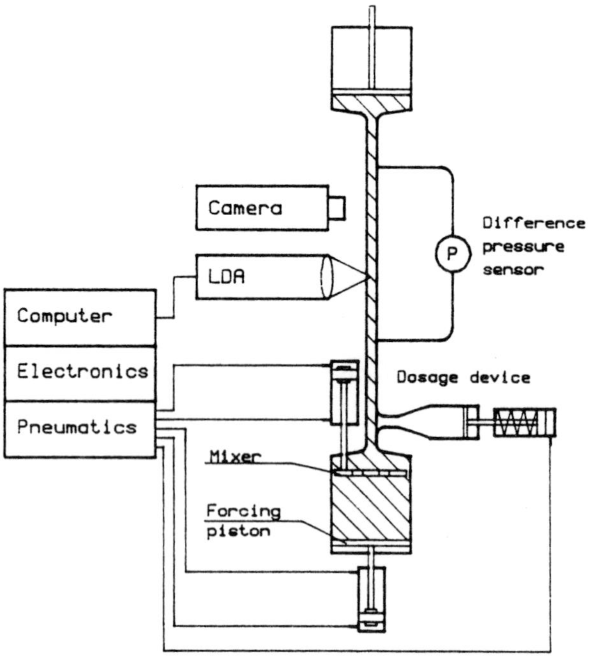

Figure 1: Experimental Set-Up

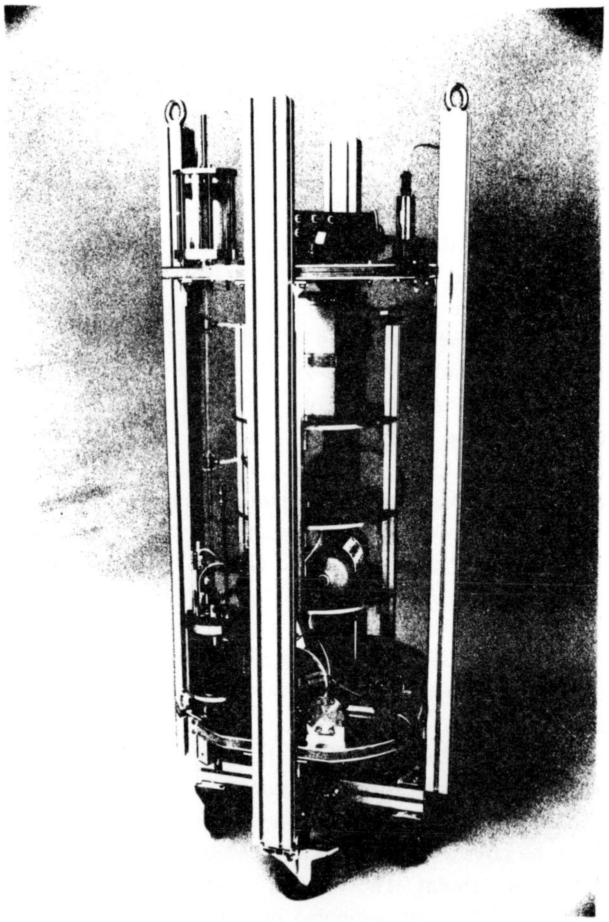

Figure 2: Photo of the Experimental Set-Up for the Drop Tower Bremen

2. EXPERIMENTAL SET-UP

The measurements are being carried out inside the drop tower Bremen. The drop tower Bremen enable experiments under Microgravity for 4.7s with a residual acceleration of 10^{-6} g_o. The experimental set-up mounted inside the drop capsule is shown in figure 1.
A suspension of glass spheres in water is chosen as model suspension, because experimental data for 1 g_o conditions is abundant and the glass spheres sediment strongly.

First the pressure drop for different flow rates and concentrations in the turbulent region was measured. The friction factor of the suspension depending on the volumen concentration c_v was then determined.

The suspension is pumped up pneumatically with a piston from a storage container through the measurement tube into a holding container, see figure 1 and figure 2.

The particles were suspended under microgravity inside the storage container. Mixing was accomplished by moving a perforated plate up and down axially. A second possibility is to inject the particles into the channel flow via a dosage apparatus.

The flow conditions can be measured by the pressure, temperature and position sensors. The flow rate is calculated from position signals of the pumping pistons.

Furthermore mixing of the flow is photographed to observe the flow structure.

Process measuring and control is accomplished by a shock-proof computer.

3. THEORETICAL BACKGROUND

There are many empirical equations which describe the pressure drop of heterogeneous suspension under $1g_o$. In general they have the form:

$$\frac{\Delta p}{l} = \frac{\rho_F}{d_R} \frac{\bar{c}_G^{\,2}}{2} (\lambda_F + \lambda_Z) \qquad (1)$$

In Table 1 you see some empirical equations for the pressure drop in horizontal pipes. Approximations of the additional pressure drop for g approaching to 0 leads in these equations to a wide range of dependence on g, this means that this equations predict the influence of g ambiguously. Therefore, it is very interesting to measure the pressure drop in the complete region from $1g_o$ to $0g_o$. Furthermore, there is no equation which describes the behaviour for $0g_o$ adequately.

Author	Equation	Behaviour g -> 0
Durand, Concolios	$\frac{\Delta p_G}{l} = \lambda_F \frac{1}{d_R}\frac{\varrho_F}{2} v_G^2 \left\{1 + N c_T \left[\frac{g d_R}{v_G^2} v_S \sqrt{\frac{3}{4}\left(\frac{\varrho_M}{\varrho_F}-1\right)\frac{1}{g d_{Km}}}\right]^{3/2}\right\}$	$\frac{\Delta P_z}{l} \sim g^{\frac{9}{4}}$
Jufin	$\frac{\Delta p_G}{l} = \lambda_F \frac{1}{d_R}\frac{\varrho_F}{2} v_G^2 \left[1 + 332{,}7 \frac{\sqrt{d_R c_T}}{v_G^3} \frac{v_S^{3/4}}{(g d_{Km})^{3/6}}\right]$	$\frac{\Delta P_z}{l} \sim g^{\frac{1}{4}}$
Newitt, Richardson, Abbot	$\frac{\Delta p_G}{l} = \lambda_F \frac{1}{d_R}\frac{\varrho_F}{2} v_G^2 \left[1 + 1100 \frac{v_S}{v_G}\frac{g d_R}{v_G^2} c_T \left(\frac{\varrho_M}{\varrho_F}-1\right)\right]$	$\frac{\Delta P_z}{l} \sim g^{2}$
Kriegel, Brauer	$\frac{\Delta p_G}{l} = \frac{1}{d_R}\frac{\varrho_F}{2} v_G^2 \left\{\lambda_F + 0{,}282\, c_T \left(\frac{\varrho_M}{\varrho_F}-1\right) k_f v_{SKm} \left(\frac{1}{g v_F}\right)^{1/3} \cdot \left(\frac{g d_R}{v_G^2}\right)^{4/3}\left[1 + 2{,}7\left(\frac{c_T}{c_{Tmax}}\right)^4\right]\right\}$	$\frac{\Delta P_z}{l} \sim g^{2}$
Silin, Witoskin, Karasik	$\frac{\Delta p_G}{l} = \lambda_F \frac{1}{d_R}\frac{\varrho_F}{2} v_G^2 \left[1 + 153\, c_T \left(\frac{\varrho_M}{\varrho_F}-1\right)^2 \sqrt{Fr_{Km}} \frac{\sqrt{g d_R}}{v_G^3}\right]$ $Fr_{Km} = \frac{\sum \frac{v_{SKi}}{g d_{Ki}} \cdot D_i}{100}$	$\frac{\Delta P_z}{l} \sim g^{\frac{1}{2}}$
Newitt, Richardson, Abbot	$\frac{\Delta p_G}{l} = \lambda_F \frac{1}{d_R}\frac{\varrho_F}{2} v_G^2 \left[1 + 66 \frac{g d_R}{v_S^2} c_T \left(\frac{\varrho_M}{\varrho_F}-1\right)\right]$	$\frac{\Delta P_z}{l} \sim g^{-1}$
Buhrke, Kecke, Richter	$\frac{\Delta p_G}{l} = \lambda_F \frac{1}{d_R}\frac{\varrho_F}{2} v_G^2 \left[\lambda_F + 3{,}5 \mu c_T \left(\frac{\varrho_M}{\varrho_F}-1\right)\frac{d_{Km}}{d_{Kg}}\left(\frac{g d_R}{v_G^2}\right)^{1,1}\left(\frac{v_{krit}}{v_G}\right)^{1/6}\right]$	$\frac{\Delta P_z}{l} \sim g^{1.18}$

Table 1: Approximation of some empirical formula* for the additional pressure drop for heterogeneous suspensions in horizontal pipes for g → 0.

(* additional symbols see [6])

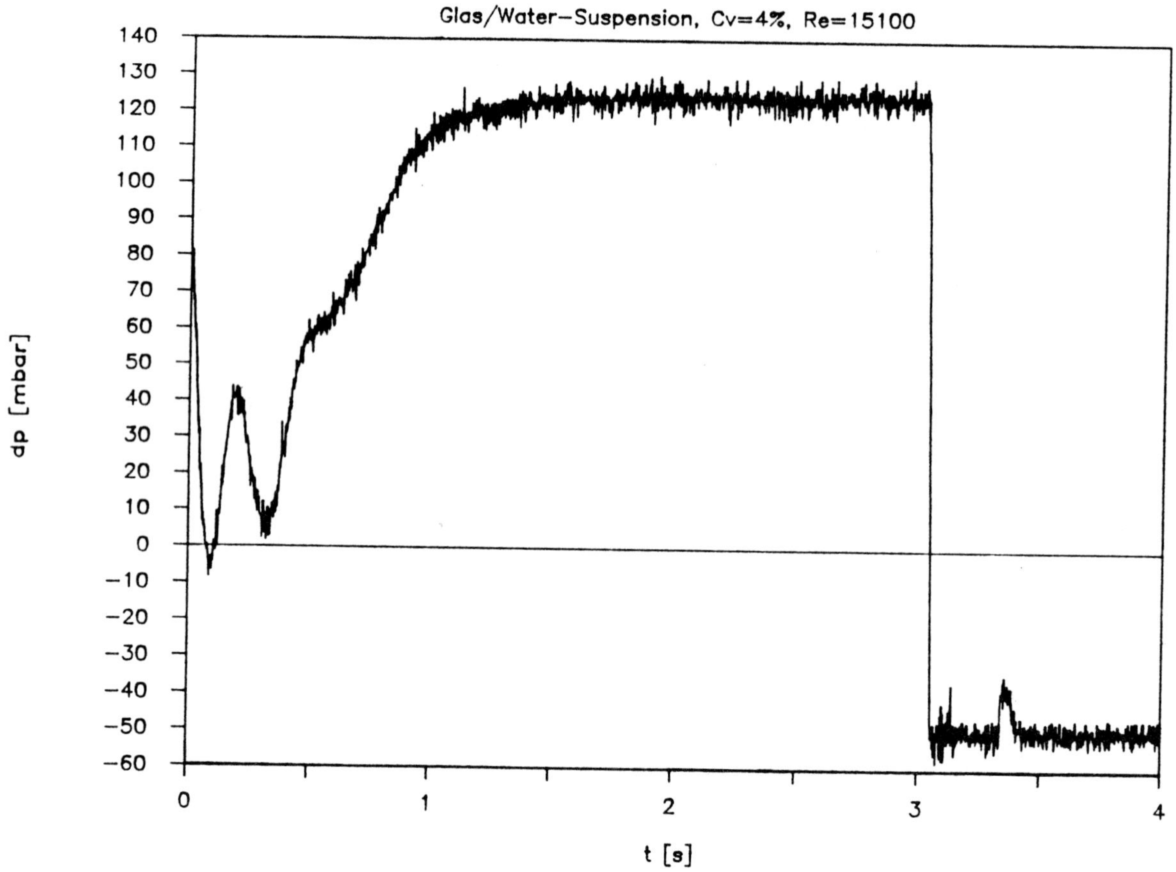

Figure 3: Typical Pressure Drop Signal during the Pumping Time

Now let us look at results for homogeneous suspensions without sedimentation, i.e. the densities of the continous phase and the particles are equal.

The following equation can be used for the pressure drop of a homogeneous suspension:

$$\frac{\Delta p}{l} = \lambda_{sus} \frac{\rho_{sus}}{d_R} \frac{\bar{c}_G^{\,2}}{2} \qquad (2)$$

According to measurements of Schröder [7] for polystyren cylinders in water or to measurements of Daily and Hardison [8] for plastic spheres in water, the friction factor decreases with the volume concentration c_v (see Figure 4). Schröder gets the behaviour:

$$\lambda_{sus} = (1-c_v)\,\lambda_F \qquad (3)$$

According to Daily and Hardison this effect for spherical particle is not so strong.

The reason for the decrease of the friction factor is a decrease of the turbulent shear stress caused by the fluctuating velocity components.

4. RESULTS

From a scientific point of view the following questions can be answered by the measurements under microgravity at the drop tower Bremen :

- Are the particle homogeneously suspended ?
- Does the presure drop become constant within the short experimental time of 4.5 s ?
- Are the friction factors within the expected range and which is the behaviour for variation of c_v ?

The glass spheres were suspended by a pneumatic mixing plate under microgravity and this process was photographed with a time intervals of 0.3s. These pictures show that the suspension is homogeneous after about 1s.

Figure 3 shows the pressure drop signal during the pumping interval. For Reynolds numbers between 11000 to 14000 the pressure drop is constant after 1.0 - 1.5 s. This means that there is enough time to measure a static pressure drop inside the drop tower Bremen. The pressure drop increases with the concentration of glass particles (20 % for c_v = 0.13).

The ratio λ_{sus}/λ_F follows from the experimental values:

$$\frac{\lambda_{sus}}{\lambda_F} = \frac{\Delta p}{l} \frac{1}{\frac{0{,}316}{\sqrt[4]{\frac{d_R \bar{c}}{\nu_{H_2O}}}} \frac{\rho_{H_2O}\bar{c}^2}{2 d_R}} \cdot \frac{1}{\left(\frac{\rho_{sus}}{\rho_{H_2O}}\right)^{3/4} \left(\frac{\eta_{sus}}{\eta_{H_2O}}\right)^{1/4}} \quad (4)$$

with:

$$\frac{\rho_{sus}}{\rho_{H_2O}} = 1 + \frac{\rho_{glass} - \rho_{H_2O}}{\rho_{H_2O}} c_V \quad (5)$$

$$\frac{\eta_{sus}}{\eta_{H_2O}} = 1 + 2{,}5\, c_V + 7{,}17\, c_V^2 \quad (6)$$

The viscosity is calculated using an equation given by Vand [9]. Later it will be measured during parabolic flights under microgravity.

First results of the relative friction factor are shown in Figure 4. They are in the expected range and decrease with higher concentration c_v.

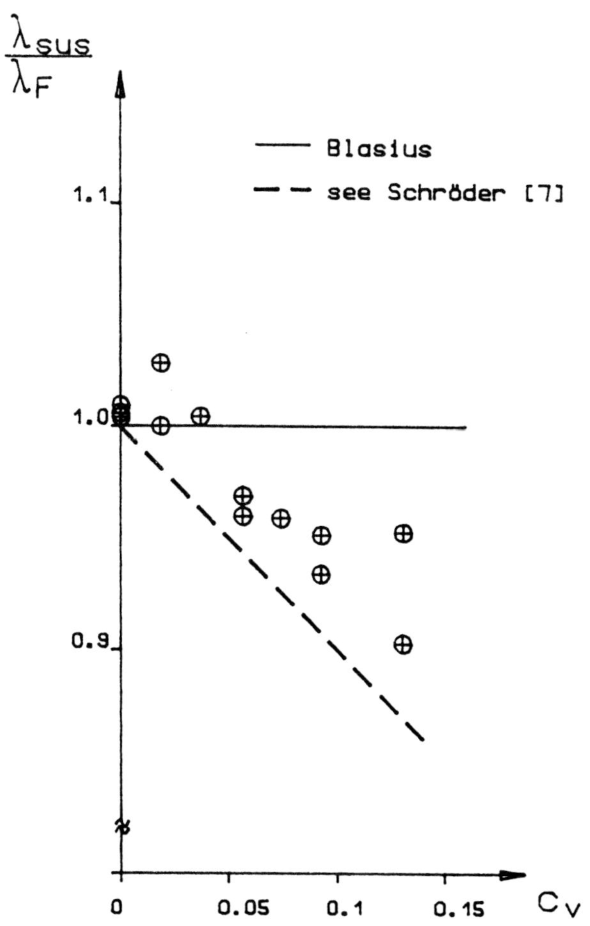

Figure 4: Reduced Friction Factor of Glass-Water Suspensions under Microgravity

5. CONCLUSION AND OUTLOOK

A method to measure the friction factor of a heterogeneous suspension without sedimentation is successfully tested. The behaviour is similar to the behaviour of suspensions with equal density of both phases. The exact material parameters can be measured.

Further measurements with a varitions in particle concentration c_v and in Reynolds number are planned.

The method of injection of the spheres into the flow will be tested.

On the other hand a shock-proof Laser-Doppler-Anemometer is under development to measure the concentration and velocity profile of both phases. Once this apparatus is developed, a microscopic view in the flow structure will be possible.

6. ACKNOWLEDGEMENT

The financial support by the Bundesminister für Forschung und Technologie and the Senator für Bildung, Wissenschaft und Kunst des Landes Bremen is gratefully acknowledged.

7. REFERENCES

1. Durand,R.;Concolis,E.: Etude experimentale du refoulement des matériaux en conduite. 2.émes Journées de l'Hydraulique, Grenoble 1952
2. Jufin,A.P.: Giromechanisazija. Moskva: Stroisdat 1974
3. Newitt,D.M.;Richardson,H.F.; Abbot,M., et al: Hydraulic Conveying of Solids in Horizontal Pipes. Trans. Inst. Chem. Engrs. 33, page 93-113, London 1955
4. Kriegel, E.;Brauer,H.: Hydraulischer Transport körniger Feststoffe durch waagrechte Rohrleitungen. VDI - Forschungsheft 515, 1966
5. Silin,N.A.;Witoskin,J.;Karasik,W.M;Oceretko,W.F.: Girotransport, woprosy gidravlika. Isdatelstvo "Naukova dumka", Kiev, 1971
6. Burke,H.;Keck,H.J.;Richter,H.: Strömungsförderer; Friedr. Vieweg & Sohn, Braunschweig/Wiesbaden 1989
7. Schröder,V.: Experimentelle Untersuchungen der Strömungsverluste einer homogenen Suspension im Rohr, Krümmer und Diffusor.

Diss. TH Darmstadt 1982

8. Daily,J.W.;Hardison,R.L.: Rigid Particle Suspensions in Turbulent Shear Flow: Measurement of Total Head, Velocity and Turbulence with Impact Probes. Massachusetts Institute of Technology, Hydrodynamics Laboratory, Report No. 67, 1964

9. Vand,V.: Viscosity of Solutions and Suspensions,I,II, J. Phys. Colloid. Chem.,52/1948

CALIBRATION OF THERMAL ANEMOMETERS AT VERY LOW VELOCITIES UNDER MICROGRAVITY

F. R. Stengele, A. Delgado, H. J. Rath

Center of Applied Space Technology and Microgravity, ZARM
University of Bremen, 2800 Bremen, West Germany

1. INTRODUCTION

Hot-wire and hot-film anemometry are widely used measuring techniques in experimental fluid mechanics and in practical processes for determining mean and fluctuating quantities of velocity, temperature and shear stress.

In order to measure flows with the help of hot-wire or hot-film probes these have to be calibrated in advance. In the literature concerning this subject many different calibration techniques are described [1]. At speeds below 0.5 m/s considerable difficulties occur during calibration.

Most traditional static calibration techniques involve the calibration of the hot-wire probe against another velocity-measuring device such as a Pitot tube. At low velocities, difficulties arise in this procedure, since the differential pressures that are registered become very low.

Further the heat transfer from the heated wire is at such low velocities a very complex process as it is composed of several physical processes taking place at the same time. They are conduction, radiation, free convection and forced convection. In most cases in gases, where the difference of the sensor temperature T_w and that of the medium T_∞ is less than 300°C and the velocity is higher than 0.5 m/s radiation, conduction and free convection can be neglected.

But at very low velocities free convection is significant and has to be taken into account in the relationship describing the heat transfer from the wire. A number of investigators have considered the effects of free and forced convection in the Reynolds and Grashof number regions that are relevant here [2,3,4,5]. For example, McAdams [2] suggested that the heat transfers be calculated separately and the higher value used. On the other hand, van der Hegge Zijnen [3] proposed that the vectorial sum of the Nusselt numbers with the two modes of convection be used. However, both these methods can result in considerable error.

Free convection is strongly reduced during microgravity. Therefore thermal anemometers can be calibrated adequately at very low velocities e.g. in drop towers and in Parabolic flights.

2. HEAT BALANCE

The total heat loss of the hot wire probe depends on the velocity of the medium, the temperature difference between the wire and the medium, the physical properties of the medium and also the dimension of the wire and its material.

The dimensionless parameter, that describes the heat transfer of the hot wire, is the Nusselt number Nu which can be defined as

$$Nu = I^2 \cdot R_w / (\pi \cdot l \cdot \lambda_m (T_w - T_\infty)) \qquad (1)$$

where I is the electrical current, R_w the electrical resistance of the hot wire, and l the length of the sensor.

King [8] was the first, who studied theoretically and experimentally thermal anemometers. King's law, equation (2), which relates the Nusselt number Nu to the Reynolds number Re of the flow around the wire, describes the basic response of the hot-wire anemometer.

$$Nu = A + B\ Re^{1/2} \qquad (2)$$

Here, A and B are constants, Re is the Reynolds number $u \cdot d/\nu$, u is representing the fluid velocity and d is the wire diameter. Later studies have shown that an exponent between 0.33 and 0.5 describes the measurements better than the exponent 0.5 of King [3,4,5].

But the heat loss of the hot wire probe at very low velocities is not only caused by the forced convection but also by the heat transfer into the prongs and the free convection. Radiation losses can be neglected in this case.

Because the free convection is strongly reduced under microgravity thermal anemometers are examined and calibrated here at very low velocities in the Bremen drop tower and in an additional small drop tower facility.

At very low velocities the Nusselt number is a function of the following parameters :

$$Nu=f(Re,Gr,Pr,\frac{d}{l}) \quad (3)$$

where Gr is the Grashof $\beta \cdot g \cdot d^3 \cdot (T_w - T_\infty)/\nu^2$, Pr is the Prandtl number ν/a, a is the thermal accomodation coeffizient and d/l is the aspect ratio of the hot wire.

3. EXPERIMENTAL APPARATUS

The basic idea of the calibration module is to move the wire at steady, accurately controlled velocities in a quiescent fluid.

There are two methods of moving a probe through a quiescent fluid. The probe travels either along a circular path or in a straight line. The latter possibility has been realized on board an aircraft. The results obtained hereby will be published in [9]. The experiments presented here have been carried out in a small drop tower facility, problems occur because of the restriction in the space available and, therefore, the possibility at which the probe is moved along a circular path has been selected.

Fig. 1 shows the principal design of the experiment. The hot-wire probe is attached to a rotating disk, which is driven by a centrally positioned stepping motor. A ring shaped measuring tunnel is placed below the rotating disk, that is until then filled with ambient air. In order to ensure a high flexibility regarding the use of other media like water it was arranged to seal the tunnel with a labyrinth packing. The probe is driven along a path which lies in the middle of the tunnel. The available measuring distance is 1.2 m while the tunnel's inner dimenions are 100 mm x 100 mm.

With this kind of calibration system it is possible to investigate thermal anemometers under microgravity in a velocity range from 0 to 1.5 m/s. As the temperature of the surrounding air has a very large influence on the measuring results, especially at very low velocities, a latent heat reservoir is placed around the circular measuring tunnel. The latent heat reservoir is filled with wax, before the experiment it is heated with the help of a heating coil. The hot wax guarantees a constant temperature inside the tunnel over the whole period of the experiment. By using different types of wax the temperature in the tunnel can be varied between 5 and 50°C. The complete system is further insulated and then integrated in the capsula.

The hot wire and hot film probes were triggered by a TSI constant-temperature-anemometer of the 1150 series, the used probes are TSI hot wire probes 1210-T1.5, 1211-T1.5 with a diameter of 4 μm and an aspect ratio l/d = 312.5. Hot film probes of the type 1210-20, 1211-20 with a diameter of 51 μm and an aspect ratio l/d = 24.5 were used as well. The temperature of the wires is in all cases held constant at 250°C.

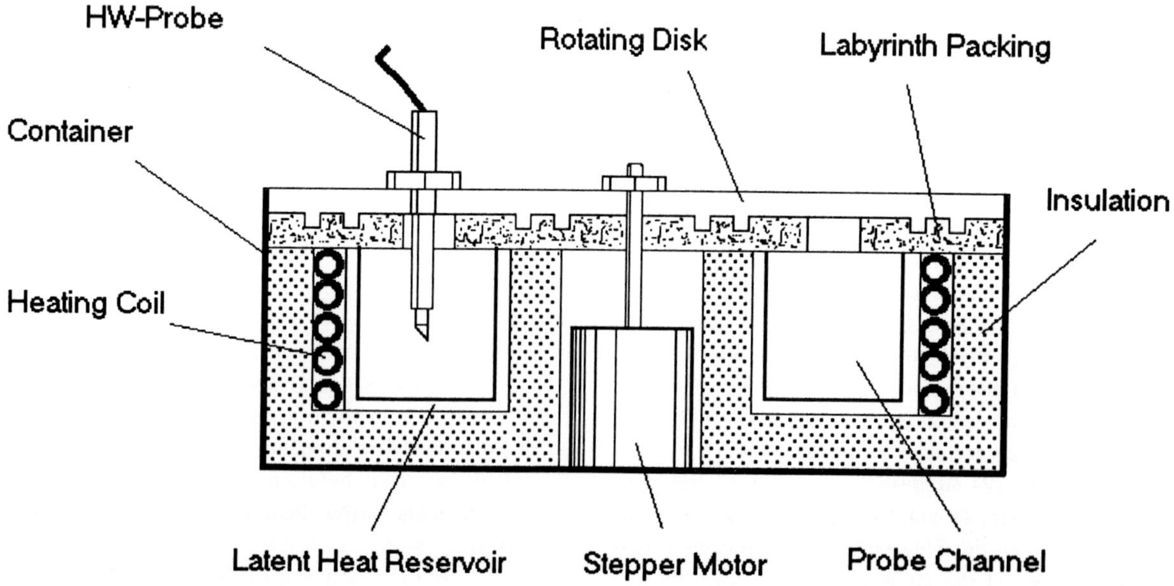

Figure 1: Experimental apparatus

Storing and analysing of the obtained data is done by a computer.

4. RESULTS

Experiments in the Bremen drop tower are being carried out currently. The results obtained will be published elsewhere.

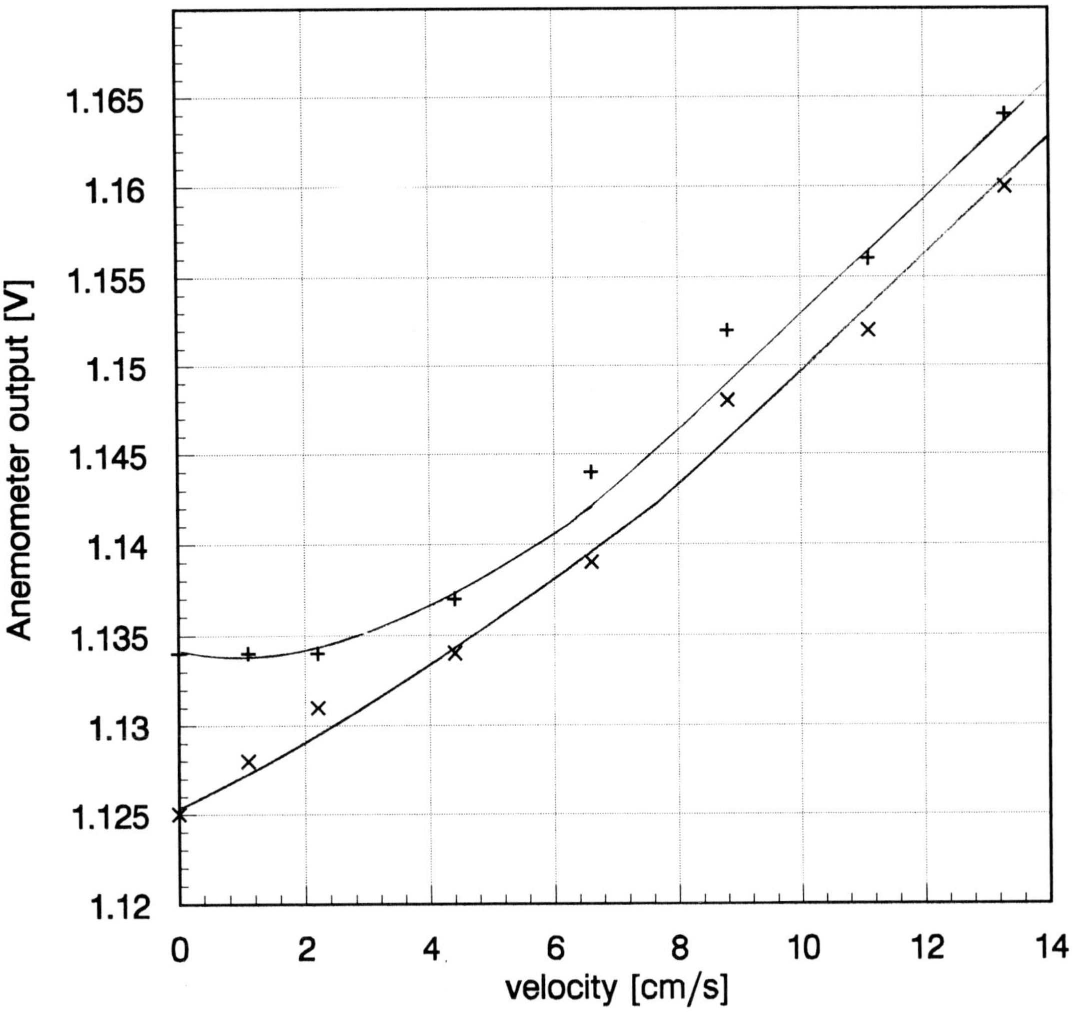

Figure 2: Comparison between μg and terrestrial measurements
x μg experiments, + terrestrial experiments

Here, the results of tests which were performed in another drop facility (drop height: 18 m) are presented. The μg-quality in the smaller drop tower facility in Bremen is 0.01 g and the experimentation time is 1.5 sec. These experiments have been performed without the latent heat reservoir and therefore the surrounding air temperature was far from constant. A DC-motor was used to drive the disk.

Fig. 2 illustrates the influence of gravity on the anemometer output voltage for different velocities of the probe. It is evident that the values for μg lie below that obtained in the earth laboratory.

The results of the terrestrial measurements show for velocities $v \rightarrow 0$ that the curve appears as a straight line, the determination of the velocity from the anemometer output is ambiguous in this range.

In μg-tests this is not the case, even so the slope of the curve decreases here, too. At higher velocities a steady difference between the two curves is shown in fig. 2. Here it is still necessary to determine the range of influence of the free convection more exactly with the help of higher velocities.

In further experiments planned in the Bremen drop tower higher velocities will be realized. From the experimental conditions available there (μg-time of 4,7 sec; residual acceleration better than 10^{-5}g) a significant improvement of the results is expected.

ACKNOWLEDGEMENT

The financial support by the Bundesminister für Forschung und Technologie and the Senator für Bildung, Wissenschaft und Kunst des Landes Bremen is gratefully acknowledged.

REFERENCES

1. Andrews, G.E., Bradley, D. and Hundy, G.F., *Hot wire anemometer calibration for measurements of small gas velocities, J. Heat Mass Transfer,* Vol. **15**. pp. 1765-1786 (1972).
2. MC-Adams, W.H., *Heat Transmission*, third edition. Chap. X McGraw-Hill, New York (1954).
3. Van der Hegge Zijnen, B.G., *Modified correlation formulae for the heat transfer by natural and forced convection from horizontal cylinders*, Appl. Sci. Res. A6, 129-140 (1956).
4. Collis, D.C. and Williams, M.J., *Two-dimensional convection from heated wires at low Reynolds numbers, J. Fluid Mech.* **6**, 357-384 (1959).
5. Hatton, A.P., James, D.D., and Swire, H.W., *Combined forced and natural convection with lowspeed air flow over horizontal cylinders, J. Fluid Mech.*, vol. **42**, part 1, pp. 17-31 (1970).
6. Joergensen, F. E., *An Omnidirectional thin-film probe for indoor climate research,* DISA Info., **24**, 24-29 (1980).
7. Rasmussen, C.G., *Das Luft-strahl-Hitzdraht-Mikrophon*, DISA Info., **2**,5-13 (1965).
8. King, L. V., *On the convective heat transfer from small cylinders in a stream of fluid. Determination of convective constants of a small platinum wires with applications to hot-wire anemometry*, Phil. Trans. R. Soc. 214A, 373 - 432 (1914).
9. Stengele, F.R., Delgado, A, Rath, H.J., *Calibration of thermal anemometers at very low Reynolds numbers under microgravity.* To be published in the Proceedings of the IUTAM - Symposium on Microgravity Fluid Mechanics (1991).

Chemistry, Structure Formation and Biotechnology

GELS, GELS, GELS AND THEN, PERHAPS, GELS

Pier Giorgio Righetti, Cecilia Gelfi and Marcella Chiari

Chair of Biochemistry and Department of Biomedical Sciences and Technologies, University of Milano, Via Celoria 2, Milano 20133, Italy

ABSTRACT

The chemistry of acrylamides is here reviewed. Both neutral and charged monomers are dealt with. The latter ones (acrylamido weak acids and bases) are used for generating immobilized pH gradients, for isoelectric focusing separation of amphoteric macromolecules. A series of cross-linkers is investigated and the use of highly-cross linked gels critically examined. New, N-substituted monomers are described, and data on their hydrophobicity and hydrolytic resistance in both, monomeric and polymeric state, presented. A unique monomer appears to be trisacryl, which is extremely hydrophilic but highly prone to hydrolysis. However, in the polymer, it is quite resistant to alkaline hydrolysis, suggesting that the gel phase could be stereo-regular.

Keywords: *electrophoresis, isoelectric focusing, immobilized pH gradients, polyacrylamide.*

1. INTRODUCTION

While most of the participants at this meeting are involved with hydrodynamics and hydromechanics (disciples of Leonardo da Vinci, we might say, since the maestro started studying these phenomena already in 1490, depicting whirls of liquid in rivers flowing against bridge pillars) we are obsessed just by the opposite problem: how to prevent them. Convective liquid flows are anathema in electrophoresis, as they quickly destroy the separation. Biochemists were fully aware of that and in fact gel matrices (*i.e.* a capillary system able to co-ordinate and trap the solvent in the separation space) were introduced already in 1953 by Grabar and Williams (Ref. 1) (agar) and in 1955 by Smithies (Ref. 2) (starch). But the big revolution in protein separation came with polyacrylamide gels, first reported by Raymond and Weintraub (Ref. 3), Ornstein (Ref. 4) and Davis (Ref. 5).

2. CHEMISTRY OF POLYACRYLAMIDES

Acrylamide polymers are most peculiar: if one looks up at an encyclopedia, one finds that their principal uses are in paper, water treating, mining and oil recovery. They are also used in food processing, surface coatings (e.g. latex paints), in adhesives, even in the medical field (contact lenses) and, finally, in research (electrophoresis and chromatography). Table 1 summarizes the important physico-chemical properties of acrylamide (2-propenamide), the mono-functional monomer which, together with a cross-linker (typically N,N'-methylene bisacrylamide), originates a polyacrylamide matrix.

Polyacrylamide matrices (PAG) are quite unique for protein separations: they can be prepared at any range of %T and %C so that they can effectively sieve proteins (which means that the average pore size of the gel is of the same order of magnitude as the average protein diameter) (Ref. 6). The great resolution of PAG electrophoresis is due to this simultaneous mechanism of mass and charge fractionation. Over the years, a sort of dichotomy has developed: polyacrylamides have been confined mostly to protein separations and agaroses to nucleic acid fractionations (Ref. 7). Yet, years ago, when working with NASA (MSFC, Huntsville, Ala) we became intrigued by the possibility of obtaining polyacrylamide gels of very large pore size, by using relatively high %C (>20%) (Ref. 8). Table 2 lists the series of cross-linkers we had explored. It turned out that we could produce gels of extremely high pore size (>600 nm) but useless, since they were

TABLE 1 Physico-chemical properties of acrylamide[1]

Property	Value
Molecular mass	71.08
Melting point	84.5
Vapor pressure (solid, Pa):	
25°C	0.9
40°C	4.4
50°C	9.3
Vapor pressure (liquid, kPa):	
87°C	0.27
103°C	0.67
116.5°C	1.5
136°C	3.3
Density (g/mL, 30°C)	1.12
Equilibrium moisture content at 22.8°C	1.7 g water/Kg dry acrylamide
Optical properties:	
crystal system	monoclinic or triclinic
crystal habit	thin tabular to lamellar
Refractive indexes:	
N_x	1.46
N_y	1.55
N_z	1.58
Optical sign	(-)
Solubilities (g/100 mL, 30°C)	
acetonitrile	39.6
acetone	63.1
benzene	0.346
ethylene glycol monobutyl ether	31
chloroform	2.66
1,2-dichloroethane	1.50
dimethyl formamide	119
dimethyl sulphoxide	124
dioxane	30
ethanol	86.2
ethyl acetate	12.6
heptane	0.0068
methanol	155
pyridine	61.9
water	215.5
carbon tetrachloride	0.038

[1]By permission from W.M. Thomas and D.W. Wang, in *Encyclopedia of Polymer Science and Engineering*, Wiley, New York (1985) 169-211.

opaque and would collapse at 1 g exuding water. Even when using highly hydrophilic cross-linkers (such as DHEBA) the situation was not ameliorated: it would thus appear that at high %C the matrix becomes 'hydrophobic' not in the chemical sense (because the monomers are hydrophobic) but more in a physical sense (because the matrix is so entagled and 'knotted' that solvent cannot penetrate any longer).

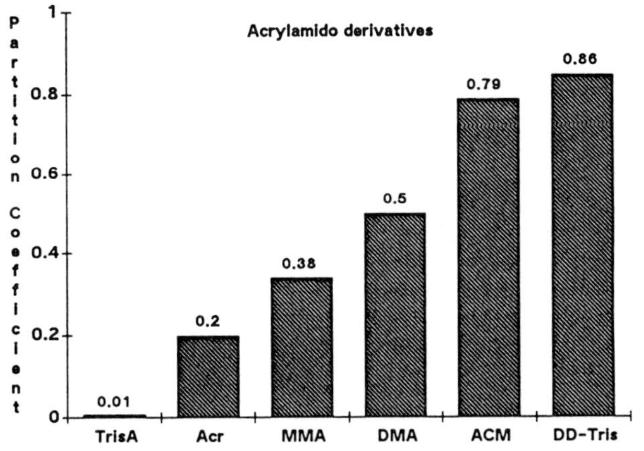

Figure 1: Hydrophobicity scale of 6 acrylamide monomers. It has been obtained by partitioning in water/n-octanol and quantifying the concentration in the two phases by CZE. TrisA: trisacryl; Acr: acrylamide; MMA: monomethyl acrylamide; DMA: dimethyl acrylamide; ACM: acryloyl morpholine; DD-Tris: di-deoxy trisacryl.

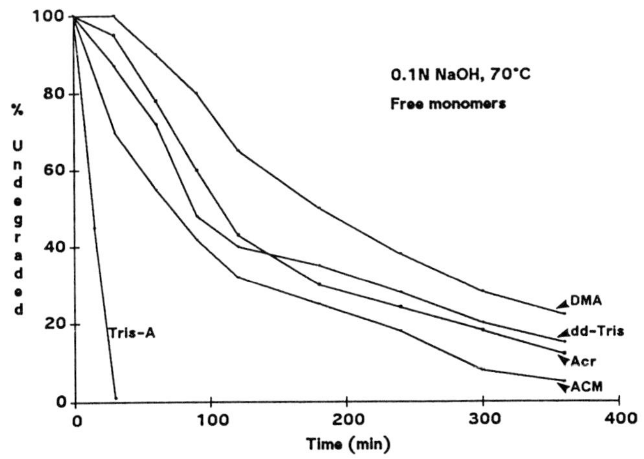

Figure 2: Kinetics of hydrolysis of different acrylamide monomers. Hydrolysis was performed in 0.1 N NaOH at 70°C for the times indicated. The amounts were assessed by harvesting triplicates at each point, neutralizing and injecting in a CZE instrument (Beckman P/ACE). Abbreviations as in Fig. 1. Note that, whereas all other monomers exhibit first order kinetics, TrisA follows a zero-order degradation kinetic.

3. ISOELECTRIC FOCUSING AND IMMOBILIZED pH GRADIENTS

The project of high %C gels was abandoned because after 1980 we started developing immobilized pH gradients (IPG) (Ref. 9). This is a new, most powerful variant of isoelectric focusing, a protein separation technique based on driving amphoteric species to their isoelectric point (pI) value along a stationary pH gradient (Ref. 10). The gradient was kept stationary by a most complex mixture of perhaps thousands of amphoteric chemicals, called carrier ampholytes, having M_r values in the range 600 to 1000 Da. It turned out that the pH gradient would indeed drift with time, and this was associated with all sorts of other problems. Thus followed the IPG breakthrough. Here our knowledge of acrylamide chemistry came very handy: IPGs are based on grafting (immobilizing) onto a neutral polyacrylamide backbone a series of acrylamido weak acids and bases, in such ratios as to create along the matrix a pH gradient of any desired size (from ultranarrow, barely 0.1 pH unit, up to very extended, a pH 8.5 interval) and shape. Tables 3 to 5 list 14 of these chemicals, which exhibit pK values fairly well distributed along the pH 1 to 12 scale. This is the fundamental characteristic of IPGs: by using a simple set of crystalline chemicals, well defined and of known structure, but with pKs covering the useful values of the pH scale for protein fractionation, one can engineer at whim any desired pH gradient. At the beginning, we have proposed a series of formulations (obtained by extensive computer modelling) covering any pH interval but rigourously linear (Ref. 11). We then became aware that in fractionation of highly complex samples and in two-dimensional separations non-linear pH gradients would be very useful. We have thus recently proposed concave exponential, convex (logarithmic) and even sinusoidal pH gradients (Ref. 12, 13).

4. NEW ACRYLAMIDE MONOMERS

More recently, as the IPG technique has now been fully developed, we came back to the chemistry of neutral acrylamide monomers. We have concentrated on a series of N-substituted acrylamides (given in Figure 1), in the hope of obtaining gels of peculiar characteristics (such as larger pore size and more resistant to hydrolysis). We have now made extensive physico-chemical studies on these compounds, by exploiting the unique separation power and quantitation capability offered by capillary zone

TABLE 2 Structural data on cross linkers[1]

Name	Abbr.	Chemical formula	M_r	Chain length
N,N'-methylene bisacrylamide	Bis (MBA)	$CH_2{=}CH{-}C({=}O){-}NH{-}CH_2{-}NH{-}C({=}O){-}CH{=}CH_2$	154	9
Ethylene diacrylate	EDA	$CH_2{=}CH{-}C({=}O){-}O{-}CH_2{-}CH_2{-}O{-}C({=}O){-}CH{=}CH_2$	170	10
N,N'-(1,2-Dihydroxyethylene) bisacrylamide	DHEBA	$CH_2{=}CH{-}C({=}O){-}NH{-}CH(OH){-}CH(OH){-}NH{-}C({=}O){-}CH{=}CH_2$	200	10
N,N'-Diallyltartardiamide	DATD	$CH_2{=}CH{-}CH_2{-}NH{-}C({=}O){-}CH(OH){-}CH(OH){-}C({=}O){-}NH{-}CH_2{-}CH{=}CH_2$	288	12
N,N',N''-Triallylcitric triamide	TACT	$CH_2{=}CH{-}CH_2{-}NH{-}C({=}O){-}CH_2{-}C(OH)(C({=}O)NH{-}CH_2{-}CH{=}CH_2){-}CH_2{-}C({=}O){-}NH{-}CH_2{-}CH{=}CH_2$	292	12—13
Poly(ethylene glycol) diacrylate 200	$PEGDA_{200}$	$CH_2{=}CH{-}C({=}O){-}O{-}CH_2CH_2OCH_2CH_2O{-}C({=}O){-}CH{=}CH_2$	214	13
N,N'-Bisacrylylcystamine	BAC	$CH_2{=}CH{-}C({=}O){-}NH{-}CH_2{-}CH_2{-}S{-}S{-}CH_2CH_2NHC({=}O)CH{=}CH_2$	260	14
Poly(ethylene glycol) diacrylate 400	$PEGDA_{400}$	$CH_2{=}CH{-}C({=}O){-}O{-}CH_2CH_2OCH_2CH_2OCH_2CH_2{-}O{-}CH_2CH_2OCH_2CH_2OCH_2CH_2{-}O{-}C({=}O){-}CH{=}CH_2$	400	25

[1]From Ref. 8, by permission.

electrophoresis (CZE; Beckman P/ACE 2000 instrument). We have explored the following parameters: a) hydrophobicity of the different monomers; b) hydrolytic stability in NaOH solutions in the monomeric state; c) hydrolytic stability in the polymer and d) efficiency of incorporation. Figure 1 shows the hydrophobicity scale we have built by partitioning in water/n-octanol phases: it would appear that the limit to the hydrophobicity of the monomer is around P = 0.8. At higher P values (P = 0.86, such as in di-deoxy Trisacryl, DD-Tris) the polymer becomes an opaque plastic, unable to swell in water. Thus, we have focused our attention on three most promising chemicals: trisacryl (TrisA), dimethyl acrylamide (DMA) and acryloyl morpholine (ACM). Some interesting findings were obtained when studying the hydrolytic stability of these compounds: as shown in Figure 2, most of them exhibit first order degradation kinetics in 0.1 N NaOH at 70°C, with the following half lifes ($T_{1/2}$, min): DMA: 185; dd-Tris: 130; Acr: 111; ACM: 80 and TrisA: 15. However, most strikingly, TrisA exhibits a zero-order kinetic of degradation, which means that probably the monomer is intrinsically unstable, possibly due to its electronic structure.

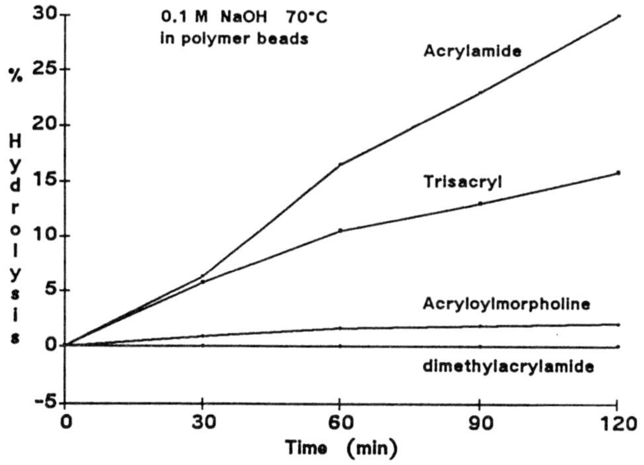

Figure 3: Degradation kinetics of the monomers in the polymeric gel. The different monomers were polymerized (by emulsion polymerization) as beads, subjected to hydrolysis in 0.1 N NaOH and then analyzed for hydrolytic products. Hydrolysis was assessed in the beads by titrating free acrylic acid residues by frontal analysis. Note the much increased stability of triasacryl in the polymer as compared to its behaviour as free monomer (Figure 2).

This is quite unfortunate, since TrisA appeared to be the most likely candidate for new gel matrices: it is in fact extremely hydrophilic (see Figure 1) and it should give an open pore structure, in view of its large size as a monomer and its high M_r. However, a completely different picture emerges when investigating the stability of the same monomers in the polymeric gel structure. Now, as shown in Figure 3, the situation is reversed: the most unstable monomer (TrisA) is decidedly more stable than acrylamide, although less stable than a polyacryloyl morpholine or a polydimethylacrylamide gel. This opens up interesting perspectives and speculations: it suggests to us that perhaps, in the polymer, TrisA gives highly oriented and structured fibres, able to shield the amido bond from incoming $-OH^-$ ions in an hydrolytic process. In other words, while in the monomer the stability (or better the high instability) is driven by electronic factors, in the polymer the greatly added stability should be driven by steric factors. If this is so, it might turn out that a polytrisacryl could be a rather stereo-regular polymer (it should be remembered that polyacrylamides are thought to be a 'random meshwork of fibres', totally disordered) and it could be a promising candidate for structural studies both in earth and microgravity. We have in fact started a collaboration with Drs. T.P. Lyubimova and V.A. Briskman (Perm, USSR) for decoding the structure (if any) of such new polymers. This has been quite an evolution: the original approach to polyacrylamides has been to wear a miner helmet (due to the important applications of this polymer in mining), but now we will have to wear the more subtle tools of X-ray diffraction and soft-neutron scattering for the final drive.

5. ACKNOWLEDGMENTS

PGR thanks ESA-ESTEC (Noordwijk, The Netherlands) for a grant (No. 9393/91/NL/JSC) allowing this investigation on new polymer matrices.

6. REFERENCES

1. P. Grabar and C.A. Williams, *Biochim. Biophys. Acta* 10 (1953) 193-194.

2. O. Smithies, *Biochem. J.* 61 (1955) 629-641.

3. S. Raymond and L. Weintraub, *Science* 130 (1959) 711-713.

TABLE 3 Acidic acrylamido buffers

pK^d	FORMULA	NAME	M_r	SOURCE
1.0	$CH_2{=}CH{-}CO{-}NH{-}C(CH_3)(CH_3){-}CH_2{-}SO_3H$	2-acrylamido-2-methylpropane sulfonic acid	207	a
3.1	$CH_2{=}CH{-}CO{-}NH{-}CH(OH){-}COOH$	2-acrylamido glycolic acid	145	b
3.6	$CH_2{=}CH{-}CO{-}NH{-}CH_2{-}COOH$	N-acryloyl glycine	129	c
4.4	$CH_2{=}CH{-}CO{-}NH{-}(CH_2)_2{-}COOH$	3-acrylamido propanoic acid	143	c
4.6	$CH_2{=}CH{-}CO{-}NH{-}(CH_2)_3{-}COOH$	4-acrylamido butyric acid	157	c

(a) Polysciences Inc., Warrington, PA. 18976, USA.
(b) P.G., Righetti, M. Chiari, P.K. Sinha and E. Santaniello, *J. Biochem. Biophys. Methods* 16 (1988) 185-192.
(c) Pharmacia-LKB Biotechnology, Uppsala, Sweden.
(d) The pK values for the three Immobilines and for 2-acrylamido glycolic acid are given at 25°C; for AMPS (pK 1.0) the temperature of pK measurement is not reported.

TABLE 4 Basic acrylamido buffers

pK[e]	FORMULA	NAME	M_r	SOURCE
6.2	$CH_2=CH-CO-NH-(CH_2)_2-N\bigcirc O$	2-morpholino ethylacrylamide	184	a
7.0	$CH_2=CH-CO-NH-(CH_2)_3-N\bigcirc O$	3-morpholino propylacrylamide	198	a
8.5	$CH_2=CH-CO-NH-(CH_2)_2-N(CH_3)_2$	N,N-dimethyl aminoethyl acrylamide	142	a
9.3	$CH_2=CH-CO-NH-(CH_2)_3-N(CH_3)_2$	N,N-dimethyl aminopropyl acrylamide	156	a
10.3	$CH_2=CH-CO-NH-(CH_2)_3-N(C_2H_5)_2$	N,N-diethyl aminopropyl acrylamide	184	b
>12[d]	$CH_2=CH-CO-NH-(CH_2)_2-\overset{+}{N}(C_2H_5)_3$	N,N,N-triethyl aminoethyl acrylamide	198	c
>12	$CH_2=CH-CO-NH-(CH_2)_2-\overset{+}{N}(CH_3)_3$	N,N,N-trimethyl aminoethyl acrylamide	153	c

(a) Pharmacia-LKB Biotechnology, Uppsala, Sweden.
(b) P.K. Sinha and P.G. Righetti, J. Biochem. Biophys. Methods 15 (1987) 199-206.
(c) IBF, Villeneuve La Garenne, France.
(d) QAE(quaternary amino ethyl)-acrylamide.
(e) All pK values (except for pK 10.3) measured at 25°C. The value of pK 10.3 refers to 10°C.

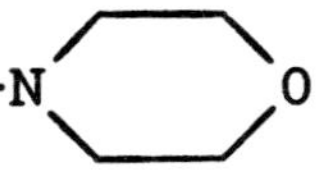

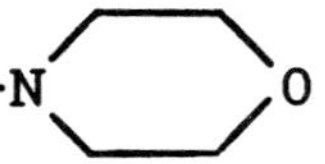

TABLE 5 New basic acrylamido buffers

pK^e	FORMULA	NAME	M_r	Ref
6.6	$CH_2=CH-CO-NH-(CH_2)_2-N\underset{S}{\bigcirc}$	2-thiomorpholino ethylacrylamide	200	a
6.85	$CH_2=CH-CO-N\bigcirc N-CH_3$	1-acryloyl-4-methylpiperazine	154	b
7.0	$CH_2=CH-CO-NH-(CH_2)_2-\text{(imidazole)}$	2-(4-imidazolyl) ethylamine, 2-acrylamide	165	c
7.4	$CH_2=CH-CO-NH-(CH_2)_3-N\underset{S}{\bigcirc}$	3-thiomorpholino propylacrylamide	214	a
8.05	$CH_2=CH-CO-NH-(CH_2)_3-N(CH_2CH_2OH)_2$	N,N-bis(2-hydroxyethyl)-N'-acryloyl-1,3-diaminopropane	200	d

a) M. Chiari, P.G. Righetti, P. Ferraboschi, T. Jain and R. Shorr, *Electrophoresis* 11 (1990) 617-620.
b) M. Chiari, C. Ettori, A. Manzocchi and P.G. Righetti, *J. Chromatogr.* 548 (1991) 381-392.
c) M. Chiari, M. Giacomini, C. Micheletti and P.G. Righetti, *J. Chromatogr.* 558 (1991) 285-295.
d) M. Chiari, L. Pagani, P.G. Righetti, T. Jain, R. Shorr and T. Rabilloud, *J. Biochem. Biophys. Methods* 21 (1990) 165-172.
e) All pK values measured at 25°C.

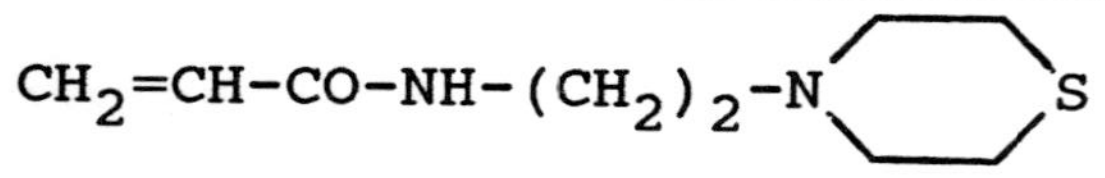

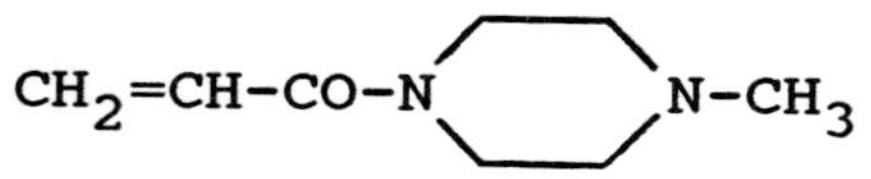

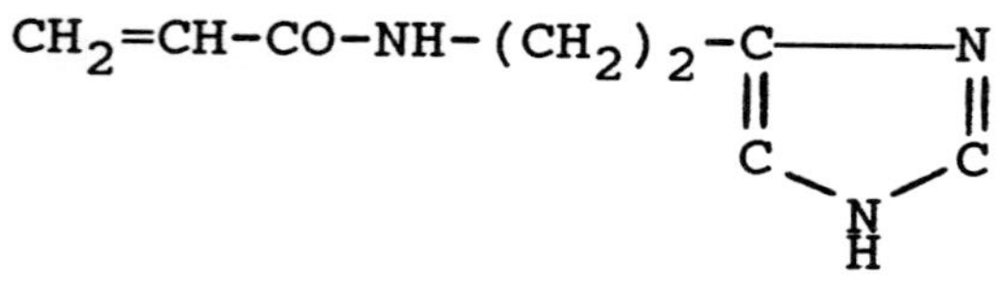

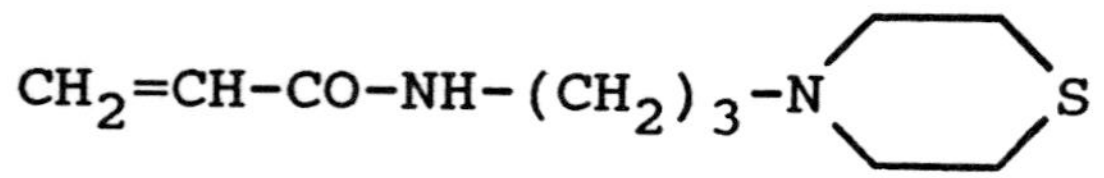

4. L. Ornstein, *Ann. N. Y. Acad. Sci.* 121 (1964) 321-349.

5. B.J. Davis, *Ann. N. Y. Acad. Sci.* 121 (1964) 404-427.

6. P.G. Righetti, *J. Biochem. Biophys. Methods* 19 (1989) 1-20.

7. P.G. Righetti, *J. Chromatogr.* 516 (1990) 3-22.

8. P.G. Righetti, B.C.W. Brost and R.S. Snyder, *J. Biochem. Biophys. Methods* 4 (1981) 347-363.

9. P.G. Righetti, *Immobilized pH Gradients: Theory and Methodology*, Elsevier, Amsterdam 1990.

10. P.G. Righetti, *Isoelectric Focusing: Theory, Methodology and Applications*, Elsevier, Amsterdam, 1983.

11. E. Gianazza, F. Celentano, G. Dossi, B. Bjellqvist and P.G. Righetti, *Electrophoresis* 5 (1984) 88-97.

12. C. Tonani and P.G. Righetti, *Electrophoresis* 12, 1991, in press.

13. P.G. Righetti and C. Tonani, *Electrophoresis* 12, 1991, in press.

POLYMERIZATION UNDER TERRESTRIAL AND ORBITAL CONDITIONS. COMPARATIVE STUDY.

T.P.Lyubimova

Institute of Continuous Media Mechanics, UB Acad. Sci.USSR, 1, Akad. Korolyov Street, Perm, 614061, USSR

ABSTRACT

The role of gravity dependent mechanisms in the polymerization is studied on the base of comparative analysis of polymer processing under terrestrial and orbital conditions. Two processes are considered. First of them is photoinitiated crosslinking polymerization of acrylamide. Another one is homopolymerization of methyl methacrylate with the initiator concentrated before the onset of the process in the polymer underlayer. Both processes occur as the frontal ones. Gravity dependent heat and mass transfer mechanisms of the polymerization connected with the reaction exothermality and the appearance of a new more dense phase are investigated. The significant influence of the convection on the polymerization processes characteristics and the structure of obtained samples has been demonstrated.

Keywords: *Polymer Processing, Frontal Polymerization, Gravity Dependent Heat and Mass Transfer Mechanisms, Buoyancy Convection, Polyacrylamide Gel, Methyl Methacrylate, Gel Structure*

INTRODUCTION

Gravity dependent heat and mass transfer mechanisms often play an important role in the material processing. It gives the possibility to use the unique conditions of space for the technological purposes. Wide usage of microgravity conditions for the experiments on the processing of semiconductors is well known. Comparatively new field of application is the synthesis of polymers in space (Refs.1,2). It has been found out recently (Ref.3) that the gravity dependent mechanisms are at work during the processing of polyacrylamide gel (PAAG) and leave the traces of their activity in the structure of final product. The structure of gel samples obtained under terrestrial and orbital conditions was found to be different. It is displayed in the difference of electrophoretic separation when using these samples. Depending on the molecular weight and other protein properties the electrophoretic separation was improved by a factor of 2 - 6 when the orbital gels were used.

The difference of the structure of gel samples produced on the Earth and in orbital conditions indicates that the gravity dependent phenomena play an active role in the polymerization. In this paper the acrylamide crosslinking photopolymerization under terrestrial and orbital conditions has been analyzed with the aim to detect gravity dependent mechanisms and to determine their influence on the polymerization process characteristics and the resulting gel structure. Another polymerization process which is the methyl methacrylate homopolymerization has been considered as well and the role of convective phenomena in this case has been studied.

POLYACRYLAMIDE GEL PROCESSING UNDER TERRESTRIAL AND ORBITAL CONDITIONS

The PAAG production is usually carried out by crosslinking polymerization of acrylamide in aqueous solution. The product being obtained is a porous matrix. Its structure strongly depends on the total concentration of both comonomers and the crosslinking monomer percentage. Among the other factors are the solvent viscosity, the temperature and the initiator concentration. Detailed investigation of the initial reactional mixture composition and the temperature effects on the polymerization kinetics as well as on the structure of obtained gels has been made in Refs.4-6. In Refs.7,8 the influence of the initiation rate on the permeability of PAAG has been studied and two-phase model has been developed for the permeability. We have investigated (Ref.9-13) the role of convective flows caused by the polymerization process itself.

The aqueous solutions of acrylamide and crosslinking monomer NN'-methylen bisacrylamide were used in the experiments. Initial reactional mixtures were placed into the vertically placed rectangular glass cavities. Polymerization was carried out at room temperature. The process was initiated by the reaction photosensibilizator - riboflavin. Monochromatic light beam source of the wave length 445 nm was used for the illumination of the cavity through one of the transparent walls.

Real polymerization process might occur differently at different points of the vessel. In the case of photopolymerization spatial non-uniformity of irradiation due to the absorption of light beam by the solution might be the reason of this non-uniformity. If the thickness of the layer is more than one mm, the difference of light beam intensities at different distances from irradiated surface might be large enough. Hence, at zero gravity polymerization process should be realized as the frontal one, with the polymerization front gradually moving from irradiated surface.

If the polymerization is carried out in gravity, then the gravity dependent phenomena might play an active role. PAAG polymerization is the exothermal process. In the experiments (Refs. 9-13) heating due to the reaction exothermality was about several degrees. This internal non- uniform heating can cause thermal buoyancy convection. Another factor is the appearance of a new more dense phase. If the reaction develops non-uniformly, it might lead to the marked density gradients and the appearance of the flows.

Transport phenomena bear direct influence on the macrostructure of resulting product. They cause the non-uniformities of density, index of refraction and other properties, all with the characteristic size above tenth of millimeter. However the forming of macro- and microstructure are interconnected. It is evident, that reciprocal influence exists too, since the molecular characteristics determine such macroscopic gel parameters, as density, elasticity, permeability and optical properties.

With the aim to detect the convection the experiments were carried out for the processing of polyacrylamide gel with the simultaneous visualization of the flows with the help of light scattering particles (Refs.9-11). The inlet of ultra-dispersed chemically neutral amber crumb were injected into the initial solution. Perpendicularly to the wide sidewall the cavity was dissected by the light blade from the special collimator. The visual observation and photographing of the process were made in the light blade plane.

The dependencies of polymerization process characteristics and resulting gel structure on the heat and mass transfer were studied with the help of shadowing apparatus with the point-wise source (Refs.9-12) and by interferometry method (Ref.13). In the last case the cavity was confined from its wide sidewalls by the flat semi-transparent mirrors. If these mirrors are parallel each other they form a working cell of an interferometer. The transformations of the fields of refraction index gradients, which are connected with the properties of the reacting mixture and the temperature were observed.

Mathematical modelling of the process was developed on the base of the equations of buoyancy convective heat and mass transfer in polymerizing mixture:

$$\frac{\partial \vec{v}}{\partial t} + (\vec{v}\nabla)\vec{v} = -\nabla p + \mathrm{Div}\,\hat{\sigma} + \left(Gr_T\theta + Gr_\eta\eta\right)\vec{\gamma} \qquad (1)$$

$$\frac{\partial \theta}{\partial t} + \vec{v}\nabla\theta = \frac{1}{Pr}\Delta\theta + \frac{Fk}{Pr}(1-\eta)^n \times F(I,H)\exp\{\theta/(1+\beta\theta)\} \qquad (2)$$

$$\frac{\partial \eta}{\partial t} + \vec{v}\nabla\eta = \frac{1}{Sc}\Delta\eta + \delta\frac{Fk}{Pr}(1-\eta)^n \times F(I,H)\exp\{\theta/(1+\beta\theta)\} \qquad (3)$$

$$\mathrm{div}\,\vec{v} = 0\ ,\quad \hat{\sigma} = H\hat{e} \qquad (4)$$

Here $\vec{v}$, p and θ are the dimensionless velosity, convective part of the pressure and temperature, $\hat{\sigma}$ and $\hat{e}$ are the viscous stresses and shear rates tensors, η characterizes the conversion of the monomer into a polymer, n stands the order of the reaction, F(I,H) describes the dependence of the reaction rate on the intensity of irradiation I and the effective viscosity H. Dimensionless parameters are Grasshof numbers Gr_T and Gr_η, Prandtl number Pr, Schmidt number Sc, parameters δ and β and Frank - Kamenetzky parameter Fk :

$$Gr_T = \rho^2 g\beta_T R T_0^2 L^3/\mu^2 E_{ef},$$

$$Gr_\eta = \rho^2 g\beta_\eta L^3/\mu^2,$$

$$Pr = \mu/\rho\chi,\quad Sc = \mu/\rho D_f,$$

$$\delta = \rho C_p R T_0^2/QE_{ef},\ \beta = RT_0/E_{ef},$$

$$Fk = QE_{ef}L^2K_{ef}^0\exp(-E_{ef}/RT_0^2)/(\kappa R T_0^2).$$

It is known that the effective viscosity of polymerizing mixture is strongly changing during polymerization. Figure 1 shows the example of the effective viscosity time dependence during PAAG polymerization obtained with the help of rotational viscosimeter (Refs.9-11). Sharp increase of the effective viscosity rate is clearly observed near the gelation point. After that the system loses fluidity almost instantly.

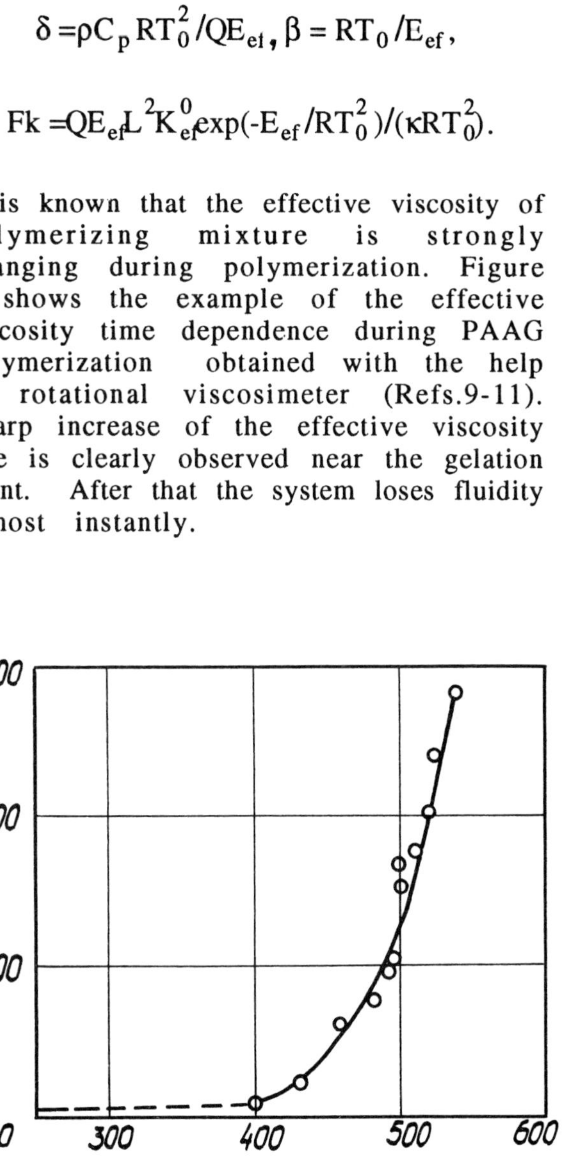

Figure 1. The example of the effective viscosity time dependence during PAAG polymerization.

In the calculations the effective viscosity of polymerizing mixture was supposed to be depending on the conversion and the temperature:

$$H = (1 + A\eta)^{m} \exp(-\gamma\theta) \qquad (5)$$

The term (5) is the final non-dimensional formula obtained after the expansion in the exponent of the effective viscosity temperature dependence.

The dependence of the effective reaction rate constant on the intensity of irradiation and the effective viscosity was taken in the form:

$$F(I,H) = \begin{cases} I_d^{1/2} & \text{if } H<H_c \\ I_d^{1/2}(H/H_c)^{1/2} & \text{if } H\geq H_c \end{cases} \qquad (6)$$

In these terms I_d is the local intensity of irradiation at the distance d from irradiated surface defined by the formula:

$$I_d = \exp(-B\,d) \qquad (7)$$

Here B is the dimensionless parameter, characterizing the absorption of irradiation by the solution.

The problem was solved for the rectangular parallelepiped with one of the sizes being essentially shorter than the others (the Hele-Shaw cell).

On the rigid sidewalls of the cell no-slip conditions, linear law for the heat transfer and impermeability condition applyed, while the external heating was absent:

$$\vec{v}|_{\Gamma} = 0, \quad \left.\frac{\partial\theta}{\partial n}\right|_{\Gamma} = -\alpha\,\theta, \quad \left.\frac{\partial\eta}{\partial n}\right|_{\Gamma} = 0 \qquad (8)$$

In the case of Hele-Show cell it is possible to use the approximation of plane trajectories and to solve the problem in two-dimensional formulation. We used the following approximation, when reducing the problem to 2D form:

$$v(x,y,z,t) = (1 - z^2/D^2)\,u(x,y,t)$$
$$\theta(x,y,z,t) = \vartheta(x,y,t) + (1 - z^2/D^2)\,a(x,y,t) \qquad (9)$$

In (1) - (8) polymerization kinetics is described with the help of phenomenological approach. It is simple enough and contains a few number of the parameters. More detailed description of the process can be obtained with the help of kinetic approach. We used the stationary state hypothesis for the radicals and long chain approximation and neglected by the physical crosslinking and termination by disproportionation and have derived the kinetic equations

$$\frac{d[A]}{dt} = -k_i[R^*][A] - k_p([\dot{P}]+[\dot{V}])[A] \qquad (10)$$

$$\frac{d[B]}{dt} = -2k_i[R^*][B] - 2k_p([\dot{P}]+[\dot{V}])[B] \qquad (11)$$

$$\frac{d[U]}{dt} = 2k_i[R^*][B] + 2k_p([\dot{P}]+[\dot{V}])[B] - k_i[R^*][U] - k_p([\dot{P}]+[\dot{V}])[U] \qquad (12)$$

$$\frac{d[S]}{dt} = k_p([\dot{P}]+[\dot{V}])[U] \qquad (13)$$

$$[\dot{P}]+[\dot{V}] = \left(\frac{k_i[R^*]}{k_t}([A]+2[B]+[U])\right)^{1/2} \qquad (14)$$

Here [A] and [B] stand the concentrations of comonomers, [S] is the crosslinks concentration, $[\dot{P}]$ and $[\dot{V}]$ are the concentrations of polymer radicals, k_i, k_p and k_t are the rate constants for the initiation, propagation and termination. The termination rate constant k_t is the function of the effective viscosity defined in accordance with (6). $[R^*]$ is the function of the local intensity of irradiation defined by (6)-(7). Temperature dependencies of the

rate constants are given by the Arrenius law.
The compatible solution of the equations (10)-(14) with (1),(2),(4)-(8) was carried out when the kinetic approach was applied.

Finite-difference method was used in the calculations. The numerical algorithm included the decomposition on different processes, the combination of the Lagrange and Euler approaches and the predictor-corrector method with the variable time step.

It is well known that the intensity of irradiation is one of the main factor of the photopolymerization processes control. Our experiments and mathematical modelling have shown that the orientation of photoinitiative light beam with respect to the cavity is important factor as well varying of which one can observe qualitatively different behavoir.

If one illuminates the cavity from above then the essential non-uniformity of illumination along the vertical direction occurs because of the absorption of light beam by the solution. As far as the duration of the induction period decreases with the growth of the irradiation intensity and the reaction rate is increasing function of I then the gelation starts from above and develops much faster at the upper part of the cavity. Therefore the unstable stratification is created in the region: the more dense phase is placed beyond the lesser dense one. It leads to the non-uniformity of polymerization front due to the Rayleigh-Taylor instability of an interphase surface and the development of the flows.

In the Figure 2 the series of experimental photographs obtained in the terrestrial experiments with this direction of irradiation is presented. The pictures illustrate the emergence of gel germs near the upper surface and the motion of this germs down through the fluid volume. The structure of the final product is strongly non-uniform.

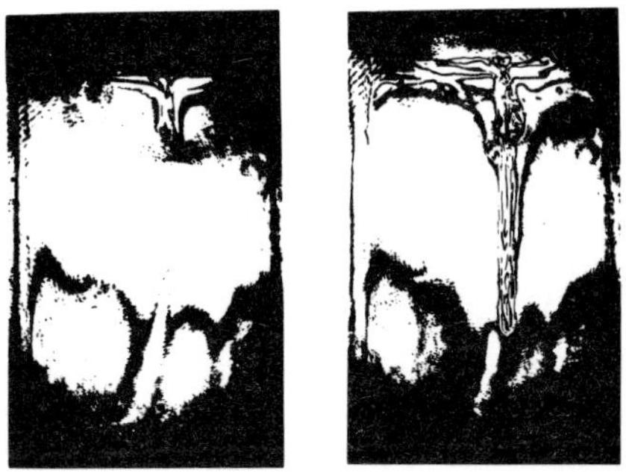

t = 265 sec t = 300 sec

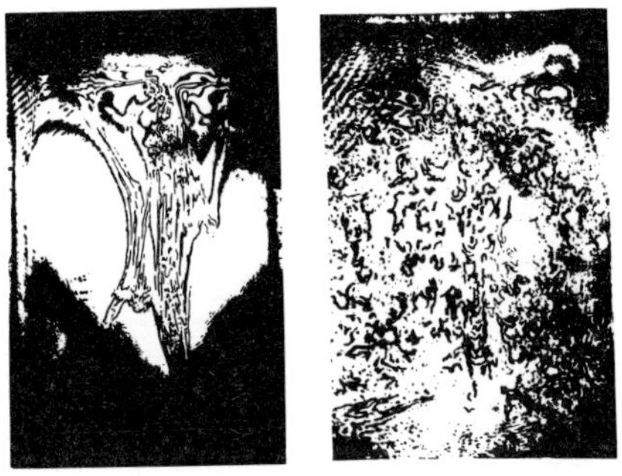

t = 330 sec t = 780 sec

Figure 2. Experimental photographs obtained in the case of photoinitiation from above with the help of interferometry method.

Figures 3 and 4 show the numerical results obtained in the calculations with this direction of irradiation. The left pictures are the stream function lines, the right ones are the lines of constant value of the conversion. As one can see, with the decrease of the Grasshof numbers Gr_T, and Gr_η , that corresponds to the decrease either of the cavity size or of the gravity level, the contribution of convective phenomena into the total heat and mass transfer decreases. In the pictures of the Figure 4 the frontal character of the process displays more clearly than in the Figure 3.

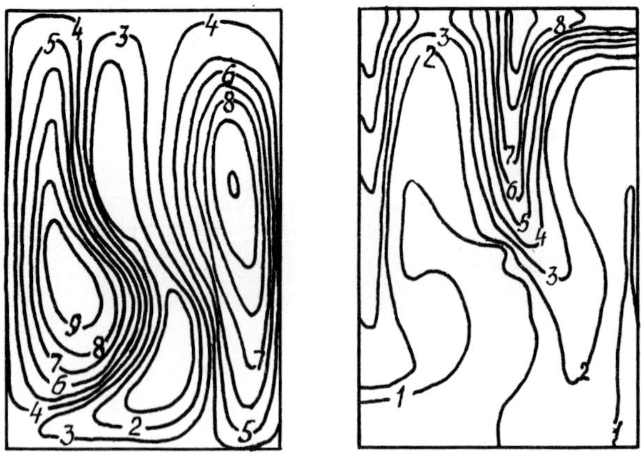

Figure 3. Numerical results obtained in the case of photoinitiation from above at $Gr_T = 10^6$, $Gr_\eta = -2 \cdot 10^7$, Fk= 5, $\delta = 0.5$.

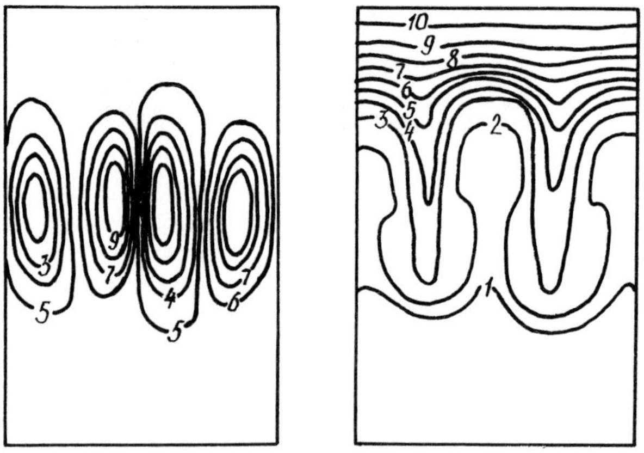

Figure 4. Numerical results obtained in the case of photoinitiation from above at $Gr_T = 10^5$, $Gr_\eta = -10^6$, Fk= 5, $\delta = 0.5$.

The results obtained in the experiments and with the help of numerical methods are similar. The structure of the final product keeps the traces of convective motions, existed before the formation of macrogel.

When one illuminates the cavity from the bottom, the process evolves differently. In this case the gelation starts from below and the Rayleigh-Taylor instability of the polymerization front is impossible - the more dense phase is placed lower than the lesser dense one. But the temperature gradient emerging due to the reaction exothermality creates potentially unstable stratification: it corresponds to the heating from below. It means, that in this case thermal convection should be the main mechanism of heat and mass transfer.

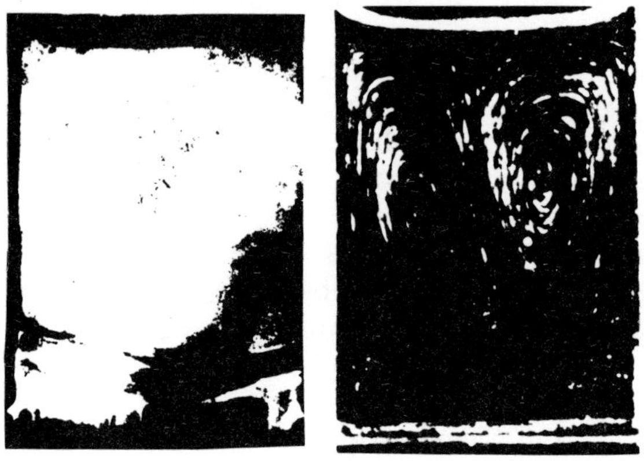

t = 90 min

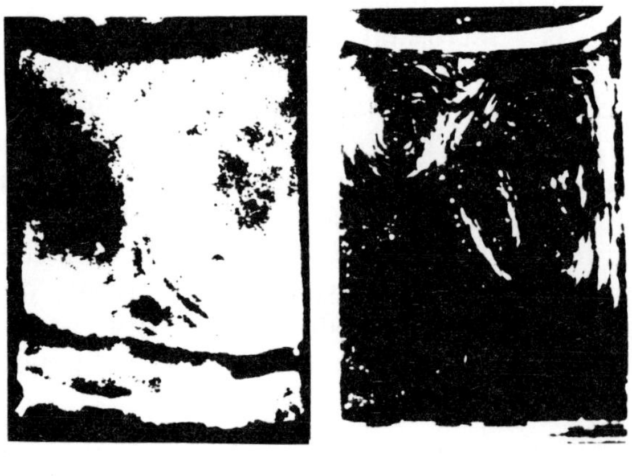

t = 110 min

Figure 5. Experimental photographs obtained in the case of the photoinitiation from below: a - the shadow photographs of the structure, b - the convective motions visualized with the help of light scattering particles

The experiments with this direction of irradiation (Refs.9-11) were including both visualization of convective motions and the investigation of the structure with the help of shadow apparatus. The total duration of the polymerization process was equal to 115 minutes. The obtained experimental photographs are presented in the Figure 5. The left photographs are the pictures of the structure, the right ones are the visualizations of convective motions.

As one can see the polymerization front remains plane during the first stages of the process. Then it is distorted, but does not lose continuity, because of the absence of the mechanism, causing its instability. At the same time in the region, where the liquid phase is placed, the thermal buoyancy convection is developing. It has the form of two almost symmetrical vortices, separated by the ascending fluid flow. The velocity of convective motion may be determined using the length of tracks on the photographs and the exposure time. In this experiment it is of the order of 1 mm per sec. The boundary between the vortices displays slow random motions. The convective motion takes place at comparatively small values of the effective viscosity, i.e. corresponds to the starting segment of the rheological curve. During the final stage the random moving of the boundary between the vortices leads to the decrease of convection and speeding up of the gelation in the left part of the cavity. With the approach to the point of macrogel formation the convection is dumped.

The series of experimental photographs obtained at the photoinitiation from below with the help of interferometry method is presented in the Figure 6. The creation and the development of a convective flare over the irradiated wall is clearly observed. The final product structure is strongly non-uniform. It keeps the traces of the flows observed before the macrogel formation.

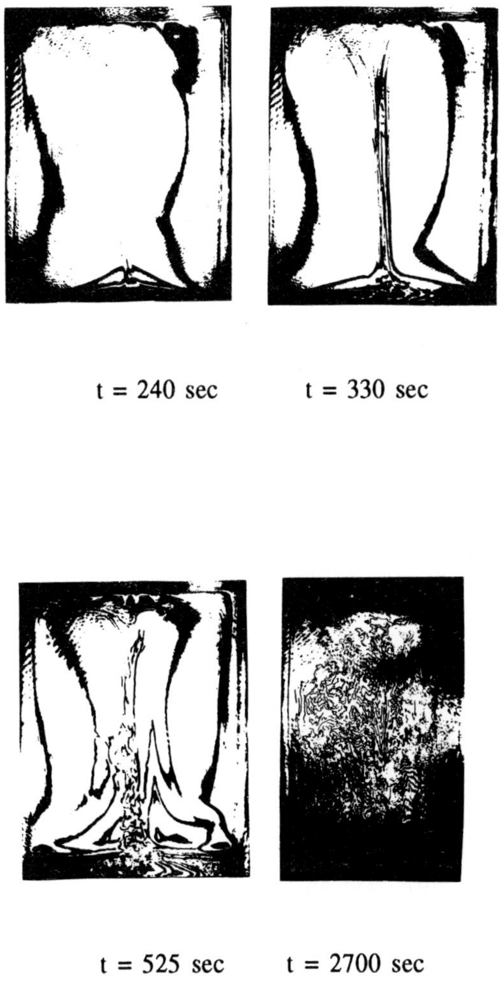

Figure 6. Experimental photographs obtained in the case of photoinitiation from below with the help of interferometry method.

The experiments and numerical modelling with the illumination from the narrow sidewall have detected the important role of the both mechanisms of heat and mass transfer in the structuration. At the first stages of the process the new more dense phase, which appears near the irradiated wall, by the jets moves from above to the bottom. At this time the convective motion is one-cellular. It is displaced to another part of the cavity. Later on the gel jets fill out step-by-step the whole volume of the cavity. The main role of buoyancy convection in this case results in twisting and tangling of gel filaments.

The most favorable conditions for producing of uniform gel are created, when the photoinitiative light beam is directed to the wide sidewall of the cavity. In this case the non- uniformity of irradiation inside the cavity is less than at the other directions of irradiation. However under the laboratory conditions the uniform gel samples were obtained only in the cavities of the thickness 1 mm. The terrestrial experiments for the containers of the thickness more than 3 mm have demonstrated that the polymerization process was accompanied with the convective motions, leading to the deterioration of the final structure of gel. In the orbital conditions the intensity of convective motions should be markedly less. It can be supposed that this difference resuts in the distinction between the structures of orbital and control samples of gels obtained in Ref.3.

Thus the experimental and numerical results demonstrate that the gravity dependent heat and mass transfer mechanisms can play role in the PAAG polymerization important. If we need in sufficiently lengthy samples of gel with more uniform or prescribed structure then the space production might be perspective. It must be noted however, that really the inertia forces are at work at space installations which cause the convection. Hence, with the aim of obtaining the gels with prescribed structure in orbital conditions it is necessary to carry out the preliminary investigations of heat and mass transfer during polymerization under real space flight conditions.

METHYL METHACRYLATE POLYMERIZATION GRAVITY DEPENDENT MECHANISMS

The role of gravity dependent phenomena in the methyl methacrylate homopolymerization has also been studied (Ref.13). The process was activated by an initiator, small amount of which was concentrated at the beginning of the process in the polymer underlayer.

The underlayer is a polymethyl methacrylate sample polymerized in advance up to a glassy state and containing unconsumed initiator. It is placed near one of the cavity walls. Polymerization is carried out at room temperature under argon atmosphere.

At the beginning of the process the strong boundary exists between the underlayer and the monomer. As the process develops, the condensed polymer from the underlayer is dissolved in a monomer and the boundary is spreading into an intermediate layer. The initiator is distributing from this layer all over the volume. Then the frontal and background polymerizations are simultaneously developing in the volume.

The background polymerization is connected with the distribution of the initiator all over the volume. As the measurements show, the conversion of a monomer into a polymer in the region near the upper surface is about 10-20 percent. In a control sample containing no underlayer this value constitutes 1-2 percent.

The frontal polymerization mechanism is associated with the initiator presence in a monomer, availability of condensed polymer phase capable of dissolution in a monomer and creation of a peculiar microreactors near the interphase surface. By means of those microreactors it is possible to carry out polymerization under the gel effect conditions.

There are different viewpoints on the mechanisms of initiator spreading all over the volume and its role in the considered polymerization process. To our mind the distribution of the initiator occurs due to the diffusion and buoyancy convection. And besides the most contribution is given by convection. The last is caused by the reaction exothermality and the appearance of a new more dense phase.

With the aim to detect the convection we have investigated polymerization process by the interferometry method. In the Figure 7 series of experimental photographs illustrating considered process is presented.

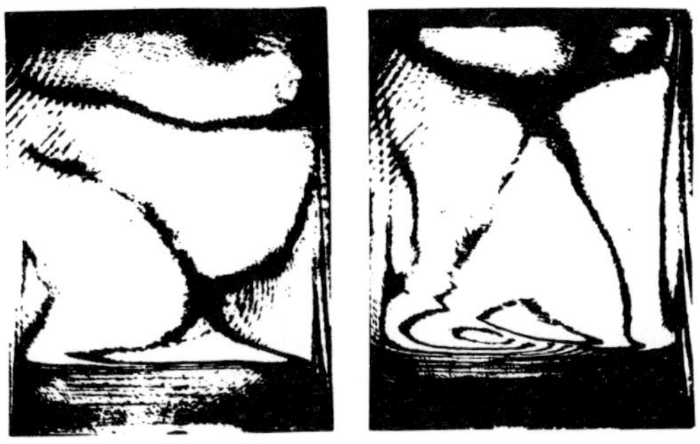

t = 15 min t = 93 min

Figure 7. Experimental photographs obtained during polymerization of methyl methacrylate with the help of interferometry method.

As one can see the convective phenomena are of the same type as in the case of PAAG polymerization at the photoinitiation from below (see Figure 6). But the intensity of the motion is sufficiently less than in the above mentioned case. The creation of a convective flare over the polymer underlayer is observed. The convective motion looks like two vortexes of the opposite direction separated by the rising thermal flow. It exists at the first stage of a process. With the increase of the effective viscosity the convective motion is dumped. From our point of view this motion is the main factor promoting the initiator spreading all over the volume and polymerization process developing.

Thus the buoyancy convection plays the positive role in this process. Under microgravity conditions such a process can not be developing without the additional forces of non-gravitational origin such as inertia forces or electromagnetic effects.

CONCLUSION

I. The considered examples demonstrate that the gravity dependent phenomena can play an important role in the polymerization.

II. Gravity dependent heat and mass transfer mechanisms of the polymerization are caused by the reaction exothermality and the appearance of a new more dense phase.

III. The same gravity dependent phenomena may exist and play important role at the other processes in multiphase and multicomponent media.

IV. Gels display very high gravitational sensitivity. They keep the traces of the buoyancy convective flows, taking place during the gelation, in the structure of resulting product. This gels property make them useful for the modelling investigations of the structuring media behaviour.

V. The low intensity of convection under microgravity conditions might provide the possibility to synthesize the polymers with new properties. However the space production of polymers with the prescribed properties should be elaborated on the base of preliminary investigation of polymerization under the real space flight conditions.

REFERENCES

1. *Microgravity Polymers, (Pamphlet). NASA, Conf. Publ. NASA CP-2392, June 1986* (1986)

2. J.W.Vanderhoff, M.S.El-Aasser, F.J. Micale, E.D.Sudol, C.M.Tseng, A.Silwanowicz and H.R.Sheu In *Proc. Amer. Chem. Soc., Div. Polym. Mater. Sci. Eng.*, 54 (1986) 587-592.

3. A.S.Sadykov, V.B. Leontyev, Yu.S. Mangutova, G.M.Grechko, G.S. Nechitailo and A.L. Mashinsky *Akademiia Nauk SSSR, Doklady*, 303 (1988) 1004-1007.

4. P.G. Righetti, B.C.W. Brost, and R.S. Shyder, *J. Biochem. Biophys. Methods*, 4 (1981) 347.

5. C. Gelfi and P.G. Righetti *Electrophoresis*, 2 (1981) 213.

6. P.G. Righetti, C. Gelfi and A.Bianchi Bosisio *Electrophoresis*, 2 (1981) 291.

7. N. Weiss, T. Van Vliet and A. Silberberg *J. Polym. Sci.* 17 (1979) 2229.

8. N. Weiss, T. Van Vliet and A. Silberberg *J. Polym. Sci.* 19 (1981) 1505.

9. Sh.D. Abdurakhmanov, L.G. Bogatyreva, V.A. Briskman, M.G. Levkovich, V.B. Leontyev, T.P. Lyubimova, A.L. Mashinskii and G.S. Nechitailo In *Numerical and Experimental Modelling of Hydrodynamic Phenomena under Weightlessness* Sverdlovsk (1988) 120.

10. L.G. Bogatyreva, V.A. Briskman, M.G. Levkovich, V.B. Leontyev, T.P. Lyubimova and G.S. Nechitailo In *Space Science and Technology* Kiev (1989)

11. Sh.D. Abdurakhmanov, V.G. Babskii, L.G. Bogatyreva, V.A. Briskman, M.G. Levkovich, V.B. Leontyev, T.P. Lyubimova, A.L. Mashinskii and G.S. Nechitailo In *The XIX Gagarin Scientific Readings on the Astronautics and Aeronautics*, 1989, Moscow (1990) 219.

12. V.B. Leontyev, Sh.D. Abdurakhmanov and M.G. Levkovich *In Proceedings of AIAA/IKI Micro-gravity Science Symposium Moscow, USSR, May 13-17, 1991* Moskow (1991) 273-280

13. V.B.Golubev, B.A.Korolyov, K.G.Kostarev and T.P.Lyubimova In *Hydromechanics and Heat/Mass Transfer in Microgravity, Perm - Moscow, USSR, July 6-14, 1991*, Perm (1991) 202.

THERMAL CONVECTION IN CONTINUOUS FLOW ELECTROPHORESIS

M. S. Bello
Institute of Macromolecular Compounds Academy of Sci.
199004, St. Petersburg, Bolshoi pr., 31, USSR

ABSTRACT

An incompressible viscous fluid flow with a step-wise distributed heat sources in the vertical infinite plane channel is considered as a two-dimensional model of a continuous flow electrophoresis (CFE) chamber entrance. Combined forced and thermal convection are studied numerically for the ranges of Reynolds and Grashof numbers which are characteristic of the CFE devices.

Key Words:

Continuous flow electrophoresis, Thermal convection, Internal heat sources, Vertical plane channel, Entrance region.

1. INTRODUCTION

Absence of thermal convection in microgravity was a prerequisite for Space experiments with continuous flow electrophoresis (CFE). These experiments stimulated research in CFE fluid dynamics and, in particular, studies of thermal convection in the electrophoresis chamber. To estimate the influence of thermal convection on the fluid flow, the previous studies[1-4] used critical Rayleigh numbers calculated for similar hydrodynamical systems. Thus, Ostrach has drawn an analogy between the CFE chamber and a parallelepiped enclosure[1], whereas Saville has used critical Rayleigh numbers for an infinite vertical channel of rectangular cross-section with uniform vertical temperature gradient imposed on the walls. An origin of the temperature gradient was assumed to be heating of the fluid at the entrance[3,4]. Naumann and Rhodes[4] have calculated the temperature field at the chamber entrance assuming that details of the fluid flow had negligible effect on the temperature distribution. The energy balance equation was solved for uniform plug flow and Poiseuille flow. Temperature gradients potentially able to cause instability were found. Moreover, Rayleigh numbers corresponding to these gradients agreed well with those estimated[2] and experimentally measured[4]. Nevertheless, it seems useful to investigate this hydrodynamic system on the basis of the full Navier-Stokes equations and to examine whether or not the

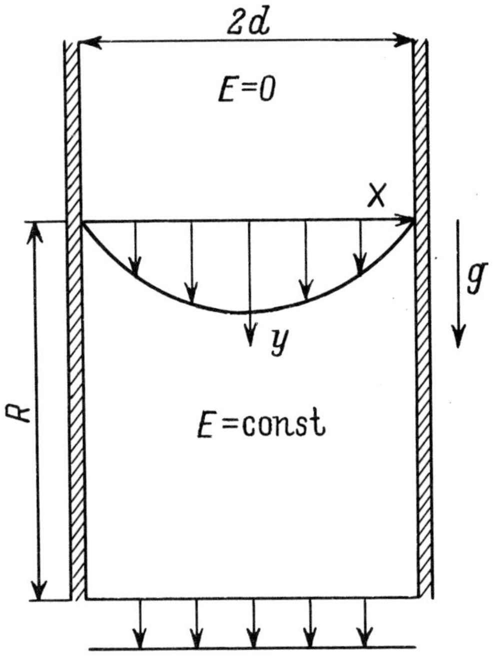

Figure 1 Vertical channel with step-wise distributed heat sources.

instability occurs at the entrance.
This paper numerically investigates an incompressible viscous fluid flow with step-wise distributed heat sources in the vertical infinite plane channel with the walls maintained at a constant temperature (Figure 1). This system is considered as a two-dimensional model of the chamber entrance. Fluid flow and heat transfer are governed by Navier-Stokes equations in Boussinesq's approximation. Numerical studies were carried out in the ranges of Reynolds and Grashof numbers which are characteristic of the CFE.
These results are discussed in view of the recent studies[5,6], considering thermal convection in the front plane of the CFE chamber.

2. MATHEMATICAL MODEL

When thermal convection in a CFE chamber is studied, the following is assumed: the fluid is homogeneous, Newtonian and incompressible. In this paper all its properties except density are considered constant. Additional assumptions are the following:

1. the electric field $\bar{E}$ is uniform when $y > 0$ and vanishes when $y < 0$;
2. the fluid flow and the temperature field in the domain $y > 0$ do not influence those in the upper part of the channel $y < 0$.

The steady fluid flow and the temperature field in the lower part of the channel are governed by the two-dimensional Navier-Stokes equations in Boussinesq's approximation:

$$v_y \frac{\partial \omega}{\partial y} + v_x \frac{\partial \omega}{\partial x} = \frac{1}{Re}\left(\frac{\partial^2 \omega}{\partial y^2} + \frac{\partial^2 \omega}{\partial x^2}\right) + \frac{Gr}{Re^2}\frac{\partial \vartheta}{\partial x}, \quad (1)$$

$$\omega = \frac{\partial^2 \psi}{\partial y^2} + \frac{\partial^2 \psi}{\partial x^2}, \quad (2)$$

$$v_y = -\frac{\partial \psi}{\partial x}, \quad v_x = \frac{\partial \psi}{\partial y},$$

$$v_y \frac{\partial \vartheta}{\partial y} + v_x \frac{\partial \vartheta}{\partial x} = \frac{1}{Re\ Pr}\left(\frac{\partial^2 \vartheta}{\partial y^2} + \frac{\partial^2 \vartheta}{\partial x^2} + 1\right) \quad (3)$$

$$\vartheta = (T - T_w)/T_c, \quad T_c = \frac{\sigma E^2 d^2}{\chi},$$

$$Re = \frac{\upsilon d}{\nu}, \quad Gr = \frac{g\beta T_c d^3}{\nu^2}, \quad Pr = \frac{\eta}{\chi} C_p.$$

ω being the dimensionless vorticity, ψ the dimensionless streamfunction, ϑ the dimensionless temperature, x and y the dimensionless coordinates measured in the units of the channel half width d; v_x and v_y the dimensionless velocities along the x and y coordinates (measured in the units of the average velocity); T the dimensional temperature of the fluid, T_w the wall temperature, T_c the characteristic temperature difference due to internal heating, ν and η the kinematic and dynamic fluid viscosities; σ, χ, β and C_p the fluid electric conductivity, thermal conductivity, coefficient of thermal expansion and specific heat, respectively; υ the average fluid velocity, g the acceleration due to gravity; Re, Gr and Pr the Reynolds, Grashof and Prandtl numbers, respectively.

The boundary conditions are given by the Poiseuille profile and constant temperature at the entrance of the computational domain:

$$v_y = \frac{3}{2}(1 - x^2), \quad \vartheta = \vartheta_0,$$
$$\psi = \frac{3}{2}(x^3/3 - x), \quad (y = 0); \quad (4)$$

The no-slip and the constant temperature conditions at the rigid walls are given by:

$$v_{y,z} = 0,\ \vartheta = 0,\ \psi = \mp 1, (x = \pm 1) \quad (5)$$

At the outlet boundary of the computational domain, which is chosen at a distance y = R (see Figure 1), the boundary conditions are given by:

$$v_y = 1,\ \frac{\partial\vartheta}{\partial y} = 0,\ \psi = -x,\ (y = R). \quad (6)$$

The value of R is large enough for the fluid flow and the temperature field to become one-dimensional. The one-dimensional temperature ϑ and vertical velocity v_y are given by

$$\vartheta = \frac{1}{2}(1 - x^2),$$

$$v_y = \frac{3}{2}(1-x^2) - \frac{Gr}{120\ Re}(1-6x^2+5x^4) \quad (7)$$

A numerical solution for eqs. (1-6) was found by a finite difference method[7].
The fluid flow and the temperature field were calculated on a uniform finite difference grid 21×65.

3. RESULTS AND DISCUSSION

The calculations were performed for the following values of the Grashof to Reynolds numbers ratio: Gr/Re = 100, 150, 180; Reynolds numbers ranged from 1 < Re < 3. Although the Prandtl number in most cases was 5, other values were also used to check the results. The length R of the computational domain was 12.8. This value ensures the development of the one-dimensional temperature and velocity fields within the domain.
Pictures of the fluid motion and distribution of the temperature field are shown in Figures 2 and 3. Figure 2 shows smooth rearrangement of the flow for Gr/Re = 180, ϑ_0 = 0.

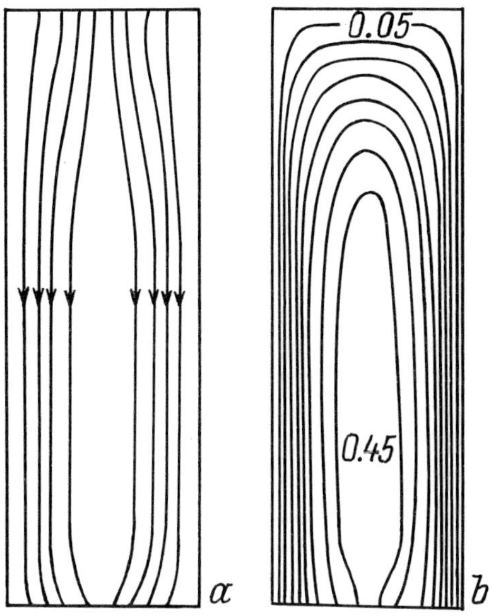

Figure 2. Streamlines (a) and isotherms(b) for Gr/Re=180, ϑ_0=0.

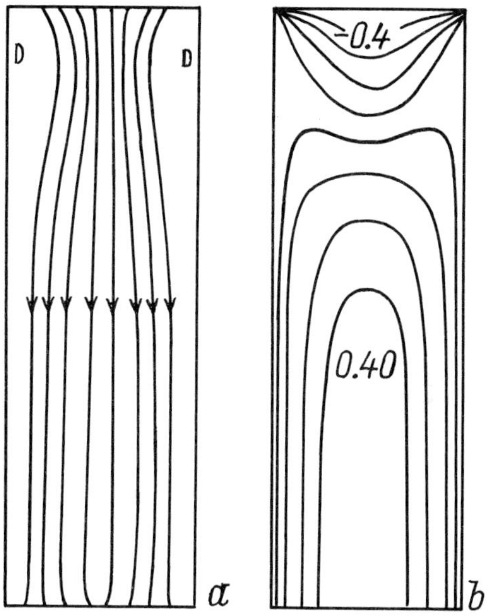

Figure 3. Streamlines(a) and isotherms (b) for Gr/e = 100, ϑ_0=-0.5.

This value of Gr/Re corresponds to the most "dangerous" case when the vertical velocity vanishes at the axis of the channel (see eq. 7). Reverse flows at the entrance occur when the inlet temperature of the fluid is less then that of the walls (Figure 3, ϑ_0 = -0.5).

These reverse flows are due to the increase in the axis fluid velocity caused by the influence of the buoyancy on the cooler and, therefore, heavier fluid in the core of the flow. Behaviors of the vertical velocity and the temperature at the channel axis are shown in Figure 4 which illustrates the most common case ($\vartheta_0 = 0$).

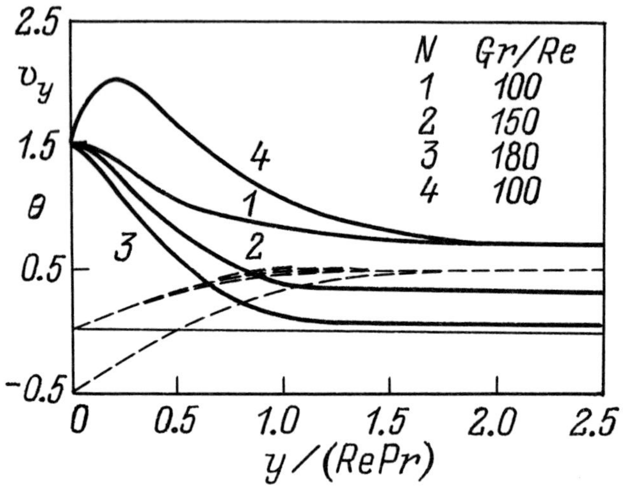

Figure 4. Dependence of the vertical velocity (solid) and the temperature (dashed) at x=0 on the axial coordinate. Three close dashed curves correspond to the different values of *Gr/Re*. The lowest dashed curve corresponds to the case (4).

The temperature is weakly dependent on the details of the flow field in agreement with Naumann and Rhodes. This fact also confirms the present assumption (2). All four velocity and four temperature curves approach their asymptotic values determined by eqs. (7).

Velocity curve (4) shows an increase in the vertical velocity discussed above. Fig. 4 also shows that the hydrodynamic entrance region is controlled by the thermal entrance region, which is approximately equal to *RePr* (in dimensionless units). Thus one can estimate the vertical temperature gradient A, which can potentially lead to instability:

$$A \approx T_c/(2Re\cdot Pr\cdot d).$$

The Rayleigh number corresponding to the temperature gradient A is

$$Ra = \frac{g\beta A d^4}{\nu^2} Pr = \frac{Gr}{2Re},$$

where Gr has been defined above. Therefore, the Rayleigh numbers corresponding to the flows shown in figs. 2-4 are approximately equal to 50, 75 and 90. These values considerably exceed the critical Rayleigh number for the onset of instability in a vertical plane channel[8], which was used as an estimation[2,4]. No instability, however, was revealed in these calculations. This may be because the region of potential instability (Figure 2, b) has a wedge-like shape and the viscous forces prevent the instability more effectively than would a rectangular region.

It was predicted[5,6] that large-scale reverse flows can occur at the front chamber plane even when the inlet fluid temperature is equal to the wall temperature. Therefore, results of this paper also confirm previous studies[5,6] in that the front-plane temperature distribution and fluid flow play decisive role in the CFE hydrodynamics.

REFERENCES

1. S. Ostrach *J. Cromatogr.*, **140**, **3**, 187 (1977)
2. D.A. Saville *(COSPAR) Space Research, ed. .J. Rycroft* (Oxford and New York: Pergamon Press 1979)
3. E.D.Lynch and D.A. Saville *Chem. Eng.Commun.*, **9**, 201 (1981)
4. R.J. Naumann and P.H. Rhodes, *Sep. Sci. and Techn.*, **19**, 1, 51 (1984)
5. M.S. Bello and V.I. Polezhaev, *Microgravity sci. technol.*, **3**, **4**, 231 (1991)
6. M.S. Bello and V.I. Polezhaev, *Hydrodynamics, Gravitational Sensitivity and Transport Phenomena*

in Continuous Flow Electrophoresis (AIAA paper No 91-0112 1991)
7. V.I Polezhaev, A.V. Bune, N.A. Verezub, G.S.Glushko, V.L.Gryaznov, K.G. Dubovik, S.A.Nikitin, A.I.Prostomolotov, A.I. Fedoseev, S.G.Cherkasov, *Mathematical Modelling of Convective Heat and Mass Transfer on the Basis of Navier-Stokes equations* (Moscow, Nauka 1987 in Russian).
8. G.Z. Gershuni and E.M. Zhukhovitskii, *Convective Stability of Incompressible Fluid* (Moscow, Nauka 1972 in Russian)

PLANT RESEARCH IN SPACE

A.L.Mashinsky[1] and G.S.Nechitailo[2]

1) Institute of Medico-Biological Problems, USSR Ministry of Health, Moscow, USSR
2) NPO 'Energiya', Kaliningrad, Moscow Region, USSR

ABSTRACT

The data obtained indicate a possible impairment of chemical reactions at zero gravity and the opportunity of obtaining biosynthesis products other than the terrestial ones. To investigate seeds, seedlings and vegetating higher plants, the technical means were worked out with due regard for the in-flight exposures. The effects connected with microgravity were demonstrated. The plants that may develop through all stages in-flight have substantial metabolism disorders. The latter, however, are not the obstacle to consider the plants as a component of the life support systems in long-term space missions. The results are based on the authors' original studies at Salyut and Mir orbital stations.

1. INTRODUCTION

The symposium's topic is confined to problems related to the effects of weightlessness on the processes of hydromechanics and heat mass exchange. While not trying to reduce the complicated processes occurring in a cell and in a plant organism on the whole only to physico-chemical terms, we call attention to the currently revealed spaceflight effects on plants. The goal is to consider the most important aspects of plant physiology in weightlessness from the standpoints of physical chemistry and biophysics.

Without entering in all details of spaceflight exposures, we shall recall here the factors that cannot be reproduced in laboratory on Earth, namely galaxic cosmic radiation (GCR) and weightlessness.

2. METHODS AND REQUIREMENTS

Special research has been carried out on weightless conditions at Salyut spacecraft, and peculiarities in the processes of some chemical reactions have been studied. Absolute weight-forces acting on the body is equal to zero, does not actually exist due to unbalanced perturbations of various origin constantly present at spacecraft. Numerical assessment that led us to this conclusion has been obtained earlier (Ref. 1). It was demonstrated that aerodynamic forces induce the microgravity from $8\cdot10^{-7}$ to $1\cdot10^{-5}$ g, depending on orbit height. Solar pressure also exerts on an orbiting object a microgravity of about 10^{-9} g. The effect of executive organs of orientation system varies from 10^{-7} to 10^{-9} g. Rotational manoeuvers induce the microgravity of $2\cdot10^{-1}$ – $2\cdot10^{-2}$ g, and in docking process the latter exceeds $5\cdot10^{-1}$ g.

The most significant contribution to the magnitude of 'gravitational noise' is provided by movements and physical training of the crew and may reach $8 \cdot 10^{-3} - 1 \cdot 10^{-2}$ g. Thus, there exist various in-flight perturbations that lead to biological subject overloads in a wide range commensurate with the gravitational sensitivity threshold.

Theoretically it was shown that the weightlessness effect on intracellular processes is manifested by redistribution of particles that are responsible for gravitational perception. The mass forces influence only the cells whose size exceeds 10 μm (Ref. 2,3) This made grounds for some researchers to conclude that it is impossible to obtain any effect on cellular organism in weightlessness. Parfenov (Ref. 4) infers that the elementary biological processes at cellular level are gravitation-independent. However, in this case the chemical reactions are not evaluated in the field of forces whose total sum diminishes to zero. Moreover, a cell with size exceeding 10 μm is not considered although it is often observed in a plant cell of the eukaryotic organism.

Our research demonstrated the gravitational dependence of chemical processes that take place in phase transition or mass transfer in a porous medium. It was shown that spaceflight exposures modify the form and properties of molecules of biopolymers such as gelatin and agar solutions (Ref. 5). In particular, the increment in characteristic viscosity, depending on agar concentration, was revealed as well as the change in electrophoretic properties of polyacrylamide gel obtained from monomer under in-flight conditions. The physico-chemical properties of gel were also changed: NMR-spectra parameters, pattern of X-ray angular dispersion, etc. (Ref. 6).

The potentials were studied to synthesize in outer space biologically active substances such as nucleotides and nucleosides. New compounds were found to have a molecular weight larger than in natural analogues (Ref. 7).

The data obtained indicate a possible impairment of chemical reactions at zero gravity and the opportunity of obtaining reaction products other than the terrestrial ones. We are convinced that this cannot but tell on classical aspects of plant physiology in space such as water regimen, mineral nutrition, metabolism, photosynthesis, growth and development. Suffice it to say that in weightlessness the cytoplasm can somewhat change its structure along with the plasma membrane surrounding it that changes its selective permeability to control cell metabolism. The water moves along the conducting xylemal elements. The water plant regimen is mostly characterized by the value of water potential

$$\Psi = P + \rho \cdot g \cdot h - \pi ,$$

where

- Ψ – water potential of plant tissue,
- P – hydraulic pressure in water solution,
- π – osmotic pressure,
- ρ – water density,
- g – acceleration of gravity,
- h – vertical distance,

and gravitational member $g\,h$ diminishes to zero in-flight. Ions and other soluble substances participating in the root-to-leave movement of water are transported through the cell plasmalemma, whereas organic compounds extend along the phloem to the area of lower concentrations. This selected examples of plant transport finally depend on the properties of the cell membranes which, like biopolymers formed in weightless conditions, change their characteristics. Earlier studies and materials presented at the symposium make it clear. The alteration in the rate and pattern of cell metabolism can, therefore, be expected. This can, probably, account for changes in plants that were studied in-flight.

To obtain the results, a special vegetation equipment was elaborated and approbated with due regard for space technology requirements, which allowed:

1) to answer the question on the possibility of plant growth in-flight, features of their morphogenesis and spatial orientation, type of metabolism, productivity;

2) to provide the required moistening regimen in the root layer for prolonged experiments with pea and wheat that were carried out in the updated Oasis unit at Salyut-4 station where the first results on biochemical composition of these plants were obtained;

3) to improve illumination and mineral nutrition, to perform plant electrostimulation and aeration of the root area, etc.

4) to ensure plant growth from seed to seed at spacecraft. This was accomplished in the Phyton unit where arabidopsis seeds were obtained in-flight (Ref. 8);

5) to grow plants in artificial gravity, magnetic field gradient or by damping the external influences;

6) to fix plant material directly under the spaceflight conditions;

7) to study in-flight the fluid transfer in granulated artificial substrate as well as to solve other specific problems of plant cultivation and examination.

At present the Soviet-Bulgarian greenhouse Svet is in action at Mir station, where microprocessors control cultivation parameters, such as humidity and temperature of the root layer and air, illumination level, operation of executive organs. The information is periodically transmitted to Earth by telemetry. Radish root crops were first produced using this apparatus.

3. RESULTS

Now the main results of study will be presented. The experiments with seed seemed to be most simple to carry out. However, the results obtained by different authors during the short-time flights (up to a few days) varied greatly. Nevertheless, in prolonged seed exposure to spaceflight (827 days) an almost complete destruction of all arabidopsis seed was shocking (0.41% for test and 43.7% for control germination), while the number of chromosomal aberrations in Welsh onion exceeded 50%. There are also other data on impaired structure and function of seed. Therefore, we evaluated the stability of seed biochemical characteristics and determined their free radical state. The content of nitrogen substances and mineral acid-soluble phosphates decreased in lettuce and aneth seeds which were exposed to spaceflight for 240 days; total amount of carbohydrates also dropped through the consumption of easily hydrolysed polysaccharides, starch in particular. The level of unsaturated fatty acids, especially the polyunsaturated class, rose to increase the ratio of saturated to unsaturated fatty acids. This is characteristic of organism's response to stressful exposures that are provoked by free radical accumulation. The amount of free radicals in seeds was determined by the EPR method in two 2-month space experiments prior to and following the mission. The intensity of EPR spectra was found to increase by 10-30% in 8 out of 10 types of test seed compared to controls. This may activate the mutagenic process because it is known that free radicals can impair the function of enzymes responsible for replication or correction of the genetic apparatus. Most developing mutations are harmful to the organism and population on the whole. Therefore, protection of the genetic apparatus from mutagens is a burning problem. One of possible solutions is the use of antimutagens. Hexamidine, α-tocopherol, auxin, and kinetin proved to decrease in-flight the incidence of chromosomal aberrations in Welsh onion whose example demonstrated the adverse effect of spaceflight exposures on seeds and the possibility to decrease and even eliminate undesirable sequelae.

One of the questions still open in space biology is the feasibility to attain during mission a complete ontogenetic cycle including its reproductive phase. Polarization of the plant body starts with initial cell division, and the force of gravity plays a part in these processes. Therefore, the research on plant growth and development at microgravity or weightless conditions presents both practical and theoretical interest.

The study on potato tubers (18 days in spaceflight), gladiolus and grape hyacinth bulbs (60 days in-flight) demonstrated mild metabolic disorders coupled to changes in structure and ultrastructure of cells. The latter concerned for the most part the electron-dense zones that were apparently constructed of lipids (in grape hyacinth) (Ref. 9). Morphological changes related to spatial disorientation were at the same time revealed in potato seedlings. Short-term experiments with pea, corn and barley seeds showed slight changes in germination and energy of growth but a significant decrease in the mitotic activity of root cells and a change in the ratio of mitotic phases. Trinucleate cells in the meristem zone, the increase in nuclear sizes and binucleate cell count were observed. Larger number of nuclei was, probably, associated with impaired mitotic division. The anatomy pattern during space mission and in controls is as follows: shoots started growing, their internodes became longer, the test and control number of cells did not noticeably change. A fundamental result is a conclusion that shoot germination, growth and morphogenesis in-flight are normal, the initial growth phases being determined by the organism's embryonic development on Earth, i.e. early plant is tolerant of gravity. However, this is valid under the conditions with a constant acceleration of 10^{-5} to 10^{-6} g and occasional gravity varying from 10^{-2} to 10^{-4} g. The experiments with damped disturbances demonstrated a considerable lag in seedling growth compared to flight and ground-based controls.

The research on plant growth and development during spaceflight includes all most important steps from germination to fertilization and completion of the developmental pathway. The first experiment on cultivation from seed to adult plant was made in 1971 at Salyut orbital station in the Oasis-1 unit. Crew's observations, plant exterior indicated that the form and size of aerial organs in Khibin cabbage, flax and crepis grown in-flight were oriented to light source, the stalk was straight and the foliar color normal. However, in this and subsequent experiments the plant growth was inhibited.

The development of the root area in test peas was obviously impaired, which was manifested by diminution in the main and lateral root lengths. The dividing cells were not revealed in the meristem of the test pea roots, while they made 2.8% in controls.

The evaluation of cell ultrastructure showed the presence of the major organelles with typical structure: mitochondria contained the whole cristae and double membranes, and well-developed lamellar structure was observed in chloroplasts. But there were organelles with the altered structure in the test: disrupted outer membrane, destructed intra-organoid structures. A change in the content of pigments (chlorophyll 'a' and 'b', carotenoids), carbohydrates and mineral elements was noted. The results were confirmed for other biological subjects cultivated during space missions for different time intervals.

The potentials for morphogenesis were studied on radish plants cultivated in the greenhouse Svet and on potato culture in model experiments on agar-agar and organic nutrient medium. The experience demonstrated possible formation of root-crops in radish and storing tubers on the ground potato shoots developed in-flight. However, growth rates were markedly depressed in test radish that developed dwarfism like in permafrost. This evidence, along with other similar results, made us study the initial growth phases in more detail.

The examination of major lipids in the test wheat membranes revealed their considerable decrease to indicate possible destruction of some lipids. The data agree with 230-232 nm absorption in the differential absorption spectra of test lipid extracts. This suggests the presence of conjugated dienes, that is, the intermediary peroxidation products of unsaturated fatty acids. Peroxidation of lipids (POL) is known to be universal and reacts to any strong environmental influence within a short time. POL activation lowers the protective biomembrane function enhancing its permeability to micro- and macromolecules. Besides, test lipid extracts showed the presence of fluorescing products that accumulate in lipid peroxidation (Ref. 10).

4. CONCLUSIONS

Thus, a decrease in the total amount of pigments, lipids and polyunsaturated fatty acids, and also the occurrence of typical oxidation products suggest the activation of oxidative free radical processes in plants exposed to spaceflight. At the same time, the changes may result from pigment reparation in plant ontogenesis, which is the adaptation response to destruction provoked by peroxidation.

In summary, the plants that may develop through all stages inflight have substantial metabolic disorders. The latter, however, are not an obstacle to consider the plants as a component of the life support systems in long-term space missions to supply the crew with food, atmosphere and water.

5. REFERENCES

1. Mashinsky A.L., Mitichkin O.V., Grechko G.M. In: The Organism and Gravity. Institute of Botany, Lithuanian Academy of Science, Vilnyus, 1976, 228-238.

2. Kondo S. Jap. J. Genet.,1968, 43, 472-478.

3. Pollard E. J. Theoret. Biol.,

1965, 8, 113-123.

4. Parfenov G.P. Problems of Space Biology, 1988, 57, 272.

5. Khenokh M.A., Kuzicheva E.A., Mashinsky A.L., Konshin N.E., Pershina V.P. In: Molecular Fundamentals of Cellular Structure and Functional Activity. USSR Academy of Science, Leningrad, 1977,195-199

6. Sadykov A.S., Leontyev V.B., Nechitailo G.S., Abdurakhmanov Sh.D., Mashinsky A.L. Systems of Chemical Processes Depending on Gravity. DAN SSSR, 1989, 303, 4, 1004-1007.

7. Khenokh M.A., Kuzicheva E.A., Mashinsky A.L., Nechitailo G.S., Semenov Yu.P. In: Biological Studies at Salyut Orbital Stations. Nauka, Moscow, 1984, 21-25.

8. Merkis E.I., Laurinavichyus R.S. Complete Cycle of Individual Development of Arabidopsis thaliana (L) Heynh. at Salyut 7 Orbital Station. DAN SSSR, 1983, 271, 2, 509-512.

9. Kordyum E.L., Mashinsky A.L., Popova A.F., Uvarova S.A., Khristenko L.A. Kosmicheskie issledovaniya na Ukraine, 1978, 12, 42-49.

10. Rumyantseva V.B., Merzlyak M.N., Mashinsky A.L., Nechitailo G.S. Kosmicheskaya biologiya i aviatsionnaya meditsina, 1990, 1, 53-55.

ON NUMERICAL SIMULATION OF FREE FLUID ELECTROPHORESIS

S.V.Ermakov, O.S.Mazhorova, Yu.P.Popov

Keldysh Institute of Applied Mathematics USSR Academy of Sciences
Miusskaya sq.4, Moscow, 125047, USSR.

ABSTRACT

The results of 1-D and 2-D numerical simulation of free fluid zone electrophoresis corresponding to capillary and column techniques are presented. Three sample transport mechanisms convection, electrophoretic migration and diffusion are accounted for mathematical treatment. The action of the first one is illustrated for column electrophoresis, when the origin of bulk fluid motion is the electroosmotic slipping on the column walls. For capillary electrophoresis two mathematical approaches are compared. It is showed that their predictions are similar only for the case when buffer concentration is much more than the sample one.

Key Words: Zone Electrophoresis, Phenomenological Approach, Capillary, Sample, Buffer.

INTRODUCTION

Space experiments on electrophoretic fractionation of biological samples, earth preparation and post flight analysis have stimulated the mathematical modeling of electrophoresis. Currently two approaches are used to describe this process. The first (purely phenomenological) considers the electrophoretic migration only for species which constitute the sample (Ref.1,2). The buffer is treated as a passive medium, that transports the sample by means of convection. Other interactions between buffer and sample species are not regarded (Ref.1) or they are based on simplified formulas, e.g. the linear dependence of solution conductivity on sample concentration (Ref.2).

The second approach treats the electrolyte mixture in the electrophoretic column as a multicomponent medium. Time evolution of sample and buffer compounds is described using mass and charge conservation laws and chemical equilibrium equations (Ref.3,4). Although this approach yields more complex mathematical models, it is sufficiently general to simulate all electrophoretic techniques, while the approximation mentioned earlier may be used only for zone electrophoresis (ZE). In our report we have attempted to compare these two approaches.

MATHEMATICAL MODEL

We shall study electrophoretic separation of organic acids with ZE technique. Assume that an electrophoretic column is filled with an aqueous solution of **n** univalent acids and one univalent base. The last together with one acid forms the buffer medium. Dissociation-association reactions for acids, base and water have schemes:

$$HA_i \rightleftarrows A_i^- + H^+ \; (i=1,\dots n), \quad H^+B \rightleftarrows B + H^+,$$
$$H_2O \rightleftarrows H^+ + OH^- \qquad (1)$$

where HA_i, B - are neutral acid and base molecules, while A_i^-, H^+B, H^+, OH^- - are acid, base, hydrogen and hydroxyl ions, respectively. Since the rate of chemical reactions is much higher compared to the transport processes (convection, electrophoretic migration, diffusion), we can use the local chemical equilibrium hypothesis. It implies that a chemical equilibrium exists at every point in space and time. Hence it is convenient to use analytical concentrations a_i, b_i, $[H^+]$, specified as:

$$a_i = [HA_i] + [A_i^-] \quad , \quad b = [H^+B] + [B],$$

$$\alpha_i = \frac{[A_i^-]}{a_i} = \frac{K_i}{K_i + [H^+]}, \qquad (2)$$

$$\beta = \frac{[H^+B]}{b} = \frac{[H^+]}{K_0 + [H^+]},$$

where [] – molar concentration, K_i(i = 0,...n), – equilibrium constant for i–th reaction, α_i i=1,...n) and β – ionization degrees.

Generally, sample transport is composed of:

– convection, when it has traveled captured by the bulk fluid motion with a velocity $\overline{V}$;

– electrophoretic migration under the action of the electric field with intensity $\overline{E}$;

– diffusion .

Let's present the equations, describing evolution of concentration fields for every specimen and solution behavior. We assume that absolute concentration of each compound is small compared to that of water, so the difference in properties between solution and water is negligible. It allows to regard solution as incompressible fluid. Thus conservation relations of individual species, momentum, energy and net charge for solution as well as neutrality condition are written as:

$$\frac{\partial a_i}{\partial t} + \nabla\left[(\overline{V} - \alpha_i \cdot \gamma_i \cdot \overline{E}) \cdot a_i - D_i \cdot \nabla a_i\right] = 0, \qquad (3)$$

$$\{i = 1,\dots n\}$$

$$\frac{\partial b}{\partial t} + \nabla\left[(\overline{V} + \beta \cdot \gamma_0 \cdot \overline{E}) \cdot b - D_0 \cdot \nabla b\right] = 0, \qquad (4)$$

$$\frac{\partial \overline{V}}{\partial t} + (\overline{V} \cdot \nabla)\overline{V} = -\nabla\frac{p}{\rho} + \nu \cdot \Delta\overline{V} - \zeta \cdot \overline{g} \cdot T, \qquad (5)$$

$$\nabla\overline{V} = 0, \qquad (6)$$

$$\frac{\partial T}{\partial t} + \overline{V} \cdot \nabla T = -\chi \cdot \Delta T + \sigma \cdot (\nabla\Phi)^2, \qquad (7)$$

$$\overline{E} = -\nabla\Phi \ , \ \nabla\overline{j} = 0 \ , \qquad (8)$$

$$\frac{\overline{j}}{F} = \sum_{i=1}^{n}\left[D_i \cdot \nabla(\alpha_i \cdot a_i) + \alpha_i \cdot \gamma_i \cdot a \cdot \overline{E}\right] - D_0 \cdot \nabla(\beta \cdot b) +$$

$$+\beta \cdot \gamma_0 \cdot b \cdot \overline{E} - D_{n+1} \cdot \nabla[H^+] + \gamma_{n+1} \cdot [H^+] \cdot \overline{E} +$$

$$+D_{n+2} \cdot \nabla[OH^-] + \gamma_{n+2} \cdot [OH^-] \cdot \overline{E}, \qquad (9)$$

$$\sum_{i=1}^{n} \alpha_i \cdot a_i + [OH^-] = \beta \cdot b + [H^+] \qquad (10)$$

Here D_i (i = 0,...n+2), – diffusion coefficients, γ_i(i = 0,...n+2), –ion mobilities; Φ – electric field potential; $\overline{j}$ – electric current density; σ – conductivity; F – Faraday constant; p – pressure; ρ,ν –mean density and kinematic viscosity; ζ, χ – thermal expansion coefficient and thermal diffusivity; T – temperature; $\overline{g}$ – acceleration of mass force. Subscripts i = 1,2,..n corresponds to acids, i = 0 – to base, i = (n+1) and i = (n+2) denote hydrogen and hydroxyl ions. The equation set (3)–(10) includes both approaches previously discussed. It is too complicated to be solved without further simplification.

PHENOMENOLOGICAL APPROACH

First we pay attention to the phenomenological approach. Suppose that the buffer consists of one acid and one base solution, while the sample has two compounds (n = 3). The phenomenological approach implies that the buffer concentration is so large, in comparison with that of the sample, that sample addition doesn't change the solution acidity (pH). Hydrogen and buffer concentration is believed to be constant, hence equation (3) is solved only for the sample components. Diffusion current terms (eq. (9)) are usually omitted because of their small value as well as terms, which account for hydrogen and hydroxyl contribution. After these assumptions eq. (9) looks as follows:

$$\bar{j} = F\cdot\left[\beta\cdot\gamma_o\cdot b + \sum_{i=1}^{3}\alpha_i\cdot\gamma_i\cdot a_i\right]\cdot\bar{E} =$$

$$= F\cdot\left[(\alpha_1\cdot\gamma_1\cdot a_1 + \beta\cdot\gamma\cdot b) + \sum_{i=2}^{3}\alpha_i\cdot\gamma_i\cdot a_i\right]\cdot\bar{E} =$$

$$= (\sigma^{*} + \sum_{i=2}^{3}\sigma_i)\cdot\bar{E} \qquad (11)$$

where σ^{*}– represents the pure buffer conductivity. Subscript i = 1 corresponds to acid in buffer, i = 2,3 – sample acids. By transforming this equation the dependence of electrophoretic velocity on the sample constituents concentration is easily derived.

Two problems were considered using the above approach. The first concerned ZE in a thin isothermal capillary where the solution is in a mechanical equilibrium ($\bar{V}$ = 0) and the electric current density along the capillary is constant ($\bar{j}$ = const). These restrictions enable us to study 1–D problem. The set of equations (3)–(10) is reduced to relation (11) and two equations (3) for each sample constituent.

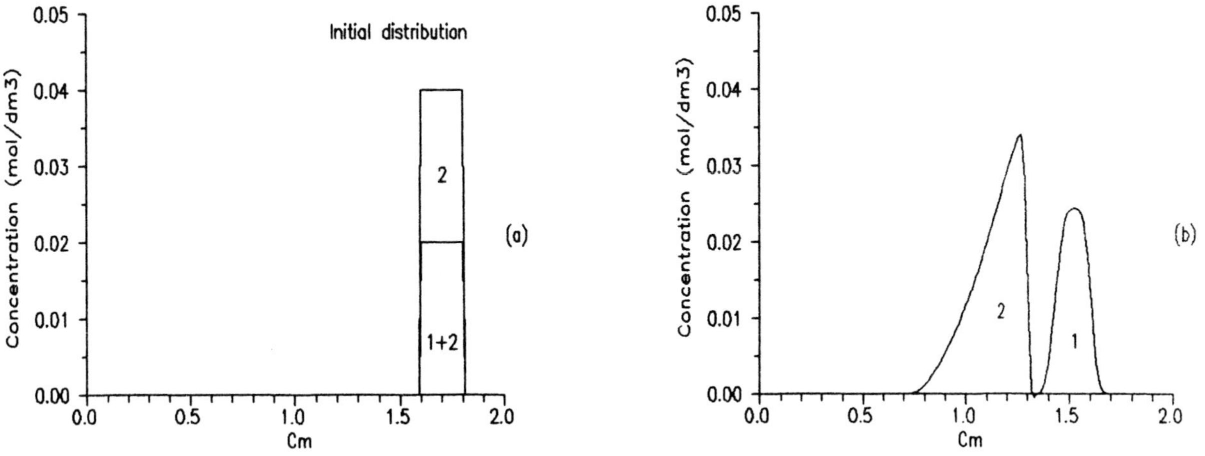

Figure 1: 1–D concentration {mol/dm^3} profiles along capillary at t=0.0 (a) and t=0.1 (b), obtained by using phenomenological model

The results of such a simulation are presented in figure 1 (a,b). The concentration profiles of the two sample specimens are plotted for initial moment {t = 0} and for final state {t = 0.1}, when they are separated. For cases in which the concentration of buffer constituents is not so big compared to that of the sample species, the buffer – sample interaction leads to sample spreading in addition to diffusion. When we increase the buffer concentration, such additional sample spreading is reduced.

Another problem considers the situation where the buffer ($\bar{V} \neq 0$) moves under the action of electroosmotic or thermal convection. This case may occur when high electric current densities are used or the electrophoretic column cooling system is not effective. Now together with equation (3), Navier–Stokes (5)–(6), heat transfer (7) and field potential (8) equations are solved. As an example, two variants, when electroosmotic slipping distorts the sample and when the buffer is stationary, are presented here. For this case concentration contour lines are shown at the initial {t = 0, figure 2 (a)} and final {t = 0.1, figure 2 (b , c)} states for each specimen. By implementing simulations with different slipping velocities (or Raleigh numbers for thermal convection), one may establish criteria by which the buffer motion doesn't affect considerably the separation process. It's very important for preparing space experiments.

MULTICOMPONENT MODEL

We will now elaborate on the alternative electrophoresis model (Ref.3,4) i.e. the 1–D capillary ZE problem as described earlier. For comparison we shall use similar conditions. Here four mass transfer equations (3)–(4) have to be solved for buffer and sample specimens. The hydrogen ion concentration (i.e. the pH) varies from one zone to the other and its distribution is calculated along the capillary by using eq.(10). No terms in equation (9) are omitted.

The simulation results are presented in figure 3 (a–b) where the concentration profiles along the capillary are plotted for initial and final states. The time moments correspond to those of figure 1, but, unlike the latter the buffer species profiles are also plotted. A detailed comparison of the results obtained in both approaches shows that they have remarkable discrepancies. The phenomeno-

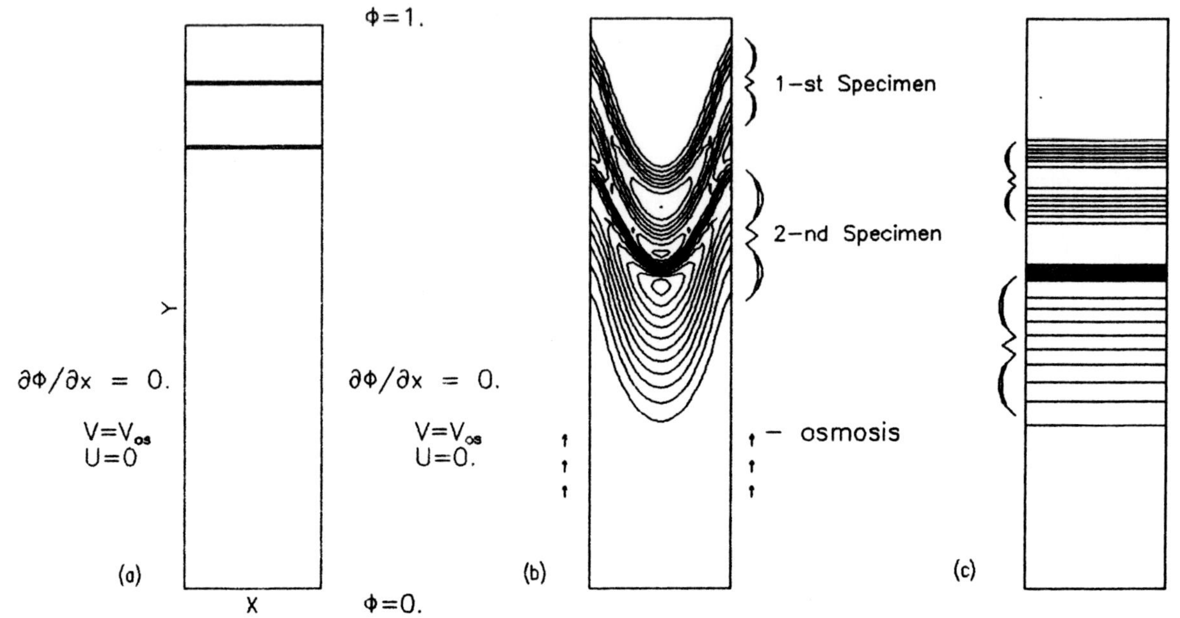

Figure 2: 2–D concentration contour lines: (a) initial zone location {t=0}, (b) when buffer moves under the action of electroosmosis {t=0.1}, (c) when buffer is stationary {t=0.125}.

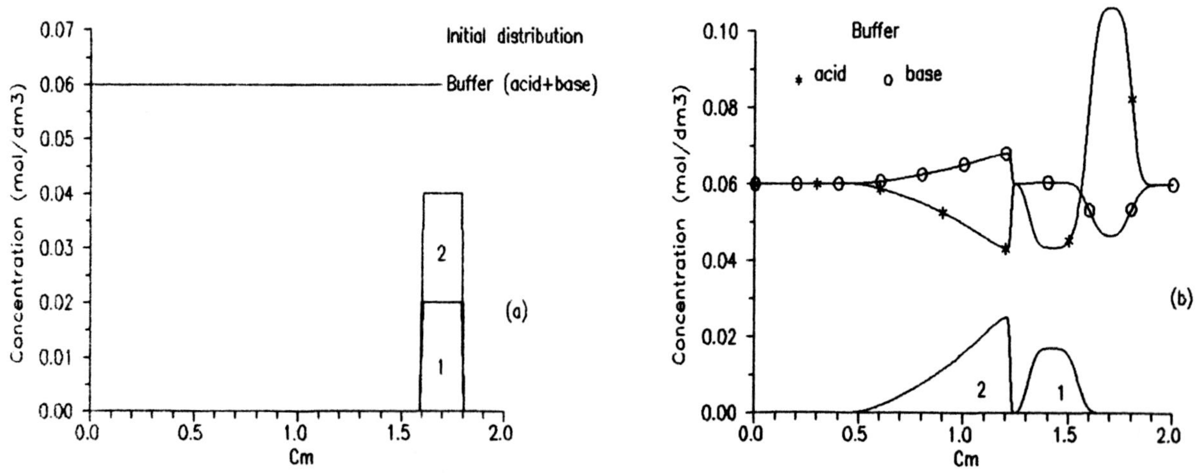

Figure 3: 1–D concentration {mol/dm^3} profiles along capillary at t=0.0 (a) and t=0.1 (b), obtained by using alternative multicomponent model

logical model predicts less sample distortion and shorter distance passed by every zone. This is explained by the fact that suggestion about constant buffer concentration and pH value isn't justified, that is clearly seen in figure 3 (b). Both approaches give approximately equal results only when the buffer concentration is much more than that of the sample. It is a trivial situation, when the sample is distorted mainly by diffusion, so the use of more simple models is reasonable.

Thus we may conclude that when the buffer and sample concentrations have comparable values the phenomenological approach gives incorrect results, because the buffer species concentration and pH value do not remain constant. They vary in space and time. Only in case in which the sample concentration is negligible compared to the buffer, we may take into account only diffusion transport while the conductivity may be considered constant.

REFERENCES

1. **O.S.Mazhorova, Yu.P.Popov, V.I.Pohilko, A.I.Feonychev.** *Fluid Dynamics.* (1988), #3, 333.
2. **A.A.Aksenov, A.V.Gudzovskii, T.V.Kondranin, A.A.Serebrov.** in *The modern problems in hydrodynamics, aerophysics and applied mechanics.* Moscow MPhTI (1986), 122. (in Russian).
3. **V.G.Babskii, M.Yu.Zhukov, V.I.Yudovich** in *Mathematical Theory of Electrophoresis,* Consultants Bureau ,New York, (1989), p.241.
4. **D.A.Saville, O.A.Palusinsky.** *AIChE Journal.* **32**, (1986), 207.

INSTABILITY AND SELF-OSCILLATIONS IN PROCESSES OF COMBINED POLYMERIZATION AND CRYSTALLIZATION.

Yu.A. Buyevich and I.A. Natalukha.

Department of Mathematical Physics, Ural State University, Sverdlovsk 620083, U.S.S.R.

ABSTRACT

Results are presented revealing the physical mechanism of the instability of steady states of continuous crystallization in polydisperse assemblages of polymeric particles forming by means of anionic polymerization. Self-oscillating regimes occurring due to the instability of steady ones are investigated.

Keywords: Crystallization, Anionic Polymerization, Distribution Function, Instability, Neutral Stability Surface, Self-oscillations

1. INTRODUCTION

This paper deals with physical and mathematical modelling of a combined process of anionic polymerization and crystallization at the particulate level. An adequate description, an improvement of technological operation of polymerization reactors and an appropriate choice of optimal regimes are inconceivable without creating a comprehensive theory of heat and mass transfer between a continuous phase and a polydisperse system of particles taking into account the kinetics of polymerization and crystallization.

Crystallization from a supercooled polymer melt is frequently simultaneous with the process of polymerization from a monomer, some temperature interval existing within which heat effects of crystallization and polymerization are comparable. Under such conditions a tendency at developing various instabilities and waves has been repeatedly indicated [1-4].

2. THE MODEL AND ANALYSIS

The following assumptions are made to develop a mathematical model. In a polymer melt the polymerization of a monomer is proceeding at the same time as crystallization of a polymer. The physical properties of the melt and crystals are assumed to be homogeneous throughout the whole volume considered, which is achieved by means of intensive mixing of a suspension. The concentration of suspension, hydrodynamic conditions and temperature levels are supposed to prevent particle agglomeration and breakage. The heat balance and kinetic equations governing the polymerization degree (i.e. the ratio of a polymerized portion of a monomer to the whole of its mass in the volume) and the crystal size distribution can be written in the following way:

$$\rho c\frac{dT}{dt} = \alpha(T_m-T) + \rho_1 L_1 \Phi(w,T) + \rho_2 L_2\int_{r_*}^{\infty}\frac{d}{dt}\left(\frac{4}{3}\pi r^3\right)f(t,r)dr, \qquad (1)$$

$$\frac{dw}{dt} + \gamma_* w = \Phi(w,T), \qquad (2)$$

$$\frac{\partial f}{\partial t} + \frac{\partial}{\partial r}\left(\frac{dr}{dt}f\right) + \gamma(r)f = 0,$$
$$\left.\frac{dr}{dt}f\right|_{r=r_*} = J, \quad \frac{dr}{dt} = B(T_0-T)F(r). \qquad (3)$$

Here ρ, ρ_1 and ρ_2 are the densities of the suspension, polymer and crystals, respectively, w is the polymerization degree, Φ - the polymerization rate, γ_* - the withdrawal rate of the polymer melt, $\gamma(r)$- the crystal withdrawal rate, $J(w,T_0-T)$ - the nucleation rate, c - the specific heat capacity of the suspension, L_1, L_2 are the latent heats of polymerization and crystallization, respectively, T_m - the temperature of external media, T_0- thermodynamically equilibrium temperature, α- the effective heat transfer coefficient per unit volume and $f(t,r)$ is the crystal size distribution function.

By using eqns. (1)-(3) and applying the technique suggested in [5], we obtain the following system of equations governing the dimensionless temperature and polymerization degree:

$$\frac{B(u)}{B_s}\frac{d(x+Kv)}{d\tau} + (St_m-St)x - St + K(v+1) + \frac{St-K}{\Omega}B(u)\int_0^{\infty} r^2(s)F(s)\times$$
$$\times\frac{J(\tau-s)}{J_s B(\tau-s)}\exp\left(-B_s\int_0^s\frac{G(s-z)dz}{B(\tau-z)}\right)ds=0, \qquad (4)$$

$$\frac{B(u)}{B_s}\frac{dv}{d\tau} + v + 1 = \frac{\Phi(u,w)}{\Phi_s}, \qquad (5)$$

where the following dimensionless variables and parameters are introduced:

$$u=(T_0-T)/T_0,\quad x=(u-u_s)/u_s,$$
$$v=(w-w_s)/w_s,\quad G=\gamma(r)/\gamma_*, \qquad (6)$$
$$\tau=\gamma_*\int_0^t\frac{B(u)}{B_s}dt,\quad s=\gamma_*\int_0^r\frac{dr}{B_s F},$$

$$K=\rho_1 L_1 w_s/\rho c T_0 u_s,\quad u_m=(T_0-T_m)/T_0,$$

$$St=\alpha(u_m-u_s)/\rho c\gamma_* u_s,$$

$$St_m=\alpha u_m/\rho c\gamma_* u_s,$$

$$\Omega=\int_0^\infty r^2(s)F(s)\exp\left(-\int_0^s G(z)dz\right)ds,$$

where the subscript s indicates steady state conditions. It is worth noting that K=St and K=0, $J'_w=0$ correspond to pure polymerization and crystallization, respectively.

Assuming $u=u_s$ and $w=w_s$, one can easily obtain a stationary solution of eqns. (1)-(3) defining the steady states of the process with constant nucleation and polymerization rates and crystal size distribution:

$$\alpha(T_m-T_s)=\rho_1 L_1\Phi(u_s,w_s)+$$
$$+4\pi\rho_2 L_2 J_s\Omega,\ \gamma_* w_s=\Phi(u_s,w_s),$$
$$f_s(s)=\frac{J_s}{B_s}\exp\left(-\int_0^s G(z)dz\right). \qquad (7)$$

Let us analyse the stability of solution (7) with respect to random small disturbances of temperature. Linearization of eqns. (4) and (5) with respect to x and v and their representation in the form

$$\begin{pmatrix}x\\ v\end{pmatrix}=\begin{pmatrix}x_0\\ v_0\end{pmatrix}\exp\lambda\tau\ ,\ \lambda=\vartheta+i\omega \qquad (8)$$

gives the following characteristic equation:

$$\left[\lambda+St_m-St+B_u(St-K)+(St-K)I_2B_u+\right.$$
$$\left.+(St-K)(J_u-B_u)I_1\right](\lambda+1-\Phi_{01})+$$
$$+\Phi_{10}\left[K+I_1(St-K)\right]=0, \qquad (9)$$

where

$$I_1=\frac{1}{\Omega}\int_0^\infty r^2(s)F(s)e^{-\lambda s}\times$$
$$\times\exp\left(-\int_0^s G(z)dz\right)ds,$$

$$I_2=\frac{1}{\Omega}\int_0^\infty\left(\int_0^s e^{-\lambda z}G(s-z)dz\right)r^2(s)F(s)\times$$
$$\times\exp\left(-\int_0^s G(z)dz\right)ds,$$

$$B_u=\left.\frac{u_s d\ln B}{du}\right|_{u=u_s},\quad J_u=\left.\frac{u_s\partial\ln J}{\partial u}\right|_{u=u_s},$$

$$\Phi_{ij}=\left.\frac{u_s^i w_s^j}{i!j!}\frac{\partial^{i+j}\Phi}{\partial u^i dw^j}\right|_{\substack{u=u_s\\ w=w_s}}.$$

Note that eqn. (9) has been derived by using the modified time in expression (8). However, considering that a transition to the real time causes an appearance of additional terms in the linearized eqns. (4) and (5), having order higher than one with respect to x and v, this does not influence the linear stability.

If $\vartheta=0$, eqn. (9) gives the neutral stability surface in the space of parameters J_u, B_u, K, St, St_m and Φ_{ij} for an arbitrary kinetics of polymerization, nucleation, crystal growth and removal. Equation (9) has the root $\omega=0$ (provided that $\vartheta=0$), to which the following instability condition corresponds

$$J_u < \frac{St-St_m}{St-K} - \frac{B_u\Omega_1}{\Omega} - \frac{\Phi_{10}St}{(1-\Phi_{01})(St-K)}, \tag{10}$$

where

$$\Omega_1 = \int_0^\infty\left(\int_0^s G(z)dz\right)r^2(s)F(s)\times$$

$$\times \exp\left(-\int_0^s G(z)dz\right)ds.$$

Condition (10) corresponds to the so-called aperiodic bifurcation, when the instability of a steady process arises initially with respect to perturbations of a zero frequency. Since $St_m > St > K > 0$ and $\Phi_{10} < 0$, $\Phi_{01} < 0$, condition (10) can be accomplished for both the positive and negative Gibbs numbers. The latter case can be observed under high supercooling - the so-called Tamman effect [6] - when a viscisity of a melt increases greatly. This is undesirable on technological considerations. Thus further analysis of this case has not been carried out, in view of the fact that polymerization could scarcely proceed at such a supercooling. Note that the Tamman effect does not take place in crystallization from supersaturated solutions [7].

By means of separating real and imaginary parts of eqn. (9), one can obtain also non-zero roots ω (i.e. the oscillation frequency with respect to which the stability breaks) and the corresponding neutral stability surface $J_u=S$. The traces of the neutral stability surface in the plane of J_u and St_m are shown in Figure 1. The instability region is defined by the inequality $J_u > S$. Figure 1 shows that, in

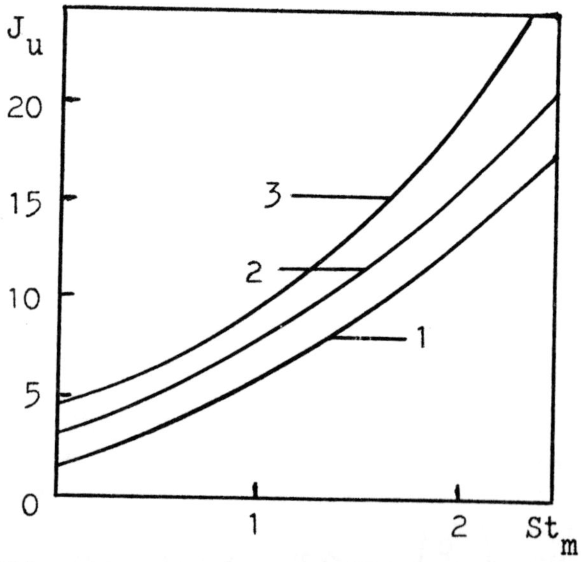

Figure 1: The traces of the neutral stability surface in the plane J_u, St_m; $\Phi_{01}=-1$, $\Phi_{10}=-2$, $St=St_m$, 1: K=0, 2: K=1, 3: K=10

spite of high supercooling, provided that $J_u < S$ (since Gibbs number under this condition is small), stationary regimes in this region are stable with respect to non-zero frequency perturbations. This means that small fluctuations of the temperature and the polymerization degree diminish exponentially with time, and the stability breaks only at the attainment of a sufficiently large slope of the nucleation rate as a function of supercooling (parameter J_u is proportional to the derivative $J'(u_s)$). When the image point in the parametric space passes over the neutral stability boundary in the vertical direction at constant Stanton number, Gibbs number increases, i.e. the mean supercooling drops, which is caused by increasing of $J'(u_s)$ and is not a result of variations of the nucleation rate $J(u_s)$. Thus the physical reason of the instability is the strongly non-linear dependence of the nucleation rate on the supercooling. The instability is caused by a competition between external heat sources, polymerization and nucleation rates, crystal growth and withdrawal of crystals. An increase of the polymerization reaction heat (which is characterized by K) results in stabilization of the system. An increase of Stanton number St_m stabilizes the process.

Nonlinear analysis of the system (4),(5) has shown that perturbations of the steady process can assume both the form of slightly nonlinear auto-oscillations and pulsating ones of a rather wide frequency spectrum. Figure 2 shows the dependence of the

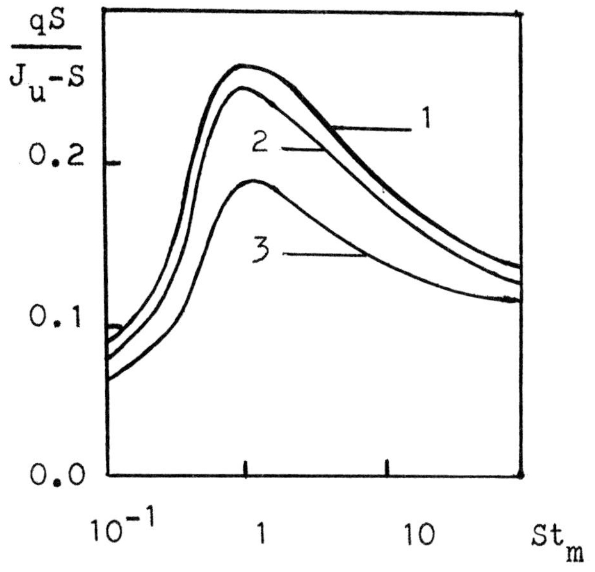

Figure 2: Squared auto-oscillation amplitude for the dimensionless temperature x as a function of supercriticality J_u-S and St_m; St=0.2; 1: K=0,012: K=0.01, 3: K=0.1

squared amplitude q of slightly nonlinear auto-oscillations of the dimensionless temperature x on the supercriticality and Stanton number.

3. CONCLUSION

The theory developed provides one with a convenient tool to determine conditions under which stationary regimes of the process of combined polymerization and crystallization lose their stability with respect to occasional small disturbances. The neutral stability surface and the corresponding oscillation period can be expressed in terms of relevant physical, chemical and processing parameters for various kinetics of polymerization, nucleation, crystal growth as well as for different in- and output conditions. Essential properties of slightly nonlinear auto-oscillating regimes occurring on the threshold of instability are calculated.

4. REFERENCES

1. A.G.Merzhanov, Fiz. Gor. Vzryva, 9, 4 (1973) (in Russian).
2. V.A.Volpert and S.P.Davtyan, Dokl. AN SSSR, 268, 62 (1983) (in Russian).
3. V.A.Volpert, I.N.Megrabova, S.P.Davtyan and V.P.Begishev, Fiz. Gor. Vzryva, 21, 69 (1985) (in Russian).
4. L.Yu.Artyukh, A.T.Luk'yanov and S.E.Nysanbayeva, Fiz. Gor. Vzryva, 21, 57 (1985) (in Russian).
5. Yu.A.Buyevich, V.V.Mansurov and I.A.Natalukha, Inzh.-Fiz. Zhurn., 55, 275 (1988) (in Russian).
6. Ja.I.Frenkel, The Kinetic Theory of Fluids (Nauka, Moscow 1975) (inRussian).
7. A.D.Randolph and M.A.Larson, Theory of Particulate Processes (Academic Press, New York 1988).

Crystallization under Microgravity Conditions

CELLULAR PATTERNS AND DENDRITIC TRANSITION IN DIRECTIONAL SOLIDIFICATION

Bernard Billia
Laboratoire de Physique Cristalline (U.R.A n°797), Université d'Aix-Marseille III
Faculté des Sciences de St Jérôme, Case 151, 13397 Marseille Cedex 13, France

ABSTRACT

Cellular patterns and the cell-dendrite transition in directional solidification of a binary alloy are investigated. It is shown that the analogy between cells and viscous fingers is not validated by experimental data in the succinonitrile-acetone system. A phenomenological improvement of the tip undercooling relation is yet feasible in the physical limit. Based on the analysis of the local dendritic transition that is observed in standard experiments, a method is proposed to estimate the onset of dendritic growth, which might be useful in the conception and preparation of microgravity experiments.

Keywords: *Directional Solidification, Alloys, Cellular Patterns, Cell-Dendrite Transition.*

1. INTRODUCTION

As the physical properties of engineering components are strongly influenced by the microstructure which is formed in the solid phase during processing, the prediction of the correlation between the elaboration conditions and the microstructure is crucial. Besides, the current understanding of the stability and selection of nonlinear structures in pattern-forming instabilities has very much advanced in recent years and cells and dendrites are since receiving further attention as archetypes.

The major goal of this paper is to develop, from the analysis of cellular solidification and dendritic transition in real experiments, a coherent approach to improve the conception and preparation of space studies. First, directional solidification of a binary alloy is described, the control parameters and microstructure characteristics introduced and the specificities of cellular growth and cell - dendrite transition in the succinonitrile-acetone sytem given. In the main section, a multiple-scale analysis of cellular solidification, with negligible surface tension effects, is carried out, in two mathematical limits and in the physical limit. Then, the Local Dendritic Transition, that is common in standard experiments in which the growth rate is switched on after thermal stabilization, is examined. Finally, a method for the choice of the growth parameters in microgravity experiments is proposed.

2. CELLULAR SOLIDIFICATION

2.1. Control parameters and characteristics of cellular shape

In directional solidification of a binary alloy (Figure 1), the liquid phase of initial solute concentration C_0 is solidified at velocity V in a temperature gradient G.

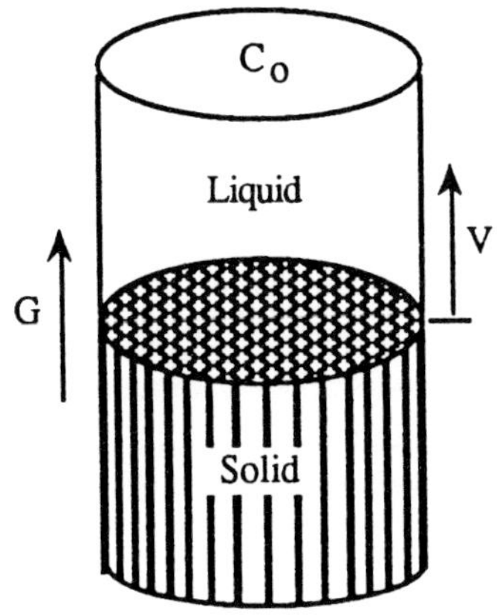

Figure 1 : Schematic representation of cellular upward solidification of a binary alloy.

From these dimensional control parameters and the physico-chemical coefficients of the alloy system, characteristic lengths are constructed : a solutal length

$l_s = D/V$, with D the solute diffusion coefficient, a thermal length $l_t = \Delta T_0/G$, where ΔT_0 is the freezing range, and a capillary length $d_0 = \Upsilon/\Delta T_0$ in which Υ is proportional to the solid-liquid surface tension.

In dimensionless form, directional solidification only depends on two dimensionless control parameters, for instance the level of morphological instability $\nu = l_t/l_s$ and Sekerka's capillary number $A = Kd_0/l_s$, in which K is the segregation coefficient of solute.

In general, when a series of experiments are carried out, C_0 and G are kept constant and the growth velocity V is varied. Starting from a small value, the solidification front first is planar. Then , cells are observed (finite amplitude cells and deep cells) that are followed by dendrites with sidebranches[1].

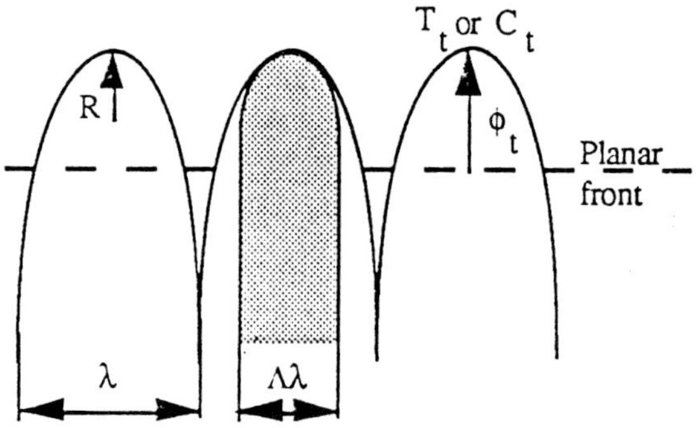

Figure 2 : Cellular shape characteristics.

Figure 2 shows the relevant cellular shape characteristics : primary spacing λ, tip radius R and tip shift ϕ_t from the planar front position. Under equilibrium solidification, ϕ_t can be assimilated to the temperature T_t, or solute concentration, C_t, at the tip as, in the cellular range, the lowering of the melting point due to the local curvature is usually very small. Dimensionless Peclet numbers are commonly introduced : a spacing Peclet number $P_\lambda = \lambda/l_s$, a radius Peclet number $P_R = R/l_s$ and a shift Peclet number $P_t = \phi_t/l_s$.

From the mathematically self-consistent analysis developed in the recent years[2,3], it follows that the cellular relative width Λ is a key parameter, although it cannot be directly measured. Therefore, Λ has to be evaluated from its geometrical definition

$$\Lambda = 2 / [1 + (1 + 4\lambda / \pi R)^{1/2}] \qquad (1)$$

which shows that it is equivalent to the ratio of the primary spacing to the tip radius. Furthermore, rather than T_t, it is convenient to consider the tip undercooling ΔT, that is equal to the difference between the liquidus temperature for the initial solute concentration, T_L, and T_t (Figure 3). In dimensionless form, $\Delta T^* = \Delta T/\Delta T_0$ is obtained by dividing ΔT by the freezing range ΔT_0 and C_t replaced by $C_t^* = KC_t/C_0$.

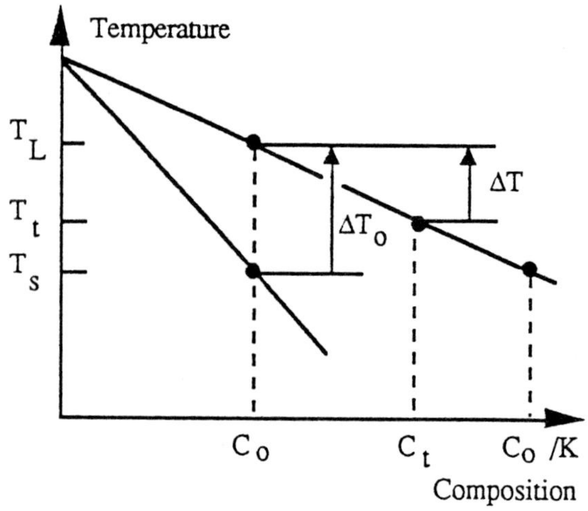

Figure 3 : Solid-liquid phase diagram of a binary alloy, with the tip undercooling ΔT and the freezing range ΔT_0.

2.2. Specificities of cellular growth and cell-dendrite transition

For the present analysis of cellular growth and cell-dendrite transition, the succinonitrile-acetone system (K=0.1) will be adopted as the reference. Indeed, the physico-chemical properties of this alloy system has been completely determined[4] and extensive data are available for samples grown in a Hele-Shaw cell[5-7], i.e. between two parallel glass plates with a space of about 100 µm, so that convection in the liquid phase is negligible.

The variation of the major shape parameters with ν is given in Figure 4. The critical results for the following are that : the spacing Peclet number and constitutional supercooling at the tip, $\nu C_t^* -1$, are small quantities, of order ε , for cells and become of order unity only at the dendritic transition.

3. MULTIPLE - SCALE ANALYSIS OF CELLULAR SOLIDIFICATION

For cellular growth, a multiple-scale analysis in the small spacing Peclet number limit was first carried out by Pelcé and Pumir[2], neglecting capillary effects. The liquid phase has to be divided into three zones (Figure 5). In the outer zone, the growth axis x is scaled by the solutal length and the y-axis by λ. At that scale, the cellular front looks planar and the problem is one dimensional. In the tip region, or inner zone, both x and y are scaled by the primary spacing, $X=x/l_s$ and $Y=y/l_s$. For deep cells, there exists a tail region in which the liquid is homogeneous in the y-direction and the interface

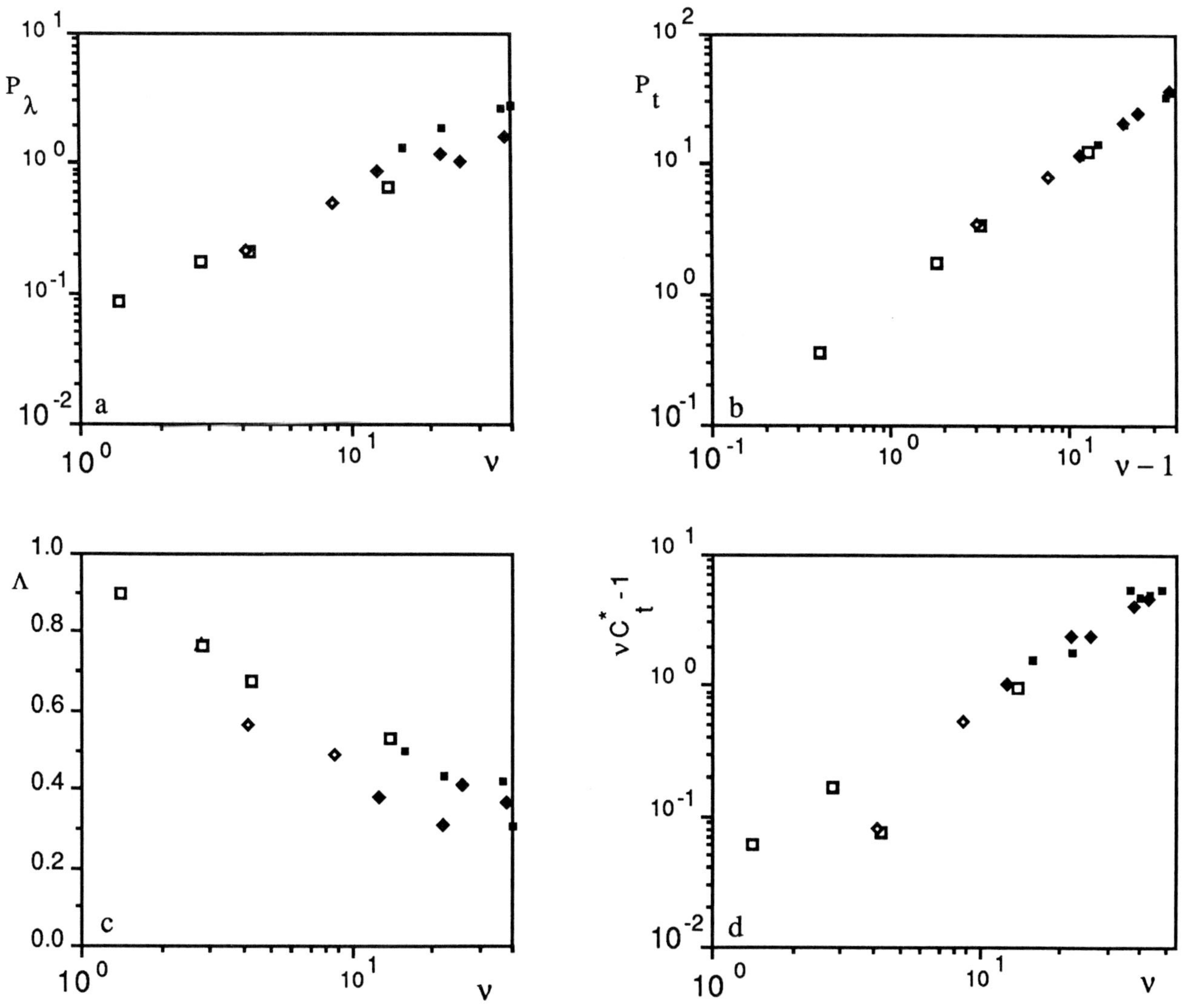

Figure 4 : Variation with the level ν of morphological instability of - a) the spacing Peclet number, - b) the shift Peclet number, - c) the relative width and - d) the constitutional supercooling at the tip, in the cellular range and about the dendritic transition. Succinonitrile-acetone alloys, data from Ref. 7. The open symbols correspond to cells and the full symbols to dendrites.

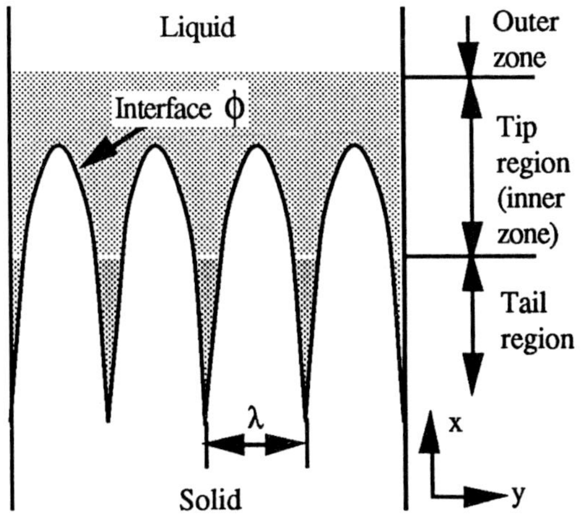

Figure 5 : The various regions in deep-cellular growth.

shape given by the Scheil equation[8]. Then, solutions should be found for the outer liquid and the tip region that are asymptotically matched.

The interesting part is the inner zone. Seeking for some analogy with viscous fingering[9], it is worth to substract the effect of the temperature gradient and replace C by the solute concentration

$$\omega = C - G\,(x - x_t)\,/\,m \qquad (2)$$

where m is the liquidus slope. Then, the solidification front ϕ becomes an isoconcentration line, $\omega_\phi = C_t$. For small $P_\lambda (= \varepsilon)$, the diffusion equation of solute reduces to the Laplace equation

$$\Delta C + P_\lambda \partial C/\partial X = 0 \Rightarrow \partial^2 C/\partial X^2 + \partial^2 C/\partial Y^2 = 0 \quad (3)$$

so that $\Delta\omega = 0$.

It will be seen that the difficulties will result from the square bracket in the normal balance of solute concentration ω

$$\frac{\partial \omega}{\partial N}\Big|_{\phi} = P_\lambda C_t (K - 1) \left[1 - \frac{1}{\nu C_t^*} - \frac{P_\lambda (1 - K)(\Phi - \Phi_t)}{\nu C_t^*}\right] \cos\theta \quad (4)$$

where N is the normal to the front directed towards the liquid and θ the angle between N and the growth axis.

Different mathematical limits are a priori conceivable but which ones are physically meaningful ? The last term in the bracket always is small, of order ε, as it is linked to the depth of the inner region. In the Pelcé-Pumir[2] and modified Pelcé-Pumir limit[3],the constitutional supercooling at the tip, νC_t^*-1, is assumed of order unity but experiments definitely show that it is actually small in the physical limit (see Figure 4d).

3.1. Pelcé - Pumir theory

In the Pelcé-Pumir (PP) limit, there is a direct analogy between the equations of cellular solidification and the equations of viscous fingering, that have been solved by Saffman and Taylor[9]. For discussion, it is convenient to consider the relation obtained for the tip undercooling

$$\nu \Delta T^* = 1 + \Lambda (\nu C_t^* - 1) \quad (5)$$

In Figure 6a, the experimental points (open squares for cells, full squares for dendrites) are obtained from the direct measurement of the tip temperature and the theoretical points (x and + symbols) are given by the RHS of the PP-relation. For the tip undercooling, the agreement can be considered as satisfactory for cells. Nevertheless, it should be noticed that the crosses are above unity, which has a nonphysical consequence. Indeed, the PP-relation can be combined with the relation of equilibrium solidification to give a new expression, indexed ΔT^*, for the relative width

$$\Lambda_{\Delta T^*} = 1 - \frac{K\nu}{1 - K} \frac{1 - C_t^*}{\nu C_t^* - 1} \quad (6)$$

of which a first expression was given by the geometrical definition (Eq.1). In the variation of the ratio $\Lambda_{\Delta T^*}$ over Λ versus ν (Figure 6b), a negative peak is observed in the cellular range that is nonphysical as the relative width must be a positive quantity, in between zero and unity. Therefore, it should be concluded that the relative width Λ cannot be predicted from the PP-relation, although it is mathematically self-consistent.

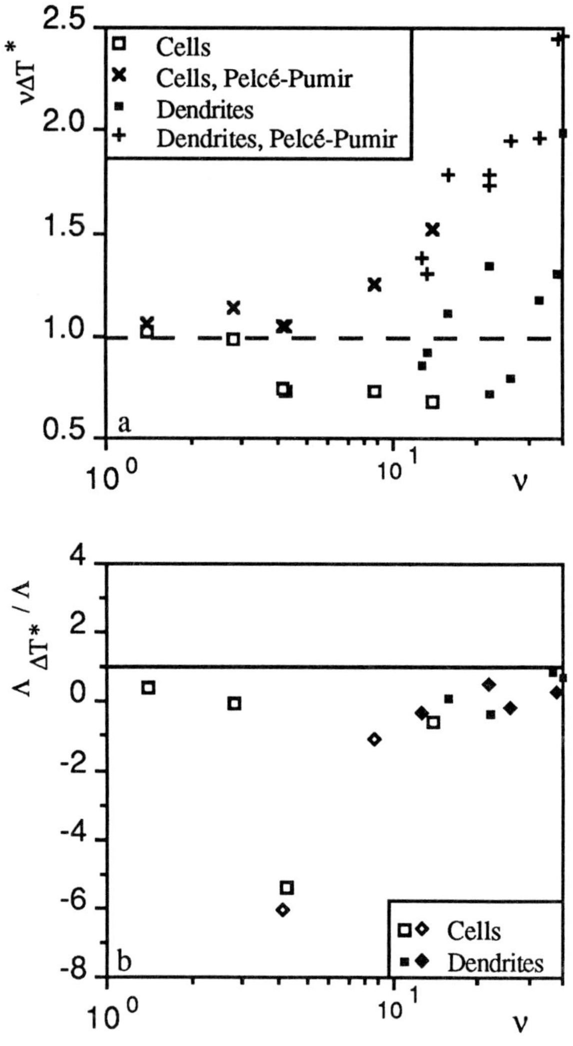

Figure 6 : - a) Experimental test of the Pelcé-Pumir relation for the tip undercooling, - b) Variation in the Pelcé-Pumir limit of $\Lambda_{\Delta T^*} / \Lambda$ with ν.

To clarify the origin of the discrepancy, it is worth to focuss on the two points that deviate the most to negative values in Figure 6b. From Figure 4a, it is obvious that, for these points, as assumed by Pelcé and Pumir, P_λ is effectively small but Figure 4d shows that the constitutional supercooling at the tip also is of order ε and not of order unity. Consequently, the analogy between cells and Saffman-Taylor fingers is not validated.

3.2. Modified Pelcé - Pumir theory

In the modified Pelcé-Pumir (mPP) limit, the constitutional supercooling at the tip is replaced by the distance from the threshold of morphological instability, ν-1, which ensures the analogy with viscous fingering for $\nu \geq 2$.

In the tip undercooling relation

$$\nu \Delta T^* = 1 + \Lambda (\nu C_t^* - 1) + (\Lambda - 1)(1 - C_t^*) \quad (7)$$

there is now an additional negative term which, for cells, leads to an excellent agreement with experiment (Figure 7a). Concomitantly, the ratio $\Lambda_{\Delta T*}/\Lambda$ is always very close to unity (Figure 7b).

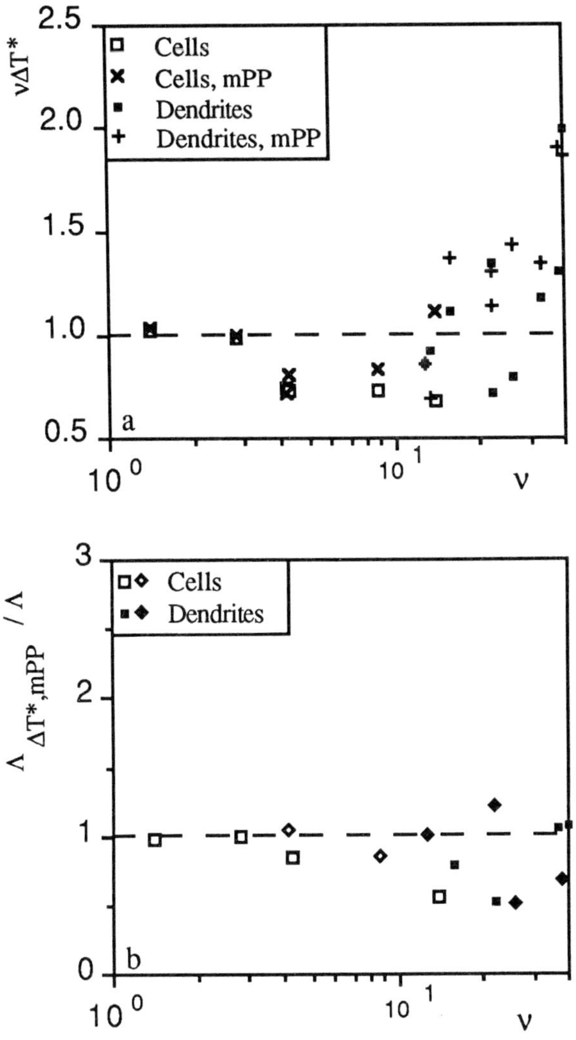

Figure 7 : - a) Experimental test of the mPP-relation for the tip undercooling, - b) Variation in the mPP-limit of $\Lambda_{\Delta T*}$ / Λ with ν.

In some sense, these results show that, sometimes, it is possible to derive a good relation out of the real physical limit as experiments say that constitutional supercooling at the cell tip is small. Thus, one is founded to seek for what is implied when reducing $\nu C_t^* - 1$ to $\nu - 1$. By using equilibrium solidification, it follows that

$$\nu C_t^* = \nu - P_t (1 - K) \ . \quad (8)$$

For the last term to be negligible, P_t should be small, of the order of the spacing Péclet number, which is not verified in experiments (see Figure 4a,b). For instance, P_t equals 20 times P_λ for the two points selected in Figure 6b.

3.3. Physical limit

In the physical limit, constitutional supercooling at the tip is of order ε so that, to the first order in P_λ, the normal balance of the solute concentration is zero. Then, the problem is a Neumann problem[3], whose solution is simply

$$\nu \Delta T^* = 1 \quad (9)$$

This solution is long known as the Bower-Brody-Flemings (BBF) solution. This expression, that was derived in 1966 under ad hoc assumptions, actually is the leading order term of the exact solution, which explains why the departure from the BBF relation is always small for cells.

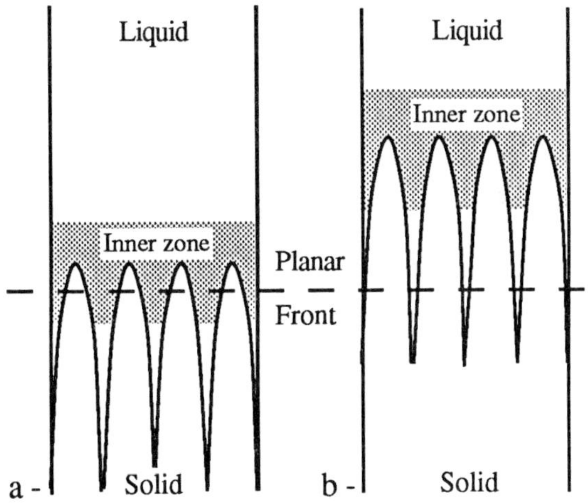

Figure 8 : Schematic representation of cellular solidification showing the difference between - a) mPP cells and - b) real cells.

At this point, it should be stressed that a cell is not only characterized by its shape but also by its position in the temperature field. As shown in Figure 8, the difference between the mPP-limit and the physical limit results solely from the vertical position of the tips. In the former case, the tips remain close to the position of the

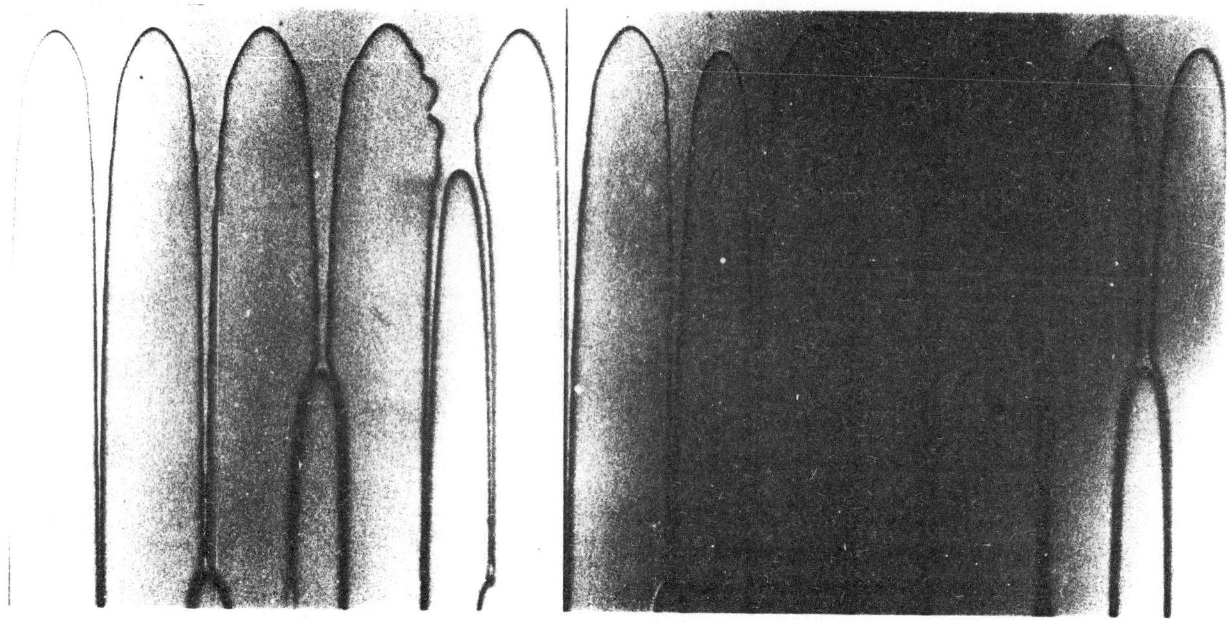

Figure 9 : Local Dendritic Transition as observed in the building of a one-dimensional array. Succinonitrile-1.3 wt % acetone - V = 2.5 μm/s - G = 72.3 °C/cm. (Courtesy of W. Kurz). x60.

planar front (Figure 8a) whereas, in the physical limit, the tips are protuding far in the liquid in order to reduce constitutional supercooling (Figure 8b).

3.4. Phenomenological approach

In the physical limit, it is also possible to use a phenomenological approach[11], whose bases are : - i) the tip undercooling relation should tend to the PP-relation (Eq.5) as the solute miscibility gap, $\Delta C = (C - C_s)_\phi$, approaches a constant value so that the departure from the Pelcé-Pumir relation results from the variation of ΔC along the front (non parallel solidus and liquidus lines), - ii) the difference with the Pelcé-Pumir relation is small for cells so that additivity can be assumed and the tip undercooling, ΔT^*, written as the PP-contribution, ΔT^*_{PP}, plus a ΔC-contribution, $\Delta T^*_{\Delta C}$, which naturally must go to zero as the solidus and liquidus lines tend to be parallel.

Then, a relation is sought for that expresses the tip undercooling as a function of the tip concentration and cell shape, $\Delta T^* = f(C_t^*, \text{shape})$. As for cells, constitutional supercooling at the tip is small and Λ close to unity, one rather seeks for $\Delta T^* = f(\nu C_t^* - 1, \Lambda - 1)$ that can be twice linearized

$$\Delta T^* = A_o|_{\Lambda=1} + \frac{\partial A_o}{\partial \Lambda}\Big|_{\Lambda=1} (\Lambda - 1)$$

$$+ (\nu C_t^* - 1)\left[A_1|_{\Lambda=1} + \frac{\partial A_1}{\partial \Lambda}\Big|_{\Lambda=1} (\Lambda - 1)\right] \qquad (10)$$

where A_o and A_1 have to be determined. As the PP-relation can be put in a similar form, the ΔC-contribution also has the same form. Besides, $\Delta T^*_{\Delta C}$, that should vanish for a constant miscibility gap, must be proportional to $1 - C_t^*$ which characterizes the variation of ΔC from its value at the tip. Finally, one gets

$$\nu \Delta T^* = 1 + \Lambda (\nu C_t^* - 1) + (\Lambda - 1)(1 - C_t^*)$$

$$- \beta \nu (\Lambda - 1)(1 - C_t^*) \qquad (11)$$

The first two terms in the RHS correspond to the PP-relation, with a correction to the leading order term in the wrong direction (see Figure 6a). Including the third term, the mPP-relation is obtained and the global change from unity is now in the right direction (see Figure 7a). In addition, there is a new negative term that, although it is small (a fit with the experimental data gives β=0.014, i.e. about the square of the solute segregation coefficient), has an important physical meaning. Indeed, it is this term that expresses the difference between supercritical and subcritical bifurcations. At the threshold (ν=1), it is zero for a supercritical bifurcation, $\Lambda = C_t^* = 1$, but not for a subcritical bifurcation for which Λ and C_t^* are less than unity as the pattern has a finite amplitude.

4. LOCAL DENDRITIC TRANSITION

The building of a 1D-cellular array after the growth velocity has been switched on is shown in Figure 9. In the beginning, there is a coarsening of the primary spacing due to a cell elimination process. After a while, sidebrances locally begin to develop towards the eliminated cells so that the dendritic transition is highly local and nonsymmetric. It is this kind of Local Dendritic Transition (LDT), that is common in standard experiments, in which the growth velocity is suddenly switched on after thermal stabilization, that we have analyzed.

4.1. Criterion for local dendritic transition

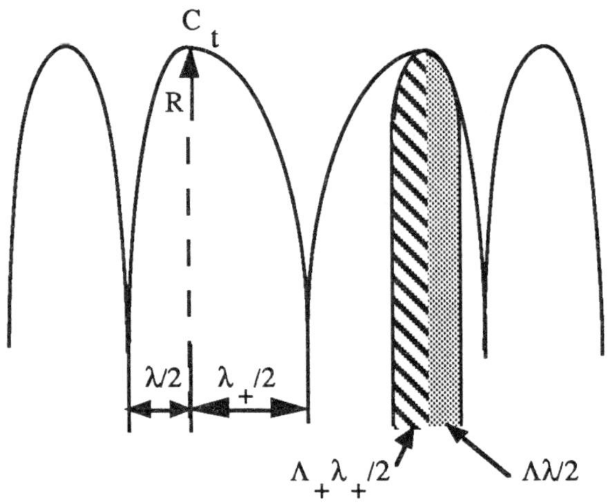

Figure 10 : Sketch of the changes induced by local dendritic transition in a 1D-cellular array.

Local dendritic transition is sketched in Figure 10. When a cell is eliminated, it is overgrown by its two neighbours. Basically, the half-primary spacing is increased and the tip radius unchanged so that the half relative width is decreased. Tip concentration and constitutional supercooling are unchanged. It follows that in the microsolvability condition[12]

$$\nu C_t^* - 1 = \frac{\nu A}{K}\frac{F_\Lambda}{P_\lambda^2} \Rightarrow \nu C_t^* - 1 = \frac{\nu A}{K}\frac{F_{\Lambda+}}{P_{\lambda+}^2} \qquad (12)$$

in which constitutional supercooling at the tip is balanced by a capillary term[13], only the R.H.S. is altered.

The physical idea is the following : if the capillary term is decreased there will be a deficit of capillary compensation and some constitutional supercooling will become available to drive the local dendritic transition.

By expanding $F_{\Lambda+}/P_{\lambda+}^2$ and using Eq.1, LDT is predicted when

$$\frac{1}{F_\Lambda}\frac{d F_\Lambda}{d\lambda} - 2\frac{\Lambda - 2}{\Lambda(1-\Lambda)} > 0 \qquad (13)$$

It is obvious that the increase in spacing Peclet number leads to a deficit. The change in the capillary function F_Λ can be estimated from experiments

$$\frac{1}{F_\Lambda}\frac{d F_\Lambda}{d\lambda} = -\frac{2.5}{\Lambda(1-\Lambda)} \qquad (14)$$

It leads to a surplus[14,15]. Therefore, there is a critical value of the relative width, Λ_D = 0.75, below which LDT will occur. Figure 4c shows that cells actually are observed above a critical value which yet is about 0.5. Owing to the roughness of the model, the qualitative fit is rather gratifying.

4.2. Practical estimation of the cell-dendrite transition

Combining results from the study of the cellular tip undercooling and from the local dendritic transition it is possible to estimate the value of the level ν of morphological instability at the dendritic transition, which might be useful to choose the values of the control parameters in microgravity experiments.

Indeed, from the definition of the tip undercooling, the BBF and the mPP relations, one can express the relative width as a function of the level of morphological instability and segregation coefficient

$$\Lambda = 1 - \frac{K\nu}{1 + K(\nu - 1)} \qquad (15)$$

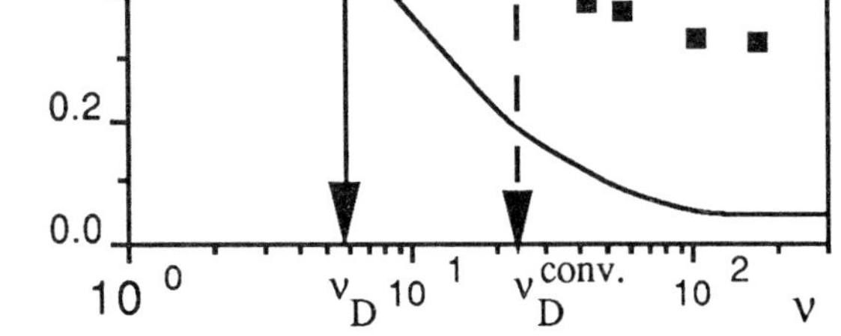

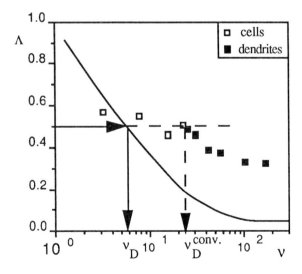

Figure 11 : Practical estimation of the dendritic transition. Al - 4.1 % Cu alloys. Data points from Ref.16.

Microgravity experiments are carried out on bulk samples. For bulk samples, the dendritic transition (ν_D^{conv}) also occurs about Λ_D=0.5 even when there is convection in the liquid phase (Figure 11). The comparison with Figure 4c shows that only the variation of Λ with ν is changed. Using relation 15 for the relative width, which corresponds to the full line in Figure 11, the dendritic transition under diffusive conditions is estimated at about $\nu_D = 6$.

5. CONCLUSION

Such predictions about the onset of dendritic growth, as well as the tip undercooling relation, will be checked in the microgravity experiments that, in collaboration with other groups, we are presently preparing for the D2, IML2 and USML2 missions.

For the future, it should be stressed that cellular experiments are highly desirable in order to span in K, which would allow a definite test of theories. For a complete interpretation of the observations, it is mandatory to measure λ, R *and* C_t or T_t, i.e. to determine the tip position. Dynamical studies would improve the understanding of the formation of cellular arrays and local dendritic transition. If only because the dynamics is richer in 3D than in 2D, reference microgravity experiments on bulk samples, with a diffusive transport in the melt, will still be required.

6. ACKNOWLEDGEMENTS

This paper gets together recent results obtained in the "Spatial Structuration and Convection in Directional Solidification" group, in Marseille. M. Hennenberg, H. Jamgotchian, H. Nguyen Thi and R. Trivedi have each contributed to one section, or more. Financial supports from the Centre National d'Etudes Spatiales, Centre National de la Recherche Scientifique, European Space Agency and NATO Scientific Exchanges Programme are gratefully acknowledged.

7. REFERENCES

1. R. Trivedi, *Met. Trans. A* **15** (1984) 977.
2. P. Pelcé and A. Pumir, *J. Cryst. Growth* **73** (1985) 337.
3. H. Hennenberg and B. Billia, *J. Phys. I France* **1** (1991) 79.
4. M. A. Chopra, *PhD Thesis*. Rensselaer Polytechnic Institute (1983).
5. K. Somboonsuk, J. T. Mason and R. Trivedi, *Met. Trans. A* **15** (1984) 967.
6. R. Trivedi and K. Somboonsuk, *Mat. Sci. Eng.* **65** (1984) 65.
7. H. Esaka, *PhD Thesis*. Ecole Polytechnique Fédérale de Lausanne (1986).
8. J. D. Weeks, W. Van Saarloos and M. Grant, *J. Cryst. Growth* **112** (1991) 244.
9. P. G. Saffman and G. I. Taylor, *Proc. Roy. Soc. London A* **245** (1958) 312.
10. T. F. Bower, H. D. Brody and M. C. Flemings, *Trans. AIME* **236** (1966) 624.
11. B. Billia, H. Jamgotchian and R. Trivedi, *J. Cryst. Growth*, submitted.
12. J. S. Langer, in *Chance and Matter*, J. Souletie, J. Vannimenus and R. Stora Eds, North-Holland, Amsterdam (1987), P. 629.
13. R. Trivedi, *J. Cryst. Growth* **49** (1980) 219.
14. B. Billia, H. Jamgotchian and R. Trivedi, in *Materials and Fluid Sciences in Microgravity*, ESA SP-295, ESA, Noordwijk (1990), P. 189.
15. B. Billia, H. Jamgotchian and H. Nguyen Thi, in *Nonlinear Phenomena Related to Growth and Form*, M. Ben Amar, P. Pelcé and P. Tabeling Eds, Plenum, to appear.
16. Y. Miyata, T. Suzuki and J. I. Uno, *Met. Trans. A* **16** (1985) 1799.

ON THE POSSIBILITY OF HEAT AND MASS TRANSFER AND INTERFACE SHAPE CONTROL BY ELECTROMAGNETIC EFFECT ON MELT DURING UNIDIRECTIONAL SOLIDIFICATION

Yu.M.Gelfgat, M.Z.Sorkin, J.Priede, O.Mozgirs
Institute of Physics, Latvian Academy of Sciences
Salaspils, Riga, Latvia, 229021

Abstract *A possibility to control both a heat/mass transfer in melt and a crystal-melt interface shape by applying forced convection to unidirectional solidification process is considered. Some general conditions providing desirable shape of crystal-melt interface as well as the effect of rotating magnetic field on heat transfer are predicted. The ability to maintain interface shape concave by means of the rotating magnetic field is also showed experimentally. Some further developments of electromagnetic means to control transfer processes in melt are discussed.*

Key Words: Unidirectional solidification, Crystal-melt interface, forced convection, Rotating magnetic field.

1. INTRODUCTION

Unidirectional solidification under microgravity shows such negative feature as an increase of non-uniformity in impurity distribution when the intensity of convective mass transfer is comparable with the diffusional one (Ref.1-4) and low mass transfer velocity at the containerless crystallization refining. Considering the possibility to use the forced MagnetoHydroDynamic convection to eliminate those disadvantages, we shall try: a) to find thermal conditions providing stable phase interface shape; b) to choose a concrete type of MHD device meeting required conditions; c) to estimate the possibility of MHD usage for flow pattern control in melt.

2. FORCED CONVECTION EFFECT ON THE CRYSTALLIZATION FRONT SHAPE AND HEAT TRANSFER IN THE BRIDGMAN TECHNIQUE.

Let us examine the general effects of the forced convection in melt, on both the phase interface shape and the temperature field. We shall examine the heat transfer in the following simplified growth system, shown in figure 1., incorporating regions of liquid phase G_l, solid phase G_s and container G_a. Due to both melt flow and phase interface translation, the temperature distribution in all regions under

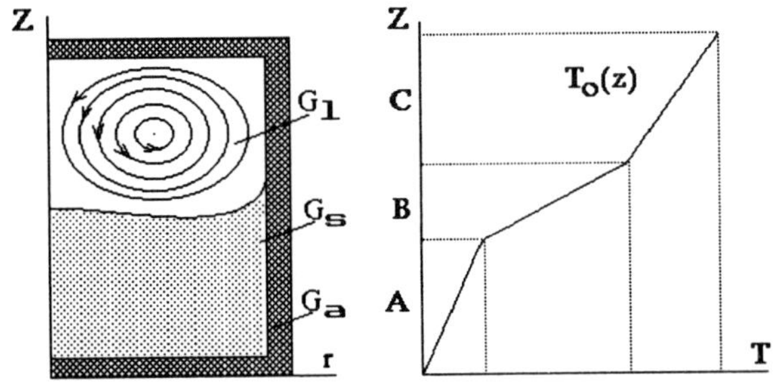

Figure 1. *Sketch to the formulation of the problem.*

consideration is governed by heat advection-diffusion equation, considered in a frame of reference moving uniformly with the velocity of solidification front

$$k\frac{\partial T}{\partial t}+\vec{\nabla}(-\lambda\vec{\nabla}T+k\vec{v})=-\mathbf{Tk}\,\vec{w}\cdot\vec{\delta}\,(\vec{r}-\vec{r_f}) \quad (1.1)$$

The equation (1.1) is represented in dimensionless form by choosing as being characteristic thermal parameters of the melt and introducing as scales the following quantities: velocity of temperature diffusion $v=a/R$, time of temperature diffusion $\tau=a/R$, where a is thermal diffusivity of melt and R is a radius of a container. Coefficients k and λ are respectively dimensionless relative specific heat and thermal conductivity, varying depending on region and being equal to unity in the region of the melt. $\mathbf{Tk}=\frac{\sigma\rho}{kT^*}$ is dimensionless parameter, where σ is a specific latent heat and T^* is a characteristic temperatue; ρ is a density of melt; $\vec{w}=w\vec{e_z}$ is dimensionless velocity of phase interface translation. The term responsible for latent heat is included in equation as source by making use of Dirac vector delta-function, being defined as $\vec{\delta}(\vec{e_i}x_i)=\vec{e_i}\,\delta(x_i)$. The velocity field $\vec{v}$ is given by

$$\vec{v}=\vec{w}+\frac{1}{r}\vec{\nabla}\Psi\times\vec{e_\alpha} \quad (1.2)$$

where Ψ is a simplified Stokes stream function approximating real melt motion by obeying only kinematic condions (boundary and symmetry conditions) but dynamic ones (Navier-Stokes equation)

$$\Psi(r,z)=\mathbf{Pe}\cdot r^2(r^3-1.5\cdot r^2+0.5)\cdot \left[1-cos(2\pi\cdot\frac{\mathbf{G}+\Delta_\mathbf{a}-z}{\mathbf{G}+\Delta_\mathbf{a}-z_f(r)})\right] \quad (1.3)$$

Dimensionless parameter **Pe** controls intensity of melt flow and is refered as Peclet number, **G** is aspect ratio of container interior, $\Delta_\mathbf{a}$ is the thickness of container wall relative to crystal radius, $z_f(r)$ is the function of radius defining the shape of phase interface. The boundary conditions imposed on temperature field at the container surface is following

$$\left[T+\mathbf{Rs}\cdot\frac{\partial T}{\partial r}\right]_{r=1+\Delta_a}=T_0(z) \quad (1.4)$$

where $T_0(z)$ is a fixed temperatue distribution along a heater, **Rs** is the parameter of a thermal resistance between the heater and container. At the end walls of container far-field like conditions are imposed.

Now we shall examine an interface shape in terms of its deflection δ_f defined as follows $\delta_f=z_f(0)-z_f(1)$. Further interface will be refered as concave if the deflection value is negative $\delta_f<0$ (deflection is directed to the solid phase, provided a liquid phase is above solid one) and convex - in opposite case.

Provided the function $T_0(z)$ is linear along the side wall of container, and melt convection as well as phase interface translation is absent and both liquid and solid phase has the same heat conductivity, then the resulting phase interface is flat ($\delta_f=0$).

Increasing of the parameter **Tk** results in the change of the initially flat phase interface to a concave one. The forced convection at **Pe**>0, corresponding to melt ascending along the container walls and descending to the crystallization front in the central part, leads to the interface concavety increase. At **Pe**<0 (a central jet is ascending and returning down along the container walls) the deflection value is practically independent on **Pe**, but has a negative value, although the central part of a profile becomes flatter. For a linear temperature profile imposed on side wall, the possibilities of the crystallization front shape control by forced convective flows are rather limited. It is known that the shape and location of crystallization front can be controlled by imposing a more complicated, than linear one, temperature profile along the side wall of the container (Ref.5). For example, Fig.2 shows the interface profiles at different **Tk** values when $T_0(z)$ is the piecewise-linear function consisting from two sections having diffe-

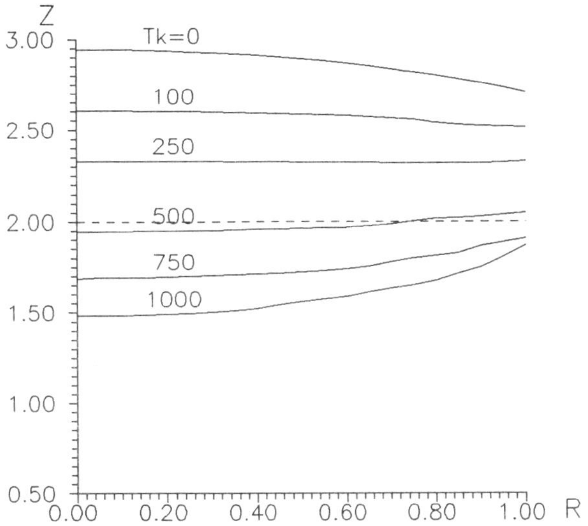

Figure 2. *Phase interface shape dependence on the latent heat parameter Tk*

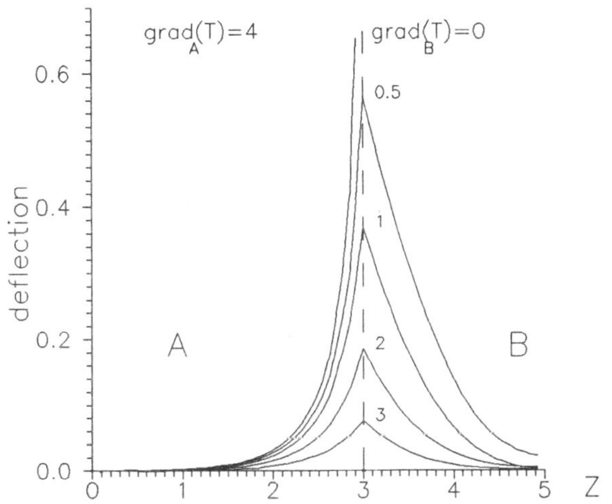

Figure 3. *Crystalization front defection value versus position of crystalization point on heater profile at different temperature gradiends in section B*

rent temperature gradients - section **A**: $\left.\frac{\partial T_0}{\partial z}\right|_{z=0\div2} = 4$, section **B**: $\left.\frac{\partial T_0}{\partial z}\right|_{z=2\div5} = 0.5$. As it is seen form Fig.2, the front is convex, while **Tk**$<$ 250. As **Tk** increases, the interface descends and its shape changes into a concave one. Let us examine dependence of the parameter of interface shape δ_f on the front displacement relative to the deflection point of boundary temperature profile T_0 at fixed temperature profiles having sections with different gradients. It is seen from Fig. 3, that δ_f has a sharply pronounced maximum when the position of point having temperature equal to phase transition one coincides with the deflection point of heater temperature profile. The deflection of the interface δ_f decreases as the distance between phase transition point z^*, located on the heater surface, and the deflection point of heater temperature profile $T_0(z)$ increases. There is more rapid variation of δ_f on z^* in section **A**, where the temperature gradient is higher than in the section **B**. The data shown on Fig.3 allow to predict how the parameters of crystallization will effect the front shape. Provided that **Tk**= **Pe**= w= 0 and the front is located above the deflection point of temperature profile in section **B** , then by increasing any of the parameters **Tk** or **Pe** the interface tries to become more concave, but at the same time it shifts downward and approaches to the deflection point of boundary temperature profile giving rise to to opposite trend of increasing of convexity of the interface. In such a way, the self-compensation process takes a place, maintaining flatness of the interface. Note, that this is the case only when interface is remaining above the deflection point. When front shifts below the deflection point, the processes considered begin to work in the same direction and concavity of the front rapidly increases. Thus, we conclude that a stable, convex or flat front, whitch as it is known is more preferable at crystal growth, can be obtained at the forced convective stirring under the following conditions: 1) meridional flows of melt are organized in such a way that in central part of the container there occurs an ascending jet with a returning flow along the container walls; 2) the temperature profile of the heater has two zones - an upper and a lower one and in the lower zone the temperature gradient is higher than in the upper one; the point of the profile with the temperature of crystallization is in the upper zone; 3) the crystallization conditions have been arranged so that during the process the front has been in the upper zone of the heater temperature profile.

3. NUMERICAL SIMULATION OF A ROTATING MAGNETIC FIELD EFFECT ON THE CRYSTALLIZATION FRONT SHAPE IN THE BRIDGMAN TECHNIQUE

A rotating magnetic field is the convenient technique to generate a forced convection in the seeded containers under conditions of unidirectional crystallization conducted at high temperatures. It has the following advantages: 1) generation of the azimuthal swirl in melt (it in turn gives rise to meridional flows); 2) high efficiency of electromagnetic effects on the melt under specific conditions of the unidirectional crystallization, when the melt diameter is significantly smaller than the diameter of the inductor gap; 3) a convenient connection of the inductor with the growth chamber. The disadvantage of the rotating magnetic field is an impossibility to control both intensitivity of meridional flow and its configuration (flow is directed to the crystallization front near the side wall and backward in the center). Note, that this configuration corresponds to the one, considered above.

We have developed a mathematical model of MHD flow generated by rotating magnetic field, yielding correct magnetic torque distribution in containers of finite length (Ref.6).

There are primary azimuthal swirl induced by electromagnetic torque due to rotating magnetic field and secondary meridional flow due to disbalance of a preasure gradient and centrifugal force at the phase interface and at the endwall of the container. It has been shown by numerical simulation, that there are different regimes of motion possible in the melt under the action of rotating magnetic field. At Stoke's regime of flow occuring when the melt motion is sufficiently slow to neglect an effect of secondary flow to primary one, the electromagnetic torque is balanced by viscous shear of azimuthal swirl in the whole volume of the melt. Increase of an induction or a frequency of rotation of the magnetic field results in transition from regime of Stoke's flow to a non-linear stable one, where boundary layers of velocity on the side walls and crystalization front appear on account of significant interaction between the azimuthal and meridional flows. A further increase of the parameters gives rise to unstable regime of flow, where velocity oscillations (periodic or stochastic) exist. Obviously, the nonlinear stable regime is more preferable for using in the unidirectional crystallization processes.

There are two scales of meridional velocity having different orders of magnitute in the boundary layer and in the bulk of the melt at the nonlinear regime of the flow. Meridional velocity in the boundary layer at the phase interface is cosiderably higher than in the bulk of the melt due to conservation of the flow. It enables to conduct the recrystallization under conditions when either a) mass transfer is convective, but heat transfer is diffusional (due to the difference in transfer coefficients) or when both mass and heat transfers are convective; b) the difference of radial velocity in the boundary layer and in the bulk of melt allows to choose a regime, when heat (or mass) transfer at the crystallization interface is convective, but in the bulk the transfer remains diffusional. The mathematical model of MHD flow in a rotating field mentioned above has been used for numerical simulation of heat transfer in the Bridgman technique. Without going into details we shall illustrate the obtained results on the example of GaAs crystallization by a vertical Bridgman technique. Let us consider a heater consisting of two radiating surfaces with constant temperatures, separated by the 100 mm wide "adiabatic" area, with melt radius 45 mm and its height 300 mm. Calculations showed that if a characteristic temperature gradient is 14 $^{o}C/cm$ and a container velocity is 2 mm/h, the phase interface is concave, if we consider only molecular heat transfer in the melt. If we take into account a free convection, the interface is flat with a slight convexity. The action of a uniform rotating magnetic field with induction 4 mT

(50 Hz) changes the interface into a concave one with deflection ~0.15 R, resulting in the increase of the process stability and gives a possibility to increase the velocity of container, if necessary. If characteristic grad T= 7°C/cm, the interface remains a convex, when B > 0, although the interface is more flatter in comparison with case B = 0.

4. PHYSICAL SIMULATION OF A ROTATING FIELD EFFECT ON THE PHASE INTERFACE GEOMETRY

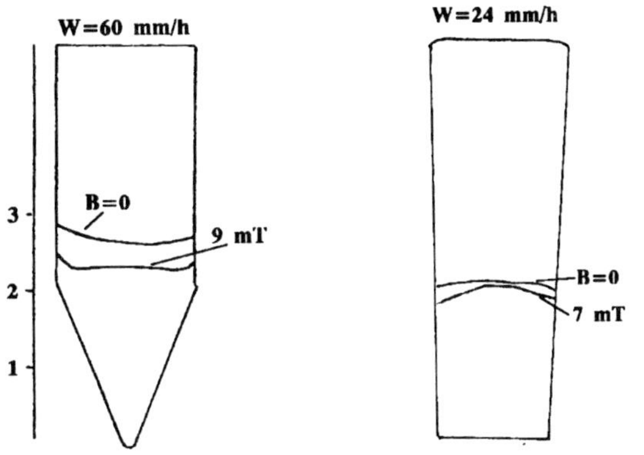

Figure 4. *Interface profiles determined experimentaly during crystallization of tin samples.*

The variation of the phase interface shape has been studied experimentally by unidirectional solidification of tin under the action of a rotating magnetic field on the melt. The input of bismuth impurity has been used to mark the interface shape during the crystallization. Without the effect of the field the interface was initially convex (the crystallization parameters in use were - dT/dz=1°C/cm, w = 60 mm/h). The field influence leads, as it has been predicted, to interface shifting downward and flating of its central part with lugs near the container side wall - Fig.4(left). When dT/dz= 3°C/cm and *w*= 24 mm/h the interface is concave without field affection and becomes more concave, if B>0 - Fig.4(right) - (note, that for tin $\lambda_s > \lambda_l$ and that promotes the concave shape of the interface).

5. MELT FLOW STRUCTURE CONTROL

As it has been already mentioned, the inductor of a uniform rotating field is the most advantageous one among induction devices generating the forced convection in melt under conditions of the unidirectional crystallization. We have stated that by using the conventional rotating field inductor it is impossible to change the direction of a meridional flow. Regardless of an azimuthal flow direction, in meridional plane melt is moving toward interface

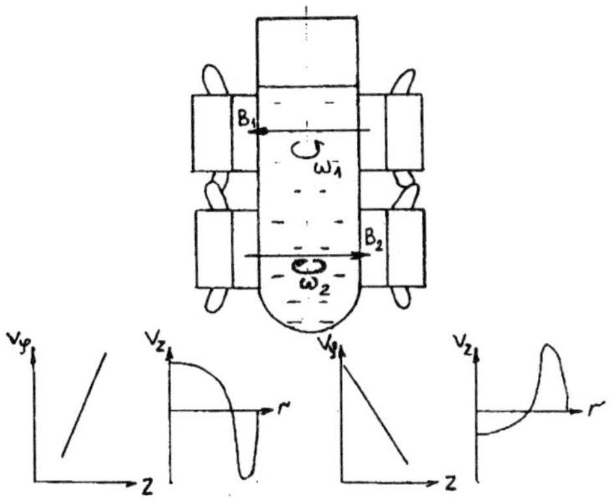

Figure 5. *Flow control by using of double inductor.*

along the side walls of the container and ascending along the symmetry axis. It is also impossible to control independently the intensity of azimuthal and meridional flows. At the same time there are situations, when the downward direction at the central region is preferable, for example when initially we have a strong concave interface and we have to make it flatter. For intensification of the crystallization refining by the floating zone technique it is most rationally to stir the melt using flow velocity oscillations generated by MHD inductors, when the oscillation amplitude is comparable with a mean velocity of the flow. Heat and mass transfer caused by the velocity oscillations does not lead to the impurity radial stratification and does not require an intensive azimuthal flow, that results in the floating zone instability. A double inductor, consisting of two located on different height levels rotating field inductors with the independent controlled power supply, have been used to solve the given problems. The different current loads of the inductors (including counterphase supply) makes possible to maintain the controlled nonuniformity of the field and azimuthal velocity distribution Fig.5. Thus, one can control the intensity and configuration of the meridinal flow. The generation of intensive velocity pulsations, spreading in the whole melt volume, can arise from a shear-layer between two differentially rotating liquid volumes (due to the hydrodynamic instability) by using the double inductor. The experimental measurements of velocity structures under the double inductor effect, proved the possibility of the meridional flow control including its reversing. The measurements

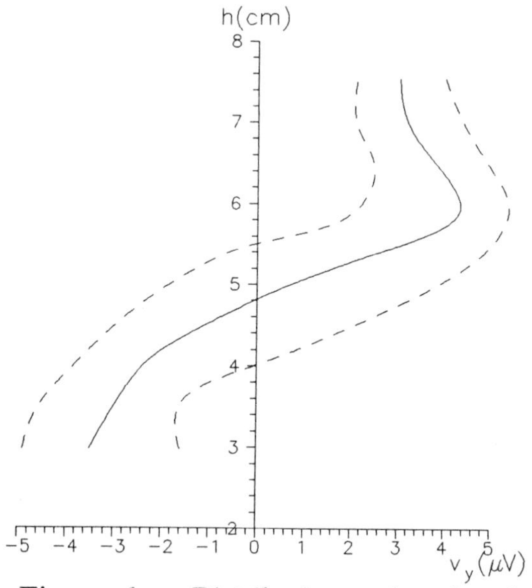

Figure 6. *Distribution of azimuthal velocity (solid line) and range of velocity pulsations (dashed lines generated by double inductor. 1μV=0,7mm/s.*

have been carried out in a cylindrical layer of a model melt (Ga-In-Sn) with diameter 40 mm, height 50 mm, by a conductive anemometer with its self magnetic field. When the current load in the upper section of the inductor was twice larger then current load in the lower section (counter-field rotation in the sections), the azimuthal velocity increased upwards and meridional flow directed downwards along the side walls; if the ratio of loads was opposite, the distribution of azimuthal velocity and configuration of meridional flow were opposite too. If the load is equivalent in the opposite cut in sections, the differential rotation in a liquid phase is worked out, followed by the velocity pulsations, the pulsation amplitude is comparable with the azimuthal velocity order of magnitude in the whole melt volume - Fig.6. The experiments on solution crystallization of bismuth-antimony alloy samples, when a liquid bismuth is saturated due to solution of floating solid antimony feed shows that the intensity of mass transfer in a "pulsating" regime is equivalent to the intensity of mass transfer caused by a mean meridional flow generated by conventional inductor at sufficiently larger current loads, but the interface deflection is less due to the decrease of impurity radial segregation.

Thus, the used numerical and physical simulation techniques and results obtained permit to carry out experiments which combine the advantages of microgravity and magnetohydrodynamic technologies.

References

1. V.S.Zemskov and M.R.Raukhman, *Gidromehanika i teplomassoobmen pri poluchenii materialov*, (Nauka, Moscow, 1990), pp.131-141.
2. V.S.Zemskov et al. *Gidromehanika i teplobmen v nevesomosti,* (Novosibirsk, 1988), pp.142-152.
3. *Material Science in Microgravity,* Springer Verlag, 1986.
4. V.S. Zemskov et al. *Gidromehanika i teplomassoobmen pri poluchenii materialov,* (Nauka, Moscow, 1990), pp.115-120.
5. A.Horowits and I.S.Horowits, *Math. Res. Bull.,* 21, 1123 (1986)
6. J.Priede, *The 13th Riga MHD Conference 1990,* vol.1, pp.127-128

PARABOLIC FLIGHT EXPERIMENT TO EVALUATE SIMULATION SOFTWARE

-UNI-DIRECTIONAL SOLIDIFICATION OF SUCCINONITRILE-

H. Kimura+, M. Shimizu, S. Ishikura* H. Nakamura** and M. Ishikawa**

National Space Development Agency of Japan, 2-5-6 Shiba, Minato-ku,Tokyo 105 Japan
**Japan Space Utilization Promotion Center, 2-21-16 Nishiwaseda, shinjyuku-ku, Tokyo 169, Japan*
***Mitsubishi Research Institute Inc., 2-3-6 Otemachi, Chiyoda-ku, Tokyo 100, Japan*

+Present Address : National Research Institute for Metals, Tsukuba, Ibaraki 305, Japan

Abstract

Low gravity experiment using an airplane for an uni-directional solidification of succinionitrile has been carried out to evaluate simulation software. As compared with normal gravity experiment, rapid nucleation was observed for the Neumann type solidification due to the high gravity before the low gravity entry and rapid solidification rate was observed for the isothermal solidification.

Keywords: Unidirectional Solidification, Succinonitrile, Parabolic Flight, Simulation Software

1.INTRODUCTION

Computer software which simulate solidification process with convection under microgravity are being developed as a supporting tool for researcher in the Space Station era.
The target of the simulation software is shown in the following.

(1) Transportation phenomena
  a.Convection induced by residual gravity
  b. Marangoni convection
(2) Solidification process
  a. Convection - solidification coupling process
  b. Isothermal solidification process

The simulation software solved moving boundary problem with convection is programmed using an algorism of the boundary fixing method, which is based on Munakata et al. (Ref.1) for 2-dimensional rectangular space. The morphological instability due to kinetic effect is neglected in their algorism, which can be solved easily by convective transport equation as shown in details in Ref.2.

In the present work, parabolic flight experiment to evaluate the simulation software for uni-directional solidification were carried out using MU-300 airplane in 10^{-2}g.

2.EXPERIMENTAL PROCEDURES

Succinonitrile was used as a model material. Since Jackson's factor of this material which shows degree of kinetic effect at solid-liquid interface is small as 2.3, the kinetic effect can

be neglected. A schematic drawing of the central part of an experiment system is shown in Figure 1. The material was filled in an acrylic resin cell 13mm in height and width , and 2mm in thickness. The top and bottom part of the cell were contacted isothermal metal blocks. A sub-heater was used to maintain the atmospheric temperature constant near the cell. The temperature of the sample was controlled using the isothermal blocks and the sub-heater. A nucleation trigger with pertier device was used.
to accelerate the nucleation. The solidification process was observed with an optical microscope in low magnification. The solidification length was measured using video image with superimposed 0.01 sec time signal. The spatial accuracy of the measurement was 32 μm.

The temperature condition of the experiment was shown in Table 1. No.1 to 4 were for the Neumann type solidification, and No.5 for the isothermal solidification. Here the melting temperature of succinonitrile was 57.2 °C. First the temperature of all part was kept 60 °C isothermally, then cooling was started.

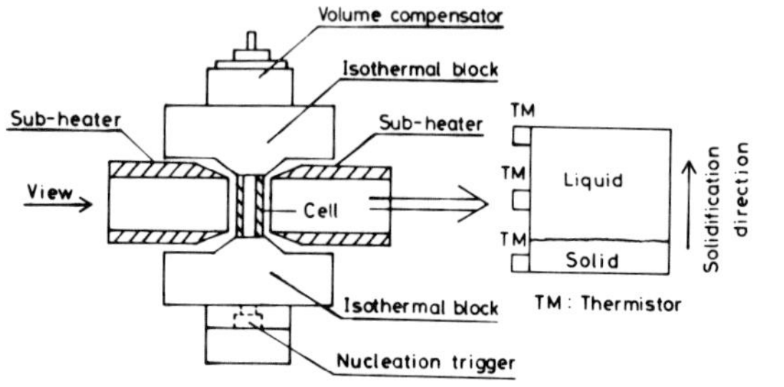

Figure 1 :Schematic drawing of central part of an experimental system.

TABLE 1 Temperature condition of the experiment for the Neumann type (1-4) and isothermal (5) solidification. Unit is °C.

No.	*Cell top*	*Cell bottom*	*Atmosphere*
1	60	43	60
2	60	45	60
3	60	47	60
4	60	49	60
5	52	52	52

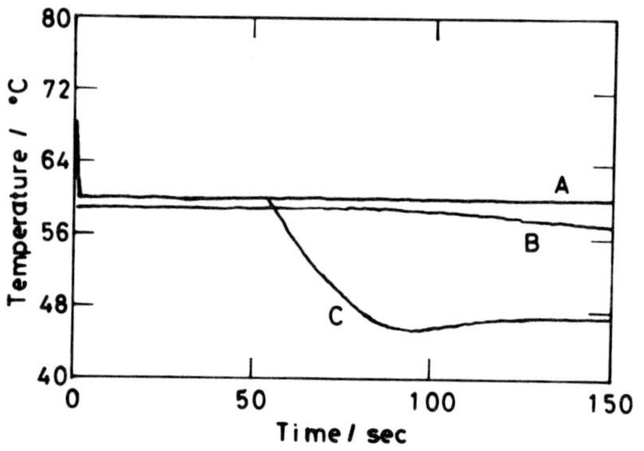

Figure 2 : Temperature profile during the experiment at cell top:(A), cell center:(B) and cell bottom:(C).

3. VALIDITY OF EXPERIMENT

Figure 2 shows temperature profile of the experiment for No.3 condition. Time origin is 90 sec before the low gravity entry. The temperature of the cell bottom deviates 0.9 °C from the set-point temperature.

3.1 Estimation of heat transfer

In the Neumann type solidification, determining factor of the solidification rate is (1) the heat transfer induced by the temperature gradient in the solid near the solid-liquid interface, (2) the latent heat of solidification at the solid-liquid interface and (3) the heat transfer induced by the temperature gradient in the liquid near the solid-liquid interface. In the experiment, it is possible to evaluate a heat transfer as one-dimensional model of solid-liquid interface, because (a) the cell is assumed adiabatic, (b) the shape of the solid-liquid interface is flat under the low gravity, (C) no convection is occurred under the low gravity. Furthermore it is possible to assume the temperature gradient in the solid constant along the solidification direction. We now consider the effect of temperature fluctuation during the solidification process resulting from effects (1) to (3) . Here the temperature of the solid-liquid interface is assumed 57.2 ° C, the melting temperature of succinonitrile.

(1) Heat transfer in the solid

The amount of heat transfer in the solid is shown in eq.(1), assuming constant temperature gradient in the solid.

$$Q_s = \lambda_s (T_m - T_c) / X, \quad (1)$$

where Q_s is the heat transferred in unit time and unit area, λ_s the thermal conductivity of the solid, T_m the melting temperature, T_c the cold edge temperature and X the solidification length.

Since the solidification length from the cold edge is 0.73 - 1.49 mm, and the thermal conductivity of succinonitrile solid is 13.438 J/m sec K, Q_s is 1.88×10^5 - 9.23×10^4 J /m^{2}sec. Namely Q_s is 1.37×10^5 J/m^{2}sec at solidification length of 1 mm.

(2) Latent heat of solidification

Heat generation per unit time and unit surface area is estimated as shown in eq.(2).

$$Q_c = v\ L, \quad (2)$$

where v is the growth rate and L the latent heat per unit volume.

Since the latent heat of succinionitrile is 4.725×10^7 J/m^3 and the solidification rate under the low gravity is 1.78×10^{-5} - 2.60×10^{-5} m/sec, heat generation is 1.23×10^3 - 8.41×10^2 J/m^{2}sec. This value is 0.01 less than Q_s.

(3) Heat transfer in the liquid

The amount of heat transfer in the liquid is shown in eq.(3).

$$Q_l = \lambda_l (\partial T_l / \partial X_l), \quad (3)$$

where Q_l is the heat transferred in the liquid near the solid-liquid interface in unit time and unit area,
$(\partial T_l / \partial X_l)$ the temperature gradient near the interface, λ_l the thermal conductivity of the liquid. The heat transfer is 0.99 Q_s from the heat conservation law. Temperature gradient in the liquid near the interface is the same as that in the solid, because the thermal conductivity of succinonitrile melt is 13.373 J/m sec K.

After all, the effect of the expelling latent heat on over all heat transfer is small as 1%. Therefore it is possible to assume that temperature gradient in the liquid is the same as that in the solid.

3.2 Effect of initial temperature fluctuation

Initial temperature gradient of 1.3 ° C/cm is observed between the cell edge and the cell center under the isothermal stage before the low gravity entry. Since temperature gradient of the experiment under the low gravity is higher as 68 - 134 ° C/cm, initial temperature gradient is neglected. Namely this effect is small less than 2%.

3.3 Effect of temperature overshoot at cold cell edge

Maximum effect on solidification rate by temperature overshoot of 0.9 ° C is determined as shown in eq.(4).

$$(T_m - T_r) / (T_m - T_c), \quad (4)$$

where T_r is the lowest temperature. In the

present work, the lowest temperature is 46.1 °C, the value of eq.(4) is 1.088. Namely the temperature gradient becomes only 9% larger than set-point temperature.

Therefore overall error is estimated less than 12%.

4. RESULTS AND DISCUSSION

Figure 3 shows solidification pattern obtained under the No.3 condition. Cooling was started 30 sec before the low gravity entry. Solid is nucleated 20 sec after the cooling at the cell bottom near the trigger, which spread immediately in the horizontal direction. Rapid nucleation is observed as compared with that in the normal gravity due to the high gravity before the low gravity entry. The solid-liquid interface is nearly flat under the low gravity. The succinonitrile is solidified stable in polycrystalline form. The shape of the solid-liquid interface becomes dendritic 60 sec after the low gravity entry. The shape transition might be caused by impurity to form binary system. A slow convection, whose rate is about 1 mm/sec, is observed in the transition period between normal gravity and low gravity, but no convection is observed in the low gravity period.

Figure 4 shows the time dependence of the solidification length at the cell edge and the cell center. The solidification rate is about 0.03 mm/sec. Figure 4 also shows the simulation results under the No.3 condition assuming initial solid part length of 0.73 mm. This value is obtained from the experiment. The initial and boundary conditions are shown in the following.

(1) Initial condition
  a. Temperature of cell top = cell side
    = sample liquid > melting temperature
  b. Temperature of cell bottom
    < melting temperature
(2) Boundary condition
  a. Cell top and bottom ---------- Isothermal
    non- slip wall
  b. Cell side -------------------- Heat insulated
    non-slip wall

Figure 5 shows the calculation model of the simulation denoted the Neumann solution. The experimental results are in good agreement with the simulation ones. The simulation can be evaluated for no convection case, but not for convection case in the present work. The same results are obtained for other temperature conditions as shown in TABLE 1.

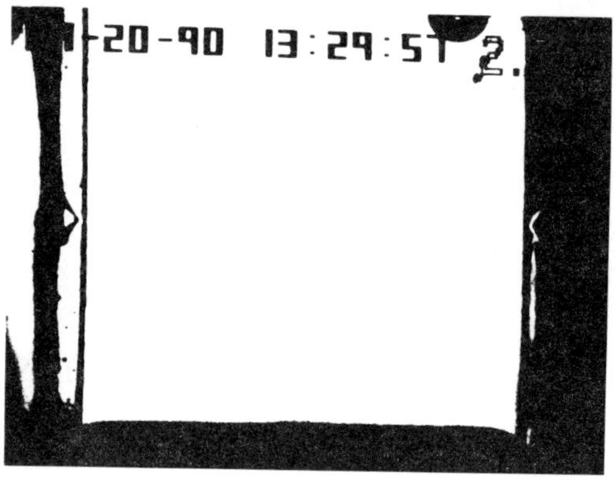

(a)

(b)

Figure 3 : Solidification pattern under the No.3 condition of the cell bottom 24 sec :(a) and 50 sec:(b) after the cooling. (b) is the pattern in the low gravity period.

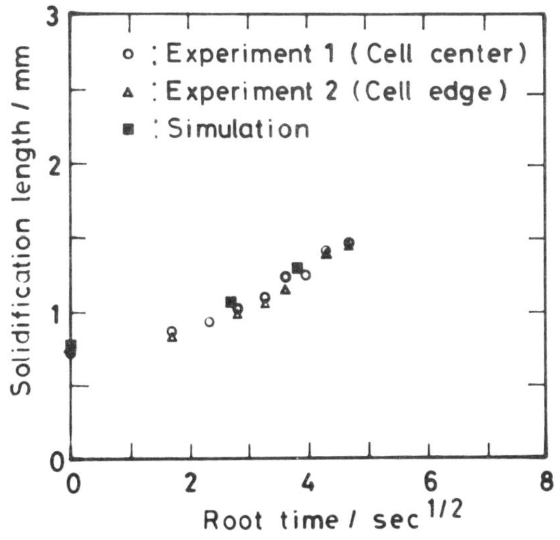

Figure 4 :Time dependence of solidification length.

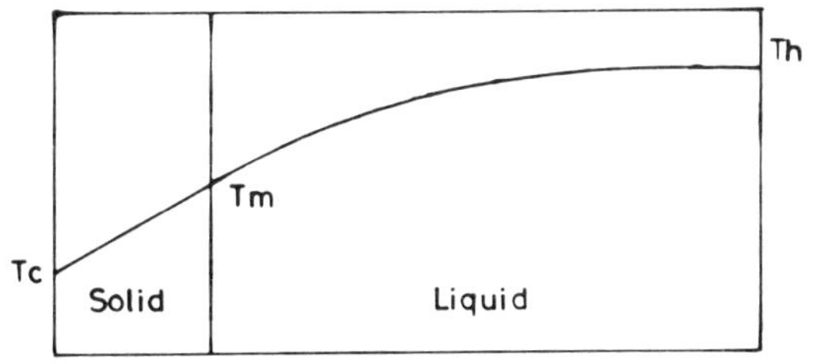

Figure 5 : The calculation model of the simulation.

Figure 6 shows solidification pattern of the isothermal condition. Higher-order axes of the dendrite are solidified massively in the low gravity as compared with the dendrite of lower-order axes in the normal gravity. It is difficult to observe details of dendrite shape. Figure 7 shows the time dependence of the solidification length. The solidification rate in the low gravity is larger than that in the normal gravity. Open circle and closed circle in the normal gravity show the results before and after the parabolic flight. This rapid solidification is against our expectation, and can not be simulated now. The reason why the rapid solidification is observed in the low gravity is expected resulting from difference of supercooling and/or effect of g-jitter. But time dependence of the degree of supercooling is not observed. The effect of g-jitter should be considered, but it is not carried out in the present work.

(a)

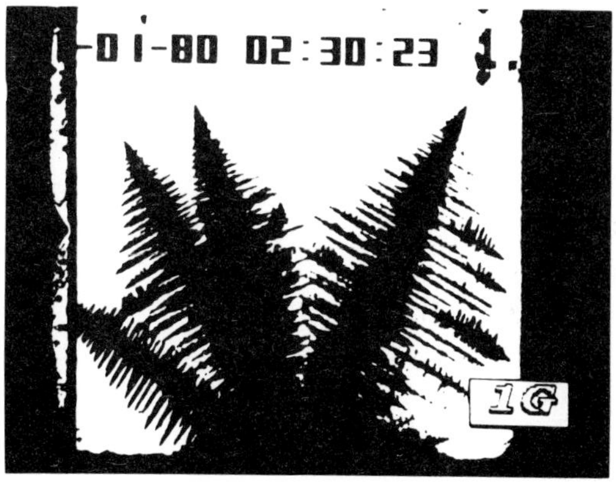

(b)

Figure 6 : Solidification pattern of isothermal condition in the low gravity:(a) and in the normal gravity:(b).

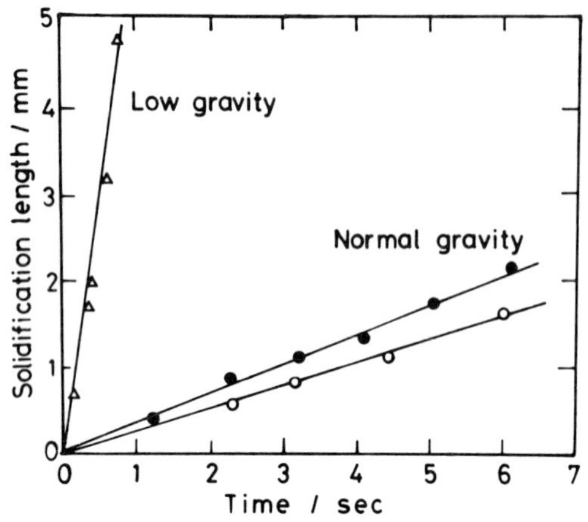

Figure 7 : Time dependence of solidification length for the isothermal condition. o and • show the results before and after the parabolic flight.

5. CONCLUSION

The experiment under the low gravity to evaluate the simulation software of unidirectional solidification of succinonitrile was carried out. For the Neumann type solidification the experimental results are in good agreement with the simulation ones, but for isothermal solidification the experimental results can not be simulated. We now reprogram the software being referred to the experimental results.

6.REFERENCES

1. T. Munakata and I.Tanasawa, *Trans. Jpn. Soc.Mech.Eng. B52(1986)876.*
2. Y. Fujimori, H. Kimura, T. Kusunose, H. Nakamura, S. Kamei and M.Ishikawa, *Proc. 17th Int. Symp. Space Tech. Sci. (Tokyo, 1991) p.2165.*

STABILITY OF STATIONARY REGIME OF DIRECTED CRYSTALLIZATION.

A.P.Gus'kov

Institute of Solid State Physics Academy of Sciences of the USSR, Moscow District, Chernogolovka, 142432,U.S.S.R.

ABSTRACT

One of the causes of spatial inhomogeneity of material structure is the interface instability at the directed crystallization. The numerical calculations of the impurity layer period coincides with the experimental results. This theory explains diversity of the periodical structure of a component and gives the classification of solidification regimes depending on the system stability value.

Keywords: *Stability of stationary solutions, temperature and concentration pertubations, growth increment δ, temporary pulsations and spatial distortions of the crystallization front.*

1. FORMULATION OF THE PROBLEM

The beginning of the interface stability investigations was founded in 1960–s (Refs. 1, 2). The problem has been recently comprehensivel reviewed (Refs.3,4). It has been shown (Ref.5) that the crystallization front instability may provide a spatial distortions as well as a temporary pulsations. It is the reason for the inhomogeneous partition of a component concentrations in the solid phase. It has been demonstrated (Ref.5) that the evaluations of the inhomogeneity scale are in good agreement with the experimental results. The paper contains results of the dispersion equation numerical solution of model problem and explanations of some phenomena as observed in experiments.

The designations are: $u(y,z,t)$ is the temperature, normalized to the phase transition temperature at the initial impurity concentration; $c(y,z,t)$ is the impurity concentration normalized to the initial one; y,z,t are nondimensional spacecoordinates and the time; D is the nondimensional diffusion coefficient in the melt, χ is thermal diffusity, e is the phase transition heat. Suppose the conditions of cooling the melt are such that the plane crystallization front in stationary conditions is moving at a constant velocity v . The initial temperature of the melt u , at the distance of z from the melt front, has a constant temperature u . Thermal conductivity in solid and liquid phases and the diffusion of impurity in the liquid phase are taken into account. To simplify the notation of the equations the coordinate x is not written out. The values referring to the solid phase are denoted by a prime. The initial system of equations in the coordinates τ , y , z , moving with the solidification front has the form :

$$\frac{\partial u'}{\partial \tau} = \chi' \Delta u' + v \frac{\partial u'}{\partial z}; z_o \leq z \leq 0 \quad (1)$$

$$\frac{\partial u}{\partial \tau} = \chi \Delta u + v \frac{\partial u}{\partial z}; 0 \leq z < \infty \quad (2)$$

$$\frac{\partial c}{\partial \tau} = D \Delta u + v \frac{\partial c}{\partial z}; 0 \leq z < \infty \quad (3)$$

$$\chi \frac{\partial u}{\partial z}\Big|_{z=+0} - \chi' \frac{\partial u}{\partial z}\Big|_{z=-0} = -\varepsilon V \tag{4}$$

$$\frac{\partial c}{\partial z}\Big|_{z=0} = V(1-\kappa)c \; ; \; c(\omega) = 1 \tag{5}$$

$$U'(z_o) = U_o ; U(\infty) = U_{oo} ; U(0) = U'(0); \; c(\omega) = 1, \tag{6}$$

$$V = V(U(y,0,\tau); C(y,0,\tau)), \tag{7}$$

here κ is the equilibrium partition coefficient.

2. ANALYSIS OF STABILITY

To study the stability of stationary solutions at low temperature and concentration perturbations the solutions are searched in the form:

$$U' = U'_S + U(z) \exp(\Omega\tau + iKy)$$

$$U = U_S + U_m(z) \exp(\Omega\tau + iKy)$$

$$C = C_S + C_m(z) \exp(\Omega\tau + iKy);$$

$$U'_m \ll U_S ; U_m \ll U_S ; C_m \ll C_S; \tag{8}$$

here U_S, U_S, C_S are the solutions of the stationary problem, and $\Omega = \Omega_1 + i\Omega_2$, where Ω_2 is the frequency of temporary pulsations.

Problem (1)—(7) was linearized in small perturbations (8), and frequency dependencies of the growth increment δ were determined. The numerical calculations of δ have shown that oscillations arise at some definite values of the governing parameters. In this case different types of impurity concentration waves may arise at the phase boundary. We consider the results of numerical calculations. Figure 1 presents the dependence of the growth increment δ of the minority component concentration on the frequency of temporary pulsations Ω_2 and of the wave number K.

Curves 1—6 correspond to different values of thermal flux Φ passing through the crystallization front,with $\Phi_1 < \Phi_2 < \ldots < \Phi_6$. According to the hypothesis of the predominating development of the pertur

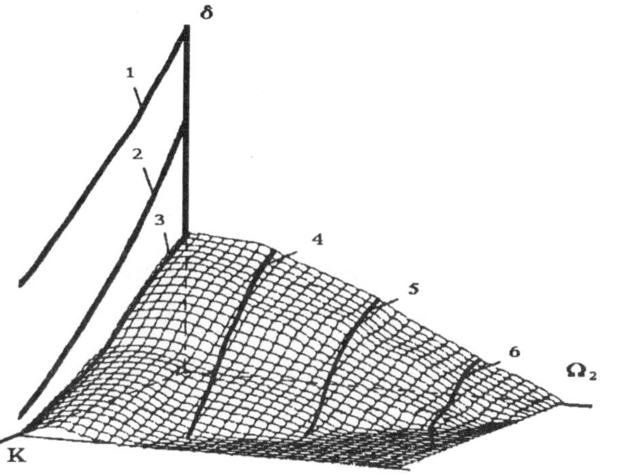

Figure 1: The dependence of the growth increment of the minority component concentration $\delta(\Omega_2, K)$.

bation mode with the maximum increment, the regimes of hardening with $K = 0$ correspond to fluxes $\Phi_1 - \Phi_6$. In the cases of Φ_4, Φ_5, Φ_6 (curves 4,5,6) the experiment shows a laminated component distribution in the solid phase.

Figure 2 shows an example of dependence $\delta(\Omega_2, K)$, in which there exist both temporary pulsations and spatial distortions of the crystallization front at Φ_3 and Φ_4 (curves 3 and 4).Under these conditions the minority component concentration distribution in the solid phase will be more complicated. The

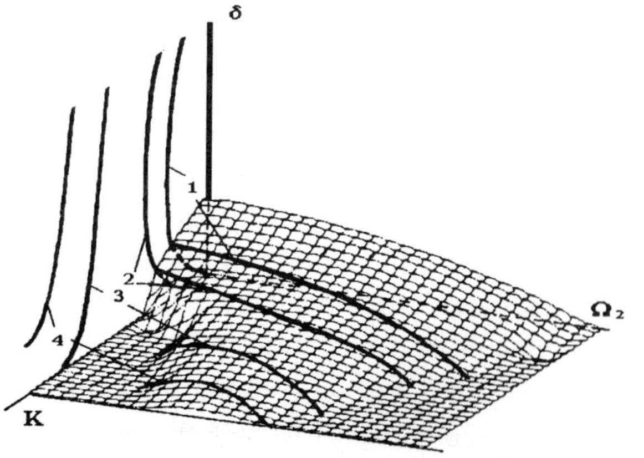

Figure 2: The dependence $\delta(\Omega_2, K)$.

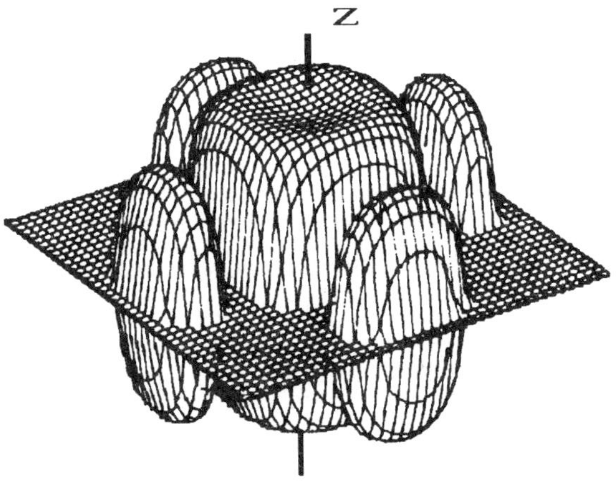

Figure 3: The isosurfaces of the impurity concentration of a two–component alloy.

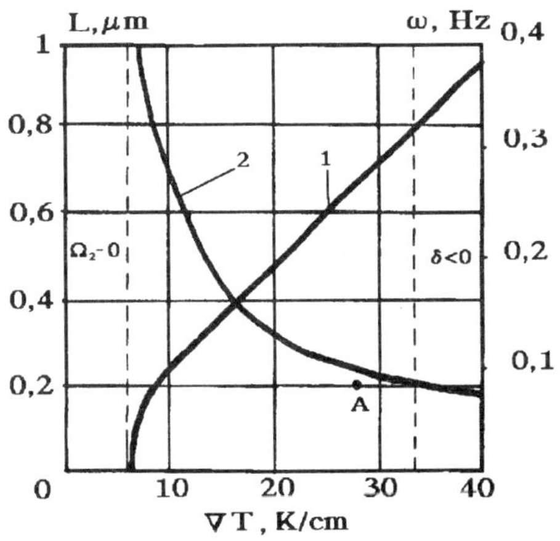

Figure 4:Dependencies of $\omega(\nabla T)$ (curve 1) and $L(\nabla T)$ (curve 2) .

distribution in this case is depicted in Figure3. Here presented an isosurface is of the impurity concentration of a two–component alloy for the case of a single–mode regime of temporary pulsations (the front is propagating in the direction of the z axis) and a two–mode regime of spatial distortions. The isosurface depicted is a of dendrite–like structure. Such formations are likely to determine the spacings between the branches of dendrites of polycrystal materials.

The results obtained make it possible to elucidate the origin of the layered structure at crystals growth, from the melt in the absence of convective transfer. In case we construct dependencies and $\omega(\nabla T)$ for $K = 0$ at the conditions shown in Figure 1, we get the curve shown in Figure 4. In the range $\Phi_{min} < \Phi < \Phi_{max}$ the system displays temporary pulsations of the minority component concentration and temperature at the crystallization front. Pulsations arise as Φ exceeding the threshold value of Φ_{min}. With increasing Φ the frequency of pulsations grows. At $\Phi > \Phi_{max}$ the pulsations vanish, since the system transforms to a stable state.

3. CONDITIONS FOR OSCILLATIONS OF THE COMPONENT CONCENTRATION

We consider the two thermal fluxes, passing through the phase front, and the associated temperature gradients. Flux $q_{\varepsilon} = \varepsilon v$ is developing due to the phase transition energy release, and flux q_S is formed by external conditions. In the general case such a separation is useful and serves as clarification of qualitative rules. However, there exist experiments, in which one of the fluxes may be neglected. One of these limiting cases is the crystallization of extremely undercooled liquids. Suppose the liquid undercooling is such that the phase transition heat release does not heat the material above the phase transition point. Then, in the ideal case of uniform under cooling, $q_S = 0$. One may easily see that in these conditions there exist only two stable stationary regimes of crystallization: one with a fixed solidification front and the other with the front displacing with some limiting velocity. Any other regime is unstable and transforms to one of the stable regimes under a small perturbation.

Another limiting case consists in slow crystallization with a fixed temperature profile, determined by external conditions. This corresponds to growing single crystal materials. Suppose the speed of the phase front displacement is sufficiently small (this can always be achieved, since the temperature profile is moving along the ingot), that is $q_\varepsilon = q_S$ .In this case the system is stable: under small perturbations it returns to the initial stationary state. In the general case the stability of the system depends on the ratio of the flux magnitudes q_ε and q_S.

The numerical estimates of the component concentration wave frequency, correlated (as was the case in (Ref.5)) by the values of inhomogeneities in the materials actually produced with thermal fluxes at the hardening front, make it possible to distinguish typical regimes of hardening as the ratio q_S/q_ε varies. Relation $q_S/q_\varepsilon = 0$ corresponds to the crystallization of an extremely undercooled melt, $q_S/q_\varepsilon = 0.5$ to fine crystalline material obtained by fast quenching, $q_S/q_\varepsilon = 5$ for the rim zone of the ingot, $q_S/q_\varepsilon = 50$ stands for laminated distribution of the impurity in growing crystals from the melt, i.e., in growing single crystals.$q_S/q_\varepsilon \to \infty$ corresponds to the solidification of amorphous materials. In the transition from fine crystalline materials under fast quenching to amorphous materials under ultra fast quenching, the ratio q_S/q_ε is likely to vary from small to large values. This may be the reason for the variety of materials structures obtained under changing quenching rates.

The experimental results of crystallization of an *Al–Cu* alloy without fracture of plane front have been given in paper (Ref.6). It has been shown, that the component concentration changing bands have been seen in the alloy in the direction of crystallization. The purpose of the study was the comprehension this effect by temperature oscillation ahead the front of crystallization occurrence at temperature gradient. However, in this case the period of oscillation of the component concentration must be by the order of magnitude lower than that we see. The dependencies $\delta(\Omega_2, K)$ shown in Figure 1 and $L(\nabla T)$, here L is the period of concentration distribution of the component, shown in Figure 4, have been constructed at the parameters under conditions of experiment (Ref.6). The point A on Figure 4 is the result given from paper (Ref.6). It may be supposed that the concentration distribution of the component observed in paper (Ref.2) has been connected with self–waves of the concentration of the component, described in this report.

4. CONCLUSIONS

i.The interface instability at the directed crystallization is one of the reasons for the spatial inhomogeneity of a material structure.Numerical calculations of the impurity partition period are coinciding with the experimental results.

ii.The theory explains the variety of component periodical structures.

iii.The theory provides the classification of solidification regimes depending on the system stability value.

5.REFERENCES

1.W.W.Mullins,R.F.Sekerka,*J.Appl.Phys.*34 (1963) 323.

2.W.W.Mullins,R.F.Sekerka,*J.Appl.Phys.*35 (1964) 444.

3.M.C.Flemings, *Solidification Processing*, N.Y.(1974).

4.R.W.Cahn and P.Haasen, *Physical Metallurgy*, Nord Holland Physics Publishing,1983.

5.А.П.Гуськов , *ТВТ* , 29 , (1991) , 275

6. Х.Утек, М.Флемингс В кн. "*Проблемы роста кристаллов*" ,"Мир" (1968).

STABILITY OF THE SOLID-LIQUID INTERFACE DURING THE DIRECTIONAL SOLIDIFICATION OF BINARY ALLOYS

G. Zimmermann

ACCESS e. V., Intzestraße 5, D-5100 Aachen, Federal Republic of Germany

ABSTRACT

Directional solidification of metallic alloys with a planar solid-liquid interface is often preferred to get well-defined material properties. This paper demonstrates, that purely diffusive conditions, which can be realized under microgravity, stabilize a planar phase-boundary. Secondly, the macroscopic curvature of the interface is given by the temperature field of the furnace which can be optimized by numerical simulation. It is shown that a nearly planar phase-boundary can be realized.

Keywords: *Directional solidification, planar interface, numerical simulation, microgravity*

1. INTRODUCTION

Directional solidification of metallic alloys is an important method to get defined process parameters. In practice the Bridgman-Stockbarger-technique is frequently used, which has the advantage of realizing a quasi one-dimensional heat flux along the sample axis in the region of the solid-liquid interface, combined with a substantial temperature gradient. To get well-defined experimental conditions and resulting homogeneous material properties, directional solidification with a planar solid-liquid interface is often preferred. This paper is focussed on two mechanisms, influencing the stability of a planar interface: thermosolutal convection in the melt in front of the interface and radial temperature gradients.

2. PLANAR INTERFACE THROUGH SUPPRESSED THERMOSOLUTAL CONVECTION IN THE MELT

During directional solidification of a binary metallic alloy vertically upwards, one component is enriched ahead of the moving solid-liquid interface due to segregation. If the less dense component piles up in front of the phase boundary, solutal convection may occur in the melt, deforming the shape of the interface. A stability criterion based on a linear theory is given by Coriell et al. (Ref. 1).

In order to check this theory, solidification experiments were performed by Rex and Sahm (Ref. 2) with an Al-Mg alloy. Samples with 0.3 wt% Mg were solidified both on earth and under microgravity during the German Spacelab Mission D1. In agreement with the predictions of the linear stability theory, the longitudinal macrosegregation profile of the samples processed on ground confirms the convective mixing in the melt. Purely diffusive conditions prevail during the solidification under microgravity. This is demonstrated in figure 1a, where the sample was remelted to position 45 mm. After starting directional solidification, the concentration of Mg piles up in front of the solid-liquid interface, leading to a minimum in the Mg concentration in the solid. After about 5 mm growth steady state conditions of solidification are reached. The interface is quenched at position 77 mm. The profile shows the expected concentration peak.

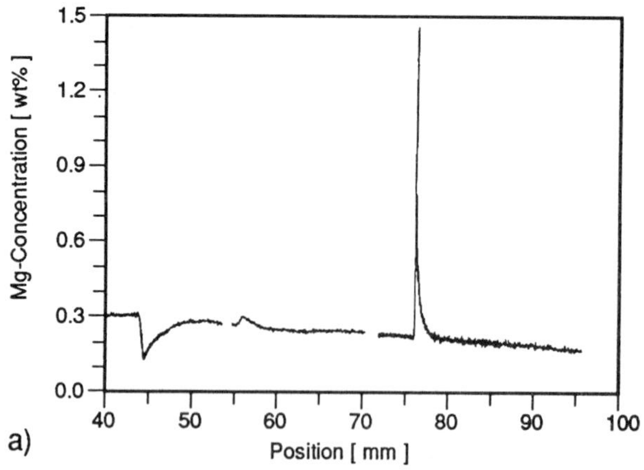

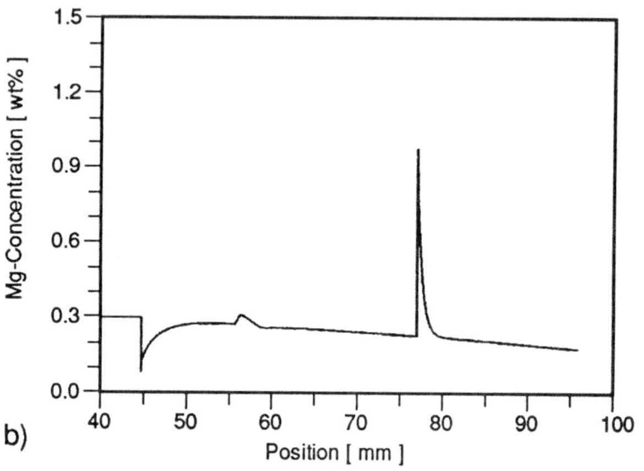

Figure 1:Longitudinal concentration profile of a directional solidified and quenched AlMg-sample; a) solidified in microgravity; b) numerical simulation

Assuming a diffusive solidification behaviour, the longitudinal concentration profile C(x) can be calculated using the one-dimensional equation of diffusive mass transport given by Smith et al. (Ref. 3):

$$\frac{\partial C}{\partial t} = D \cdot \frac{\partial^2 C}{\partial x^2} + v(t) \cdot \frac{\partial C}{\partial x},$$

with concentration C of magnesium, diffusivity coefficient D and a time dependent solidification velocity v(t). The continuity condition of the mass balance at the moving interface has to be fulfilled at every time step.

The diffusion equation is solved numerically by an explicit finite difference method. The calculated concentration profile for $D = 3.5 \cdot 10^{-9}$ m²/s, $v = 6.5 \cdot 10^{-6}$ m/s, segregation coefficient $k = 0.29$ and initial concentration $C_0 = 0.30$ wt% is given in figure 1b and fits very well to the experimental profile shown in figure 1a.

Two additional remarks should be noted:

- During the microgravity experiment an uncontrolled decrease of the heater temperature of about 20 °C took place. This leads to a locally increased solidification velocity and therefore to the small concentration peak apparent at position 57 mm, figure 1a. This effect can be simulated numerically using a time-dependent solidification velocity. A linear increase of the solidification velocity within 80 s to $8.5 \cdot 10^{-6}$ m/s, followed by a linear decrease to the initial value within 300 s reproduces the experimental behaviour.
  The level of the Mg concentration is lowered by about 2 wt%/m due to a continuous loss of Mg in the melt during the experiment. This effect is taken into consideration in the numerical simulation.

Comparing figures 1a and 1b confirms purely diffusive solidification conditions under microgravity. The micro-g level supresses thermosolutal convection in the melt and stabilizes a planar solid-liquid interface during directional solidification.

3. PLANAR INTERFACE THROUGH REDUCED RADIAL TEMPERATURE GRADIENTS

For directional solidification of metallic alloys the Bridgman-Stockbarger technique is often used. It is realized in high-temperature gradient furnaces, consisting of a heater and cooling section, separated by an adiabatic zone to establish a high gradient. The upper part of the sample is heated from the side via radiation and cooled in the lower part by conduction. In the region of the solid-liquid interface a one-dimensional heat flux along the sample axis is desirable, which results in a macroscopically planar phase-boundary. To optimize the temperature field numerous experiments and accompanying numerical simulations must be done, variing the process parameters and sample design.

Figures 2 shows the numerically simulated temperature field of the Dornier gradient heating furnace GFQ containing a Cu-Mn sample, an alloy which will be used for experiments in microgravity during the German D-2 mission. The cross-section of the whole furnace and sample region is given in figure 2a, while figure 2b shows an enlargement of the adiabatic zone with an isotherm at the melting temperature of the CuMn alloy at 880°C.The temperature distribution has been calculated with the three-dimensional finite element solver CASTS (Hediger and Hofmann (Ref. 4)). Input data are the geometry and the temperatures of the furnace, as well as the geometry and material properties of the sample consisting of nine different materials and the

a)
b)
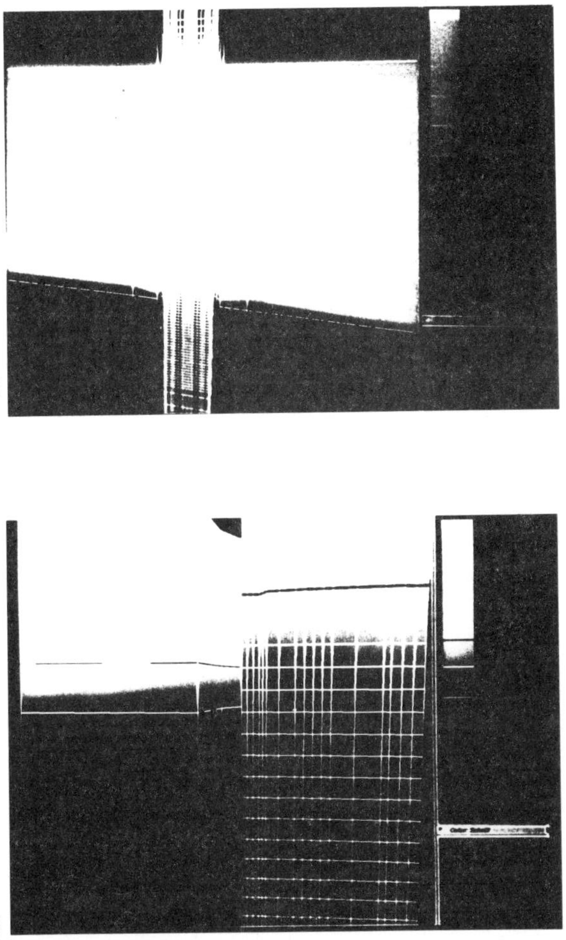

Figure 2: Numerical simulated temperature field during directional solidification, compressed in axial direction by a factor 3.3 a) within the whole GFQ gradient furnace and the sample, b) within the adiabatic region, including an isotherm at melting temperature of 880°C.

heat transfer coefficients between furnace and sample due to radiation and conduction in the gas.

To get an impression of the quality of the realized temperature distribution, figure 3 shows calculated radial temperature profiles in the sample at the melting temperature of CuMn of about 880 °C. The curves a) and b) are calculated for a furnace filled with a gas mixture of 35 mbar He and 55 mbar Ar at steady state conditions (curve a) and for a solidification velocity of $2.5 \cdot 10^{-5}$m/s (curve b). Because of the low solidification velocity and the good thermal conductivities of the sample materials there is nearly no difference between the two curves.

Curve c) of figure 3 belongs to a furnace filled with pure helium, which has a better thermal conductivity. This means a better cooling of the sample and results in a shift of the interface position of about 1 mm into the heater region. There, the radial temperature gradients of the furnace are slightly larger and produce

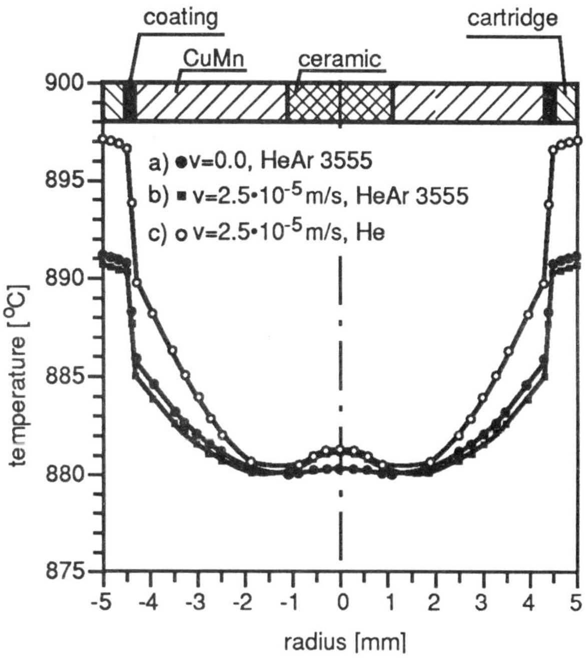

Figure 3: Numerically simulated radial temperature profiles at the melting temperature in a CuMn-sample for different solidification velocities and cooling gases in the furnace, a) v = 0, HeAr; b) v = $2.5 \cdot 10^{-5}$m/s, HeAr; c) v = $2.5 \cdot 10^{-5}$m/s, He

a greater radial temperature difference in the CuMn sample.

It is demonstrated that variations in the gas cooling shift the position of the solid-liquid interface relatively to the furnace. In the given example with a HeAr mixture the radial temperature difference in Cu-Mn can be reduced below 6 °C. Together with an axial temperature gradient of about 200 K/cm this results in a curvature of the solid-liquid interface of no more than 0.25 mm. Such small radial temperature gradients are insufficient to produce a relevant curvature of the interface by convective mass transport in the melt. For comparison with the experimental measurements, figure 4 shows the quenched solid-liquid interface of a CuMn sample, which is quite planar.

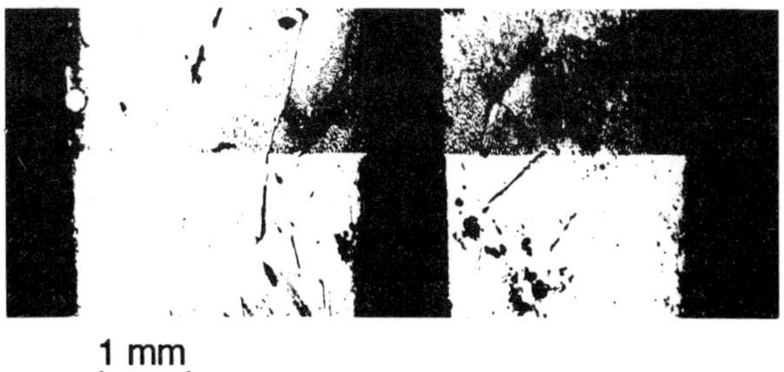

Figure 4: Picture of the nearly planar phase-boundary of a CuMn-sample, quenched during directional solidification in a gradient furnace on earth

The shape of the solid-liquid interface of a sample depends on its position in the furnace. To realize a planar phase-boundary, the temperatures of the furnace, the sample design and the heat-transfer between furnace and sample have to be optimized. A powerful tool to avoid expensive experimental test series will be the numerical simulation of the solidification process to define the parameters and ensure a stable, planar interface during directional solidification.

4. CONCLUSION

The stability of a solid-liquid interface during directional solidification with low velocities is mainly determined by the convection in the melt in front of the phase-boundary. It is demonstrated that convection can be suppressed in an microgravity environement and that thermal convection can be reduced by minimizing the radial temperature gradients at the solid-liquid interface.

5. REFERENCES

1. S. R. Coriell, M. R. Cordes, W. J. Boettinger and R. F. Sekerka, in: *J. Crystal Growth* 49, 1980, 13

2. S. Rex and P. R. Sahm, in: *Proc. Norderney Symposion on Scientific Results of the German Spacelab Mission D1,* eds.: P. R. Sahm, R. Jansen and M. H. Keller, (1987), 222

3. V. G. Smith, W. A. Tiller and J. W. Rutter, in: *Can. J. Phys.* 33, (1953), 723

4. F. Hediger and N. Hofmann, in: *Modeling of Casting, Welding and Advanced Solidification Processes V,* eds.: M.Rappaz, M.R.Ozgu, K.W.Mahin, The Minerals, Metals and Material Society, (1991), 611

THERMOELECTROMAGNETIC CONVECTION IN BULK SINGLE CRYSTAL GROWTH UNDER WEIGHTLESSNESS

L.A.Gorbunov, E.D.Lumkis
Institute of Physics of Latvian Academy of Sciences
229021, Salaspils-1, Riga, Latvia, USSR

ABSTRACT *The paper deals with single crystal growth under microgravity by the Bridgman technique and suppression of residual convective flows by steady magnetic fields. It is shown that hydrodynamics and mass transfer are affected by forced convection resulting from the interaction of an external magnetic field with thermoelectric currents in melt (the thermoelectromagnetic convection - TEMC). The thermoelectric currents can arise as a result of the thermo-e.m.f. effect near melt-single crystal or melt-conducting container boundaries. The paper presented thoroughly analyses the problem of thermo-e.m.f. and thermo-currents near the melt conducting container boundary.*

Keywords: thermoelectromagnetic convection, single crystal, magnetic field, Bridgman.

1. INTRODUCTION

To optimize the conditions for the bulk single crystal growth from the melt under gravity and to suppress various residual convective flows, it has been proposed (Ref.1,2) to use steady magnetic fields. However, (Ref.3,4) the magnetic field may stimulate additional flows, namely thermo- electromagnetic convection (the TEMC). First investigations of the TEMC have shown that at comparatively large temperature gradients in semiconductor melts interactions with an external magnetic field cause electromagnetic forces and TEMC. Nowadays these effects are studied by the Czochralski method of single crystal production from a graphite highconducting crucible (Ref.5). It is shown experimentally and theoretically that even on Earth TEMC significantly affects the melt hydrodynamics, and it can be compared to the thermogravitational convection (the TGC) and thermocapillary convection (the TCC) and dominate them. In (Ref.3) a possibility of the thermocurrent occurrence in melts has been proposed due to the varying of thermo-e.m.f. coefficient a for a solid phase, for example, at facets , thermal stress, non-uniformity of impurity and dopant distribution along a crystal, etc. But this effect has not been yet studied.

2. MATHEMATICAL MODEL AND CALCULATION TECHNIQUE

The present paper considers the numerical investigations of the TEMC effect on the melt hydrodynamics in single crystal semiconductor growth by the method of directional crystallization at $g = 0$. Let us analyze the mathematical model (Figure 1). Calculation region D consists of melt region D_1; crystal region D_2 and container region D_3. Region D is cylindrically symmetrical to the z-axis. The crystallization front is considered to be flat. The container is placed in a steady magnetic field $\vec{B}=\left(0, 0, B_z\right)$. The container is made of a conducting material with conductivity σ_c. Container wall thickness is d. Gravity vector g is directed along the z-axis (at g = 0) Liquid motion in region D_1 is described by two-dimensional equations of viscous incompressible liquid in cylindrical system of coordinates r, z in the Boussinesq's approximation in the following variables: Ψ is the current function, function $\xi = \omega/r$, where ω is the velocity vortex; angular momentum $M = V_\varphi r^2$; Θ - temperature deviation from melting temperature:

$$\frac{\partial \xi}{\partial t} + div\,(\vec{v}\xi) + r^{-4}\frac{\partial M^2}{\partial z} =$$
$$\mathrm{Re}^{-1}\left[r^{-1}\frac{\partial}{\partial r}\left(r^{-1}\frac{\partial}{\partial r}\left(r^2\xi\right)\right) + \frac{\partial^2 \xi}{\partial z^2}\right] \quad (1)$$
$$Gr\mathrm{Re}^{-2}\frac{\partial \theta}{\partial r} + N\cdot r^{-1}\frac{\partial (B_z^2 v_r)}{\partial z}$$

$$r\frac{\partial}{\partial r}\left(r^{-1}\frac{\partial \Psi}{\partial r}\right) + \frac{\partial^2 \Psi}{\partial z^2} = -r^2\xi \quad (2)$$

$$\frac{\partial M}{\partial t} + div\,(\vec{v}M) =$$
$$\mathrm{Re}^{-1}\left[r^{-1}\frac{\partial}{\partial r}\left(r\frac{\partial M}{\partial r} - 2M\right) + \frac{\partial^2 M}{\partial z^2}\right] + \quad (3)$$
$$N\left[rB_z\frac{\partial \Phi}{\partial r} - B_z^2 M\right]$$

$$\frac{\partial \Theta}{\partial t} + div\,(\vec{v}\Theta) =$$
$$\mathrm{Re}^{-1}\mathrm{Pr}^{-1}\left[r^{-1}\frac{\partial}{\partial r}\left(r\frac{\partial \Theta}{\partial r}\right) + \frac{\partial^2 \Theta}{\partial z^2}\right] \quad (4)$$

In equations (1) - (4) the function Φ is the potential of an electric field; ξ and Ψ are connected with velocity vector $\vec{V}=(V_r, V_\varphi, V_z)$ by relations

$$\xi = r^{-1}\left(\frac{\partial V_r}{\partial z} - \frac{\partial V_z}{\partial r}\right);$$
$$V_r = r^{-1}\frac{\partial \Psi}{\partial z};\ V_z = -r^{-1}\frac{\partial \Psi}{\partial r}.$$

The equations are written in dimensionless variables: the crystal radius R_k is chosen as a length scale, velocity scale V = 0.3146 cm/sec (rotational velocity of a crystal edge with radius 1.5 cm at container rotational velocity 2 rot/min), temperature scale 1K. Dimensionless parameters that characterize the problem are: the Reynolds number Re, Grashof number Gr, Prandtle number Pr, Hartmann number Ha, MHD interaction parameter $N = Ha^2/Re$.

They are determined according to the general formulae. The boundary conditions

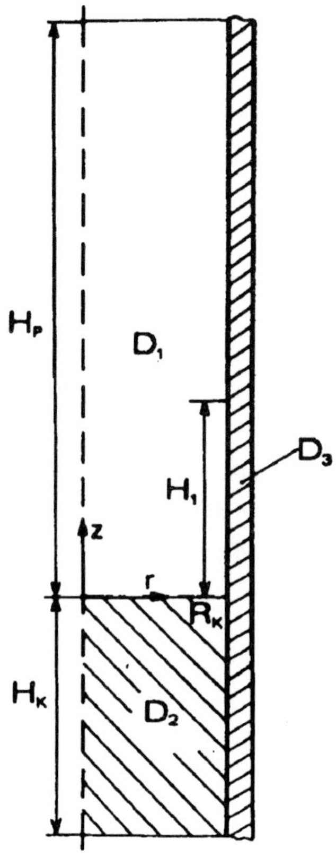

Figure 1. *Formulation of problem*

for (1) - (3) are the following ones

$$\Psi|_\Gamma = 0;\ \frac{\partial \Psi}{\partial n}|_\Gamma = 0$$
$$\xi|_\Gamma = 0;\ M|_\Gamma = 0 \quad (5)$$

where Γ is the boundary of melt with a container and crystal. The boundary conditions for equation (4) are the followingones:

$$\Theta|_{\Gamma}=0; \quad \Theta|_{z=H_p}=15;$$
$$\frac{\partial\Theta}{\partial r}|_{r=R_k}=0; \quad (0\le z\le H_1) \tag{6}$$
$$\Theta|_{r=R_k}=C_1(z-H_1)^2+C_2(z-H)+10$$
$$(H_1<z<H_p)$$

where C_1, C_2, H_p are empiric constants. The equations which determine the electric current and electric field potential Φ , associated with it, have been calculated in the D region. Considering the thermoelectric effect, the Ohm's law in a magnetic field can be written as follows:

$$\frac{\vec{j}}{\sigma}=\vec{\nabla}\Phi+\vec{V}\times\vec{B}-\alpha\vec{\nabla}\Theta \tag{7}$$

where α - the coefficient of the thermo-e.m.f. Numerical calculations have been made due to (Ref.5), which considers the thermoelectric effect on a melt-container boundary. A thermoelectric effect on a crystal-container boundary at the given boundary conditions (Ref.6) have been neglected. The effects of Peltiet, Thompson and Joule heat are ignored. The system of equations (1) - (4) with the relating appropriated boundary conditions has been numerically solved by the method of finite differences. The area was covered by a rectangular, non-uniform grid with step smaller towards the boundaries. A conservative approximation of differential equations was used, convective terms were approximated on an extended nine-point grid (Ref.6). The equations were solved step by step; the scheme for every equation due to a corresponding variable was explicit, the finite-difference equations have been solved by the ORTOMIN method (Ref.7). The calculations have been done for germanium single crystals growth in a graphite container with radius R_k = 1.5 cm, crystal height H_k = 3 cm, melt height H_p = 9cm. The temperature distribution in melt is such that at $0 < z < H_1/R_k = 2$ the thermal flow near the melt boundary equals zero ($\frac{\partial\Theta}{\partial r}=0$), but at $H_1< z <H_p$ the temperature varies from $\Theta = 10$ up to $\Theta = 15$. Thermal and physical characteristics in solid and liquid phases have been used at the basic criteria definition: Re = 465, Pr = 0.016, Gr = 1.7 . 10^5 g = 9.81 m/sec^2; $\sigma_k/\sigma = 0.11$ is the ratio of solid and liquid phases conductivities; the ratio of the container material (graphite) conductivity to the germanium conductivity in a liquid phase is $\sigma_c/\sigma = 0.07$. Since the effective temperature difference $\Theta_{max} = 15$, then the effective Grashof number,determining the motion, is $Gr_{ef} = Gr\Theta_{max}$. (Besides, at $g < g_o$ the effective Grashof number $Gr_{ef} = Gr\Theta_{max}\, g/g_o$). B varies from 0.05 up to 0.2 T,

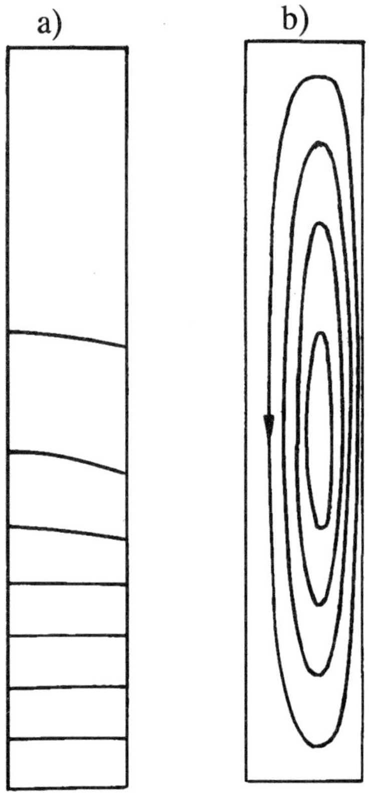

Figure 2. *Isotherms (a) and isolines (b) of the current function in melt at* $g = 10^{-2}\, g_o$ *and* $B = 0.05$ *T and without TEMC (isotherms and isolines are shown in equal intervals,* Θ_{max} = 15, Ψ_{max} = 0.00168).

Ha - from 30 up to 120. According to the fact

that the temperature of a solid phase Θ = *const,* relating coefficient of thermo-e.m.f. α has been considered on a container-melt boundary $\alpha = -1.2\ 10^{-5}$ V/K.

3. RESULTS AND DISCUSSION

Let us analyze the calculation results, first, for the case when g < < g_o (g_o = 0.0981 m/sec^2) in the absence of the TEMC (Figure 2-3). Figures 2 - 3 show the data for a temperature field in melt, Figures 2b-3b - the isolines of current function Ψ. It is seen that in this case there exist a single vortex motion in the melt with a rather small (low) intensity that is caused by a residual level of gravity. The increase of a longitudinal magnetic field leads to the flow intensivity suppression at its structure possession (Figures 2-3). The molecular heat transfer dominates in the melt already at B = 0. So, practically there is no magnetic field effect on the temperature field (see Figures 2-3). Figures 4,5 depict the analogous results for the TEMC at g = 0 (i.e. in the absence of the thermogravitational convection) for B = 0.1 and 0.2 T (d/R_k = 0.28). Figure 4a and 5a show the isolines of the melt azimuthal rotational velocity, Figures 4b-5b - the isotherms in the melt; Figures 4c-5c - the current functions of a meridional flow. The occurrence of azimuthal velocity of melt rotation V_φ should be noted. It is associated with the existence of a thermocurrent radial component in the melt and its interaction with an axial component of the magnetic field induction. V_ϕ maximal value occurs in the region of comparatively small values of B = 0.02-0.03 T. At B increase the melt rotation intensivity decreases. The mentioned effect influences the intensitivity of the

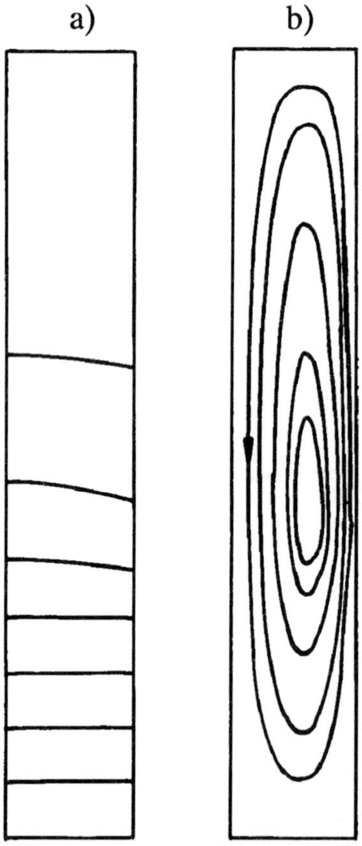

Figure 3. *Isotherms (a) and isolines (b) of the current function at g* $=10^{-2}g_o$, *B* = *0.2 T and without TEMC* (Θ_{max} = *15,* Ψ_{max} = *0.000416).*

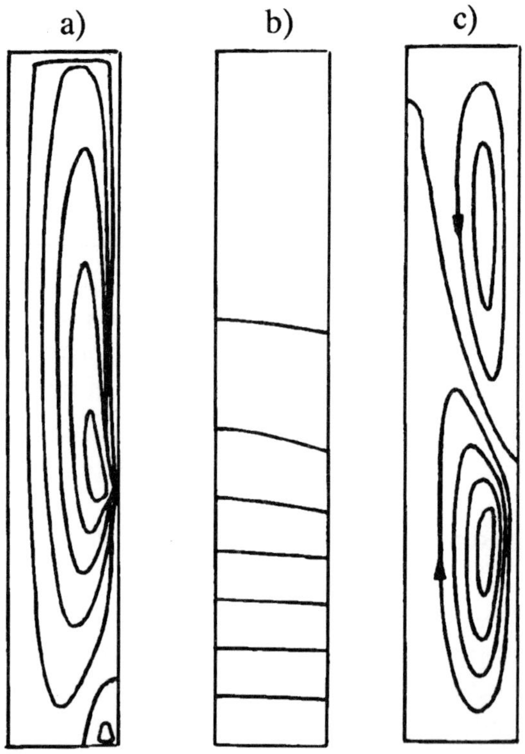

Figure 4. *Isolines of the azimuthal velocity of the melt rotation (a), isotherms (b) and isolines (c) of the current function in melt at g* = *0, B* = *0.1 T* ($V_{\varphi max}$ = *1.2;* Θ_{max} = *15;* Ψ_{max} = *0.0085;* Ψ_{min} = *-0.011).*

secondary meridional melt flow (Figures 4c,5c) as well as the temperature field structure - Figures 4b,5b. Comparison with the calculated results at small g ($g = 0.0981$ m/sec^2) (see Figures 2-4) shows that the secondary meridional flow in the TEMC presence significantly differs (30-100 times) in structure (a two-vortex structure) and intensivity from the meridional flows at $g << g_0$. The estimation proves that at an abrupt change of the graphite container wall thickness from $d/R_k = 2.8\ 10^{-1}$ down to $d/R_k = 2.8\ 10^{-3}$ the TEMC intensivity remains rather high and can be compared with the TGC. Thus, the data obtained prove that the TEMC should be accounted for in experiments and calculation, when the growth process is being carried out in graphitizied quartz ampoules even with a very thin conducting layer on the ampoule walls ($d = 4\ 10^{-2}$ mm). In conclusion we analyze the effect of the TEMC and TGC simultaneous interaction in germanium single crystal growth by the method of directional crystallization in a graphite container with wall thickness $d/R_k = 0.28$ in a vertical magnetic field. As it has been mentioned above, the highest TEMC intensivity is reached in the region of comparatively weak magnetic fields. So, Figures 6a,b,c and 7a,b,c depict the isolines of the melt azimuthal velocity rotation (a), isotherms (b) and isolines of the current function (c) of a meridional flow. It is seen that at B = 0.05 T (see Figure 6) the TEMC with a characteristic structure of a secondary flow (Figure 6c) and rather high velocity of melt rotation (6-7 rot/min (Figure 6a)) dominates in melt in comparison with the case of the diffusional heat transfer (see Figure 3a). Increasing the magnetic field induction results in the TEMC suppression in com-

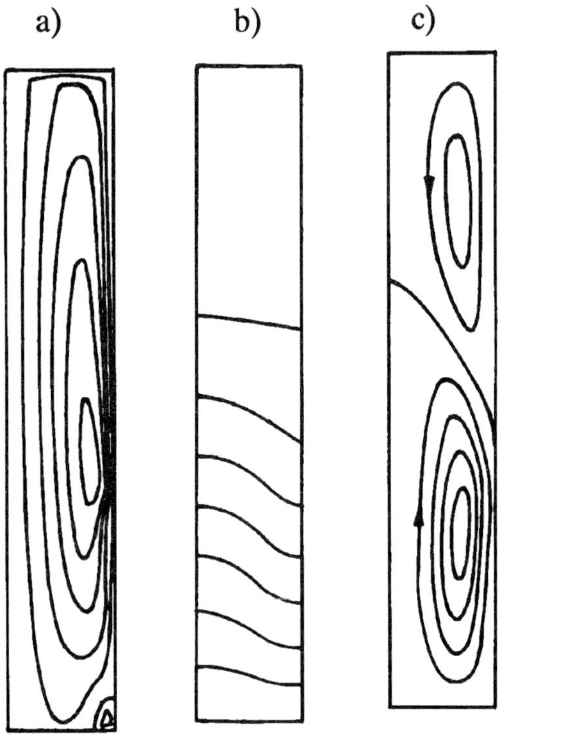

Figure 5. *Isolines (a) of the azimuthal velocity of melt rotation, isotherms (b) and isolines (c) of the current function in melt at g = 0, B = 0.2 T* ($V\varphi_{max} = 2.3$; $\Theta = 15$; $\Psi_{max} = 0.072$, $\Psi_{min} = -0.12$).

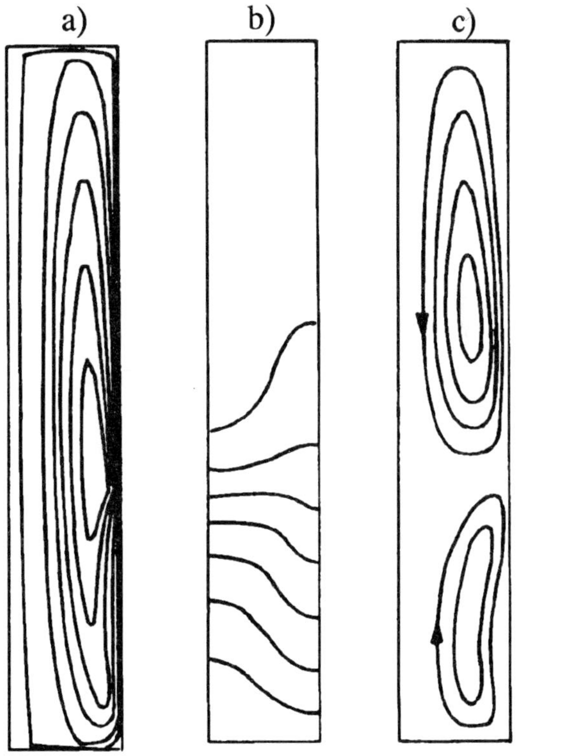

Figure 6. *Isolines (a) of the azimuthal velocity of melt rotation, isotherms (b) and isolines (c) of the current function in melt at g = g₀, B = 0.05 T* ($V\varphi_{max} = 4.6$; $\Theta_{max} = 15$; $\Psi_{max} = 0.42$; $\Psi_{min} = -0.19$).

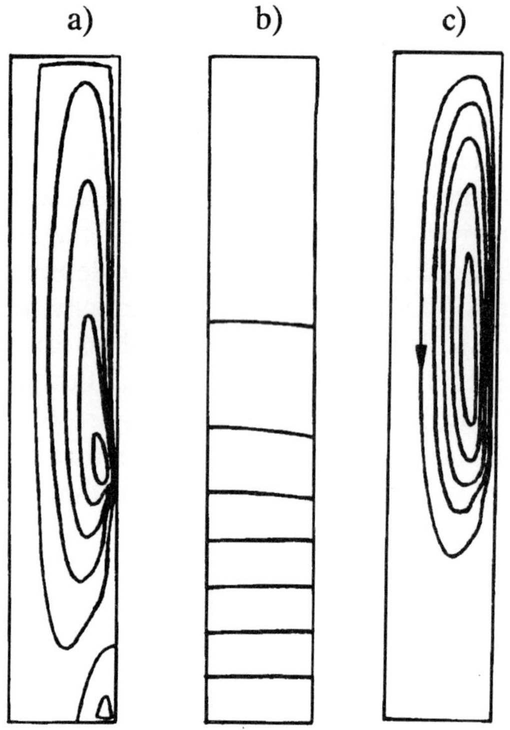

Figure 7. *Isolines (a) of the azimuthal velocity of melt rotation, (b) isoterms and isolines (c) of the current function in melt at* $g = g_0$ $B = 0.2$ *T* ($V_{\varphi max} = 1.2; Q_{max} = 15; \Psi_{max} = 0.0186$).

parison with the TGC. As a result, at B = 0.2 T the flow structure is similar to the case when B = 0. Such TEMC dependence on B stresses that the TEMC may strongly affect the process at small B and especially at small g. As the analysis shows, weak magnetic fields from e.g. heaters, or power supply can affect the process of growth. The results obtained prove that such effects should be accounted for in real processes.

References

1. G,Sarma, Space Technoligy (Moscow,1980), p.73-79.
2. A.Bojarevics and L.A.Gorbunov, Hydrodynamics and Heat Transfer at Material Production (Moscow, Nauka, 1990), p.89-102.
3. L.A.Gorbunov, Magnetohydrodynamics, 4, (1987).
4. Yu.M.Gelfgat and L.A.Gorbunov The 4th USSR Seminar on Hydrodynamics and Mass Transfer under Microgravity (Novosibirsk, 1987), p.51-52.
5. L.A.Gorbunov and E.D.Lumkis, Magnetohydrodynamics, 2, (1990).
6. L.A.Goncharov and I.V.Fryazinov, USSR Academy of Sciences (Moscow, 1986), p.93.
7. L.Heigeman and D.Yang, Applied Itteration Methods (Moscow, Nauka, 1986, 445 p.).

MATHEMATICAL MODELLING OF CONVECTION DURING CRYSTAL GROWTH BY THE THM

Alexander S. Senchenkov

SPLAV Technical Center, 9 Baikalskaya St., 107497, Moscow, Russia

Igor V. Friazinov and Marina P. Zabelina

Keldysh Institute of Applied Mathematics, Russian Academy of Sciences, 4 Miusskaya Sq., 126047, Moscow, Russia

ABSTRACT

The convection in a solution zone during a crystal growth by the Travelling Heater Method under microgravity and terrestrial conditions is investigated numerically. The influence of various magnetic fields (constant axial and rotating) is studied as well. Microgravity conditions and the presence of the constant magnetic field lead to the increase of the crystallization front curvature. To minimize the crystallization front curvature under microgravity conditions a rotating magnetic field could be used. Thus the magnetic field should be used for the effective control of the heat and mass transfer processes during crystal growth from solutions.

Keywords: *Crystal growth, Travelling Heater Method, Mathematical modelling, Microgravity, Magnetic field.*

1. INTRODUCTION

We consider a solution zone which is formed as a result of dissolving a compound AB in a solvent B (e.g. CdTe in Te). The solvent B is placed between a solid feed rod and a seed crystal. The solution zone is heated by a heater,which is moved along the ampoule axis with a velocity V. The relative displacement of the heater with respect to the solution zone results in the difference of the temperature ΔT in the feed interface Γ_p and the seed interface Γ_m as shown in Figure 1. Dissolution of the polycrystalline feed is caused by the undersaturation of the solution and the crystal growth is caused in the seed interface by a supercooling of the solution. This technique is known as the Travelling Heater Method (THM).

Our model describes buoyancy convection in a vertical cylindrical configuration. The influence of an external applied constant or rotating magnetic field is considered too. The problem under consideration is axisymmetric.

2. MATHEMATICAL FORMULATION

The flow is governed by Navier-Stokes equations with Boussinesq approximation. The governing equations for the temperature and solute concentration are the equations for conservation of energy and mass. The values of the magnetic forces are determined using the Maxwell equations.

It could be shown that in the case of the constant magnetic field the deviation of the disturbed magnetic field from the external field is very small. Therefore, the Lorentz force term which has to be added to the momentum equation can be

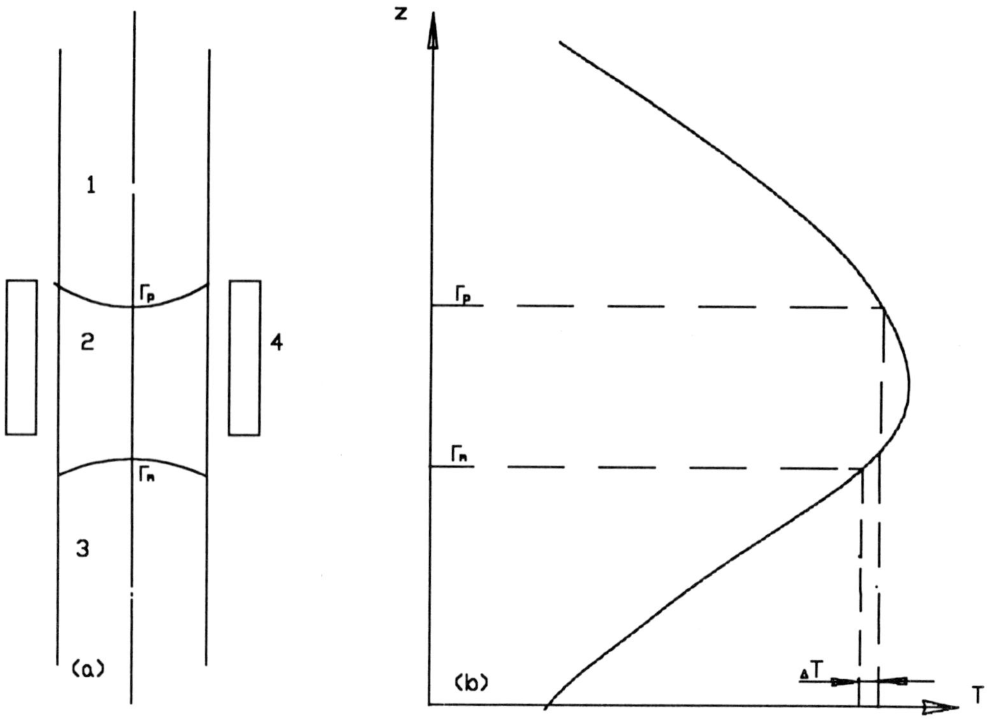

Figure 1. Schematic diagram of the THM (a) and temperature distribution (b)
1 - feed, 2 - solution, 3 - crystal, 4 - heater

written in our cases using the given external magnetic field value.

We assume that the heater velocity is small enough so all of the variables take some quasistationary values depending on the position of the heater, the external forces and so on. The problem was solved for various gravity levels and magnetic field strength.

We have employed a finite-difference method. An irregular adaptive grid was used. The implicit monotonous conservative scheme was build on the basis of balance correlation for the functions in each elemental cell (Ref.1). The solution of the implicit finite-difference equations was fulfilled by an over-relaxation method with the local-optimal choice of the iterative parameters (Ref.2).

3. CALCULATION RESULTS

The calculations were carried out for the conditions corresponding to the experiment with CdTe-solution in tellurium solvent on ZONA-4 facility (Ref.3). The main experiment parameters are given in Table 1.

TABLE.1 Calculations data

Crystal diameter, mm	15
Tellurium disk thickness, mm	8
Growth temperature, ^{o}C	750
Heater velocity, mm h^{-1}	0.2

On earth the crystallization front is almost plane and the solute concentration is almost constant along the crystallization front due to a strong convective flow. There is one toroidal vortex in the fluid zone. The fluid is moved up at the periphery and down at the axis. The moving results in a temperature increase at the crystallization front in the central part of the sample and the temperature decrease near the ampoule wall. The fluid flow makes the solution composition more homogeneous at the growing interface and the crystallization front coincides practically with an isotherm. The crystal-solution interface has a concave shape in the central part of the sample and a convex one in the periphery.

Microgravity conditions ($g/g_0=10^{-3}$) lead to an appreciable change of the crystallization front shape. The vortex becomes weaker and the flow velocity decreases by two orders of magnitude. The temperature field is determined now largely by thermal conductivity. The zone length decreases both at the axis and at the periphery. The form of the growing interface becomes convex respective to the solution.

The effect of a constant magnetic field is qualitatively similar, but very strong magnetic fields are required for obtaining the same quantitative results. We have obtained a convex crystallization front only when the magnetic induction was more than 0.5 T. Figure 2 shows the form of the solution zone and streamlines at various values of the magnetic induction B_c under terrestrial conditions.

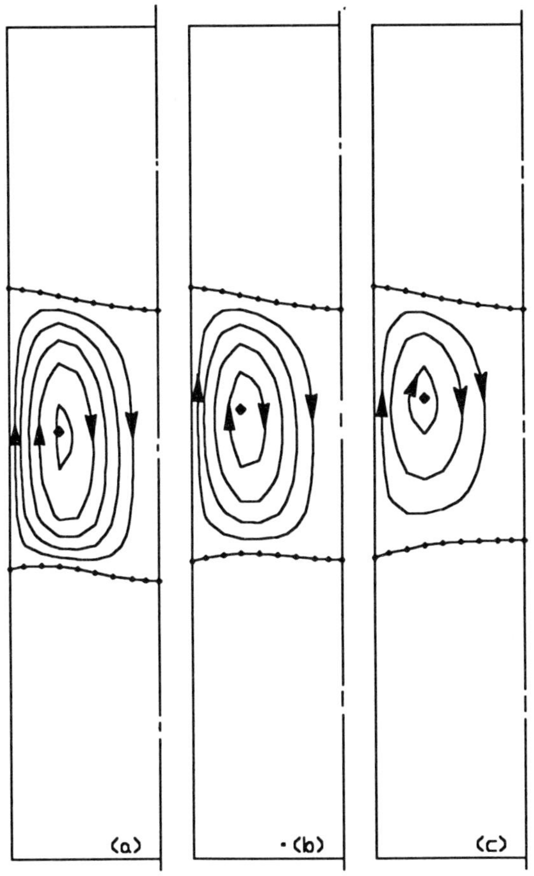

Figure 2 The solution zone and streamlines for various constant magnetic fields (a): $B_c=0$ T, (b): 0.3, (c): 0.6

It can be seen that the increase of the magnetic induction results in an increase of the crystallization front curvature and a decrease of the flow velocity. The influence of the constant magnetic field on the crystallization front shape is shown at Figure 3.

The dependence of the maximum flow velocity on the magnetic induction is given in Table 2. It can be seen that the fluid velocity is decreased by a factor of 2 in the presence of the magnetic field at induction 0.6 T as compared to the case of the magnetic field absence.

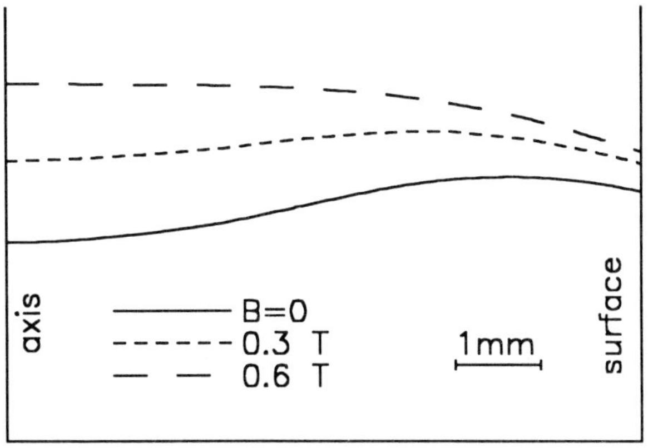

Figure 3. Variation of the crystallization front shape with the magnetic field strength.

To minimize the crystallization front curvature under microgravity conditions a rotating magnetic field could be used. In this case we can chose the magnetic induction strength such that the flow is laminar and steady.

The rotating magnetic field intensifies the flow remarkably in a meridional plane. As it can be seen in Figure 4, the application of the rotating magnetic field under microgravity conditions leads to a change of the flow pattern. At the dissolving interface the flow induced by the rotating magnetic field has the same direction as compared to the one driven by the buoyancy convection. At the crystallization front these flows are directed opposite to each other. Therefore a second vortex arises near the crystallization front at the magnetic induction B_ω of about 0.7 mT.

TABLE 2 Influence of the constant magnetic field on the flow velocity

Magnetic induction B_c ,T	Maximum velocity, cm s^{-1}	
	Radial	Axial
0	1.03	1.81
0.3	0.70	1.49
0.6	0.48	0.89

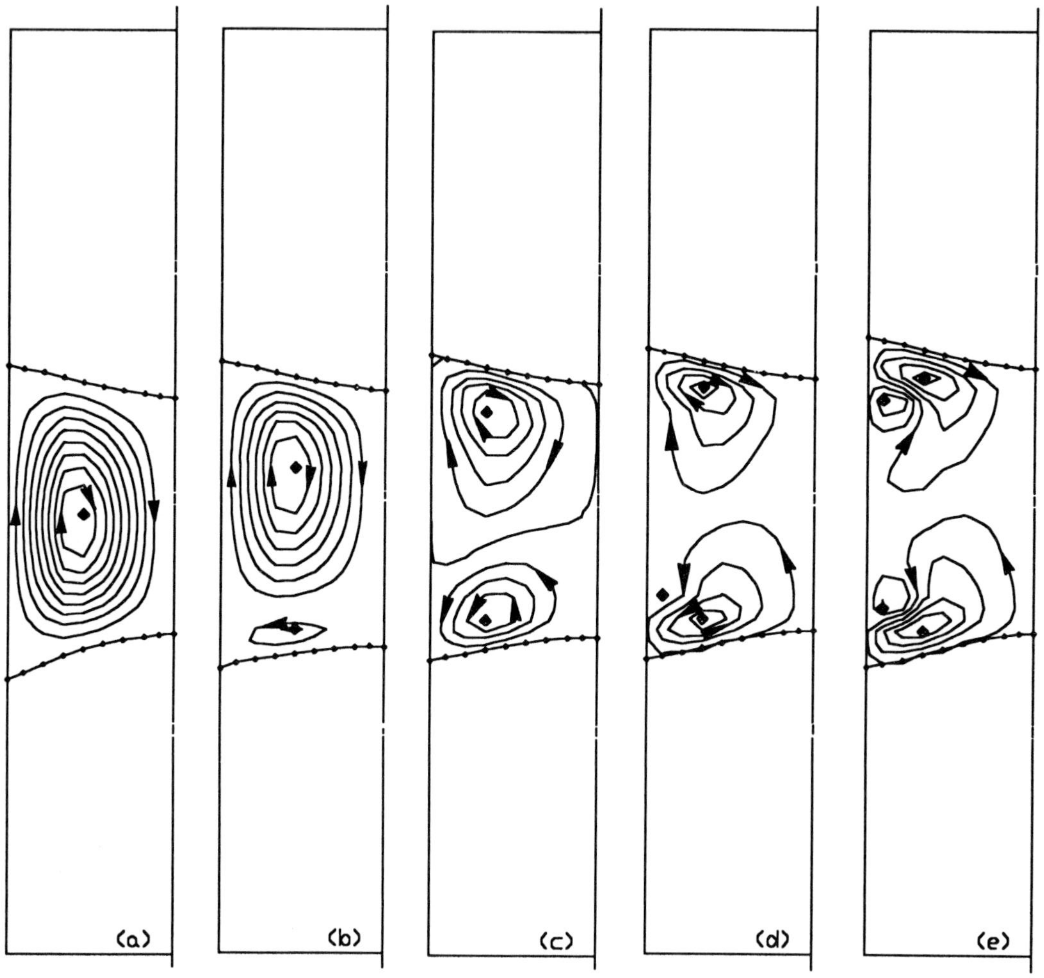

Figure 4. Flow pattern in meridional plane for the various rotating magnetic field strength at gravity level of 10^{-3} g_0

(a): B_ω =0 , (b): 0.7, (c): 2, (d): 4, (e): 8 mT

An appreciable deformation of the crystal-solution interface takes place in this case. The interface has a convex shape with small curvature (Figure 5). When the magnetic induction increase the flow induced becomes stronger as compared to the one driven by natural convection. At the crystallization front the fluid is now moved in a direction from the ampoule wall to the axis. The interface shape becomes more convex again but the curvature is essentially less than the one in the absence of a rotating magnetic field. If the magnetic induction B_ω is about 4 mT, third vortex is appeared near ampoule wall and for B_ω=8 mT we have four symmetric vortices. The influence of the buoyancy convection is negligibly small in this case. The crystal-solution interface is faintly deformed when the magnetic induction increases from 2 to 8 mT.

Table 3 shows the effect of the rotating magnetic field on the hydrodynamics of the fluid in microgravity. The data demonstrate remarkable acceleration of mass transfer in the solution zone in presence of the rotating magnetic field. The axial velocity is increased by factor 10 for B_ω=8 mT.

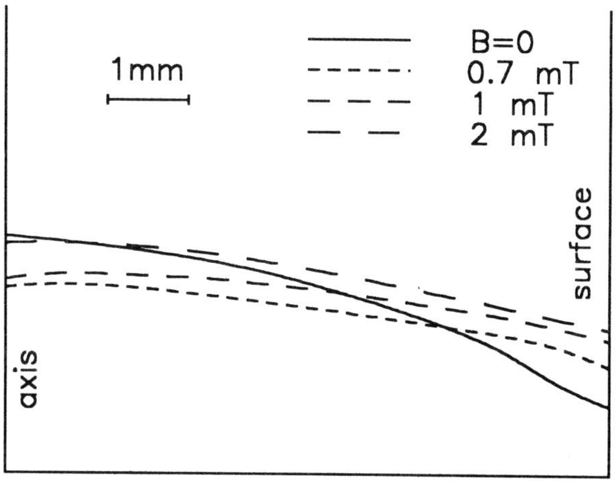

Figure 5. Variation of the crystallization front shape with the rotating magnetic field strength under microgravity conditions.

TABLE 3 Influence of the rotating magnetic field on the flow

Magnetic induction, B_ω *,mT*	*Maximal velocity,cm* s^{-1}	
	Azimuthal	*Axial*
0	0	0.0279
0.5	0.0393	0.0282
1	0.112	0.0328
2	0.323	0.0600
4	0.871	0.1180
8	2.060	0.2710

We have presented here the results in case of residual accelerations level of 10^{-3} g_0. With reduction of gravity the effect of the rotating magnetic field will be more considerable.

4. CONCLUSION

The calculations performed show a possibility to use constant and rotating magnetic fields for effective control of the crystal growth process by THM both under terrestrial and microgravity conditions.

5. REFERENCES

1. A.A.Samarsky, *Theorija raznostnykh skhem*(Nauka, Moskva 1977). In Russian.

2. M.P.Zabelina and I.V.Friazinov, *Differenzialjnye uravneniya*, **23**, 1188 (1977). In Russian.

3. I.V.Barmin, A.V.Jegorov, E.I.Gorjunov, E.S.Ivanchenko, A.V.Kotov, Ju.P.Perfiljev and A.S.Senchenkov, *Oborudovanie kosmicheskogo proizvodstva* (Mashinostroenie, Moskva 1989). In Russian.

CRYSTALLIZATION OF TWO–PHASE SYSTEMS UNDER CONDITIONS OF COMPENSATION OF GRAVITY – INDUCED LAMINATION BY ELECTROMAGNETIC FORCES

J. Yu. Chashechkina , D.B.Orlov , M.Z.Sorkin[1) , O.V.Abramov

Institute of Solid State Physics Academy of Sciences of the USSR , Moscow District , Chernogolovka , 142432 , USSR

[1) *Institute of Physics of Latvian Academy of Sciences , Riga region , Salaspils–1 , 229021, Latvija*

ABSTRACT

A magnetohydrodynamic (MHD) technique of simulating microgravity in an electroconducting medium has been developed. The technique is based on the selective effect of activated electromagnetic volume forces (EMVF) on separate components of a systema as a function of the electrical conducticonductivity. The installation for crystallizing the melts of the Al-Pb immiscible system, in the liquid state, has been constructed to realize the method. The Al-Pb system is an important component when new highperformance antifriction alloys are developed. The Al-Pb alloys, produced by means of the above mentioned installation, possess homogeneous structure and high antifriction properties. Both the method and the materials obtained are recommendable for realization in industry.

Keywords: *Gravity–Induced Lamination, Electromagnetic Forces, Magnetohydrodynamic Method, Quasi-Weightlessness, Immiscible, Structure Homogeneity, Al-Pb.*

1. INTRODUCTION

At present the most essential objective of materials science is the development of new composite materials, consisting of components, laminating substantially under ordinary ("earth") conditions. That is, the components are immiscible in the liquid state. The perspective to create such materials, possessing unique properties, initiated the performance of a number of experiments for obtaining compositions of the "fixed" emulsion type, such as Zn–Pb, Al–Pb, Al–In, Au–Rh, Al–Bi during space–flight (Refs.1-3). It was supposed that under the conditions of microgravity one could manage to decrease lamination when alloys with different component densities are crystallized and it would lead to their more homogeneous distribution in volume. However the results of the performed experiments were far from successful: spatially homogeneous distribution of components in the material was not achieved. Further development of space investigations is restricted because of complexity and high cost. It convinces one of the actuality to develop methods enabling to simulate microgravity under earth conditions.

2. EXPERIMENTAL RESULTS

The magnetohydrodynamic (MHD) method of simulating weightlessness conditions

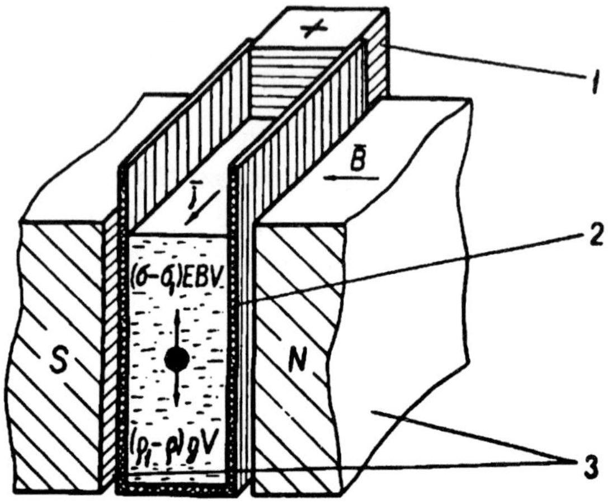

Figure 1: The principal scheme of the MHD-method realization to control the distribution of phases in separating liquid metallic systems: 1 — electrode, 2 — crucible with melt, 3 — poles of electromagnet.

(quasi-weightlessness) in electroconducting media (Ref.4) is of a significant interest. The method is based on the selective effect of electromagnetic volume forces (EMVF) in the melt on separate components of a system as a function of the electrical conductivity and on the possibilities of the EMVF arbitrary orientation in space. In this case the opportunity of realizing the simulated state of indifferent equilibrium of the components arises under earth conditions like as it is in space. To estimate the behaviour of the second phase under these conditions the Frenkel–Zeldovitch equation and the Smolukhovsky formula for coagulation of colloid solutions were used (Ref.5). The analysis reveals that the precipitation of particles is described by the equation.

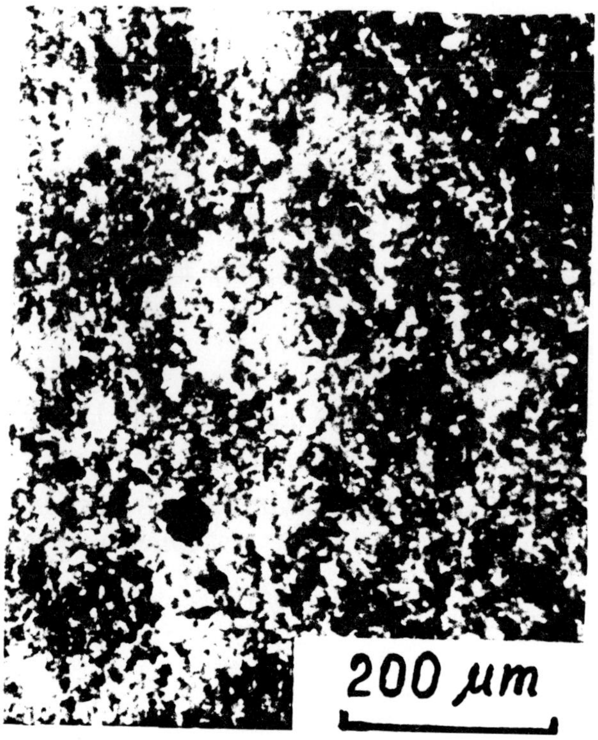

Figure 2: The microstructure of an alloy Al – 13% Pb (in weight).

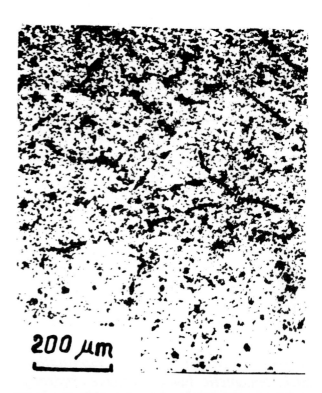

Figure 3: The microstructure of a cast alloy Al – 10% Pb – 1% Sn – 1% Cu.

$$\frac{8}{9}\frac{\rho}{\rho_1}\frac{dR_e}{d} = -R_e + \frac{32}{9}\frac{R^3 g}{\nu_1^2}\frac{\rho_1 - \rho}{\rho_1}, \quad (1)$$

where $R_e = R\nu/\nu_1$, $\tau = t\nu_1/2R^2$; g is the acceleration of gravity, ν is the velocity of the second phase particle, ρ_1 is the liquid density, R is the particle radius of the second phase, ν_1 is the viscosity of liquid, t is the time of precipitation of a particle.

From the analysis of the above mentioned expression the conditions of the electromagnetic treatment of melt, avoiding the sedimentation processes, were determined. The principal scheme of the MHD–method is presented in Figure 1. The scheme was used both for calculations and for constructing the installation of a laboratory type.

To realize this method an experimental installation was constructed. It enables us to obtain materials of the Al–Pb system, possessing a wide range of immiscibility in the liquid state. This specific system is effective for the development of antifriction alloys. When exposing the crystallized melt to electromagnetic forces and optimizing the treatment modes, ingots containing Pb up to 15% in bulk are produced. The finely dispersed Pb inclusions ($\approx 50\mu m$) are homogeneously distributed in an ingot bulk. The alloy micro-

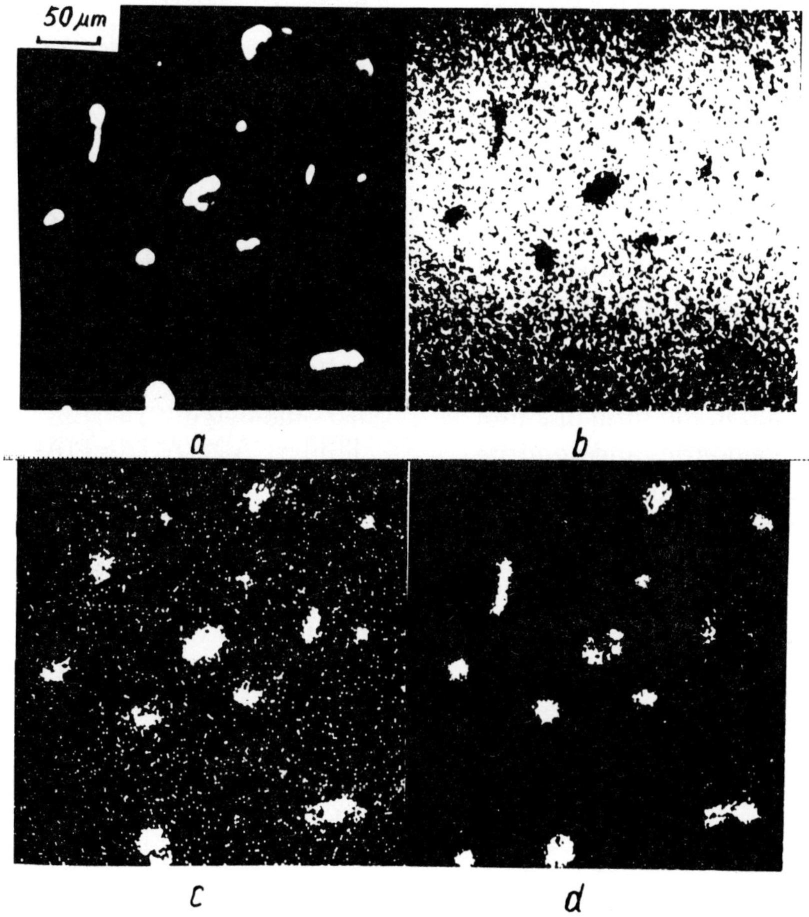

Figure 4: The microstructure of a cast alloy Al – 10% Pb – 1% Sn – 1% Cu:
a photographed in reflected electrons;
b — "— in X–ray AlK_α;
c — "— in X–ray PbM_α;
d — "— in X–ray SnL_α.

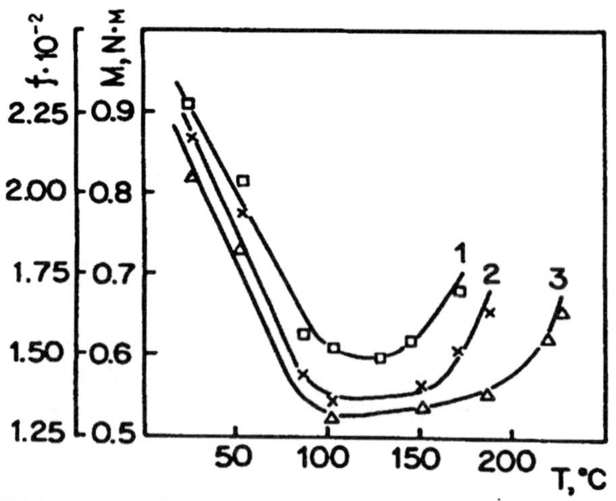

Figure 5: The change of a moment of friction (*M*) and of a coefficient of friction (*f*) as a function of engine lubricating oil–temperature at the friction of alloys coupled with steel.
1—Alloy AO20–1 (Al–20%Sn –1%Cu);
2—Alloy Al –10%Pb –1%Sn –1%Cu;
3—Alloy Al –12%Pb –10%Sn –1%Cu.

structure is presented in Figure 2.

Figure 3 displays the alloy microstructure on the basis of Al–Pb system with tin, copper and antimony components. An ingot alloy is an aluminum matrix in which the easily melted component is distributed in the form of a thin network along grain boundaries and dendrite cells of a basic aluminum phase. There are homogeneously distributed dispersed inclusions of $\approx 5 \div 30 \mu m$ in size.

According to the data of the local X–ray spectral analysis (Figure 4), the small inclusions, contained in an aluminum matrix, are enriched in lead and tin. It should be noted that alloying of Al–Pb system does not cause principal changes of Pb distribution.

The obtained aluminum–lead alloys show high antifriction quality due to structure homogeneity (Ref.6).

Figure 5 shows the change in the moment of friction and the friction coefficient *f*, calculated on the basis of the moment of friction values, as a function of engine lubricating

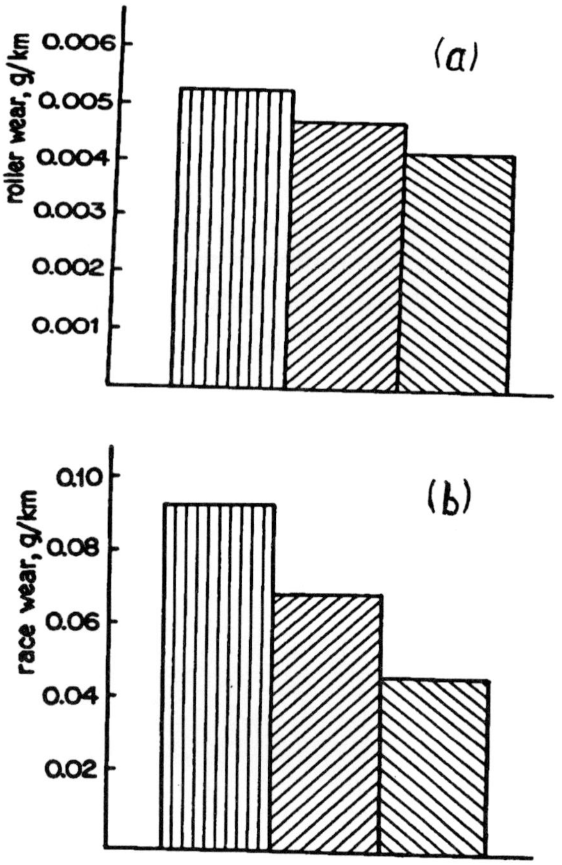

Figure 6: The diagrams of wear: **(a)** the steel roller wear; **(b)** the wear of race made of investigated alloys.
(vertical hatching) – Alloy AO20–1 (Al–20% Sn–1% Cu)
(diagonal hatching) – Alloy Al–10%Pb–1%Sn–1%Cu
(reverse diagonal hatching) – Alloy Al–12%Pb–10%Sn–1%Cu

oil–temperature under trials in the mode of liquid friction of experimental Al–Pb alloys and industrial ones on the basis of aluminum and tin.

The highest critical temperature of the transition to boundary friction is observed for aluminum alloys, containing 12% of lead (in weight).

Figure 6 presents the results of trials in the mode of dry friction of a couple, consisting of a race, produced of industrial and experimental alloys and a steel roller. The trial results demonstrate the fact that the Al–Pb alloys are more durable than the Al–Sn ones.

3. CONCLUSIONS

The Al–Pb materials produced according to industrial technology were used to manufacture sliding bearings for Diesel engines. The results of the trials were successful, so that both the method of production and the produced materials can be recommended for realization in industry.

4. REFERENCES

1. J.L.Reger in *Interim Report to NASA Marshall Space Flight Center, Contract NASA — 28267*, May (1973).
2. G.H.Otto in *Final Report to NASA, Contract NASA — 27809*, January (1976).
3. J.B.Andrews, C.J.Briggs and M.B.Robinson in *Proceedings VIIth European Symposium on Materials and Fluid Sciences in Microgravity*, Oxford, UK, September (1989), ESA SP-295 January (1990).
4. O.V.Abramov, Yu.M.Gel'fgat, S.I.Semin, M.Z.Sorkin and J.Yu.Chashechkina, *Fiz. Khim. Obrab. Mater.* 3 (1981) 47.
5. Ya.I.Frenkel in *Vvedenie v teoriyu metallov*, L.: Nauka (1972).
6. O.V.Abramov, J.Yu.Chashechkina and D.B. Orlov in *6th Int. Symp. Compos. Metal., Vysoke Tatry Stara Lesna*, October (1986) 2, Bratislava (1986) 361.

NUMERICAL SIMULATION OF SOLUTION CONVECTION ABOVE THE SURFACE OF A GROWING CRYSTAL AT VARYING GRAVITY

V.A.Brailovskaya, V.V.Zil'berberg and L.V.Feoktistova

Institut of Applied Physics AS USSR, Nizhny Novgorod

ABSTRACT

In order to way searching to improve of quality of aqueous-soluble monocrystals numerical simulation of solution hydrodinamic and mass exchanges above the surface of growing crystal are investigated.

To create the homogeneity of concentration in boundary layer near crystal surface the regime with low gravity or suppresing of natural convection by forced one are used.

Keywords: Aqueous-soluble Crystal. Numerical Simulation, Low Gravity, Natural and Forced Convection, Crystal Growth Velosity.

1.INTRODUCTION

Numerical simulation of the solution dynamic above the surface of a growing water-soluble crystal is of great applied importance. It allows for the investigation of mechanisms regulating the velocity of the crystal growth and its quality. The quality of a grown crystal is determined, in particular, by homogeneity of a concentration boundary layer above the surface of a growing crystal. Such an homogeneity can be found in crystals grown under conditions of low gravity (Ref.1) when the effect of natural convection is reduced. Another way to suppress natural convection is to create forced convection which suppresses the natural one.

Natural and forced convections in a solution above the surface of a growing crystal at different values of the gravity parameter are modeled in this paper.

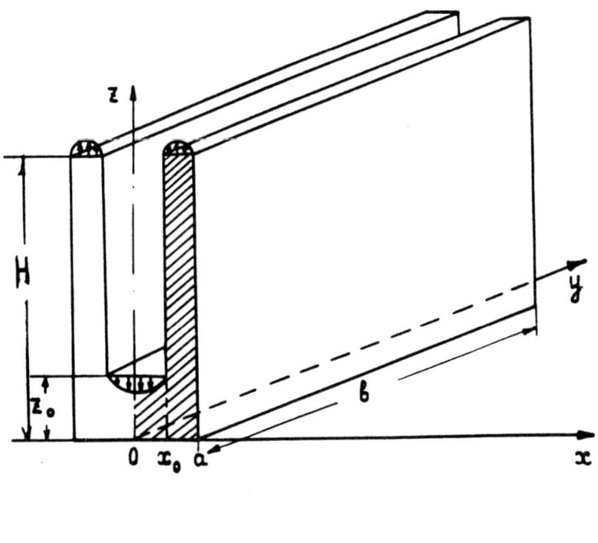

Figure 1: The calculation scheme.

2. FORMULATION OF THE PROBLEM.

We consider the solution flow and mass exchange above the crystal surface in a crystallization chamber which is schematically given in Figure 1.

At b>>a,H the solution flow can be assumed two-dimensional and is limited by the solution of the problem in (XZ) plane; here only the region x>0 (dashed region in Figure 1) is considered from the condition of symmetry with respect to (YOZ) plane.

The solution is fed into the chamber through the cross-section $z=z_o$, $0<x<x_o$ and is removed from it through the cross - section z=H, $x_o<x<a$. The state of the solution in each point is described with the variables $\psi(x,z)$ stream function, $\omega(x,z)$ vorticity, $C(x,z)$ concentration of salt in the solution, which satisfy the following equation (Ref. 2).

$$\omega_t + u\omega_x + v\omega_z = \omega_{xx} + \omega_{zz} + F \qquad (1)$$

$$\omega=\psi_{xx}+\psi_{zz} \qquad (2)$$

$$C_t+uC_x+vC_z=(C_{xx}+C_{zz})/Sc \qquad (3)$$

$$F=-Gr_D C_x \qquad (4)$$

$$u=\psi_z, \quad v=\psi_x \qquad (5)$$

In equations (1)-(5) the height of the crystallization chamber H is taken as a unit of length

$$C=(Cp-C^*(T_1))/(C^*(T_2)-C^*(T_1))$$

Cp is "dimensional" concentration (kg of salt/ kg of solution); $C^*(T_1)$ is "dimensional" concentration of salt in the solution saturated at T_1 ; T_2 is solution temperature in the crystallizer; T2 is temperature of solution saturation

$Sc=\nu/D$ - Schmidt number

$Gr_D=(g\beta(C^*(T_2)-C^*(T_1))H^3)/\nu^2$ - Grashof number

ν is dynamic viscosity, D is diffusion coefficient, β_c concentration expansion coefficient.

As initial condition we consider $\psi(x,z)=0$, $C(x,z)=1$ in entier region.

Boundary conditions.
At rigid boundaries of the region the no-slip condition is met while symmetry conditions are fulfilled on the axis.

$$\psi(x,0)=\psi(x,1)=\psi(0,z)=\psi(a,z)=0$$

$$\psi_z(x,0)=\psi_z(x,1)=\psi_x(a,z)=0, \quad w(0,z)=0$$

In the section $z=z_0$, $0<x<x_0$, where the solution is fed into the chamber,

$$\psi(x,z_0)=-(V_0/2x_0)x^2-V_0x, \quad \psi(x,z)=0,$$

where V_0 is maximum velocity of the solution in flow. At $z=1,(x_0\le x\le a)$ is is solution outflow from the chamber.

$$\psi(x,1)=V_0x_0/(a-x_0)^2\{(x_0-x)^2-(x_0-a)^2/2\}$$

$$\psi(x,1)=0, \quad x_0\le x\le (x_0+a)/2$$

Boundary conditions for concentration C (x,z) are:

$$C_x(0,z)=C_x(a,z)=0$$

and at the solution-crystal boundary (z=0):

$$C_z(x,0)=\text{const}, \qquad C(x_c,0)= RG/\beta,$$

where x_c is coordinate of the growth center (further we assume $x_c=0$), RG is normal growth velocity,

$$RG=2D\rho_{H_2O}/a\,\rho_0 \int^a \partial/\partial z(C_P/(1-C_P))\big|_{z=0}\,dx$$

β is a temperature-dependent constant which caharacterizes the power of the growth center on the crystal surface. Here the proximity of C(x,0) to unity characterizes a degree of "kineticity" of the crystal growth regime.

3. METHOD OF THE NUMERICAL SOLUTION

To numerically solve the initial system of equations (1)-(5) using the method of finite differences we turned to a scheme of variable directions. Poisson equation was solved with fast Fourier transform. Time integration was carried out according to an implicit two-layer scheme of the first order accuracy. We also used monotonous approximation of spatial derivatives on an nonuniform mesh with Samarsky mesh scheme viscosity compensation. Nonuniform mesh with appreciable thickening near the boundary layer at solution-crystal boundary were applied (minimum mesh spacing along z is 0.001)

4. RESULTS

Calculations were made with those parameters of D, ν, β, β_c which are typical for KDP (KH_2PO_4) crystals. Figure 2 gives the basic results of numerical study of the solution dynamics above the surface of a growing crystal. Figure 2a corresponds to the case when $g=0, V_0=0.001$ cm/c, while two other figures are done for terrestrial g and the velocities V_0: 0.001 cm/c (2b), 0.4 cm/c (2c). The upper picture in each figure is isoline of the stream function in all the chamber while isolines of ψ and concentration C in a 0.05*H - deep boundary layer are shown in lower pictures. In Figure 2a natural convection is absent while in Figure 2b it dominates above the forced one (in the absence of forced convection the calculations in this region were obtained in (Ref. 4)).

Figure 2c shows a situation contrary to that in Figure 2b (that is, the forced convection dominates above the natural one). Each figure also gives rate velocity RG of the crystal growth and solution concentration near the growth center C(0,0). Further increase of forced convection velocity (as compared with Figure 2c) is useless, i. e., it does not result in essential

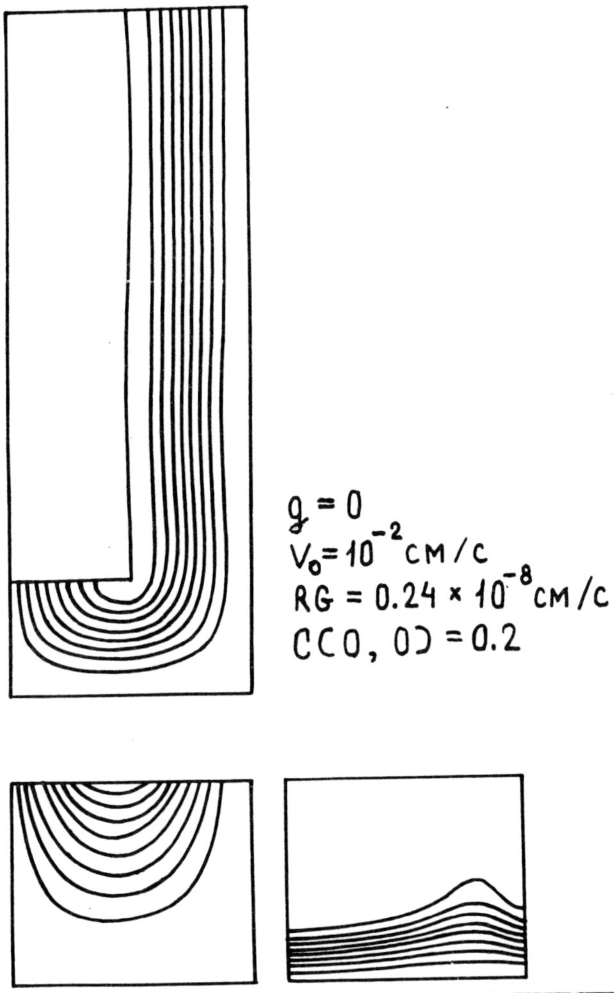

Figure 2a

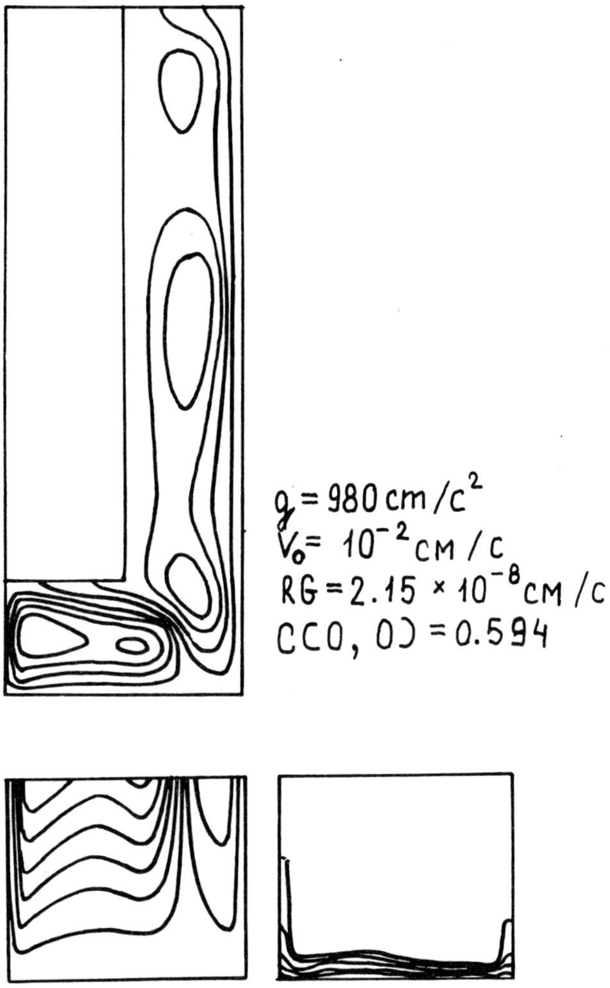

Figure 2b

enhancement of the crystal growth rate but increases inhomogeneity of the boundary layer.

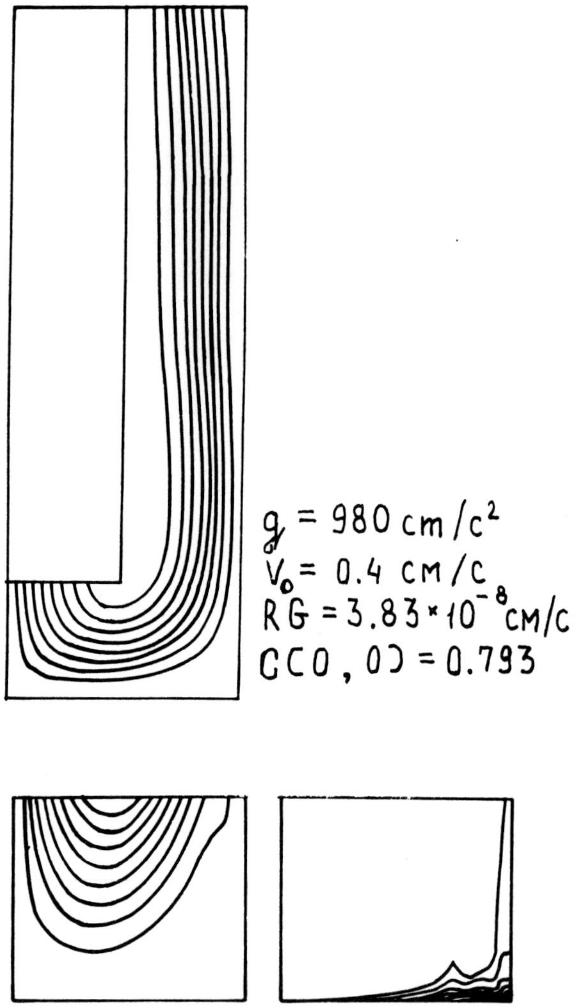

Figure 2c.

5. CONCLUSIONS

i It is evident that both the decrease of gravity and intensification of forced solution flow essentially change homogeneity of the boundary layer (as compared to the case of natural convection).

ii Investigations directed to optimization of the regime of solution inflow into the crystallization chamber seem rather promising.

REFERENCES

1. A.Authier, K.W.Benz, M.C.Robert, F.Wallrafen, Crystal Growth from Solution.//In "Fluid Sciences and Materials Sciences in Space" by edition H.U.Walter, Springer-Verlag (1988) 436

2. V.I.Polezhaev, Hydromechanics, Heat and Mass Exchange in Crystal Growth. //Itogi Nauki i Techniki, Mehanika Zhidkosti i Gasa, M., VINITI,18 (1984) 198 (in Russian)

3. V.A.Brailovskaya,G.A.Galushkina, V.V.Zil'berberg and L.V.Feoktistova, Simulation of Solution Natural Convection above the Surface of a Growing Crystal, // In "Hydromechanics and Heat Exchange in Material Production" M.,Science (1990) 120 (in Russian).

4. V.A.Brailovskaya, V.V.Zil'berberg and L.V.Feoktistova, Numerical Simulation Mass Exchange Processingin Regeneration of Crystal Growing from Aqueous Solution, These of 11 All-Union Conference

HEAT TRANSFER SIMULATION DURING THE GROWTH OF A HGI2 CRYSTAL FROM VAPOR PHASE AT LOW TEMPERATURES.

A. ROUX*, P. BONTOUX*, A. FEDOSEYEV** and R. SANI***

*Institut de Mécanique des Fluides, Marseille, FRANCE.
**Institute for Problems in Mechanics, Moscow, USSR.
***Center for Low Gravity Fluid Mechanics, Boulder, USA.

Abstract

We consider the modeling of heat flux and temperature fields in a sealed growth ampoule in order to control the growth of a mercuric iodide (HgI_2) crystal from vapor phase and at low temperatures. Under usual growth conditions, the pressure inside the ampoule is very small. Then, most of the computations concern radiative and conductive heat transfers. The computations are performed using finite element methods. The influence of several parameters such as the geometries of the source and the crystal, the thermal boundary conditions, the undercooling, the conductivity anisotropy of the crystal and the possible transparency or opacity of the source and the crystal, is discussed. Numerical simulations are relevant of experimental situations investigated in the team of Prof. Kaldis at ETH-Zürich.

Key Words: HgI2, Crystal Growth, Heat Transfer, Radiation, Numerical Simulation.

1. INTRODUCTION

Single crystals of mercuric iodide have high technological interest as detectors of X- and gamma-rays at room temperature. HgI2 crystals are grown from a vapor phase in closed, evacuated ampoules, at approximately 120°C. Here, the crystal growing ampoule has a cylindrical shape. The bottom surface of the ampoule is indented so that internally it forms a pedestal on which the crystal grows. The indentation fits over a metal support in copper. The thermal profile around the ampoule necessary for crystal growth is provided by six independently controlled heaters. The toroidal source is slowly heated and a cristallization is obtained on the pedestal. A range of ground experiments show that for a given undercooling the crystal size increases during few hours then stops and that a new undercooling is necessary to restart the growth process. These steps probably correspond to an equilibrium between the temperatures at the crystal/gas and source/gas interfaces, leading to an equilibrium of the partial pressures inside the ampoule and to the vanishing of the sursaturation necessary to the growth. The purpose of our numerical simulation is to help for a better understanding of this HgI2 crystal growth experiment.

2. ASSUMPTIONS

We consider a purely axisymmetric configuration with the following assumptions: uniform physical properties over the surfaces, gas transparent to radiation (non-participating), opaque internal pyrex walls, grey and diffuse radiating surfaces, transparent or opaque source and crystal. Former tests based on comparison with experimental temperature measurements in HgI2 vapor filled ampoule have shown that, for guessed internal pressure (p=0.2 torrs), convection was negligible.

3. EQUATION AND BOUNDARY CONDITIONS

Then, for the following tests, the convection is neglected and the problem is governed by the energy equation only. We have:

$$0 = \frac{1}{r}\frac{\partial}{\partial r}\left(kr\frac{\partial T}{\partial r}\right) + \frac{\partial}{\partial z}\left(k\frac{\partial T}{\partial z}\right) + q \qquad (1)$$

where q is a source term.
We have defined six major regions along the walls of the ampoule connected to the six

independently controlled heaters: the upper zone (u), the source zone (s), the lower zone (l), the bottom zone (b), the pedestal zone (p) and the cold finger zone (cf). In each region, we dispose of one temperature measurement. As an example, the following temperatures have been measured by Piechotka[1]:

$$Tu=120.6°C,\ Ts=117°C,\ Tl=120.2°C,\ Tb=122°C,\ Tp=116.1°C,\ Tcf=115.3°C \quad (2)$$

We have decided to set these six temperature values as boundary conditions at the locations where they are measured experimentally. The walls are assumed to be insulated elsewhere.

For tests taking into account radiation, the heat exchange relationship between radiating boundaries is written as follows:

$$\sum_{j=1}^{N}\left(\frac{\delta_{ij}}{\varepsilon_j}-F_{ij}\frac{1-\varepsilon_j}{\varepsilon_j}\right)q_{rj}=\sum_{j=1}^{N}(\delta_{ij}-F_{ij})\sigma T_j^4 \quad (3)$$

where δ_{ij} is the Kronecker symbol, ε_j is the emissivity of the surface A_j, q_{rj} is a part of the term q of the energy equation (1) and where F_{i-j} is a viewfactor which represents the fraction of diffuse radiant energy leaving a surface A_i and which directely falls upon a surface A_j (see Figure 1). By definition, we have:

$$F_{ij}=\frac{1}{A_i}\int_{A_i}\int_{A_j}\frac{\cos\phi_i\cos\phi_j}{\pi r^2}dA_i dA_j \quad (4)$$

As we know, the emissive power W of a non black surface at temperature T radiating to the hemispherical region above it, is written:

$$W=\varepsilon W_b=\varepsilon\sigma T^4 \quad (5)$$

with ε, the emissivity; $\sigma=5.5598\ W/m^2K^4$, the Boltzman constant and Wb, the emissive power of a black surface. When incident energy (Q_i) falls on a surface, it can be absorbed (Q_a) or reflected (Q_r) or transmitted (Q_t) through the material. We can write:

$$Q_i = Q_r + Q_a + Q_t \quad (6)$$

and divide by Q_i to obtain:

$$1=\rho+\alpha+\tau \quad (7)$$

where ρ, α, τ are respectively, the reflectivity, absorptivity and transmitivity.

For an opaque material:

$$\tau=0 \Longrightarrow \rho=1-\alpha \quad (8)$$

For a diffusive grey surface:

$$\varepsilon=\alpha \quad (9)$$

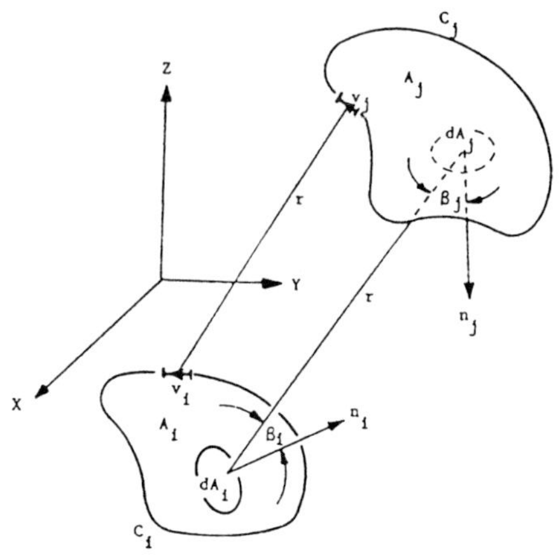

Figure 1. View factor calculation (from ref. 3).

4. NUMERICAL RESULTS

The preliminary computations are performed using two codes FEMINA[2] and FIDAP[3], both based on a finite element method (see reference manuals for the details on the modeling and on the numerical method). Associated computational domains, built with triangular or quadrilateral elements, are shown on Figure 2.

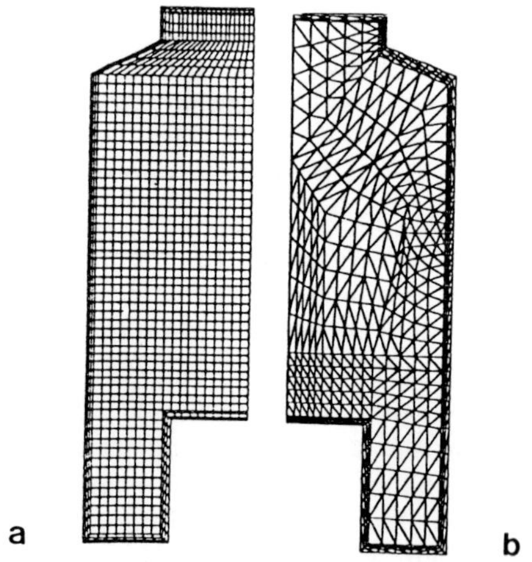

Figure 2. Computational domains with quadrilateral or triangular elements. (a) FIDAP code, (b) FEMINA code.

All the results presented in this section have been obtained with FIDAP code, which allow to solve radiative exchanges inside enclosures of complex geometry (hidden walls, etc...). These results concern computed temperature fields which are represented by plotting isotherms, with thirty levels. With FEMINA code, validation has been made.

Three different models have been considered. The first one concerns the *pure conduction* (PC), the second one the *conduction combined with radiation for transparent source and crystal* (CRT), the third one concerns the *conduction combined with radiation for opaque source and crystal* (CRO). As the actual emissivities of the source and the crystal are unknown, we test several values from 0 to 1.

The purpose of the study is to test these three models and their results on the thermal field inside the ampoule; particularly, check the occurence (and determine the position) of an "equilibrium" isotherm between the source/gas and crystal/gas interfaces that should be associated with the stop of the growing process.

In this approach, three idealized shapes of the source and the crystal are considered (rectangular sections) in order to display three growth stages (see Figure 3): the beginning of growth (source only, no crystal), the midterm of growth (half source, half crystal) and the end of growth (no source, crystal only). The rectangular models satisfied globally the mass conservation during the process. At each stage, the isotherm pattern is displayed and an "equilibrium" isotherm is identified.

This "equilibrium" isotherm in an actual experiment would correspond to the interfaces of the source and the crystal. Indeed, all the part under this isotherm should tend to crystallize and all the part upper this isotherm should tend to evaporate. At least, a crystal with a shape delimited by this isotherm is expected. Shaded areas are displayed (see Figures 5,6,7) in order to give an idea of the actual shapes. The influence of the anisotropic crystal thermal conductivity is also studied.

4.1. The PC model

The PC model leads to a large difference between temperatures at the source/gas and crystal/gas interfaces compared to other models taking into account the radiation (see Figures 4a,5a,6a). In fact, it is shown by decreasing the temperature value of the cold finger, noted Tcf (simulation of several undercoolings), that this temperature difference never tends to zero considering the experimental conditions. So, with a PC model, it's impossible to reach an equilibrium between temperatures at the source/gas and crystal/gas interfaces and the growth should never stop.

4.2. The CRT model

With the CRT model, a radiation coming from a pyrex/gas interface can go through the source or the crystal and be reflected at an other pyrex/gas interface. For this model, the difference between temperatures at the source/gas and crystal/gas interfaces is smaller (see Figures 4b,5b,6b) than for the PC model.

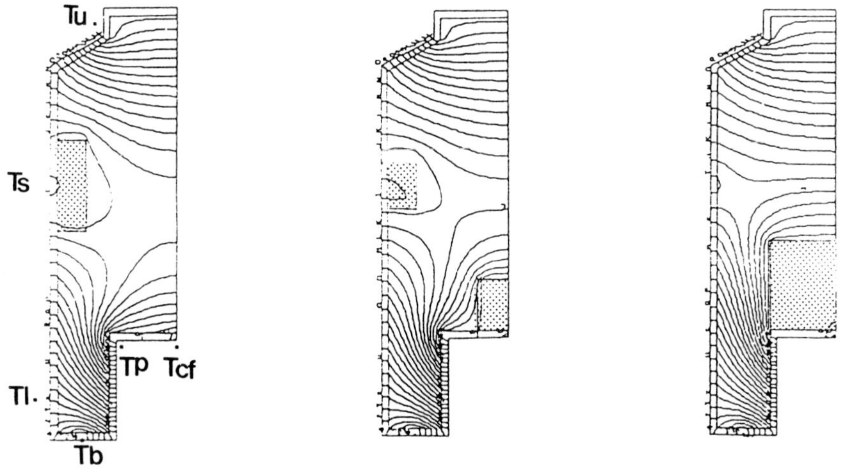

Figure 3. Different source and crystal geometries. (a) begin of the growth, (b) midterm of the growth, (c) end of the growth.

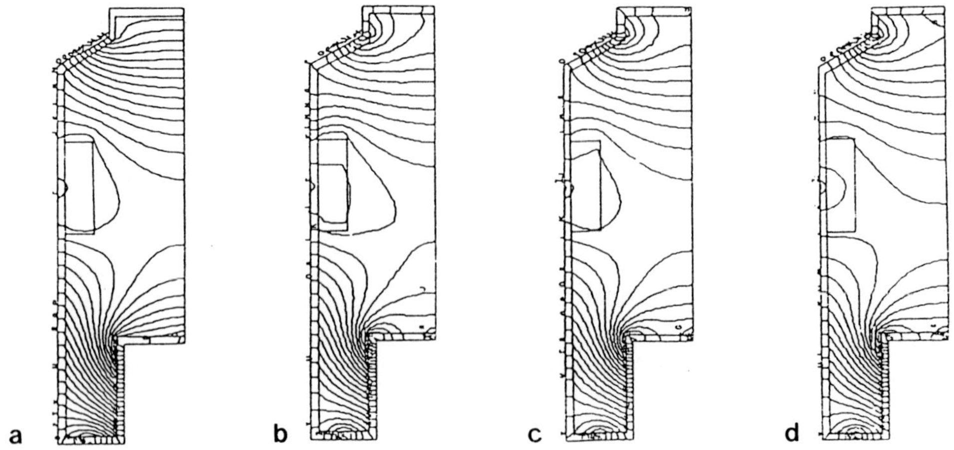

Figure 4. Influence of the three models on the thermal field at the beginning of the growth. (a) pure conduction, (b) transparent source, (c) opaque source with εs=0, (d) opaque source with εs=1.

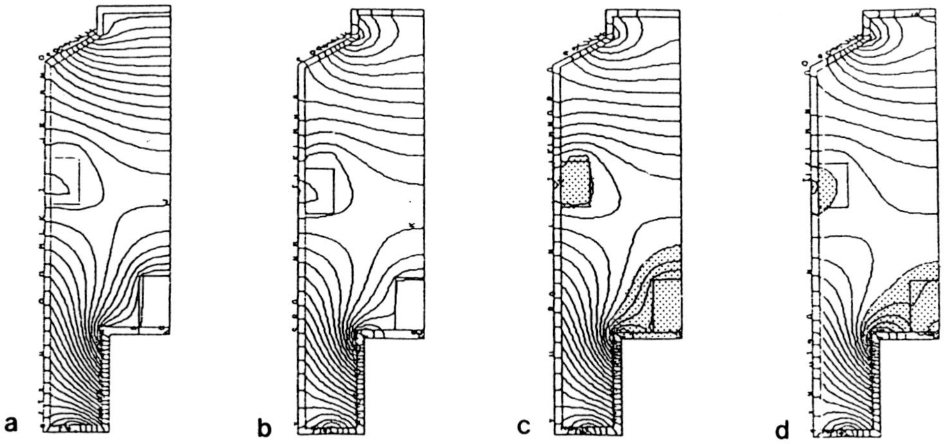

Figure 5. Influence of the three models on the thermal field at the midterm of the growth. (a) pure conduction, (b) transparent source and crystal, (c) opaque source and crystal with εs=εc=0, (d) opaque source and crystal with εs=εc=1.

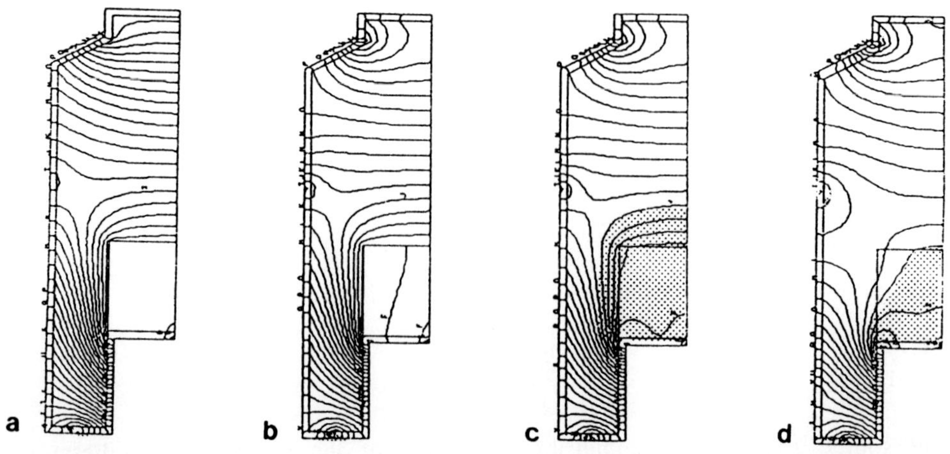

Figure 6. Influence of the three models on the thermal field at the end of the growth. (a) pure conduction, (b) transparent crystal, (c) opaque crystal with εc=0, (d) opaque crystal with εc=1.

4.3. The CR0 model

But, the most interesting results are obtained for the CRO model. In this model, a radiation coming from a pyrex/gas interface can't go through the source or the crystal and is reflected at the source/gas or the crystal/gas interface.

4.3.1. Influence of the emissivity

When the emissivities of the source or the crystal are increased from 0 to 1 (i.e. when their absorbtivities are increased, see relation (9)), the temperatures along the source/gas and crystal/gas interfaces become hotter. Indeed, the crystal region become hotter and the "equilibrium" isotherm moves inside the crystal (see Figures 5c,d and 6c,d). We observe that for $\varepsilon c=0$ (see Figure 6c), the shaded area would lead to a volume of the crystal larger than the assumed initial rectangular one. Differently, for $\varepsilon c=1$ (see Figure 6d), the "equilibrium" isotherm limits an area of volume roughly equal to the initial one. The (shaded) form displayed in Figure 6d is actually found in the experiments performed on ground. This suggests that the crystal emissivity should be closer to $\varepsilon c=1$ than to $\varepsilon c=0$.

4.3.2. Crystal conductivity's effect

As the crystal has an anisotropic conductivity, the effect of the radial conductivity of the crystal on the thermal field is investigated for the situation of the end of the growth. The axial and radial conductivities, respectively noted kz and kr, are assumed to be constant. The two cases kr=kz and kr=kz/5 are considered. With the PC model, the CRT model and the CRO model when $\varepsilon c=0$, no significant effect is observed. But, the CRO model with $\varepsilon c=1$ exhibits strong effects for a high emissivity (see Figure 7). Indeed the "equilibrium" isotherms have very different but very realistic shapes. This suggests that the more realistic results would be obtained with a more real 2D axisymetric case (kr constant in the plane perpendicular to z-axis and when kz=kr/5.

5. CONCLUSION

The PC and CRT models are shown to be insufficient for modeling correctly the major feature of the vapor growth technique. However, CRO model bring more consistent results comparable with experimental observations made in the team of Prof. Kaldis at ETH-Zürich. The limitation of the analysis results from the lack of physical data; in particular the values of the source and the crystal emissivities are not given in the literature. Then, the effect of the emissivity was investigated parametrically. The results tend to suggest that the realistic values for the crystal emissivity would be about 0.5 to 1. The imposition of the thermal boundary conditions should be improved. Presently, six temperature measurements only are used. In the future, a more complete modeling of the temperature distribution is scheduled for pedestal region. Also, we will consider the influence of parameters as the position of the growth axis, the cold finger radius, the emissivity...

6. ACKNOWLEDGEMENT

This research is supported by ESA, contract n° 8431/89/F/BZ. The authors thank E. Kaldis, M. Piechotka, M.Zha (ETH Zürich), P. Behrmann and O.Minster (ESTEC) for fruitful discussions.

7. REFERENCES

1.M. PIECHOTKA, M. ZHA and E. KALDIS, 1991, private communication.
2.A.FEDOSEYEV,FEMINA code, IPM AS USSR, Moscow,1986,65P(VITINI N1447-B86).
3.M.S. ENGELMAN, FIDAP Users Manual, Fluid Dynamics, Evanston, IL, 1986.

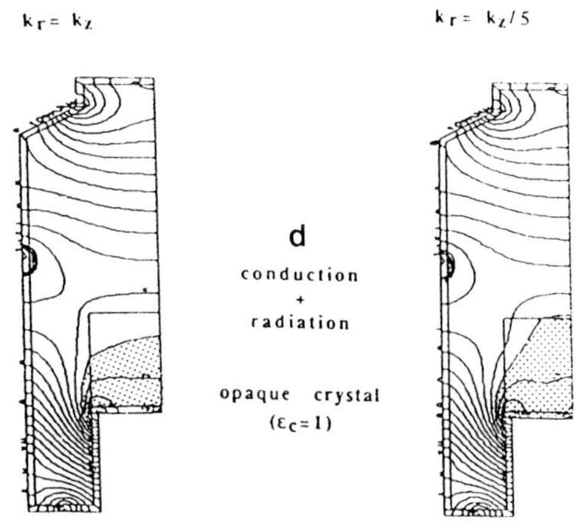

Figure 7. Influence of the radial crystal conductivity for an opaque crystal with $\varepsilon c=1$.

COMPUTATION OF THERMAL FIELDS IN THE SPACE FURNACE "CRYSTALLIZATOR" (ČSK-1)

A.I. Fedoseyev, V.I. Polezhaev, A.I. Prostomolotov
Institute for Problems in Mechanics AS USSR, Vernadsky 101, 117526 Moscow, USSR
E.V.Chernyaev, I.I.Petrenko, S.V.Purtov
Institute for High Energy Physics, 142284 Protvino, Moscow region, USSR
Č.Barta, A.Triska
Institute of Physics ČS AS, 180 40 Praha 8 - Libeň, NaSlovance 2, ČSFR

Abstract

The problems arising in the mathematical simulation of furnaces are discussed. Multilayer furnace design with materials of different thermal properties and temperature-dependent thermal conductivities is treated. Requirements on the software for the solution of this class of problems are analyzed. The mathematical problem is presented together with the solution technique based on a finite element method. The results of a numerical simulation of the thermal furnace ČSK-1 for crystal growth under microgravity are presented and compared with the experimental data obtained in the stationary mode. The influence of the design features and thermal convection flows on the process are analyzed.

The main characteristics of the program package MGD2 developed and used for numerical simulation are presented. The discussion of the efficiency of multigrid domain decomposition technique used in MGD2 for the solution of nonlinear problems is also presented.

Keywords: *Space Furnace Design, Heat Transfer, Finite Element Method, Multigrid Method, Numerical simulation, Efficient Software for PC.*

1 INTRODUCTION

The main requirements imposed on the space-furnace design and development are:

- to provide a required space-time thermal operational mode;
- to determine and take into account factors affecting temperature field in the furnace;
- to provide an uniform temperature gradient and isothermal zones [3, 4];
- to provide remote control and safe service;
- to minimize the furnace volume, its weight and power.

These requirements are analysed by numerical simulation and computer calculations to optimize the furnace configuration.

The available refs.[1, 2, 3, 4, 5, 6] on microgravity furnaces have not addressed numerical methods of heat transfer calculations. The most recent papers [7, 8, 9] study the crystal growth under earth gravity conditions in Czochralski and Bridgman with forced convection and heat radiation and compare the calculated thermal profiles with experimental observations.

The furnace design is impossible without an appropriate mathematical model which includes different ways of heat transfer. Usually the computational model is rather complicated and bulky. So, the choice of suitable computational techniques and application programs is important.

To calculate the heat transfer in furnaces we use a numerical method based on the finite element technique. The efficiency of Multigrid Method and the transferability of the MGD2-program package [11] allow an engineer to use them at personal computers of the IBM PC/AT family.

This paper presents the computations of the furnace "Crystallizator/ČSK-1 "[1, 4, 10, 12]. This furnace has been employed for crystal growth experiments under microgravity by the Physical Institute of Čhescoslovensko AS, Praha.

A specific feature of the ČSK-1 furnace consists in multilayer thermal shielding made from materials with thermal parameters differing by three orders of magnitude and, moreover, being temperature-dependent.

The contributions of convective and radiative heat flows are estimated also.

2 PROBLEM FORMULATION

2.1 The Construction of Crystallizer ČSK-1

The major units of the furnace which determine the temperature field are shown in Figure 1. Five resistive heaters A-E are used whose power and locations are listed in Table 1. Four materials are

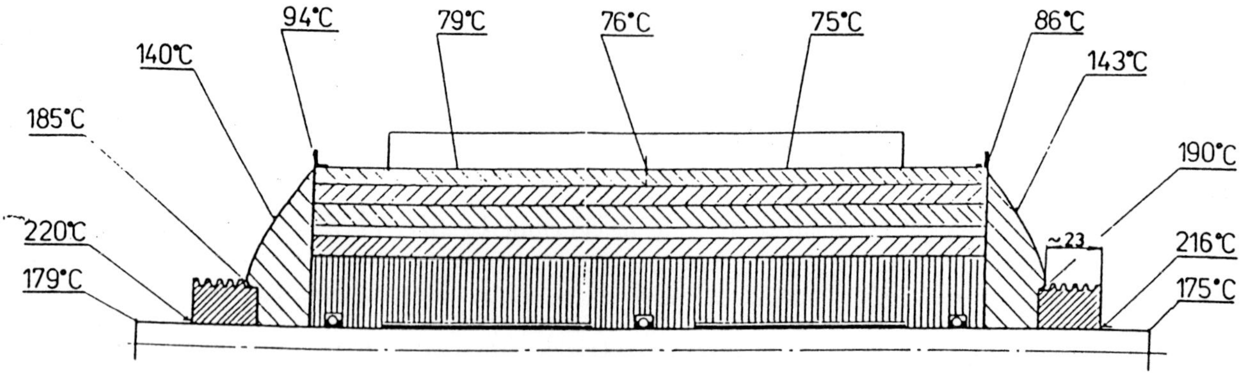

Figure 1: The temperature values on the surface of furnace ČSK-1

used: steel, corundum, air, felt-like Al_2O_3. Temperature dependence of thermal conductivity $k(\theta)$ is linearly interpolated between points listed in Table 2.

Table 1: Power of heaters at the temperature 850 C^o

Heater	Power [W]	Position z, r [mm]	
A	53.1	-124 ÷-116	11 ÷ 15
B	27.1	-100 ÷ -20	11 ÷ 11.5
C	16.0	-4 ÷ 4	11 ÷ 15
D	28.7	20 ÷ 100	11 ÷ 11.5
E	53.9	116 ÷ 124	11 ÷ 15

Table 2: Temperature-dependent thermal conductivity of materials is used

Material	Thermal conductivity $k(\theta)$ [W/M/K]				
T, C^o	200	400	600	800	1000
Al_2O_3	.050	.078	.108	.148	.195
Air	.039	.052	.062	.072	.081
Steel	12.	14.	16.	18.	20.
Corundum	2.40	2.33	2.26	2.19	2.12

2.2 Mathematical Model

The equation governing the axisymmetric temperature field $\theta(r,z)$ is the steady-state heat conduction Equation

$$div(k(\theta)grad\ \theta) + q = 0, \qquad (1)$$

where $q(r,z)$ is the heat source term. To complete the mathematical model, we still need the appropriate boundary conditions (2). There are three possible types of boundary conditions at the surface S:

$$\begin{aligned} \theta\Big|_{\Gamma_1} &= \bar\theta(S), && \text{with given temperature} \\ -k(\theta)\frac{\partial\theta}{\partial n}\Big|_{\Gamma_2} &= \bar q(S), && \text{with given heat flux} \quad (2) \\ -k(\theta)\frac{\partial\theta}{\partial n}\Big|_{\Gamma_3} &= h(\theta-\theta_0), && \end{aligned}$$

where h is heat-exchange coefficient, θ_0 is environment temperature. Boundary condition $\bar\theta(S)$ is imposed on the external surface according to the measurements [10], (Figure 1). Due to balanced load, equipment and boundary conditions symmetry the computations are reduced to one half a furnace. The condition $\partial\theta/\partial n = 0$ is imposed on the symmetry plane z=0.

3 NUMERICAL METHODS

Numerical methods used for temperature field calculations are outlined below.

Finite Element Method (FEM). To solve equations (1-2), the region G is divided into bilinear finite elements $\Phi^i(x,y)$. Galerkin finite element formulation constructed for the field variables (3)[13] gives an algebraic system (4)

$$\theta = \theta^i\Phi^i \qquad (3)$$

$$A(\theta)\ \theta = f, \quad \text{where} \qquad (4)$$

$$A^{ij} = \int_G k(\theta)\nabla\Phi^i\nabla\Phi^j dG + \int_{\Gamma_3} h\Phi^i\Phi^j dS \qquad (5)$$

$$f^i = \int_G q\nabla\Phi^i dG + \int_{\Gamma_3} h\theta_0\Phi^i dS - \int_{\Gamma_2} \bar q\Phi^i dS \qquad (6)$$

Simple iterations. The entire set of nonlinear algebraic equations (4-6) for the field variables θ^i is solved by means of Picard's iterations $A(\theta_{n-1})\theta_n = f$ with simultaneous updating of matrix A until condition (7) is satisfied.

$$\begin{aligned} \|\ \theta_n - \theta_{n-1}\ \| &< \epsilon_1 \\ r_n = \|\ f - A(\theta_n)\theta_n\ \| &< \epsilon_2 \qquad (7) \end{aligned}$$

Multigrid Method (MM). There are many methods to solve algebraic system (4). We use the Multigrid Method proposed by Fedorenko in 1961 [14] to solve finite-difference systems. The efficiency of MM for FEM schemes on different types of finite elements have been studied in many papers (e.g., see [15]), where the optimal efficiency O(N) of this method is proved.

Domain Decomposition. We suggest an approach which unites both Multigrid and Domain Decomposition Methods. The region is subdivided into a set of domains topologically equivalent to a rectangle. This approach has some advantages, as listed below:

i) the possibility of program execution on small-memory computers;

ii) parallel computations on a multiprocessor computer;

iii) simple and clear programming of MM stages.

4 THE MGD2 PACKAGE

The program package MGD2 can be used for large-scale computations of electrostatic, temperature and concentration fields in two-dimensional and axisymmetric regions of arbitrary shapes. The package is based on the effective version of multigrid method [11] applied to finite-element formulation of the problem. The comparison of this approach with traditional methods shows 50 - 500 times gain in computational time. This gain is proportional to the size of the discrete problem. This is especially important in case of using personal computers. The package MGD2 is running on the IBM, VAX and PC families of computers. The organization of dynamic memory allows the solution of problems up to 200,000 nodes using 450 Kbytes of computer memory. A typical problem with 10,000 nodes requires about 10 mins of running time on an IBM PC/AT-386 computer.

5 NUMERICAL SIMULATION

The objective of current study is as follows:

i) to investigate Multigrid applicability to nonlinear problems;

ii) to analyze various factors affecting the temperature profile in the furnace tube;

iii) to calculate the steady-state temperature field for the ČSK-1 furnace.

5.1 Discussion of Primary Results

Effects of the following factors have been studied:

1) material choice for the heat loading region;

2) presence of a steel shield inside furnace;

3) heat cartridge size and location;

4) joint size between an insulating cover and the furnace tube;

5) variation of steel brand;

6) presence of air in the furnace tube;

7) decreasing of heat consumption under stationary condition;

8) change of spacer material;

9) gap allowance for detachable joint;

10) nonuniformity of heat release to the cartridge.

The temperature profiles in crystallizer furnace tube are graphically represented in Figure 2 for all the models calculated.

Table 3 compares the calculated and experimentally measured profiles for different model parameters.

Table 3: The models calculated by MGD2.
Runtime 10 mins on IBM PC AT/386

Model	Design factor	$\Delta T, C^{o}$	compare
Ex	*Experimental data*	$840 \div 865$	with
1B	Felt + corundum capsules and insulator spacers (h=0.25mm)	$350 \div 230$	*Ex*
2B	Steel shield within furnace	$150 \div 170$	*Ex*
3B	The thickness of A,C,E heaters (h=4mm)	$-40 \div -25$	2B
4A	Extension of joint radius from 11.3 upto 25 mm	$5 \div 15$	3B
5A	Change of steel brand	$5 \div 15$	3B
6A	The presence of air in tube (r=8.125mm)	$-15 \div -25$	3B
7A	Change of mode for 6A $Q_{stat} = 70\% Q_{heat}$	$-60 \div -30$	*Ex*
7D	Gap between cover and tube (h=0.4mm)	$-35 \div -25$	*Ex*
7E	Change of mode for 7D $Q_{stat} = 75\% Q_{heat}$	$-10 \div +10$	*Ex*

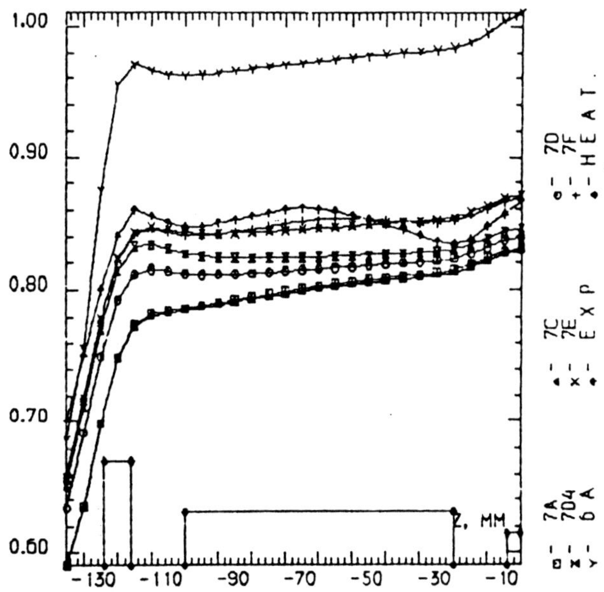

Figure 2: Temperature profiles on the axis of the ČSK-1 furnace tube after power decrease

Let us comment some of the results obtained:

The steel edge of h=0.8mm thickness (see Figure 3) in felt-like heat insulation in model 2A is, as it should be expected, the dominant heat conductor in the furnace.

The extension of thickness from h=0.25mm to 4mm of spiral-shaped heaters A,C and E enclosed in corundum capsules resulted in a steeper temperature profile lowered by 15-35 C^o(model 3B).

A 2mm-shift of heat source A from edge of the body has resulted in a 10-25 C^otemperature rise which emphasizes the need for a accurate positioning of cartridges (model 3C).

A significant temperature drop occurs at the joint of body cover and furnace tube. An extension of joint radius from 11.3 mm up to 25 mm allowed to increase temperature in furnace tube by 5-15 C^o(model 4A).

In all the previous models the thermal conductivity was taken for heat-resistance steel brand [16] and was supposed to vary as $k = 10 + 10 * \theta$. This steel brand being replaced by the stainless one with conductivity behavior $k = 12 + 5 * \theta$ (model 5A), both a slight 5-15 C^otemperature rise and temperature profile steepening occur.

The modification 6A is a more involved one due to account of air thermal conductivity in the furnace tube. The temperature difference between axis and inner surface of the furnace tube rises up to 15-25 C^o. Heat power removed through a furnace tube edge by the air conductivity is 1.8 W.

5.2 Comparison with Measurement Data

On the completion of the design stage of an acceptable furnace model the primary results were discussed with ČSK-1 designers. The heat power supply and material characteristics in use were revised.

Thus, on heating to the required temperature, power is reduced roughly by 30%. The radius of joint body cover and furnace tube is 22 mm. For model 7A temperature profile has lowered, compared to that of the experimentally measured one by 60-30 C^o.

In model 7C the felt brand "Pyrostol" was replaced by "Sybral" insulator spacers. Thus, due to a higher thermal conductivity the temperature in the region of interest decreases only by 1-3 C^o. On the other hand, when spacers are removed and the circular gap is filled by air the temperature increases by 4-9 C^o.

On inserting the 0.4mm circular gap between the body cover and the furnace tube (see Figure 4), the temperature profile acquires some qualitative changes. This is due to the increase in the thermal resistance of heat removal through the body cover near the heater A. The differences from the measured values are in the range of 25-30 C^o(model 7D).

It should be noted that two-fold grid refinement (model 7D4 has 20,000 finite elements) still causes an appreciable change of up to 10 C^oor, in other words, up to 1%, which can be explained by strong nonlinearity, high gradients, heat load discontinuity and insufficient finite-element density in the heater region.

The best agreement with the measured curve is achieved when the heater consumes 75 % of the maximum heating power available (model 7E). In this variant the discrepancy is in the range of 10 C^o. Figures 2, 5 present the curves discussed above.

The bends of isotherms point to such peculiarities as well-conducting steel shields and furnace tube gap. The isotherms are separated by steps of 20 C^o, see Figures 3,4.

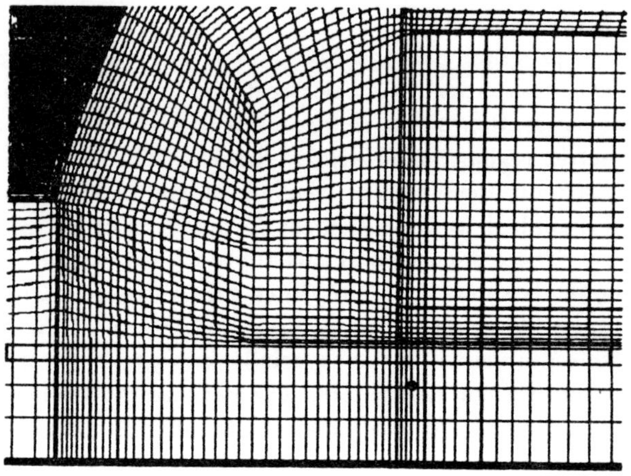

Figure 3: Grid fragment for model 7E

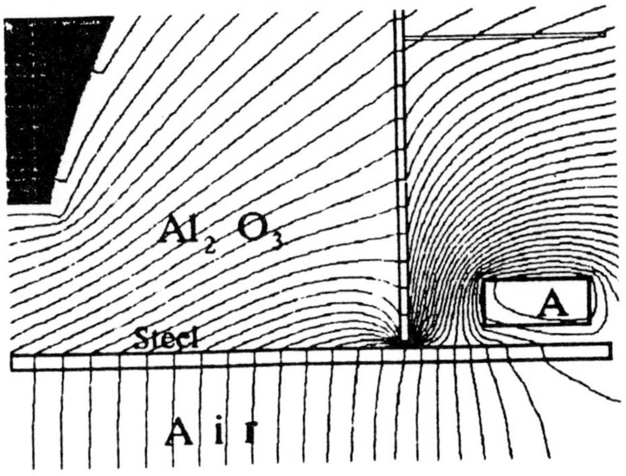

Figure 4: Isotherms with step 20 C^o T=76(20)911 C^o. Model 7E. The heater A and steel layer contours are shown by bolder lines

The calculations have proved the possibility of easy modifications of the MM algorithm so as to solve nonlinear problems as well, the residual vanishing being kept stable (see Table 4).

Table 4: The convergence of MM-iterations and solution. r_n as defined in (7). Interpolation from coarse grid is as Iteration 0. Model T7D4.

Stage	r_n	$\frac{r_n}{r_{n-1}}$	T(r=0,z=0),C^o
Initial	1.6 E+3		0.0
Iteration 0	2.4 E-2		844.0
Iteration 1	1.9 E-3	13.3	846.0
Iteration 2	2.9 E-4	6.4	845.26
Iteration 3	4.2 E-5	6.8	845.47
Iteration 4	6.7 E-6	6.3	845.43

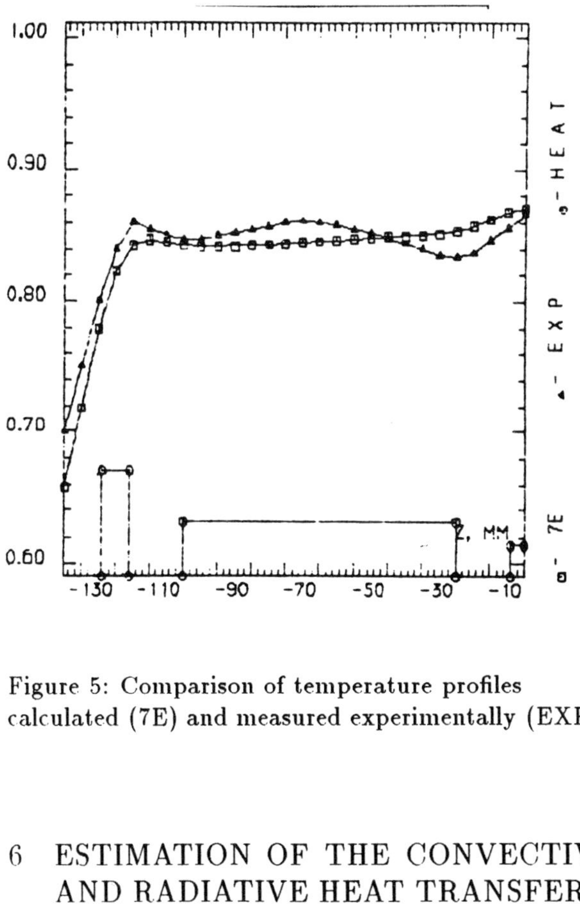

Figure 5: Comparison of temperature profiles calculated (7E) and measured experimentally (EXP)

6 ESTIMATION OF THE CONVECTIVE AND RADIATIVE HEAT TRANSFER

Practically, the heat transfer in the whole furnace occurs simultaneously via thermal conductivity, convection and heat radiation. The contributions of convective and radiative heat flows are estimated. It is shown that the contribution of these to heat exchange is negligible due to a small diameter of furnace tube and low working temperature range of 100-970 C^o[1]. This complicated heat exchange takes place only in the furnace tube. However, in the remainder of the furnace the heat is spread by heat conduction. For the time being, these processes are not included in ČSK-1 mathematical model.

6.1 Convective Heat Transfer

As an approximation, the equivalent thermal conductivity in enclosed volumes (plane, cylindrical, spherical gaseous or liquid gaps) can be treated by formula (8) [17], where ε_c is correction factor for equivalent conductivity due to convection.

$$\varepsilon_c = \frac{k_{eq}}{k} = 0.18(GrPr)^{\frac{1}{4}}, \left(\varepsilon_c = 1 \,\Big|\, PrGr < 10^3\right) \tag{8}$$

The k, Pr and Gr values are taken from reference book[16] at average temperature $T = 0.5(T_h + T_c)$. In case of Rayleigh number $Ra = PrGr < 10^3$, the convection factor is assumed to be $\varepsilon_c = 1$.

As an example, when $T = 950C^o, \Delta T = 20^o$, the typical scaling length d=16.25mm (model 6A), air has a reference Prandtl number of Pr=0.718, Gr=110, so Ra=79 < 1000 and $\varepsilon_c = 1$.

The temperature field calculated under the assumption of zero convection does not cause the convection which could vary heat exchange. It can be expected that the measurements made under earth conditions would not differ significantly from those under microgravity, i.e. without convection.

6.2 Radiative Heat Transfer

In the crystallizer the radiative exchange could take place in the furnace camera-tube only. In order to simplify the estimation of radiation heat flux we use the following hypothesis: furnace-tube surface S_a is diffuse-gray (emissivity ε =0.8, oxidized steel), tube edges S_b are black-bodies and air is transparent.

$$S_a = \{|\, z \,| < L = 200mm;\ |\, \varphi \,| < \pi;\ r = R\}$$
$$S_b = \{z = \mp L;\ |\, \varphi \,| < \pi;\ 0 < r < R = 8.125mm\}$$

Two cases of radiative exchange allow the easy analysis:

a) between pieces of surface S_a;

b) between edge S_b and tube surface S_a.

The difference of heat emission and absorption between two surface differentials dS_1 and dS_2 is given by formula (9) [17].

$$d^2Q = \varepsilon c_0(\theta_1^4 - \theta_2^4)\frac{cos\varphi_1 cos\varphi_2}{\pi r_{12}^2} dS_1 dS_2 \tag{9}$$

where $c_0 = \sigma_0 10^6 = 5.6710^{-2} W/mm^2/T^4, 1T = 10^3 K$. The detailed description of other variables of (9) for cases a) and b) is presented in [18].

On analytical integration over $dr, d\varphi$ of Eq.(9) and substitution of $\theta(z)$ by results of numerical calculation, one derives the expressions for the density of radiated heat flux (10) and (11).

$$d^2Q(z_1, z_2) = A\left[1 - \frac{t^4(t^2+6)}{(t^4+4t^2)^{3/2}}\right] dz_1 dz_2 \tag{10}$$

$$d^2Q(z_1, \mp L) = A\left[1 - \frac{t^2+2}{(t^4+4t^2)^{1/2}}\right] tR\, dz_1 \tag{11}$$

where $A = \pi\varepsilon c_0(\theta_{z1}^4 - \theta_{z2}^4)$ and $t = | z_1 - z_2 | /R$, θ is reffered to absolute zero.

Numerical integration of (11) over the tube length gives the value of radiation loss through open tube edge $Q_r = 0.44W$. For the radiation heat flux across tube element dz_1 to be found, the expression (10) is integrated over z_2 and added to (11). The maximum of this sum over tube is achieved near source A and equals to $q = 7mW/mm^2$. An accurate account of radiative exchange will result in a small local ripple of temperature profile. However, the average temperature in the operating region would change insignificantly because the power losses by radiation constitute only about 0.6% of the overall heating power.

7 CONCLUSION

It has been shown that the computation of a space furnace designed for materials research can be done efficiently with the program MGD2.

The entire set of ČSK-1 ' models has been computed. Effects of the following factors on the furnace performance have been analysed: material choice for the heat loading region, presence of a steel shield and choise of steel brand, the heat cartridges size and location, joint size between an insulating cover and the furnace tube, and besides that the presence of air in the furnace tube, gap for detachable joint and nonuniformity of heat release to the cartridge.

The package MGD2 was tested for a wide set of problems having analytical solutions. A numerical solution achives 0.01-1% accuracy for the range of parameters overlaping the ČSK-1 ones.

The application of the effective numerical procedures allows one to compute large scale problems on IBM PC computers. Typical problem with 10,000 unknowns requires about 10 mins on PC/AT-386.

A new version of the MGD2 package including the elaborated account of radiative exchange mechanism is under development now. The solution techniques for time-dependent heat transfer problems on a personal computer are included as well.

References

[1] W.Shtainborn, The Furnaces. *Space material science.* Mir, Moscow 1989 (in Russian).

[2] L.L.Regel, Space material science. *Itogi nauki i techniki. The Space Research.* V.**21** Moscow 1989 (in Russian)

[3] J.Chi, R.Zyedensticker, S.Duncan, The multipurpose furnace. *In: Space technology,* Mir, Moscow 1980 (in Russian)

[4] I.V.Barmin, A.W.Egorov, E.I.Goryunov et al. *The equipment for space industry.* Mashinostroenie, Moscow 1988 (in Russian)

[5] I.Jasinski, A.F.Witt, On the control of the crystal-melt interface shape during growth in a vertical Bridgman configuration. *J. Crystal Growth,* 1985, v.**71**, **2**, p.295.

[6] B.N.Rosental, C.R.Krolikowski, Programmable Multi-zone Furnace for Microgravity Research. *Proc. 29th Aerospace Sciences Meeting.* 1991 Reno, Nevada.

[7] M.J.Crochet, F.Dupret, Y.Ryckmans, Numerical simulation of crystal growth in a vertical Bridgman furnace. *J. Crystal Growth,* 1989, v.**97**, p.173.

[8] R.A. Brown, T.A. Kinney, P.A. Sackinger, D.E. Bornside, Toward an integrated analysis of Czochralski Growth. *J.Crystal Growth,* 1989, v.**97**, p. 99.

[9] S.Ch.Atabaev, W.S.Khenkin, A.I.Prostomolotov, S.A.Sidel'nikov The complex heat exchange and hydrodinamics for Czochralski modifications. *Prep. 427,* IPM AS USSR, Moscow 1989 (in Russian)

[10] C.Barta, A.Triska et al., The technical description of ČSK-1 . Phys.Inst.of the CS AS, Praha 1989.

[11] A.I.Fedoseyev, I.I.Petrenko, S.W.Purtov, Solution of large scale FEM problem in complex-shape regions by the Multigrid technique. *Prep 364,* IPM AS USSR, Moscow 1988 (in Russian)

[12] C.Barta, A.Triska, J.Trnka, L.L.Regel, Experimental facility for materials research in space/CSK-1. *Proc. 5th Europ Symp. on Material Sciences under Microgravity,* SchlossElmau, FRG 1984.

[13] O.C.Zienkiewicz, *The Finite Element Method.* Mc Graw-Hill 1977.

[14] R.P.Fedorenko. Relaxation Method for finite-differential elliptic equations. *Soviet Journal for Computational Methods and Mathematical Physics* 1961 V.**1**, **5** p.922 (in Russian)

[15] R.A.Nicolaides, On the l^2 Convergence of an Algorithm for Solving Finite Element Equation. *Math.Comp.,* 1977, v.**31**, p.892.

[16] W.S.Chirkin. *The Thermophysical Properties of Materials.* Gosfizmatizdat, Moscow 1959 (in Russian)

[17] M.A.Mikheev, I.M.Mikheeva, *The Basis of Heat Transfer.* Energiya, Moscow 1973 (in Russian)

[18] A.I.Fedoseyev, I.I.Petrenko, S.V.Purtov et al., Computation of Thermal Fields in Space Furnace "Crystallizator" (ČSK-1). *Prep 496,* IPM AS USSR, Moscow 1991 (in Russian)

Containerless Undercooling of Metals

Berndt Feuerbacher, Ivan Egry, and Dieter M. Herlach
Institute of Space Simulation, DLR, Cologne, Germany

Abstract

Solidification is an activated process requiring a certain amount of undercooling. This fact may be used to access the interesting regime of the metastable, undercooled melt, which allows for non-equilibrium solidification into metastable solids. In metals, heterogeneous nucleation generally leads to a lowering of the activation threshold, thus limiting the undercooling range. This is avoided through containerless processing techniques in an ultraclean environment. Experiments are presently performed in the laboratory on Earth but will be extended in future into space, where the reduced positioning forces allow for a substantially extended undercooling range and induced flow effects are largely suppressed.

Deep undercooling of a metallic melt opens access to an increasing number of metastable solid phases. In addition, the large driving force for solidification leads to an extraordinary rapid crystallization process. Measurements of the solidification velocity confirm recent theories on dendritic growth for moderate undercooling, while discrepancies are identified at extreme undercooling ranges. Phase selection phenomena can be observed in situ and are correlated to nucleation and growth processes. In addition, possibilities are demonstrated to actively influence the phase selection in order to obtain predetermined metastable metallic materials with novel properties.

Key Words: Containerless Processing, Metals, undercooling, metastable phases.

1. INTRODUCTION

The undercooling of a melt is an ubiquitous effect that occurs prior to any solidification process. It arises from the fact that an activation barrier is present, separating the molten state from the energetically preferred solid state at temperatures below the melting point. Usually this activation barrier is small, so only minor degrees of undercooling occur.

It is, however, attractive both from a scientific and an application oriented point of view to increase the level of undercooling. This is because a deeply undercooled melt represents a metastable thermodynamic system, which is hardly explored due to obvious experimental difficulties. The observation of the rapid crystallization process provides new insight into the physical nature of solidification[1] On the other hand, solidification from a deeply undercooled state gives access to a variety of metastable solid phases with novel material properties[2]. This is illustrated in Figure 1, which represents a schematic free-energy diagram of a metal having several metastable phases (β, γ...) in addition to a stable phase α. The stable state of the system corresponds to the lowest, heavy line, with an equilibrium melting point T_f^{α}.Undercooling occurs along the liquid line and gives access to other crystalline phases as the respective metastable melting points T_f^{β}, T_f^{γ} are reached.

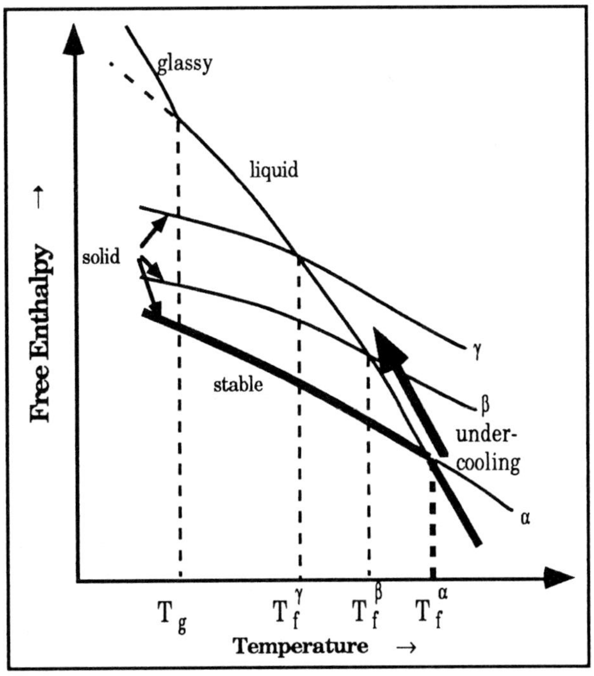

Figure 1: Free enthalpy as a function of temperature, indicating stability of the various phases melt, solid β,γ, and glassy state. The heavy line gives the stable phases. With increasing melt undercooling, additional metastable phases become accessible.

Ultimately glassy solidification is possible provided the melt can be cooled below the glass point T_g.

A deep undercooling level can be achieved by rapid quenching methods such as melt spinning, splat cooling or atomization. The disadvantage of such techniques is twofold. Firstly, the rapid heat extraction requires samples that are small (typically tens of microns in size) in at least one dimension. Secondly, processing geometry and timescales do not allow for direct diagnostics of the metastable liquid nor the solidification process itself.

An alternative, avoiding these problems, is offered by techniques of containerless processing. Here the effect of heterogeneous nucleation, which usually initiates solidification and such limits further undercooling, is suppressed by avoiding any contact with solid surfaces or surface contaminants. Therefore bulk liquid samples (typically several millimeters in size) can be undercooled slowly to low temperatures. In a suitable geometry, such as offered by the electromagnetic levitation technique, the undercooled sample is accessible to various contactless diagnostic tools to investigate its thermodynamic properties[3] and its solidification behaviour. In addition, active stimulation by external means is possible, guiding the early stages of crystallization process into predetermined directions and such allowing to voluntarily influence the solidification path as well as the metastable solid product.

2. Experimental Techniques

The requirement of an efficient suppression of heterogeneous nucleation events implies the avoidance of contact with any solid crystalline material - such as containers - and a high degree of surface cleanliness. Metals in particular are sensitive to surface contamination, e.g. by oxidation, so vacuum techniques are usually necessary[4]. Levitation under vacuum can be achieved by electrostatic, magnetic or electromagnetic forces. Here the latter proves to be the most useful technique in the laboratory, as it provides stable levitation configurations for conductive materials.

Electromagnetic levitation has been used for contamination free melting of metals since a long time. In the present context, where low temperatures are required to achieve undercooling, the Joulean heat produced by the very same eddy currents that provide levitation is a problem. A special technique therefore is required to achieve large temperature differences [5]. Here the levitation coil is designed in such a way that a zone with a high, but rather uniform field is found in the lower part, while a high field gradient prevails in the upper part of the coil. In this way, a spatial separation exists between a heating and a levitation zone. The sample is melted in the lower part and then lifted against gravity by increasing RF power, leading to cooling in the high-gradient zone. The achievable temperature span between melt overheating and lowest undercooling temperature is limited, so specific coils have to be carefully tailored for each sample size and material.

From the above discussion it becomes apparent that earth gravitation imposes stringent limits on the applicability of electromagnetic levitation to undercooling experiments, which call for an extension towards space experimentation. The most important aspects are:

- **Lower temperature limit**. As levitation and heating effects cannot be decoupled completely, considerable sample heating is inevitable. Therefore gas cooling is usually required, and even here temperatures below 1000° cannot be achieved.
- **Accessible temperature span**. The difference between the highest achievable melt temperature and the lowest undercooling temperature is essential for the possibilities of melt overheating before solidification and therefore for the accessibility of metastable products.

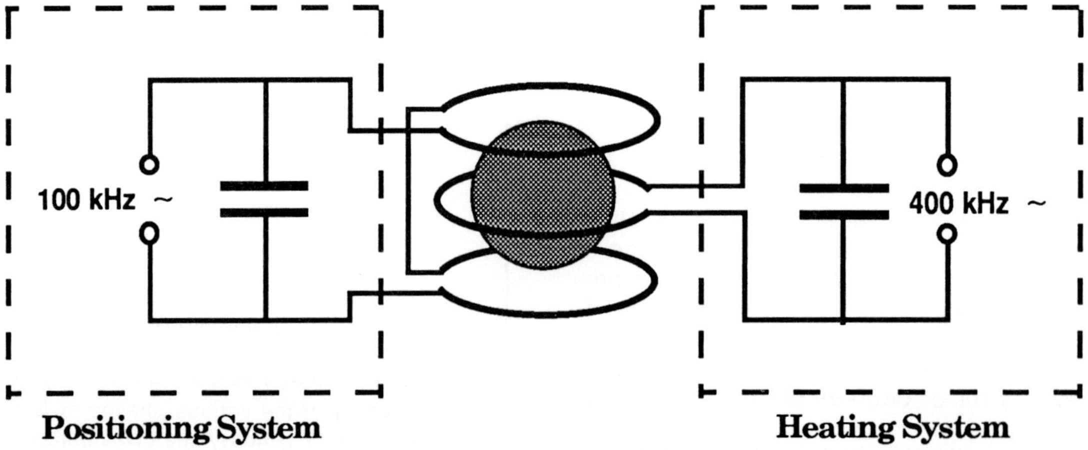

Figure 2: Schematic of the two-coil system used in the TEMPUS -instrument for containerless processing in space.

- **Cleanliness**. As pointed out above, gas cooling still is a prerequisite to achieve substantial undercooling in laboratory levitators. Even the best gas cleaning systems show impurity levels of reactive gases (O_2, H_2O etc.) which are 4-5 orders of magnitude above those attainable easily through ultrahigh vacuum techniques.
- **Dynamic effects**. The levitation forces required to support a sample against earth gravity inevitably induce dynamic motions inside the molten metal. These may give rise to dynamic nucleation and thus limit the undercooling range achievable. The same forces induce deformations of the sample shape and such hamper measurements of surface tension and viscosity.

In a space microgravity environment, the required positioning forces are reduced about four orders of magnitude compared to the earth laboratory. This gives access to new materials classes such as easy glass forming metals or semiconductors. Some fundamental thermodynamic measurements such as viscosity or heat capacity in the metastable state of the undercooled melt[3], are probably possible only in the low gravity environment of space.

As a result of the laboratory measurements reported here, a space facility has been conceived under the name "TEMPUS", which is presently in the process of industrial development and construction in the Dornier company[7]. Here full advantage is taken of the low gravity environment by using a coil arrangement that largely separates the functions of positioning and heating. This is achieved by a double-coil system, using a quadrupole field for positioning and a dipole field with negligible positioning forces for heating. Additional optimization is achieved by different RF frequencies for the two coils, as shown in a schematic way in Figure 2. With this coil design, heating efficiencies of more than 20 % have been measured, about an order of magnitude higher than achievable on ground and therefore drastically reducing the RF generator power and size requirements.

3. EXPERIMENTAL RESULTS

In the following, a number of ground-based laboratory results will be presented, which are typical for the capabilities of the techniques used and illustrate the potential of future space experiments.

3.1. Melt Undercooling

The result of a typical undercooling experiment as performed in a ground levitation system is shown

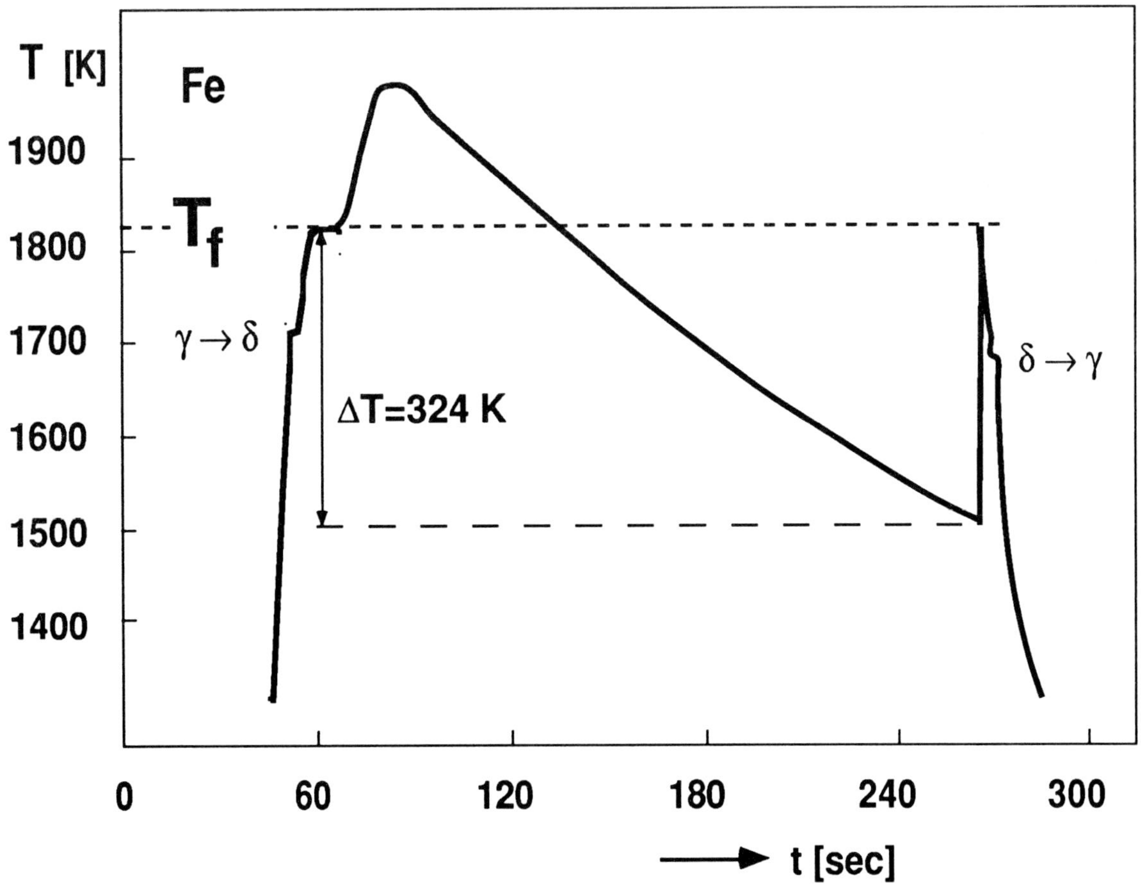

Fig. 3: Temperature-time profile as obtained from a bulk Fe melt solidified in containerless state.

in Figure 3. Initially, the temperature of the solid sample is raised to the melting point T_f, which appears as a plateau in the time-temperature plot and serves as an absolute calibration point. After heating the melt well above the melting point, the temperature is lowered without a phase change well below the equilibrium melting point. Upon nucleation, which occurs spontaneously at $\Delta T = 324$ K in this example, the temperature rises rapidly due to the release of the latent heat of crystallization in a typical recalescence event. Also visible in the figure are the γ-δ solid state phase transformations both during heating and cooling.

Experiments on undercooling give fundamental information on nucleation and phase formation. Studies performed so far have indicated a dominant influence of heterogeneous nucleation sources[8]. Space experiments, allowing for deep undercooling in an ultraclean environment, hold promise of reducing heterogeneous effects to such an extent that the homogeneous undercooling limit may be reached[9].

The deeper undercooling attainable under reduced gravity may also allow to surpass the hypercooling level, which implies a recalescence to temperatures below the equilibrium melting point. This reduces possible transformations of phases formed in the nucleation process and thus gives direct access to novel metastable materials.

3.2. Solidification Velocity

In a levitation system it is possible to initiate nucleation intentionally by touching the undercooled melt with a solid trigger needle. Under these circumstances, the geometric position of the nucleation point is known. This is achieved by imaging the sample sphere onto a fast silicon photodiode. In addition, the temperature at the time of triggering may be chosen freely over a wide range. In this way a direct measurement of the solidification velocity as a function of undercooling is possible[10].

A typical result as obtained from a pure nickel sample is shown in Figure 4. The growth is extremely rapid, reaching up to 70 m/sec for the highest undercooling. Two regions of different temperature behaviour can be distinguished. Below a critical value ΔT^* the data follow a power law $V \sim \Delta T^{\beta}$ with $\beta \approx 3$. In this regime, the data may be interpreted following predictions of recent theoretical work on dendritic solidification[11] (solid line), including non-equilibrium effects at the solid-liquid interface (dashed line). On the other hand, for undercooling levels beyond ΔT^* the solidification velocity is found to increase linearly, an observation that is presently not understood in terms of theoretical models. The temperature ΔT^* has been shown to coincide with the critical undercooling for grain refinement[12].

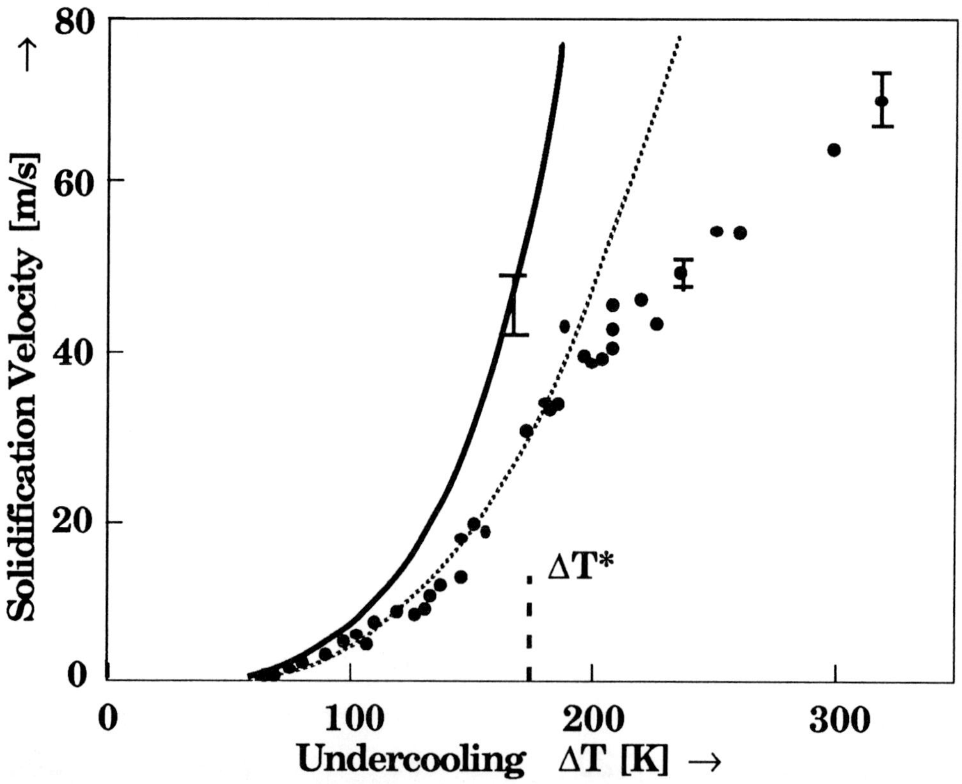

Figure 4: Growth velocity as a function of undercooling for solidification of a pure Ni sample. Full dots, measured values; full line, theoretical prediction according to Ref. 11, dashed line, theory including non-equilibrium effects at the solid-liquid interface.

3.3. Triggered Metastable Nucleation

In crystallization from a deeply undercooled melt, an increasing number of metastable phases may become accessible. It is, however, difficult to predict which of the phases will be selected in the solidification process, so a method to intentionally select a certain metastable phase is desirable.

In a recent experiment, controlled external phase triggering was demonstrated in the Fe-Ni alloy system[13]. For Ni concentrations above 4 %, this alloy solidifies into a stable fcc (γ) phase. A metastable bcc (δ) phase becomes accessible for undercooling intervals below 200 K, which can be easily achieved experimentally.

In order to initiate metastable bcc solidification, an external trigger made of a Fe-Mo alloy was used, which has a stable bcc crystal phase at the desired nucleation triggering temperatures. In contrast to the experiments discussed above, the Fe-Ni sample was not vacuum levitated but embedded into a protective glass skin, a method that also reduces heterogeneous nucleation for certain materials combinations. After melting and undercooling to a predetermined temperature, the solid trigger needle was heated to the same temperature and brought in contact with to liquid, which immediately led to solidification and recalescence.

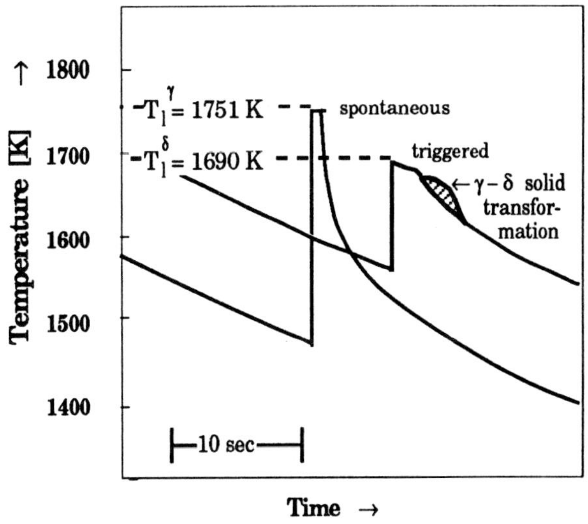

Figure 5: Two temperature-time profiles obtained during solidification of undercooled Fe-24 at.% Ni melts. Left-hand side, spontaneous crystallization in a fcc structure-at an undercooling of 278 K. Right-hand side, solidification upon triggering at an undercooling of 194 K into metastable bcc phase.

The detection of metastable solidification makes use of the fact that the melting point of a metastable phase is always lower than that of the corresponding stable phase. An example for such a measurement is given in Figure 6. The left-hand

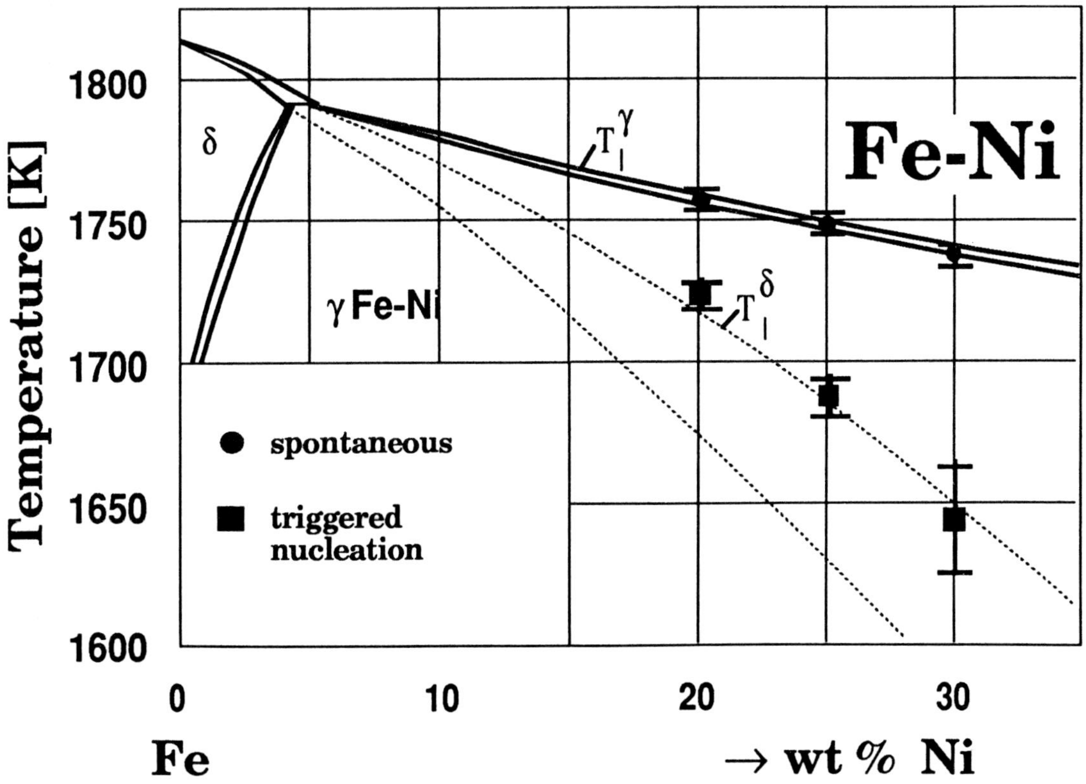

Figure 6: Phase diagram of Fe-Ni, including metastable extensions for the bcc phase (dashed lines). Experimental points represent recalescence temperatures of the corresponding temperature-time profiles. In case of spontaneous crystallization of stable fcc phase they coincide with the liquidus line of the equilibrium phase diagram (dots), while in case of triggered bcc solidification (squares) they fall on the liquidus line of the metastable bcc phase. Error bars indicate experimental scattering.

trace corresponds to spontaneous nucleation, which happened - in this case - at an undercooling of 278 K. Upon nucleation, the temperature rises rapidly to the equilibrium liquidus line at 1751 K. Completion of solidification occurs during a temperature plateau, followed by cooling of the solid sample. In the right-hand trace, nucleation was triggered intentionally at 194 K undercooling using a Fe-Mo needle. Recalescence now is observed to end at the metastable melting point of the bcc (δ) phase. A further indication for metastable solidification is the observed solid-state transformation into the stable fcc (γ) phase, which is shown in Figure 5 by a cross hatched area.

Similar measurements have been performed for three different alloy concentrations in order to confirm the interpretation in terms of metastable solidification. A summary of the results is presented in Figure 7. Here the maximum temperatures of the recalescence peaks have been plotted in the Fe-Ni phase diagram, which is augmented by a metastable extension for the bcc phase[14]. All measured points for spontaneous nucleation (full dots) closely coincide with the liquidus line of the equilibrium phase diagram. On the other hand the recalescence temperatures measured following triggered solidification are found close to the calculated metastable liquidus line for the bcc phase. This observation leads to the conclusion that indeed metastable bcc solidification has been introduced through the Fe-Mo trigger material. At the same time, it is apparent that metastable phase diagrams can be constructed experimentally in this way.

CONCLUSIONS

Containerless processing of metals has been demonstrated to provide a valuable tool in the investigation of undercooled metallic melts. In contrast to methods relying on rapid cooling, deep undercooling in a contactless environment allows for detailed diagnostic measurements during the process of solidification and even for intentional interference into the phase selection process. In this way, novel insight is possible into the solidification phenomenon and into metastable thermodynamics.

This is illustrated using examples of laboratory results. Deep undercooling to a level previously not achievable in bulk melts is demonstrated. The direct observation of solidification velocities as a function of undercooling allows a comparison to recent theoretical predictions, free of adjustable parameters. Finally, the possibility of externally influencing the phase selection by using a suitable trigger material is demonstrated.

The limitations of containerless processing under the influence of earth gravity are pointed out. Space experiments are shown to open new avenues for fundamental investigations and give access to new materials classes.

5. References

1. J.C. Baker and J.W. Cahn in *"Thermodynamics of Solidification"*, ASM, Metal Park 1971
2. B. Feuerbacher, *Mat. Sci. Rep.* 4, 1 (1989)
3. I. Egry and B. Feuerbacher, this volume
4. For some metals, melt fluxing is an alternative method
5. R. Willnecker, Thesis, University Bochum 1988
6. D.M. Herlach, F. Gillessen, *J. Phys. F.* 17, 1635 (1987)
7 J. Piller, R. Knauf, P. Preu, G. Lohöfer and D.M. Herlach, ESA SP 256, 437 (1987)
8. R. Willnecker, D.M. Herlach, and B. Feuerbacher *Appl. Phys. Letters* 49, 1339 (1986)
9. D.M. Herlach, B. Feuerbacher, and A.L. Greer Proposal to AO OSSA 4-88 (1988)
10. R. Willnecker, D.M. Herlach, and B. Feuerbacher, *Phys. Rev. Letters* 62, 2707 (1990)
11. J. Lipton, W. Kurz, and R. Trivedi, *Acta Met.* 35, 957 (1987) and 35, 965 (1987)
12. R. Willnecker, D.M. Herlach, and B. Feuerbacher, *Appl. Phys. Letters* 56, 324 (1990)
13. E. Schleip, D.M. Herlach, and B. Feuerbacher *Europhys. Letters* 11, 751 (1990)
14. Y.-Y. Chuang, K.-C. Hsieh, and Y.A. Chang *Metal Trans. A* 17, 1373 (1986)

METHOD OF LARGE SINGLE CRYSTALS GROWTH FROM MELT WITH A GIVEN SHAPE OF MELT CRYSTAL INTERFACE

Golyshev V.D., Gonik M.A.,

Institute for Synthesis of Minerals, Institutskaya Str., 1, Alexandrov, Vladimir Region, 601600, USSR.

ABSTRACT

The apparatus and methods for large size single crystals growth, that permit to obtain similar thermal conditions both in space and on the earth, are described. The method that permits to study sources of nonreproductivity of the thermal conditions while growing crystals of semitransparent materials is discussed. The results of testing of technique for crystal growth of NaCl, KCl and $NaNO_3$ are given.

Key words

Thermal Conditions, Thermal Model, Computer Controlling, Plane Crystal-Melt Interface, NaCL, KCl.

1. INTRODUCTION

While working out a crystal growth technology for its outer space application, it seems to be very important a problem to grow shorter the period of the technology devising.

It is possible, if one would have a method of crystal growth that satisfy the following demands:

- while realizing the method, thermal conditions at the interface of crystal-melt would be the same on the earth and in space;
- we can guarantly reproduct the wanted thermal conditions at the interface (temperature gradient in the crystal and in the melt, superheating of the melt, crystal growth rate);
- thermal conditions at the interface do not depend on cross section of growing single crystal.

To satisfy the first demand, it is necessary that natural convection does not appear on the earth. To satisfy the second one, it is necessary that we shall be able, while growing crystals, to determine precisely temperature fields. It is more complicated to do it, while growing single crystals of semitransparent materials. It takes place because of complicated character of equations of radiative-conductive heat transfer and of absence of data on thermophysical properties for such materials at high temperatures. So, there is a problem of studying a possible determination and reproductivity of thermal conditions at interface of semitransparent materials. Without these investigations we can't say how realy is a solution of the above raised problem. For these purposes, the modelling method of crystal growth is helpful.

2. MODELLING METHOD OF CRYSTAL GROWTH

In the modelling method, an idea of creating conditions that permit to simplify the problem of heat transfer calculation at crystal growth by making one-dimension thermal field and eliminating convective heat transfer have been realized.

The thermal model of the method is shown in figure 1. According to this model the growing crystal and the melt are the infinite plane layers. Because cooling is made from below, the heat transfer by convection in the melt is eliminated and heat problem is reduced to relatively simple one-dimension one of the radiative-conductive heat transfer, with crystal growth occuring with plane shape of the interface.

A scheme of the crystal growth method, called by us the modelling method, in which this thermal model was realized, is shown in figure 2. The background heater consists of two sections and serves to make initial value of temperature. The main heater serves to make a temperature drop on the two-layer melt-crystal system. It consists of two sections, with help of those, according to data of thermocouples 1 and 2, one makes the surface I to be isothermal. Isothermality of the surface II is controlled by thermocouples 3 and 4. The ratio of a height of the crucible H to its diameter is equal to ~0,1. The surfaces II and I of the crucible are placed horizontally.

The measures have been taken not to permit lifting of a seed, while melting, by forces of surface tension. For the contact of the melt with the surface I to be constant during all the growing process, a volume III was foreseen in the crucible, the value of which is defined by ratio of specific weights of the melt and the furnace charge. Measuring thermocouples 1-4 have been put into pyrometric canals of 1,7x1,7 mm size, close to a bottom and a lid, and were separated from the melt by a wall of 0,2-0,5 mm thickness.

Crystal growth is carried out by computer controlling temperatures in points 1 and 3 (figure 2) with help of the thermal model, so that crystal growth rate would be equal to the given values in the given moments of the time.

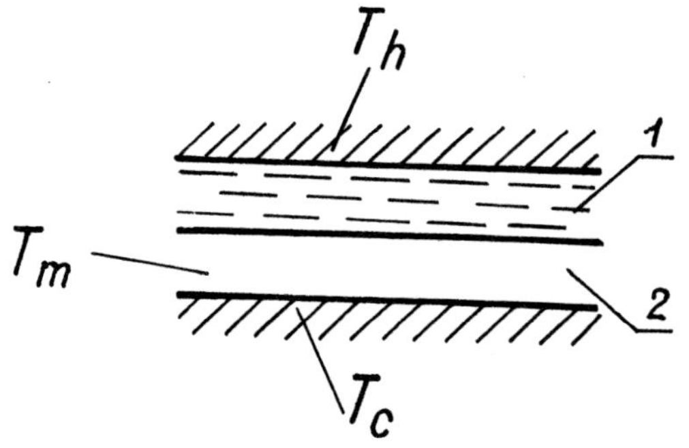

Figure 1: Thermal model of the method: 1-melt; 2-crystal; T_C : temperature of cold surface; T_h : hot surface one.

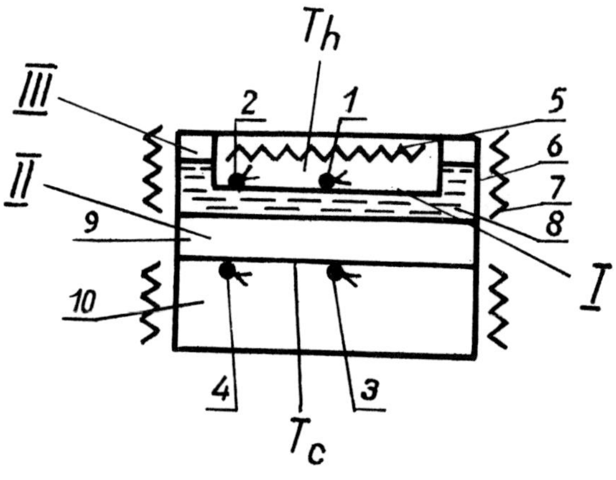

Figure 2: A scheme of modelling method: 1-4-thermocouples;5-main heater; 6-crucible; 7-heater; 8-melt;9-crystal; 10-metal block.

Testing of the constructive decisions and quality of the model have been made, while growing single crystals of NaCl on a seed. For these purposes, marks of a position and a shape of interface were made in the chosen moments of the time. Marks were realized by switching off, for a some time, the main heater. It causes an increase of crystal growth rate and capture of the KCl admixture, put earlier in the melt.

A light beam is dispersed on layers with great amount of admixture. It makes possible to measure the position of the interface in moments of starting and finishing of crystal growth with great rate. To measure the positions of boundaries of regions with great and little rate of crystal growth the samples were examined with laser light and photographed with a microscope.

To make calculation of the interface position and to control on the thermal model the data on thermophisical properties are necessary. The thermal conductivity of the melt, Λ_m, have been determined by a steady-state method of coaxial cylinders with solving the inverse problem of the radiative-conductive heat transfer. A cell for measurements of the thermal conductivity is shown in figure 3. The spectrum absorptivity, K, for melts have been determined according to the scheme shown in figure 4. The method of all the measurements is described, in details, in preliminary work (Ref. 1).

The comparison of the calculated positions of the interface with found in experiment ones at given moments of the time (figure 5) shows that the crystal growth occurs with a plane shape of the interface, that is we, indeed, have the one-dimension problem. Within the accuracy of the experiment the position of the interface for NaCl corresponds to calculated one from solution of stationary problem, with neglecting the latent heat of solidification and

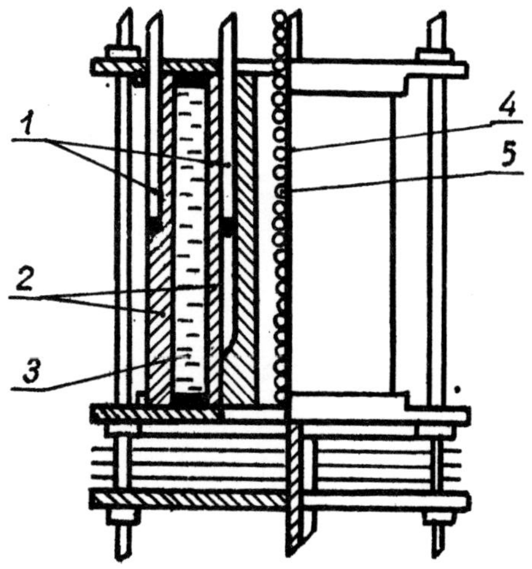

Figure 3: Experimental cell.
1-thermocouples; 2-metalic tube; 3-melt; 4-guard heater; 5-main heater; 6-alumina tube.

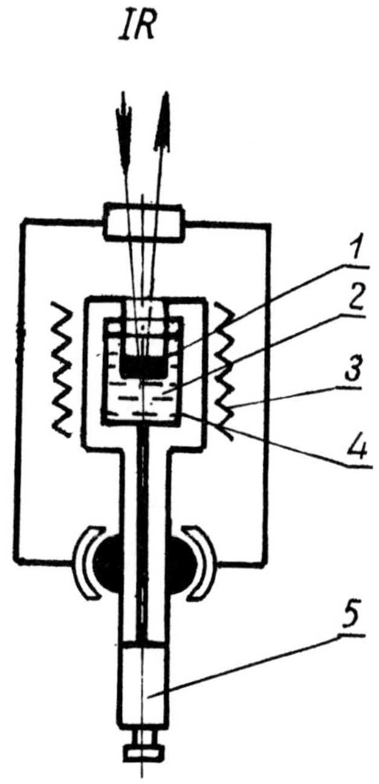

Figure 4: Experimental cell for optical properties studies: 1-mirror; 2-melt; 3-heater; 4-crucible; 5-micrometer.

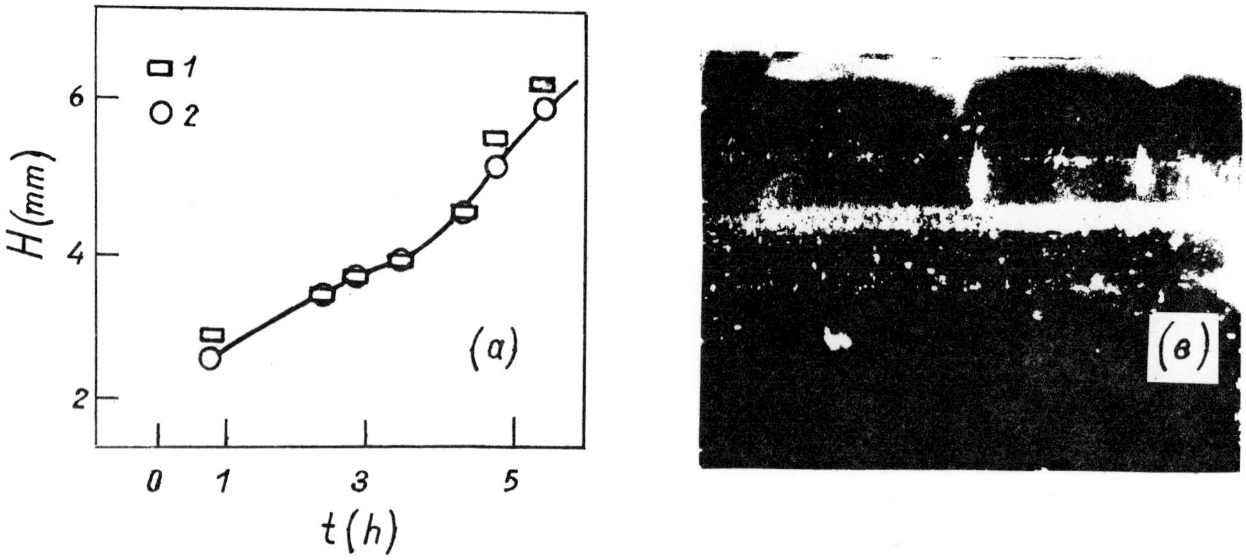

Figure 5: A comparison of calculated and experimental determination of the interface position: a) diagram; b) photo of the NaCl crystal.

radiative-conductive character of heat transfer. The coincidence of calculated and real positions of the interface is in range of 4-5%. The accuracy of calculations is so good that it allowed us not to melt completely the seed with 80 mm in diameter and 2,5 mm of thickness. An absolute error of the position calculations was equal to 0,1-0,2 mm.

The simplicity of the thermal model and knowledge of thermophysical properties permit us to analize the sources of nonreproductivity of the thermal conditions at the interface.

In particular we have studied the effect of the found by ourselves dependence of the thermophysical properties from experimental conditions (Ref.2) on the character of the crystal growth process. It is revealed that, even for the same materials, we may have qualitatively different dependences of growing rate on time at just the same conditions in a crystal growth chamber (figure 6). Therefore, the knowledge of thermophysical properties in concrete conditions of growing becomes a matter of principle for a guaranted reproduction of the thermal conditions.

In detail, the modelling method, the procedure of testing and result of it, and, also, the effect of thermophysical properties values are considered in (Ref.3). Here it should be noted that realization of the crystal growth in conditions of one-dimension thermal field with cooling from below and controlling on the thermal model with thermophysical properties, obtained for the given concrete conditions, results to reproduction of thermal conditions at the interface with satisfactory, for a technology practice, accuracy. It should be also said that crystal growth, in the one-dimension thermal field, and, consequently, with the plane interface, results to independence of the thermal conditions on cross section of the growing single crystal.

So, with account of guaranted reproduction of thermal conditions at crystal growth in the

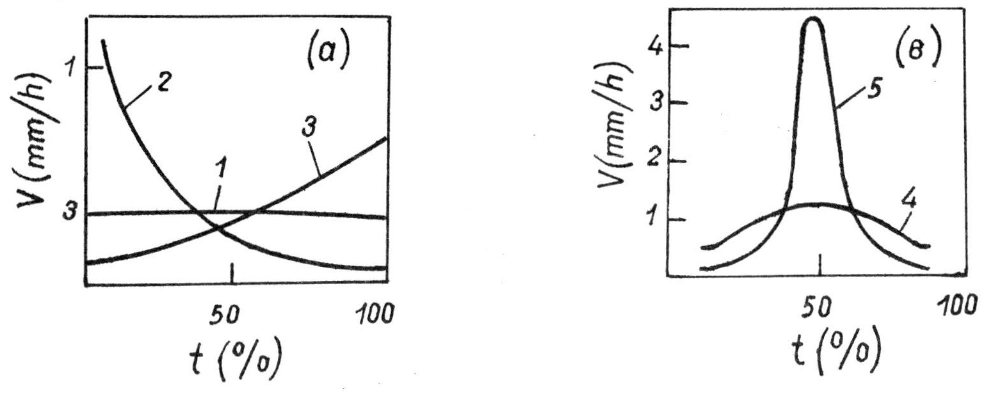

Figure 6: Dependence of the growth rate of crystal by cooling with constant rate for different values of thermophysical properties: 1-$\Lambda_m/\Lambda_c = 1$; 2-$\Lambda_m/\Lambda_c < 1$; 3-$\Lambda_m/\Lambda_c > 1$; 4-K=5 cm^{-1}; 5-K=0,3 cm^{-1}; Λ_c : thermal conductivity of the crystal, t-time of crystalization.

one dimension thermal field, a problem to work out a method that permits, in contrary to the modelling one, to obtain one-dimension thermal field close to the interface with an arbitrary ratio of the height of the single crystal to its diameter (in modelling method this value is less than 0,1) becomes of importance.

3. METHOD OF LARGE SINGLE CRYSTAL GROWTH

A problem to make the one-dimension thermal field close to the interface is reduced to the problem to obtain the plane shape of the interface for any cross section of growing crystal and any ratio of the height to the diameter of the single crystals.

To solve this problem we used a heater, called the main heater, placed close to and along a surface of the crystal-melt interface. In figure 7 the scheme of the method of crystal growth in a crucible and in figure 8 without crucible method of growing crystals are shown. The crystal growth occurs owing to relative shift of the main heater and the crystal. The plane shape of the interface is attained by maintaining the plane surface of the bottom of a hermetic container isothermal on thermpcouples T_1 and T_2 , (so as in figure 2) owing to redistribution of a power between the two sections of the main heater.

The testing of possibility to use one-dimension thermal model have been made, while growing KCl single crystals of 80 mm in diameter and 60 mm in height. CsI of the 0,1 %

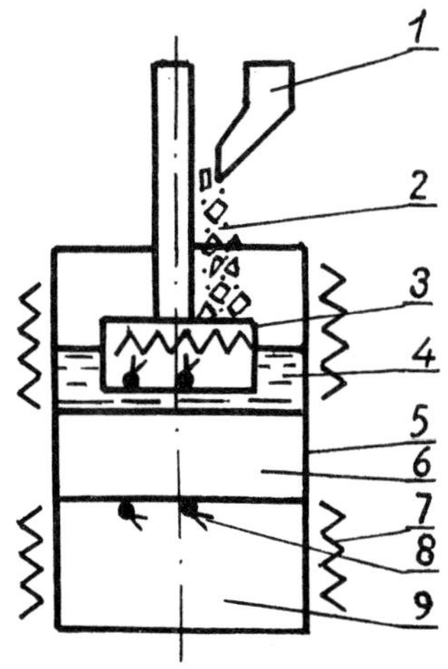

Figure 7: A scheme of the method:
1-bunker; 2-furnace charge; 3-container with heater; 4-melt; 5-crucible; 6-crystal; 7-heater; 8-thermocouples; 9-metal block.

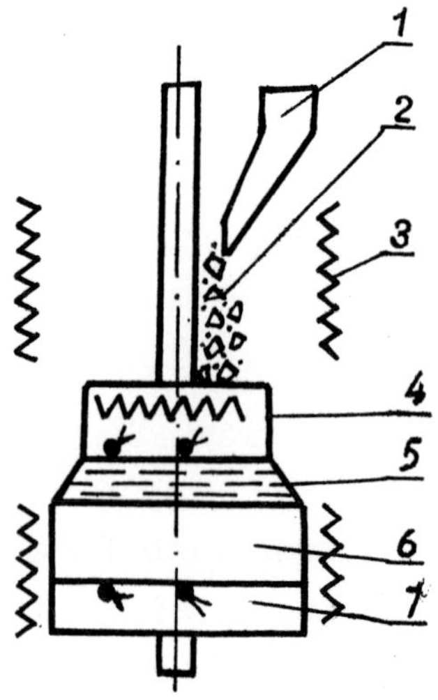

Figure 8: A scheme of the method:
1-bunker; 2-furnace charge; 3- heater; 4-container with heater; 5-melt; 6-crystal; 7-thermocouples.

in weight was the admixture for analysis of the position and shape of the interface, as it was above described. To make the interface visible γ -radiation was used. As we found, the interface was plane all over the height of crystal. The position of the interface was determined with an accuracy of 4-5%.

Because in crystallization process there is a shift of the main heater relatively of the growing crystal, we have studied conditions of an appearance of the convective movement. The research has been carried out for $NaNO_3$ solidification. The melt was lit by incandenscent lamp and laser beam. The convection was observed on moving of special put particles that disperse radiation.

It was found that in crucible method, if the hermetic container is in horizontal plane with accuracy to $\pm 30'$, there is no convective mass transfer. While using the without crucible method, the Marangoni convection was observed. The moving rate of tracer, for the melt layer of 1 mm thickness and temperature drop ~20 K, was about 1 mm/s.

4. REFERENCES

1. V.D. Golyshev, M.A. Gonik: *Izvestiya Akademii Nauk SSSR, ser. Physics, 52,* 1896, (1988).
2. V.D. Golyshev, M.A. Gonik: *High Temperatures-High Pressures,* (1991), (to be published).
3. V.D. Golyshev, M.A. Gonik: *Crystal properties and preparation,* Vol. 36-38, (1991), pp. 623-630.

PROBLEMS OF GROUND-BASED REFERENCE EXPERIMENTS FOR SOLUTION CRYSTALLIZATION, EXEMPLIFIED BY THE CALCIUM PHOSPHATE SYSTEM.

H. E. Lundager Madsen and F. Christensson.

Chemistry Department, RVA University, Thorvaldsensvej 40, DK-1871 Frederiksberg C, Denmark.

ABSTRACT

It is shown by a series of crystallization experiments in silica gel that 1) interactions between silicic acid and the reacting ions may strongly influence mass transport and 2) osmosis may under certain conditions lead to gel shrinkage. The importance of convection in the initial stage of a free-diffusion experiment is also demonstrated. The results raise the question of how to optimize the design of both ground-based and space experiments.

Key Words: Diffusion crystallization, gel growth, osmosis, calcium phosphate.

1. INTRODUCTION

Solution crystallization in space is most often carried out by a two-way diffusion method, i.e. by reaction crystallization or, less frequently, solvent modification. The two main effects eliminated by microgravity are sedimentation and convection. On earth, the former results in uneven growth and a high degree of aggregation and the latter in a large number of nuclei - due to rapid mixing of reagents - as well as a high defect concentration in the crystals (Ref.1). Under microgravity conditions, transport of matter may be purely diffusive; this has the additional advantage that a theoretical treatment is simpler.

It is well known that both sedimentation and convection can be eliminated by crystallization in a gel. This is the closest possible approach to microgravity conditions in a ground-based crystallization experiment. The most frequently employed gelling substance is silicic acid, formed either by acidifying an alkali silicate solution - the classical method (Ref.2) - or by hydrolysis of a silicate ester (Ref.3); the latter has the advantage that it is much easier to control pH of the gel. The unknown factor is the influence of the gelling substance on crystallization; very few systematic studies of this problem have been made. On the other hand, diffusion does not appear to be much influenced by the gel (Ref.4).

In ordinary solution diffusion experiments sedimentation

of crystals can hardly be avoided, but in the initial stage of the process convection may be minimized by matching the solution densities. As soon as crystallization starts, this condition can, of course, no longer be fulfilled. Nevertheless, there is no doubt that this method yields a relatively close approach to microgravity conditions as well.

The aim of the present work is to study the problems encountered with the methods of ground-based reference experiments outlined above and to show, how the results influence the choice of conditions for a microgravity experiment. We are presently preparing a crystallization experiment in the Solution Growth Facility of the EURECA space mission scheduled for 1992. The system studied is calcium phosphate crystallizing from aqueous solution at 40°C, selected in part for its biological relevance. It has the advantage, in the actual connection, that several different crystal phases - in addition to an amorphous phase - may be formed, depending on precipitation conditions (concentrations and pH), and that both structure and morphology of the precipitate are highly sensitive to variations in these conditions (Refs.5,6). This makes the system well suited for a study like the present one.

2. EXPERIMENTAL

The gel growth experiments were carried out in test tubes of 20 cm^3 capacity or U-tubes of up to 250 cm^3 capacity. The ordinary solution diffusion experiments were carried out in the engineering models of the SGF reactors for the EURECA mission. These have a total volume of 4.5 liters and are divided in two reservoirs of 2 liters each and a mixing chamber of 0.5 liter. The valves between the chambers are opened and closed synchroneously with very low speed.

All chemicals were reagent grade. The water used as solvent was ordinary demineralized water further purified by passing through an activated carbon filter and a second mixed-bed ion exchange column. The specific conductivity of the water never exceeded twice the literature value for pure water.

Silicic acid gel was prepared by hydrolysis of tetramethyl silicate (Fluka). 5 cm^3 of ester was added per 100 cm^3 of aqueous solution, yielding a silicic acid concentration of a little less than 5 %. The tube with the gel mixture was kept at 40°C until the gel was set (at least overnight).

All experiments were carried out at 40°C. When an experiment in a SGF reactor was finished, the valves were closed, and the contents of each chamber were withdrawn, filtered and analyzed. The precipitate in the mixing chamber was examined by microscopy and X-ray diffraction. In the former case a Zeiss Jenapol Interphako microscope was used. Calcium in solutions was determined by complexometric titration with EDTA, using eriochrome black T as indicator. Phosphate was determined by colorimetry, using the molybdenum blue method (reduction by ascorbic acid with Sb(III) catalyst); the absorbance was measured at 890 nm with a Zeiss PM QII spectrophotometer.

3. RESULTS

Figure 1 shows two experiments of gel growth in test-tubes. In one the gel contained $NH_4H_2PO_4$ and $(NH_4)_2HPO_4$; it was 0.05 M with respect to either. Above this a small amount of gel without salts was placed. The

Figure 1: Test-tube experiments of calcium phosphate crystallization in silica gel. Left: calcium in gel, right: phosphate in gel.

Figure 2: U-tube experiment of calcium phosphate crystallization in silica gel. Calcium in left limb, phosphate in right.

solution on top was 0.1 M $CaCl_2$. The volumes were 8, 2 and 8 ml, respectively. The second was similar with the exception that calcium chloride was in the gel and ammonium phosphate in the top solution. It is obvious that calcium diffuses a longer distance in the gel than does phosphate. This is also seen in the U-tube experiment in Figure 2. The initial precipitate was observed near the middle of the gel column, but as the photograph shows, further precipitation occurred in the part nearest to the phosphate solution.

Figure 3 shows what may happen in a U-tube if the water activity is higher in the gel than in the solutions. Water flows out of the gel by osmosis, the gel shrinks, and the two solutions are mixed directly by flowing freely along the wall of

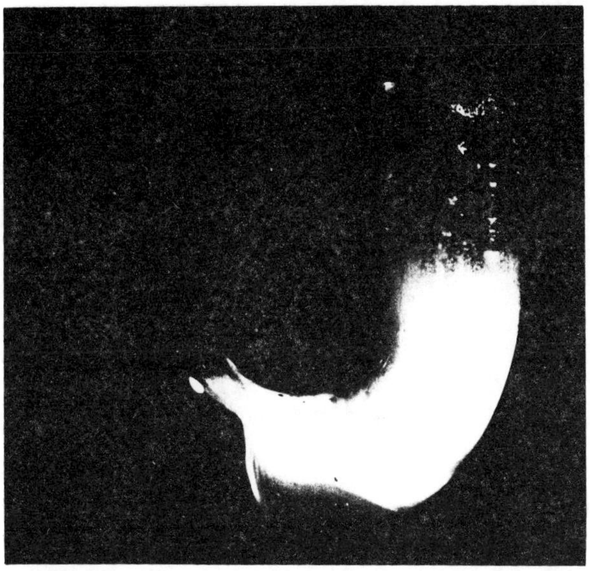

Figure 3: Cryptocrystalline precipitate along tube wall as a result of gel shrinkage due to osmosis from salt-free gel.

the tube; this leads to a cryptocrystalline precipitate as in an experiment with simple mixing. No salt was added to the gel; in the experiment in Figure 2, it was 0.15 M with respect to NH_4Cl.

The importance of density matching of solutions in ground-based diffusion experiments without gel is obvious from the results of experiments with the

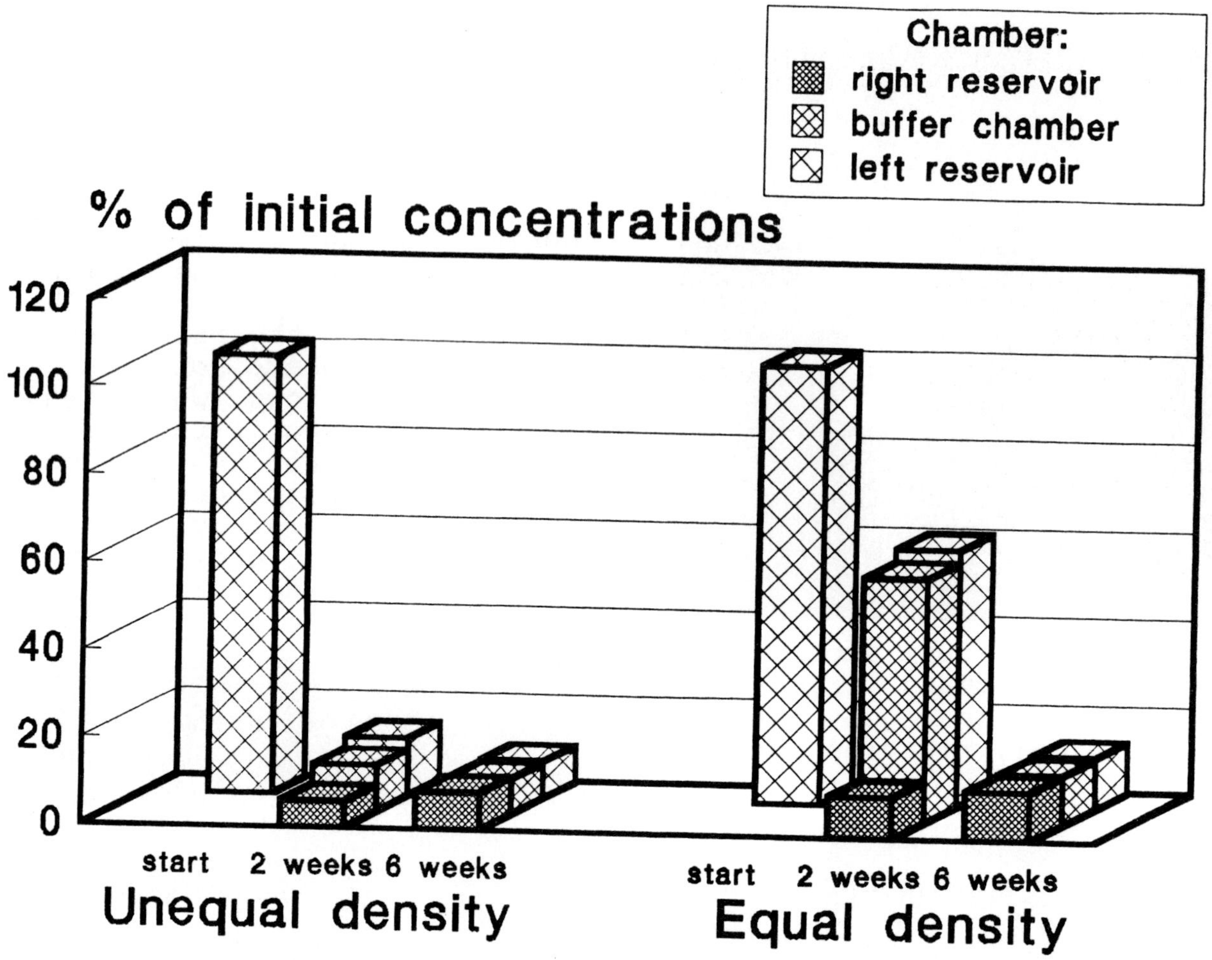

Figure 4: Solution concentrations of calcium in the three chambers of the SGF reactor in two series of experiments. Phosphate showed a similar distribution from the opposite end.

SGF reactors shown in Figure 4. In these experiments we had to replace ammonium in phosphate and chloride solutions by potassium in order to match the density of the calcium chloride solution. In the experiments with unequal densities, the water activities were matched instead. Figure 5 show microphotographs of the products of the two series of experiments. There was significantly less precipitate after 2 weeks in the reactor with equal initial densities than in the other.

4. DISCUSSION

The above results clearly demonstrate that there is no single optimal method for ground-based reference experiments in solution crystallization. The cost of eliminating sedimentation and convection in the gel is the strong interaction of phosphate ions with the gelling substance. This is not unexpected, considering the chemical similarities of silicon and phosphorus, which are neighbours in the periodic system. The consequence is, however, that gel experiments do not match microgravity very well in this case. The problem of osmosis means that if the gel method is used in reference experiments, then the mixing chamber in the reactor for the space experiment must be filled with a salt solution, not with pure water. This is also the case when the densities of the

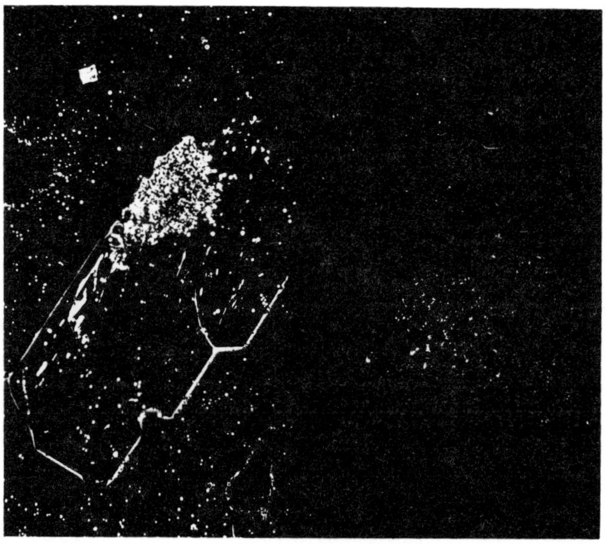

Figure 5: Positive phase contrast microphotographs of products from SGF reactor experiments. Left: Unequal density. Large crystal of $CaHPO_4,2H_2O$, small crystals of $CaHPO_4$ and agglomerates of submicroscopic $Ca_5OH(PO_4)_3$. Right: Equal density. Only submicroscopic $Ca_5OH(PO_4)_3$.

reacting solutions are matched. The latter method is probably the one which most closely simulates a microgravity experiment with the present system.

Since the number of space experiments which may be carried out is strongly restricted, it is of utmost importance that the maximal imformation is obtained from each. This implies that results of both space and ground experiments should be as simple to interpret as possible. The present study exposes some of the problems to be expected.

5. ACKNOWLEDGEMENTS

This work was supported financially by the European Space Agency, the Danish National Scientific Research Council and the Danish Space Board. We wish to thank Miss Marianne Petersen for technical assistance.

6. REFERENCES

1. A. A. Chernov, *Contemp. Phys.* **30**, 251 (1989).
2. K. H. Henisch, *Crystal Growth in Gels* (Pennsylvania State University Press, University Park, Pennsylvania 1970).
3. H. Arend and J. J. Connelly, *J. Crystal Growth* **56**, 642 (1982).
4. Y. Bernard, F. Lefaucheux, S. Gits and M.-C. Robert-Picard, *Compt. Rend. Acad. Sci. Paris* **295**, 1065 (1982).
5. H. E. Lundager Madsen and G. Thorvardarson, *J. Crystal Growth* **66**, 369 (1984).
6. H. E. Lundager Madsen and F. Christensson, *J. Crystal Growth, in press.*

Gravity Related Effects on Transport Processes During the Solidification of a Nondilute Semiconductor Alloy

J. Schilz, G. Mahr von Staszewski
DLR, Institute for Materials Research, P.O. Box 906058, D-5000 Köln 90, FRG

and A. Chait
NASA Lewis Research Center, Cleveland, Ohio 44135, USA

ABSTRACT

This paper introduces a microgravity experiment that will be performed on the Advanced Gradient Heater Facility (AGHF). A homogenous ingot of a binary semiconductor compound will be partly remolten in space and then, under microgravity conditions, directionally solidified again. Aim is the investigation of the occuring mass transport mechanisms during the solidification process. For this purpose, the residual gravity vector, temperature profiles, and the shape of the solid/liquid interface will be recorded.

As a model substance to study residual effects of gravity, the nondilute $Ge_{1-x}Si_x$ *semiconductor alloy has been selected. There are a lot of technical applications for this compound and, for the time being, no satisfactory method has been developed to grow a homogenous crystal. This is because Si segregates highly during crystal growth, and there is a lack of understanding of such segregation effects.*

Our microgravity experiments on the AGHF facility together with extensive investigations in our institutes, are intended to shed light on the phenomena appearing during growth of such a system. The parameters recorded during solidification under microgravity conditions are fundamental data to be obtained in addition to subsequent crystal characterisation. These data are input for a full numerical analysis, modelling the convectional and segregational effects during solidification.

1. INTRODUCTION

Microgravity provides a unique environment for the study of the interaction between thermosolutal convection and important aspects of solidification from the melt. However, a growing body of knowledge derived from experience gained in previous space experiments, and from the results of numerical modeling, shows that the residual gravity vector aboard the Shuttle carrier cannot be overlooked if a deep insight in the related physics is required [1, 2].

A space experiment on the Advanced Gradient Heater Facility (AGHF) endeavours to shed light on key open questions in semiconductor processing, as well as to provide a comprehensive experimental database for future theoretical studies on understanding gravity-related effects on the growth of nondilute semiconductor alloys. The influence of convection on axial and radial segregation for semiconductor alloys is of paramount fundamental and technological importance. In particular, the material system chosen, $Ge_{1-x}Si_x$, has important applications in several fields and can also serve as a model alloy of pseudobinary systems for which space experiments would otherwise be prohibitively dangerous or complicated (e. g. $Hg_{1-x}Cd_xTe$).

2. THE BINARY SYSTEM GeSi

Because of its adjustable lattice constant germanium-silicon alloys may be considered to become an important material for substrates used for other compound semiconductors such as GaAs. Besides, heavily doped $Ge_{1-x}Si_x$ is the only available thermoelectric alloy for use at high temperatures in excess of 1000 K [3]. Pure Ge and Si have been studied extensively, but a thorough study on the properties of the $Ge_{1-x}Si_x$ alloy is still lacking. For example, the ionization energy of doping impurities is known to depend on the alloy composition. This phenomenon forms a base for applications of $Ge_{1-x}Si_x$ alloys as extrinsic infrared radiation detectors [4].

The most interesting composition has a Si content of 15 at%. This is because the band gap minima in Ge and Si are located at different k-points in the Brillouin zone. Therefore the curve of the mixed crystals gap as a function of composition shows a kink, which is located at an x-value of 15 at% Si [5]. (See Figure 1.) At lower x-values the conduction band ordering remains Ge-like with the minimum at the L-point (along the [111]-direction). At higher Si-contents the lowest energy in the conduction band is located near the X-point ([100]-direction).

As the carriers are redistributed at this composition, a number of electronic properties such as conductivity, mobility, thermoelectric power, etc. show a discontinuity [3, 6, 9]. As an example the thermoelectric power dependence is shown in Figure 2. It exhibits a maximum at an x-value of 0.15.

Despite of the enormous potential of applications, up to now $Ge_{1-x}Si_x$ has not

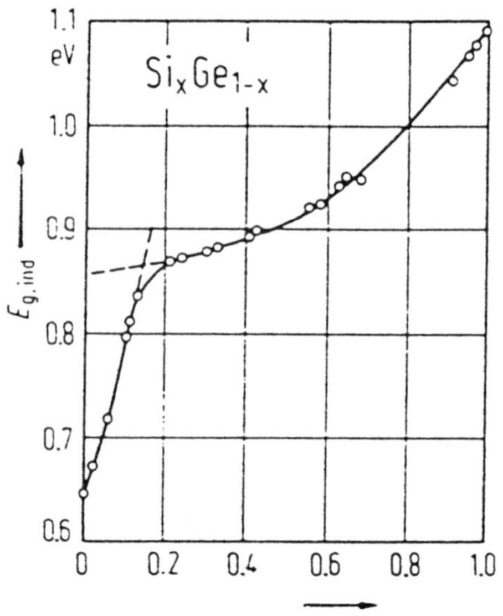

Fig. 1. **Bandgap** of $Ge_{1-x}Si_x$ as a function of composition x [5].

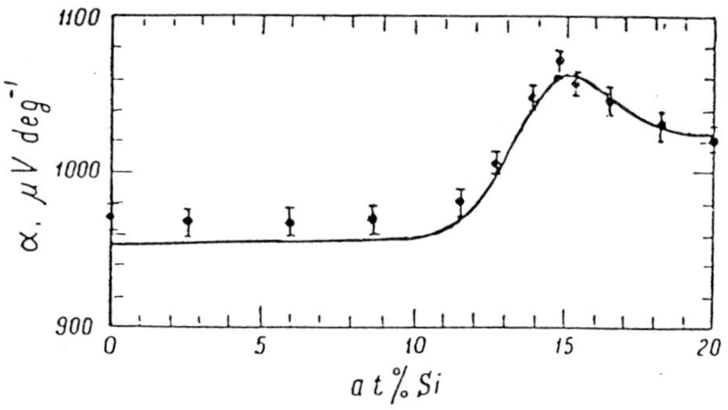

Fig. 2. **Thermoelectric power change** in heavily doped n-type $Ge_{1-x}Si_x$ at 300 K [9].

been used as a standard material in technical applications. This is mainly due to the difficulties arising in growing homogenous compounds or even single crystals.

2.1 Growth of GeSi

The main difficulties associated with the growth of homogenous alloys are the wide separation of the solidus/liquidus lines (Figure 3) [7], the low diffusion coefficient of the component elements in the solid ($\leq 10^{-11} m^2/s$) [8], and the lattice mismatch of more than 4 %.

There are a number of methods developed to grow $Ge_{1-x}Si_x$, which all try to overcome the above mentioned difficulties by

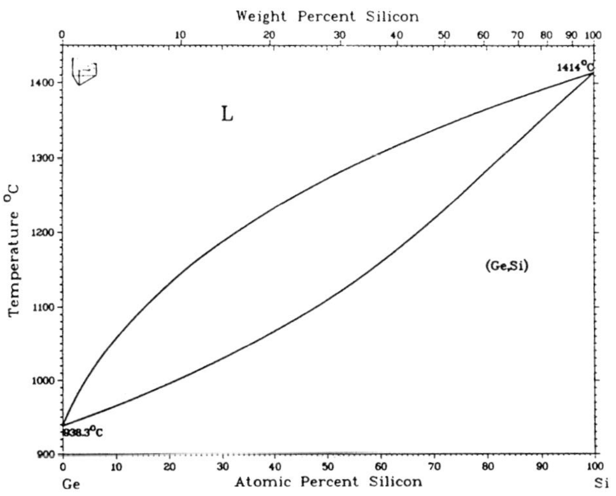

Fig. 3. Phase diagram of the $Ge_{1-x}Si_x$ binary compound semiconductor [7].

often unconventional ideas. However, a systematic knowledge in the formation of the Si segregations has not been derived.

Investigations on the solidification are the main purpose of the proposed flight project [10]. We are preparing $Ge_{1-x}Si_x$ samples by two different methods:
One is the growth of the alloy from the melt by a modified Czochralski method, where Si is continuously fed to the melt, and alternatively the formation of a binary system by means of powder metallurgy.

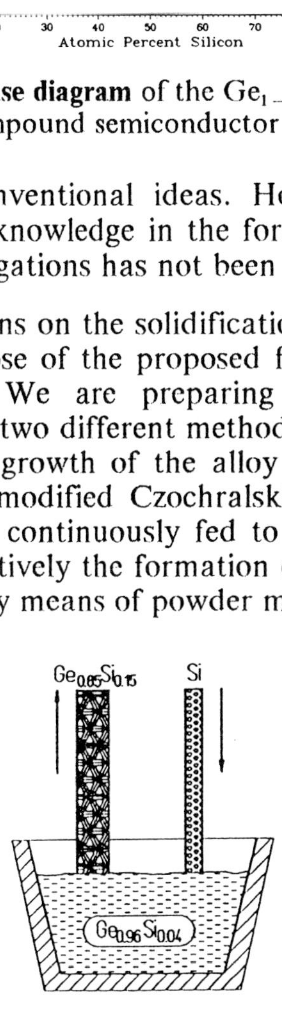

Fig. 4. Modified Czochralski method for growing homogenous $Ge_{1-x}Si_x$ crystals from the melt [13].

Figure 4. shows the principle of growing crystals from the melt [13]. Because of the high segregation coefficient of Si, the molten substance continuously looses that component. To compensate, a second bar consisting of pure Si is dissolved into the melt. The whole process is controlled by a computer measuring the growth rate to make sure that the GeSi-melt always has the same composition.

The second method is the formation of a homogenous polycrystal by a mechanical procedure. Si and Ge in the desired ratio are milled together in a ball mill under Ar atmosphere. It is possible to achieve grains having a size below 1 μm. After milling, the powder is hot pressed at a temperature of 900°C and a pressure of 200 MPa and then sintered during several weeks at 900°C. All steps are carried out in a clean atmosphere containing no oxygen [14].

Both methods yield homogenous polycrystals or even single crystals. Such $Ge_{1-x}Si_x$-ingots of a composition of 15 at% Si will be used as the starting material for the flight experiment.

3. FLIGHT EXPERIMENT

Figure 5 shows a sketch of one the cartridges used for the AGHF furnace. The sample inside the quartz ampoule has a diameter of 15 mm. The experiment will be carried out as follows: Two third of the $Ge_{1-x}Si_x$ ingot will be heated to a temperature above 1100 °C and thus remolten. The AGHF provides a thermal gradient of 100 °C/cm within the melt at the solidification interface. During the directional solidification the solid-melt interface will be periodically marked by electrical current pulses to allow observation of interface shape and location. Such a pulse demarcation can either be provided by electrodes connected to the crystal and the melt or by periodically rising the heater's temperature 2 or 3 degrees. 5 thermocouples between the crucible made of quarz or boron nitride and the cartridge made of niobum or tantalum, and four thermocouples placed at the ampoule's axis will record the temperature distribution within the liquid. One cartridge will have 4 additional thermocouples within the melt.

Solidification of a compound having a highly segregating component leads to a

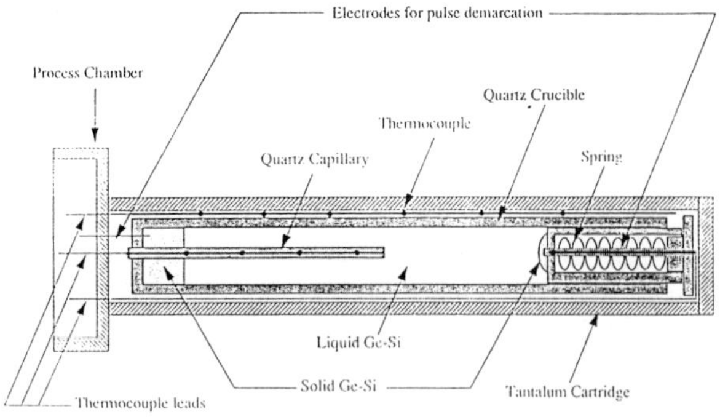

Fig. 5. Sketch of crucible for the AGHF with electrodes for pulse demarcation, thermocouples and high temperature spring.

density gradient in the melt in front of the phase boundary. Such an inhomogenous, diffusion governed concentration profile leads, under gravity conditions, to solutal damping. By increasing the growth rate, one can thus change the mass transport properties from convective to diffusion controlled flow [15, 16]. In a low gravity environment only diffusion will be the driving force for the melt components.

Marangoni convection is eliminated by covering any free surfaces of the melt with a quartz stopper. Because of the volume change during phase transition one has to make sure the surface is always covered. This is achieved using a spring, which has already been developed [17].

4. EVALUATION OF GROWTH PARAMETERS

To investigate the influence of the residual gravity, two different types of samples will be processed during the flight experiment: One sample will be solidified under quiet microgravity conditions and a second one under μg-conditions, where strong variations of the gravity vector occur. A time dependent recording of magnitude and direction of these g-jitters by accelerometers together with the measurement of temperature fields and shape of the solidification front and subsequent comparison of the data to those of samples processed on earth, will help to clarify the role of g-variations on solidification.

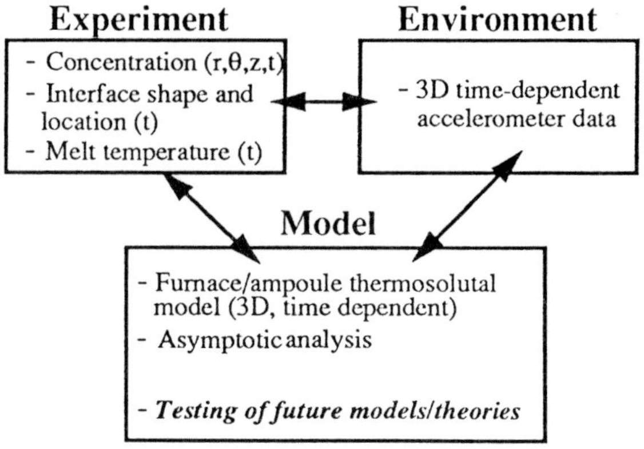

Fig. 6. Relationship among proposed experiment, space environment, and modeling.

The post-flight analysis of the samples contains the observation of concentration profiles in radial and longitudinal direction and the analysis of the time dependent solid/liquid interface shape. Together with the data recorded during the flight experiment, this will form a database for a numerical study of the effects of the transport phenomena in the melt. Such a computational study intends to couple the simultaneous measurements of the temperature field, the solid/liquid interface location through demarcation, and the three-dimensional gravitational field through accelerometers. This data will be used to propose models which predict the causal relationship between residual gravitational acceleration, thermal and concentration fields. Figure 6 illustrates this concept of such a joint physical/computational experiment.

Special attention has to be paid to the direct start of recrystallisation: Here the Si contents will show a discontinuity and thus a high intercrystalline strain occurs. The amount of strain can be observed by measuring the band gap across this interface location.

5. CONCLUSIONS

Microgravity experiments to be carried out on the AGHF facility will help to understand the formation of compound crystals of the technological important $Ge_{1-x}Si_x$ system. For the first series of experiments the material system will have a composition of 15 at% Si. This is a potentially good model system for other semiconductor compounds, such as $Hg_{1-x}Cd_xTe$ which is presently difficult to study in detail. In addition, $Ge_{1-x}Si_x$ is a semiconductor with its own applications, i.e. in thermoelectric generators, infrared detectors, solar cells, substrate material, etc.

The data collected during the flight experiment, as well as post flight crystal and electrical characterisations will be used as base for a comprehensive modeling approach of the solidification process. As a whole, the μg-experiment and the investigations on the $Ge_{1-x}Si_x$ systems prepared under different conditions propose to give new results for the use of $Ge_{1-x}Si_x$ in technical applications. Gscheidner , Jr., F. C. Laabs,

6. BIBLIOGRAPHY

[1] J. I. D. Alexander, J. Quazzani, F. Rosenberger, *Analysis of Low Gravity Tolerance of Model Experiments for Space Station: Bridgman Technique* NASA Contract NAG8-684, First Semi-Annual Report (1988)

[2] E. S. Nelson, *An Examination of Anticipated g-Jitter on Space Station and Its Effects on Materials Processes,* NASA Technical Memorandum 103775, (1991)

[3] C. M. Bhandari, D. M. Rowe, *Silicon-Germanium Alloys as High-Temperature Thermoelectric Materials,* Contemp. Phys. 21 (1980) 219

[4] M. L. Schultz, *Ionization Energies of Some Impurities in Ge-Si Alloys,* Bull. Amer. Phys. Soc. 2 (1957) 135

[5] R. Braunstein, A. R. Moore, F. Herman, *Intrinsic Optical Absorption in GeSi;* Phys. Rev. 109 (1958) 695

[6] G. Busch, O. Vogt, *Elektrische Leitfähigkeit und Halleffekt von Ge-Si-Legierungen,* Helvetica Phys. Acta 33 (1960) 437

[7] T. B. Massalski, Am. Soc. for Metals (1986) p. 1248

[8] Y. Xu, B. J. Beaudry, K. A. *Pressure-assisted reaction bonding between a W sheet and a SiGe alloy,* J. Appl. Phys. 68(9) (1990) 4846

[9] E. V. Khoutsishvili, M. G. Kekoua, *Electrical Properties of Ge-Si Alloy System,* Scripta Fac. Sci. Nat. Univ. Purk. Brun. 16 (1986) 265

[10] J. P. Dismukes, L. Ekstrom, *Homogenous Solidification of Ge-Si Alloys,* Transact. of the Metallurgical Soc. of AIME, 233 (1965) 672

[11] M. S. Saidov, R. S. Umerov, *Growing single crystals of silicon-germanium solid solutions by electron-beam zone melting without a crucible,* Sov. Phys. Crystallogr., 23 (1978) 130

[12] R. M. Davis, C. C. Koch, *Mechanical Alloying of Brittle Components: Silicon and Germanium,* Scripta Metallurgica, 21 (1987) 305

[13] S. I. Tairov, V. I. Tagirov, M. G. Shakhtakhtinskii, A. A. Kuliev, Report of the Acad. Sci. USSR 176 (1967) 851

[14] J. Schilz, M. Langenbach, *Formation of GeSi Solid Solution by Powder Metallurgy,* (1991) to be published

[15] E. Karthaus, Dissertation TH Aachen, Mat. Nat. Fak., FRG, 1991

[16] G. Mahr v. Staszewski, *Solidification of the GeSi (15 at% Si) Solid Solution,* J. Mat. Sci. Lett. 10 (1991) 451

[17] H. Buhl, *High-Temperature Spring for use in excess of 1000°C,* (1991) (to be published)

Programme of Space Experiments

THE MICROGRAVITY RESEARCH PROGRAMME OF THE EUROPEAN SPACE AGENCY (ESA)

H.U. Walter, ESA Headquarters, Paris (France)

Abstract

The first milestone for microgravity research in Europe was the First Spacelab Mission in 1983 followed by the German Spacelab Mission D-1 in 1985. Future Spacelab flights with ESA participation are NASA's IML-1 (Jan. 1992), IML-2 (1994) and the German D-2 mission (1993). ESA is planning a further European Spacelab flight dedicated to microgravity in 1996 (E 1). In addition, ESA's Retrievable Carrier EURECA, scheduled for launch in July 1992 with reflights in 1995 and 1997 will greatly expand the opportunities for Europeans to conduct research under microgravity conditions. Continuity in microgravity research in Europe is provided by various Sounding rocket projects (TEXUS, MASER, MAXUS) and three to four payloads carrying some 40 experiments are flown annually. Caravelle aircraft flying parabolic trajectories and drop tower (Univ. Bremen) and drop tube (CEA Grenoble) facilities are used for short duration experiments. With ESA's participation in the International Space Station FREEDOM and its contribution COLUMBUS (Man Tended Free Flyer and Attached Laboratory) there will be ample opportunities for microgravity research towards the end of the century.

A wide variety of flight experiment facilities for research in Materials and Fluid Scineces as well as Life Sciences has been developed. Besides operation by payload specialists and automation one can employ robotics, expert systems and experiments can also be controlled by the scientists from the ground directly (telescience).

Results obtained demonstrate the potential of microgravity as a tool for research. Commercial applications may develop in the long term, however, they are presently not emphasized in ESA's programme. Progress in microgravity research depends on the efficient use of the existing resources and the concerted effort of the international scientific community. ESA is therefore fostering cooperation not only within Europe but also with the USA, the CIS and Japan.

1. Introduction

The ESA Council at ministerial level held in January 1985 and in November 1987, affirmed the commitment to expand Europe's autonomous capability and competitiveness in all sectors of civil space activities. Following that resolution, the member states of ESA have embarked upon a major expansion of their space programmes. This encompasses not only the traditional space activities but also the establishment of permanent manned systems in space, in order to exploit and to utilise the unique features of earth orbiting laboratories. The observation of the earth and its environment, cosmic observations and telecommunications are major issues. In addition there is the microgravity condition in orbiting spacecraft, which is of major interest for research in life sciences, materials and fluid sciences.

Microgravity opens a new avenue for experimental research in Life Sciences, Fluid and Materials Sciences. The essential features of microgravity environment are the elimination of gravity-driven convection, the lack of sedimentation and buoyancy and the absence of fluidstatic pressure. Furthermore the synergistic effects of near weightlessness and radiation are of major interest in life sciences.

In the Materials Sciences, one can thus investigate phase transitions, solidification, crystal growth and chemical reactions under conditions of diffusion-controlled heat and mass transfer. There is furthermore the feasibility of containerless processing, since samples can be easily manipulated in a levitated or a freely floating state. In Fluid Sciences the focus is on hydrodynamics in the absence of gravity-driven convection, capillarity, the stability of liquid bridges, near critical point phenomena, the precise determination of physicochemical constants as well as model experiments on combustion etc.

In the Life Sciences, microgravity provides unique experimental conditions that cannot be achieved on Earth. The evolution of life on earth has been responding to the gravitational field in many respects and microgravity studies on the underlying mechanisms in cellular processes may add a new dimension to our basic knowledge in biology. Studyinghigher organised life forms such as plants, small animals or even man himself should reveal gravity-dependent processes and responses to gravity, which cannot be observed on the ground.

Opportunities to conduct research in microgravity have been very limited so far. Within the next decade, this situation should change radically. From a few tens of hours of microgravity research per year, the European user community will have permanently manned orbiting laboratories available. This sudden increase in opportunities will change the pace of microgravity research demanding a coherent strategy, a well defined research programme, the necessary facilities and infrastructure, and an enthusiastic, competent and sufficiently large community of users.

Presently microgravity research is still in a phase of exploration. It is therefore difficult to predict how it will evolve and on which topics should be emphasized. It is even more difficult to predict areas of interest for industrial applications. The best way to proceed may be to follow in a first approximation the requests by the scientific community and to ascertain a detailed evaluation of proposals by high level peer groups. In addition external advisors (e.g. the Microgravity Advisory Committee, MAC) could make recommendations on topics which should be emphasized in ESA's microgravity programme in order to develop "focused science" projects in parallel to the diversity of individual projects.

This paper reviews the microgravity programme of ESA. Following a brief overview of current research activities, the major multi-user experiment facilities as well as the flight opportunities and experiment carriers are described. Finally, some general considerations and concerns are presented.

2. Current topics in ESA's Microgravity programme :

ESA's Microgravity programme concentrates on two major disciplines :
Materials and Fluid Sciences, and
Life Sciences

These disciplines are subdivided as follows :

Materials and Fluids Sciences :

Fluid statics and capillarity
Fluid dynamics
Near critical point phenomena and phase transitions
Chemical pattern formation
Physical chemistry
Mass transport and diffusion
Combustion

Crystal growth from the melt/solidification
Crystal growth from the vapour phase
Crystal growth from solutions

Protein crystallization
Metals, alloys and composites
Metastable phases and glasses

Life Sciences :

Human Physiology and Space Medicine
Developmental biology
Plant biology
Cell biology
Radiation biology
Exobiology.

The main features of interest of Microgravity are the elimination of gravity-driven convection, sedimentation and fluidstatic pressure. The feasibility of containerless processing is another advantage of this unique environment.

Density gradients in liquids and gases arise wherever temperature or concentration gradients exist. This is the case with any heat treatment and any process involving phase transitions or chemical reactions. Density-gradients and their corollary, buoyancy-driven convection, are thus modifying numerous processes involving fluid systems on Earth. The effects of transient convective flow may dominate heat and mass transfer and may thus limit the control of numerous processes such as crystallisation. It prohibits the execution of defined model experiments for testing theoretical models. Under microgravity conditions gravity-driven convection is virtually eliminated. Consequently, a new generation of experimental research in Fluid and Materials sciences becomes feasible.

Phenomena of fluid dynamics can be investigated in space without the interference of gravity-driven convection. Results from such studies are expected to provide important feedback for the testing and development of theories describing three-dimensional laminar, oscillatory and turbulent flow regimes and flow phenomena generated by various other driving forces.

In space, diffusion-controlled transport conditions in fluids can be achieved. This has been demonstrated with a variety of mass diffusion experiments, whereby exceptionally precise data on self-diffusion, hetero-diffusion and thermo-transport (Soret effect) have been obtained.

In crystallization, the control of heat and mass transfer presents a major problem, and microgravity is therefore of great interest for the crystallization of semiconductors, the solidification of metals and alloys and for crystal growth in general. Virtually all crystal growth processes from melts, solutions and vapours involve thermal and concentration gradients driving buoyancy flow on earth. The crystallisation under conditions of diffusion-controlled heat and mass transfer in microgravity should enable one to optimise these processes and to control the product. Chemically homogeneous semiconductors have been obtained and crystals grown from the vapour phase and from solutions have greatly reduced defect densities and mosaic spread.

One of the very promising projects is the crystallization of proteins. It is estimated that there exist about one million different proteins, yet the molecular structure of only about 200 have been unravelled. The difficulty exists in obtaining sufficiently large (0.1 to 1 mm) and sufficiently perfect crystals for X-ray diffraction analysis. For neutron diffraction even larger crystals (order of some mm) are needed. There are clear experimental indications that the crystallisation of proteins is enhanced in space. Evewn though the reasons for this improvement is not understood yet, very recent experiments have yielded a variety of protein crystals of a quality superior to that which could be obtained on the ground.

In space the behaviour of mixtures of liquids with gases and solids can be investigated without sedimentation causing separation of the components according to their densities. The elaboration of composites such as high porosity alloys or particle and fibre dispersions which cannot be obtained on the ground can be investigated in space.

The absence of fluidstatic pressure together with the lack of gravity-driven convection provides optimal conditions for the investigation of capillarity phenomena, phase transitions, and critical point phenomena in fluids. As the critical point is approached, the compressibility of fluids increases by orders of magnitude and on earth a fluid is therefore compressed under its own weight. The fluid is stratified according to density, and homogeneous bulk samples needed for investigating phenomena near the critical point can only be obtained in space.

In the absence of gravity melts can be positioned and manipulated without containers. Containerless processing is of interest since crucibles can be the source of contamination, heterogeneous nucleation, and mechanical stress. So far very few microgravity experiments on glass formation have been undertaken. There is the possibility of studying undercooling, nucleation, and the formation of metastable phases in the containerless mode.

Combustion is an example of a process where all of the features of microgravity come to bear. Combustion involves complex chemical reactions with large temperature and concentration gradients and, therefore, strong convection. Microgravity offers the opportunity to investigate combustion phenomena under diffusion-controlled conditions. In addition samples can be investigated while freely floating. The combustion of levitated droplets or clouds of particles or droplets (Diesel engine) and the burning of supercritical fluids are examples demonstrating the relevance of the microgravity environment for conducting unique experiments.

Gravity effects on living systems are difficult to isolate. This is primarily due to the built-in regulatory and compensatory mechanisms which are inherent to any life form, which respond to the new environment in complex ways. Exposing the system to weightlessness may reveal scientifically important underlying processes which are hitherto unknown and cannot be studied on earth. Therefore, life sciences in space encompass research areas which are related to gravity detecting and gravity responding systems. Experiments in cell and developmental biology give strong evidence that gravity affects biological systems. In particular changes in cell differentiation and cell proliferation were observed. Experiments in human physiology gave new insights into the functioning of the vestibular system and the homeostatic regulation of the cardiovascular system.

Provided that the conditions for conducting fundamental research will be favorable one can expect microgravity research to mature to a level where industrial applications in the widest sense will be derived. At present this component is barely visible at the horizon and it should not prohibit the scientific cooperation on an international level in microgravity research.

3. Flight opportunities/Microgravity experiment carriers

The first experiments under microgravity conditions were carried out in the early seventies during the flyback of Apollo spacecraft from the moon. These early demonstrations were followed by investigations on Skylab in 1973/1974 and ASTP in 1975. Microgravity research activities in the USSR were intensified with the availability of Salyut-7 in 1982 and with the setting up of a permanently manned laboratory "MIR" in 1984. The first flight of the SACELAB built by ESA was in November 1983 (ESA mission SL-1) followed by the German Spacelab mission D-1 in November 1985. A Spacelab mission by NASA, which was also dedicated to microgravity research was conducted in 1984 (SL-3). The Challenger disaster put a halt to these activities for several years and the technical difficulties encountered during 1990 caused further delays. The latest Shuttle manifest issued in December 1990 can be viewed with a certain confidence. The microgravity missions that are included are shown in the following table:

The most important mission coming up for European researchers is the first mission of the EUROPEAN RETRIEVABLE CARRIER, EURECA-1, in 1992. This mission has been delayed for several years and the experimenters are waiting impatiently. As an average per year at least one microgravity-dedicated mission of Spacelab is on the NASA manifest up to 1998, when the International Space Station FREEDOM should become operational. ESA is preparing for two Spacelab flights, E-1 and E-2 in 1995 and 1996 as well as for two EURECA flights in 1994 and 1997 (COLUMBUS Precursor Flights). A call for proposals has been issued recently and the letters of intent that were received reflect the considerable interest of the science community in Europe in the opportunity offered. More then 500 proposals were received, approx. 70% are concerned with microgravity.

A certain continuity in microgravity research in Europe could be maintained with the Sounding Rocket programmes TEXUS and MASER.
27 payloads have been launched successfully with the German TEXUS and 4 payloads with the Swedish MASER. ESA has been participating in these flights at a level between 50 and 90 %. The duration of free fall conditions is about 6 minutes with experiment payloads of approx. 250 kg. A new project, MAXUS, has been initiated , the first flight was in May 1991. The payload amountsa to about 300 kg, which is to be carried to an altitude of almost 900 km, which corresponds to a duration of free fall conditions of approx. 13 minutes. Unfortunately this first flight was a failure due to technical reasons. It will be repeated in May 1992.

In addition simple look and see experiments and simple tests can be carried out on board of Caravelle aircraft flying parabolic trajectories. Somwe 20 to 25 seconds of low gravity conditions can be achieved and there is considerable interest by scientists and engeneers in this opportunity.
The Drop Tower in Bremen has become operational in September 1990, large experiment packages can be exposed to free fall conditions during 4.7 seconds (heigth 110 m). An ultrahigh vacuum drop tube which is 47,1 m high, allows for free fall experiments particularly on containerless solidification during 3.1 seconds.

Despite of these activities, one must realize that the total experiment time logged for microgravity research in the western world amounts to approximately 2000 hours at most. This is marginal in demonstrating in any detail the potential of the microgravity environment as a tool for research in experimental physics and it is totally insufficient to demonstrate the potential of microgravity for industrial R&D not to mention industrial production. Although numerous investigations were conducted in the former USSR on Salyut Space Stations and on MIR, there is rather little information about these and the present Conference was the first occasion to meet the scientists involved and to obtain first hand information.

The forthcoming International Space Station FREEDOM with the European element COLUMBUS will provide for a permanently manned orbiting laboratory with ample opportunities for Microgravity Research.

"FREEDOM" consists of a cluster of three laboratories which are manned permanently : the U.S. laboratory by NASA, the Japanese Experiment Module (JEM) with an exposure facility and the COLUMBUS attached laboratory by ESA. The station will be equipped with solar panels for electrical power, and Canada will provide a robotic servicing system. Further elements will be the Man Tended Free Flyer pressurized laboratory and a Polar Platform, both by ESA. In addition a European Space Shuttle, HERMES, is foreseen.

The framework within which ESA provides its elements is called the Columbus Programme.

The Columbus Attached Laboratory is planned to be launched in 1997, the Polar Platform a year later, and the Columbus Free Flying Laboratory in 1999. The Columbus Attached Laboratory will of course be complemented by the U.S. and Japanese equivalents. It will rely on the U.S. Space Station system for resources and for the crew habitation and maintenance.

A payload mass of up to 10.000 kg should be available in the Columbus Attached Laboratory, which can be accommodated in some 47 single racks, each having a size of about 2/3 of a Spacelab rack. This payload includes not only the experiment systems but also experiment storage and general purpose user support.

The crew time available per rack and day, will be substantially lower than for Spacelab. There, some 20 racks were supported by at least 2 operators. For the Columbus Attached Laboratory only one astronaut from a total of eight available, would normally be allocated. Therefore, experiments will have to be automated and the system operations will differ from those used on Spacelab.

The overall characteristics of the Columbus Attached Laboratory are summarized in the following table :

Overall dimensions	
length :	12.8 m
diameter	4.0 m
launch mass	14700 kg (depending on Space Shuttle performance)
Operational mass	up to 27900 kg
Payload mass	up to 3000 kg at launch, up to 10000 kg operational
Payload racks	47 single racks equivalent
Volume	
Payload dedicated	38 m^3
Crew compartment	52 m^3
Total volume	115 m^3
Power	20 kW total 10 kW average to payload, 120 V DC
Communications	100 Mbps downlink 25 Mbps uplink
Operational orbit	450 km circular 28.5° inclination
Design life	30 years

The performance specifications project a gravity level lower than 10^{-5}g.

The Columbus Free-Flying Laboratory is about half the size of the Attached Laboratory. Gravity levels are expected to be substantially lower than for the Attached Laboratory, namely 10^{-6}g. Servicing will occur in 6 months intervals. Consequently, the operations will be fully automated. Power levels will be around 5 kW to the payload. Data rates will be 100 Mbps downlink and 25 Mbps uplink.

4. Flight Experiment Facilities

Besides the facilities provided by ESA various national Space Agencies, especially the German DLR/DARA, the French CNES and the Italian ASI have developed microgravity research facilities in their national programmes.

The most important ones are:

ITALY:

Fluid Physics Module (FPM)

FRANCE

Holographic Interferometer (HOLIDDO)
Electrophoretic Separator (RAMSES)
Solidification Fundamental Research System (MEPHISTO)
Echograph, Ergometer, Portable Doppler
Instrumentation for Neurophysiology

GERMANY

Holographic Optics Laboratory (HOLOP)
High Precision Thermostat (HPT)
Mirror Furnace (ELLI)
Gradient Furnace with Quenching Device (GFQ)
Zone Melting Furnace (ZMF)
Isothermal Heating Facility (IHF)
Electromagnetic Levitator (TEMPUS)
Biolaboratory
Zentrifuge (NIZEMI)

This list includes only the major developments. The variety of facilities is rather impressive and one should be aware of the fact that the coordination is sometimes difficult and duplication and overlap could not always be avoided. Worldwide this situation is even more complex, there are for example 12 different facilities for the crystallization of proteins and one can ask oneself whether this is required for the advancement of microgravity research. The international cooperation on board of the International Space Station FREEDOM will require a concerted effort in order to achieve progress in a cost efficient way. Microgravity research is expensive and duplication should be avoided on all levels.

The International Space Station FREEDOM once it becomes operational will be a major challenge to the International Microgravity Research Community. The opportunities offered will be orders of magnitude larger then what we are used to. ESA is presently initiating the following laboratory developments:

COLUMBUS ATTACHED LABORATORY

BIOLAB	4,5 RACKS
ANTHROLAB	5 RACKS
FLUID SCIENCES LAB	4 RACKS
HIGH TEMP.MAT. PROCESSING LAB	5 RACKS
CRYSTALLIZATION LAB	4 RACKS
CONTAINERLESS PROCESSING LAB	2,5 RACKS
TOTAL	25 RACKS

COLUMBUS FREE FLYING LABORATORY

AUTOMATED CRYSTAL LAB	3 RACKS
AUTOMATED BIOLAB	3 RACKS
AUTOMATED HIGH TEMP. FAC.	2 RACKS
AUTOMATED FLUID SCIENCES LAB	2 RACKS
TOTAL	10 RACKS

For more detailed information see the experiment catalogues for MS/FS facilities (1,2,3).

CONCLUSIONS

The first probing steps by researchers engaged in materials and fluid sciences as well as life sciences to elucidate the role of the body force of gravity on various phenomena and processes were made during the last decade. A wide variety of topics relevant for microgravity research have been identified. However, very few investigations have been conclusive so far.

The permanently manned capability and the payload capacity of the International Space Station FREEDOM will enable researchers to conduct large series of long duration experiments to investigate a given problem in all the details needed. Presently the user community is not prepared for this opportunity. The number of scientists actively engaged in research on gravity-dependent phenomena is rather limited. Well prepared and high quality flight experiments can only be expected if there is sufficient activity in the research laboratories. This calls for a considerable increase in the funding for ground-based research not only in order to increase the user community but also in order to exploit all the cheaper possibilities of investigation on the ground prior to endeavouring in expensive experimentation on board of orbiting laboratories.

Depending on the type of experiments, appropriate ways of operation have to be elaborated, spanning from experimentation and operations conducted by astronauts to full automatisation. Teleoperations, where the experimenter can observe on the ground and modify the settings of his experiment, should become a very important mode conducting investigations. This requires the possibility of observing in real time during the critical phases of an experiment. Continuous data links with sufficiently high data rates are therefore required, which is presently not the case. The use of expert systems to control and monitor experiments is another possibility, which will need to be developed. Eventually a combination of these different modes of operation will have to be employed in order to render the operations as efficient as possible and to optimize the scientific return.

The complexity of the task both from a technical and operational point of view and also regarding the coordination between the various participants of the Space Station FREEDOM project is very high indeed. Considerable efforts will be needed by space agencies and industry, and the scientific community will have to play a major role in assuring that the best possible use will be made of these systems.

REFERENCES

1. H.U.Walter, Flight Experiment Hardware for Sounding Rocket Investigations. ESA-SP-1116, August 1989

2. H.U.Walter, Facilities for Materials and Fluid Sciences Experiments
embarked on Spacelab. ESA SP-1120

3. H.U.Walter, W.Riesselmann, Facilities for Materials and Fluid Sciences Experiments embarked on EURECA, ESA SP-1118

OUTLINE OF THE JAPANESE SPACE ACTIVITIES IN THE FIELD OF MICROGRAVITY

Kazuo Ishida

Space Technology Corporation

1-4-5 Kudankita, Chiyoda-ku, Tokyo, 102, Japan

ABSTRACT

Outline of the Japanese space activities in the field of microgravity is described in this paper. Most of the microgravity experiments are dependent on those of USA as well as Europe so far. In addition to those, various domestic experiment opportunities are becoming available recently.
The major microgravity experiment program in Japan now in progress consists of the following projects.
They are grouped as follows;
a) joint programs with other countries developing systems and/or experiment facilities calling for potential users.
b) domestic programs developing systems and experiment facilities calling for potential users.
c) programs supporting experimenters to use established missions.

1. Japanese experiment module(JEM) to be attached to the international space station Freedom. (a) ('98-)
2. Space Lab.
   * FMPT(first material processing test)(a)('92)
   * IML-1('92),IML-2('94)(a)
   * D-2(c)('93)
3. Space flyer unit (SFU)('94)(a)
4. Parabolic flight using airplane (MU-300)(c) ('90-)
5. Sounding Rocket (TR-1A)(b)('91-)
6. Reusable capsule, joint project with Germany (Express)(a)('94)
7. Drop shaft (JAMIC, MGLAB)(b)('91-)

The outline of the above individual projects will be explained.

Keywords: Japanese microgravity program. Review.

1. Introduction

Space related policy in Japan is decided by Space Activities Commission(SAC), which secretariat is the Science and Technology Agency(STA).
The microgravity activities consist of hierarchical part having space station(JEM) in view and flexible part having stimulation of science and technology in view.
There are three centers for microgravity utilization in JEM.
a) Institute of Space and Astronautical Science (ISAS): for universities.
b) National Space Development Agency of Japan (NASDA): for national institutes.
c) Space Utilization Promotion center (JSUP): for industries.

The microgravity experiment program for the preparation of JEM in these three years has been established by NASDA/STA. The program for '95-'98 is now under planning. Ministry of Education(MOE) is supporting ground based research of universities via ISAS. Ministry of International Trade & Industry(MITI) encourages industries for this field in a flexible way via Society of Japanese Aerospace Companies,Inc.(SJAC), Key Technology Center of Japan(JKTC), Institute for Unmanned Space Experiments Free flyer(USEF) and New Energy and industrial technology Development Organization(NEDO). Thus, microgravity researchers are encouraged in various ways, resulting in active state with some confusion. In addition to participating in the developing program, users are often called for performing experiments flexibly in the established foreign microgravity opportunities such as;
space shuttle(GAS), sounding rocket(TEXUS, MASER) airplane(KC-135, Caravelle), reusable capsule (Photon), drop tower(ZARM). Activities of this kind will not be involved in this paper.
Semiconductor materials research of Space Technology Corporation will be explained more in detail as an example.

2. JEM

Japanese Experiment Module for the international space station is being developed by NASDA/STA. It consists of a pressurized module(PM), an experiment logistic module(ELM) and an exposed facility(EF) as shown in Figure 1.

PM: multipurpose laboratory in which material processing experiments, life science experiments, etc. are performed.
ELM: It is used to store and resupply experimental samples, gases and to transport materials between JEM and the earth.
It consists of a pressurized section and an exposed section.

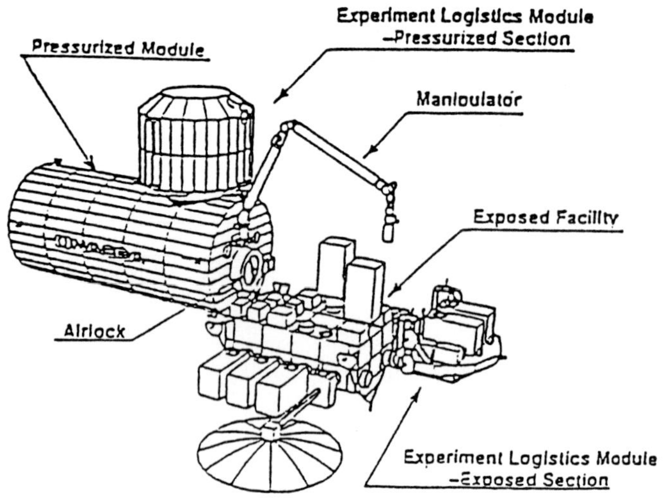

Figure 1. JEM configuration

EF: a working station for scientific observation, communication experiments, science and engineering experiments, material experiments, etc. They are performed mainly by remote manipulation.

Announcement of opportunities for experimenters are under preparation and will be made soon. Public invitation for candidates of the astronauts for JEM will be made from July to August this year. The systems to support potential experimenters are under discussion.

3. FMPT

The FMPT program, which is in charge of NASDA(STA), is now ready for flight by NASA space shuttle/ spacelab. The experiment facilities to be installed in three double-racks have been developed in Japan. A Japanese payload specialist will be on board for 7 days to perform following 34 experiments (22: material processing, 12: life science).

Material processing:
* Large high-quality compound semiconductor crystal.
* Homogenous amorphous semiconductors.
* Glasses tranparent for infrared light.
* Samarskite crystal.
* Organo-metallic crystal.
* Superconducting composite.
* Particle dispersion alloys.
* Carbon fibre hybridized composites.
* Improved metal deoxidation process.
* Improved liguid-phase metal sintering.
* Improved gas evaporation in metals.
* Sphere Si crystals .
* Behavior of molten glasses.
* Interdiffusion of molten metals.
* Solidification of immiscible alloys.
* Behavior of liquid drops.
* Behavior of bubbles.
* Marangoni convection.

Life science:
* Elecrophoresis of biomaterials.
* Crystal growth of enzymes.
* Cosmic radiation protection technology.
* Radiation biology.
* Physiological effects.
* Biological rhythm.
* Metabolism.
* Vestibular function.
* Visual stability.
* Perceptual-motor function

This program has been delayed for several years due to space shuttle accident in 1986.

4. IML(International Microgravity Laboratory)

NASDA has joined the IML program of NASA using space shuttle/spacelab. In IML-1, 16 experimental equipments developed by 13 countries will be used for 39 themes. Specific onboard facilities has been and will be developed in Japan for the following experimental themes;

IML-1
* Crystal growth of organic superconductors.
* High energy cosmic radiation monitoring and analysis of biological specimens.

IML-2
* Mechanism of vestibular adaptation of fish.
* Fertilization and embryonic development of Japanese newt.(Pending final approval by NASA)
* Mating behavior of the fish(medaka) and development of their eggs. (Pending final approval by NASA)
* Realtime dosimetry and biological experiments for heavy charged particles.
* Growth and differentiation of cultured bone-derived cells.
* Differentiation of dictyostelium discoideum. (Pending final approval by NASA)
* Relation between behavior and proliferation in protista.(Pending final approval by NASA)
* Separation of chromosome DNA of a nematode, C. elegans, by electrophoresis.
* Experiments of separating the culture solution of animal cells in high concentration.
* Microstructure and strength of ordered TiAl intermetallic alloys.
* Mixing of melt of multicomponent compound semiconductor.
* Influence of G-jitter on natural convection and diffusive transport.
* Study on thermally driven flow.
* Crystallization of food protein. (Pending final approval by NASA)
* Crystallization of bacteriorhodopsin. (Pending final approval by NASA)

5. SFU

An unmanned free flyer unit is being developed as a joint project of MITI, MOE and STA under management by USEF. It will be launched by the Jpanese H-2 rocket and be retrieved by NASA space shuttle after 6 months' operation in orbit. Six payload boxes of ca. 1 ton in total weight are integrated on it with solar cell panel extending to a length of 24 m as shown in Figure 2.

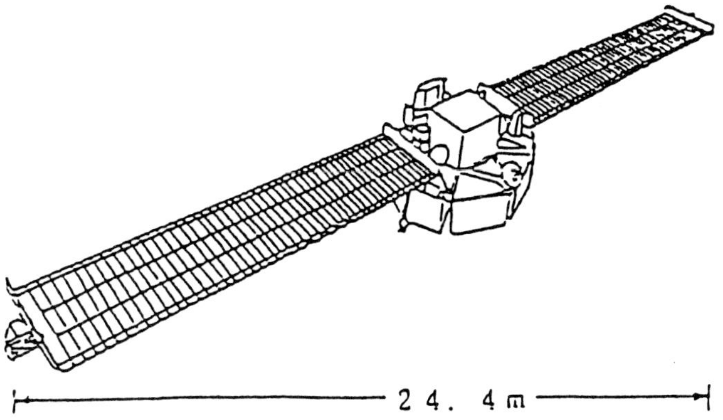

Figure. 2

SFU flight configuration

Experiments on the first flight are under preparation as follows;

* Crystal growth of GaAs by Bridgman method.
* Non contact crystal growth of CdTe by Bridgman method.
* Thin film crystal growth of CdS from solution.
* Crystal growth of AlGaAs from solution.
* Crystal growth of InP by Bridgman method.
* Thin film crystal growth of InGaP from vapor.
* Crystal growth of InP by travelling-heater method.
* Crystal growth of InGaAs by travelling-heater method.
* In situ observation of crystal growth of $Ba(NO_3)_2$
* Hibernation, fertilization and development of newt.
* Crystal growth of diamond from vapor.

6. MU-300

Microgravity time of 20 seconds has been provided by parabolic fight of airplane. Recently also in Japan, the service has been started by Diamonnd Air Service using MU-300 airplane. The service is convenient for experimenters to participate in onboard experiments using their own facilities.

7. TR-1A

The mission has been developed by NASDA(STA) and is ready for the first flight (Sep.'91). The payload part is 750 kg in weight, 0.85 m in dia., 2.4 m in length. It will be launched every summer at Tanegashima island in the southern part of Japan and be recovered on sea. The experiments to be performed in these 3 years are as follows;

* In-situ observation of solid-liquid interfaces during crystal growth.
* Marangoni convection and its control.
* Bubble generation, dynamics and handling.
* Melting and solidification of dispersed alloys.
* Melting and solidification of ceramics.
* Crystal growth from liquid phase.
* Interface parameters between solid and liquid.
* Melting and solidification of oxide superconductor.
* Melting and solidification of glasses.

8. Unmanned recoverable satellite (Express)

The study of this mission has started by a joint program between German(DARA) and Japanese(MITI) governments. Launching will be from Kagoshima Space Center in the southern part of Japan under reponsibility of ISAS using M-3SII rocket and will be recovered in Autsralia after several days flight in orbit under responsibility of DARA(MBB/ERNO). Mission experiment weight is 165 kg for the first one to be launched in January 1994. Preparation of catalytic materials such as zeolites are being planned for the mission.

9. Drop shaft

Microgravity time of 10 seconds has been achieved by the Japan Microgravity Center(JAMIC) by using an existing former coal mine shaft located at Kamisunagawa, Hokkaido, the northern part of Japan. The facility utilizes 720 m in total of the mine shaft, 500 m for free fall, 200 m for braking, and 20 m for emergency braking.
A rocket-shaped evacuated capsule, 1.8 m in diameter, has a inner floating payload capsule and falls through the drop shaft along non-contact guide rails in atmospheric condition.
To compensate the air drag in falling, optimum thrust is added to the capsule by gas jet thrusters. The drop shaft is schematically shown in Figure 3.
The operation will start in September 1991. After several demonstration experiments, a variety of material science and life science experiments will be performed.

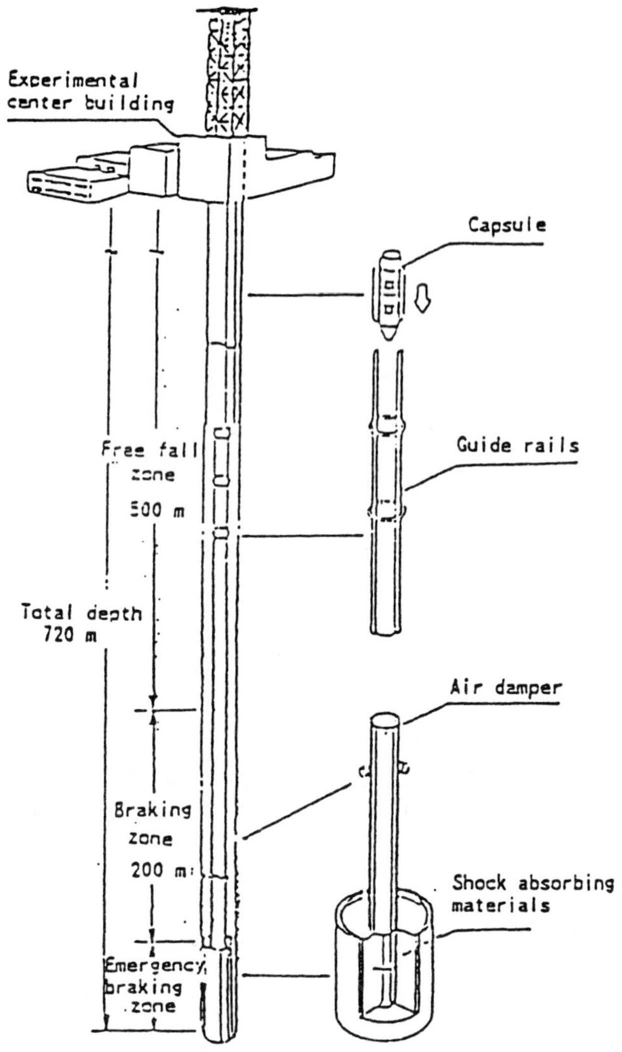

Figure 3. JAMIC drop shaft

Another drop shaft for microgravity time of 4.5 seconds, is under construction by Micro-Gravity Labratory of Japan(MGLAB) at Gifu prefecture, a middle part of Japan. It also uses an existing mine shaft of 150 m. A capusule, 0.9 m in diameter, falls along an evacuated tube. The operation will start in 1993.

10. Space Technology Corporation(STC)

Space Technology Corporation(STC) is a kind of research consortium (1986.4-1993.3) for microgravity experiments invested 70% by government(Japan Key Technology Center, JKTC), 30 % by industries. Six companies(IHI, Toshiba, NEC, Hitachi, Fujitsu, MELCO) invested 5 % each and transiently transfer to STC their researchers, four in average, working as one group in the investor laboratory building. The subject of research is related to processing of semiconnducting materials.The sub-themes in individual laboratory are listed in Table 1 with corresponding microgravity missions.

Table 1 STC research program

Sub-theme	Investor	Mission
Materials diffusion in PbSnTe melt	IHI	TEXUS('88,'90)
Impurity diffusion in In(Ga)Sb melt	Toshiba	TEXUS('88),D-2('93)
Thermal conductivity of InSb melt	NEC	TEXUS('89),MU300('90) JAMIC('91)
Microstructure in Ge-GaAs eutectic	Hitachi	D-2('93)
Crystal growth of GaAs from Ga solution	Fujutsu	TEXUS('88), D-2('93)
Crystal growth of InP from vapor	MELCO	TEXUS('88), D-2('93)
Measurements of wettability etc.	IHI	

The category of research in STC can be classified into three groups.

1) Measurement of intrinsic parameters of melts such as materials diffision coefficients in InSb, PbSnTe and thermal conductivities in InSb. These are the most important parameters controlling actual processes in the manufacturing of semiconducting devices. Diffusion under low gravity will reveal the essential features.
2) Study of crystal growth processes having practical importance such as GaAs and InP. These materials comprise two of the most important for industrial application in compound semiconductors and the knowledge obtained in various environments will deepen the understanding of the related processes.
3) Preparation of unique materials such as Germanium-Gallium Arsenide eutectic crystals having micro-structures. The forming process of micro-structure has an attractive application potential and this experiment will hopefully be the first step in scouting around in the field.

11. Concluding remark

Japan is slow in coming into the field of microgravity. In these several years, however, several projects, as briefly reviewed in this paper, has started. A number of researchers in the field has increased to ca.200 due to various supporting measures of the government. Participation of industries seems to be more remarkable than that in other countries.
In future, international coorporation will become more important, by which we shall achieve the best combination of research subjects and experiment tools in the world wide scale.

FLUID DYNAMICS AT ZARM: AN OVERVIEW

H.J. Rath and H.C. Kuhlmann

Center of Applied Space Technology and Microgravity (ZARM)
University of Bremen, 2800 Bremen 33, Hochschulring/Am Fallturm, Germany

ABSTRACT

The fluid dynamics projects of the Center of Applied Space Technology and Microgravity are reviewed. After an introduction to the drop tower experiment facility a brief overview over the diverse research subjects is given. Projects concerned with viscosity measurements, the flow between concentric spheres, and the diffusion flame around spherical fuels in slightly buoyant atmospheres are described in more detail.

Keywords: *Elongational viscosity, spherical gap flow, diffusion flames.*

1. INSTITUTE AND EXPERIMENT FACILITY

During the last decade microgravity research has become an independent branch of science and has established its own interdisciplinary community. The Center of Applied Space Technology and Microgravity (ZARM) was founded in 1985 as a result of an increased microgravity research activity and demand for experiment facilities. Since that time a broad spectrum of projects is being carried out at ZARM. The emphasis of these projects is on fundamental research in fluid dynamics. The two other major departments are concerned with satellite technology and supersonic wind tunnel technology. The key experimental facility of ZARM is the drop tower (see figure 1), a laboratory unike in Europe.

The ZARM drop tower offers a very "clean" microgravity environment during 4.74 s of free fall. The heart of the drop tower

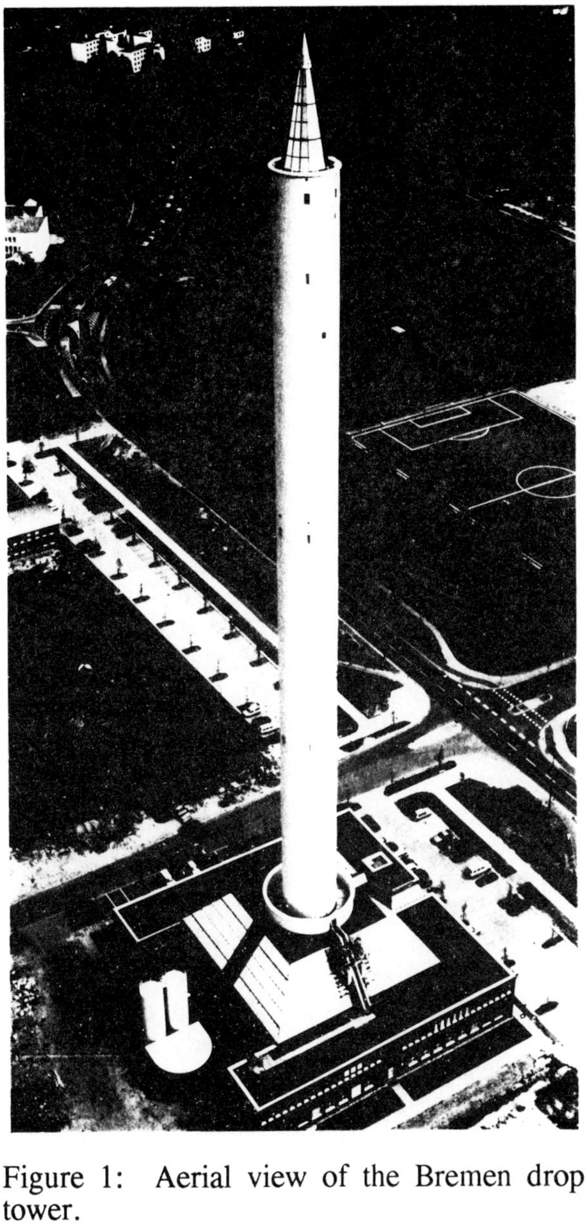

Figure 1: Aerial view of the Bremen drop tower.

system is a 110 m vertical tube of diameter 3.5 m that is placed excentrically in a 145.5 m concrete tower. The tube is merely supported at its lower end and is otherwise standing freely. Typically, experiments are mounted on a rack that fits into a drop capsule. After dropping this capsule it is decelerated in a container located in the deceleration chamber below the tube supporting structure. The container can be moved horizontally to allow for future implementation of a vertical launch device to be installed even below the deceleration chamber. This way, the duration of the "free fall" can be extended to 9.5 s. To reduce atmospheric drag the whole system has been designed as a large vacuum chamber. Residual accelerations less than 10^{-6} g can be obtained if the tube is evacuated to a pressure of 1 Pa.

During the free fall period the control center is in contact with the capsule /experiment via TV and remote control enabled by a laser telemetry system the main purpose being data acquisition. To prevent the dropped equipment from being damaged, the capsule is gently decelerated (<30g) in the 8.5 m tall deceleration container filled with expanded polystyrene. Due to a great degree of automatization of the required technical operations 3 to 4 drop cycles are feasible during one day. This performance even allows for series experiments within a reasonable time span. Depending on space demand and communication channels required it is possible to place several experiments in the capsule which is essentially cylindrical of size 0.4 x 1.5 m (radius x height, see figure 2).

The utilization of the ZARM drop tower is open to all scientists worldwide. Interesting experiments range from physics of fluids, physical chemistry, materials science, and engineering to biology. Depending on the typical time scales involved in the respective experiment it can be successfully performed in the drop tower. To obtain stationary conditions, e.g., in fluid dynamics, within the given drop time t_D the typical length scale of the experiment must be less than $\sqrt{\nu t_D}$, $\sqrt{\kappa t_D}$, or $2\pi\sqrt{2\nu t_D}$, if equilibrium is obtained by momentum diffusion, thermal diffusion or capillary waves, respectively (ν: kinematic viscosity, κ: thermal diffusivity). These limits apply only if steady states are to be investigated and the above processes determine the characteristic time scales. Thus investigation of stationary flows close to transition points is prohibited. However, information can also be gained from the transient behavior near phase transitions (see, e.g., Kaukler[1]). These problems are generally not encountered when studying phenomena which, apart from being intrinsically transient, occur on a small length scale. Examples are a hot wire in a fluid with small relative velocity perpendicular to the wire's axis or diffusion flames around small particles. Problems of this type are excellent candidates for drop tower experiments.

Figure 2: Open drop capsule.

2. PROJECT OVERVIEW

Fluid phases are present in many technological processes. Some of these, especially in materials sciences and biotechnology, can possibly be optimized by proper utilization of the microgravity environment to gain basic information about the underlying physics. Since microgravity research is a very young discipline, it is clear that much of its research effort has to be fundamental in character.

This need is reflected in the more than twenty projects being carried out at ZARM (table 1). Half of them are concerned with

TABLE 1 Project Overview

	Fundamental Problems	Tool Development/Measurement Technique	Technological Projects
Experimental	Diffusion flames Thermocapillary convection in spherical drops Spherical gap flow with radial body force Suspension flows Weissenberg effect	Laser Doppler velocimetry Laser Speckle velocimentry Low velocity thermal anenometers Elongational viscosity measurement Heat conductivity measurement Surface tension measurement	Microgravity tank systems Capillary Transport Polyreactions
Numerical/ Theoretical	2-D diffusion flames Spherical gap flows Two-phase flow Thermocapillary convection in liquid bridges Rotating disk flow Through flow effects	Elongational viscosity Multigrid methods	

fundamental fluid physics. The activities in this section cover many areas such as capillary effects, multiphase flows, combustion, and non-isothermal flows.

Besides these fundamental hydrodynamic problems, ZARM is developing tools supporting the physical problem experiments or numerics, respectively. Some of these developments are projects on their own. They are listed in the second column of table 1. The most important area in this respect is the velocity measurement technique. Into the same category also fall projects that are mainly devoted to the improvement of known or development of new measurement techniques utilizing microgravity.

Examples are the measurement of surface tension by use of well defined capillary waves, the determination of the heat conductivity of very low viscosity fluids, or the investigation of elongational viscosity coefficients of non-Newtonian fluids.

In the third column of figure 3 projects are listed, results of which have the best potential for being utilized in technological processes or satellite technologies in the near future.

Many problems, especially the fundamental ones, are tackled both experimentally and numerically. The numerical methods used range from finite element to multigrid methods and are generally selected according to the nature of a given problem. It should be mentioned that there are many cross links between all projects listed; for instance rotating disk flows are fundamental to microgravity tank systems in which a liquid is to be positioned by rotation.

A number of experiments related to the projects listed in table 1 can be carried out in the drop tower. However, for those experiments that require longer durations of weightlessness, KC-135 flights and other commercial carriers are used. TEXUS and MICROBA flights are in preparation.

3. SUMMARY OF SOME SELECTED PROJECTS

An overview over the microgravity research program of ZARM has been given before[2]. Here, we present some projects in more detail.

Measurement of the Elongational Viscosity

For viscosity measurements one has to determine the stresses in the fluid produced by a given strain. Apart from compression the motion near a point can be decomposed into a pure straining motion and a rigid body

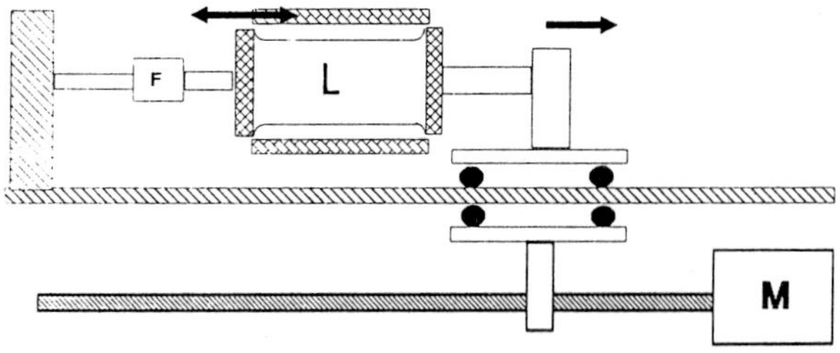

Figure 3: Sketch of the elongational viscosimeter. L: liquid bridge, M: motor, F: force measurement.

rotation. In non-Newtonian fluids the elongational viscosity associated with the pure straining motion depends on the rate of change of the strain as well as on the strain history. For a measurement of this viscosity it is desirable to create a pure straining motion of the fluid. In practice, however, this is very hard to achieve due to a number of difficulties. For most existing measurement techniques the fluid volume under investigation is already deformed before it enters the elongational flow field. Moreover, the elongational rate cannot be kept constant in space and time. As a result, different methods can lead to considerably different results[3]. Techniques in which the fluid is fixed between two coaxial disks avoid the problem of interface distortions mentioned above before it is stretched. To keep the geometrical shape of the liquid bridge close to cylindrical during the stretching phase a measurement technique has been developed utilizing the microgravity environment[4]. The main advantage of weightlessness is that large liquid volumes can be used which facilitates the observations. Forces that tend to deform the cylindrical liquid bridge are inertia, capillary break up, and end effects. To minimize the two former disturbance sources a proper elongation rate $\dot{\epsilon}$ has to be chosen since small (large) values of $\dot{\epsilon}$ suppress inertia effects (growth of capillary induced deformations). End effects are being minimized in the present set up (see figure 3) by using rubber-like supporting disks, which radii can be varied externally during the elongation period to satisfy the volume constraint. In drop tower experiments elongation rates of 0.01 - 2.0 s^{-1} have been used. The maximum length of the liquid bridge is at present 20 cm and elongation times of up to 3 s can be achieved.

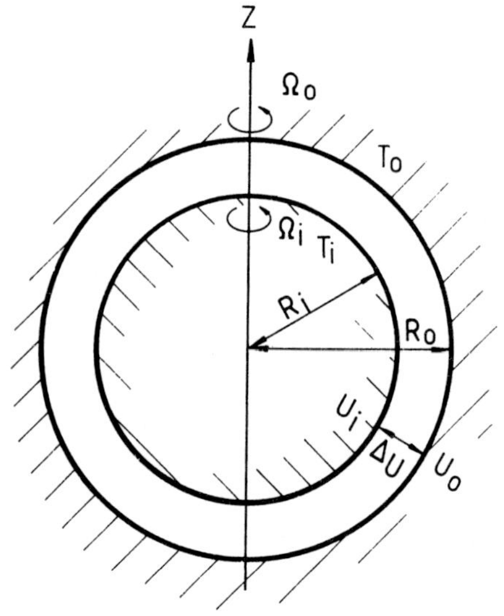

Figure 4: Geometry and boundary conditions for the flow between concentric spheres. Ω, T, R, and U denote circular frequency, temperature, radius, and electrical potential, respectively.

Spherical Gap Flow with Radial Body Force

This project is concerned with the flow between rigid concentric spheres of different but constant temperature, that rotate independently around a common axis. The dielectric fluid is subjected to a radial body force by application of a low frequency AC potential difference across the spherical gap. This model is basic to the understanding of more complicated atmospheric flows, yet it is sufficiently involved. The experimental modelling is essentially the same as in ref. 5 and is also described in ref. 6. Figure 4 shows the basic geometry. For small gap size the strength of both the electrically induced body force[5] $(\sim r^{-5})$ and Newton's gravity force field $(\sim r^{-2})$ can be approximated by a linear function of the radius r, thus having a very similar influence on the flow. During on ground experiments buoyant flows usually mask the weak flow due to the radial electrostatic 'gravity'. Therefore, weightlessness is required for serious modelling. Since the radial diffusion time across the gap is of the order of minutes in the present realization, long durations of microgravity are required. However, flow and temperature measurement techniques are currently developed and tested in the terrestrial laboratory.

Parallel to the experiment preparations computations are being carried out using a

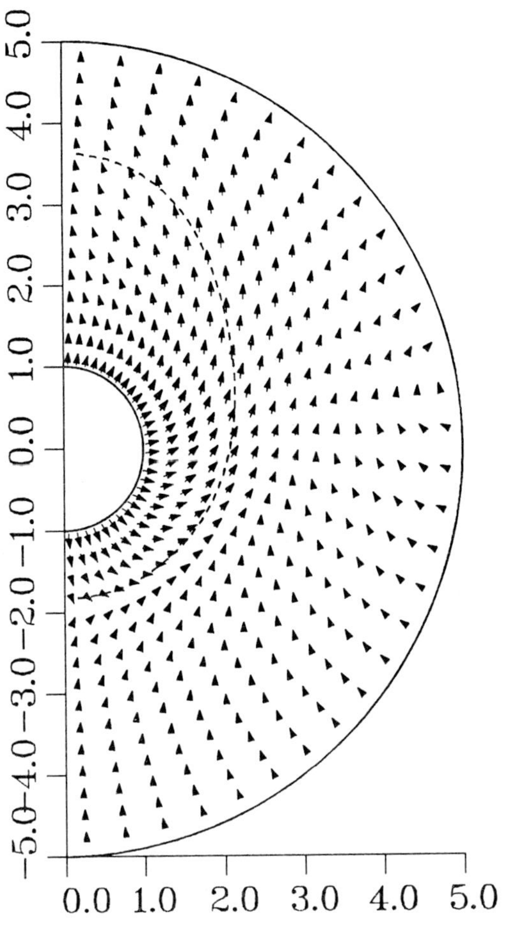

Figure 5: Numerically calculated flame location (dashed line) and rescaled velocity field (arrows) around a spherical particle for small buoyant forces. Shown is a vertical cut through the axisymmetric fields.

time-dependent two-dimensional finite difference code. This code has already been validated[7] by reproducing the dynamics and boundary layer scaling[8] of flows induced by a rapidly rotating outer shpere started from rest.

Weakly Buoyant Flow around a Burning Sphere

The combustion of spherical condensed fuels can be advantageously studied in drop towers.

Under microgravity the whole combustion process is generally radial symmetric, whereas under terrestrial gravity conditions buoyancy effects lead to a strong distortion of the otherwise spherical diffusion flame. This loss of symmetry significantly influences the combustion rates and different power laws have been proposed to describe the rate of burning as function of the Grashof number[9,10,11]. It is not quite clear, in which

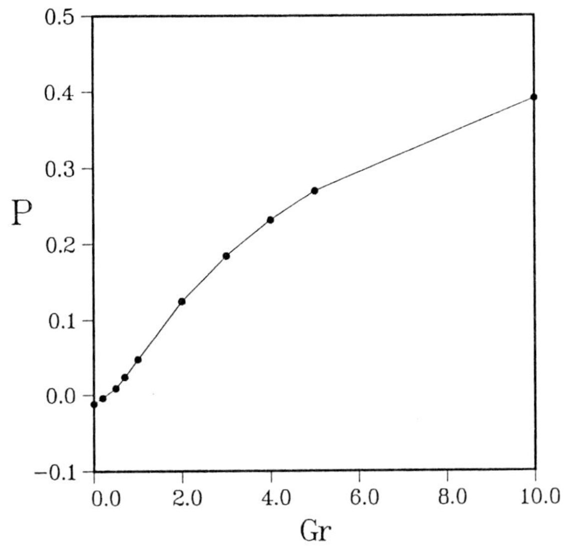

Figure 6: Reduced burning rate P of a spherical particle as function of the Grashof number Gr. The offset from zero indicates the magnitude of the numerical error.

way the burning rate depends on the Grashof number, if buoyancy is weak. For an experimental investigation of this problem under normal gravity the fuel spheres would have to be very small making quantitative measurements increasingly difficult. Therefore, this problem is investigated numerically by a time-dependent two-dimensional calculation of the flow, concentration, and temperature fields in a spherical annulus. The method is briefly described in ref.12. The simplifying assumptions are

- the reaction is one step irreversible,
- Soret and Dufour effects can be neglected,
- the condensed fuel is at constant temperature,
- the regression rate of the drop radius is small,
- the reaction rate is large (Burke-Schumann limit),
- radiative heat transfer is insignificant,
- the Boussinesq approximation is valid.

The resulting set of model equations are solved on a staggered equidistant grid with a scheme based on the MAC method and a modified SOLA code. The position of the reaction front is determined by a Shortly-Weller technique. A typical flame shape and flow field indicated by arrows is shown in figure 5. Figure 6 shows the reduced quasi-steady state burning rate as function of the Grashof number. Obviously the character of the curve changes as $Gr \to 0$. For symmetry reasons, the burning rate must be $\sim Gr^2$. To determine the precise curvature, however, more accurate calculations have to be done.

4. ACKNOWLEDGEMENTS

The authors acknowledge discussions with C. Egbers, R. Kröger, and T. Coordes.

5. REFERENCES

1. W. F. Kaukler, *Metallurgical Trans. A* **19**, 2625 (1988).
2. H. J. Rath, in *International Symposium on Applications of Microgravity - Possibilities of Ten Seconds -*, Tokyo 1991, p. 21.
3. for results on the fluid M1, see: *J. Non-Newtonian Fluid Mechanics* **35** (1990).
4. S. Berg, A. Delgado, R. Kröger, and H. J. Rath in *Proceedings of the 1st ASME/JSME Fluid Engineering Conference, Forum on Microgravity Flows,* Ed.: A. Hashemi, B. N. Antar, and I. Tanasawa, Portland 1991, p. 47.
5. J. E. Hart, G. A. Glatzmaier, and J. Toomre, *J.Fluid Mech.* **173**, 519 (1986).
6. C. Egbers, A. Delgado, and H. J. Rath in *Proceedings of the 1st ASME/JSME Fluid Engineering Conference, Forum on Microgravity Flows,* Ed.: A. Hashemi, B. N. Antar, and I. Tanasawa, Portland 1991, p. 41.
7. M. Liu, A. Delgado, and H. J. Rath in *Proceedings of the 7th International Conference on Numerical Methods in Laminar and Turbulent Flow,* Ed.: C. Taylor, J. H. Chin, and G. M. Homsy, Pineridge Press, Swansea (1991), p. 23.
8. C. E. Pearson, *J. Fluid Mech.* **28**, 323 (1967).
9. D. B. Spalding in *Proceedings of the 4th Symposium (International) on Combustion*, The Combustion Institute, Pittsburgh (1953), p. 847.
10. T. Saitoh and N. Sasaki, *Trans. Jap. Soc. Mech. Engineers* **50**, 1397 (1984).
11. J. Sato, M. Tsue, M. Niwa, and M. Kono, *Combust. Flame* **82**, 142 (1990).
12. H. C. Kuhlmann, T. Coordes, and H. J. Rath in *Proceedings of the 7th International Conference on Numerical Methods in Laminar and Turbulent Flow,* Ed.: C. Taylor, J. H. Chin, and G. M. Homsy, Pineridge Press, Swansea (1991), p. 1174.

THE PROGRAMME OF INVESTIGATION INTO DYNAMICAL PROCESSES OF SPACE OBJECTS WITH PARTIALLY FILLED FUEL TANKS IN MICROGRAVITY.

E.L.Kalyazin, V.N.Kulikov, A.G.Mednov

Moscow Aviation Institute, 125871, Moscow, USSR.

ABSTRACT

Complex approoach to investigate dynamical problem is proposed. The approach is based on the mathematical models parameters identification according to the results of experiments. Experiments programe on the earth installations andon the orbital station is worked out. The programme include the stady of liquid reorientation transitional processes, nonlinear liquid oscillation in microgravity and these processes influence on spacecraft dynamics.

Keywords: *Microgravity, Spacecraft Dynamics, Experiment.*

Space engineering development, creation of large orbital and interplanetary stations, interorbital space tugboats and systems such as Space Shuttle on liquid propellants demand new methods to study dynamical interaction of spacecraft structure with partially filled fuel tanks. This problem is very important for liquid-fueled upper stages and orbital transfer vehicles, that have relative liquid mass as high as 80%. When spacecraft performs manoeuvres in orbit, and systems of orientation and stabilization have worked, fluid motion in tanks with various devices is significantly nonlinear. The perturbation acting from fluid to spacecraft become considerable. It is necessary to take into account when designing the tanks and angle stabilization systems to satisfy stringent operating requirements.

The importance of this phenomenon is reflected in research efforts of the engineering and science communities (Refs.1,2,3,4). The research works show that correct analytical solutions of the dynamic problems of partially filled tanks don't exist now. Most of experimental methods allow to define hydrodynamic parameters for given objects and the type of perturbation.

A complex approach to the investigation of dynamical problems is proposed. The approach is based on the mathematical models identification according to the results of the experiments. Various levels of the mathematical models are used. The methods of linear and nonlinear dynamics of partially filled tanks are made use of.

The deformation of the spacecraft frame is not considered. It is assumed, that when spacecraftr is rotated, the control system creates constant correcting moment. The external perturbation forces and moments are absent. The linear equations of spacecraft rotation motion are considered as first approximation. The angle velocities and accelerations are small values. Dynamical coefficients and spacecraft mass-inertial characters are changed depending on time.

The experimental part of the programme is considered in detail.

The experimental part includes:

- design of experiments implementation methods;
- design and creation of the expe-rimental modules and installations;
- implementation of the experiments on the drop tower, on the laboratories plane and on board the space station.

The basic problem of the experimental researches is definition of dynamical characteristics of spacecraft-fluid system (mode's natural frequencies and shapes of oscillations, damping ratios, add mass and inertia moment from fluid, reaction slosh forces and moments) in microgravity and various accelerations. Spacecraft's rotation, harmonic excitation and reorientation of fluid are considered.

Full scale testing of spacecraft-fluid system is neither practical nor efficient for determining the dynamical characteristics of spacecraft-fluid system. The cost associated with full scale testing and the question of the validity of extrapolating the full-scale results to other fluid/spacecraft configurations, are the main drawbacks. Sub-scale model testing is an attractive alternative. This-scale testing is cheaper and can be tested to model many fluid/spacecraft configurations. The size and mass of sub-scale models are also suitable for testing on board the orbital station.

To determine model parameters of sub-scale testing scaling conditions are used:

$$Fr=idem;\ Re=idem; \qquad (1)$$
$$Bo=idem;\ k_m=idem;\ k_f=idem,$$

where
$Fr=V^2(ga)^{-1}$ - Froud Number;
$Bo=pga^2s^{-1}$ - Bond Number;
$Re=Van^{-1}$ - Reynolds Number;
$k_m=M_{fluid}(M_{fluid})^{-1}$ - mass ratio;
$k_f=f_{slosh}(f_{sp.cr})^{-1}$-frequency ratio;

V - fluid velocity; g - apparent acceleration component at a tank location; a - tank diameter; p - fluid density; s - surface tension of the fluid; n - kinematic viscosity of the fluid.

The following system spacecraft-fluid variants of motion must be simulated on the experimental installation:

A - turn relative to transverse axis at various angles with various angle velocities and accelerations ;

B - turn relative to lengthwise axis at various angles with various angle velocities and accelerations ;

C - fluid reorientation when tran-sition from microgravity to motion with acceleration is occurs;

D - harmonic oscillation of the spacecraft-fluid system in transverse plane.

The design features peculiarities of experimental installations are as following. The testing model of the tank is placed into experimental module. The principle scheme of the module is identical to the drop tower and installations on the laboratories plane and orbital station. The experimental module includes the moving frame into which tank model is

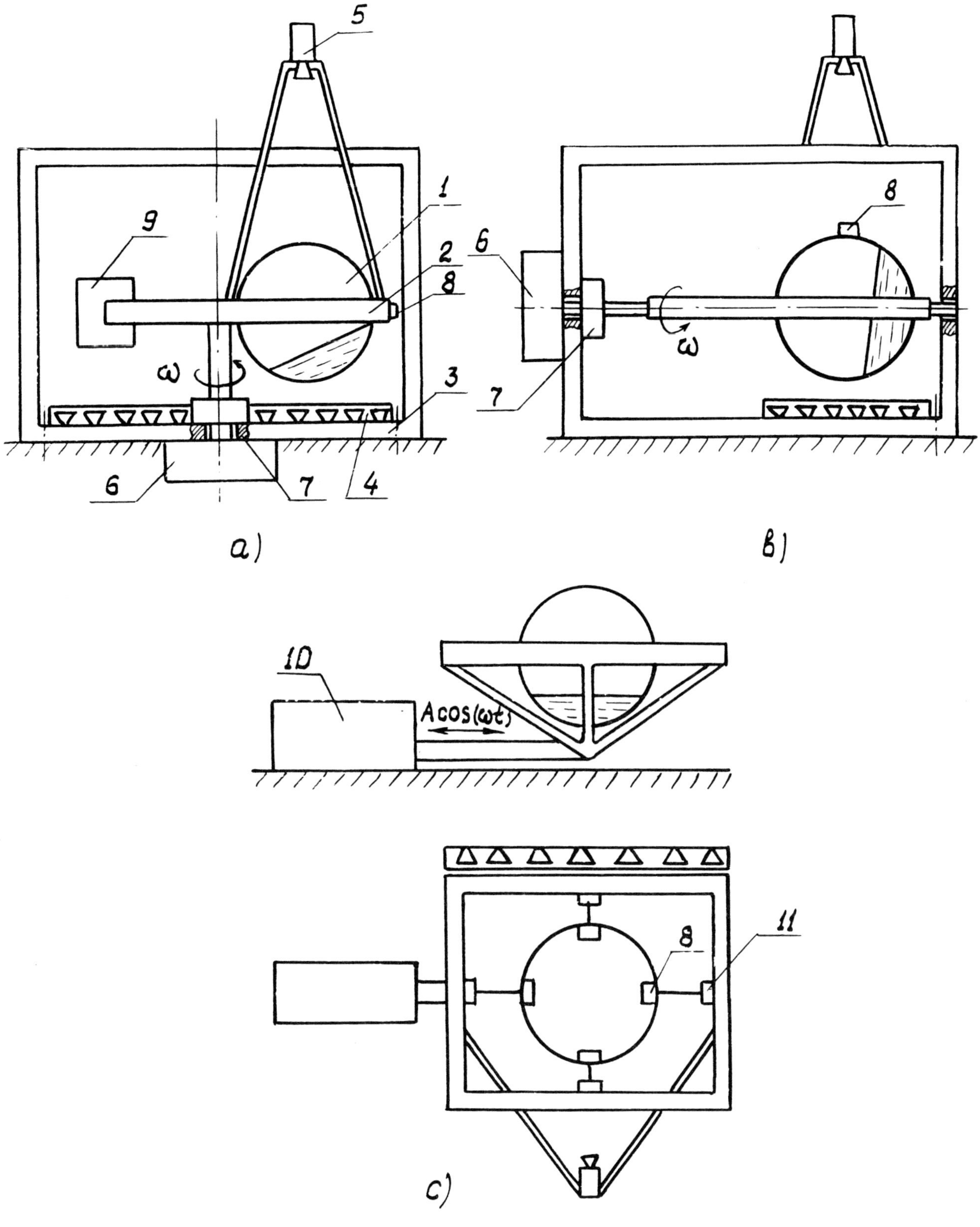

Figure 1: Shemes of the experimental module. 1 - tank model; 2 - rotating frame; 3 - force frame; 4 - sorhits; 5 - photo camera; 6 - motor; 7 - moment sensor; 8 - accelerometer; 9 - balans load; 10 - electro-mechanical shaker; 11 - force sensor.

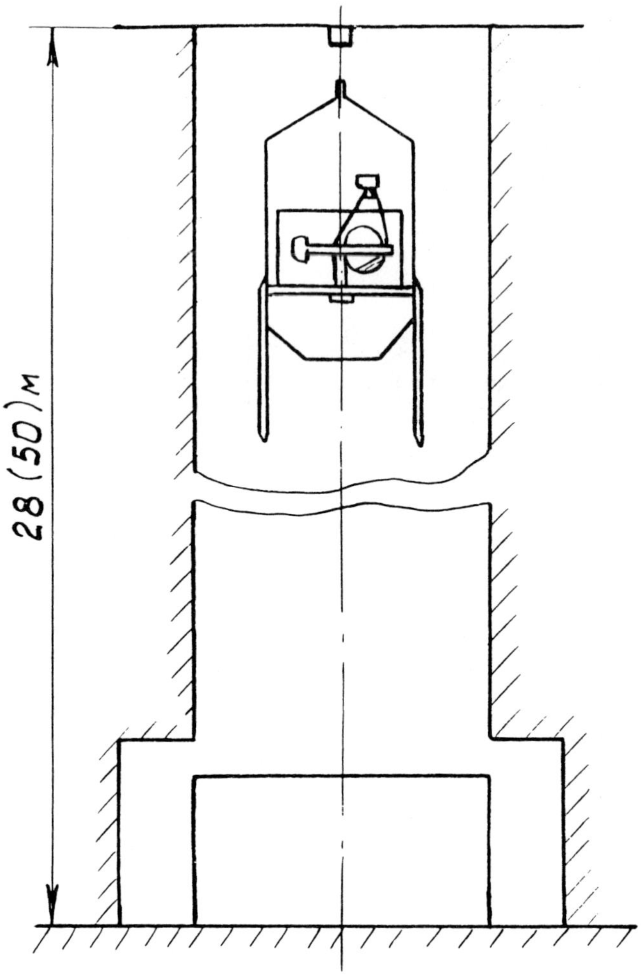

Figure 2: Experimental module in the Drop Tower.

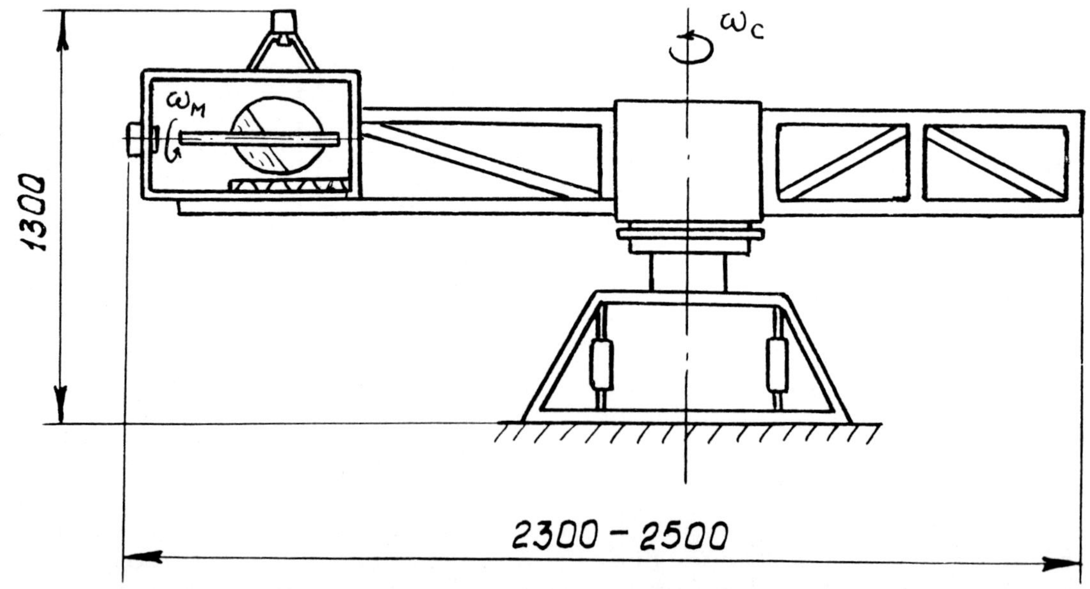

Figure 3: Experimental module on the centrifuge.

mounted, the force frame, the units of rotation or movement, the system of photo(video) recording, motors and sensors of the registering system. Figure 1 shows schemes of the module for simulation of variants A,B,D.

System of recording consists of the photo(video) camera and sophits. When the spacecraft turns are simulated, the spring motor is used to move the module. The electro-mechanical shaker moves the module when harmonic oscillations are studied. The control of the simulation processes parameters is realized by angle velocity, angle acceleration, linear velocity and angle sensors. The dynamical parameters of fluid-spacecraft system are defined using the special methods. The measurements of moment sensors and accelerometers or force and acceleration sensors are used. They allow to obtain the value of the forces and moments, that act from fluid to tank, and add mass and inertia moment from fluid. Acquisition and recording of experimental data and work control of module operation are realized by personal computer.

Figure 2 shows module in the Drop Tower. Figure 3 shows the module on the experimental installation called "centrifuge". This installation operates on the laboratories plane. The frame of the installation is able to revolve in its axis with necessary angle velocity. It is provided initial conditions for tests. The time of microgravity is about 20 seconds, The simulation of all variants (A,B,C,D) of spacecraft-fluid motion is possible. However only preliminary data are possible to obtain for variant D. This data can be made more accurate by increasing the time of microgravity.

The tests on the drop tower and on the plane are necessary to check the methods of experiments in operation that are planned to be realized on board the orbital station. High cost of experiments in space demand that an installation should be designed in such a may would solve various problems. The installation designed for the orbital station, allows to study the problems of nonlinear spacecraft-fluid systems as well as the problem of structure dynamics,

biotechnology and materials production in space.

The realization of this programme allows to reduce time and material expenses considerably when new space objects will be designed and tested in operation.

REFERENCES

1. H.N. Abramson, W. Chu, L.R. Garza, Liquid Sloshing in Spherical tanks", *AIAA Journal*, Vol.1, No.2, p.384-389,Feb.(1963).

2. A.D. Myshkis, V.G. Babskii, N.D. Kopachevskii, L.A. Slobozhanin, L.A. Tyuptsov, *Low-Gravity Fluid Mechanics*, Springler-Verlag, New York, (1987).

3. G.S. Narimanov , L.V. Dokuchaev, I.A. Lukovskii, *Nonlinear dynamics of flying apparatus with fluid*. Moscow,(USSR),(1977).

4. K.A. Abgarian, E.L. Kalyazin, V.P. Mishin, I.M. Rapoport, *Dynamics of rockets*. Moscow,(USSR)(1990).

DIFFUSIVE INSTABILITY DURING TERNARY ISOTHERMAL DIFFUSION IN THE ABSENCE OF GRAVITATION

N.D. Kosov, Yu.I. Zhavrin, V.N. Kosov

Kazakh State University, Timiryazev 46, Alma-Ata, 480121, USSR

ABSTRACT

It has been shown that under isothermal conditions in some three-component gas mixtures the diffusion leads to the stratification of the boundary into two sublayers enriched with heavy and light components correspondingly. Under the influence of gravitation or gas mixture flow caused by the diffusive baroeffect some convective structures may appear which makes the process of mass transfer unstable.

Keywords: Diffusion Instability, Convective Structure, Diffusive Baroeffect, Gravitation

1. INTRODUCTION

The study of a non-homogeneous gaseous mixture motion in comparison to that of the incompressible liquid dynamics has a number of peculiarities in both the mathematical simulation of the phenomenon and the analysis of its physical mechanism. By analogy with the effects of "buoyancy" and "double diffusion", the formation of density-stratified regions in gaseous systems is explained as the result of the competing processes of heat conduction and diffusion /Refs. 1,2/. It has been generally believed that convection may appear only under the conditions of the hydrostatic mixture stability disturbance, the stability description be-

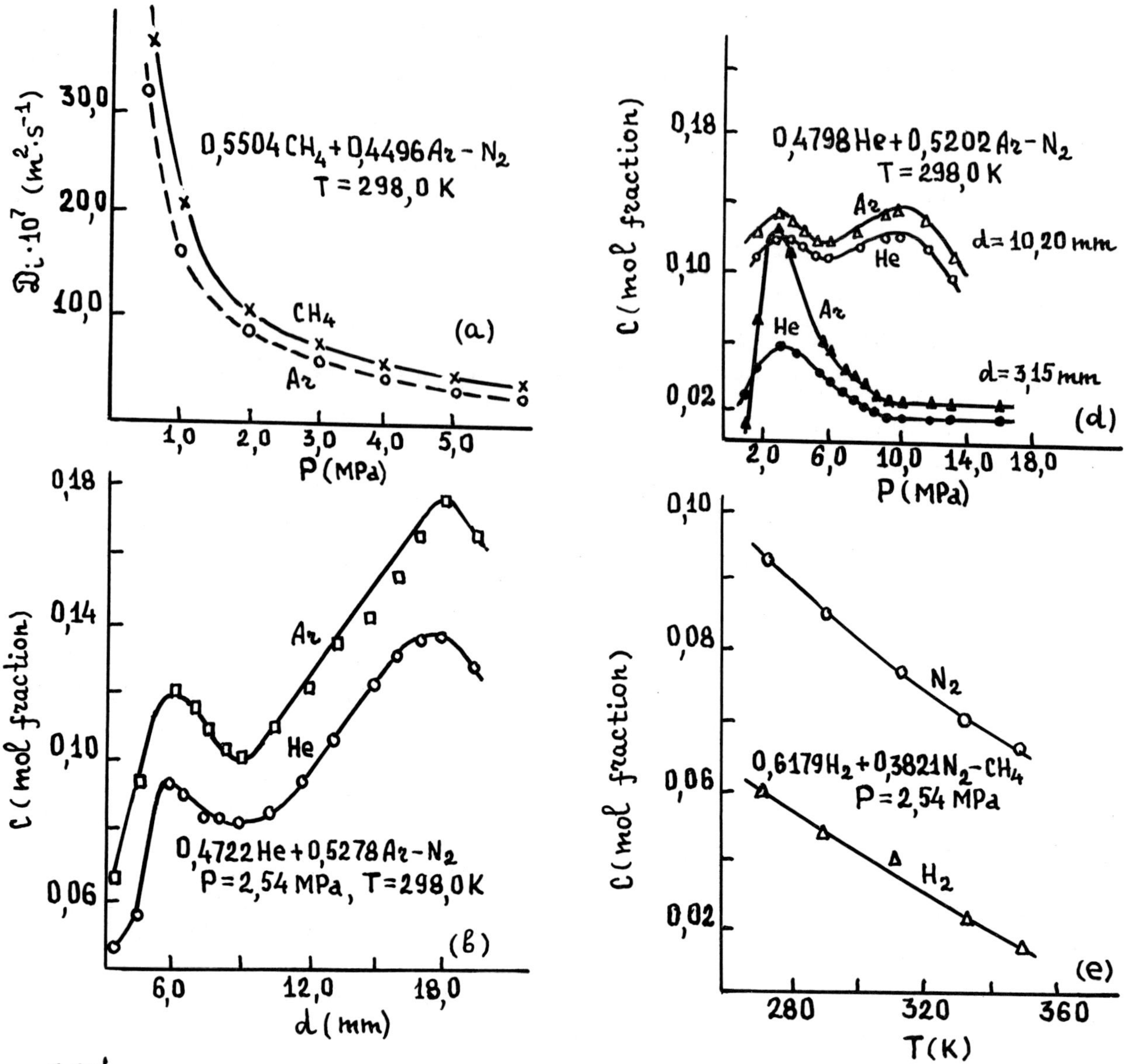

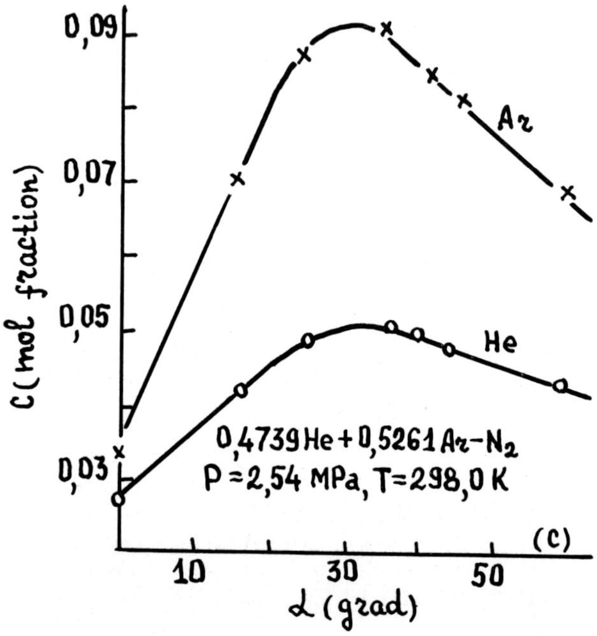

Figure 1: Diffused amounts of components in the unstable diffusion in some of the threefold systems as a function of a) the ratio of the diffusion coefficients (the process- is stable); b) diffusion channel diameter; c) angle of slope of the diffusion channel; d) pressure; e) tempreture. In all the cases, excert "a", experimental time was as long as 1 hour.

ing similar to that of the standard heat convection in a uniform medium. However, the experimental data showed /Refs. 3-5/ that for a number of three-fold hydrostatic stable gaseous systems diffusion led to the boundary separation into two diffusive sublavers enriched with heavy and light components respectively thus resulting in the local stratification of the mixture density and the appearance of convective flows in the gravity field. The superposition of the convective flows on the molecular transfer makes the diffusion unstable. As it is shown below, under the isothermal conditions diffusion in multi-component mixtures causes the density inversion. Under the action of gravitation (other mass forces) or the gaseous mixture flowing due to the diffusive baroeffect, the density inversion may lead to the appearance of convection cells in the absence of gravitation thus making the process unstable.

2. DIFFUSION INSTABILITY IN THE GRAVITATION FIELD

Experiments were carried out using the method of two flask apparatus /Ref. 4/, temperature and pressure ranging from 276°K to 473°K and from 0.1 to 17.0 MPa, respectively. Construction peculiarities of the experimental devices allowed to visualize the process dynamics and take photos of the convective streams, to change the geometrical characteristics of the diffusional channel as well as to use the anemometrics method of investigation /Ref. 2/. The diffusion process in the system is shown to be stable, provided the diffusion coefficients of the components are comparable with each other (Fig. 1a). Fig. 1b shows experimental results for the transferred He and Ar vs pressure and the diameter of the diffusion capillar, extremums being clearly seen. Occurance of the maximums in concentration may be explained by interaction of structures formed. The structures have been observed to move in opposite directions. The unstable process takes place for a definite values of the diffusive channel diameter. Influence of pressure on the transporting qualities of the components is of non-linear character (Fig. 1d), occurs at certain values of the diffusional channel diameter which is due to the same reasons as the occurance of extremums in Fig. 1b. If in the system there is an unstable diffusion process at a given temperature, then increase in temperature promotes the stabilization of the process (Fig. 1d). The intensity of unstable process essentially depends on the position of diffusion channel with reference to vertical line, which confirms the convective character of the fluxes obtained (Fig. 1c). Experimental data analysis showed that calculation methods widely applied in thermo-convection can be used for the description of diffusional instability.

The model under consideration is applied to an ideal incompressible isothermic ternary gas mixture limited within two parallel planes. According to the independent diffusion condition mass transfer of the component is described by its diffusion coefficient $\mathcal{D}_i$ /Ref. 3/, $\mathcal{D}_i$ being constant for given mixture composition (see perturbation method in /Ref. 2/. Having linearized by the Boussinesq method one obtains the system of hydrodynamic equations:

$$
\begin{aligned}
&\frac{\partial c_1}{\partial t} + \vec{u}\nabla c_1 = \mathcal{D}_1 \nabla^2 c_1, \\
&\frac{\partial c_2}{\partial t} + \vec{u}\nabla c_2 = \mathcal{D}_2 \nabla^2 c_2, \\
&\frac{\partial \vec{u}}{\partial t} + (\vec{u}\nabla)\vec{u} = \\
&= -\frac{1}{\rho_0}\nabla P + \nu\Delta\vec{u} + g(\beta_1 c_1 + \beta_2 c_2)\vec{\gamma},
\end{aligned}
\qquad (1)
$$

where ρ_0 is average density, and $\beta_k = \frac{1}{\rho_0}\left(\frac{\partial\rho}{\partial c_k}\right)$. The unit vertical vector is denoted by $\vec{\gamma}$. The other quantities are denoted conventionally. Making the system of equations dimensionless and introducing a streamfunction $\psi(x,z)$ one obtains as a result of some mathematical transformations the following system of equations for dimensionless perturbations C_1 and C_2:

$$
\begin{aligned}
&\frac{\partial c_1}{\partial t} - \tau_1 \nabla^2 c_1 = -\frac{\partial \psi}{\partial x}, \\
&\frac{\partial c_2}{\partial t} - \tau_2 \nabla^2 c_2 = -\frac{\partial \psi}{\partial x}, \\
&\left(\frac{1}{P_r}\frac{\partial}{\partial t} - \nabla^2\right)\nabla^2\psi = \\
&= \tau_1 R_1 \frac{\partial c_1}{\partial x} + \tau_2 R_2 \frac{\partial c_2}{\partial x},
\end{aligned}
\qquad (2)
$$

where $z = 0,1;\ c_1 = c_2 = \psi = 0.$

The dimensionless parameters of similarity are as follow: $P_r = \frac{\nu}{\mathcal{D}_3};\ R_i = g\beta_i \Delta c_i d^3/\nu\mathcal{D}_i;\ \tau_i = \frac{\mathcal{D}_i}{\mathcal{D}_3};\ i = 1,2.$ The solution to the boundary problem

$$
\begin{aligned}
&\psi = \psi_1 \sin(\pi a x)\sin(\pi n z)\exp[-\lambda t], \\
&\begin{Bmatrix} c_1 \\ c_2 \end{Bmatrix} = \begin{Bmatrix} c_1^1 \\ c_2^1 \end{Bmatrix} \cos(\pi a x)\sin(\pi n z)\exp[-\lambda t], \\
&K^2 = \pi^2(n^2 + a^2),\ n = 1,\ a = \frac{1}{\sqrt{2}}.
\end{aligned}
\qquad (3)
$$

Using method /Ref. 2/ we obtain the expressions which characterize the monotonous and oscillating instability as well as the frequency of neutral oscillations of   on the border of stability in the form:

$$R_1 = -\frac{k^6}{(\pi a)^2} - R_2, \qquad (4)$$

$$R_1 = -\frac{k^6\{(\tau_1+\tau_2+P_2)[\tau_1\tau_2+P_2(\tau_1+\tau_2)]-\tau_1\tau_2P_2\}}{(\pi a)^2(\tau_1+P_2)\tau_1P_2} - \frac{\tau_2(\tau_2+P_2)}{\tau_1(\tau_1+P_2)}R_2.$$

The ternary gas mixture stability analysis carried out with the help of (4), may be illustrated by the example of the ideal system Ar (1)- N_2 (2)- He (3) (indexation of the component is shown in brackets), which may be found in any of the three areas (Fig. 2), classified as it is shown in

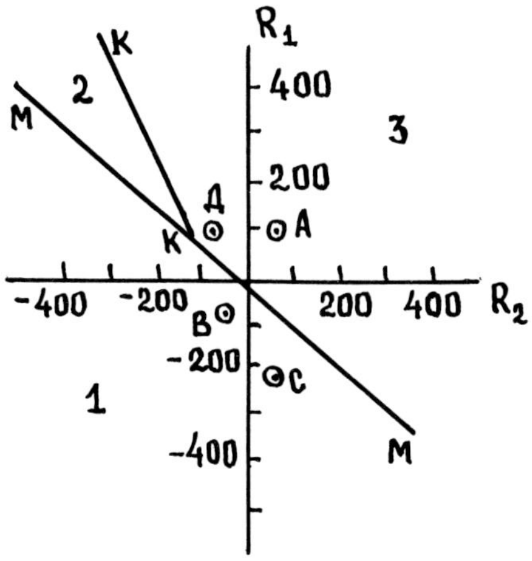

Figure 2: Neutral lines and areas of existing perturbations MM and KK are neutral lines of monotonus and oscillatory perturbations. The points correspond to the /Refs. 3-5/.

/Ref. 2/ in the following way: 1- is the area of stable diffusion; 2,3- are the areas of monotonous and oscillating perturbations. Obviously, for the case R1 > 0 and R2 < 0, the gas mixture He (3)+ Ar (1)- N_2 (2) will be diffusionally unstable. Begining with certain critical parametres (pressure, temperature, diameter of the diffusional channel) the system passes from stable to unstable condition in which the complication regime of oscillation is observed. The characteristics size of convective cells, the time of establishment of unstable regime were also determined.

3. DIFFUSION INSTABILITY IN THE ABSENCE OF GRAVITATION

It is obvious that the experimental features characteristic of the unstable process will be similar for the zero-gravity state as well. The numerical experiment based on the solution of the diffusion equations under corresponding concentration boundary conditions and using the finite difference approach for the cases of one and two independent variables showed the appearance and development of inhomogeneities, with their subsequent blurring. The convective structure formation in such systems is possible under the condition that the laminar flow caused by the diffusive baroeffect is superimposed on the standard diffusive separation of the gaseous mixture. The flow velocity in the framework of the chosen diffusion model given by the following expression:

$$\vec{v}=(\mathcal{D}_1-\mathcal{D}_3)\nabla C_1+(\mathcal{D}_2-\mathcal{D}_3)\nabla C_2. \quad (5)$$

Solutions in dimensionless variables to the diffusion equations of the system (1) for which the velocity is determined according to (5) are found in the form (3). The performed linear stability analysis allows to obtain the condition under which the following oscillating disturbances take place

$$\frac{\{L[N+S]-K^2(\tau_1+\tau_2)\}^2}{\{\tau_1\tau_2K^2-L(\tau_1N+\tau_2S)\}}<4K^2, \quad (6)$$

$$L=(\pi a)tg(\pi a x),\ N=\Delta C_2(\tau_2-1),$$

$$S=\Delta C_1(\tau_1-1).$$

The results obtained may be extended to liquid mixtures as well because the physical principles and mathematical models used in the paper are common to both gaseous and liquid mixtures.

Thus, the diffusive baroeffect both in the presence and in the absence of gravitation may cause the convection structure formation in diffusion in the multicomponent media. This fact should be necessarity taken account of when measuring the heat and mass transfer constants as well as dealing with the problems of chrystallization and melting in the absence of gravitation.

4. REFERENCES

1. J.S. Turner, Buoyancy effects in fluids.(Univ.Press. Cambridge, 1973)

2. G.Z.Gershuny,E.M.Zhukhovitsky. Convective stability of an incompressible liquid.(Nauka, Moscow,1972) p.392.

3. Yu.I.Zhavrin,N.D.Kosov et al. JTPh.,54,943(1982).

4. Yu.I.Zhavrin,V.N.Kosov,JEPh., 55,92(1988).

5. I.V.Bolotov,Yu.I.Zhavrin,V. N.Kosov,Problems of heat and mass transfer(Alma-Ata,1989)p.7.

PROBLEMS OF EQUILIBRIUM CONVECTIVE STABILITY CONTROL

I.O. Keller and E.L. Tarunin

Perm State University, 15, Bukirev str., 614600, Perm, USSR

ABSTRACT

The stabilization of convective equilibrium with the help of single-loop feedback control is considered. It's shown that analysis of bifurcation allows to construct control function. Both linear and nonlinear feedback control were used for stabilization. A circle thermosyphon and rectangular cavity filled with fluid served as convective systems.

Keywords:Convective Instability, Feedback Control, Thermosyphon

1. INTRODUCTION

The theoretical question of the stability control of the systems with distributed parameters using feedback was considered in Ref.1. The possibility of restraining the convective system near the equilibrium state with the help of the temperature change on one of the bounds of the cavity is shown (Ref.2) by calculation experiments. In this paper the inclination of the cavity towards vertical position was chosen as control influence. The efficiency of the proposed methods of controlling is shown by various calculations. There are some convective systems in which proposed methods of controlling is effective. The simplest of them is well known convective loop (Ref.3).

2. BRANCHING ANALYSIS

One can consider that the problem of the stability control is solved if the dependence of inclination angle $\varphi(t)$ (see Figure 1) on the time allowing to restrain the system near the equilibrium is defined. Limiting the consideration to the single-

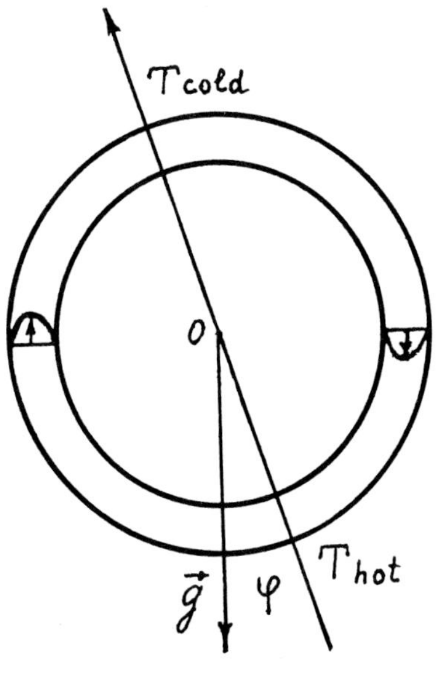

Figure 1: The circular loop.

loop control system with feedback one must obtain the dependence of angle $\varphi(U)$ on any characteristics of deviation U from the equilibrium. As U(t) the average velocity of circulation or convective heat flux may be chosen as examples. The dependence $\varphi(U)$ can be constructed with the help of the analysis of steady solutions branching of the convective system with $\varphi \neq 0$. The picture of branching for the case of circular loop is presented in Figure 2. The change of the sign of φ changes the upper and the lower parts of the branching. With sufficiently large values of the Rayleigh number $Ra > Ra_*$ and $\varphi \neq 0$ three steady solutions are possible. Two of them are stable and one is unstable. With the increasing of the inclination angle, the critical point Ra_* separating the regions of existing one ($Ra < Ra_*$) or three ($Ra > Ra_*$) solutions, moves to the right. Bifurcation point in which the stable and unstable branches coalesce is an important characteristic of the branching. In Figure 2 it's seen that by the certain choice of angle φ (just to have $Ra < Ra_*$) one can make the system move towards the stable branch which is on the contrary side of axes Ra.

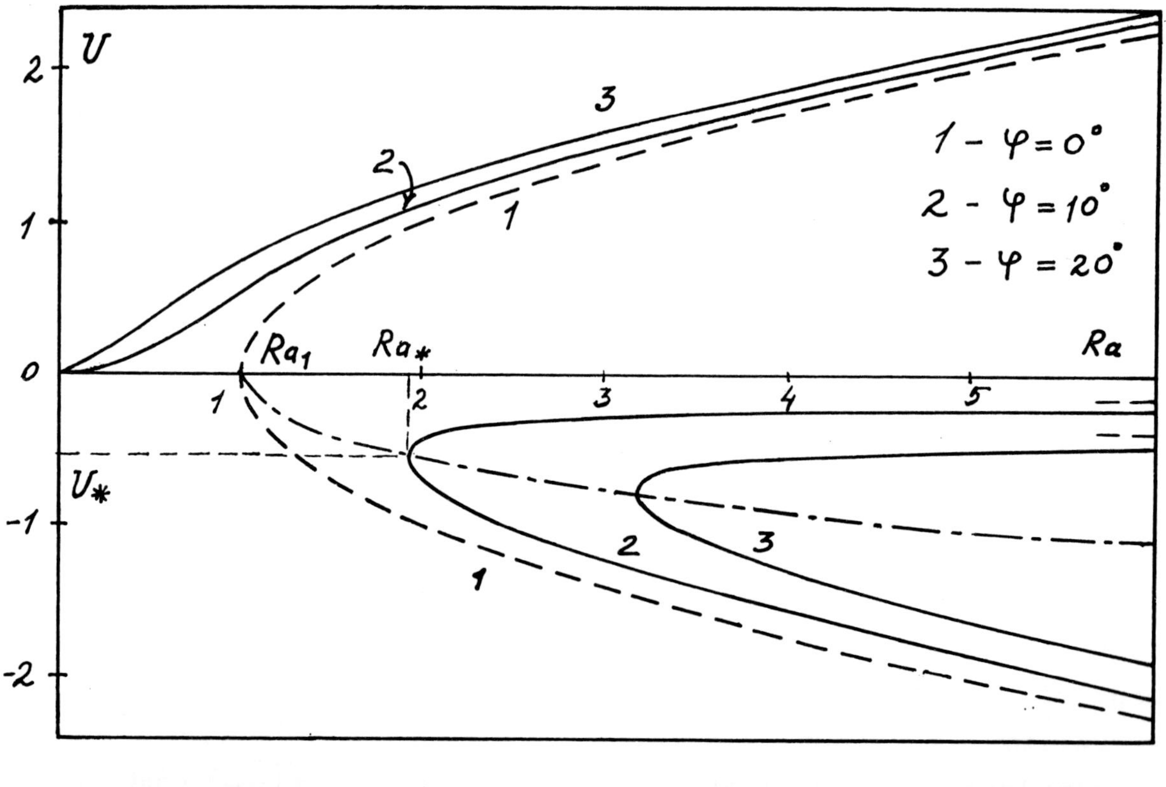

Figure 2. The fluid flux – U in the circular loop on Ra *for angles* $\varphi = 0^\circ, 10^\circ, 20^\circ$.

Let's call angle φ_1 as the one with which $Ra_*(\varphi) = Ra$. In the case of sufficiently small amplitude of the flow and $Ra < Ra_*$ one can choose angle φ to make the system move towards equilibrium. Such choice is possible when the state of the system (being characterized by values U,Ra) is between the axes Ra and line connecting the points of bifurcation (shade dotted line in Figure 2). The angle corresponding to the unstable branch coming through the point (Ra,U) will be called φ_2. The typical time of the amplitude suppressing in the cases described before depends on the closeness of the system to the stable state. For the definition of the controller $\varphi(U)$ supplying the finite time of suppression obtained angle φ was increased by coefficient $(\xi+1)$ where ξ $(\xi>0)$ is the control 'reliability' parameter.

Coming from the point of view the construction of controller $\varphi(U)$, restraining the system around the equilibrium is possible by the following choice:

$$\varphi = \begin{cases} (1+\xi)\cdot\varphi_1, & |U| > |U_*| \\ (1+\xi)\cdot\varphi_2, & |U| < |U_*| \end{cases} \tag{1}$$

As the controller one can use, for example, simply φ_1, instead of the eqn (1). However it will lead to the heightened values of the angle and fast amplitude changes. Controlling function (1) with small ξ gives the slow change of the controlling parameter and the flow amplitude; theoretically with $\xi \rightarrow 0$ the characteristic time is coming up to infinity. The slow change of the influence leads to the less energetic expense and therefore may be considered as a positive feature of the control system.

Below the linear analysis of the stability control (which is valid in the case of small amplitudes) in a rectangular region and convective loop is presented. As the controlling function the more simple linear (compared to eqn (1)) connection between the flow amplitude and the inclination angle is used.

3. RECTANGULAR CAVITY

The cavity is heated from below. Two side walls of the cavity are thermo-isolated. The equations of free convection and the boundary conditions in a dimensionless form are

$$\frac{\partial\Delta\psi}{\partial t} + J(\psi,\Delta\psi) = \Delta\Delta\psi - G[\frac{\partial T}{\partial x}\cos(\varphi) - \frac{\partial T}{\partial z}\sin(\varphi)], \tag{2a}$$

$$\frac{\partial T}{\partial t} + J(\psi, T) = \frac{1}{P}\,\Delta T, \qquad (2b)$$

$$T\Big|_{z=1} = 0, \quad T\Big|_{z=0} = 1,$$

$$\frac{\partial T}{\partial x}\Big|_{x=0,\,l} = 0, \quad \psi\Big|_b = \frac{\partial \psi}{\partial n}\Big|_b = 0. \qquad (2c)$$

Here T is the temperature, ψ–the stream function, J – Jakobian. There are three dimensionless parameters: Grashof and Prandtl numbers G, P and l. Angle φ corresponds to the controlling influence. The simplest feedback connection has the form

$$\varphi = (1+\xi)\cdot k_*\cdot \varphi_e, \qquad \varphi_e = \underset{x,z}{\mathrm{extr}}\,\varphi(x,z), \qquad (3)$$

where k_* is a critical gain value, separating the regions of stable and unstable control with ξ =0. Applying the known Galerkin-procedure of the analysis of stability to system (2) (without nonlinear terms), one can obtain the equation for the decrement:

$$p_1\lambda^2 + p_2\lambda + p_3 = 0. \qquad (4)$$

Here p_i – functions on parameters l, P, G, k. The equation (4) was obtained for two basis functions describing the one-vortex flow. Analysis (4) showed that for all the parameter's values the regime of the attenuation (amplification) of the disturbances is monotonous. Value k_* is being defined from condition p_3=0 and has the form:

$$k_* = f_1(l) - f_2(l)/Ra. \qquad (5)$$

Here Ra = G*P is the Rayleigh number and functions f_1 and f_2 are known. Critical value k_* was obtained also with the help of the numerical calculation (net method) using the full equations (2). These numerical experiments showed the good accordance with the linear theory results. The investigated control in the rectangular cavity with the help of asymmetric heating showed the existence of the oscillating regimes. On our opinion this can be explained by the existence of a controlling influence lagging connected with limited velocity heat wave spreading. In the case of the cavity inclination such lagging is absent and this fact results in the absence of the oscillating mode.

4. CONVECTIVE LOOP

A circular convective loop is shown in Figure 1. The dynamic system describing approximately convection in such loop has form

$$\dot{x} = \sigma(-x + y\cos(\varphi) + (z-r)\sin(\varphi)), \qquad (6a)$$

$$\dot{y} = rx - y - xz, \qquad (6b)$$

$$\dot{z} = -z + xy. \qquad (6c)$$

Here r and σ are the parameters of similarity (analogous to the Rayleigh and Prandtl numbers respectively), x is the fluid flow, y- the deviation of the temperature from the thermal-conductivity distribution, z- the inverse temperature difference. In the case of φ=0 eqn (6) is the well known Lorenz triplet (Ref.6).

Figure 4 shows the steady solutions branching of eqn (6) with various φ. The branching is given by following implicit dependence

$$x^3 + x(1-r\cos(\varphi)) + r\sin(\varphi) = 0. \qquad (7)$$

Equation (7) according to the above-stated point of view allowed to get the formulas for the controlling function φ(x,r,σ) in an explicit form. Because of the bulky size they are not presented here. With the use of the obtained controlling function the control process was simulated with the help of numerical solution of the system (6) by Runghe-Kutta's method. Figure 3 shows the dependence x(t) for the various values of parameter ξ and fixed r=3, σ=1. It's apparent that the control is monotonous (the control has been switched on at $t=t_0$).

The linear analysis of the

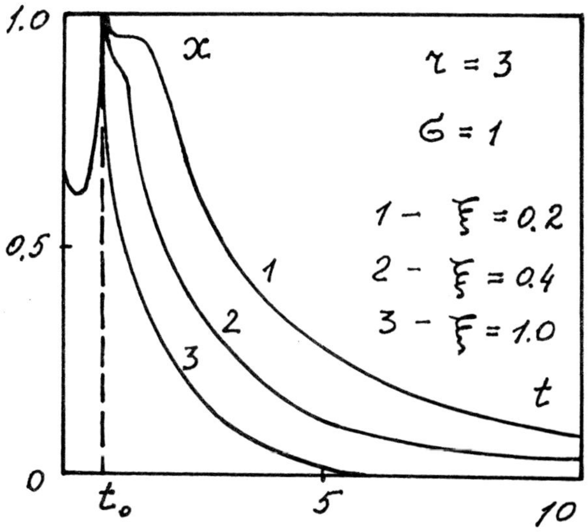

Figure 3: The process of controlling with using of eqn (1).

control with controller like (3) allows to get the value of the control stability bound

$$k_* = 1 - \frac{1}{r} \qquad (8)$$

The results of the calculation experiments with using the (3) and k from eqn (8) are presented in Figure 4. It's seen that the control is impossible for the value ξ <0.

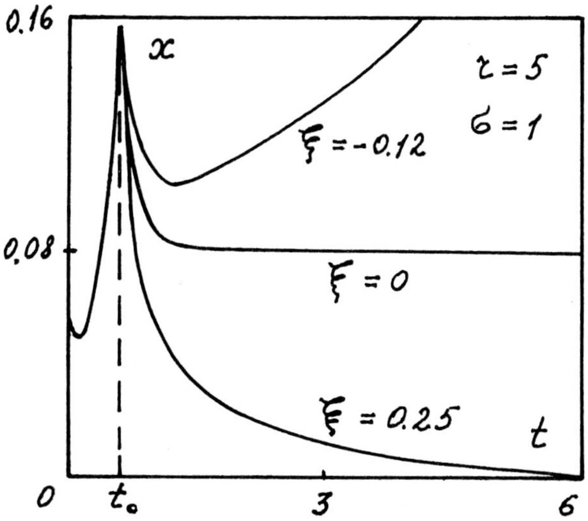

Figure 4: Controlling in the case of the linear controller from eqn (3).

5. CONCLUSIONS

It's shown that we can use the linear theory of stability for a small deviation of the system from equilibrium. In the region of finite deviations the study of branching is very useful. The efficiency of the proposed methods of controlling was demonstrated by various calculations.

6. REFERENCES

1. V.A. Buchin in Mechanics and Scince-Tecnique Progress.v.2. Mechanics of Fluid and Gas. Nauka, Moskow (1987),p.164
2. I.O. Keller and E.L.Tarunin Izv. AN SSSR, Mech. Zhidk. Gaza 4(1990)p.6.
3. G.Z. Gershuni, E.M. Zhukhovitsky and A.A. Nepomnjashchy *Stability of Convective Flows* (Nauka, Moskow 1989).

ON INSTABILITY MECHANISMS OF BINARY MIXTURE ADVECTIVE FLOW

V.M. Myznikov

Perm State Pedagogical Institute,
24, K. Marx Str., 614600, Perm, USSR

ABSTRACT

The stationary advective flow of a binary mixture in a plane horizontal layer with fixed longitudinal temperature gradient on its boundaries is considered. Besides the existence of a constant mean longitudinal concentration gradient is assumed within the cavity. The behaviour of two limiting types of small normal disturbances - planar and helical - is investigated. There were found four kinds of instability mechanisms: nonviscous hydrodynamical; Rayleigh's stratificational; inner gravitation waves; thermoconcentration.

Keywords: *Plane Horizontal Layer, Binary Mixture, Advective Flow, Plane and Helical Normal Disturbances.*

1. INTRODUCTION

The problem of stability of advective flows in thin layers is interesting in particular in connection with its application to monocrystal production in a horizontal variant of the directional crystallization method. The temperature stratification arising in the layer under the influence of longitudinal temperature gradients results in the appearance of flow instability mechanisms of different physical nature. These types of instability were studied in detail for the case of a single-component liquid in (Ref. 1-6).

In the present paper the linear stability of advective binary mixture flow in a plane horizontal layer with longitudinal temperature and concentration gradients is investigated.

2. THE BASIC EQUATIONS. THE STATIONARY SOLUTION.

We shall consider the plane horizontal liquid binary mixture layer of infinite extent limited from below and above by solid parallel planes $y = \pm h$ (coordinate axes are shown in Figure 1).

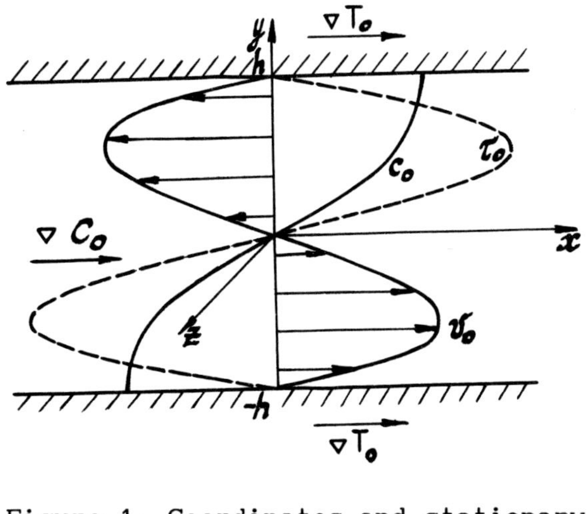

Figure 1: Coordinates and stationary distributions of velocity, temperature and concentration of the basic flow

On both planes of high heat conductivity the temperature is maintained changing lineary as a function of the horizontal coordinate x :

$$y = \pm h: \qquad T = Ax. \qquad (1)$$

Here A is a constant temperature gradient. The constant longitudinal concentration gradient B is also kept in the layer:

$$\lim_{L\to\infty} \frac{1}{2L} \int_{-L}^{L} \frac{\partial C}{\partial x}\, dx = B. \qquad (2)$$

The linear dependence of density on temperature and concentration is assumed:

$$\rho = \rho_0 (1-\beta_1 T-\beta_2 C). \qquad (3)$$

The study of the flow stability is carried out on the basis of binary mixture free convection equations in Boussinesq approximation without the thermodiffusive effect (Ref. 1):

$$\frac{\partial \vec{v}}{\partial t} + (\vec{v}\,\nabla)\vec{v} = -\frac{1}{\rho}\nabla p + \nu\,\Delta\vec{v} + g(\beta_1 T+\beta_2 C)\vec{\gamma}, \qquad (4)$$

$$\frac{\partial T}{\partial t} + \vec{v}\,\nabla T = \chi\,\Delta T, \qquad (5)$$

$$\frac{\partial C}{\partial t} + \vec{v}\,\nabla C = D\,\Delta C, \qquad (6)$$

$$\mathrm{div}\,\vec{v} = 0. \qquad (7)$$

Here the following signs are introduced: $\vec{v}$ - velocity, p - pressure (reference point is hydrostatic corresponding to the average density ρ), T - temperature referred to some arbitrary zero, C - concentration of light component, g - gravity acceleration, $\vec{\gamma}$ - unit vector directed vertically upwards, ν, χ, D, β_1, β_2 - coefficients of kinematic viscosity, heat diffusivity, diffusion, thermal expansion, concentration coefficient of density correspondingly.

The equations (4)-(7) and the condition (2) may be presented in the dimensionless form on the basis of the following unit choice: length - h, time - h^2/ν, velocity $g\beta_1 Ah^3/\nu$, temperature - Ah, pressure - $\rho g\beta_1 Ah^2$, concentration - $\beta_1 Ah/\beta_2$:

$$\frac{\partial \vec{v}}{\partial t} + Gr(\vec{v}\,\nabla)\vec{v} = -\nabla p + \Delta\vec{v} + (T+C)\vec{\gamma}, \qquad (8)$$

$$\frac{\partial T}{\partial t} + Gr\,\vec{v}\,\nabla T = \frac{1}{Pr}\,\Delta T, \qquad (9)$$

$$\frac{\partial C}{\partial t} + Gr\,\vec{v}\,\nabla C = \frac{1}{Sc}\,\Delta C, \qquad (10)$$

$$\mathrm{div}\,\vec{v} = 0, \qquad (11)$$

$$\lim_{l\to\infty} \frac{1}{2l} \int_{-l}^{l} \frac{\partial C}{\partial x}\, dx = \frac{Ra_d}{Gr\ Sc}\,. \qquad (12)$$

The equations contain four dimensionless parameters - Grasshof number Gr , Prandtl number Pr, Schmidt number Sc and concentration Rayleigh number Ra_d (or concentration Grasshof number $Gr_d = Ra_d/Sc$):

$$Gr = \frac{g\beta_1 Ah^4}{\nu^2}\,, \quad Pr = \frac{\nu}{\chi}\,, \quad Sc = \frac{\nu}{D}\,, \quad Ra_d = \frac{g\beta_2 Bh^4}{\nu^2}\,.$$

The boundary conditions of no-slip, temperature distribution on the boundaries, impenetrability and close flow condition are formulated as follows:

$$y = \pm 1:\ v=0,\ T=x,\ \frac{\partial C}{\partial y}=0,\ \int_{-1}^{1} v_0\,dy=0 \qquad (13)$$

Equations (8)-(12) with conditions (13) have the stationary solution (base flow)

$$v_x = v_0 = \frac{1}{6}\,Q\,(y^3 - y), \qquad (14)$$

$$T_0 = x+\tau_0(y) = x+\frac{1}{360}GrPrQ(3y^5-10y^3+7y), \qquad (15)$$

$$C_0 = (Ra_d/Gr\ Sc)x+\frac{1}{360}Ra_d\,Q(3y^5 -10y^3 + 15y), \qquad (16)$$

where $Q = 1 + Ra_d/GrSc$.

3. FORMULATION OF THE LINEAR STABILITY PROBLEM

We shall introduce small disturbances of the base flow. Let us consider disturbed velocity, temperature, concentration and pressure fields:

$$\vec{v}=\vec{v}_0+\vec{v}', T=T_0+T', C=C_0+C', p=p_0+p', \quad (17)$$

where $\vec{v}'$, T', C', p' are the small disturbances. After substituting (17) into (8)-(11) and linearization in the vicinity of the principal state we shall obtain a system of linear equations for disturbances (primes of $\vec{v}'$, T', C', p' are dropped):

$$\frac{\partial \vec{v}}{\partial t} + Gr[(\vec{v}_0\nabla)\vec{v}+(\vec{v}\,\nabla)\vec{v}_0] = -\nabla p + \Delta\vec{v} + (T+C)\vec{\gamma}, \quad (18)$$

$$\frac{\partial T}{\partial t} + Gr(\vec{v}_0\nabla T + \vec{v}\,\nabla T_0) = \frac{1}{Pr}\Delta T, \quad (19)$$

$$\frac{\partial C}{\partial t} + Gr(\vec{v}_0\nabla C + \vec{v}\,\nabla C_0) = \frac{1}{Sc}\Delta C, \quad (20)$$

$$\operatorname{div}\vec{v} = 0, \quad (21)$$

with boundary conditions

$$y=\pm 1: \quad \vec{v}=0, \quad T=0, \quad \frac{\partial C}{\partial y}=0. \quad (22)$$

In the general case of spatial normal disturbances all the variables are proportional $\exp[-\lambda t+i(k_x x+k_z z)]$.

There are no Squire's transformations in the problem. So we shall discuss two limiting cases - plane disturbances ($k_z= 0$) and spatial helical disturbances ($k_x= 0$). For the case of plane disturbances the stream function $\psi(x,y,t)$ is introduced:

$$v_y = -\frac{\partial\psi}{\partial x}, \quad v_x = \frac{\partial\psi}{\partial y}.$$

Eliminating pressure we obtain the following amplitude equations:

$$\varphi^{IV}- 2k_x^2\varphi''+ k_x^4\varphi + ik_x Gr[v_0''\varphi-v_0(\varphi''-k_x^2\varphi)]-ik_x(\theta+\xi)=-\lambda(\varphi''-k_x^2\varphi), \quad (23)$$

$$\frac{1}{Pr}(\theta''-k_x^2\theta)+ik_x Gr(\tau_0'\varphi-v_0\theta)-Gr\varphi' = -\lambda\theta, \quad (24)$$

$$\frac{1}{Sc}(\xi''-k_x^2\xi)+ik_x Gr(C_0'\varphi-v_0\xi)-Gr_d\varphi' = -\lambda\xi, \quad (25)$$

with boundary conditions

$$y=\pm 1: \quad \varphi=\varphi'=0, \quad \theta=0, \quad \xi'=0. \quad (26)$$

(here and further the prime denotes differentiation with respect to y). $\varphi(y)$, $\theta(y)$, $\xi(y)$ are the amplitudes of stream function, temperature and concentration.

For the case of helical disturbances the stream function is defined as follows:

$$v_z = \frac{\partial\psi}{\partial y}, \quad v_y = -\frac{\partial\psi}{\partial z},$$

and the amplitude problem has the form:

$$\varphi^{IV}-2k_z^2\varphi''+k_z^4\varphi-k_z^2(\theta+\xi)=-\lambda\Delta\varphi, \quad (27)$$

$$u_x''-k_z^2 u_x -Gr\, v_0'\varphi=-\lambda u_x, \quad (28)$$

$$\frac{1}{Pr}(\theta''-k_z^2\theta)-Gr\,\tau_0'\varphi+Gr\,u_x=-\lambda\theta, \quad (29)$$

$$\frac{1}{Sc}(\xi''-k_x^2\xi)-Gr\,C_0'\varphi+Gr_d u_x=-\lambda\xi, \quad (30)$$

$$y=\pm 1: \quad u_x=0, \quad \varphi=\varphi'=0, \quad \theta=0, \quad \xi'=0. \quad (31)$$

The boundary problems (23) - (26) and (27)-(31) define the eigenvalues spectrum λ. The Galerkin technique was applied to aproximate solving of spectral problems. There were used up to 80 basic functions.

4. RESULTS

The calculations are carried out for the following values of parameters: Gr_d= 0.1, 1.0, 10; Pr=0.1, 1.0, 10; Sc= 0.1, 1.0, 10. First of all the results indicate the presence of three mechanisms of instability which occur in the case of single-component liquid advective flow. As is evident from Figure 1 the velocity profile of the stationary solution (14) possesses the inflection point occurring in the centre of the layer section $y = 0$. Due to occurence of the inflection point the base flow is displaied the nonviscous hydrodynamical instability mode. This mode is related to the formation of stationary vortexes system on the boundary partition of opposing flows.

Then as (15) and Figure 1 shows undisturbed temperature profile has zones of potentially unstable stratification adjoining the upper and lower horizontal boundaries. In this zones the appearance of Rayleigh instability stratification caused by heating from below is going on. This instability leads to development of plane or helical structures. The central part of the liquid layer is stratificated in potentially stable manner and generation of disturbances of the inner gravitation waves type is observed.

But so long as the concentration distribution (16) is potentially stable the critical values of Grasshof number grows by increasing of diffusion Grasshof number and Schmidt number (Table 1).

The analogous growth of the critical Grasshof number is indicated with increasing of the Schmidt number. For Sc=10 (Gr=0.1,Pr=1.0) the monotonous helical Rayleigh instability transforms into the oscillatory form. The stream function amplitude of the helical vortexes in unstable temperature stratification zones starts to change periodically. Besides the enumerating instability types the interaction of the temperature and concentration fields in the layer causes the new thermoconcentration mechanism of instability which displays in the formation of the oscillatory plane vortex chain. The critical wave number of such disturbances is $k_x \approx 1.2 \div 1.4$.

TABLE 1. The critical Grasshof numbers for the hydrodynamical (HD), Rayleigh helical (RH) and oscillatory helical (OH) modes; Pr = 0.1, Sc = 0.1 .

Modes	*Diffusion Grasshof number*			
	$Gr_d = 0$	$Gr_d = 0.1$	$Gr_d = 1.0$	$Gr_d = 10$
HD_*	788	797	816	2162
RH_{**}	5716	5735	5813	7031
RH	5911	5926	5958	7192
OH	1248	1558	6421	$>5x10^4$

* - even mode; ** - odd mode.

5. REFERENCES

1. G.Z. Gershuni, E.M. Zhukhovitsky and A.A. Nepomnjashy, *Stability of convective flows* (Nauka, Moscow 1989).

2. G.Z. Gershuni, E.M. Zhukhovitsky and V.M. Myznikov, *J. Appl. Mech. and Techn. Phys.* 1(1974)95.

3. G.Z. Gershuni, E.M.Zhykhovitsky and V.M. Myznikov, *J. Appl. Mech. and Techn. Phys.* 5(1974)145.

4. V.M. Myznikov in *Hydrodynamics,* Perm 7(1974),p.33.

5. V.M. Myznikov in *Convective Flows,* Perm (1989),p.28.

6. H.P. Kuo, S.A. Korpela, A. Chait and P.S. Marcus in *8th Heat Transf. Conf.,* San Francisco 3(1986),p.1539.

Authors Index

List of Participants

Professor V S Afraimovich
Institute of Applied Physics
Russian Academy of Sciences
46 Ulyanov Street
606600 Nizhni Novgorod
Russia

Dr I M Aliev
Moscow Technical University
5, 2 Baumanskaya Street
107005 Moscow
Russia

Dr Gustav Amberg
Hydromekanik KTH
S-10044 Stockholm
Sweden

Dr V K Andreev
Computing Centre
Siberian Branch of Russian Academy of Sciences
Academgorodok
660036 Krasnoyarsk
Russia

Mr Arquis
ENSCPB Master Laboratoire
351 cours de la Liberation
F-33405 Talence Cedex
France

V K Artemyev
Institute of Physics and Power Engineering
1 Bondarenko Sq
249020 Obninsk
Kaluga Region
Russia

Dr N A Avdonin
Research Institute of Mathematics and Computer Science
Latvian State University
29 Rainis Boulevard
226050 Riga
Latvia

Professor V S Avduevsky
Machine Science Institute
Russian Academy of Sciences
4 Griboedova Street
101830 Moscow
Russia

Dr A Babiano
LMD-ENS
24 rue Lhomond
Paris 75231 Cedex 05
France

I A Babushkin
Perm State University
5 Bukirev Street
614005 Perm
Russia

Professor O M Barabash
Institute of Metallophysics
Ukrainian Academy of Sciences
36 Vernadsky Prospect
252680 Kiev
Ukraine

R I Barabash
Kiev Polytechnical Institute
37 Prospect Pobedy
KP1-1150
252056 Kiev
Ukraine

Professor I V Barmin
SPLAV Technical Center
9 Baikalskaya Street
107497 Moscow
Russia

Professor A Barrero
Industriales Universisdad De Sevilla
ETSI Industriales
Avda Reina Mercedes
S/n 41012 Sevilla
Spain

N Baturin
Russian Academy of Sciences
88 Profsoyuznaya Street
117810 Moscow
Russia

Professor Dr Helmut F Bauer
University of the German Armed Forces
Werner Heisenberg Weg 39
8014 Neubiberg
Munich
Germany

A A Bauere
Institute of Physics
Latvian Academy of Sciences
32 Miera Street
Salaspils-1 Riga District 229021
Latvia

Dr M S Bello
Institute of Macromolecular Compounds
Russian Academy of Sciences
31 Bolyshoy Prospect VO
199004 Shankt-Peterburg
Russia

Dr A G Belonogov
Institute of Physics and Applied
Mathematics
Urals State University
51 Lenin Avenue
620083 Ekaterinburg
Russia

A Bennema
Center of Microelectronics
Latvian Academy of Sciences
19 Turgenev Street
Riga
Latvia

Dr Daniel Beysens
Centre d'Etudes Nucleaires de Saclay
Service de Physique du Solide et de
Resonance Magnetique
Orme des Merisiers
F-91191 Gif-sur-Yvette Cedex
France

Dr B A Bezuglyi
Tyumen State University
10 Syemakov Street
625003 Tyuman
Russia

Dr Bernard Billia
University Aix-Marseille III
Faculte de St Jerome
Case 151
F-13397 Marseille Cedex 13
France

Dr R V Birikh
Perm State Pedagogical Institute
24 Karl Marx Street
614600 Perm
Russia

Dr V M Biryukov
Research Institute "Nauchny Tsentr"
103460 Moscow
Russia

Dr V N Blinkov
Kharkov Aviation Institute
17 Chkalova Street
310084 Kharkov
Ukraine

A F Bogdanova
Kiev Polytechnical Institute
37 Prospect Pobedy KP1-1150
252056 Kiev
Ukraine

Dr N E Boitsun
Dnepropetrovsk State University
72 Gagarin Prospect
320625 Dnepropetrovsk
Ukraine

Dr V A Brailovskaya
Institute of Applied Physics
Russian Academy of Sciences
46 Ulyanov Street
606600 Nizhni Novgorod
Russia

Dr V A Briskman
Institute of Continuous Media Mechanics
Russian Academy of Sciences
1 Korolyov Street
614061 Perm
Russia

Dr Ing Camel
Center d'Etudes Nucleaires
DEM/SESC
85 X 38041 Grenoble Cedex
France

Dr Antonio Castellanos-Mata
University of Sevilla
DPTO Electronica y Electromagnetismo
Avda Reina Mercedes S/N
41012 Sevilla
Spain

Professor Hsueh-Chia Chang
Department of Chemical Engineering
University of Notre Dame
Notre Dame
Indiana 46556
USA

Dr J Yu Chashechkina
Institute of Solid State Physics
Russian Academy of Sciences
Chernogolovka
Noginsky District
Moscow Region 142432
Russia

Dr Christian Chauveau
Centre de Recherches sur la Chimie de la Combustion et des Hautes Temperatures
CNRS-CCHT
1C Av de la Recherche Scientifique
45071 Orleans Cedex 2
France

V P Chegnov
Research Institute "Nauchny Tsentr"
103460 Moscow
Russia

Professor T A Cherepanova
Center of Microelectronics
Latvian Academy of Sciences
19 Turgenev Street
Riga
Latvia

Dr V I Chernatinskii
Perm State Pedagogical Institute
24 Karl Marx Street
614600 Perm
Russia

S I Chuckalov
Institute of Applied Physics
Academy of Science Republic of Moldova
5 Acad. Grosul Street
277028 Kishinev
Republic of Moldova

Dr Arne Croll
Kristallographisches Institut
der Universitat
Hebelstr 25
D-7800 Freiburg
Germany

Professor G Cubiotti
Instuto di Fisca Teorica
PO Box 50 - 98166
S. Agata di Messina
Italy

Dr Andreas N Danilewsky
Kristallographisches Institut
der Universitat
Hebelstr 25
D-7800 Freiburg
Germany

Dr Ye A Demekhin
Department of Applied Mathematics
Krasnodar Polytechnical Institute
2 Moscowskaya Street
350072 Krasnodar
Russia

A V Demin
SPA "Kompozit"
Kaliningrad
Moscow Region 141070
Russia

Dr Yu Derbenev
Research Institute "Biotechnics"
Russian Ministry of Medical Industry
38 Kropotkinskaya Street
119034 Moscow
Russia

Dipl. Ing M Dreyer
University of Bremen
ZARM
D-2800 Bremen 33
Germany

Dr A S Drobyshev
Physics Department
Kazakh State University
96 Komsomolskaya Street
480012 Alma-Ata
Kazakhstan

Dr K G Dubovik
Institute for Problems in Mechanics
Russian Academy of Sciences
101 Prosp Vernadskogo
117526 Moscow
Russia

Andreas Ecker
Dept RVM
PO Box 14 20
7990 Friedrichshafen
Germany

Professor Dr Ivan Egry
Institut für Raumsimulation
DRL Linder Hohe
5000 Koln-Porz 90
Germany

Dr Ing Peter Ehrhard
Kernforschungszentrum Karlsruhe GmbH
Institut für Reaktorbauelemente
Postfach 3640
D-7500 Karlsruhe 1
Germany

Dr M K Ermakov
Institute for Problems in Mechanics
Russian Academy of Sciences
101 Prospect Vernadskogo
117526 Moscow
Russia

Dr S V Ermakov
Keldysh Institute of Applied Mathematics
Russian Academy of Sciences
4 Miusskya Square
125047 Moscow
Russia

L Yu Erokhin
Institute for Problems in Mechanics
Russian Academy of Sciences
101 Prospect Vernadskogo
117526 Moscow
Russia

Dr Rierre Evesque
LCPC
Anrenne d'Orly BP
Orly Sud n 155
F-94396 Orly Aerogare Cedex
France

Professor H Fahr
Extraterrestriche
Forschung Universitat
Bonn auf dem Hugel 71
D-5300 Bonn
Germany

Dr Ing Piergiuseppe Falciani
Officine Galileo SpA
35 via Einstein
50013 Campi Bisenzio
Italy

Dr Yadwiga Fangrat
Building Research Institute
Fire Research Division
Filtrowa 1
00-950 Warsaw
Poland

Dr M Farge
LMD-CNRS
Ecole Normale Supereur
24 rue Lhomond
75231 Paris Cedex
France

Dr Jean-Jacques Favier
Cerem/DEM/SESC
Laboratoire d'Etude de la Solidification
85 X CENG
F-38041 Grenoble Cedex
France

Dr A I Fedoseev
Institute for Problems in Mechanics
Russian Academy of Sciences
101 Prospect Vernadskogo
117526 Moscow
Russia

Dr A I Fedyushkin
Institute for Problems in Mechanics
Russian Academy of Sciences
101 Prospect Vernadskogo
117526 Moscow
Russia

L V Feokistova
Institute of Applied Physics
Russian Academy of Sciences
46 Ulyanov Street
606600 Nizhni Novgorod
Russia

Dr A I Feonychev
Cosmonautics and Automatical
Spacecraft
Faculty
Moscow Aviation Institute
4 Volokolamskoye Shosse
125871 Moscow
Russia

Professor Dr Berndt Feuerbacher
Institut für Raumsimulation
DLR Linder Hohe
D-5000 Koln 90
Germany

Professor Robert Finn
Department of Mathematics
Stanford University
Building 380
Stanford
California 94305
USA

Ass. Professor Amalia E Finzi
Aerospace Dept
V Golgi 40-20133 Milano
Italy

Professor Yasushi Fuhuzawa
1603-1 Kamitomioka
Nagaoka 940-21
Japan

Dr Alfonso Ganan
Industriales Universidad De Sevilla
ETSI Industriales
Avda Reina Mersedes
S/n 41005 Sevilla
Spain

M Garcia-Rviz
Instituto Andaws de Geologia
mediterronea
Av Fuentenueva s/n
Granada 18002
Spain

Dr A Yu Gelfgat
Research Institute of Mathematics and
Computer Science
University of Latvia
29 Rainis Boulevard
226250 Riga
Latvia

Professor Yu M Gelfgat
Institute of Physics
Latvian Academy of Sciences
32 Miera Street
Salaspils-1 Riga District 229021
Latvia

Dr Philippe Georis
Laboratore de Chimie Physique
Ep Cp 165
Unisersite Libre de Brusseles
50 Avenue F D Roosevelt
1050 Brussels
Belgium

Professor G Z Gershuni
Perm State University
15 Bukirev Street
614005 Perm
Russia

Dr Jose Antonio Nicolas Gimeno
Departamento de Fundamentos
Matematicos ETSI Aeronauticos
Universsidad Politecnica
Plaza del Cardenal Cisneros, 3
Cuidad Universitatria
28040 Madrid
Spain

V N Glotov
Institute of Bioorganic Chemistry
Russian Academy of Sciences
16/10 Miklukho-Maklai Street
117871 Moscow
Russia

Dr Iskender Gokalp
Centre de Rechercehes sur la Chimie de la Combustion et des Hautes Temperatures
CNRS-CCHT
1C, Av. de La Recherce Scientifique
45071 Orleans Cedex 2
France

Dr V D Golyshev
Research Institute for Mineral Resources
601600 Aleksandrov
Vladimir Region
Russia

Dr V A Goncharov
Research Institute "Nauchny Tsentr"
103460 Moscow
Russia

Dr M A Gonik
Research Institute for Mineral Resources
601600 Aleksandrov
Vladimir District
Russia

Dr O N Gontcharova
Altay State University
66 Dimitrov Street
Barnaul
Russia

Dr G A Gorbenko
Kharkov Aviation Institute
17 Chkalova Street
310084 Kharkov
Ukraine

Dr L A Gorbunov
Institute of Physics
Latvian Academy of Sciences
32 Miera Street
Salaspils-1, Riga District 229021
Latvia

Professor G Gouesbet
Lesp, Ura. CNRS 230
Insa. Rouen
BP08-76131
Mont-Saint-Aignan
Cedex
France

Dr Patrick Guenoun
SPSRM/CEN Saclay
91191 GIF/Yvette Cedex
France

Dr V L Gryaznov
Institute for Problems in Mechanics
Russian Academy of Sciences
101 Prospect Vernadskogo
117526 Moscow
Russia

A P Gus'kov
Institute of Solid State Physics
Russian Academy of Sciences
Chernogolovka
Noginsky District
Moscow Region 142432
Russia

Mark Hellemans
American Consulate General
Nationalestraat 5
B-2000 Antwerpen 1
Belgium

PhD Hennenberg
Department de Chimie Physique
Faculty des Sciences
Campus Plaine, CP 231, ULB
Blord du Triomphe
1050 Brussels
Belgium

Dr Daniel Henry
Laboratorie de Mecanique des Fluides et
d'Acoustique Ecole Centrale de Lyon
69131 Eculty
France

Dr Taketoshi Hibiya
Kudan New Central Building
Kudan-Kita 1-chome
Chiyoda-Ku
Tokyo 102
Japan

Professor R H Hung
University of Alabama in Huntsville
Huntsville
Alabama 35899
USA

Dr G I Ilyin
Design Bureau "Yuzhnoye"
3 Krivorozhskaya Street
320059 Dnepropetrovsk
Ukraine

A I Ivanov
SPA "Energia"
Kaliningrad
Moscow District 141070
Russia

N N Ivanov
SPA "Lavochkin"
24 Leningradtsev Street
141400 Khimki
Moscow
Russia

Dr A A Ivanova
Perm State Pedagogical Institute
24 Karl Marx Street
614600 Perm
Russia

Professor P Joos
University Antwerp
Department Chemistry UIA
Universiteitsplein 1
B-2610 Wilrijk
Belgium

Dr I S Kalachinskaya
Moscow State University
Leninskye Mauntains
117234 Moscow
Russia

Dr E L Kalyazin
Cosmonautics and Automatical
Spacecraft Faculty
Moscow Aviation Institute
4 Volocolamskoe shosse
125871 Moscow
Russia

Dr A E Kazakova
Central Specialized Design Bureau
18 Pskovskaya Street
443009 Samara
Russia

I O Keller
Perm State University
15 Bukirev Street
614005 Perm
Russia

Eng. Hideo Kimura
Shibaryoshin Building
2-5-6- Shiba
Minato-ku
Tokyo 105
Japan

Dr A G Kirdyashkin
Institute of Geology and Geophysics
Siberian Branch of Russian Academy of Sciences
3 Universitetsky Avenue
630090 Novosibirsk
Russia

Dr V M Kiseev
Department of General and Molecular Physics
Urals State University
51 Lenin Avenue
620083 Ekaterinburg
Russia

Reiner Klett
Kayzer-Thread GmbH
Wolfratshauser 48
D-8 Munchen 70
Germany

Dr S M Klimov
Institute of Applied Physics
Academy of Sciences Republic Moldova
5 Academy Grosul Street
277028 Kishinev
Republic Moldova

Dr A K Kolesnikov
Perm State Pedagogical Institute
24 Karl Marx Street
614600 Perm
Russia

Professor N D Kosov
Kazakh State University
46 Timiryazev Street
480121 Alma-Ata
Kazakhstan

Dr V N Kosov
Kazakh State University
46 Timiryasev Street
480121 Alma-Ata
Kazakhstan

K G Kostarev
Institute of Continuous Media Mechanics
Russian Academy of Sciences
1 Korolyov Street
614061 Perm
Russia

Dr V G Kosuskhin
Research Institute of the Materials for Electrotechnics
1 Gagarina Street
248650 Kaluga
Russia

S R Kosvintsev
Perm State University
15 Bukirev Street
614005 Perm
Russia

Dr I A Kojukhari
Institute of Applied Physics
Academy Science Republic Moldova
5 Academy Grosul Street
277028 Kishinev
Republic of Moldova

A E Kovalev
Institute of Continuous Media Mechanics
Russian Academy of Science
1 Korolyov Street
614061 Perm
Russia

A V Kozlov
Urals Branch
Russian Academy of Science
91 Pervomaisckaya Street
620219 Ekaterinburg
Russia

Dr V G Kozlov
Perm State Pedagogical Institute
24 Karl Marx Street
614600 Perm
Russia

V M Kulikov
Cosmonautics and Automatical
Spacecraft
Faculty
Moscow Aviation Institute
4 Volocolamskoe Shosse
125871 Moscow
Russia

Dr V N Kurdyumov
Institute for Problems in Mechanics
Russian Academy of Sciences
101 Prospect Vernadskogo
117526 Moscow
Russia

V V Kuznetsov
Institute of Applied Physics
Russian Academy of Sciences
46 Ulyanov Street
606600 Nizhni Novgorod
Russia

Professor Dr Dieter Langbein
Dipl. Physiker
Battelle-Institute e.V.
Postschliessfach 900160
6000 Frankfurt/Main 90
Germany

A Yu Lapin
Institute of Continuous Media Mechanics
Russian Academy of Sciences
1 Korolyov Street
614061 Perm
Russia

Dr Jean-Claude Launay
Prame-Aerospatiale, BP 11
33165 St Medard en Jalles
Cedex
France

Dr O M Lavrentyeva
Lavryentyev Institute of Hydrodynamics
Siberian Branch of Russian Academy of
Sciences
630090 Novoskibirsk
Russia

A P Lazarev
International Connection Committee
Perm Regional Administration
23 Lenin Street
614000 Perm
Russia

Professor Jean-Cleaude Legros
ULB, CP165
Chemie Physique EP
50 Avenue F D Roosevelt
B-1050 Brussels
Belgium

Dr Klaus Leonartz
Aachener Centrum fur Erstarrung unter Schwerelosigkeit
Access e.V. Intzestrasse 5
D-5100 Aachen
Germany

Civ. Ing. Marten Levenstam
Hydromekanik, KTH
S-10044 Stockholm
Sweden

Dr Liggieri Libero
ICFAM-CNR
Lungoblsagno Istria 34
16141 Genova
Italy

Dr D V Lyubimov
Perm State Univesrsity
15 Bukirev Street
614005 Perm
Russia

Dr T P Lyubimova
Institute of Continuous Media Mechanics
Russian Academy of Sciences
1 Korolyov Street
614061 Perm
Russia

Yu N Macagon
Kiev Polytechnical Institute
37 Prospect Pobedy
KPI-1150
252056 Kiev
Ukraine

Dr A S Makarova
Dnepropetrovsk State University
72 Gagarin Prospect
320625 Dnepropetrovsk
Ukraine

AS Eng. Alessandra Malino
CISI Itàlia SPA
Piazza Della Republica 32
20124 Milano
Italy

Dr V V Mansurov
Department of Mathematical Physics
Urals State University
51 Lenin Avenue
620083 Ekaterinburg
Russia

Dr E V Markov
Research Instiatute "Nauchny Tsentr"
103460 Moscow
Russia

Dr B J Martuzans
Research Institute of Mathematics and Computer Sciences
University of Latvia
29 Rainis Boulevard
226250 Riga
Latvia

Dr A L Maschinsky
Institute of Biomedical Problems
Russian Ministry of Health
76a Khoroshyevskoye Shosse
123007 Moscow
Russia

Dr L N Maurin
Ivanovo State University
Ivanovo
Russia

A G Mednov
Cosmonautics and Automatical
Spacecraft
Faculty
Moscow Aviation Institute
4 Volocolamskoe Shosse
125871 Moscow
Russia

Professor Jose Meseguer
Laboratorio de Aerodinamica
ETSI Aeronauticos
28040 Madrid
Spain

Professoresor V B Molodkin
Institute of Metallophysics
Ukrainian Academy of Sciences
36 Vernadsky Prospect
252680 Kiev
Ukraine

Professor V V Moshev
Institute of Continuous Media Mechanics
Russian Academy of Science
1 Korolyov Street
514061 Perm
Russia

Dr Yu I Mosknenko
Design Bureau "Yuzhnoye"
3 Krivorozhskaya Street
320059 Dnepropetrovsk
Ukraine

Dr M Z Mukhoyan
SPA "Kompozit"
Kaliningrad
Moscow Region 141070
Russia

Professor A D Myshkis
Institute for Railway Transport Engineers
103055 Moscow
Russia

Dr V M Myznikov
Perm State Pedagogical Institute
24 Karl Marx Street
614600 Perm
Russia

Dr B I Myznikova
Institute of Continuous Media Mechanics
Russian Academy of Sciences
1 Korolyov Street
614061 Perm
Russia

Professor Luigi G Napolitano
Microgravity Advanced Research and
Support (MARS) Center
Via Diocleziano
328-80125 Naples
Italy

Professor Robert J Naumann
Office of the Dean
College of Science
University of Alabama in Huntsville
Huntsville
Alabama 35899
USA

Dr G S Nechitailo
SPA "Energia"
Kaliningrad
Moscow District 141070
Russia

Dr Alexander Nepomnyashchy
Department of Mathematics
Technion
32000 Haifa
Israel

Professor V A Niculin
Institute of Scientific Applied Research
PO Box 3011
426008 Izhevsk
Russia

Mr K F Nielsen
Technical University of Denmark
2800 Lyngby
Denmark

Dipl. Phys. Stefan Odenbach
Universitat Munchen
Sektion Physik
Schellingstrabe 4
D-8000 Munchen 40
Germany

Dr Y M Orlov
Perm Polytechnical Institute
29a Komsomolsky Avenue
614600 Perm
Russia

E V Ostrovskii
Research Institute "Biotechnics"
Russian Ministry of Medical Industry
38 Kropotkinskaya Street
119034 Moscow
Russia

Dr A S Ovcharova
Lavrentyev Institute of Hydrodynamics
Siberian Branch of Russian Academy of Sciences
630090 Novosibirsk
Russia

I L Ozernyih
Department of Design Bureau "Lavochkin"
17 Oktyabrskaya
248600 Kaluga
Russia

Dr Alberto Passerone
ICFAM - CNR
Lungoblsagno Istria, 34
16141 Genova
Italy

D S Pavlovsky
Institute for Problems in Mechanics
Russian Academy of Sciences
101 Prospect Vernadskogo
117526 Moscow
Russia

PhD Jose M Perales
Laboratorio de Aerodinamica
ETSI Aeronauticos
28040 Madrid
Spain

V A Pilgoon
Research Institute "Nauchny Tsentr"
103460 Moscow
Russia

Dr I G Podolsky
Institute of Biomedical Problems
Russian Ministry of Health
76a Khoroskyevskoye Shosse
123007 Moscow
Russia

V I Pokhilko
Research Institute of Applied Mathematics and Electronics
Moscow Aviation Institute
4 Volokolamskoye Shosse
125871 Moscow
Russia

Dr V K Polevikov
Byelorussian State University
4 Lenin Avenue
220080 Minsk
Byelorrusia

Professor V I Polezhaev
Institute for Problems in Mechanics
Russian Academy of Sciences
101 Propect Vernadskogo
117526 Moscow
Russia

Professor V P Polischuk
Department of Magnetohydrodynamics
Ukrainian Academy of Sciences
34/1 Vernadsky Avenue
252660 Kiev
Ukraine

V V Polischuk
Department of Magnetohydrodynamics
Ukrainian Academy of Sciences
34/1 Vernadsky Avenue
252660 Kiev
Ukraine

Professor V V Pukhnachov
Lavryentyev Institute of Hydrodynamics
Siberian Branch of Russian Academy of Sciences
15 Lavryentyev Avenue
630090 Novosibirsk
Russia

Dr G F Putin
Perm State University
15 Bukirev Street
614005 Perm
Russia

Professor Hans J Rath
University of Bremen
ZARM
D-2800 Bremen 33
Germany

Professor Pier Georgio Righetti
University of Milano
Department Biomedical Sciences and Technologies
via Celoria 2
Milano 20133
Italy

Professor F P J Rimrott
Department of Mechanical Engineering
University of Toronto
Toronto
Ontario
M5S 1A4
Canada

Dr Bernard Roux
Institute Mecanique des Fluides
1 rue Honnorat
F-13003 Marseille
France

N N Rusakova
Ivanovo Power Institute
34 Rabfakovskaya Street
153548 Ivanovo
Russia

E A Ryabitsky
Computing Centre
Siberian Branch of Russian Academy of Sciences
Akademgorodok
660033 Krasnoyarsk
Russia

Dr Ziad Saghir
Canadian Space Agency
National Research Council
100 Sussex Drive
Ottawa
Ontario
K1A 0R
Canada

Professor Hiroshi Sakuta
Kamitomioka
Nagaoaka 940-21
Japan

S N Salghin
Research Institute "Nauchny Tsentr"
103460 Moscow
Russia

Professor V V Sazonov
Kedysh Institute of Applied Mathematics
Russian Academy of Sciences
4 Miusskaya Square
125047 Moscow
Russia

Dr Jurgen Schilz
DLR
Institute of Material Research
WB-WF Postfach 906058
D-5000 Koln 90
Germany

Dipl.-Phys B Schwark
University of Bremen
ZARM
D-2800 Bremen 33
Germany

Professor Robert F Sekerka
Mellon College of Science
Scaife Hall 115
Carnegie Mellon University
Pittsburgh
Pennsylvania 15213
USA

Dr V A Semenov
Perm State University
15 Bukirev Street
614005 Perm
Russia

Dr A S Senchenkov
SPLAV Technical Center
9 Baikalskaya Street
107497 Moscow
Russia

Dr V N Serebryakov
SPA "Energia"
Kaliningrad
Moscow District 141070
Russia

Professor V P Shalimov
Space Research Institute of Russian
Academy of Sciences
88 Profsoyuznaya Street
117810 Moscow
Russia

V I Sharamkin
Insitute of Physics
Latvian Academy of Sciences
32 Miera Street
Salaspils-1 Riga District 228021
Latvia

S B Shatunov
Perm State Pedagogical Institute
24 Karl Mark Street
614600 Perm
Russia

Dr M V Shevtsova
Institute for Problems in Mechanics
Russian Academy of Sciences
101 Prospect Vernadskogo
117526 Moscow
Russia

Professor S I Sidorenko
Kiev Polytehnical Institute
37 Prospect Pobedy
KPI-1150
252056 Kiev
Ukraine

Dr I B Simanovskii
Perm State Pedagogical Institute
24 Karl Marx Street
614600 Perm
Russia

Dr I E Sinitsina
Institute of Scientific Applied Research
PO Box 3011
426008 Izhevsk
Russia

PdD Slavcho G Slavchev
Institute of Mechanics and Biomechanics
Bulgarian Academy of Sciences
St Acad. G Bontchev Block 4
Sofia 1113
Bulgaria

Professor L A Slobozhanin
Institute for Low Temperature Physics
and Engineering
Ukrainian Academy of Sciences
47 Lenin Avenue
310164 Kharkov
Ukraine

Dr Robert S Sokolowski
Chief Solidification Physics Branch
Space Science Laboratory
NASA HQ, Code SN
Washington DC 20546
USA

Dr Kurt Sommer
OHB-System
Universitatsallee 27
D-2800 Bremen
Germany

Dr M Z Sorkin
Institute of Physics
Latvian Academy of Sciences
Salaspils-1
Riga District 229021
Latvia

A F Spivak
Center of Microelectronics
Latvian Academy of Sciences
19 Turgenev Street
Riga
Latvia

Dr G Mahr V Staszewski
Insitut fur Werkstoff-Forschung
DLR
Postfach 90 60 58
D-5000 Koln 90
Germany

Dipl. Ing F R Stengele
University of Bremen
ZARM
D-2800 Bremen 33
Germany

Dr A T Sukhodolsky
Institute of General Physics
Russian Akademy of Sciences
38 Vavilov Street
117942 Moscow
Russia

Dr A V Svido
SPA "Electroterm"
143500 Istra-2
Moscow Region
Russia

Professor E L Tarunin
Perm State University
15 Bukirev Street
614005 Perm
Russia

Professor V A Tatarchenko
Institute for High Temperature
Russian Academy of Sciences
1 Korovinskoye Shosse
Moscow
Russia

Dr Ing Tisson
Centre d'Etudes Nucleaires
DEM/SESC
85 X, 38041 Grenoble Cedex
France

PhD Paul W Todd
US Department of Commerce
NIST
325 Broadway
Boulder
Colorado 80303-3328
USA

Professor Ales Triska
Institute of Physics
Czechoslovakian Academy of Sciences
Na Slovance 2
CS-18040 Prague 8
Czechoslovakia

Dr Stefan Van Vaerenbergh
Chemical Physics Department, EP
CP 165 ULB
50 Avenue F D Roosevelt
1050 Bruxelles
Belgium

Dr A N Vereshchaga
Institute of Continuous Media Mechanics
Russian Academy of Sciences
1 Korolyov Street
614061 Perm
Russia

Dr I I Vertgeim
Institute of Continuous Media Mechanics
Russian Academy of Sciences
1 Korolyov Street
614061 Perm
Russia

Professor V M Vinokur
Perm Polytechnical Institute
29a Komsomolsky Avenue
614600 Perm
Russia

Dr P K Volkov
Institute of Thermophysics
Siberian Branch of Russian Academy of Sciences
1 Lavryentyev Avenue
630090 Novosibirsk
Russia

Dr J P B Vreeburg
National Aerospace Laboratory
PO Box 90 502
NL-1006 BM Amsterdam
The Netherlands

Dr H U Walter
European Space Agency
6-10 rue Mario Nikis
75015 Paris
France

Ass. Professor Xueli Wei
Institute of Metal Research
Academia Ainica
Wenhua Road
Shenyang 110015
People's Republic of China

AS Eng Mark M Weislogel
NASA Lewis Research Center
21000 Brookpark Road
Mail Stop 500-217
Cleveland
Ohio 44135
USA

Dr V I Yakushin
Perm State Pedagogical Institute
24 Karl Marx Street
614600 Perm
Russia

M P Zabelina
Keldysh Institute of Applied Mathematics
Russia Academy of Science
4 Miusskaya Square
125047 Moscow
Russia

I V Zakharov
Institute of Continuous Media Mechanics
Russian Acadmy of Science
1 Korolyov Street
614061 Perm
Russia

Dr Bernard Zappoli
Direction de la Recherce
Sous-Direction Microgravite et Sciences de la Vie
Division Fluides et Materiaux en Microgravite
Centre Spatial de Toulouse
18 avenue Edouard-Belin
31055 Toulouse Cedex
France

M P Zavarykin
Perm State University
15 Bukirev Street
614005 Perm
Russia

Dr Yu F Zavialov
Research Institute "Biotechnics"
Russian Ministry of Medical Industry
38 Kropotkinskaya Street
119034 Moscow
Russia

Dr A I Zhakin
Kaharkov State University
4 Dzerzhinky Square
310077 Kharkov
Ukraine

Dr Yu I Zhavrin
Kasakh State University
46 Timiryazev Street
480121 Alma-Ata
Kazakhstan

Dr A L Zheleznyak
Institute of Applied Physics
Russian Academy of Sciences
46 Ulyanov Street
606600 Nizhni Novgorod
Russia

Dr Ing Gerhard Zimmerman
Intzestrabe 5
D-5100 Aachen
Germany

V B Zinovyev
Department of Design Bureau "Lavochkin"
17 Oktyabrskaya
248600 Kaluga
Russia

Dr S V Zorin
Perm State Pedagogical Institute
24 Karl Marx Street
614600 Perm
Russia

Dr A L Zuev
Institute of Continuous Media Mechanics
Russian Academy of Sciences
1 Korolyov Street
614061 Perm
Russia

S G Zykov
SPA "Energia"
Kaliningrad
Moscow District 141070
Russia